Physical Sciences

General

Brief Calculus with Applications

Brief Calculus with Applications

Roland E. Larson and **Robert P. Hostetler**
Exercises by **David E. Heyd**

The Pennsylvania State University, The Behrend College

D. C. HEATH AND COMPANY
Lexington, Massachusetts   Toronto

Photo Credits

Title page

B. A. King

Chapter-opening photos

0 Peter Vandermark/Stock, Boston
1 Frank Wing/Stock, Boston
2 John Zoiner/Peter Arnold, Inc.
3 Roy Pinney/Monkmeyer Press Photo Service
4 National Archives
5 Peter Menzel/Stock, Boston
6 Hugh Rogers/Monkmeyer Press Photo Service
7 W. B. Finch/Stock, Boston
8 Joe Ofria/Global Focus
9 Peter Menzel/Stock, Boston
10 Frank Siteman/Stock, Boston.

Published simultaneously in Canada.

Printed in the United States of America.

International Standard Book Numbers: 0-669-04803-8

Library of Congress Catalog Card Number: 82-81606

Preface

Brief Calculus with Applications is designed for use in a beginning calculus course for students in the management, social, and life sciences. In writing this book, we were guided by two primary objectives that have crystallized over many years of teaching calculus. For the student, our objective was to write in a precise and readable manner, with the basic concepts and rules of calculus clearly defined and demonstrated. For the instructor, our objective was to design a comprehensive teaching instrument that employs proven pedagogical techniques, thus freeing the instructor to make the most efficient use of classroom time.

As full-time calculus instructors with combined experience of more than thirty years, and as authors for the past ten years, we have found the following features to be valuable aids to the teaching and learning of calculus.

Introductory Examples Each section in Chapters 1–10 begins with a one-page motivational example designed to show the applicability of the material in the section. These examples are to be read for enjoyment and not for complete mastery, since many of the concepts and techniques involved will be discussed in the material that follows. We expect that, after reading an Introductory Example, students will want to learn the mathematics in that section because it has interesting and practical uses.

Prerequisites Chapter 0 is designed as a quick review of the algebra needed to study calculus. An instructor may elect to skip Chapter 0 and begin the course with Chapter 1, depending upon the mathematical background of a particular class. In such cases, Chapter 0 can serve as an algebraic reference for students.

Section Objectives Each section begins with a list of the major objectives to be covered in the section.

Theorems and Definitions Special care has been taken to state the theorems and definitions simply, without sacrificing accuracy. The theorems and definitions are boxed to emphasize their importance.

Examples The text contains over 425 examples. Each example has been carefully chosen to illustrate a particular concept or problem-solving technique.

Exercises We learn mathematics by doing mathematics. Thus, a calculus text should include enough exercises to build competence, skill, and understanding. This text has over 2250 exercises that have been graded for difficulty and comprehensiveness.

The answers to the odd-numbered exercises are given in the back of the text. The even-numbered exercise answers are given in the *Answer Key*. Every effort has been made to provide *correct* answers. Each exercise was independently solved by at least four people.

Many of the exercise sets contain problems identified as calculator problems. Although most of these problems can be solved without a calculator, we find that students can benefit from using calculators, and we recommend that they use calculators for these specially-marked exercises.

Graphics The ability to visualize a calculus problem goes hand in hand with the ability to solve the problem. We have included over 825 figures in the body of the text and in the exercise sets to nurture the student's ability to visualize problems.

Applications For easy reference, an index of applications has been included on the inside covers of the text. We have included in the text over 425 applications taken from a variety of fields, with special emphasis on applications in business and economics. A summary of basic business terms and formulas is provided on pages 237–238.

Order of Coverage For curriculum planning, certain dependency guidelines should be noted: Chapter 0 is optional, Chapters 1–6 should be covered in order, and Chapters 7–10 can be covered in any order.

Acknowledgements We would like to thank the many people who have helped us at various stages of this project. Their encouragement, criticisms, and suggestions have been invaluable to us. Over the past several years, many users of our texts have taken the time to write to us with comments, praise, and suggestions for improvement. We treasure such input.

Special thanks go to:

Our reviewers: Bruce Edwards, University of Florida-Gainesville; Miriam E. Connellan, Marquette University; DeWitt Sumners, Florida State University; Stephen Rodi, Austin Community College; Robert Yawin, Springfield Technical Community College; Maurice L. Monahan, South Dakota State University; Norbert Lerner, SUNY-Cortland; W. Cary Huffman, Loyola University of Chicago; and Earl H. McKinney, Ball State University.

Our colleagues: David E. Heyd, who created and solved the exercise sets; the staff at D. C. Heath, who worked on the book; our dean, John M. Lilley, who provided the space and schedules necessary for writing; Norman Patterson, whose philosophy of teaching we value; Dianna Zook, who proofread the galleys and solved the exercises; Deanna Larson, who typed the entire manuscript; our student assistant, Wilbur Walls, who read the entire manuscript for accuracy; and our students, whose input has helped form our teaching and writing styles.

Our wives, Deanna Larson and Eloise Hostetler, whose love and understanding is without price.

Roland E. Larson

Robert P. Hostetler

The Pennsylvania State University
The Behrend College

Contents

Brief Calculus with Applications

Chapter 0

A Precalculus Review

Section 0.1

The Real Number Line and Order

Section Objectives: ***To review the real number line and the properties of real numbers ▪ To discuss the notation for intervals on the real number line ▪ To discuss the properties of inequalities and demonstrate their uses.***

The coordinate system we use to represent the real numbers is called the **real line** or x-axis. The **positive direction** (to the right) is the direction of increasing values of x. The real number corresponding to a particular point on the real line is called the **coordinate** of the point. As shown in Figure 0.1, it is customary to label those points whose coordinates are integers.

The importance of the real line lies in the fact that it provides us with a perfect picture of the real numbers. That is, each point on the real line corresponds to one and only one real number, and each real number corresponds to one and only one point on the real line. This type of relationship is called a **one-to-one correspondence.** (See Figure 0.2.)

Each of the four points in Figure 0.2 corresponds to a real number that can be expressed as the ratio of two integers. (Note that $1.85 = \frac{37}{20}$ and $-2.6 = -\frac{13}{5}$.) We call such numbers **rational.** Rational numbers have either terminating or infinite repeating decimal representations:

Terminating Decimals	*Infinite Repeating Decimals*
$\frac{2}{5} = 0.4$	$\frac{1}{3} = 0.333\ldots = 0.\overline{3}$*
$\frac{7}{8} = 0.875$	$\frac{12}{7} = 1.714285714285\ldots = 1.\overline{714285}$

Real numbers that are not rational are called **irrational**, and they cannot be represented as the ratio of two integers (or as terminating or infinite repeating decimals). To represent an irrational number, we usually resort to a decimal approximation. Of course, some irrational numbers occur so frequently in applications that we have invented special symbols to represent them. For example, the symbols $\sqrt{2}$, π, and e represent the irrational numbers whose decimal approximations are as follows:

$$\sqrt{2} \approx 1.4142135623$$

$$\pi \approx 3.1415926535$$

$$e \approx 2.7182818284$$

* The bar indicates which digits repeat.

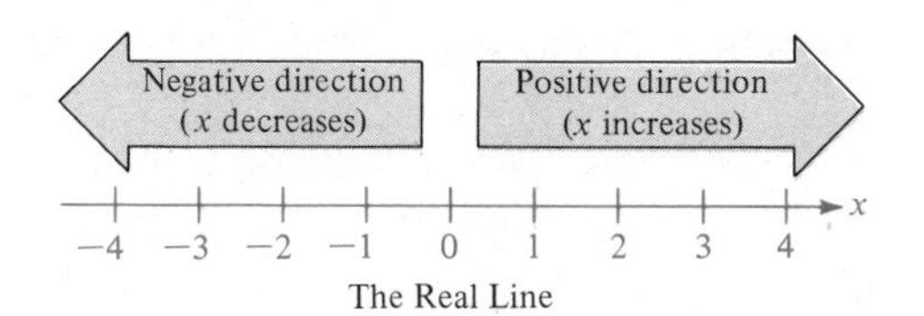

Figure 0.1

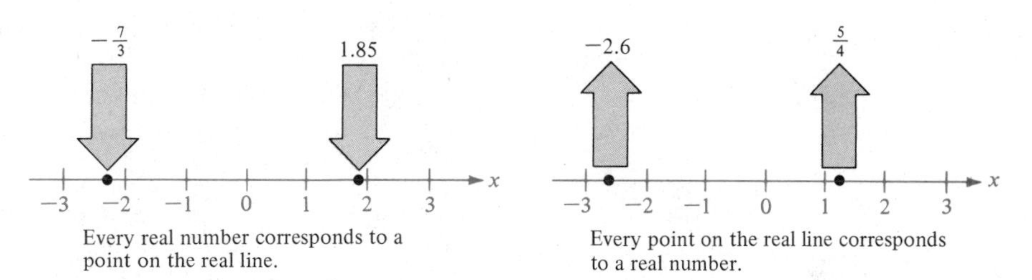

Figure 0.2

(Note that we use $\approx$ to mean *approximately equal to*.) Remember that, even though we cannot represent irrational numbers *exactly* as (finite) decimals, we can represent them *exactly* by points on the real line, as shown in Figure 0.3.

One important property of the real numbers is that they are **ordered;** 0 is less than 1, -3 is less than -2.5, π is less than $\frac{22}{7}$, and so on. We can visualize this property on the real line by observing that a is less than b if and only if a lies to the left of b. Symbolically, we denote "a is less than b" by the inequality

$$a < b$$

For example, the inequality $\frac{3}{4} < \frac{7}{5}$ follows from the fact that $\frac{3}{4}$ lies to the left of $\frac{7}{5}$ on the real line, as shown in Figure 0.4.

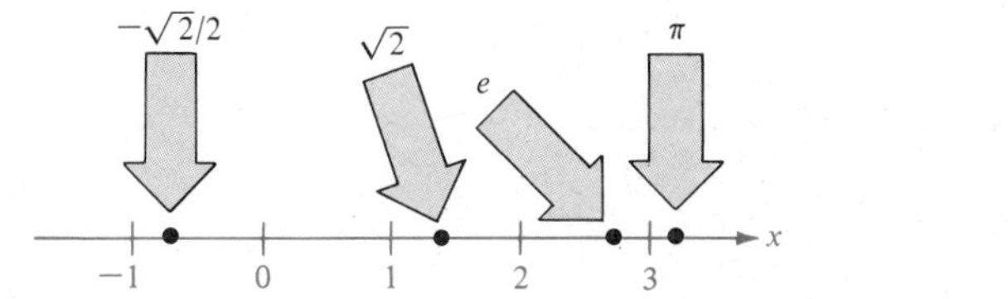

Figure 0.3

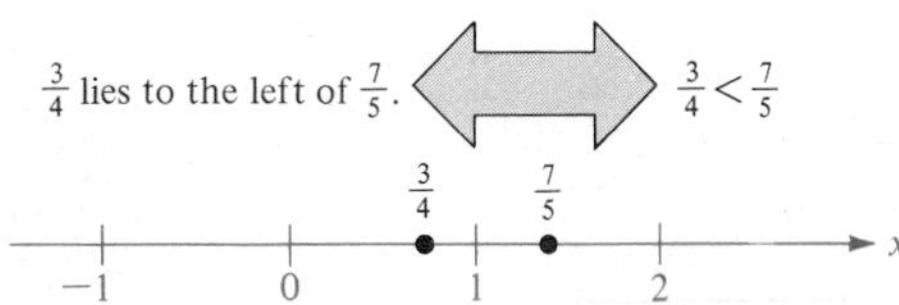

Figure 0.4

In practice, we often work with subsets of the real line rather than with the entire real line. For example, water is normally in a liquid state between the temperatures of 32° and 212° Fahrenheit, as pictured in Figure 0.5. We denote this subset of the real numbers by the inequality

$$32 < x < 212$$

or by the **interval**

$$(32, 212)$$

Similarly, the **infinite interval**

$$[212, \infty)$$

which corresponds to the inequality

$$212 \leq x$$

can be used to denote the temperature range of steam, as pictured in Figure 0.6. Note that a square bracket is used to denote "less than or equal to" ($\leq$). Furthermore, we use the symbols ∞ and $-\infty$ to denote positive and negative infinity. These symbols do not denote real numbers; they merely enable us to describe certain unbounded conditions more concisely.

Figure 0.5

Figure 0.6

An interval of the form (a, b) does not contain the "endpoints" a and b, and it is called an **open interval.** The interval $[a, b]$ does contain its endpoints, and it is called a **closed interval.** Intervals of the form $[a, b)$ and $(a, b]$ are called **half-open intervals.** Table 0.1 pictures the nine types of intervals on the real line.

Table 0.1
Intervals on the Real Line

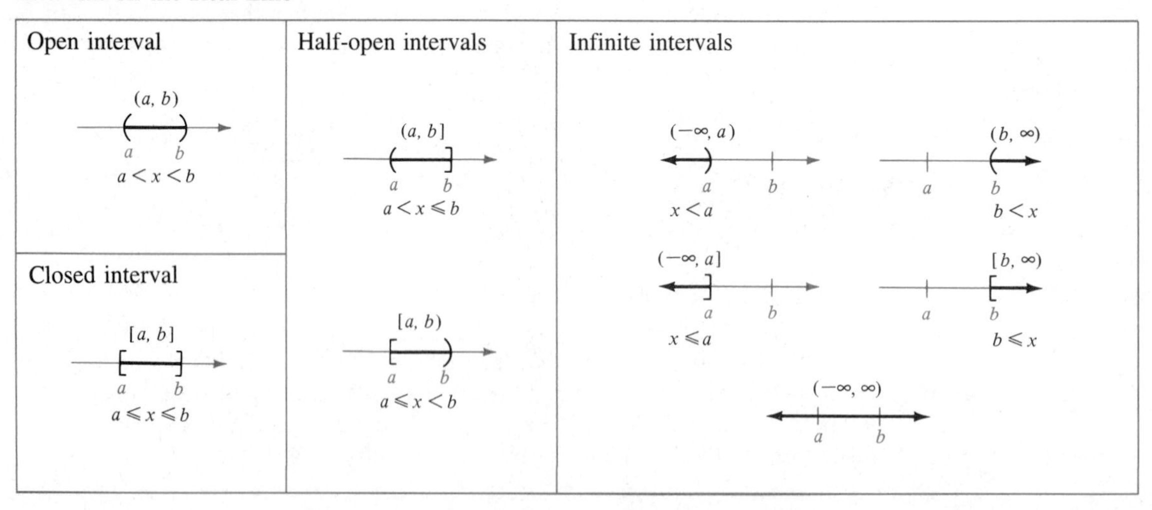

The following rules are often used when working with inequalities. These same rules hold if $<$ is replaced by $\leq$.

Operations with Inequalities

Transitive Property

$a < b$ and $b < c \Rightarrow a < c$

Multiplying by a (Positive) Constant

$a < b \Rightarrow ac < bc$

Multiplying by a (Negative) Constant

$a < b \Rightarrow ac > bc$

Adding Inequalities

$a < b$ and $c < d \Rightarrow a + c < b + d$

Adding a Constant

$a < b \Rightarrow a + c < b + c$

Subtracting a Constant

$a < b \Rightarrow a - c < b - c$

Remark

Note that we *reverse the inequality* when we multiply by a negative number. This principle also applies to division; that is, the inequality is preserved when dividing by a positive number, but it is reversed when dividing by a negative number. Study the following three examples carefully to see how these operations can be used to "solve an inequality."

Example 1

Solving an Inequality

Find the interval corresponding to the set of x-values that satisfy the inequality

$$2x - 5 < 7$$

Solution By adding 5 to both sides of the inequality, we have

$$2x - 5 + 5 < 7 + 5$$
$$2x < 12$$

Now, multiplying by $\frac{1}{2}$ gives us

$$\tfrac{1}{2}(2x) < \tfrac{1}{2}(12)$$
$$x < 6$$

Thus, the interval representing the solution is $(-\infty, 6)$, as shown in Figure 0.7.

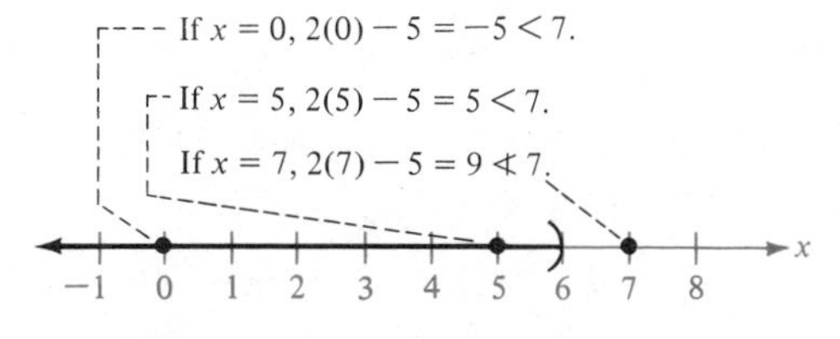

Figure 0.7

□

Once you have solved an inequality, it is a good idea to check some x-values in your solution interval to see if they satisfy the original inequality. You also might check some values outside your solution interval to verify that they do not satisfy the inequality. For example, in Figure 0.7 we see that when $x = 0$ or $x = 5$ the inequality is satisfied, but when $x = 7$ the inequality is not satisfied.

Example 2

Solving a Double Inequality

Find the interval corresponding to the set of x-values that satisfy the inequality

$$-3 \le 2 - 5x \le 12$$

Note that this double inequality means that $-3 \le 2 - 5x$ and $2 - 5x \le 12$.

Solution Although two inequalities are involved in this problem, we can work with both simultaneously. We begin by subtracting 2 from all three expressions to obtain

$$-3 - 2 \le 2 - 5x - 2 \le 12 - 2$$
$$-5 \le -5x \le 10$$

Now, we divide all three expressions by -5 (making sure to reverse both inequalities) to obtain

$$\frac{-5}{-5} \ge \frac{-5x}{-5} \ge \frac{10}{-5}$$
$$1 \ge x \ge -2$$

Thus, the interval representing the solution is $[-2, 1]$. □

Example 3

A Business Application

A manufacturer determines that (in addition to fixed overhead costs of $500 per day) the unit cost of producing a certain item is $2.50. The total daily cost of production during a given month varied between $1200 and $1325. Find the high and low production levels during the month.

Solution Since it costs $2.50 to produce one unit, it will cost $2.5x$ to produce x units. Furthermore, since the fixed cost per day is $500, the total daily cost of producing x units is

$$C = 2.5x + 500$$

Now, since the cost ranges between $1200 and $1325, we have

$$1200 \le 2.5x + 500 \le 1325$$
$$1200 - 500 \le 2.5x + 500 - 500 \le 1325 - 500$$
$$700 \le 2.5x \le 825$$
$$\frac{700}{2.5} \le \frac{2.5x}{2.5} \le \frac{825}{2.5}$$
$$280 \le x \le 330$$

Thus, we see that the daily production levels during the month varied between a low of 280 units and a high of 330 units, as pictured in Figure 0.8.

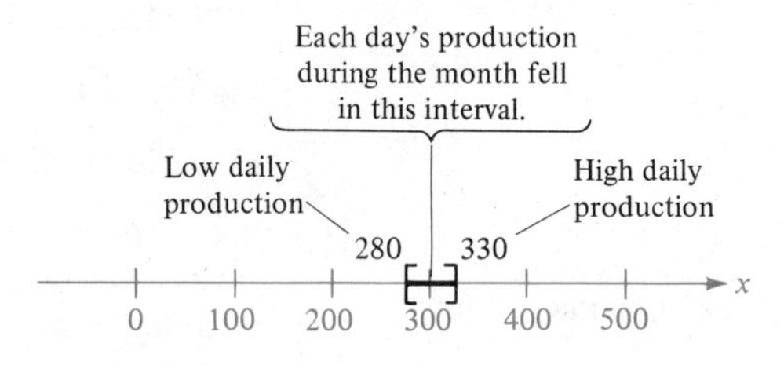

Figure 0.8

□

Section Exercises (0.1)

In Exercises 1–8, complete the two missing descriptions of the given interval.

	Interval Notation	Inequality Notation	Graph
1.			−3 −2 1 0
2.	$(-\infty, -4]$		
3.		$3 \leq x \leq \frac{11}{2}$	
4.	$(-1, 7)$		
5.			98 99 100 101 102
6.		$10 < x < \infty$	
7.	$(\sqrt{2}, 8]$		
8.		$\frac{1}{3} < x \leq \frac{22}{7}$	

In Exercises 9–20, solve the inequality and graph the solution on the real number line.

9. $x - 5 \geq 7$
10. $2x > 3$
11. $4x + 1 < 2x$
12. $2x + 7 < 3$
13. $2x - 1 \geq 0$
14. $3x + 1 \geq 2$
15. $4 - 2x < 3$
16. $x - 4 \leq 2$
17. $-4 < 2x - 3 < 4$
18. $0 \leq x + 3 < 5$
19. $\frac{3}{4} > x + 1 > \frac{1}{4}$
20. $-1 < -\dfrac{x}{3} < 1$

In Exercises 21–24, determine whether or not the given value of x satisfies the inequality.

21. $5x - 12 > 0$
(a) $x = 3$ (b) $x = -3$
(c) $x = \frac{5}{2}$ (d) $x = \frac{3}{2}$

22. $x + 1 < \dfrac{2x}{3}$
(a) $x = 0$ (b) $x = 4$
(c) $x = -4$ (d) $x = -3$

23. $0 < \dfrac{x-2}{4} < 2$

(a) $x = 4$ (b) $x = 10$
(c) $x = 0$ (d) $x = \frac{7}{2}$

24. $-1 < \dfrac{3-x}{2} \leq 1$

(a) $x = 0$ (b) $x = \sqrt{5}$
(c) $x = 1$ (d) $x = 5$

C **25.** P dollars invested at r percent simple interest for t years grows to an amount

$$A = P + Prt$$

If an investment of \$1000 is to grow to an amount greater than \$1250 in two years, then the interest rate must be greater than what percentage?

C **26.** A family establishes a business selling mini-donuts at a shopping mall. The cost of making a dozen donuts is \$1.45, and the donuts sell for \$2.95 per dozen. In addition to the cost of ingredients, the business must pay \$25 per day for rent and utilities. If the daily profit varies between \$50 and \$200, between what levels (in dozens) do the daily sales vary?

C **27.** In the manufacture and sale of a certain product, the revenue for selling x units is

$$R = 115.95x$$

and the cost of producing x units is

$$C = 95x + 750$$

In order to realize a profit, it is necessary that R be greater than C. For what values of x will this product return a profit?

C **28.** A utility company has a fleet of vans for which the annual operating cost per van is estimated to be

$$C = 0.32m + 2300$$

where C is measured in dollars and m is measured in miles. If the company wants the annual operating cost per van to be less than \$10,000, then m must be less than what value?

In Exercises 29 and 30, determine which of the two given real numbers is greater.

C **29.** (a) π or $\frac{355}{113}$ (b) π or $\frac{22}{7}$

C **30.** (a) $\frac{224}{151}$ or $\frac{144}{97}$ (b) $\frac{73}{81}$ or $\frac{6427}{7132}$

C Calculator may be helpful.

Section 0.2

Distance on the Real Line and Absolute Value

Section Objectives: *To find the distance between two points on the real line ▪ To discuss the use of absolute value to describe intervals on the real line ▪ To find the midpoint of an interval.*

Given two distinct points a and b on the real line we will, in this text, make use of each of the distances pictured in Figure 0.9:

1. the **directed distance from** a **to** b denoted by $b - a$
2. the **directed distance from** b **to** a denoted by $a - b$
3. the **distance between** a **and** b denoted by $|a - b|$ or $|b - a|$

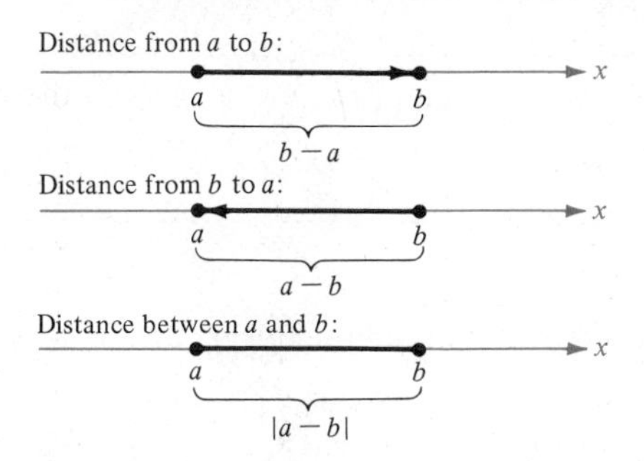

Figure 0.9

The symbol $|a - b|$ is called the **absolute value** of $a - b$, and it denotes the *magnitude* of $a - b$. The precise definition of the absolute value of a real number is as follows.

Definition of Absolute Value

If x is any real number, then the **absolute value** of x, denoted by $|x|$, is determined by

$$|x| = \begin{cases} x, & \text{if } x \geq 0 \\ -x, & \text{if } x < 0 \end{cases}$$

At first glance it may appear from this definition that an absolute value can be negative. Such is not the case. For instance, let $x = -4$. Then, by definition,

we have

$$|-4| = -(-4) \qquad \text{since} \qquad -4 < 0$$

and therefore,

$$|-4| = +4$$

Similarly,

$$|2 - 7| = |-5| = -(-5) = +5$$

and, of course,

$$|7 - 2| = |5| = 5$$

The following alternative definition of absolute value avoids potential confusion with signs, and it fits well with the definition of distance between two points. We can define absolute value as

$$|x| = \sqrt{x^2}$$

since the square root symbol $\sqrt{\ }$ denotes only the *nonnegative* root. Thus, the distance between two points on the real line can be defined as follows.

Distance between Two Points on the Real Line

The distance d between points x_1 and x_2 on the real line is given by

$$d = |x_2 - x_1| = \sqrt{(x_2 - x_1)^2}$$

Note that the order of subtracting x_1 and x_2 does not matter since

$$|x_2 - x_1| = |x_1 - x_2| \qquad \text{and} \qquad (x_2 - x_1)^2 = (x_1 - x_2)^2$$

Example 1

Finding the Distance between Two Points on the Real Line

Determine the distance between -3 and 4 on the real line in Figure 0.10. What is the directed distance from -3 to 4? From 4 to -3?

Solution The distance between -3 and 4 is given by

$$|4 - (-3)| = |7| = 7 \qquad \text{or by} \qquad |-3 - 4| = |-7| = 7$$

The directed distance from -3 to 4 is $4 - (-3) = 7$. The directed distance from 4 to -3 is $-3 - 4 = -7$.

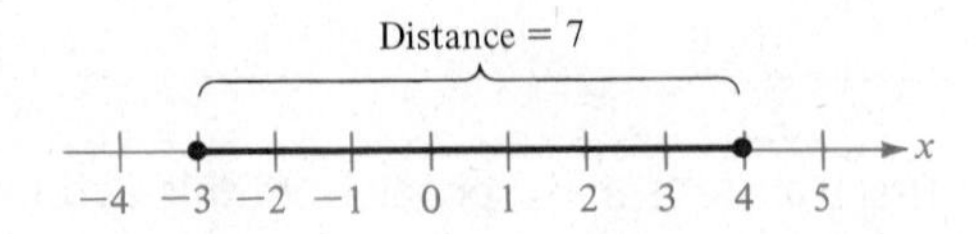

Figure 0.10

□

Example 2

Using Absolute Value to Define an Interval on the Real Line

What interval on the real line contains all numbers that lie within two units of 3?

Solution Let x be any point in this interval. Then we wish to find all x such that the distance between x and 3 is less than or equal to 2. We write this symbolically as

$$|x - 3| \leq 2$$

Requiring the absolute value of $x - 3$ to be less than or equal to 2 means that $x - 3$ must lie between 2 and -2, and hence we write

$$-2 \leq x - 3 \leq 2$$

Solving this pair of inequalities, we have

$$-2 + 3 \leq x - 3 + 3 \leq 2 + 3$$
$$1 \leq x \leq 5$$

Therefore, the desired interval is [1, 5], as shown in Figure 0.11.

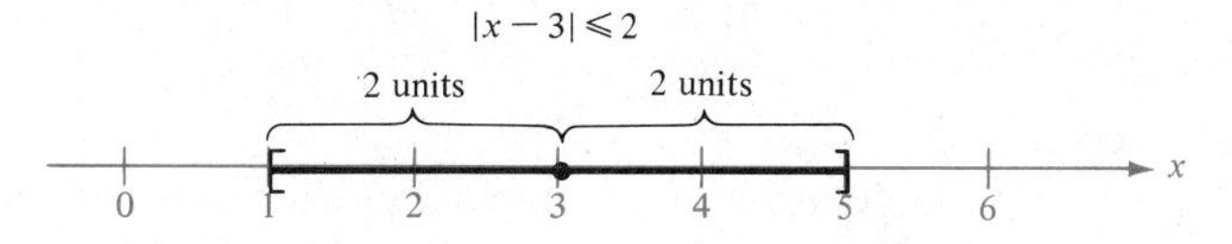

Figure 0.11 □

Example 3

Solving an Inequality That Results in Two Intervals

Determine the intervals on the real line that contain all numbers that lie more than three units from -2.

Solution Refer to Figure 0.12. Since the distance between x and -2 is given by $|x - (-2)|$, we write

$$|x - (-2)| > 3$$
$$|x + 2| > 3$$

This means that $x + 2$ must be numerically greater than 3. Hence we seek x-values that satisfy *either one* of the inequalities

$$x + 2 < -3 \qquad \text{or} \qquad x + 2 > 3$$

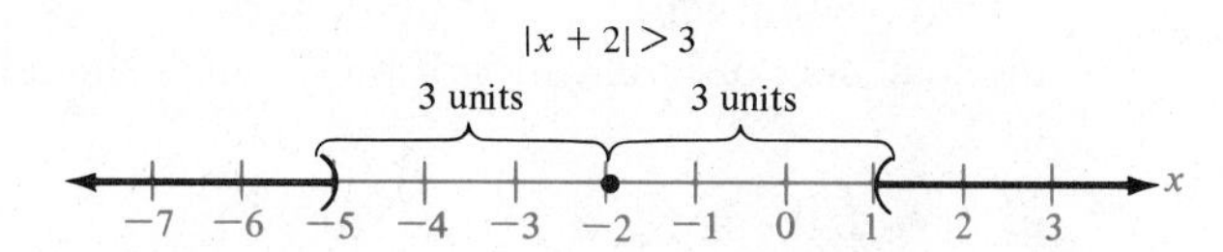

Figure 0.12

which implies that

$$x < -5 \qquad \text{or} \qquad x > 1$$

Therefore, x can lie in either of the intervals

$$(-\infty, -5) \qquad \text{or} \qquad (1, \infty)$$

□

Examples 2 and 3 suggest the following general results for absolute value and inequalities.

Intervals Defined by Absolute Value

Single Interval

$$|x - a| \leq d \Rightarrow a - d \leq x \leq a + d$$

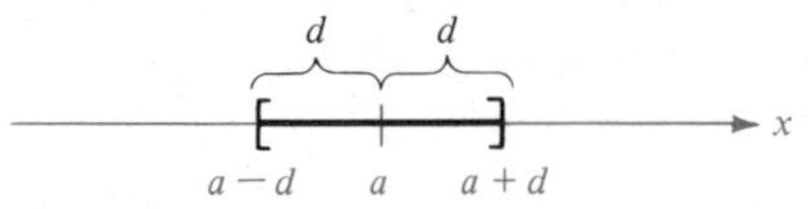

Two Intervals

$$|x - a| \geq d \Rightarrow x \leq a - d$$

or $\quad x \geq a + d$

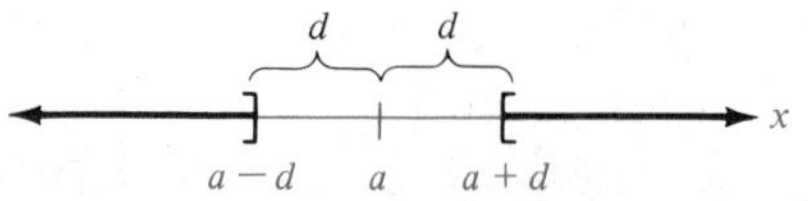

Example 4

A Business Application

A large manufacturer hired a quality control firm to determine the reliability of a certain product. Using statistical methods, the firm reported that the manufacturer could expect $0.35\% \pm 0.17\%$ of the units to be defective. If the manufacturer offers a money-back guarantee on this product, how much should be budgeted to cover the refunds on 100,000 units? (Assume that the retail price is \$8.95.)

Solution If we let r represent the percentage of defective units, we know that r will differ from 0.0035 by at most 0.0017. Using inequalities, we can write

$$0.0035 - 0.0017 \leq r \leq 0.0035 + 0.0017$$
$$0.0018 \leq r \leq 0.0052$$

Now, letting x be the number of defective units out of 100,000, we know that $x = 100{,}000r$ and we have

$$0.0018(100{,}000) \leq 100{,}000r \leq 0.0052(100{,}000)$$
$$180 \leq x \leq 520$$

Finally, letting C be the cost of refunds, we have $C = 8.95x$, and it follows that the total cost of refunds for 100,000 units should fall within the interval given by

$$180(8.95) \leq 8.95x \leq 520(8.95)$$
$$\$1611 \leq C \leq \$4654$$

(See Figure 0.13.)

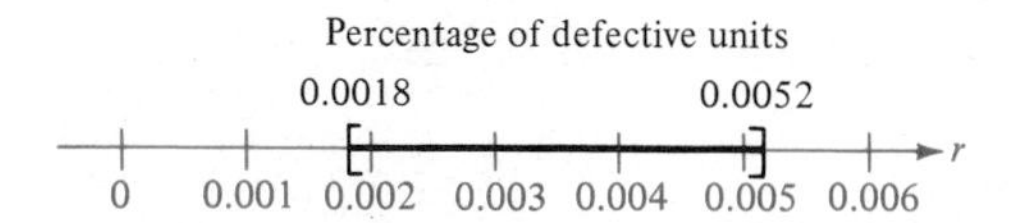

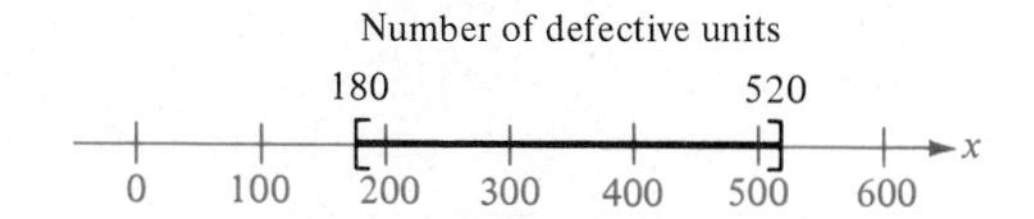

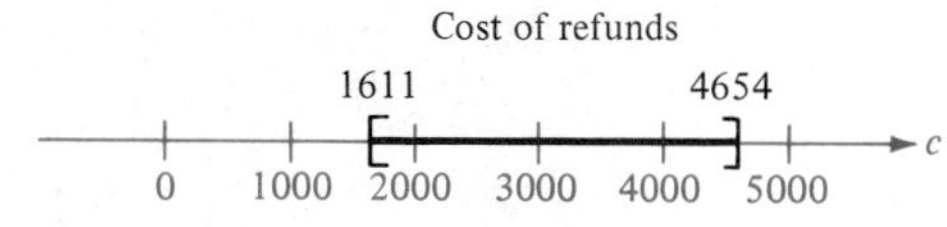

Figure 0.13 □

In Example 4, we concluded that the manufacturer should expect to spend between \$1611 and \$4654 for refunds. Obviously, the safest budget figure for refunds would be the higher of these estimates. However, from a strictly mathematical point of view, the most representative estimate would be the average of these two extremes. We see that this average is geometrically the midpoint of the interval that has these two numbers as its endpoints. (See Figure 0.14.)

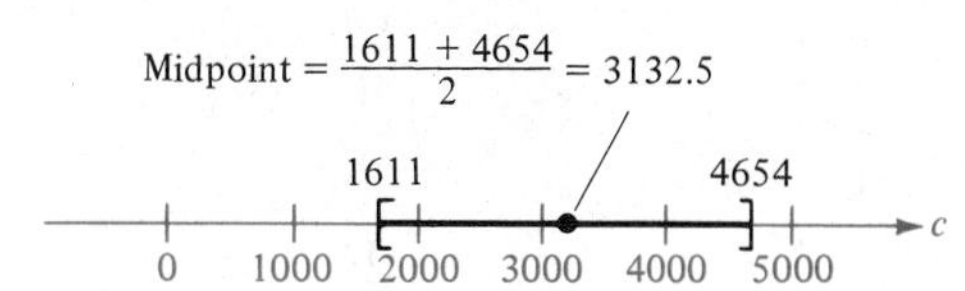

Figure 0.14

Midpoint of an Interval

The **midpoint** of the interval with endpoints a and b is

$$\text{midpoint} = \frac{a + b}{2}$$

Example 5

Finding the Midpoint of an Interval

Find the midpoint of each of the following intervals:

(a) $[-5, 7]$ (b) $(-12, -1)$ (c) $(3, 21]$

Solution

(a) $\text{midpoint} = \dfrac{-5 + 7}{2} = \dfrac{2}{2} = 1$

$$\textbf{(b)}\ \text{midpoint} = \frac{-12 + (-1)}{2} = -\frac{13}{2}$$

$$\textbf{(c)}\ \text{midpoint} = \frac{3 + 21}{2} = \frac{24}{2} = 12$$

□

Section Exercises (0.2)

In Exercises 1–8, find (a) the directed distance from a to b, (b) the directed distance from b to a, and (c) the distance between a and b.

1.

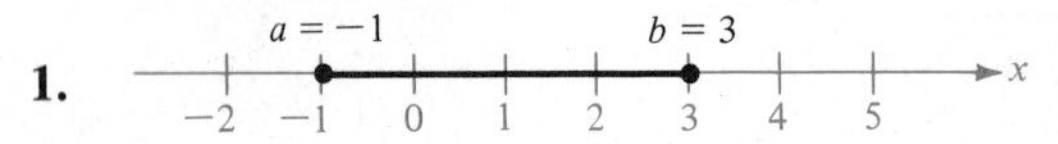

2. $a = \frac{1}{4}$, $b = \frac{11}{4}$

3. $a = -\frac{5}{2}$, $b = \frac{13}{4}$

4. $a = -4$, $b = -\frac{3}{2}$

5. $a = 126$, $b = 75$

6. $a = -126$, $b = -75$

7. $a = 9.34$, $b = -5.65$

8. $a = \frac{16}{5}$, $b = \frac{112}{75}$

In Exercises 9–16, find the midpoint of the given interval.

9. $a = -1$, $b = 3$

10. $a = -6$, $b = -1$

11. $a = -5$, $b = -\frac{3}{2}$

12. $a = \frac{3}{4}$, $b = \frac{11}{2}$

13. $[7, 21]$

ⓒ**14.** $[8.6, 11.4]$

ⓒ**15.** $[-6.85, 9.35]$

ⓒ**16.** $[-4.6, -1.3]$

In Exercises 17–26, solve the given inequality and graph the solution on the real line.

17. $|x| < 5$

18. $|2x| < 6$

19. $\left|\frac{x}{2}\right| > 3$

20. $|5x| > 10$

21. $|x + 2| < 5$

22. $|3x + 1| \geq 4$

23. $\left|\frac{x - 3}{2}\right| \geq 5$

24. $|2x + 1| < 5$

25. $|9 - 2x| < 1$

26. $\left|1 - \frac{2x}{3}\right| < 1$

In Exercises 27–36, use absolute values to define each interval (or pair of intervals) on the real line.

27. $a = -2$, $b = 2$

28. $a = -3$, $b = 3$

29. $a = -2$, $b = 2$

30. $a = -3$, $b = 3$

31. $a = 2$, $b = 6$

ⓒ Calculator may be helpful.

32. $a = -7$ $b = -1$

−8 −7 −6 −5 −4 −3 −2 −1 0 x

33. $a = 0$ $b = 4$

−2 −1 0 1 2 3 4 5 6 x

34. $a = 20$ $b = 24$

18 19 20 21 22 23 24 25 26 x

35. All real numbers within 10 units of 12.

36. All real numbers lying more than 6 units from 3.

Ⓒ**37.** The heights, h, of two-thirds of the members of a certain population satisfy the inequality

$$\left| \frac{h - 68.5}{2.7} \right| \le 1$$

where h is measured in inches. Determine the interval on the real line in which these heights lie.

Ⓒ**38.** To determine if a coin is fair, an experimenter tosses it 100 times and records the number of heads, x. Using statistical theory, the coin is declared unfair if

$$\left| \frac{x - 50}{5} \right| \ge 1.645$$

For what values of x will the coin be declared unfair?

39. The estimated daily production, p, at a refinery is given by

$$|p - 2{,}250{,}000| < 125{,}000$$

where p is measured in barrels of oil. Determine the high and low production levels.

Ⓒ Calculator may be helpful.

Section 0.3

Exponents and Radicals

Section Objectives: ***To review the basic properties of exponents ▪ To simplify and evaluate expressions involving exponents.***

Much of the work in this text deals with algebraic expressions such as

$$2x^3 + 3, \qquad \sqrt{x^2 + 5x - 2}, \qquad \text{and} \qquad (x + 3)(2x - 5)^{-1}$$

It is essential that you be familiar with the algebraic methods of simplifying and evaluating such expressions, and we devote this and the next two sections to reviewing these procedures. We begin by summarizing the basic properties of exponents.

Properties of Exponents

Whole Number Exponents

$$x^n = \underbrace{x \cdot x \cdot x \cdots x}_{n \text{ terms}}$$

Zero Exponent ($x \neq 0$)

$$x^0 = 1$$

Negative Exponents

$$x^{-n} = \frac{1}{x^n}$$

Radicals (nth Roots)

$$\sqrt[n]{x} = a \Rightarrow x = a^n$$

Special Conventions with Radicals

$$\sqrt[2]{x} = \sqrt{x} \quad \text{(Positive Square Root)}$$

Rational Notation for Radicals

$$\sqrt[n]{x} = x^{1/n}$$

Rational Exponents

$$x^{m/n} = (x^{1/n})^m = (x^m)^{1/n} = (\sqrt[n]{x})^m = \sqrt[n]{x^m}$$

Operations with Exponents

Multiplying Like Bases (add exponents)

$$x^n x^m = x^{n+m}$$

Dividing Like Bases (subtract exponents)

$$\frac{x^n}{x^m} = x^{n-m}$$

Removing Parentheses

$$(xy)^n = x^n y^n \qquad \left(\frac{x}{y}\right)^n = \frac{x^n}{y^n} \qquad (x^n)^m = x^{nm}$$

Special Conventions Involving Parentheses

$$-x^n = -(x^n) \neq (-x)^n \qquad cx^n = c(x^n) \neq (cx)^n \qquad x^{n^m} = x^{(n^m)} \neq (x^n)^m$$

Example 1

Evaluating Expressions with Exponents

Evaluate each expression for the given value of x.

Expression	*x-value*
(a) $y = -2x^2$	$x = 4$
(b) $y = 3x^{-3}$	$x = -1$
(c) $y = (-x)^2$	$x = \frac{1}{2}$
(d) $y = 2x^{1/2}$	$x = 4$
(e) $y = (2x)^{1/2}$	$x = 4$
(f) $y = \sqrt[3]{x^2}$	$x = 8$
(g) $y = (x^2 - 9)^{-3/2}$	$x = 5$

Solution

(a) $y = -2(4^2) = -2(16) = -32$

(b) $y = 3(-1)^{-3} = \dfrac{3}{(-1)^3} = \dfrac{3}{-1} = -3$

(c) $y = \left(-\dfrac{1}{2}\right)^2 = \dfrac{1}{4}$

(d) $y = 2\sqrt{4} = 2(2) = 4$

(e) $y = \sqrt{2(4)} = \sqrt{2}\sqrt{4} = 2\sqrt{2}$

(f) $y = 8^{2/3} = (8^{1/3})^2 = 2^2 = 4$

(g) $y = (5^2 - 9)^{-3/2} = (16)^{-3/2} = \dfrac{1}{(\sqrt{16})^3} = \dfrac{1}{4^3} = \dfrac{1}{64}$ □

Remark

In part (b) of Example 1, we converted to positive exponent form before evaluating the expression. This is a good practice to follow, and we suggest that you always convert to positive exponent form before evaluating an expression.

Example 2

Simplifying Expressions with Exponents

Simplify each of the following expressions.

(a) $2x^2(x^3)$ (b) $(3x)^2\sqrt[3]{x}$

(c) $\dfrac{3x^2}{(\sqrt{x})^3}$ (d) $\dfrac{5x^4}{(x^2)^3}$

(e) $x^{-1}(2x^2)$ (f) $\dfrac{-\sqrt{x}}{5x^{-1}}$

Solution

(a) $2x^2(x^3) = 2x^{2+3} = 2x^5$

(b) $(3x)^2\sqrt[3]{x} = 9x^2x^{1/3} = 9x^{2+(1/3)} = 9x^{7/3}$

(c) $\dfrac{3x^2}{(\sqrt{x})^3} = 3\dfrac{x^2}{(x^{1/2})^3} = 3\dfrac{x^2}{x^{3/2}} = 3x^{2-(3/2)} = 3x^{1/2}$

(d) $\dfrac{5x^4}{(x^2)^3} = \dfrac{5x^4}{x^6} = \dfrac{5}{x^{6-4}} = \dfrac{5}{x^2}$

(e) $x^{-1}(2x^2) = 2x^{-1}x^2 = 2x^{2-1} = 2x$

(f) $\dfrac{-\sqrt{x}}{5x^{-1}} = -\dfrac{1}{5}\dfrac{x^{1/2}}{x^{-1}} = -\dfrac{1}{5}x^{(1/2)+1} = -\dfrac{1}{5}x^{3/2}$ □

Remark

Note in Example 2 that one characteristic of expressions considered to be in simplified form is the absence of negative exponents. Another characteristic of simplified expressions is that we usually prefer to write sums and differences in *factored form*. To do this, we can use the **distributive property:**

$$abx^n + ax^{n+m} = ax^n(b + x^m)$$

We demonstrate this procedure in the following example. Study this example carefully to ensure that you understand the concepts involved in the factoring process.

Example 3

Factoring Sums and Differences of Exponential Expressions

Simplify the following expressions by factoring out common terms.

(a) $2x^2 - x^3$ (b) $2x^3 + x^2$
(c) $2x^{1/2} + 4x^{5/2}$ (d) $2x^{-1/2} + 3x^{5/2}$
(e) $3(x+1)^{1/2}(2x-3)^{5/2} - 6(x+1)^{3/2}(2x-3)^{3/2}$
(f) $(x+1)^{-1/2}(2x-3)^{5/2} + 3(x+1)^{1/2}(2x-3)^{3/2}$

Solution

(a) $2x^2 - x^3 = x^2(2 - x)$

(b) $2x^3 + x^2 = x^2(2x + 1)$

(c) $2x^{1/2} + 4x^{5/2} = 2x^{1/2}(1 + 2x^2)$

(d) $2x^{-1/2} + 3x^{5/2} = x^{-1/2}(2 + 3x^3) = \dfrac{2 + 3x^3}{\sqrt{x}}$

(e) $3(x+1)^{1/2}(2x-3)^{5/2} - 6(x+1)^{3/2}(2x-3)^{3/2}$

$$= 3(x+1)^{1/2}(2x-3)^{3/2}[(2x-3) - 2(x+1)]$$
$$= 3(x+1)^{1/2}(2x-3)^{3/2}(2x-3-2x-2)$$
$$= 3(x+1)^{1/2}(2x-3)^{3/2}(-5)$$
$$= -15(x+1)^{1/2}(2x-3)^{3/2}$$

(f) $(x+1)^{-1/2}(2x-3)^{5/2} + 3(x+1)^{1/2}(2x-3)^{3/2}$

$$= (x+1)^{-1/2}(2x-3)^{3/2}[(2x-3) + 3(x+1)]$$
$$= (x+1)^{-1/2}(2x-3)^{3/2}(2x-3+3x+3)$$
$$= (x+1)^{-1/2}(2x-3)^{3/2}(5x)$$
$$= \frac{5x(2x-3)^{3/2}}{(x+1)^{1/2}}$$

□

Remark

Be sure you see that in Example 3 we subtracted exponents when factoring—even if the exponents were negative. For example, in part (d), we subtracted the exponent $-\frac{1}{2}$ as follows:

$$2x^{-1/2} + 3x^{5/2} = x^{-1/2}(2x^{\overbrace{0}^{-(1/2)-(-1/2)}} + 3x^{\overbrace{3}^{(5/2)-(-1/2)}}) = x^{-1/2}(2+3x^3) = \frac{2+3x^3}{\sqrt{x}}$$

Example 4

Factors Involving Quotients

Simplify the following expressions by factoring out common terms.

(a) $\dfrac{3x^2+x^4}{2x}$ (b) $\dfrac{\sqrt{x}+x^{3/2}}{x}$

(c) $\frac{3}{5}(x+1)^{5/3} + \frac{3}{4}(x+1)^{8/3}$

(d) $\dfrac{3(x+2)^2(x-1)^3 - 3(x+2)^3(x-1)^2}{[(x-1)^3]^2}$

Solution

(a) $\dfrac{3x^2+x^4}{2x} = \dfrac{x^2(3+x^2)}{2x} = \dfrac{x^{2-1}(3+x^2)}{2} = \dfrac{x(3+x^2)}{2}$

(b) $\dfrac{\sqrt{x}+x^{3/2}}{x} = \dfrac{x^{1/2}(1+x)}{x} = \dfrac{1+x}{x^{1-(1/2)}} = \dfrac{1+x}{\sqrt{x}}$

(c) $\frac{3}{5}(x+1)^{5/3} + \frac{3}{4}(x+1)^{8/3} = \frac{12}{20}(x+1)^{5/3} + \frac{15}{20}(x+1)^{8/3}$

$$= \tfrac{3}{20}(x+1)^{5/3}[4 + 5(x+1)]$$
$$= \tfrac{3}{20}(x+1)^{5/3}(4+5x+5)$$
$$= \tfrac{3}{20}(x+1)^{5/3}(5x+9)$$

(d) $$\frac{3(x+2)^2(x-1)^3 - 3(x+2)^3(x-1)^2}{[(x-1)^3]^2}$$

$$= \frac{3(x+2)^2(x-1)^2[(x-1)-(x+2)]}{(x-1)^6}$$

$$= \frac{3(x+2)^2(x-1-x-2)}{(x-1)^{6-2}}$$

$$= \frac{-9(x+2)^2}{(x-1)^4}$$ □

When working with radical expressions such as

$$\sqrt[n]{ax+b}$$

we face the potential difficulty of substituting a value of x for which the expression is not defined. For example, the expression

$$\sqrt{2x+3}$$

is *not defined* when $x = -2$ since

$$\sqrt{2(-2)+3} = \sqrt{-4+3} = \sqrt{-1} \neq \textit{real number}$$

In order that $\sqrt{2x+3}$ represent a real number, it is necessary that

$$2x + 3 \geq 0$$
$$2x \geq -3$$
$$x \geq -\tfrac{3}{2}$$

In other words, $\sqrt{2x+3}$ is defined only for those values of x that lie in the interval $[-\frac{3}{2}, \infty)$, as shown in Figure 0.15.

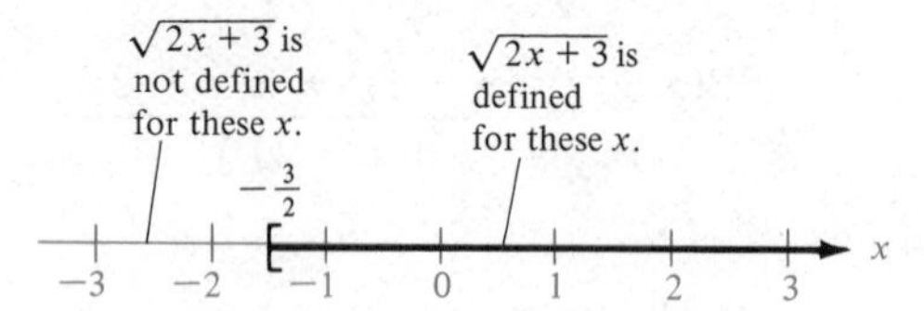

Figure 0.15

Example 5

Finding Intervals on Which Radical Expressions are Defined

Find the intervals on which the following expressions are defined.

(a) $\sqrt{3x-2}$ (b) $\dfrac{1}{\sqrt{3x-2}}$

(c) $\sqrt[3]{9x+1}$ (d) $\sqrt{2+x} + \sqrt{3-x}$

Solution

(a) $3x - 2 \geq 0$
$3x \geq 2$
$x \geq \frac{2}{3}$

(b) $3x - 2 > 0$
$3x > 2$
$x > \frac{2}{3}$

(c) $\sqrt[3]{9x + 1}$ is defined for *all* real numbers.

(d) $2 + x \geq 0$ $\quad$ $3 - x \geq 0$
$x \geq -2$ $\quad$ $3 \geq x$

Thus, the interval on which this sum is defined is $[-2, 3]$, as shown in Figure 0.16.

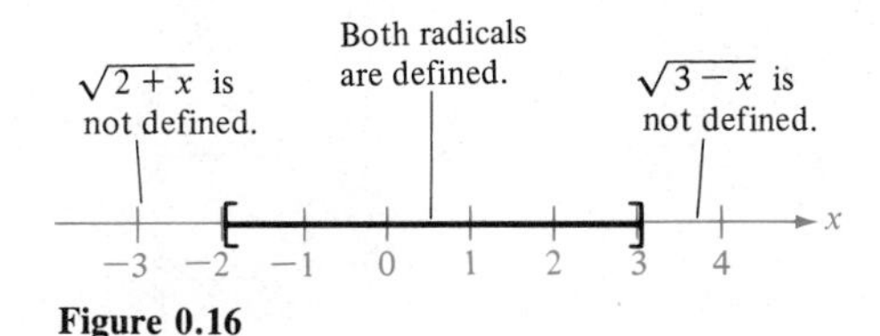

Figure 0.16

□

Note in Example 5 that the only difference between the intervals found in parts (a) and (b) is that $x = \frac{2}{3}$ is not a legitimate x-value in part (b) since it would make the denominator zero.

Section Exercises (0.3)

In Exercises 1–20, evaluate each expression for the given value of x.

1. $-3x^3$; $x = 2$
2. $\dfrac{x^2}{2}$; $x = 6$
3. $4x^{-3}$; $x = 2$
4. $7x^{-2}$; $x = 4$
5. $\dfrac{1 + x^{-1}}{x^{-1}}$; $x = 2$
6. $x - 4x^{-2}$; $x = 3$
7. $3x^2 - 4x^3$; $x = -2$
8. $5(-x)^3$; $x = 3$
9. $6x^0 - (6x)^0$; $x = 10$
10. $\dfrac{1}{(-x)^{-3}}$; $x = 4$
11. $\sqrt[3]{x^2}$; $x = 27$
12. $\sqrt{x^3}$; $x = \frac{1}{9}$
13. $x^{-1/2}$; $x = 4$
14. $x^{-3/4}$; $x = 16$
15. $x^{-2/5}$; $x = -32$
16. $(x^{2/3})^3$; $x = 10$
17. Ⓒ $500x^{60}$; $x = 1.01$
18. Ⓒ $\dfrac{10{,}000}{x^{120}}$; $x = 1.075$
19. Ⓒ $\sqrt[3]{x}$; $x = -154$
20. Ⓒ $\sqrt[6]{x}$; $x = 325$

In Exercises 21–34, simplify the given expression.

21. $5x^4(x^2)$
22. $(8x^4)(2x^3)$
23. $6y^2(2y^4)^2$
24. $z^{-3}(3z^4)$
25. $10(x^2)^2$
26. $(4x^3)^2$
27. $\dfrac{7x^2}{x^{-3}}$

Ⓒ Calculator may be helpful.

28. $\dfrac{r^4}{r^6}$

29. $\dfrac{12(x + y)^3}{9(x + y)}$

30. $\left(\dfrac{12s^2}{9s}\right)^3$

31. $\dfrac{3x\sqrt{x}}{x^{1/2}}$

32. $(\sqrt[3]{x^2})^3$

33. $\left(\dfrac{2^{1/2}x^{3/2}}{x^{1/2}}\right)^4$

34. $4x^{1/4} - 3(x^{-1/2})^{-1/2}$

In Exercises 35–50, factor the given expression and simplify your answer.

35. $\dfrac{y^4 - 3y^2}{y}$

36. $\dfrac{z^3 + 2z}{z}$

37. $4x^2 - 6x\sqrt{x}$

38. $x\sqrt{y + 1} + \sqrt{y + 1}$

39. $3x^{1/2} + 4x^{3/2}$

40. $5x^{1/3} - 4x^{4/3}$

41. $3x^{-1/2} + 4x^{3/2}$

42. $5x^{-1/3} - 4x^{5/3}$

43. $\frac{1}{2}x(x + 1)^{-1/2} + (x + 1)^{1/2}$

44. $\frac{3}{2}x(x + 1)^{1/2} + (x + 1)^{3/2}$

45. $\frac{2}{3}x^2(x^2 + 1)^{-2/3} + (x^2 + 1)^{1/3}$

46. $\dfrac{4x(x + 3)^{1/3}}{3} + (x + 3)^{4/3}$

47. $\dfrac{\frac{x}{2}(x - 1)^{-1/2} - (x - 1)^{1/2}}{x^2}$

48. $\dfrac{2(x - 1)^3(x + 2) - 3(x + 2)^2(x - 1)^2}{[(x - 1)^3]^2}$

49. $\dfrac{4x(x - 1)(x^2 + 1) - (x^2 + 1)}{(x + 1)^2}$

50. $\dfrac{12x(x + 5)^{1/2} - 3x^2(x + 5)^{-1/2}}{x + 5}$

In Exercises 51–60, find the interval on which the given expression is defined.

51. $\sqrt{x - 1}$

52. $\sqrt{5 - 2x}$

53. $\sqrt{x^2 + 3}$

54. $\sqrt[5]{1 - x}$

55. $\dfrac{1}{\sqrt[3]{x - 1}}$

56. $\dfrac{1}{\sqrt{x + 4}}$

57. $\dfrac{1}{\sqrt[4]{2x - 6}}$

58. $\dfrac{\sqrt{x - 1}}{x - 4}$

59. $\sqrt{x - 1} + \sqrt{5 - x}$

60. $\dfrac{1}{\sqrt{2x + 3}} + \sqrt{6 - 4x}$

Section 0.4

Factoring Polynomials

Section Objective: ***To review the common factorization techniques for finding the roots of polynomial equations.***

Roots and Factors

The Fundamental Theorem of Algebra states that every nth-degree polynomial equation

$$a_n x^n + a_{n-1}x^{n-1} + \cdots + a_1 x + a_0 = 0$$

has precisely n roots.* As important as this theorem is, it is of little computational value since it does not tell us how to find these n roots. The problem of finding the roots to a polynomial equation is equivalent to the problem of factoring the polynomial into linear factors. For example, the factorization

$$x^3 - 3x + 2 = (x - 1)(x - 1)(x + 2) = 0$$

implies that $x = 1$ is a (repeated) root and $x = -2$ is a root. We can check this by substitution. That is, if we let $x = 1$, then

$$(1)^3 - 3(1) + 2 = 1 - 3 + 2 = 0$$

and if we let $x = -2$, then

$$(-2)^3 - 3(-2) + 2 = -8 + 6 + 2 = 0$$

Polynomial Equation		Linear Factors		Real Roots	
$x^3 - 3x + 2 = 0$		$(x - 1)(x - 1)(x + 2) = 0$		$x - 1 = 0$	$x + 2 = 0$
				$x = 1$	$x = -2$

Similarly, the factorization

$$x^5 - x = x(x - 1)(x + 1)(x^2 + 1)$$

implies that $x = 0$, 1, and -1 are the three real roots. [The factor $(x^2 + 1)$ yields the two imaginary roots $x = \pm\sqrt{-1}$. However, we will not be concerned with such imaginary roots in this text.]

* These roots may be repeated or imaginary.

Factorization is *not* a simple problem, and there are a wide variety of techniques that we use to factor polynomials. Many of these techniques depend upon recognition of special factors, and we suggest that you study the following summary carefully.

Summary of Factorization Techniques

Quadratic Formula	*Example*
$ax^2 + bx + c = 0 \Rightarrow x = \dfrac{-b \pm \sqrt{b^2 - 4ac}}{2a}$	$x^2 + 3x - 1 = 0 \Rightarrow x = \dfrac{-3 \pm \sqrt{13}}{2}$

Special Factors	*Examples*
$x^2 - a^2 = (x - a)(x + a)$	$x^2 - 9 = (x - 3)(x + 3)$
$x^3 - a^3 = (x - a)(x^2 + ax + a^2)$	$x^3 - 8 = (x - 2)(x^2 + 2x + 4)$
$x^3 + a^3 = (x + a)(x^2 - ax + a^2)$	$x^3 + 64 = (x + 4)(x^2 - 4x + 16)$
$x^4 - a^4 = (x - a)(x + a)(x^2 + a^2)$	$x^4 - 16 = (x - 2)(x + 2)(x^2 + 4)$

Binomial Theorem	*Examples*
$(x + a)^2 = x^2 + 2ax + a^2$	$(x + 3)^2 = x^2 + 6x + 9$
$(x - a)^2 = x^2 - 2ax + a^2$	$(x^2 - 5)^2 = x^4 - 10x^2 + 25$
$(x + a)^3 = x^3 + 3ax^2 + 3a^2x + a^3$	$(x + 2)^3 = x^3 + 6x^2 + 12x + 8$
$(x - a)^3 = x^3 - 3ax^2 + 3a^2x - a^3$	$(x - 1)^3 = x^3 - 3x^2 + 3x - 1$
$(x + a)^4 = x^4 + 4ax^3 + 6a^2x^2 + 4a^3x + a^4$	$(x + 2)^4 = x^4 + 8x^3 + 24x^2 + 32x + 16$
$(x - a)^4 = x^4 - 4ax^3 + 6a^2x^2 - 4a^3x + a^4$	$(x - 4)^4 = x^4 - 16x^3 + 96x^2 - 256x + 256$
$(x + a)^n = x^n + nax^{n-1} + \dfrac{n(n-1)}{2!}a^2x^{n-2} + \cdots + na^{n-1}x + a^n$	$(x + 1)^5 = x^5 + 5x^4 + 10x^3 + 10x^2 + 5x + 1$
$(x - a)^n = x^n - nax^{n-1} + \dfrac{n(n-1)}{2!}a^2x^{n-2} - \cdots \pm na^{n-1}x \mp a^n$	$(x - 1)^6 = x^6 - 6x^5 + 15x^4 - 20x^3 + 15x^2 - 6x + 1$

Rational Root Theorem	*Example*
If $a_n x^n + a_{n-1}x^{n-1} + \cdots + a_1 x + a_0 = 0$ has integer coefficients, then every *rational* root is of the form $x = r/s$, where r is a factor of a_0 and s is a factor of a_n.	If $2x^4 - 7x^3 + 5x^2 - 7x + 3 = 0$, then the only possible *rational* roots are $x = \pm 1, \pm 1/2, \pm 3$, and $\pm 3/2$. By testing, we find the two rational roots to be 1/2 and 3.
Factoring by Grouping	*Example*
$acx^3 + adx^2 + bcx + bd$ $= ax^2(cx + d) + b(cx + d)$ $= (ax^2 + b)(cx + d)$	$3x^3 - 2x^2 - 6x + 4$ $= x^2(3x - 2) - 2(3x - 2)$ $= (x^2 - 2)(3x - 2)$

Note that the quadratic formula gives *two* roots (if $b^2 - 4ac \neq 0$), which (when the roots are real) can be viewed as the endpoints of an interval. (See Figure 0.17.) The formula gives the midpoint of this interval as well as the distance to the endpoints.

When using the quadratic formula it is important to realize that the term inside the square root (the *discriminant*) can be used to classify the roots into the following three cases:

1. two (distinct) real roots: $b^2 - 4ac > 0$
2. one (repeated) real root: $b^2 - 4ac = 0$
3. two imaginary roots: $b^2 - 4ac < 0$

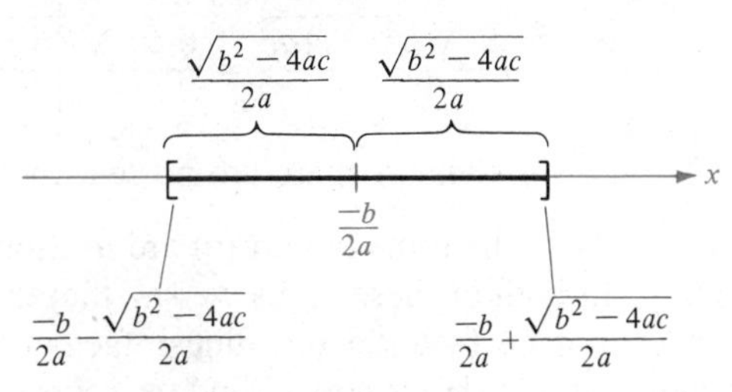

Figure 0.17

Example 1

Applying the Quadratic Formula

Use the quadratic formula to find all real roots of the following:

(a) $4x^2 + 6x + 1 = 0$
(b) $x^2 + 6x + 9 = 0$
(c) $2x^2 - 6x + 5 = 0$

Solution

(a) $a = 4,\ b = 6,\ c = 1$

$$x = \frac{-b \pm \sqrt{b^2 - 4ac}}{2a} = \frac{-6 \pm \sqrt{36 - 16}}{8}$$
$$= \frac{-6 \pm \sqrt{20}}{8}$$
$$= \frac{-6 \pm 2\sqrt{5}}{8}$$
$$= \frac{\cancel{2}(-3 \pm \sqrt{5})}{\cancel{2}(4)}$$
$$= \frac{-3 \pm \sqrt{5}}{4}$$

Thus, there are two real roots:

$$x = \frac{-3 - \sqrt{5}}{4} \approx -1.309 \qquad \text{and} \qquad x = \frac{-3 + \sqrt{5}}{4} \approx -0.191$$

(b) $a = 1, b = 6, c = 9$

$$x = \frac{-b \pm \sqrt{b^2 - 4ac}}{2a} = \frac{-6 \pm \sqrt{36 - 36}}{2} = -\frac{6}{2} = -3$$

Thus, there is one real root: $x = -3$.

(c) $a = 2, b = -6, c = 5$

$$x = \frac{-b \pm \sqrt{b^2 - 4ac}}{2a} = \frac{6 \pm \sqrt{36 - 40}}{4} = \frac{6 \pm \sqrt{-4}}{4}$$

Since $\sqrt{-4}$ is imaginary, there are no real roots. □

In Example 1, the roots in part (a) are irrational, and the roots in part (c) are imaginary. In both of these cases we say that the given quadratic is *irreducible* since it cannot be factored into linear factors with rational coefficients. The following example shows how to find the roots of *reducible* quadratic equations. Remember that we can always use the quadratic formula to find the roots of a second-degree polynomial equation. However, as we become more familiar with some of the simpler cases, we soon learn to recognize those equations whose roots are simple rational numbers. In the following example, we use factoring to find the roots of each quadratic, and we suggest that you try using the quadratic formula to obtain the same roots.

Example 2

Factoring Quadratics

Find the roots of the following quadratic equations by factoring.

(a) $x^2 - 5x + 6 = 0$
(b) $x^2 - 5x - 6 = 0$
(c) $2x^2 + 5x - 3 = 0$

Solution

(a) $x^2 - 5x + 6 = (x - 2)(x - 3) = 0$,
which implies that the roots are $x = 2$ and $x = 3$.
(b) $x^2 - 5x - 6 = (x + 1)(x - 6) = 0$,
which implies that the roots are $x = -1$ and $x = 6$.
(c) $2x^2 + 5x - 3 = (2x - 1)(x + 3) = 0$,
which implies that the roots are $x = \frac{1}{2}$ and $x = -3$. □

In this text, we will encounter several equations which have expressions of the form

$$\sqrt{ax^2 + bx + c}$$

In such cases, we are interested in finding the interval (or intervals) on the real line for which this expression is defined. To do this, it is helpful to first find the roots of the quadratic and then test its sign at points inside and outside the interval bounded by the roots. This procedure is demonstrated in the following example.

Example 3

Finding Intervals on Which Square Roots are Defined

Find the interval (or intervals) on which the following are defined:

(a) $\sqrt{x^2 - 3x + 2}$
(b) $\sqrt{-x^2 + 3x + 4}$
(c) $\sqrt{x^2 - 2x + 2}$

Solution

(a) Since

$$x^2 - 3x + 2 = (x - 1)(x - 2)$$

we know that the roots are $x = 1$ and $x = 2$, and we must test points inside and outside the interval $[1, 2]$, as shown in Figure 0.18. When $x = \frac{3}{2}$, we have

$$\left(\frac{3}{2}\right)^2 - 3\left(\frac{3}{2}\right) + 2 = \frac{9}{4} - \frac{9}{2} + 2 = \frac{9 - 18 + 8}{4} = -\frac{1}{4}$$

which implies that the quadratic is negative inside the interval and the radical is undefined there. When $x = 0$ (or when $x = 3$), the quadratic is positive,

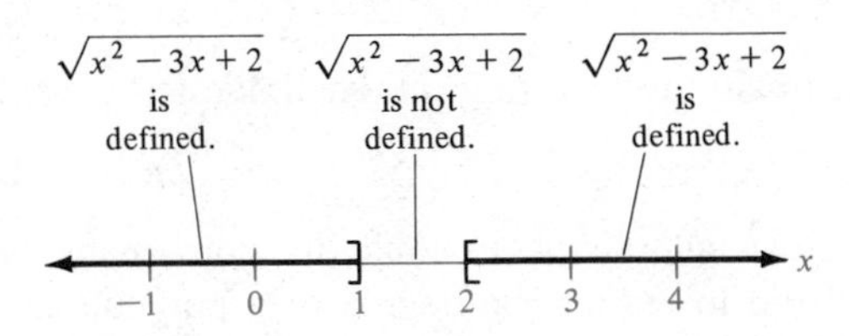

Figure 0.18

and we conclude that the radical is defined outside the interval. Thus, $\sqrt{x^2 - 3x + 2}$ is defined only on the intervals

$$(-\infty, 1] \qquad \text{and} \qquad [2, \infty)$$

(b) Since

$$-x^2 + 3x + 4 = (-x + 4)(x + 1)$$

we know that the roots are $x = -1$ and $x = 4$. By checking points inside and outside the interval $[-1, 4]$, we see that the radical is defined only inside this interval, as shown in Figure 0.19.

(c) Since $x^2 - 2x + 2$ has no real roots, we observe that the quadratic is positive when $x = 0$ and may conclude that the radical is defined for all values of x. (See Figure 0.20.)

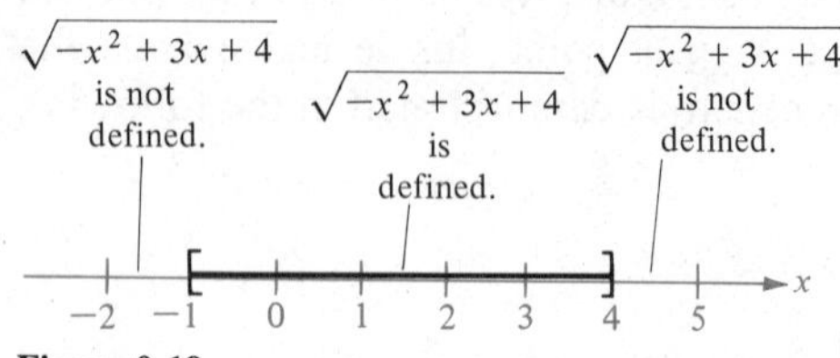

Figure 0.19

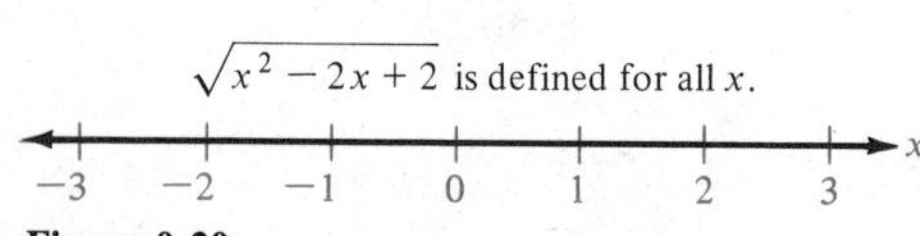

Figure 0.20

□

It is usually very difficult to find the roots of polynomials of degree three or greater unless some of the roots happen to be rational. If one or more of the roots are rational, then we can use the rational root theorem to find one root and use that root to reduce the degree of the polynomial. For example, if we know that $x = 2$ is a root of $x^3 - 4x^2 + 5x - 2$, then we know that $(x - 2)$ is a factor, and we can use long division to reduce the polynomial as follows:

$$\begin{array}{r}
x^2 - 2x + 1 \\
x - 2\overline{)\,x^3 - 4x^2 + 5x - 2} \\
\underline{x^3 - 2x^2} \\
-2x^2 + 5x \\
\underline{-2x^2 + 4x} \\
x - 2 \\
\underline{x - 2}
\end{array}$$

Thus, we know that

$$x^3 - 4x^2 + 5x - 2 = (x - 2)(x^2 - 2x + 1)$$

and finally by factoring the quadratic term, we have

$$x^3 - 4x^2 + 5x - 2 = (x - 2)(x - 1)(x - 1)$$

As an alternative to long division, many people prefer to use **synthetic division** to reduce the degree of a polynomial. We outline this procedure as follows.

Synthetic Division (for a cubic polynomial)

Given: $x = x_1$ is a root of $ax^3 + bx^2 + cx + d$

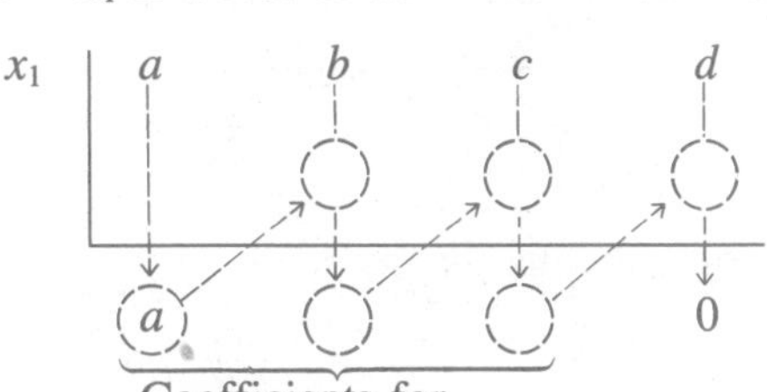

Coefficients for quadratic factor

Vertical pattern: *Add terms.*

Diagonal pattern: *Multiply by the given root.*

For example, to use synthetic division on

$$x^3 - 4x^2 + 5x - 2$$

using the given root, $x = 2$, we would have

$$\begin{array}{r|rrrr} 2 & 1 & -4 & 5 & -2 \\ & & 2 & -4 & 2 \\ \hline & 1 & -2 & 1 & 0 \end{array}$$

$$(x - 2)(x^2 - 2x + 1) = x^3 - 4x^2 + 5x - 2$$

Example 4

Using the Rational Root Theorem

Find all real roots of the following:

(a) $2x^3 + 3x^2 - 8x + 3$
(b) $x^3 - 2x^2 - 2x - 3$

Solution

(a) $2x^3 + 3x^2 - 8x + 3$

Factors of constant term: $\pm 1, \pm 3$

Factors of leading coefficient: $\pm 1, \pm 2$

The possible rational roots are the factors of the constant term divided by the factors of the leading coefficient:

$$1, \quad -1, \quad 3, \quad -3, \quad \tfrac{1}{2}, \quad -\tfrac{1}{2}, \quad \tfrac{3}{2}, \quad -\tfrac{3}{2}$$

By testing these potential roots, we see that $x = 1$ works since

$$2(1)^3 + 3(1)^2 - 8(1) + 3 = 2 + 3 - 8 + 3 = 0$$

Now, by synthetic division, we have

$$\begin{array}{r|rrrr} 1 & 2 & 3 & -8 & 3 \\ & & 2 & 5 & -3 \\ \hline & 2 & 5 & -3 & 0 \end{array}$$

Finally, by factoring the quadratic

$$2x^2 + 5x - 3 = (2x - 1)(x + 3)$$

we have

$$2x^3 + 3x^2 - 8x + 3 = (x - 1)(2x - 1)(x + 3)$$

and we conclude that the roots are

$$x = 1, \qquad x = \tfrac{1}{2}, \qquad \text{and} \qquad x = -3$$

(b) The potential rational roots of

$$x^3 - 2x^2 - 2x - 3$$

are ± 1 and ± 3. By testing these, we see that $x = 3$ works since

$$(3)^3 - 2(3)^2 - 2(3) - 3 = 27 - 18 - 6 - 3 = 0$$

By synthetic division, we have

$$\begin{array}{r|rrrr} 3 & 1 & -2 & -2 & -3 \\ & & 3 & 3 & 3 \\ \hline & 1 & 1 & 1 & 0 \end{array}$$

Finally, by the quadratic formula we can determine that

$$x^2 + x + 1$$

has no real roots, and we conclude that the only real root is $x = 3$. □

Remark Before concluding this section, we should point out that when you use synthetic division you must remember to take *all* coefficients into account — *even if some of them are zero*. For instance, if we know that $x = -2$ is a root of $x^3 + 3x + 14$, we would apply synthetic division as follows:

$$\begin{array}{r|rrrr} -2 & 1 & \boxed{0} & 3 & 14 \\ & & -2 & 4 & -14 \\ \hline & 1 & -2 & 7 & 0 \end{array}$$

Section Exercises (0.4)

In Exercises 1–14, use the quadratic formula to find all real roots of the given equation.

1. $6x^2 - x - 1 = 0$
2. $8x^2 - 2x - 1 = 0$
3. $4x^2 - 12x - 9 = 0$
4. $9x^2 + 12x + 4 = 0$
5. $y^2 + 4y + 1 = 0$
6. $x^2 + 6x - 1 = 0$
7. $3x^2 - 2x - 2 = 0$
8. $2s^2 - 7s + 3 = 0$
9. $2s^2 - 7s + 4 = 0$
10. $2s^2 - 7s + 5 = 0$
11. $x^2 - 2x + 3 = 0$
12. $x^2 + 1 = x(3 - x)$
13. $x + 1 = \dfrac{3}{x}$
14. $(z + 1)^2 = 2z^2$

In Exercises 15–30, find the roots of the given equation by factoring.

15. $x^2 + x - 2 = 0$
16. $x^2 + 5x + 6 = 0$
17. $x^2 - 5x + 6 = 0$
18. $x^2 + x - 20 = 0$
19. $2x^2 - x - 1 = 0$
20. $3x^2 - 5x + 2 = 0$
21. $x^2 - 5x = 0$
22. $2x^2 - 3x = 0$
23. $x^2 - 9 = 0$
24. $x^2 - 25 = 0$
25. $x^2 - 3 = 0$
26. $(x + 1)^2 - 5 = 0$
27. $(x - 3)^2 - 8 = 0$
28. $(x - 1)(x + 2) = 4$
29. $(x - 5)(x + 3) = 33$
30. $x + 1 = \dfrac{2}{x}$

In Exercises 31–38, find the interval (or intervals) on which the given expression is defined.

31. $\sqrt{x^2 - 7x + 12}$
32. $\sqrt{x^2 - 4}$
33. $\sqrt{4 - x^2}$
34. $\sqrt{144 - 9x^2}$
35. $\sqrt{12 - x - x^2}$
36. $\sqrt{x^2 + 4}$
37. $\sqrt{x^2 - 3x + 3}$
38. $\sqrt{-x^2 + 2x - 2}$

In Exercises 39–44, use synthetic division to complete the indicated factorization.

39. $x^3 + 8 = (x + 2)(\quad)$
40. $x^3 - 2x^2 - x + 2 = (x + 1)(\quad)$
41. $2x^3 - x^2 - 2x + 1 = (x - 1)(\quad)$
42. $x^4 - 16x^3 + 96x^2 - 256x + 256 = (x - 4)(\quad)$
43. $x^4 + 2x^3 - 6x^2 - 18x - 27 = (x - 3)(\quad)$
44. $x^5 - 243 = (x - 3)(\quad)$

In Exercises 45–56, use the Rational Root Theorem as an aid in finding all real roots of the given equation.

45. $x^3 - x^2 - x + 1 = 0$
46. $x^3 - x^2 - 4x + 4 = 0$
47. $x^3 - 6x^2 + 11x - 6 = 0$
48. $x^3 + 2x^2 - 5x - 6 = 0$
49. $4x^3 - 4x^2 - x + 1 = 0$
50. $18x^3 - 9x^2 - 8x + 4 = 0$
51. $x^3 - 3x^2 - 3x - 4 = 0$
52. $4x^3 - 6x^2 + 2x + 3 = 0$
53. $z^3 + 8z^2 + 11z - 2 = 0$
54. $3y^3 + 11y^2 - y - 1 = 0$
55. $x^4 - 13x^2 + 36 = 0$
56. $x^4 - x^3 - 7x^2 + x + 6 = 0$

In Exercises 57–60, find all real roots of the given equation by factoring by grouping.

57. $x^3 - x^2 - 4x + 4 = 0$
58. $2x^3 + x^2 + 6x + 3 = 0$
59. $(x + 2)^2(x - 1) + (x + 2)(x - 1)^2 = 0$
60. $(x + 5)(x - 2) + (x + 5)^2 = 0$

Section 0.5

Fractions and Rationalization

Section Objectives: ***To review algebraic procedures for combining and simplifying quotients ▪ To review rationalization techniques.***

In our final section of this review chapter, we look at some examples involving fractional expressions. The mathematical basis for this section is primarily the arithmetic of fractions. We summarize these rules as follows.

Arithmetic of Fractions

Adding Fractions (find the common denominator)

$$\frac{a}{b} + \frac{c}{d} = \frac{a}{b}\left(\frac{d}{d}\right) + \frac{c}{d}\left(\frac{b}{b}\right) = \frac{ad}{bd} + \frac{bc}{bd} = \frac{ad + bc}{bd}$$

Subtracting Fractions (find the common denominator)

$$\frac{a}{b} - \frac{c}{d} = \frac{a}{b}\left(\frac{d}{d}\right) - \frac{c}{d}\left(\frac{b}{b}\right) = \frac{ad}{bd} - \frac{bc}{bd} = \frac{ad - bc}{bd}$$

Multiplying Fractions

$$\left(\frac{a}{b}\right)\left(\frac{c}{d}\right) = \frac{ac}{bd}$$

Dividing Fractions (invert and multiply)

$$\frac{\frac{a}{b}}{\frac{c}{d}} = \left(\frac{a}{b}\right)\left(\frac{d}{c}\right) = \frac{ad}{bc} \qquad \frac{\frac{a}{b}}{c} = \frac{\frac{a}{b}}{\frac{c}{1}} = \left(\frac{a}{b}\right)\left(\frac{1}{c}\right) = \frac{a}{bc}$$

Cancellation

$$\frac{\cancel{a}b}{\cancel{a}c} = \frac{b}{c} \qquad \frac{ab + ac}{ad} = \frac{\cancel{a}(b + c)}{\cancel{a}d} = \frac{b + c}{d}$$

Example 1

Combining Fractions

Combine the following fractions and simplify.

(a) $x + \dfrac{1}{x}$ (b) $\dfrac{1}{x + 1} - \dfrac{2}{2x - 1}$

(c) $\dfrac{A}{x+2} + \dfrac{B}{x-3} + \dfrac{C}{x+4}$ (d) $\dfrac{A}{x+2} + \dfrac{B}{(x+2)^2} + \dfrac{C}{x-1}$

(e) $\dfrac{x}{x^2-1} + \dfrac{3}{x+1}$ (f) $\dfrac{1}{2(x^2+2x)} - \dfrac{1}{4x}$

Solution

(a) $x + \dfrac{1}{x} = \dfrac{x^2}{x} + \dfrac{1}{x} = \dfrac{x^2+1}{x}$

(b)
$$\frac{1}{x+1} - \frac{2}{2x-1} = \frac{(2x-1)}{(x+1)(2x-1)} - \frac{2(x+1)}{(x+1)(2x-1)}$$
$$= \frac{2x-1-2x-2}{2x^2+x-1}$$
$$= \frac{-3}{2x^2+x-1}$$

(c)
$$\frac{A}{x+2} + \frac{B}{x-3} + \frac{C}{x+4}$$
$$= \frac{A(x-3)(x+4)}{(x+2)(x-3)(x+4)} + \frac{B(x+2)(x+4)}{(x+2)(x-3)(x+4)} + \frac{C(x+2)(x-3)}{(x+2)(x-3)(x+4)}$$
$$= \frac{A(x-3)(x+4) + B(x+2)(x+4) + C(x+2)(x-3)}{(x+2)(x-3)(x+4)}$$
$$= \frac{A(x^2+x-12) + B(x^2+6x+8) + C(x^2-x-6)}{(x+2)(x-3)(x+4)}$$
$$= \frac{Ax^2 + Bx^2 + Cx^2 + Ax + 6Bx - Cx - 12A + 8B - 6C}{(x+2)(x-3)(x+4)}$$
$$= \frac{(A+B+C)x^2 + (A+6B-C)x + (-12A+8B-6C)}{(x+2)(x-3)(x+4)}$$

(d)
$$\frac{A}{x+2} + \frac{B}{(x+2)^2} + \frac{C}{x-1}$$
$$= \frac{A(x+2)(x-1)}{(x+2)^2(x-1)} + \frac{B(x-1)}{(x+2)^2(x-1)} + \frac{C(x+2)^2}{(x+2)^2(x-1)}$$
$$= \frac{A(x+2)(x-1) + B(x-1) + C(x+2)^2}{(x+2)^2(x-1)}$$

$$\frac{A}{x+2}+\frac{B}{(x+2)^2}+\frac{C}{x-1} = \frac{A(x^2+x-2)+B(x-1)+C(x^2+4x+4)}{(x+2)^2(x-1)}$$

$$= \frac{Ax^2+Cx^2+Ax+Bx+4Cx-2A-B+4C}{(x+2)^2(x-1)}$$

$$= \frac{(A+C)x^2+(A+B+4C)x+(-2A-B+4C)}{(x+2)^2(x-1)}$$

(e) $$\frac{x}{x^2-1}+\frac{3}{x+1} = \frac{x}{(x-1)(x+1)}+\frac{3}{x+1}$$

$$= \frac{x}{(x-1)(x+1)}+\frac{3(x-1)}{(x-1)(x+1)}$$

$$= \frac{x+3x-3}{(x-1)(x+1)}$$

$$= \frac{4x-3}{x^2-1}$$

(f) $$\frac{1}{2(x^2+2x)}-\frac{1}{4x} = \frac{1}{2x(x+2)}-\frac{1}{2(2x)}$$

$$= \frac{2}{2(2x)(x+2)}-\frac{x+2}{2(2x)(x+2)}$$

$$= \frac{2-x-2}{2(2x)(x+2)}$$

$$= \frac{-\cancel{x}}{2(2\cancel{x})(x+2)}$$

$$= \frac{-1}{4(x+2)}$$ □

Remark When adding (or subtracting) fractions whose denominators have no common factors, you may find it convenient to use the following "cross multiplication" shortcut:

$$\frac{a}{b}+\frac{c}{d} = \frac{a\bigcirc}{\textcircled{b}} \overset{\nwarrow\nearrow}{+} \frac{c\bigcirc}{\textcircled{d}} = \frac{ad+bc}{bd}$$

For instance, in part (b) of Example 1 we could have used this shortcut as follows:

$$\frac{1}{x+1}-\frac{2}{2x-1} = \frac{2x-1-2(x+1)}{(x+1)(2x-1)} = \frac{2x-1-2x+2}{(x+1)(2x-1)}$$

$$= \frac{1}{(x+1)(2x-1)}$$

Example 2

Combining Fractions Involving Radicals

Combine the following fractions and simplify.

(a) $$\frac{\sqrt{x+1} - \dfrac{x}{2\sqrt{x+1}}}{x+1}$$

(b) $$\left(\frac{1}{x+\sqrt{4+x^2}}\right)\left(\frac{2x}{2\sqrt{4+x^2}}+1\right)$$

(c) $$\left(\frac{-x\left[\dfrac{2x}{2\sqrt{x^2+1}}\right]+\sqrt{x^2+1}}{x^2}\right)+\left(\frac{1}{x+\sqrt{x^2+1}}\right)\left(\frac{2x}{2\sqrt{x^2+1}}+1\right)$$

Solution

(a) $$\frac{\sqrt{x+1} - \dfrac{x}{2\sqrt{x+1}}}{x+1} = \frac{\dfrac{2(x+1)}{2\sqrt{x+1}} - \dfrac{x}{2\sqrt{x+1}}}{x+1}$$

$$= \frac{\dfrac{2x+2-x}{2\sqrt{x+1}}}{\dfrac{x+1}{1}}$$

$$= \frac{x+2}{2\sqrt{x+1}}\left(\frac{1}{x+1}\right)$$

$$= \frac{x+2}{2(x+1)^{3/2}}$$

(b) $$\left(\frac{1}{x+\sqrt{4+x^2}}\right)\left(\frac{\cancel{2}x}{\cancel{2}\sqrt{4+x^2}}+1\right)$$

$$= \left(\frac{1}{x+\sqrt{4+x^2}}\right)\left(\frac{x}{\sqrt{4+x^2}}+1\right)$$

$$= \left(\frac{1}{x+\sqrt{4+x^2}}\right)\left(\frac{\sqrt{4+x^2}}{\sqrt{4+x^2}}+\frac{x}{\sqrt{4+x^2}}\right)$$

$$= \left(\frac{1}{\cancel{x+\sqrt{4+x^2}}}\right)\left(\frac{\cancel{x+\sqrt{4+x^2}}}{\sqrt{4+x^2}}\right)$$

$$= \frac{1}{\sqrt{4+x^2}}$$

(c)
$$\left(\frac{-x\left[\dfrac{\not{2}x}{\not{2}\sqrt{x^2+1}}\right]+\sqrt{x^2+1}}{x^2}\right)+\left(\frac{1}{x+\sqrt{x^2+1}}\right)\left(\frac{\not{2}x}{\not{2}\sqrt{x^2+1}}+1\right)$$

$$=\left(\frac{\dfrac{-x^2}{\sqrt{x^2+1}}+\sqrt{x^2+1}}{x^2}\right)+\left(\frac{1}{x+\sqrt{x^2+1}}\right)\left(\frac{x}{\sqrt{x^2+1}}+1\right)$$

$$=\left(\frac{\dfrac{-x^2}{\sqrt{x^2+1}}+\dfrac{x^2+1}{\sqrt{x^2+1}}}{x^2}\right)+\left(\frac{1}{x+\sqrt{x^2+1}}\right)\left(\frac{x}{\sqrt{x^2+1}}+\frac{\sqrt{x^2+1}}{\sqrt{x^2+1}}\right)$$

$$=\frac{\dfrac{1}{\sqrt{x^2+1}}}{x^2}+\left(\frac{1}{\cancel{x+\sqrt{x^2+1}}}\right)\left(\frac{\cancel{x+\sqrt{x^2+1}}}{\sqrt{x^2+1}}\right)$$

$$=\frac{1}{x^2\sqrt{x^2+1}}+\frac{1}{\sqrt{x^2+1}}$$

$$=\frac{1}{x^2\sqrt{x^2+1}}+\frac{x^2}{x^2\sqrt{x^2+1}}$$

$$=\frac{x^2+1}{x^2\sqrt{x^2+1}}$$

$$=\frac{\sqrt{x^2+1}}{x^2}$$

□

When working with quotients involving radicals, we often find it convenient to move the radical expression from the denominator to the numerator, or vice versa. For example, we can move $\sqrt{2}$ from the denominator to the numerator in the following quotient by multiplying by $(\sqrt{2}/\sqrt{2})$.

Radical in Denominator		*Rationalize the Denominator*		*Radical in Numerator*
$\dfrac{1}{\sqrt{2}}$	➧	$\dfrac{1}{\sqrt{2}}\left(\dfrac{\sqrt{2}}{\sqrt{2}}\right)$	➧	$\dfrac{\sqrt{2}}{2}$

We call this process **rationalizing the denominator.** (If the radical is moved from the numerator to the denominator, we call the process **rationalizing the numerator.**) We summarize the three principal techniques for rationalization as follows.

Rationalization Techniques

1. To rationalize $\sqrt{a}$, multiply by $\dfrac{\sqrt{a}}{\sqrt{a}}$.

2. To rationalize $\sqrt{a} - \sqrt{b}$, multiply by $\dfrac{\sqrt{a} + \sqrt{b}}{\sqrt{a} + \sqrt{b}}$.

3. To rationalize $\sqrt{a} + \sqrt{b}$, multiply by $\dfrac{\sqrt{a} - \sqrt{b}}{\sqrt{a} - \sqrt{b}}$.

Note that the success of the second and third techniques depends upon the following elimination of radicals:

$$\begin{aligned}(\sqrt{a} - \sqrt{b})(\sqrt{a} + \sqrt{b}) &= (\sqrt{a})^2 - \sqrt{b}\sqrt{a} + \sqrt{b}\sqrt{a} - (\sqrt{b})^2 \\ &= a - b\end{aligned}$$

Example 3

Rationalizing Single-Term Denominators and Numerators

(a) Rationalize the denominator in $\dfrac{3}{\sqrt{12}}$

(b) Rationalize the numerator in $\dfrac{\sqrt{x+1}}{2}$

Solution

(a) $\dfrac{3}{\sqrt{12}} = \dfrac{3}{2\sqrt{3}} = \dfrac{3}{2\sqrt{3}}\left(\dfrac{\sqrt{3}}{\sqrt{3}}\right) = \dfrac{\cancel{3}\sqrt{3}}{2(\cancel{3})} = \dfrac{\sqrt{3}}{2}$

(b) $\dfrac{\sqrt{x+1}}{2} = \dfrac{\sqrt{x+1}}{2}\left(\dfrac{\sqrt{x+1}}{\sqrt{x+1}}\right) = \dfrac{x+1}{2\sqrt{x+1}}$ □

Example 4

Rationalizing Double-Term Denominators and Numerators

Rationalize the numerator or denominator as indicated.

(a) $\dfrac{1}{\sqrt{5} + \sqrt{2}}$

(b) $\dfrac{1}{\sqrt{x} - \sqrt{x+1}}$

(c) $\dfrac{x - \sqrt{x^2 - x}}{3x}$

Solution

(a) $$\frac{1}{\sqrt{5}+\sqrt{2}} = \frac{1}{\sqrt{5}+\sqrt{2}}\left(\frac{\sqrt{5}-\sqrt{2}}{\sqrt{5}-\sqrt{2}}\right) = \frac{\sqrt{5}-\sqrt{2}}{5-2} = \frac{\sqrt{5}-\sqrt{2}}{3}$$

(b) $$\frac{1}{\sqrt{x}-\sqrt{x+1}} = \frac{1}{\sqrt{x}-\sqrt{x+1}}\left(\frac{\sqrt{x}+\sqrt{x+1}}{\sqrt{x}+\sqrt{x+1}}\right) = \frac{\sqrt{x}+\sqrt{x+1}}{x-(x+1)} = -(\sqrt{x}+\sqrt{x+1})$$

(c) $$\frac{x-\sqrt{x^2-x}}{3x} = \frac{x-\sqrt{x^2-x}}{3x}\left(\frac{x+\sqrt{x^2-x}}{x+\sqrt{x^2-x}}\right) = \frac{x^2-(x^2-x)}{3x(x+\sqrt{x^2-x})} = \frac{\cancel{x}}{3\cancel{x}(x+\sqrt{x^2-x})} = \frac{1}{3(x+\sqrt{x^2-x})}$$

□

Section Exercises (0.5)

In Exercises 1–30, perform the indicated operations and simplify your answer.

1. $\dfrac{5}{x-1}+\dfrac{x}{x-1}$

2. $\dfrac{2x-1}{x+3}+\dfrac{1-x}{x+3}$

3. $\dfrac{2x}{x^2+2}-\dfrac{1-3x}{x^2+2}$

4. $\dfrac{5x+10}{2x-1}-\dfrac{2x+10}{2x-1}$

5. $\dfrac{4}{x}-\dfrac{3}{x^2}$

6. $\dfrac{5}{x-1}+\dfrac{3}{x}$

7. $\dfrac{2}{x+2}-\dfrac{1}{x-2}$

8. $\dfrac{x}{x^2+x-2}-\dfrac{1}{x+2}$

9. $\dfrac{5}{x-3}+\dfrac{3}{3-x}$

10. $\dfrac{x}{2-x}+\dfrac{2}{x-2}$

11. $\dfrac{1}{x^2-x-2}-\dfrac{x}{x^2-5x+6}$

12. $\dfrac{x-1}{x^2+5x+4}+\dfrac{2}{x^2-x-2}+\dfrac{10}{x^2+2x-8}$

13. $\dfrac{A}{x-6}+\dfrac{B}{x+3}$

14. $\dfrac{A}{x+1}+\dfrac{B}{(x+1)^2}+\dfrac{C}{x-2}$

15. $\dfrac{A}{x-5}+\dfrac{B}{x+5}+\dfrac{C}{(x+5)^2}$

16. $\dfrac{Ax+B}{x^2+2}+\dfrac{C}{x-4}$

17. $-\dfrac{1}{x}+\dfrac{2}{x^2+1}$

18. $\dfrac{2}{x+1}+\dfrac{1-x}{x^2-2x+3}$

19. $\dfrac{-x}{(x+1)^{3/2}}+\dfrac{2}{(x+1)^{1/2}}$

20. $2\sqrt{x}(x-2)+\dfrac{(x-2)^2}{2\sqrt{x}}$

21. $\dfrac{2-t}{2\sqrt{1+t}}-\sqrt{1+t}$

22. $-\dfrac{\sqrt{x^2+1}}{x^2}+\dfrac{1}{\sqrt{x^2+1}}$

23. $x\left(\dfrac{1}{2}\right)\left(\dfrac{2x}{\sqrt{x^2+5}}\right)+\sqrt{x^2+5}$

24. $\dfrac{\sqrt{x^2+x+1} - \dfrac{(x+2)(2x+1)}{\sqrt{x^2+x+1}}}{x^2+x+1}$

25. $\dfrac{2x\sqrt{x^2+1} - \dfrac{x^3}{\sqrt{x^2+1}}}{x^2+1}$

26. $\left(\sqrt{x^3+1} - \dfrac{3x^3}{2\sqrt{x^3+1}}\right) \div (x^3+1)$

27. $\left[\dfrac{x}{3(x+1)^{2/3}} - (x+1)^{1/3}\right] \div x^2$

28. $\left[\dfrac{x}{5(x+1)^{4/5}} - (x+1)^{1/5}\right] \div x^2$

29. $\left(1 + \dfrac{x}{\sqrt{x^2+1}}\right) \div (x + \sqrt{x^2+1})$

30. $\dfrac{1}{x} - \left(\dfrac{1}{1+\sqrt{1-x^2}}\right)\left(\dfrac{-x}{\sqrt{1-x^2}}\right)$

In Exercises 31–50, rationalize the numerator (or denominator) and simplify.

31. $\dfrac{3}{\sqrt{21}}$ **32.** $\dfrac{5}{\sqrt{10}}$ **33.** $\dfrac{\sqrt{2}}{3}$

34. $\dfrac{\sqrt{26}}{2}$ **35.** $\dfrac{x}{\sqrt{x-4}}$ **36.** $\dfrac{4y}{\sqrt{y+8}}$

37. $\dfrac{\sqrt{y^3}}{6y}$ **38.** $\dfrac{x\sqrt{x^2+4}}{3}$ **39.** $\dfrac{49(x-3)}{\sqrt{x^2-9}}$

40. $\dfrac{10(x+2)}{\sqrt{x^2-x-6}}$ **41.** $\dfrac{1}{\sqrt{6}+\sqrt{5}}$

42. $\dfrac{10}{\sqrt{x}+\sqrt{x+5}}$ **43.** $\dfrac{x+1}{\sqrt{x^2-2}-\sqrt{x}}$

44. $\dfrac{x-1}{\sqrt{3}-\sqrt{4x-x^2}}$ **45.** $\dfrac{\sqrt{3}-\sqrt{2}}{x}$

46. $\dfrac{\sqrt{15}+3}{12}$ **47.** $\dfrac{8x}{\sqrt{17x}-1}$

48. $\dfrac{x-1}{\sqrt{x}+x}$ **49.** $\dfrac{2x-\sqrt{4x-1}}{2x-1}$

50. $\dfrac{1-x^2}{\sqrt{x}-\sqrt{x^3}}$

Chapter 1

Functions, Graphs, and Limits

Section 1.1

The Cartesian Plane and The Distance Formula

Introductory Example

Prime Interest Rate

The prime interest rate is the annual rate of interest that commercial banks charge their best customers. The prime rate is usually reserved for large loans to businesses. Loans to individuals, especially those without collateral, often carry interest rates that are higher than the prime lending rate. One of the main factors used by banks to determine the prime rate is the current rate of inflation. It stands to reason that the interest paid on a loan must exceed the inflation rate in order for the lender to make money.

Table 1.1. shows the prime rate for selected years between 1960 and 1980. This data is plotted on a **rectangular coordinate system** with appropriate units (see Figure 1.1). Note that each plotted point relates two numbers: one number corresponding to the year (the horizontal axis) and one number corresponding to the prime rate (the vertical axis).

Table 1.1

Year	Prime rate (%)
1960	4.82
1965	4.54
1970	7.91
1973	8.03
1974	10.81
1975	7.86
1976	6.84
1977	6.83
1978	9.06
1979	12.67
1980	16.36

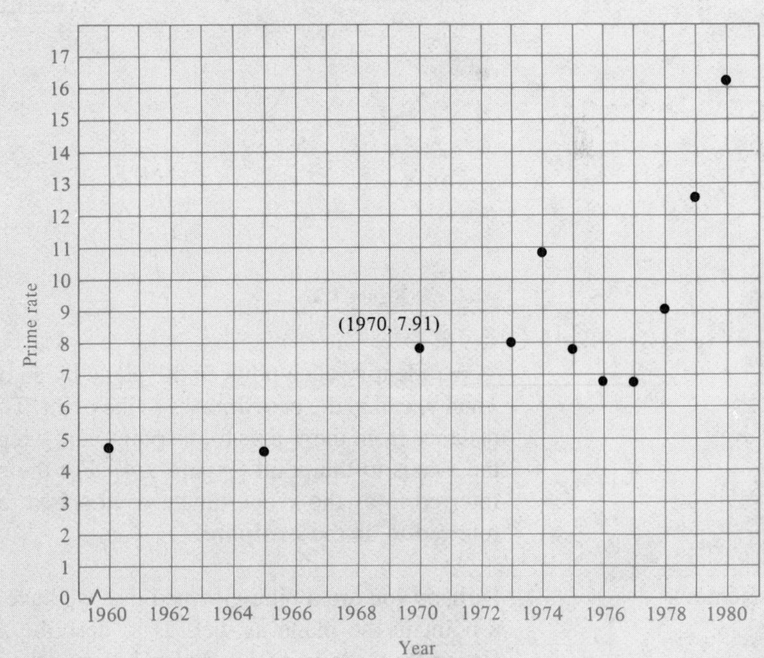

Figure 1.1 Prime Interest Rate

Section Objectives: *To review the Cartesian plane* ▪ *To develop the formula for the distance between two points in the plane* ▪ *To describe the Midpoint Rule.*

Just as we can represent the real numbers geometrically by points on the real line, we can represent ordered pairs of real numbers by points in a plane. The model we develop for representing ordered pairs of real numbers is called the **rectangular coordinate system,** or the **Cartesian plane.** We develop this model by considering two real lines intersecting at right angles (Figure 1.2).

The horizontal real line is traditionally called the ***x*-axis,** and the vertical real line is called the ***y*-axis.** Their point of intersection is called the **origin,** and the lines divide the plane into four parts called **quadrants** (Figure 1.3).

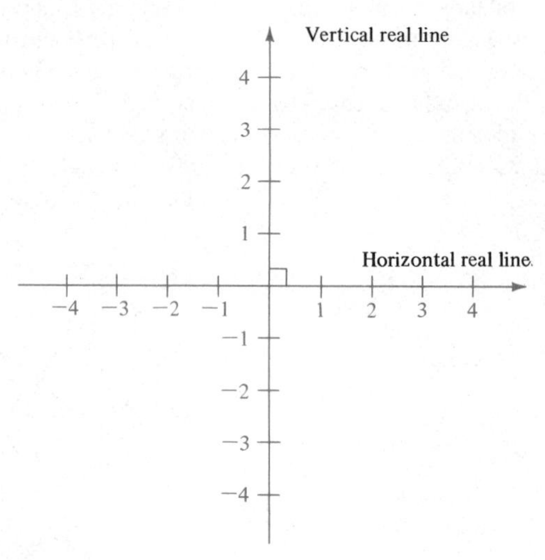

Figure 1.2

We identify each point in the plane by an ordered pair (x, y) of real numbers x and y, called the **coordinates** of the point. The number x represents the directed distance from the y-axis to the point, and y represents the directed distance from the x-axis to the point (Figure 1.4). For the point (x, y), the first coordinate is referred to as the x-coordinate or **abscissa,** and the second or y-coordinate is referred to as the **ordinate.**

Remark

Perhaps you are a bit concerned that we have used the notation (x, y) to denote a point in the plane as well as to describe an open interval on the real line. Generally there is no confusion because the nature of a specific problem will show whether we are talking about points in the plane or about intervals on the real line.

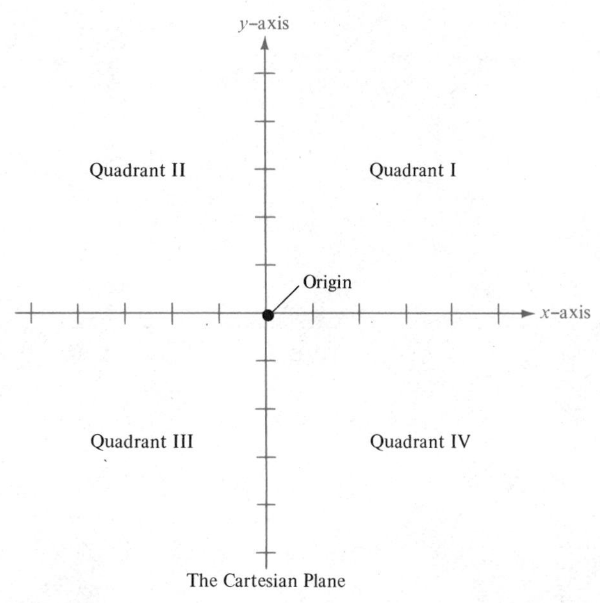

Figure 1.3

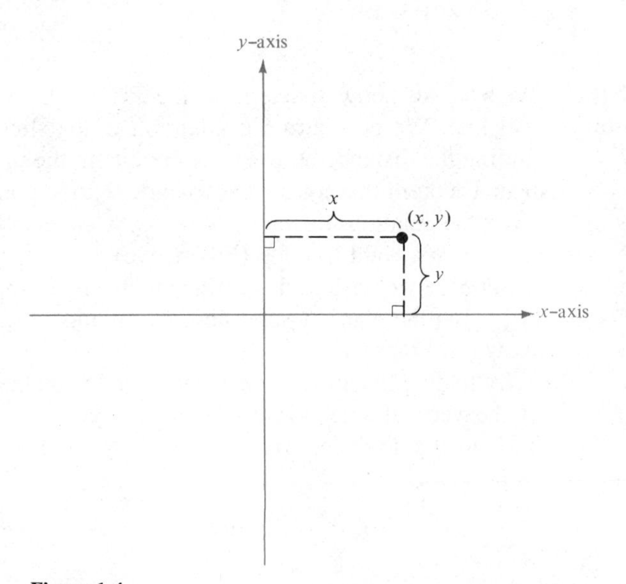

Figure 1.4

Example 1

Plotting Points in the Cartesian Plane

Locate the points $(-1, 2)$, $(3, 4)$, $(0, 0)$, $(3, 0)$, and $(-2, -3)$ in the Cartesian plane.

Solution The solution is shown in Figure 1.5.

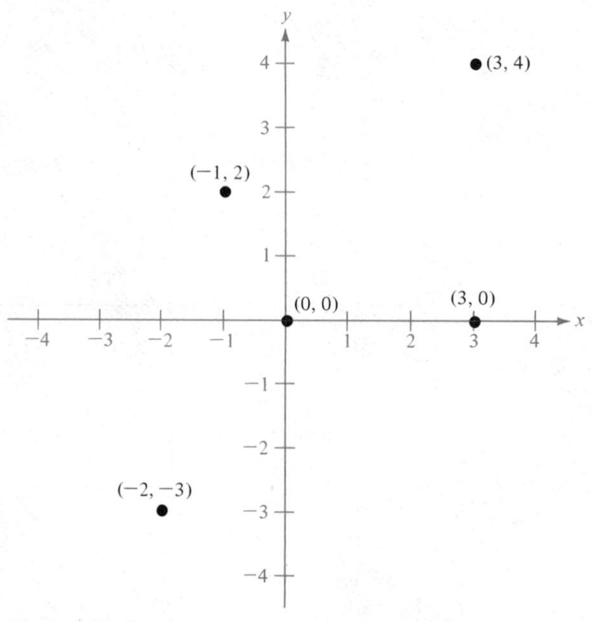

Figure 1.5

□

Development of the Distance Formula

We have seen how to determine the distance between two points x_1 and x_2 on the real line. We now turn our attention to the slightly more difficult problem of finding the distance between two points in the plane. Recall from the Pythagorean Theorem that for a right triangle with hypotenuse c and sides a and b, we have the relationship $a^2 + b^2 = c^2$. Conversely, if $a^2 + b^2 = c^2$, then the triangle is a right triangle (Figure 1.6).

Suppose we wish to determine the distance d between two points (x_1, y_1) and (x_2, y_2) in the plane. Using these two points, a right triangle can be formed, as shown in Figure 1.7.

By finding the distance between y_1 and y_2 on the y-axis, we see that the length of the vertical side of the triangle is $|y_2 - y_1|$. Similarly, the length of the horizontal side of the triangle is $|x_2 - x_1|$. By the Pythagorean Theorem, we then have

$$d^2 = |x_2 - x_1|^2 + |y_2 - y_1|^2$$

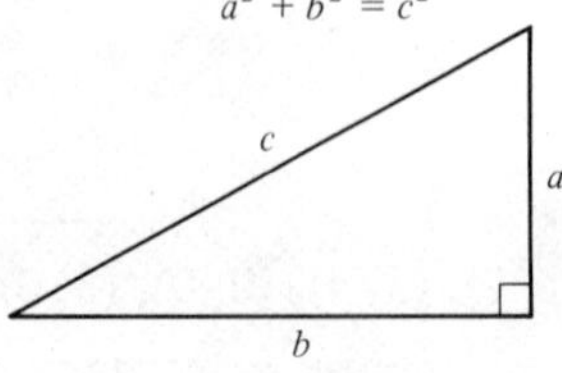

Pythagorean Theorem

Figure 1.6

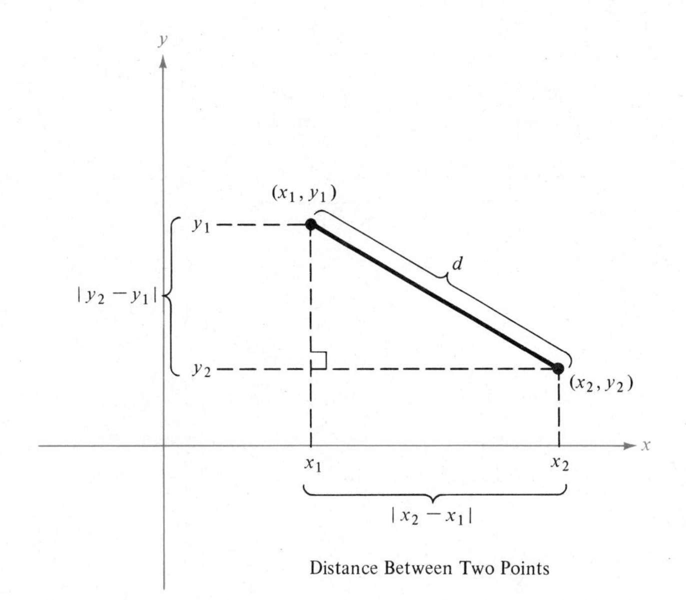

Figure 1.7

or $$d = \sqrt{|x_2 - x_1|^2 + |y_2 - y_1|^2}$$

Replacing $|x_2 - x_1|^2$ and $|y_2 - y_1|^2$ by the equivalent expressions $(x_2 - x_1)^2$ and $(y_2 - y_1)^2$, we can write

$$d = \sqrt{(x_2 - x_1)^2 + (y_2 - y_1)^2}$$

We choose the positive square root for d because the distance *between* two points is not a directed distance. Of course, we can interchange the order of subtraction and write the equivalent form

$$d = \sqrt{(x_1 - x_2)^2 + (y_1 - y_2)^2}$$

We have therefore established the following theorem.

Theorem 1.1
Distance Formula

The distance d between two points (x_1, y_1) and (x_2, y_2) in the plane is given by

$$d = \sqrt{(x_2 - x_1)^2 + (y_2 - y_1)^2}$$

Example 2

Finding the Distance Between Two Points

Find the distance between the points $(-2, 1)$ and $(3, 4)$.

Solution Applying Theorem 1.1, we have

$$d = \sqrt{[3 - (-2)]^2 + (4 - 1)^2} = \sqrt{(5)^2 + (3)^2}$$
$$= \sqrt{25 + 9} = \sqrt{34} \approx 5.83$$

(Note the use of $\approx$ to represent *approximately equal to*.) □

Example 3

An Application of the Distance Formula

Use the Distance Formula to show that the points (2, 1), (4, 0), and (5, 7) are the vertices of a right triangle.

Solution Refer to Figure 1.8. The three sides have lengths

$$d_1 = \sqrt{(5-2)^2 + (7-1)^2} = \sqrt{9+36} = \sqrt{45}$$
$$d_2 = \sqrt{(4-2)^2 + (0-1)^2} = \sqrt{4+1} = \sqrt{5}$$
$$d_3 = \sqrt{(5-4)^2 + (7-0)^2} = \sqrt{1+49} = \sqrt{50}$$

Since

$$d_1^2 = d_2^2 = 45 + 5 = 50 = d_3^2$$

we can apply the Pythagorean Theorem to conclude that the triangle must be a right triangle.

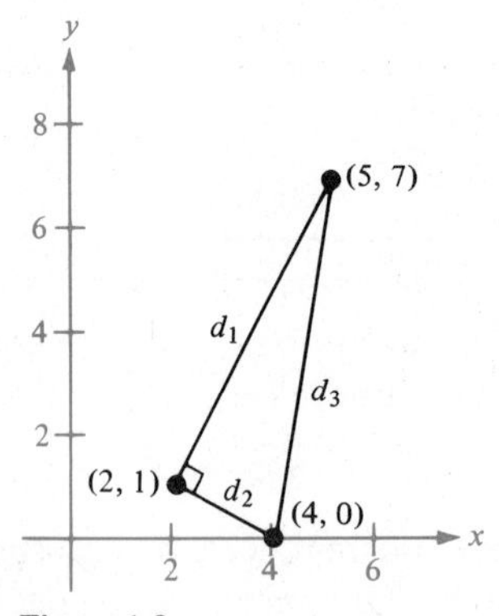

Figure 1.8

□

Remark

In Example 3, the figure provided was not really essential to the solution of the problem. *Nevertheless,* we strongly recommend that you get in the habit of including sketches with your problem solutions even if they are not specifically required. Throughout our many years of teaching calculus, we have found that students who do well with the technical aspects of calculus are very often the same students who have a good grasp of the visual aspects of the subject.

Now we introduce a rule for finding the coordinates of the midpoint of the line segment joining two points in the plane.

Theorem 1.2 Midpoint Rule

The midpoint of the line segment joining points (x_1, y_1) and (x_2, y_2) is

$$\left(\frac{x_1 + x_2}{2}, \frac{y_1 + y_2}{2}\right)$$

Proof The result of this theorem is just what we would expect. To find the midpoint of a line segment, we merely find the "average" values of the respective coordinates of the two endpoints. The formal proof requires that, in Figure 1.9, we show that

$$d_1 = d_2 \qquad \text{and} \qquad d_1 + d_2 = d_3$$

Using the Distance Formula, we obtain

$$\begin{aligned} d_1 &= \sqrt{\left(\frac{x_1 + x_2}{2} - x_1\right)^2 + \left(\frac{y_1 + y_2}{2} - y_1\right)^2} \\ &= \tfrac{1}{2}\sqrt{(x_2 - x_1)^2 + (y_2 - y_1)^2} \\ d_2 &= \sqrt{\left(x_2 - \frac{x_1 + x_2}{2}\right)^2 + \left(y_2 - \frac{y_1 + y_2}{2}\right)^2} \\ &= \tfrac{1}{2}\sqrt{(x_2 - x_1)^2 + (y_2 - y_1)^2} \\ d_3 &= \sqrt{(x_2 - x_1)^2 + (y_2 - y_1)^2} \end{aligned}$$

Thus, it follows that

$$d_1 = d_2 \qquad \text{and} \qquad d_1 + d_2 = d_3$$

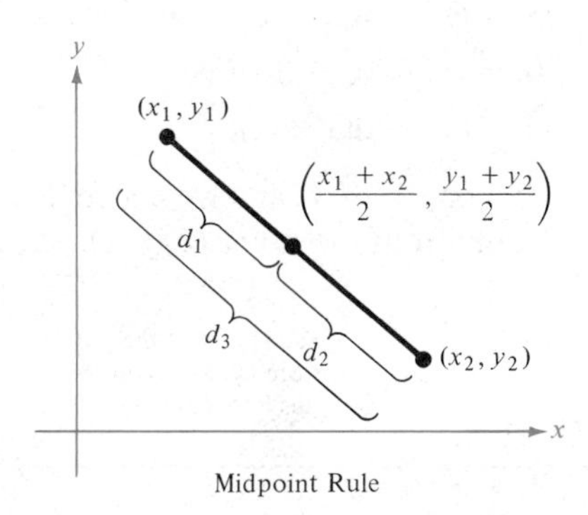

Figure 1.9

□

Example 4

Finding the Midpoint of a Line Segment

Find the midpoint of the line segment joining the points $(-3, -5)$ and $(3, 9)$.

Solution By Theorem 1.2, the midpoint is

$$\left(\frac{-3 + 3}{2}, \frac{-5 + 9}{2}\right) = (0, 2)$$

(See Figure 1.10.)

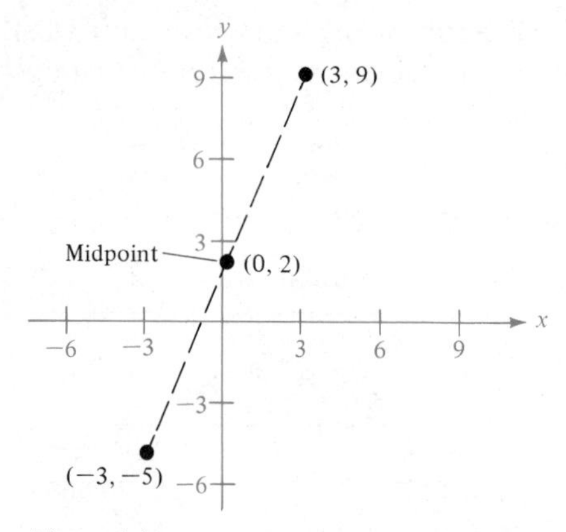

Figure 1.10

□

Example 5

Finding Points at a Specified Distance From a Given Point

Find x so that the distance between $(x, 3)$ and $(2, -1)$ is 5.

Solution Using the distance formula, we have

$$\begin{aligned}
d = 5 &= \sqrt{(x-2)^2 + (3+1)^2} \\
25 &= (x^2 - 4x + 4) + 16 \\
0 &= x^2 - 4x - 5 \\
0 &= (x-5)(x+1)
\end{aligned}$$

Therefore, $x = 5$ or $x = -1$, and we conclude that both of the points $(5, 3)$ and $(-1, 3)$ lie 5 units from the point $(2, -1)$. (See Figure 1.11.)

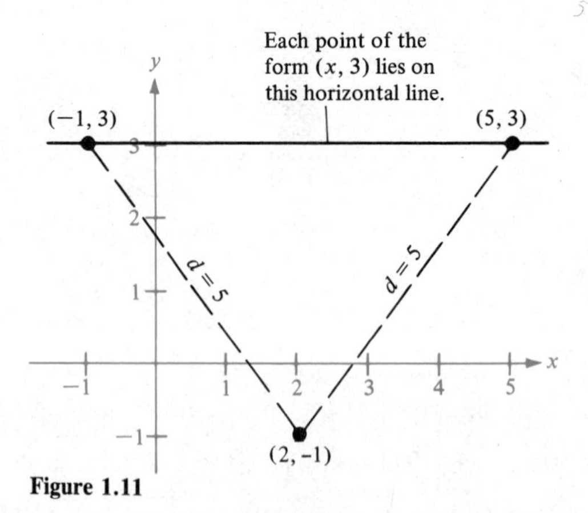

Figure 1.11

□

Figures 1.2 through 1.11 have all pictured a rectangular coordinate system whose axes cross at the point (0, 0). In practice it often occurs that the points to be plotted lie far from the origin. When this happens, we usually omit a portion of the x- or y-axis and indicate this omission by a *broken axis,* as is demonstrated in the following example.

Example 6

Using a Broken Axis

The numbers of U.S. veterans with service in Vietnam who were discharged between 1966 and 1979 are given in Table 1.2.

Table 1.2

Year	1966	1967	1968	1969	1970	1971	1972	1973	1974	1975	1976	1977	1978	1979
Number (in 1000s)	56	144	321	485	560	504	300	179	135	86	60	64	78	68

Plot these points on a rectangular coordinate system.

Solution Refer to Figure 1.12. Note that the break in the x-axis indicates that we have omitted the years between 0 and 1966.

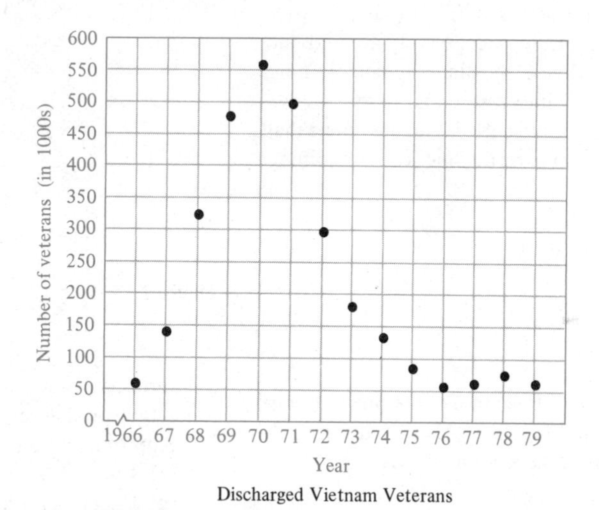

Figure 1.12

□

Section Exercises (1.1)

In Exercises 1–8, plot the points, find the distance between the points, and find the midpoint of the line segment joining the points.

1. $(2, 1), (4, 5)$
2. $(-3, 2), (3, -2)$
3. $(\frac{1}{2}, 1), (-\frac{3}{2}, -5)$
4. $(\frac{2}{3}, -\frac{1}{3}), (\frac{5}{6}, 1)$
5. $(2, 2), (4, 14)$
6. $(-3, 7), (1, -1)$
7. $(1, \sqrt{3}), (-1, 1)$
8. $(-2, 0), (0, \sqrt{2})$
9. Show that the points $(4, 0)$, $(2, 1)$, $(-1, -5)$ are vertices of a right triangle.
10. Show that the points $(1, -3)$, $(3, 2)$, $(-2, 4)$ are vertices of an isosceles triangle.
11. Show that the points $(0, 0)$, $(1, 2)$, $(2, 1)$, $(3, 3)$ are vertices of a rhombus. (A rhombus is a four-sided figure whose sides are all of the same length.)
12. Show that the points $(0, 1)$, $(3, 7)$, $(4, 4)$, $(1, -2)$ are vertices of a parallelogram.
13. Use the Distance Formula to determine if the points $(0, -4)$, $(2, 0)$, $(3, 2)$ lie on a straight line.

Ⓒ**14.** Use the Distance Formula to determine if the points (0, 4), (7, −6), (−5, 11) lie on a straight line.

Ⓒ**15.** Use the Distance Formula to determine if the points (−2, 1), (−1, 0), (2, −2) lie on a straight line.

16. Find y so that the distance from the origin to the point $(3, y)$ is 5.

17. Find x so that the distance from the origin to the point $(x, -4)$ is 5.

18. Find x so that the distance from (2, −1) to the point $(x, 2)$ is 5.

19. Find the relationship between x and y so that the point (x, y) is equidistant from (4, −1) and (−2, 3).

20. Find the relationship between x and y so that the point (x, y) is equidistant from $(3, \frac{5}{2})$ and $(-7, -1)$.

21. Use the Midpoint Rule successively to find the three points that divide the line segment joining (x_1, y_1) and (x_2, y_2) into four equal parts.

22. Use the results of Exercise 21 to find the points that divide into four equal parts the line segment joining these points:
(a) (1, −2) and (4, −1)
(b) (−2, −3) and (0, 0)

23. Prove that

$$\left(\frac{2x_1 + x_2}{3}, \frac{2y_1 + y_2}{3}\right)$$

is one of the points of trisection of the line segment joining (x_1, y_1) and (x_2, y_2). Also, find the midpoint of the line segment joining

$$\left(\frac{2x_1 + x_2}{3}, \frac{2y_1 + y_2}{3}\right) \quad \text{and} \quad (x_2, y_2)$$

to find the second point of trisection of the line segment joining (x_1, y_1) and (x_2, y_2).

24. Use the results of Exercise 23 to find the points of trisection of the line segment joining these points:
(a) (1, −2) and (4, 1)
(b) (−2, −3) and (0, 0)

In Exercises 25 and 26, use Figure 1.13 showing the average rates for new home mortgages betwen January 1980 and May 1981.

25. Approximate the average mortgage rate for:
(a) May 1980 (b) December 1980
(c) July 1980 (d) May 1981

26. Approximate the *increase* in the average mortgage rate from:
(a) July 1980 to May 1981
(b) January 1981 to May 1981

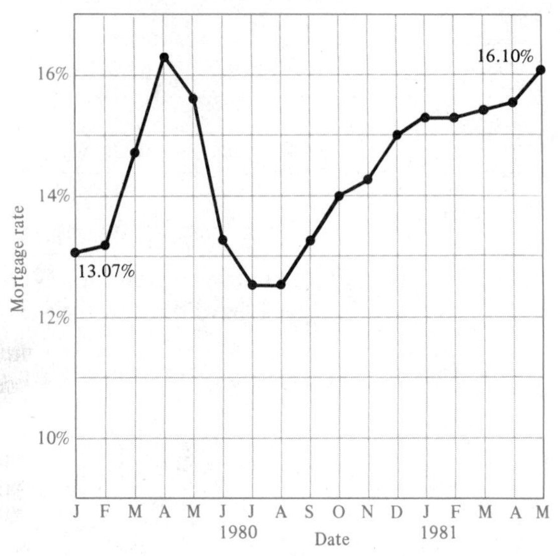

Figure 1.13

In Exercises 27–30, use Figure 1.14 showing the Dow-Jones Industrial Average (DJIA) from 1929 to 1980.

27. Approximate the DJIA for:
(a) June 1949 (b) January 1970
(c) December 1953 (d) March 1980

28. Approximate the *increase* (or *decrease*) in the DJIA from:
(a) the high of 1929 to the low of 1932
(b) the low of 1974 to the high of 1976

29. In which years did the DJIA go above 1000?

30. Approximate the *percentage increase* (or *decrease*) in the DJIA from:
(a) January 1940 to January 1950
(b) January 1973 to January 1975

Ⓒ Calculator may be helpful.

1,100
1,000
900
800
700
600
500
400
300
200
100
Dow-Jones average
1067.20
998.21
734.91
744.32
521.05
535.76
573.22
1973–1975
Recession
381.17
1929 Crash
419.79
194.40
212.50
255.49
163.12
41.22
98.95
92.92
Great Depression
1925 1930 1935 1940 1945 1950 1955 1960 1965 1970 1975 1980
Year
Dow-Jones Average of Industrial Stock Prices

Figure 1.14

Section 1.2

Graphs of Equations

Introductory Example

A Temperature Model

During a normal day the air temperature drops to a low during the night and rises to a high during the day. For example, Table 1.3 lists the air temperature in degrees Fahrenheit for a 24-hour period.

One of the goals of applied mathematics is to find equations that describe real world phenomena. We call such equations **mathematical models.** For example, a model that gives a reasonably accurate description of the temperatures in Table 1.3 is the equation*

$$y = 0.00026157t^5 - 0.01306453t^4 + 0.18365236t^3 - 0.34077620t^2 - 3.79407878t + 60.11462451$$

where $t = 1$ corresponds to 1 A.M. and $t = 13$ corresponds to 1 P.M. Figure 1.15 shows the **graph** of this model. Note the comparison between the model's temperature values and the actual temperatures.

Table 1.3

Time (AM)	1	2	3	4	5	6
Temp. (F°)	55	53	50	50	49	50
Time (AM)	7	8	9	10	11	12
Temp. (F°)	51	56	60	65	73	78
Time (PM)	1	2	3	4	5	6
Temp. (F°)	85	86	84	81	80	70
Time (PM)	7	8	9	10	11	12
Temp. (F°)	68	64	59	58	57	57

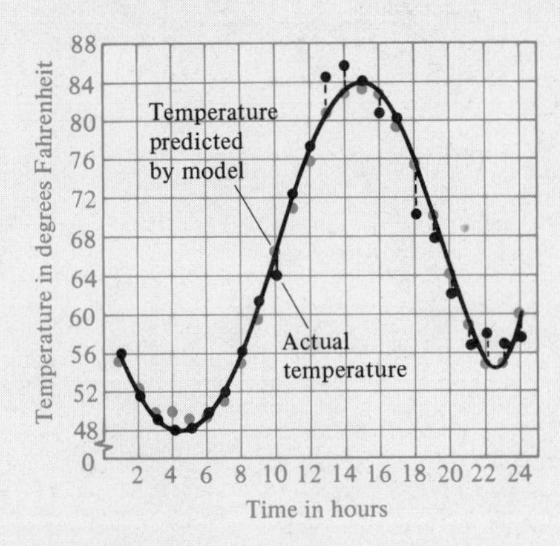

Figure 1.15

* This equation was generated by a computer by a process called least squares regression analysis.

Section Objectives: *To define the graph of an equation ▪ To introduce the point-plotting method of graphing equations ▪ To introduce the standard form of the equation of a circle ▪ To find the points of intersection of two graphs ▪ To discuss the concept of a mathematical model.*

The idea of using a graph to show how two quantities are related to each other is familiar to all of us. Newsmagazines frequently show graphs that compare the rate of inflation, the gross national product, wholesale prices, or the unemployment rate to the time of year. Industrial firms and businesses use graphs to report their monthly production and sales statistics. The value of such graphs is that they provide a simple geometrical picture of the way one quantity changes with respect to another.

Frequently, the relationship between two quantities is expressed in the form of an equation. For instance, degrees on the Fahrenheit temperature scale are related to degrees on the Celsius scale by the equation $F = \frac{9}{5}C + 32$. In this section, we introduce the basic procedure for determining the geometric picture associated with such an equation.

Consider the equation

$$3x + y = 7$$

If $x = 2$ and $y = 1$, the equation is satisfied, and we call the point (2, 1) a **solution point** of the equation. Of course, there are other solution points, such as (1, 4) and (0, 7). We can make up a table of values for x and y by choosing arbitrary values for x and determining the corresponding values for y. To determine the values for y, it is convenient to replace the equation by the equivalent form

$$y = 7 - 3x$$

x	0	1	2	3	4
y	7	4	1	−2	−5

Thus, (0, 7), (1, 4), (2, 1), (3, −2), and (4, −5) are all solution points of the equation $3x + y = 7$. We could continue this process indefinitely and obtain infinitely many solution points for the equation $3x + y = 7$. We call the collection of all such solution points the **graph** of the equation $3x + y = 7$, as shown in Figure 1.16.

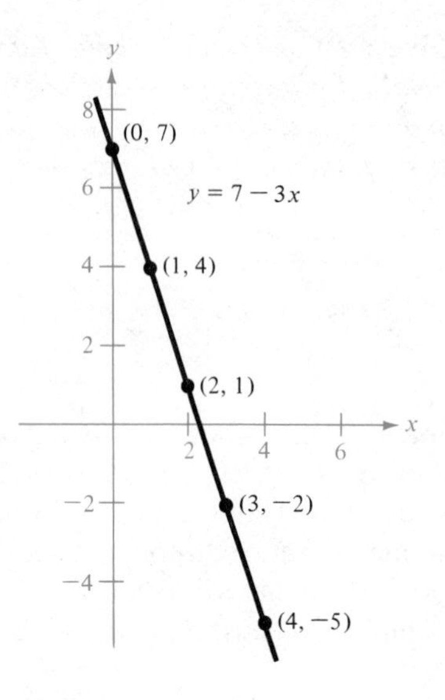

Figure 1.16

Definition of the Graph of an Equation in Two Variables

The **graph of an equation** involving two variables x and y is the collection of all points in the plane that are solution points to the equation.

Example 1

Sketching the Graph of an Equation

Sketch the graph of the equation $y = x^2 - 2$.

Solution First, we make a table of values by choosing several convenient values of x and calculating the corresponding values of y.

x	-2	-1	0	1	2	3
$y = x^2 - 2$	2	-1	-2	-1	2	7

Next, we locate these points in the plane, as in Figure 1.17.

Finally, we connect these points by a smooth curve, as in Figure 1.18.

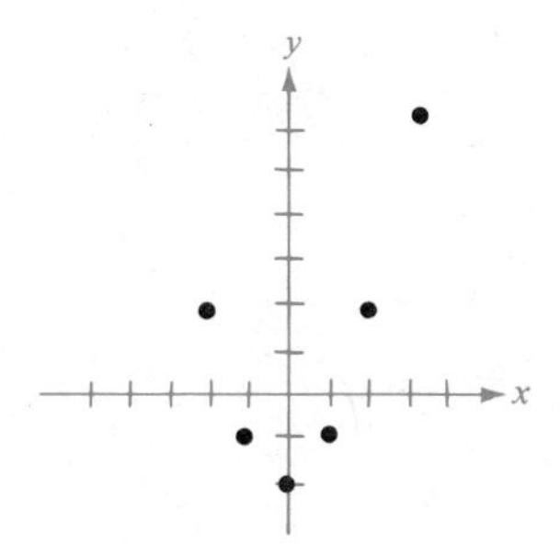

Figure 1.17

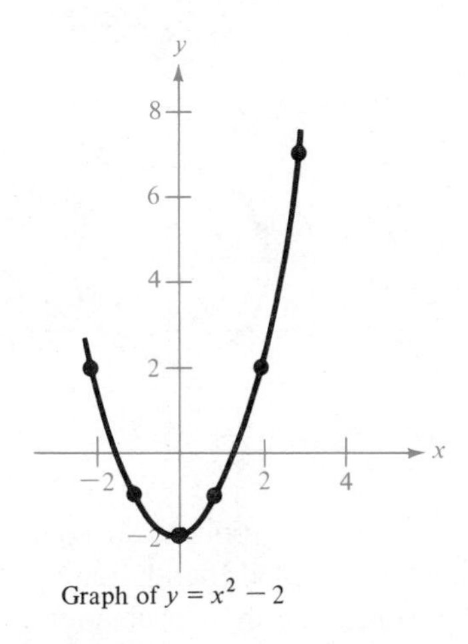

Graph of $y = x^2 - 2$

Figure 1.18

□

We call this method of sketching a graph the **point-plotting method.** Its basic features are the following:

The Point-Plotting Method of Graphing

1. Make up a table of several solution points of the equation.
2. Plot these points in the plane.
3. Connect the points with a smooth curve.

Steps 1 and 2 of the point-plotting method can usually be accomplished with ease. However, step 3 can be the source of some major difficulties. For instance, how would you connect the four points in Figure 1.19? Without additional points or further information about the equation, any one of the three graphs in Figure 1.20 would be reasonable.

Obviously, with too few solution points, we could badly misrepresent the graph of a given equation. It is hard to say just how many points should be plotted. For a straight-line graph two points are sufficient; for more complicated

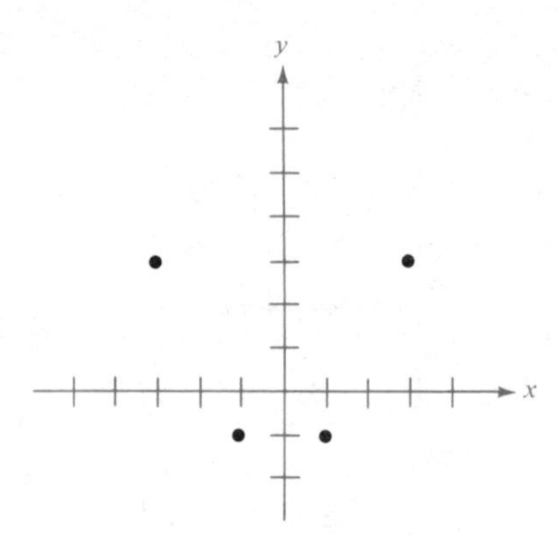

Figure 1.19

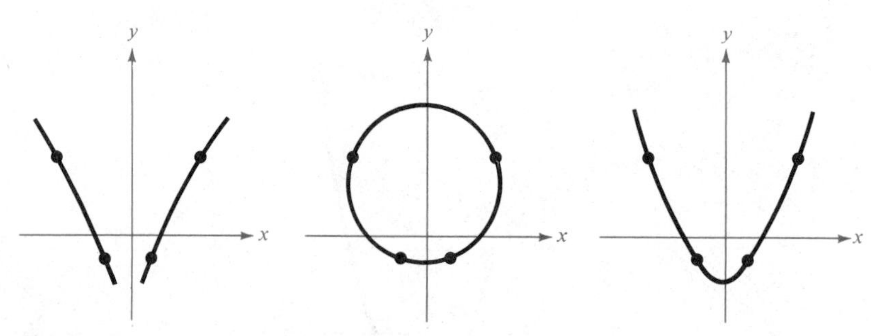

Figure 1.20

graphs we need many more points. In spite of this difficulty with the point-plotting procedure, it is a good foundation upon which to build the more sophisticated techniques discussed in later chapters. In the meantime, we suggest that you plot enough points so as to reveal the essential behavior of the graph; the more solution points you plot, the more accurate your graph will be. (A programmable calculator is a very useful device for determining the many solution points needed for an accurate graph.)

In choosing points to plot, we suggest that you start with those that are easiest to calculate. Two points that are usually easy to determine are those having zero as either their x- or y-coordinate. Such points are called **intercepts,** because they are points at which the graph intersects the x- or y-axis.

Definition of Intercepts

The point $(a, 0)$ is called an ***x*-intercept** of the graph of an equation if it is a solution point of the equation.

The point $(0, b)$ is called a ***y*-intercept** of the graph of an equation if it is a solution point of the equation.

Remark

Some authors denote the x-intercept as the x-coordinate of the point $(a, 0)$ rather than the point itself. Unless it is necessary to make a distinction, we will use "intercept" to mean either the point or the coordinate.

Of course, it is possible that a particular graph will have no intercepts, or it may have several. For instance, consider the four graphs in Figure 1.21.

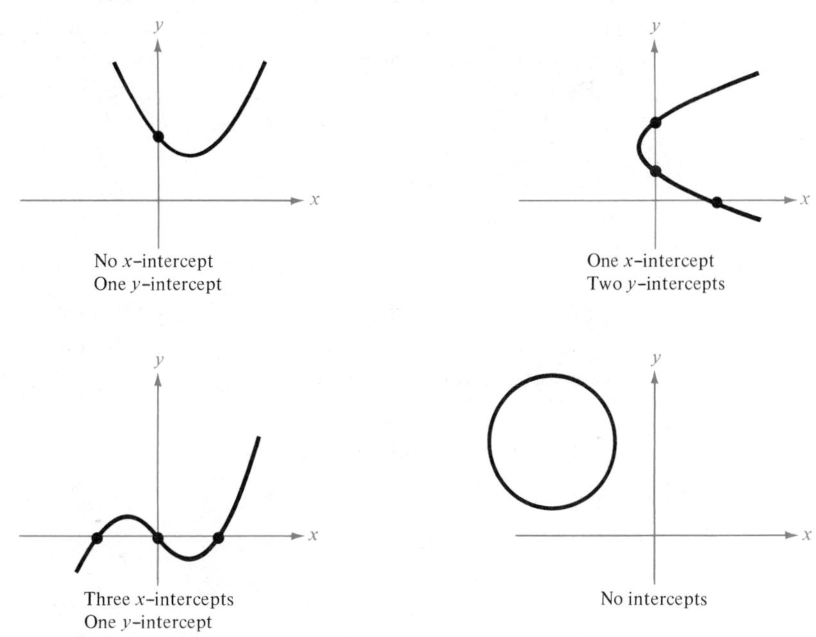

Figure 1.21

Finding Intercepts

To find the x-intercepts, let y be zero and solve the equation for x.
To find the y-intercepts, let x be zero and solve the equation for y.

Example 2

Finding x- and y-Intercepts

Find the x- and y-intercepts for the graphs of the following:

(a) $y = x^3 - 4x$
(b) $y^2 - 3 = x$

Solution

(a) Let $y = 0$; then $0 = x(x^2 - 4)$ has solutions $x = 0$ and $x = \pm 2$.

x-intercepts: $(0, 0)$, $(2, 0)$, $(-2, 0)$

Let $x = 0$; then $y = 0$.

y-intercept: $(0, 0)$

(b) Let $y = 0$; then $-3 = x$.

x-intercept: $(-3, 0)$

Let $x = 0$; then $y^2 - 3 = 0$ has solutions $y = \pm\sqrt{3}$.

y-intercepts: $(0, \sqrt{3})$, $(0, -\sqrt{3})$ □

Although useful, the point-plotting method of sketching can be tedious. For common types of equations, it is simpler to learn to recognize the general nature of the graph from the equation's form. For example, in the next section we will learn to recognize equations whose graphs are straight lines. In this section, we consider the equation of another easily recognized graph, that of a **circle.**

By referring to the circle in Figure 1.22, we see that a point (x, y) is on the circle if and only if its distance from the center (h, k) is r. This means that a circle consists of the set of all points (x, y) that are at a given positive distance r from a fixed point (h, k). Expressing this relationship in terms of the Distance Formula, we have

$$\sqrt{(x - h)^2 + (y - k)^2} = r$$

as the condition that the coordinates of (x, y) must satisfy. By squaring both sides of this equation, we obtain the **standard form of the equation of a circle,** which is part of the following theorem.

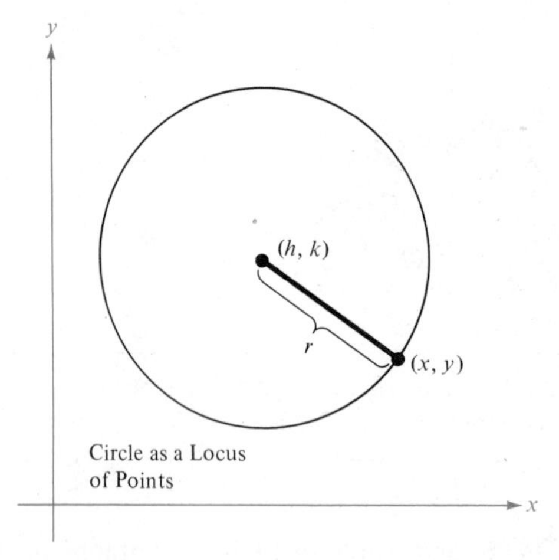

Figure 1.22

Theorem 1.3 Standard Form of the Equation of a Circle

The point (x, y) lies on the circle of radius r and center (h, k) if and only if

$$(x - h)^2 + (y - k)^2 = r^2$$

Remark

Theorem 1.3 uses the phrase "if and only if" as a way of stating two theorems in one. One theorem says that "all points on the circle satisfy the given equation." The other theorem is the *converse,* which says that "all points satisfying the given equation lie on the circle."

As a special case of Theorem 1.3, the equation of a circle with its center at the origin is simply

$$x^2 + y^2 = r^2$$

Example 3

Finding an Equation for a Circle

The point (3, 4) lies on a circle whose center is at (−1, 2) (Figure 1.23). Find an equation for the circle.

Solution The radius of the circle is the distance between (−1, 2) and (3, 4). Thus,

$$r = \sqrt{[3 - (-1)]^2 + (4 - 2)^2} = \sqrt{16 + 4} = \sqrt{20}$$

Therefore, by Theorem 1.3 the standard equation for this circle is

$$[x - (-1)]^2 + (y - 2)^2 = (\sqrt{20})^2$$
$$(x + 1)^2 + (y - 2)^2 = 20$$

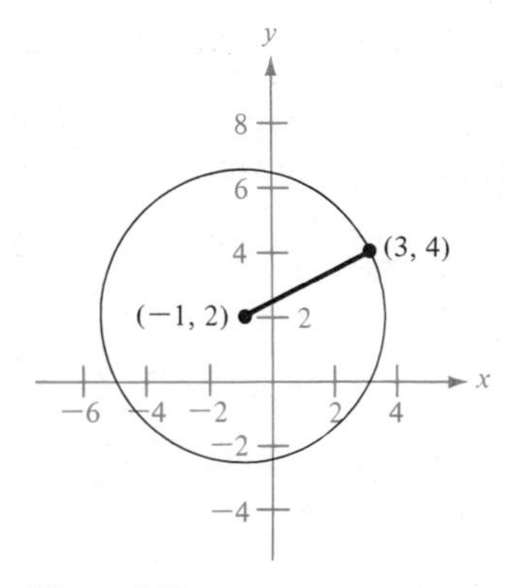

Figure 1.23

□

If we remove parentheses in the standard equation of Example 3, we obtain

$$(x + 1)^2 + (y - 2)^2 = 20$$
$$x^2 + 2x + 1 + y^2 - 4y + 4 = 20$$
$$x^2 + y^2 + 2x - 4y - 15 = 0$$

where the latter equation is in the **general form of the equation of a circle:**

$$Ax^2 + Ay^2 + Dx + Ey + F = 0, \qquad A \neq 0$$

The general form of the equation is less useful than the equivalent standard form. For instance, little is apparent about the circle with the general equation

$$x^2 + y^2 - 6x + 10y + 24 = 0$$

However, from the equivalent standard form,

$$(x - 3)^2 + (y + 5)^2 = 10$$

we can readily see that the circle is centered at (3, −5) and its radius is $\sqrt{10}$. This observation suggests that to graph the equation of a circle it is best to write the equation in standard form. This can be accomplished by using the algebraic process called **completing the square,** which we demonstrate in the following example.

Example 4

Completing the Square

Sketch the graph of the circle whose general equation is

$$4x^2 + 4y^2 + 20x - 16y + 37 = 0$$

Solution To complete the square we will first divide by 4 so that the coefficients of x^2 and y^2 are both 1. Thus, we have

$$x^2 + y^2 + 5x - 4y + \tfrac{37}{4} = 0$$

Then we write

$$(x^2 + 5x + \quad) + (y^2 - 4y + \quad) = -\tfrac{37}{4}$$

reserving space to add the square of half the coefficient of x and the square of half the coefficient of y to both sides of the equation. Thus, we obtain

$$(x^2 + 5x + \underbrace{\tfrac{25}{4}}_{(\text{half})^2}) + (y^2 - 4y + \underbrace{4}_{(\text{half})^2}) = -\tfrac{37}{4} + \tfrac{25}{4} + 4$$

$$(x + \tfrac{5}{2})^2 + (y - 2)^2 = 1$$

Therefore, the circle is centered at $(-\frac{5}{2}, 2)$, and its radius is 1 (Figure 1.24).

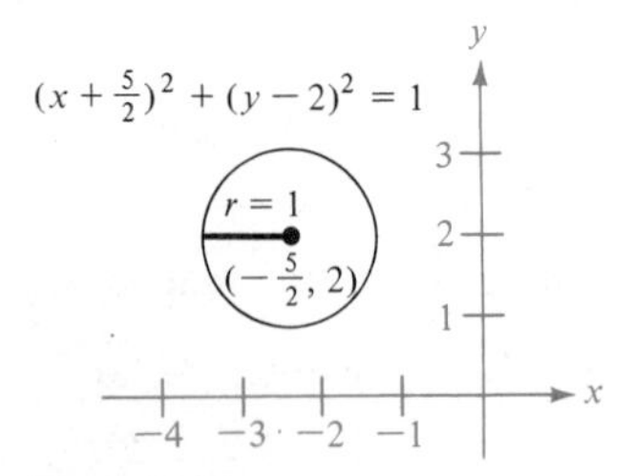

Figure 1.24

□

Since each point of a graph is a solution point of its corresponding equation, a **point of intersection** of two graphs is simply a solution point that satisfies both equations. Moreover, the points of intersection of two graphs can be found by solving the given equations simultaneously.

Example 5

Finding Points of Intersection

Find all points of intersection of the graphs of

$$x - y = 1 \qquad \text{and} \qquad x^2 - y = 3$$

Solution Applying the methods of this section, we make a sketch for each equation on the *same* coordinate plane (Figure 1.25).

From Figure 1.25, it appears that the two graphs have two points of intersection. Solving each equation for y, we obtain

$$y = x - 1 \qquad \text{and} \qquad y = x^2 - 3$$

By equating the two expressions for y, we obtain

$$x^2 - 3 = x - 1$$

or $\quad x^2 - x - 2 = 0$

Factoring yields

$$(x - 2)(x + 1) = 0$$

and thus we have $x = 2$ or $x = -1$. The corresponding values of y are obtained by substituting $x = 2$ and $x = -1$ into either of the original equations. For instance, if we choose the equation $y = x - 1$, then the values of y are 1 and -2, respectively. Therefore, the two points of intersection are

$$(2, 1) \qquad \text{and} \qquad (-1, -2)$$

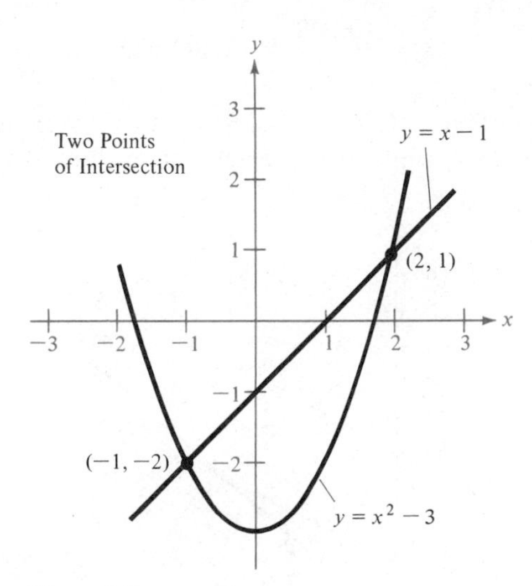

Figure 1.25 □

Many applications involve finding the point of intersection of two graphs. A common one from business is called **break-even analysis.** The marketing of a new product typically requires a substantial investment to develop and produce the product. When enough units have been sold so that the total revenue has offset the total cost, we say that the sale of the product has reached the **break-even point.** We denote the **total cost** of producing x units of a product by C, and we denote the **total revenue** received from selling x units of the product by R. Thus, we can find the break-even point by setting the cost C equal to the revenue R and solving for x.

Example 6

Break-even Analysis

Roger Fisher is setting up a small home business to manufacture and market an item he has developed. Roger has invested $10,000 in equipment and can produce each item for $0.65. If he can sell each item for $1.20, how many items must he sell before he breaks even?

Solution The cost of producing x units is

$$C = 0.65x + 10{,}000$$

and the revenue obtained by selling x units is

$$R = 1.2x$$

Since the break-even point occurs when $R = C$, we have

$$\begin{aligned} R &= C \\ 1.2x &= 0.65x + 10{,}000 \\ 0.55x &= 10{,}000 \\ x &= \frac{10{,}000}{0.55} \approx 18{,}182 \text{ units} \end{aligned}$$

Note in Figure 1.26 that sales less than the break-even point correspond to an overall loss, while sales greater than the break-even point correspond to a profit.

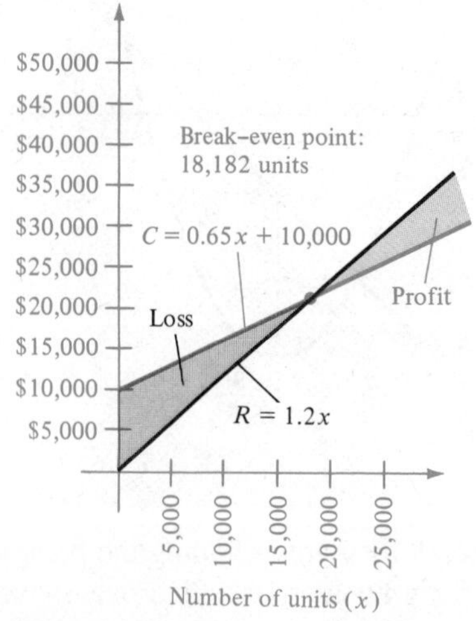

Figure 1.26

□

In this text, we will primarily be concerned with the use of equations as **mathematical models** of real world phenomena. In developing a mathematical model to represent actual data, we strive for two (often conflicting) goals: accuracy and simplicity. That is, we would like the model to be simple enough to be workable and at the same time accurate enough to produce meaningful results. Our next example describes a typical mathematical model.

Example 7

A Mathematical Model

The median income (between 1950 and 1978) for U.S. families with male householders is given in Table 1.4.

Table 1.4

Year	1950	1955	1960	1965	1970	1975	1976	1977	1978
Income (in 1000s)	3.4	4.6	5.9	7.3	10.5	14.8	16.1	17.5	19.2

A mathematical model for this data is given by

$$y = 0.02122t^2 - 0.07365t + 3.90752$$

where y represents the median income in 1000s of dollars and t represents the year, with $t = 0$ corresponding to 1950.

Using a graph, compare the data with the model and use the model to predict the median income for 1980.

Solution Table 1.5 and Figure 1.27 compare the model's values with the actual values.

Table 1.5

t	0	5	10	15	20	25	26	27	28
y	3.9	4.1	5.3	7.6	10.9	15.3	16.3	17.4	18.5
Actual income	3.4	4.6	5.9	7.3	10.5	14.8	16.1	17.5	19.2

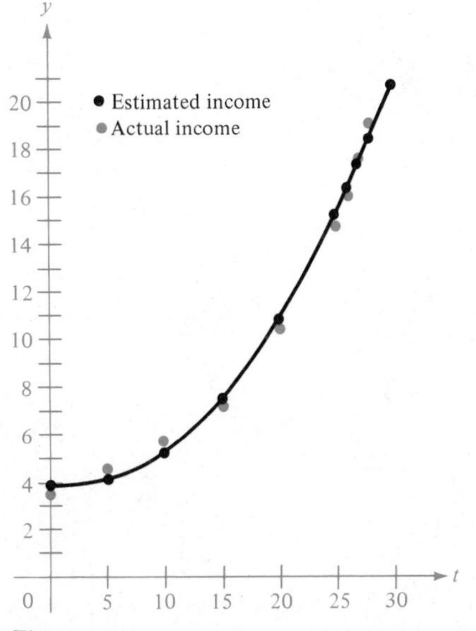

Figure 1.27

To predict the median income for 1980, we let $t = 30$ and calculate y as follows:

$$y = 0.02122(30)^2 - 0.07365(30) + 3.90752 \approx 20.8$$

Thus, we estimate the 1980 median income to be \$20,800. □

Section Exercises (1.2)

In Exercise 1–6, match the given equation with its graph. [Graphs are labeled (a), (b), (c), (d), (e), and (f).]

1. $y = x - 2$
2. $y = -\frac{1}{2}x + 2$
3. $y = x^2 + 2x$
4. $y = \sqrt{9 - x^2}$
5. $y = |x| - 2$
6. $y = x^3 - x$

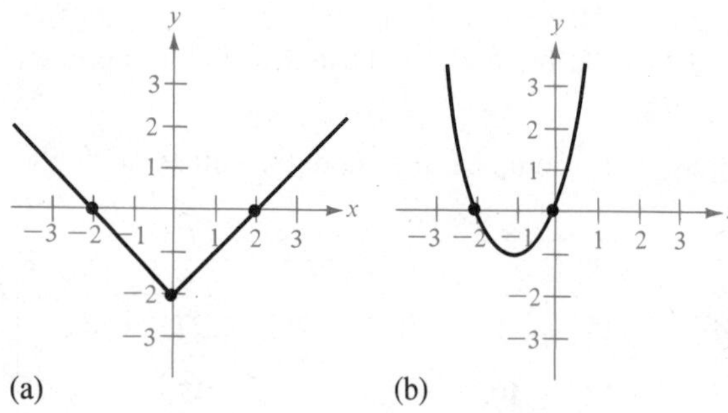

(a) (b)

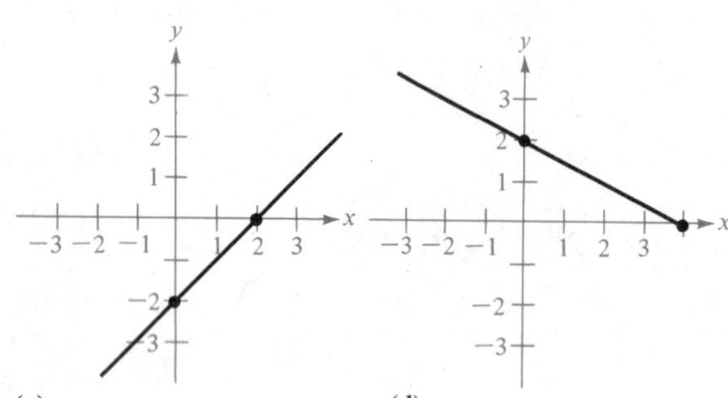

(c) (d)

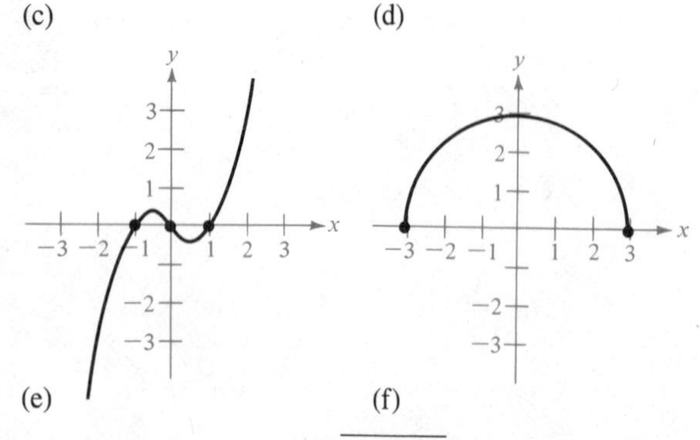

(e) (f)

Ⓒ Calculator may be helpful.

In Exercises 7–16, find the intercepts,

7. $y = 2x - 3$
8. $y = (x - 1)(x - 3)$
9. $y = x^2 + x - 2$
10. $y^2 = x^3 - 4x$
11. $y = x^2\sqrt{9 - x^2}$
12. $xy = 4$
13. $y = \dfrac{x - 1}{x - 2}$
14. $y = \dfrac{x^2 + 3x}{(3x + 1)^2}$
15. $x^2y - x^2 + 4y = 0$
16. $y = 2x - \sqrt{x^2 + 1}$

In Exercises 17–35, use the methods of this section to sketch the graph of each equation. Identify the intercepts of each graph.

17. $y = x$
18. $y = x + 3$
19. $y = -3x + 2$
20. $y = 2x - 3$
21. $y = 1 - x^2$
22. $y = x^2 + 3$
23. $y = -2x^2 + x + 1$
Ⓒ**24.** $y = x^3 - 3x$
25. $y = x^3 + 2$
26. $y = x^3 - 1$
27. $x^2 + 4y^2 = 4$
28. $x = y^2 - 4$
29. $y = (x + 2)^2$
30. $y = \dfrac{1}{x^2 + 1}$
31. $y = \dfrac{1}{x}$
32. $y = 2x^4$
33. $y = |x - 2|$
34. $y = -|x - 2|$
35. $y = \sqrt{x - 3}$

Ⓒ**36.** (a) Sketch the graph of $y = 3x^4 - 4x^3$ by completing the accompanying table and plotting the resulting points.

x	-1	0	1	2
y				

(b) Find additional points satisfying $y = 3x^4 - 4x^3$ by completing the accompanying table. Now refine the graph of part (a).

x	-0.75	-0.50	-0.25	0.25	0.5	0.75	1.33
y							

In Exercises 37–42, write the general equation of the circle with:

37. center at the origin and radius 3
38. center at the origin and radius 5
39. center at $(2, -1)$ and radius 4
40. center at $(-4, 3)$ and radius $\frac{5}{8}$
41. center at $(-1, 2)$ and passing through the origin
42. center at $(3, -2)$ and passing through $(-1, 1)$

In Exercises 43–50, write each equation in standard form and sketch its graph.

43. $x^2 + y^2 - 2x + 6y + 6 = 0$
44. $x^2 + y^2 - 2x + 6y - 15 = 0$
45. $x^2 + y^2 - 2x + 6y + 10 = 0$
46. $3x^2 + 3y^2 - 6y - 1 = 0$
47. $2x^2 + 2y^2 - 2x - 2y - 3 = 0$
48. $4x^2 + 4y^2 - 4x + 2y - 1 = 0$
49. $16x^2 + 16y^2 + 16x + 40y - 7 = 0$
50. $x^2 + y^2 - 4x + 2y + 3 = 0$

In Exercises 51–60, find the points of intersection of the graphs of the equations; check your results.

51. $x + y = 2,\ 2x - y = 1$
52. $2x - 3y = 13,\ 5x + 3y = 1$
53. $x + y = 7,\ 3x - 2y = 11$
54. $x^2 + y^2 = 25,\ 2x + y = 10$
55. $x^2 + y^2 = 5,\ x - y = 1$
56. $x^2 + y = 4,\ 2x - y = 1$
57. $y = x^3,\ y = x$
58. $y = x^4 - 2x^2 + 1,\ y = 1 - x^2$
59. $y = x^3 - 2x^2 + x - 1,\ y = -x^2 + 3x - 1$
60. $x = 3 - y^2,\ y = x - 1$
61. Determine whether the points $(1, 2)$, $(1, -1)$, $(4, 5)$ lie on the graph of $2x - y - 3 = 0$.
62. Determine whether the points $(1, -\sqrt{3})$, $(\frac{1}{2}, -1)$, $(\frac{3}{2}, \frac{7}{2})$ lie on the graph of $x^2 + y^2 = 4$.
63. Determine whether the points $(1, \frac{1}{5})$, $(2, \frac{1}{2})$, $(-1, -2)$ lie on the graph of $x^2y - x^2 + 4y = 0$.
64. Determine whether the points $(0, 2)$, $(-2, -\frac{1}{6})$, $(3, -6)$ lie on the graph of $x^2 - xy + 4y = 3$.

Ⓒ**65.** Mark Johnson is setting up a part-time marketing business in his home. After an initial investment of \$5000, Mark's unit cost is \$21.60. Mark sells each unit for \$34.10.
(a) Write an equation for the total cost (including initial investment) of x units.
(b) Write an equation for the total revenue obtained by selling x units.
(c) Find Mark's break-even point by finding the point of intersection of the cost and revenue equations.

Ⓒ**66.** A certain car model costs \$10,500 with a gasoline engine and \$11,450 with a diesel engine. The gasoline engine gets 22 miles per gallon, and the diesel gets 31. The price of both types of fuel is \$1.389 per gallon.
(a) Find the cost C_g of driving the gasoline-powered car x miles (include initial cost and fuel cost but ignore other costs).
(b) Find the cost C_d of driving the diesel-powered car x miles.
(c) Find the break-even point at which the diesel-powered car becomes more economical than the gasoline-powered car.

Ⓒ**67.** Find the sales necessary to break even if the cost of x units is

$$C = 8{,}650x + 250{,}000$$

and the revenue obtained by selling x units is

$$R = 9{,}950x$$

Ⓒ**68.** Find the sales necessary to break even if the cost of x units is

$$C = 5.5\sqrt{x} + 10{,}000$$

and the revenue obtained by selling x units is

$$R = 3.29x$$

Ⓒ**69.** The Consumer Price Index (CPI) for the 1970s is given in the following table.

Year	1970	1971	1972	1973	1974
CPI	116.3	121.3	125.3	133.1	147.7

Year	1975	1976	1977	1978	1979
CPI	161.2	170.5	181.5	195.3	211.1

A mathematical model for the CPI during this 10-year period is

$$y = 0.55t^2 + 5.85t + 114.41$$

where y represents the CPI and t represents the year, with $t = 0$ corresponding to 1970.

Ⓒ Calculator may be helpful.

(a) Use a graph to compare the CPI with the model.
(b) Use the model to predict the CPI for 1985.

70. From the model in Exercise 69, we obtain the model

$$V = \frac{100}{0.55t^2 + 5.85t + 114.41}$$

where V represents the purchasing power of the dollar (in terms of constant 1967 dollars) and t represents the year, with $t = 0$ corresponding to 1970. Use this model to complete the following table.

t	0	2	4	6	8	10	12
V							

71. The farm population in the United States as a percentage of the total population is given in the following table.

Year	1950	1955	1960	1965	1970	1975	1979
Percentage	15.3	11.6	8.7	6.4	4.8	4.2	3.4

A mathematical model for this data is given by

$$y = \frac{100}{4.90 + 0.79t}$$

where y represents the percentage and t represents the year, with $t = 0$ corresponding to 1950.
(a) Use a graph to compare the actual percentage with that given by the model.
(b) Use the model to predict the farm percentage of the population in 1990.

72. The average number of acres per farm in the United States is given in the following table.

Year	Number of acres
1950	213
1960	297
1965	340
1970	374
1975	391
1978	401

A mathematical model for this data is given by

$$y = -0.13t^2 + 10.43t + 211.3$$

where t represents the year, with $t = 0$ corresponding to 1950.
(a) Use a graph to compare the actual number of acres per farm with that given by the model.
(b) Use the model to predict the average number of acres per farm in the United States in 1985.

C Calculator may be helpful.

Section 1.3

Lines in the Plane; Slope

Introductory Example

Straight-Line Depreciation

Most business expenses are deductible for tax purposes during the year the expense occurs. An important exception to this is the cost of property with a useful life of more than one year. For example, the Internal Revenue Service does not allow businesses to deduct the entire cost of buildings, cars, furniture, or machinery in one year. Such costs must be spread out over the useful life of the property. This procedure is called depreciation. If the *same amount* is depreciated each year, the procedure is called **straight-line depreciation.** Specifically, if an item whose initial cost is C will have a salvage value of S after N years, then the amount of depreciation claimed each year is

$$D = \frac{C - S}{N}$$

Table 1.6 and Figure 1.28 illustrate the depreciation of a $12,000 machine over an eight-year period. (The salvage value at the end of eight years is $2,000.) The nondepreciated value of the machine after t years is given by the **linear equation**

$$y = -1{,}250t + 12{,}000$$

Table 1.6

t	y($)
0	12,000
1	10,750
2	9,500
3	8,250
4	7,000
5	5,750
6	4,500
7	3,250
8	2,000

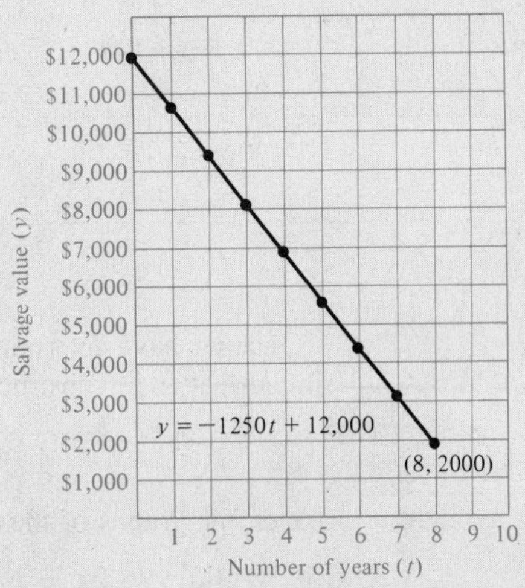

Figure 1.28

Section Objectives: ***To define the slope of a line ▪ To investigate various forms of the equation of a line.***

The simplest mathematical model for relating two variables is the **linear equation**

$$y = mx + b$$

By letting $x = 0$, we see that the line* given by this equation crosses the y-axis at $y = b$. In other words, its y-intercept is $(0, b)$. The steepness or **slope** of the line is given by m. By the slope of a line we mean the number of units the line rises (or falls) vertically for each unit of horizontal change from left to right. (See Figure 1.29.)

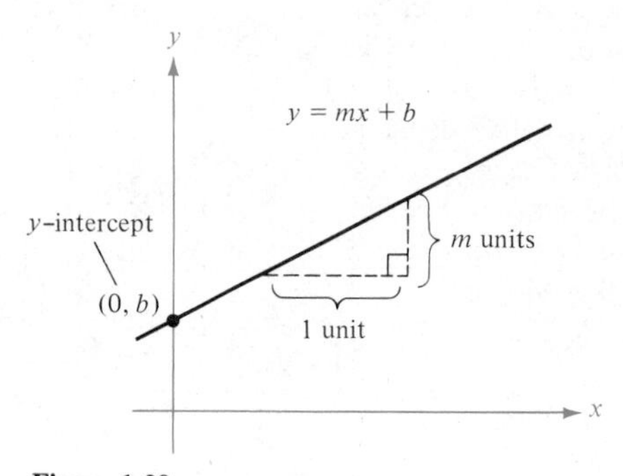

Figure 1.29

The Slope of a Line

The **slope** of the line given by $y = mx + b$ is m.

Once we have determined the y-intercept and slope of a line, it is a relatively simple matter to construct its graph.

Example 1

Graphing a Linear Equation

Sketch the graphs of the following linear equations:

(a) $y = 2x + 1$

(b) $y = 2$

(c) $y = -x + 2$

---

*Note that we use the term *line* to mean *straight line*.

Solution

(a) The y-intercept is $b = 1$, and since the slope is $m = 2$ we know that this line *rises* 2 units for each unit the line moves to the right. (See Figure 1.30.)
(b) The y-intercept is $b = 2$, and since the slope is zero we know that the line is horizontal. That is, it doesn't *rise* or *fall*. (See Figure 1.31.)

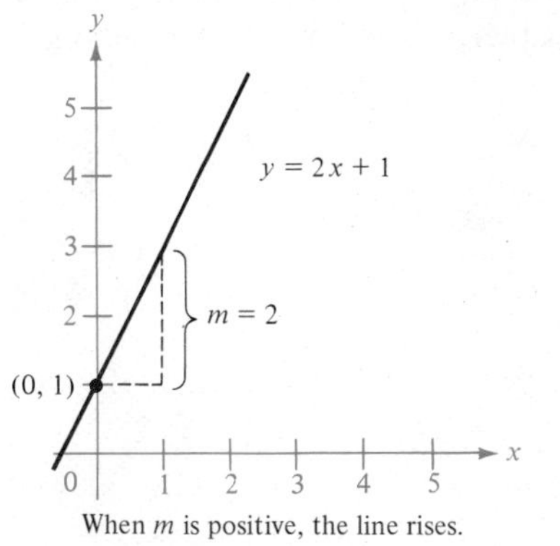

When m is positive, the line rises.

Figure 1.30

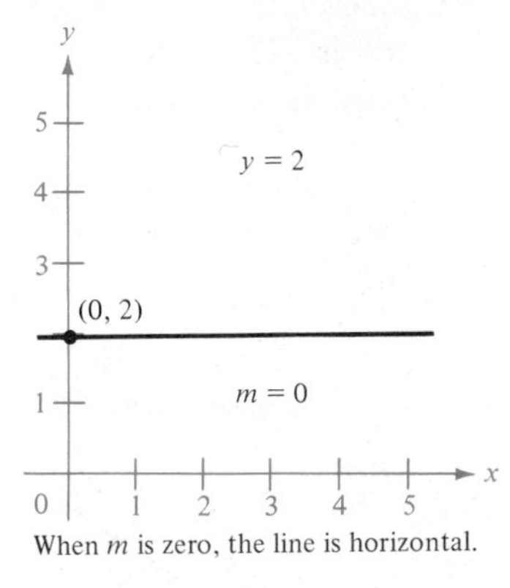

When m is zero, the line is horizontal.

Figure 1.31

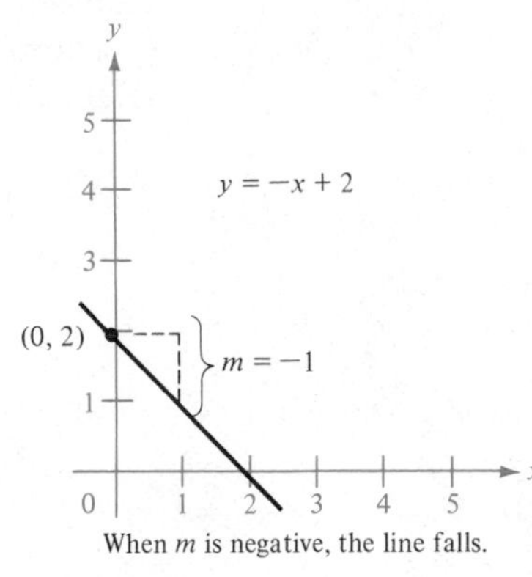

When m is negative, the line falls.

Figure 1.32

(c) The y-intercept is $b = 2$, and since the slope is $m = -1$ we know that the line *falls* 1 unit for each unit the line moves to the right. (See Figure 1.32.) □

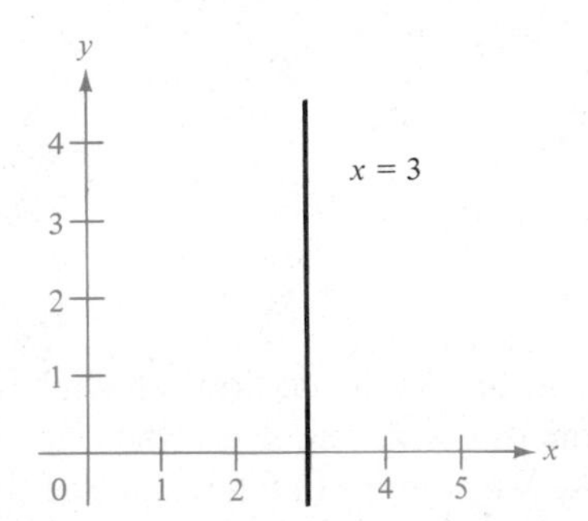

When the line is vertical, the slope is undefined.

Figure 1.33

Note in Example 1 that we gave no examples of vertical lines. Vertical lines have equations of the form $x = a$. Since such an equation cannot be written in the form $y = mx + b$, we say that the slope of a vertical line is undefined. (See Figure 1.33.)

Example 2

A Business Application

A manufacturing company determines that the total cost in dollars of producing x units of a certain product is

$$C = 25x + 3500$$

Describe the practical significance of the y-intercept and slope of the line given by this equation.

Solution The y-intercept $(0, 3500)$ tells us that the cost of producing zero units is \$3500. We call these the **fixed costs** of production, and they include costs such as product development and rent that must be paid regardless of the number of units produced. (See Figure 1.34.) The slope of $m = 25$ tells us that the cost of producing each unit is \$25.00. In the next chapter, we will see that economists refer to this unit cost as the **marginal cost.**

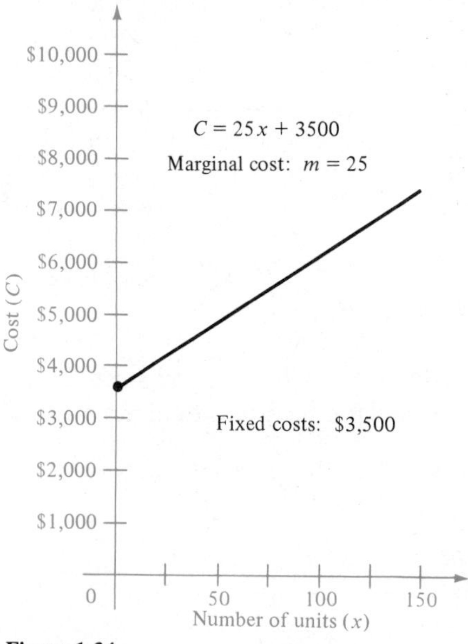

Figure 1.34

□

Now that we can determine the slope of a line from its equation, let us suppose that the equation is not given. How then can we determine the slope? For instance, suppose we want to find the slope of the line passing through the two points (x_1, y_1) and (x_2, y_2) in Figure 1.35. As we move from left to right along this line, a change of $(y_2 - y_1)$ units in the vertical direction corresponds to a change of $(x_2 - x_1)$ units in the horizontal direction. We denote these two changes by the symbols

$$\Delta y = y_2 - y_1 = \text{the change in } y$$

and $$\Delta x = x_2 - x_1 = \text{the change in } x$$

(Δ is the Greek capital letter delta, and the symbols Δy and Δx are read "delta y" and "delta x.") We use the ratio of Δy to Δx to calculate the slope of the line, and we write

$$\text{slope} = \frac{\Delta y}{\Delta x} = \frac{y_2 - y_1}{x_2 - x_1}$$

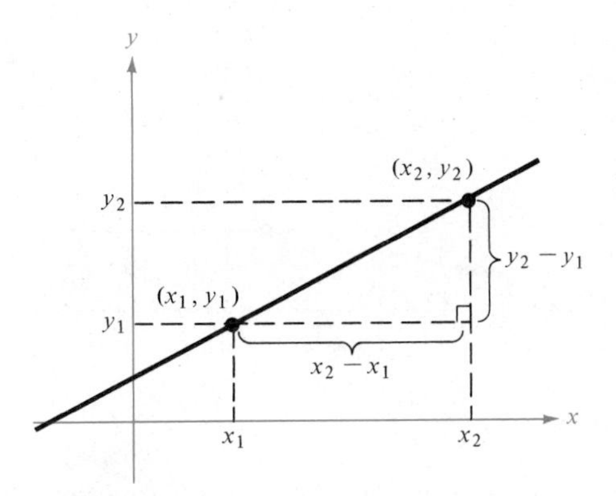

Figure 1.35

Theorem 1.4 The Slope of a Line

The **slope** m of the line passing through the points (x_1, y_1) and (x_2, y_2) is

$$m = \frac{\Delta y}{\Delta x} = \frac{y_2 - y_1}{x_2 - x_1}$$

where $x_1 \neq x_2$.

Example 3

Finding the Slope of a Line Determined by Two Points

Find the slopes of the lines containing each of the following pairs of points:

L_1: $(-2, 0)$ and $(3, 1)$
L_2: $(-1, 2)$ and $(2, 2)$
L_3: $(0, 4)$ and $(1, -1)$

Solution For L_1 the slope is

$$m_1 = \frac{1 - 0}{3 - (-2)} = \frac{1}{3 + 2} = \frac{1}{5}$$

For L_2 the slope is

$$m_2 = \frac{2 - 2}{2 - (-1)} = \frac{0}{3} = 0$$

For L_3 the slope is

$$m_3 = \frac{-1 - 4}{1 - 0} = \frac{-5}{1} = -5$$

See Figure 1.36.

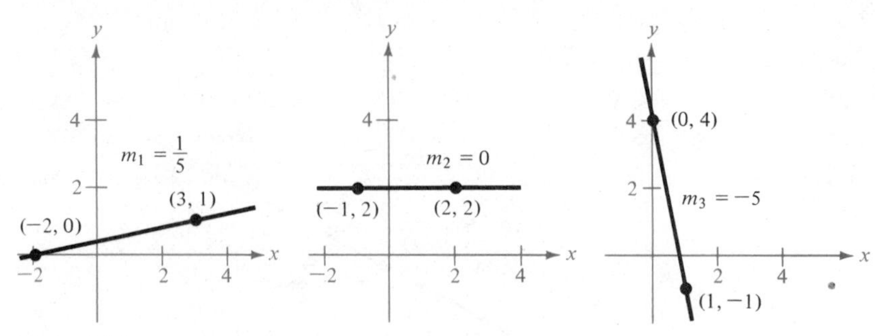

Figure 1.36

We now have enough information about linear equations and slope to look at a very important problem. If we know the slope of a line and one point on the line, how can we determine the equation of the line? Theorem 1.4 gives us the answer to this question. For, if (x_1, y_1) is a point lying on a line of slope m and (x, y) is any *other* point on the line, then

$$\frac{y - y_1}{x - x_1} = m$$

This equation, involving two variables x and y, can be rewritten in the form

$$y - y_1 = m(x - x_1)$$

which is commonly referred to as the **point-slope equation** of a line.

Point-Slope Equation of a Line

The equation of the line with slope m passing through the point (x_1, y_1) is given by

$$y - y_1 = m(x - x_1)$$

Example 4

The Point-Slope Equation of a Line

Find the equation of the line that has a slope of 3 and passes through the point $(1, -2)$.

Solution Using the point-slope form,

$$y - y_1 = m(x - x_1)$$

we have

$$y - (-2) = 3(x - 1)$$

$$y + 2 = 3x - 3$$

or

$$y = 3x - 5$$

See Figure 1.37.

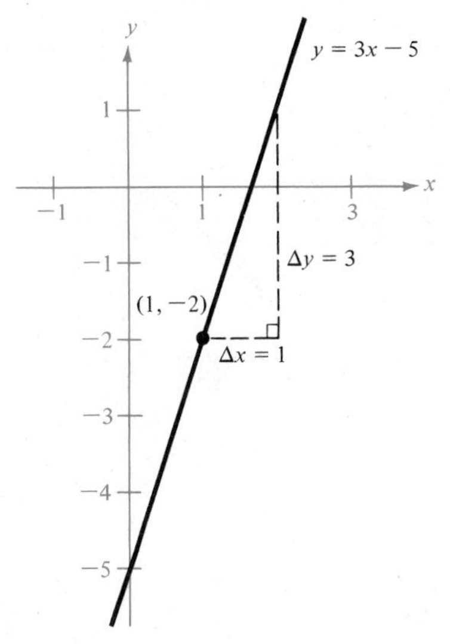

Figure 1.37

From Theorem 1.4, we know that the slope of the line passing through (x_1, y_1) and (x_2, y_2) is

$$m = \frac{y_2 - y_1}{x_2 - x_1}$$

Therefore, we can combine this result with the point-slope equation of a line to obtain the following **two-point equation** of a line.

Two-Point Equation of a Line

The equation of the line passing through the points (x_1, y_1) and (x_2, y_2) is given by

$$y - y_1 = \frac{y_2 - y_1}{x_2 - x_1}(x - x_1)$$

where $x_1 \neq x_2$.

Example 5

Using the Two-Point Equation of a Line

The total U.S. sales (including inventories) during the first two quarters of 1978 were 539.9 and 560.2 billion dollars, respectively. Assuming a *linear growth pattern,* predict the total sales during the fourth quarter of 1978.

Solution Referring to Figure 1.38, we let (1, 539.9) and (2, 560.2) be two points on the line representing the total U.S. sales. We let x represent the quarter and y represent the sales in billions of dollars. Using the two-point equation of a line, we have

$$y - y_1 = \frac{y_2 - y_1}{x_2 - x_1}(x - x_1)$$

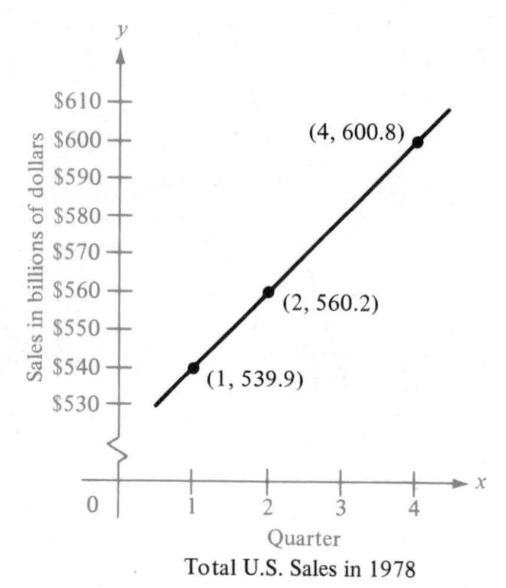

Figure 1.38

$$y - 539.9 = \frac{560.2 - 539.9}{2 - 1}(x - 1)$$

$$y = 20.3(x - 1) + 539.9$$

$$y = 20.3x + 519.6$$

Now, using this linear model, we estimate the fourth-quarter sales ($x = 4$) to be

$$y = (20.3)(4) + 519.6 = 600.8 \text{ billion}$$

(In this particular case, the estimate proved to be quite good. The actual fourth-quarter sales in 1978 were 600.5 billion dollars.) □

The prediction method illustrated in Example 5 is called **linear extrapolation.** Note that the extrapolated point does not lie between the given points. (See Figure 1.39.) When the estimated point lies between two given points, we call the procedure **linear interpolation.**

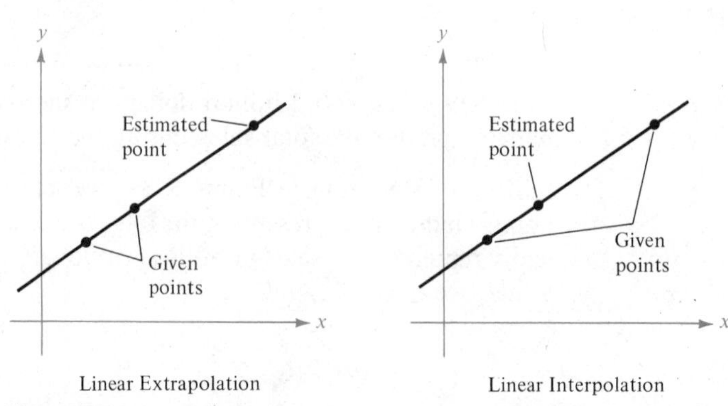

Figure 1.39

Since the slope of a vertical line is not defined, its equation cannot be written in the slope-intercept form. However, every line has an equation that can be written in the form

$$Ax + By + C = 0$$

where A and B are not both zero. For vertical and horizontal lines, the equations $x = a$ and $y = b$ can be written as

$$x - a = 0 \qquad \text{and} \qquad y - b = 0$$

Furthermore, the slope-intercept equation can be written as

$$-mx + y - b = 0$$

Therefore, the form

$$Ax + By + C = 0$$

is called the **general equation** of a line.

We now have identified the following six forms of equations of lines.

Equations of Lines

General equation: $Ax + By + C = 0$
Vertical line: $x = a$
Horizontal line: $y = b$
Slope-intercept equation: $y = mx + b$
Point-slope equation: $y - y_1 = m(x - x_1)$
Two-point equation: $y - y_1 = \dfrac{y_2 - y_1}{x_2 - x_1}(x - x_1)$.

The slope of a line is a convenient tool for determining when two lines are parallel or perpendicular. This is seen in the following two theorems.

Theorem 1.5 Parallel Lines

Two distinct nonvertical lines are parallel if and only if their slopes are equal.

Theorem 1.6 Perpendicular Lines

Two nonvertical lines are perpendicular if and only if their slopes are related by the equation

$$m_1 = -\frac{1}{m_2}$$

Example 6

Finding Parallel and Perpendicular Lines

Find the equation of the line that passes through the point $(2, -1)$ and is

(a) parallel to the line $2x - 3y = 5$
(b) perpendicular to the line $2x - 3y = 5$

Solution By writing the equation $2x - 3y = 5$ in the slope-intercept form, we have

$$3y = 2x - 5$$

or $$y = (\tfrac{2}{3})x - \tfrac{5}{3}$$

Therefore, the given line has a slope of $m = \frac{2}{3}$.

(a) Any line parallel to the given line $2x - 3y = 5$ must have a slope of $\frac{2}{3}$. Thus, the line through $(2, -1)$ that is parallel to the line $2x - 3y = 5$ has an equation of the form

$$y - (-1) = \tfrac{2}{3}(x - 2)$$
$$3(y + 1) = 2(x - 2)$$
$$3y + 3 = 2x - 4$$
$$-2x + 3y = -7$$

or $$2x - 3y = 7$$

(Note the similarity to the original equation $2x - 3y = 5$.) See Figure 1.40.

(b) Any line perpendicular to the line $2x - 3y = 5$ must have a slope of $-\frac{3}{2}$. Therefore, the line through $(2, -1)$ that is perpendicular to the line $2x - 3y = 5$ has the equation

$$y - (-1) = -\tfrac{3}{2}(x - 2)$$
$$2(y + 1) = -3(x - 2)$$

or $$3x + 2y = 4$$

See Figure 1.40.

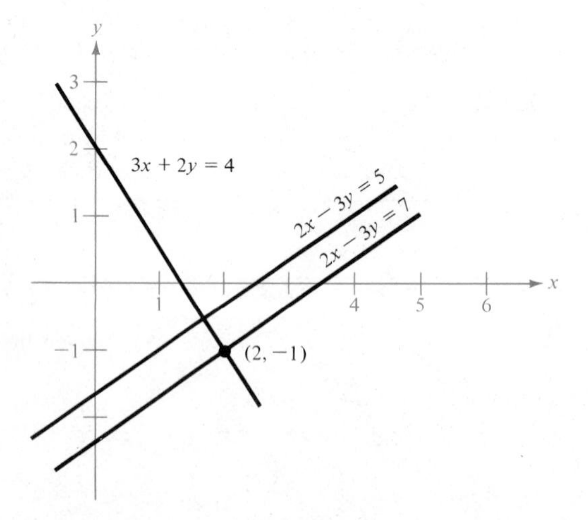

Figure 1.40 □

Section Exercises (1.3)

In Exercises 1–6, estimate the slope of the given line from its graph.

1.

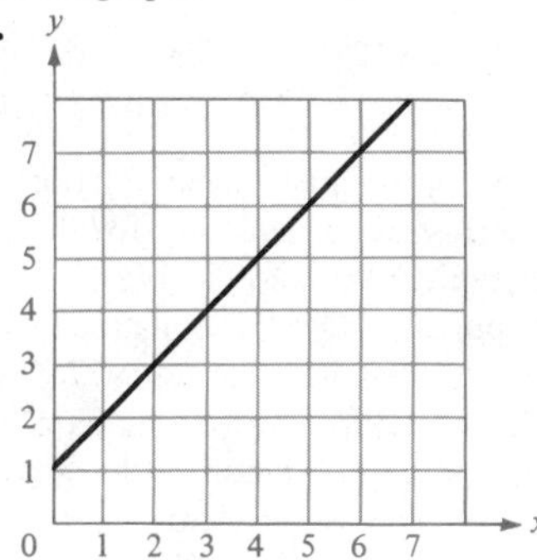

2.

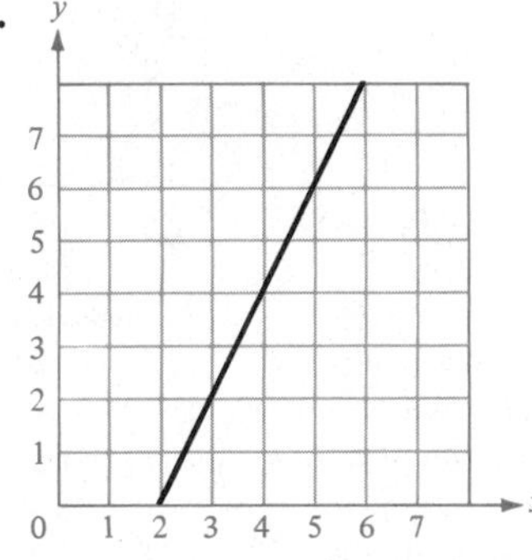

3.

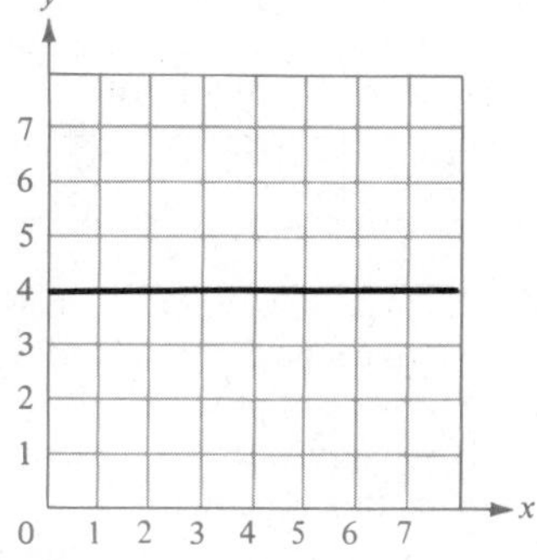

4.

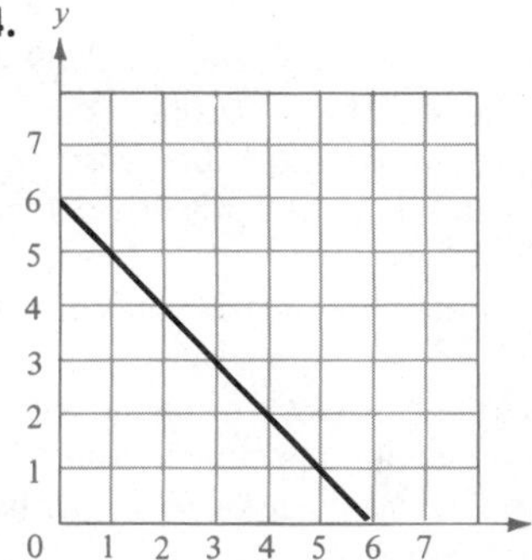

5.

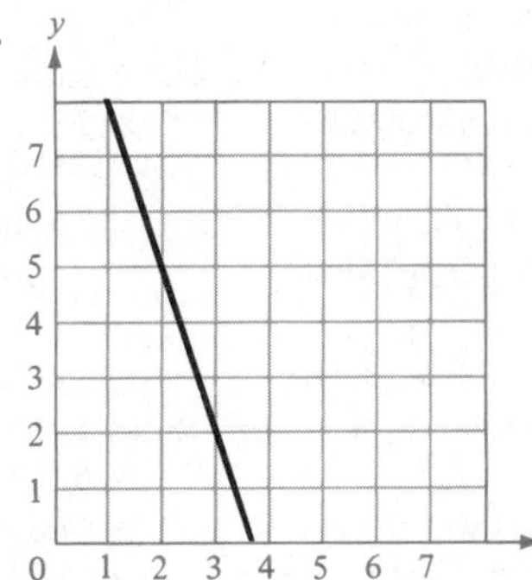

6.

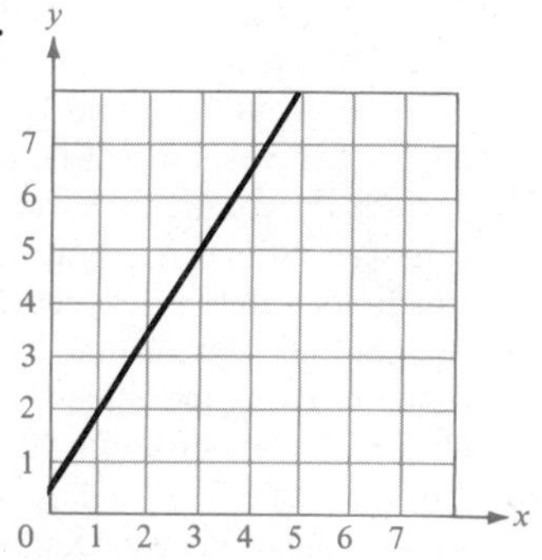

In Exercises 7–14, plot the points and find the slope of the line passing through the given points.

7. $(3, -4), (5, 2)$
8. $(-2, 1), (4, -3)$
9. $(\frac{1}{2}, 2), (6, 2)$
10. $(-\frac{3}{2}, -5), (\frac{5}{6}, 4)$
11. $(-6, -1), (-6, 4)$
12. $(2, 1), (2, 5)$
13. $(1, 2), (-2, -2)$
14. $(\frac{7}{8}, \frac{3}{4}), (\frac{5}{4}, -\frac{1}{4})$

In Exercises 15–33, find an equation for the indicated line and sketch its graph.

15. through $(2, 1)$ and $(0, -3)$
16. through $(-3, -4)$ and $(1, 4)$
17. through $(0, 0)$ and $(-1, 3)$
18. through $(-3, 6)$ and $(1, 2)$
19. through $(2, 3)$ and $(2, -2)$
20. through $(6, 1)$ and $(10, 1)$
21. through $(1, -2)$ and $(3, -2)$
22. through $(\frac{7}{8}, \frac{3}{4})$ and $(\frac{5}{4}, -\frac{1}{4})$
23. through $(0, 3)$, $m = \frac{3}{4}$
24. through $(-1, 2)$, m is undefined
25. through $(0, 0)$, $m = \frac{2}{3}$
26. through $(-1, -4)$, $m = \frac{1}{4}$
27. through $(0, 5)$, $m = -2$
28. through $(-2, 4)$, $m = -\frac{3}{5}$
29. y-intercept 2, $m = 4$
30. y-intercept $-\frac{2}{3}$, $m = \frac{1}{6}$
31. y-intercept $\frac{2}{3}$, $m = \frac{3}{4}$
32. y-intercept 4, $m = 0$
33. vertical line with x-intercept 3

In Exercises 34–39, write an equation of the line through the given point (a) parallel and (b) perpendicular to the given line.

34. $(2, 1)$; $4x - 2y = 3$
35. $(-3, 2)$; $x + y = 7$
36. $(\frac{7}{8}, \frac{3}{4})$; $5x + 3y = 0$
37. $(-6, 4)$; $3x + 4y = 1$
38. $(2, 5)$; $x = 4$
39. $(-1, 0)$; $y = -3$
40. Find the equation of the line giving the relationship between the temperature in degrees Celsius (C) and

in degrees Fahrenheit (F), knowing that water freezes at 0°C, or 32°F, and boils at 100°C, or 212°F.

Ⓒ**41.** Use the result of Exercise 40 to complete the accompanying table.

C		−10°	10°			177°
F	0°			68°	90°	

42. A manufacturer pays its assembly line workers \$4.50 per hour *plus* an additional piecework rate of 75¢ per unit produced. Write a linear equation for the hourly wages W in terms of x, the number of units produced per hour.

43. A small business purchases a piece of equipment for \$875. After five years the equipment will be outdated and will have no value. Write a linear equation giving the value V of the equipment during the five years it will be used.

Ⓒ**44.** A company constructs a warehouse for \$825,000. It has an estimated useful life of 25 years, after which its value is expected to be \$75,000. If straight-line depreciation is used, write a linear equation giving the value V of the warehouse during its 25 years of useful life.

Ⓒ**45.** A real estate office handles an apartment complex with 50 units. When the rent is \$280 per month, all 50 units are occupied. However, when the rent is \$325, the average number of occupied apartments drops to 47. Assume that the relationship between the monthly rent p and the demand x is linear.

(a) Write the equation of the line giving the demand x in terms of the rent p.

(b) (Linear Extrapolation) Use this equation to predict the number of units occupied if the rent is raised to \$355.

(c) (Linear Interpolation) Predict the number of units occupied if the rent is lowered to \$295.

Ⓒ**46.** The amount (in billions of dollars) spent by the United States for energy imports between 1975 and 1978 is given in the following table.

Year	1975	1976	1977	1978
t	0	1	2	3
Imports, y	\$96	\$121	\$148	\$172

(a) Assuming an approximate linear relation between y and t, write an equation for the line passing through (0, 96) and (3, 172).

(b) (Linear Interpolation) Use this equation to estimate the amount spent in 1976 and 1977. Compare the estimate with the actual amount.

(c) (Linear Extrapolation) Predict the amount spent on energy imports in 1980.

(d) What information is given by the slope of the line in part (a)?

Ⓒ**47.** A particular brand of wood stove sells for \$739.40, and a cord of wood sells for \$105.00.

(a) Write an equation giving the total cost, C, in terms of the number, x, of cords of wood purchased.

(b) Find the total cost of burning six cords of wood in this stove.

Ⓒ**48.** A contractor purchases a piece of equipment for \$26,500. The equipment's operator is paid \$9.50 per hour, and it uses an average of \$5.25 per hour for fuel and maintenance.

(a) Write a linear equation giving the total cost C of operating this equipment t hours.

(b) If customers are charged \$25 per hour of machine use, write an equation for the revenue R derived from t hours of use.

(c) Find the break-even point for this equipment by finding the point of intersection of the lines in parts (a) and (b).

49. A sales representative uses his own car as he travels for his company. The cost to the company is \$75.00 per day for lodging and meals *plus* 22¢ per mile driven. Write a linear equation giving the daily cost C to the company in terms of x, the number of miles driven.

Ⓒ Calculator may be helpful.

Section 1.4

Functions

Introductory Example

Federal Income Tax

The 1980 federal income tax for single taxpayers is shown in Table 1.7, where y represents the income tax (in dollars) and x represents the taxable income.

In mathematical terms, we say that y (the income tax) is a **function** of x (the taxable income). Note that the term *function* is not the same as *equation*. For example, in Table 1.7 several equations describe the single function for income tax. We can see that the graph in Figure 1.41 is made up of line segments of increasing slope. Economists call this type of tax schedule *progressive* because the rate of taxation increases as the income increases.

Table 1.7

Taxable income intervals	Income tax
$0 \le x \le 2{,}300$	$y = 0$
$2{,}300 < x \le 3{,}400$	$y = 0.14x - 322$
$3{,}400 < x \le 4{,}400$	$y = 0.16x - 390$
$4{,}400 < x \le 6{,}500$	$y = 0.18x - 478$
$6{,}500 < x \le 8{,}500$	$y = 0.19x - 543$
$8{,}500 < x \le 10{,}800$	$y = 0.21x - 713$
$10{,}800 < x \le 12{,}900$	$y = 0.24x - 1{,}037$
$12{,}900 < x \le 15{,}000$	$y = 0.26x - 1{,}295$
$15{,}000 < x \le 18{,}200$	$y = 0.30x - 1{,}895$
$18{,}200 < x \le 23{,}500$	$y = 0.34x - 2{,}623$
$23{,}500 < x \le 28{,}800$	$y = 0.39x - 3{,}798$
$28{,}800 < x \le 34{,}100$	$y = 0.44x - 5{,}238$
$34{,}100 < x \le 41{,}500$	$y = 0.49x - 6{,}943$
$41{,}500 < x \le 55{,}300$	$y = 0.55x - 9{,}433$
$55{,}300 < x \le 81{,}800$	$y = 0.63x - 13{,}857$
$81{,}800 < x \le 108{,}300$	$y = 0.68x - 17{,}947$
$108{,}300 < x$	$y = 0.70x - 20{,}113$

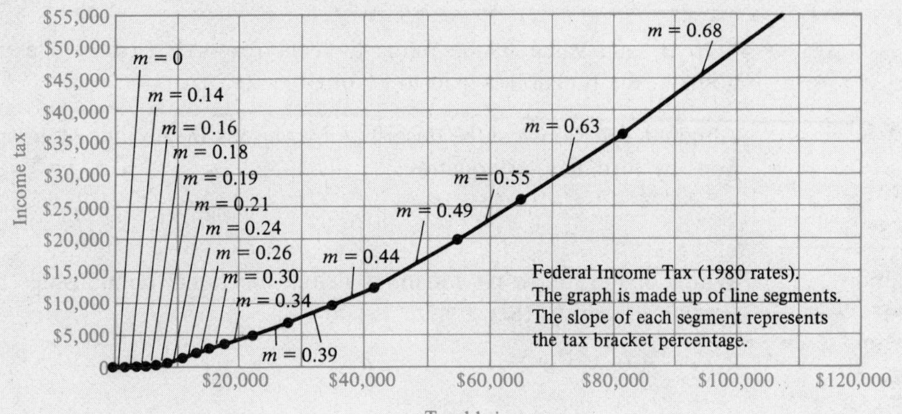

Figure 1.41

Section Objectives: ***To define the term "function" ▪ To introduce functional notation ▪ To characterize the inverse of a function.***

In many common relationships between two variables, the value of one of the variables depends on the value of the other. For example, the sales tax on an item depends on its selling price; the distance an object moves in a given time depends on its speed; the pressure of a gas in a closed container depends on its temperature; and the area of a circle depends on its radius.

Consider the relationship between the area of a circle and its radius. This relationship can be expressed by the equation

$$A = \pi r^2$$

Considering the radius of a circle to be positive, we have within the set of positive numbers a free choice for the value of r. The value of A then depends on our choice of r. Thus, we refer to A as the **dependent variable** and r as the **independent variable.**

The *type* of relationship between two variables is of extreme importance in calculus. Specifically, we are interested in relationships such that to every value of the independent variable there corresponds *one and only one* value of the dependent variable. Mathematically, we call this type of correspondence a **function.**

Definition of a Function

A **function** is a relationship between two variables such that to each value of the independent variable there corresponds exactly one value of the dependent variable.

The collection of all values assumed by the independent variable is called the **domain** of the function, and the collection of all values assumed by the dependent variable is called the **range** of the function.

If to each value in the range there corresponds exactly one value in the domain, the function is said to be **one-to-one.**

Remark Although functions can be described by various means, we often specify functions by formulas, or equations.

Example 1

Determining Functional Relationships from Equations

Which of the following equations define functional relationships between the variables x and y? (See Figure 1.42.)

(a) $x + y = 1$

(b) $x^2 + y^2 = 1$

(c) $x^2 + y = 1$

(d) $x + y^2 = 1$

Solution The standard procedure in writing an equation for a function is to isolate the dependent variable on the left-hand side. Thus, we have Table 1.8.

Note that those equations that assign two values ($\pm$) to the dependent variable for each assigned value of the independent variable do not define functions. For instance, if $y = 0$, then the equation $x = \pm\sqrt{1 - y^2}$ indicates that $x = +1$ or $x = -1$.

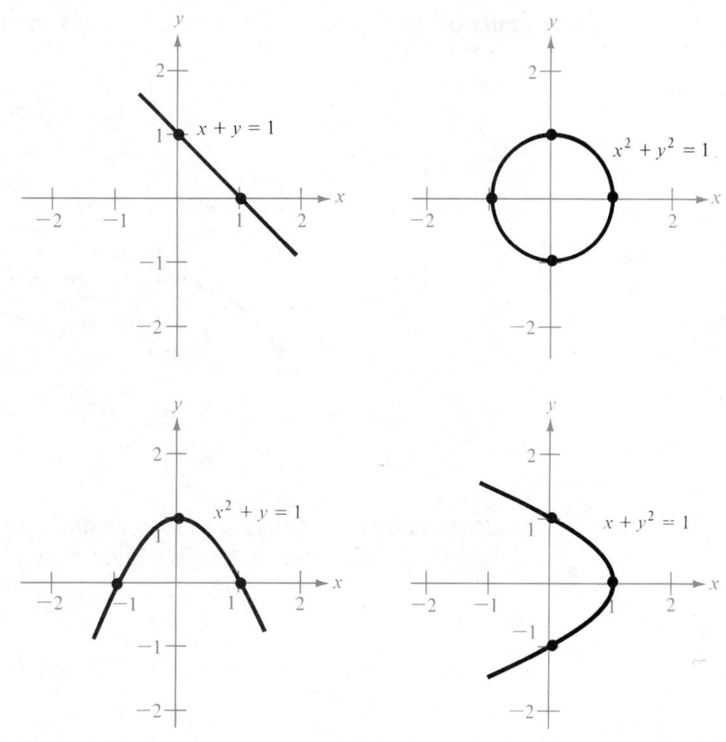

Figure 1.42

Table 1.8

Original equation	x as the dependent variable	Is x a function of y?	y as the dependent variable	Is y a function of x?
(a) $x + y = 1$	$x = 1 - y$	yes	$y = 1 - x$	yes
(b) $x^2 + y^2 = 1$	$x = \pm\sqrt{1 - y^2}$	no (two values of x for some values of y)	$y = \pm\sqrt{1 - x^2}$	no (two values of y for some values of x)
(c) $x^2 + y = 1$	$x = \pm\sqrt{1 - y}$	no	$y = 1 - x^2$	yes
(d) $x + y^2 = 1$	$x = 1 - y^2$	yes	$y = \pm\sqrt{1 - x}$	no

□

Example 2

Finding the Domain and Range of a Function

Determine the domain and range for the function of x defined by $y = \sqrt{x - 1}$.

Solution Since $\sqrt{x - 1}$ is not defined for $x - 1 < 0$ (that is, for $x < 1$), we must have $x \geq 1$. Therefore, the interval $[1, \infty)$ is the domain of the function. Since $\sqrt{x - 1}$ is always positive or zero on its domain, and since $\sqrt{x - 1}$ increases as x increases, the range of the function is the interval $[0, \infty)$. The graph of the function lends further support to our conclusions (Figure 1.43).

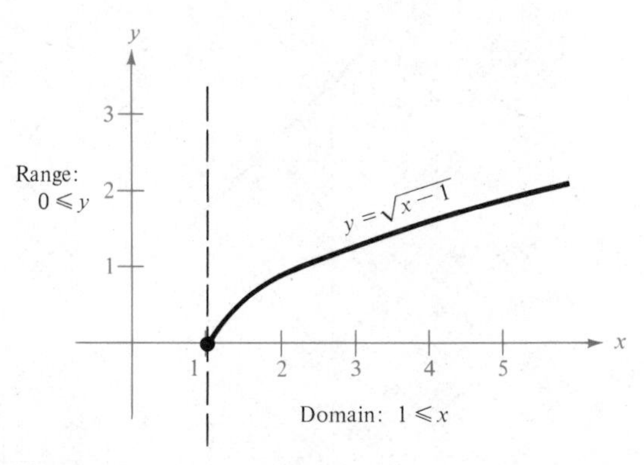

Figure 1.43

□

On occasion, we define a function using more than one equation. This procedure is demonstrated in the following example.

Example 3

Finding the Domain and Range of a Function

Determine the domain and range for the function of x given by

$$y = \begin{cases} \sqrt{x - 1}, \text{ if } x \geq 1 \\ 1 - x, \text{ if } x < 1 \end{cases}$$

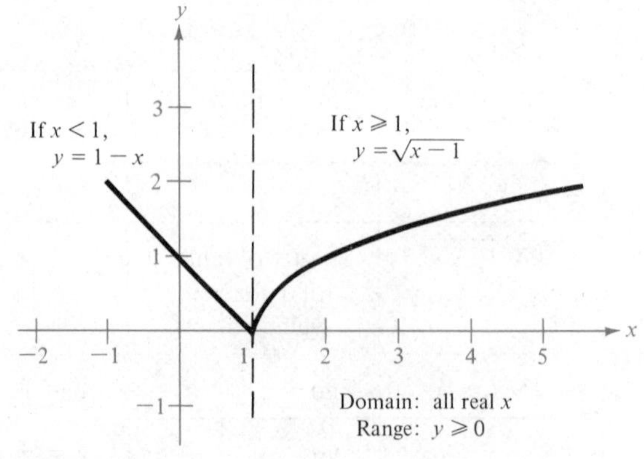

Figure 1.44

Solution Since $x \geq 1$ or $x < 1$, the domain of the function is the entire set of real numbers. On the portion of the domain for which $x \geq 1$, the function behaves as in Example 2. For $x < 1$, $1 - x$ is positive, and therefore, the range of the function is the interval $[0, \infty)$. Again, a graph of the function helps to verify our conclusions (Figure 1.44). □

Remark When equations are used to describe functions, sometimes the domain is clearly specified, as in Example 3 ($x \geq 1$ or $x < 1$). On other occasions, the domain is *implied,* as in Example 2 ($\sqrt{x - 1}$ is defined only if $x \geq 1$). On still other occasions, the physical nature of the problem may restrict the domain to a certain subset of the real numbers. For instance, the equation for the area of a circle, $A = \pi r^2$, has no specified restrictions; yet physically we always consider the radius of a circle to be positive. Thus, we list the domain of this function as all $r > 0$.

When using an equation to define a function, we generally isolate the dependent variable on the left side of the equation. For instance, writing the equation $x + 2y = 1$ in the form

$$y = \frac{1 - x}{2}$$

indicates that y is the dependent variable. In **functional notation,** this equation has the form

$$f(x) = \frac{1 - x}{2}$$

This notation has the advantage of clearly identifying the dependent variable as $f(x)$ while at the same time providing a name "f" for the function. [The symbol $f(x)$ is read "f of x."]

To denote the value of the dependent variable when $x = 3$, we use the symbol $f(3)$ as follows:

$$f(3) = \frac{1 - (3)}{2} = \frac{-2}{2} = -1$$

Similarly,

$$f(0) = \frac{1 - (0)}{2} = \frac{1}{2}$$

$$f(-2) = \frac{1 - (-2)}{2} = \frac{3}{2}$$

The values $f(3)$, $f(0)$, and $f(-2)$ are called **functional values,** and they lie in the range of f. This means that the values $f(3)$, $f(0)$, and $f(-2)$ are y-values, and thus the points $(3, f(3))$, $(0, f(0))$, and $(-2, f(-2))$ lie on the graph of f.

Example 4

Evaluating a Function

If a function f is defined by the equation $f(x) = 2x^2 - 4x + 1$, find the value of f when x is -1, 0, and 2. Is f one-to-one?

Solution When $x = -1$, the value of f is given by

$$f(-1) = 2(-1)^2 - 4(-1) + 1 = 2 + 4 + 1 = 7$$

When $x = 0$, the value of f is given by

$$f(0) = 2(0)^2 - 4(0) + 1 = 0 - 0 + 1 = 1$$

When $x = 2$, the value of f is given by

$$f(2) = 2(2)^2 - 4(2) + 1 = 8 - 8 + 1 = 1$$

Note that two different values of x may yield the same value for $f(x)$. Thus, f is *not* one-to-one. (See Figure 1.45.)

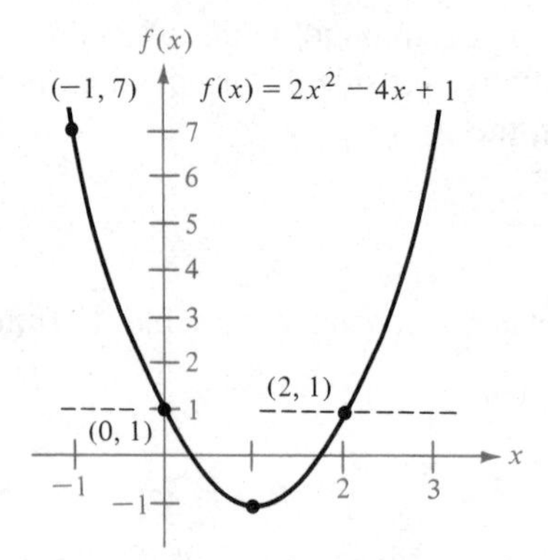

Figure 1.45

□

Remark

Graphically, we can see that the function $y = f(x)$ is not one-to-one if a horizontal line intersects the graph of the function more than once. For instance, in Example 4, the line $y = 1$ intersects the graph of the function twice, and hence the function is not one-to-one.

Example 4 suggests that the role of the variable x in the equation

$$f(x) = 2x^2 - 4x + 1$$

is simply that of a "placeholder." The same function f can be properly described by using parentheses instead of x. Thus, f can be defined by the equation

$$f(\quad) = 2(\quad)^2 - 4(\quad) + 1$$

Therefore, to evaluate $f(-2)$, we simply place -2 in each set of parentheses:

$$f(-2) = 2(-2)^2 - 4(-2) + 1 = 2(4) + 8 + 1 = 17$$

Example 5

Evaluating a Function

For the function f defined by $f(x) = x^2 - 4x + 7$, evaluate

(a) $f(3x)$ (b) $f(x - 1)$ (c) $f(x + \Delta x)$

(d) $\dfrac{f(x + \Delta x) - f(x)}{\Delta x}$

Solution We begin by writing the equation for f in the form

$$f(\quad) = (\quad)^2 - 4(\quad) + 7$$

(a)
$$\begin{aligned} f(3x) &= (3x)^2 - 4(3x) + 7 \\ &= 9x^2 - 12x + 7 \end{aligned}$$

(b)
$$\begin{aligned} f(x - 1) &= (x - 1)^2 - 4(x - 1) + 7 \\ &= x^2 - 2x + 1 - 4x + 4 + 7 \\ &= x^2 - 6x + 12 \end{aligned}$$

(c)
$$\begin{aligned} f(x + \Delta x) &= (x + \Delta x)^2 - 4(x + \Delta x) + 7 \\ &= x^2 + 2x\Delta x + (\Delta x)^2 - 4x - 4\Delta x + 7 \end{aligned}$$

(d)
$$\begin{aligned} &\frac{f(x + \Delta x) - f(x)}{\Delta x} \\ &= \frac{[(x + \Delta x)^2 - 4(x + \Delta x) + 7] - [x^2 - 4x + 7]}{\Delta x} \\ &= \frac{x^2 + 2x\Delta x + (\Delta x)^2 - 4x - 4\Delta x + 7 - x^2 + 4x - 7}{\Delta x} \\ &= \frac{2x\Delta x + (\Delta x)^2 - 4\Delta x}{\Delta x} \\ &= 2x + \Delta x - 4 \end{aligned}$$

□

Although we generally use f as a convenient function name and x as the independent variable, we can use other symbols. For instance, the following equations all define the same function:

$$f(x) = x^2 - 4x + 7$$
$$f(t) = t^2 - 4t + 7$$
$$g(s) = s^2 - 4s + 7$$

Two functions can be combined in various ways to create new functions. For example, if

$$f(x) = 2x - 3 \qquad \text{and} \qquad g(x) = x^2 + 1$$

we can form the functions

$$f(x) + g(x) = (2x - 3) + (x^2 + 1) = x^2 + 2x - 2 \qquad \text{(sum)}$$
$$f(x) - g(x) = (2x - 3) - (x^2 + 1) = -x^2 + 2x - 4 \qquad \text{(difference)}$$
$$f(x)g(x) = (2x - 3)(x^2 + 1) = 2x^3 - 3x^2 + 2x - 3 \qquad \text{(product)}$$
$$\frac{f(x)}{g(x)} = \frac{2x - 3}{x^2 + 1} \qquad \text{(quotient)}$$

We can combine two functions in yet another way called the **composition** of two functions.

Definition of Composite Function

Let f and g be functions such that the range of g is in the domain of f. Then the function whose values are given by $f(g(x))$ is called the **composite** of f with g.

It is important to realize that the composite of f *with* g may not be equal to the composite of g *with* f. This is illustrated in the following example.

Example 6

Forming Composite Functions

Given $f(x) = 2x - 3$ and $g(x) = x^2 + 1$, find

(a) $f(g(x))$ (b) $g(f(x))$

Solution

(a) Since

$$f(\quad) = 2(\quad) - 3$$

we have

$$\begin{aligned} f(g(x)) &= 2(g(x)) - 3 \\ &= 2(x^2 + 1) - 3 \\ &= 2x^2 - 1 \end{aligned}$$

(b) Since

$$g(\quad) = (\quad)^2 + 1$$

we have

$$\begin{aligned} g(f(x)) &= (f(x))^2 + 1 \\ &= (2x - 3)^2 + 1 \\ &= 4x^2 - 12x + 10 \end{aligned}$$

□

Note in Example 6 that $f(g(x)) \neq g(f(x))$, and, in general, this is the case. An important case in which these two composite functions are equal occurs when

$$f(g(x)) = g(f(x)) = x$$

We call such functions **inverses** of each other, as stated in the following definition.

Definition of Inverse Functions

Two functions f and g are **inverses** of each other if

$$f(g(x)) = x \qquad \text{for each } x \text{ in the domain of } g$$

and $$g(f(x)) = x \qquad \text{for each } x \text{ in the domain of } f$$

We denote g by f^{-1} (read "f inverse").

Remark For inverse functions f and g, the range of g must be equal to the domain of f, and vice versa.

Remark Don't be confused by the "exponential notation" for inverse functions. Whenever we write $f^{-1}(x)$, we will *always* be referring to the inverse of the function f and not to the reciprocal of f. In other words, it is generally true that

$$f^{-1}(x) \neq \frac{1}{f(x)}$$

Example 7

Demonstrating Properties of Inverse Functions

Show that the following functions are inverses of each other.

$$f(x) = 2x^3 - 1 \qquad \text{and} \qquad g(x) = \sqrt[3]{\frac{x+1}{2}}$$

Solution First, note that both composite functions exist since the domain and range of both f and g consist of the set of all real numbers.

The composite of f with g is given by

$$\begin{aligned} f(g(x)) &= 2\left(\sqrt[3]{\frac{x+1}{2}}\right)^3 - 1 \\ &= 2\left(\frac{x+1}{2}\right) - 1 = x + 1 - 1 = x \end{aligned}$$

The composite of g with f is given by

$$\begin{aligned} g(f(x)) &= \sqrt[3]{\frac{(2x^3 - 1) + 1}{2}} \\ &= \sqrt[3]{\frac{2x^3}{2}} = \sqrt[3]{x^3} = x \end{aligned}$$

Since $f(g(x)) = g(f(x)) = x$, f and g are inverses of each other. □

The following theorem suggests a geometrical interpretation of inverse functions.

Theorem 1.7 The graph of f contains the point (a, b) if and only if the graph of f^{-1} contains the point (b, a).

Theorem 1.7 can be interpreted geometrically to mean that the graph of f^{-1} can be obtained by reflecting the graph of f in the line $y = x$ (Figure 1.46).

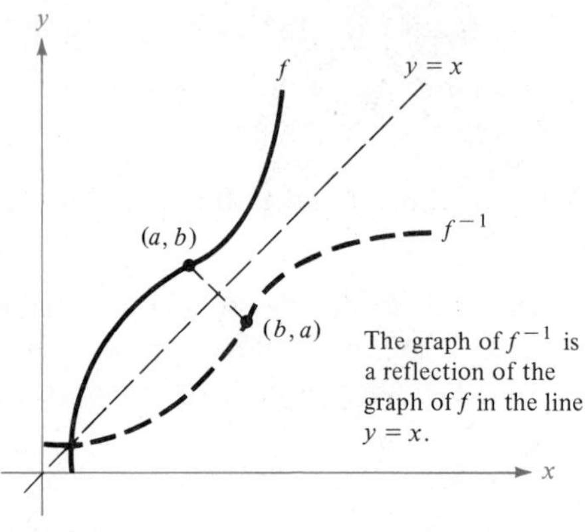

Figure 1.46

Example 8

Finding the Inverse of a Function

Find the inverse of the function given by $f(x) = \sqrt{2x - 3}$.

Solution Substituting y for $f(x)$, we have $y = \sqrt{2x - 3}$. Now to find the inverse function, we simply solve for x in terms of y. Since y is nonnegative, squaring both sides gives an equivalent equation:

$$\sqrt{2x - 3} = y$$

$$2x - 3 = y^2 \qquad y \geq 0$$

$$2x = y^2 + 3$$

$$x = \frac{y^2 + 3}{2} \qquad y \geq 0$$

Thus, the inverse function has the form

$$f^{-1}(\quad) = \frac{(\quad)^2 + 3}{2}$$

Using x as the independent variable, we write

$$f^{-1}(x) = \frac{x^2 + 3}{2} \qquad x \geq 0$$

Note from the graphs of these two functions in Figure 1.47 that the domain of f^{-1} coincides with the range of f.

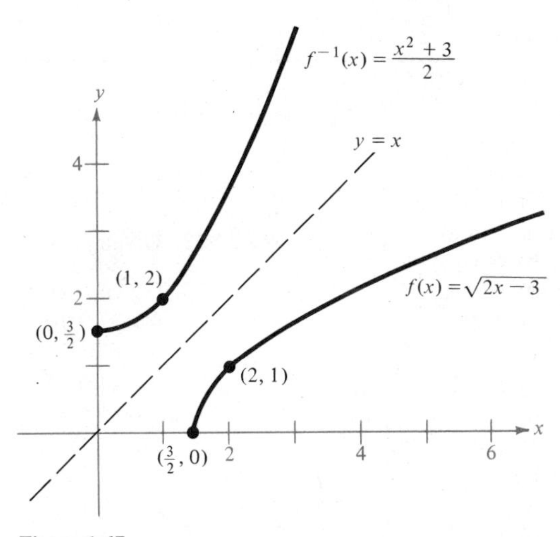

Figure 1.47

□

Not all functions possess an inverse. In fact, for a function f to have an inverse it is necessary that f be one-to-one. The next example is a case in point.

Example 9

A Function That Has No Inverse

Find the inverse (if it exists) of the function given by

$$f(x) = x^2 - 1$$

(Assume the domain of f is the set of all real numbers.)

Solution First, we note that

$$f(2) = (2)^2 - 1 = 3$$

and

$$f(-2) = (-2)^2 - 1 = 3$$

Thus, f is not one-to-one, and it has no inverse. This same conclusion can be obtained by substituting y for $f(x)$ and solving for x as follows:

$$x^2 - 1 = y$$
$$x^2 = y + 1$$
$$x = \pm\sqrt{y + 1}$$

This last equation does not define x as a function of y, and thus f has no inverse.

□

Section Exercises (1.4)

1. Given $f(x) = 2x - 3$, find:
(a) $f(1)$ (b) $f(0)$
(c) $f(-3)$ (d) $f(b)$
(e) $f(x - 1)$ (f) $f(\frac{1}{4})$

2. Given $f(x) = x^2 - 2x + 2$, find:
(a) $f(\frac{1}{2})$ (b) $f(3)$
(c) $f(-1)$ (d) $f(c)$
(e) $f(x + \Delta x)$ (f) $f(2)$

3. Given $f(x) = \sqrt{x + 3}$, find:
(a) $f(-3)$ (b) $f(-2)$
(c) $f(0)$ (d) $f(6)$
(e) $f(x + \Delta x)$ (f) $f(c)$

4. Given $f(x) = 1/\sqrt{x}$, find:
(a) $f(1)$ (b) $f(4)$
(c) $f(2)$ (d) $f(\frac{1}{4})$
(e) $f(x + \Delta x)$ (f) $f(x + \Delta x) - f(x)$

5. Given $f(x) = |x|/x$, find:
(a) $f(2)$ (b) $f(-2)$
(c) $f(-100)$ (d) $f(100)$
(e) $f(x^2)$ (f) $f(x - 1)$

6. Given $f(x) = |x| + 4$, find:
(a) $f(2)$ (b) $f(-2)$
(c) $f(3)$ (d) $f(x^2)$
(e) $f(x + \Delta x)$ (f) $f(x + \Delta x) - f(x)$

7. Given $f(x) = x^2 - x + 1$, find
$$\frac{f(2 + \Delta x) - f(2)}{\Delta x}.$$

8. Given $f(x) = \dfrac{1}{x}$, find $\dfrac{f(1 + \Delta x) - f(1)}{\Delta x}$.

9. Given $f(x) = x^3$, find $\dfrac{f(x + \Delta x) - f(x)}{\Delta x}$.

10. Given $f(x) = 3x - 1$, find $\dfrac{f(x) - f(1)}{x - 1}$.

11. Given $f(x) = \dfrac{1}{\sqrt{x - 1}}$, find $\dfrac{f(x) - f(2)}{x - 2}$.

12. Given $f(x) = x^3 - x$, find $\dfrac{f(x) - f(1)}{x - 1}$.

13. Given $f(x) = x^2 - x$ and $g(x) = 3x - 1$, find:
(a) $g(f(0))$ (b) $f(g(0))$
(c) $f(g(3))$ (d) $g(f(3))$
(e) $g(f(\frac{1}{2}))$ (f) $f(g(x))$

14. Given $f(x) = \sqrt{x}$ and $g(x) = x^2 - 1$, find:
(a) $f(g(1))$ (b) $g(f(1))$
(c) $g(f(0))$ (d) $f(g(-4))$
(e) $f(g(x))$ (f) $g(f(x))$

15. Given $f(x) = 1/x$ and $g(x) = x^2 - 1$, find:
(a) $f(g(2))$ (b) $g(f(2))$
(c) $f(g(1/\sqrt{2}))$ (d) $g(f(1/\sqrt{2}))$
(e) $g(f(x))$ (f) $f(g(x))$

16. Given $f(x) = 1/(x - 2)$ and $g(x) = \sqrt{2x + 3}$, find:
(a) $f(g(-1))$ (b) $g(f(3))$
(c) $g(f(1))$ (d) $f(g(0))$
(e) $g(f(0))$ (f) $g(f(x))$

17. Let $f(x) = x^2 - 1$. Find all real numbers x such that $f(x) = 8$.

18. Let $f(x) = x^3 - x$. Find all real numbers x such that $f(x) = 0$.

19. Let
$$f(x) = \frac{3}{x - 1} + \frac{4}{x - 2}$$
Find all real numbers x such that $f(x) = 0$.

20. Let $f(x) = a + (b/x)$. Find all real numbers x such that $f(x) = 0$.

In Exercises 21–30, find the domain and range of the given function.

21. $f(x) = \sqrt{x - 1}$

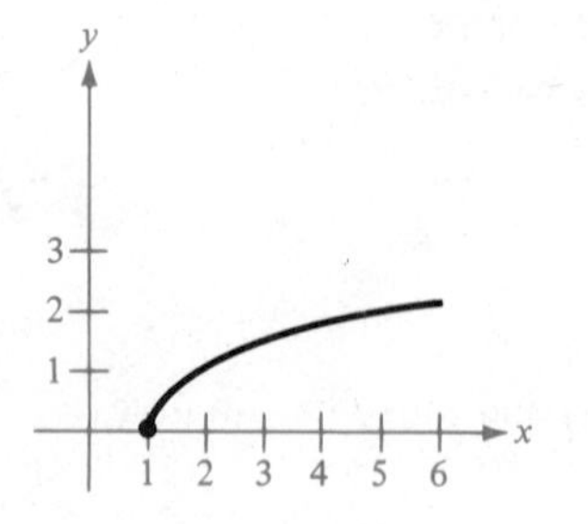

22. $f(x) = \sqrt{1 - x}$

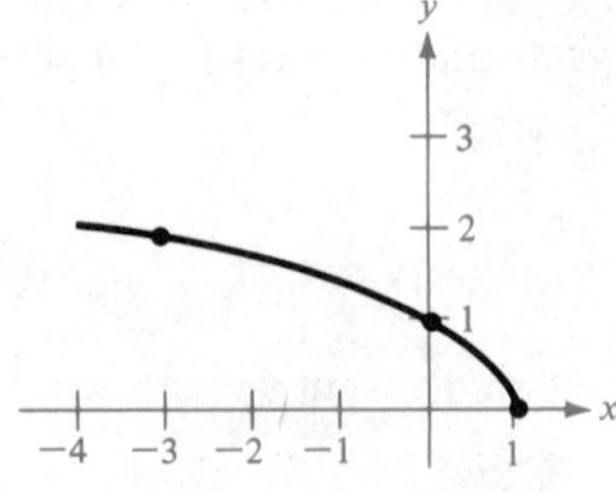

23. $f(x) = x^2$

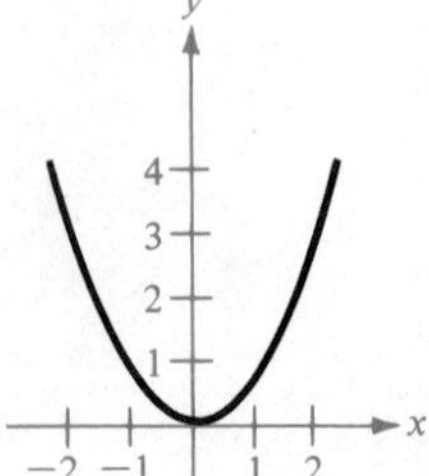

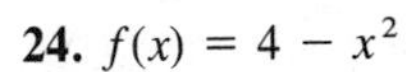

24. $f(x) = 4 - x^2$

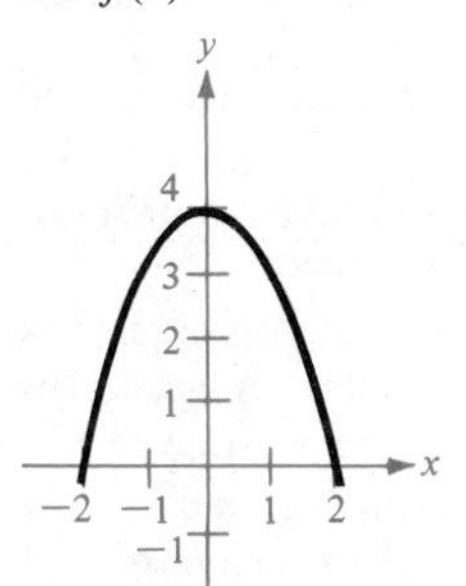

25. $f(x) = \sqrt{9 - x^2}$

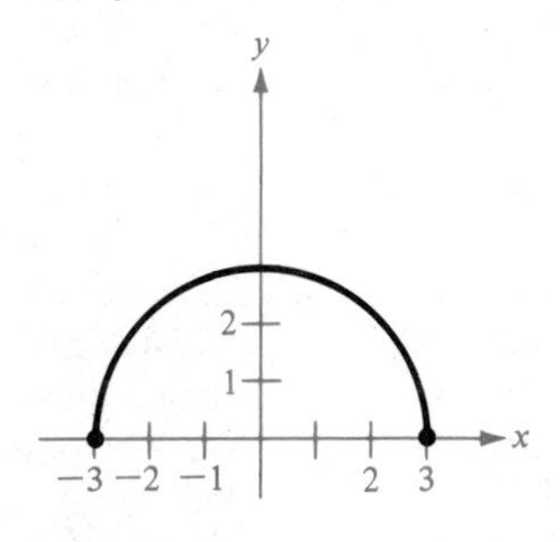

26. $f(x) = \sqrt{25 - x^2}$

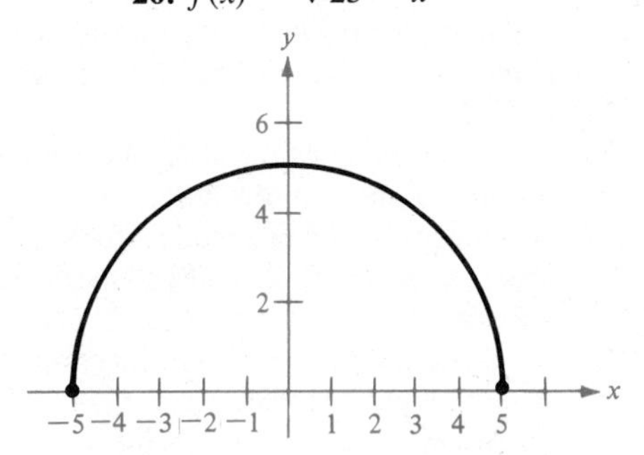

27. $f(x) = \dfrac{1}{|x|}$

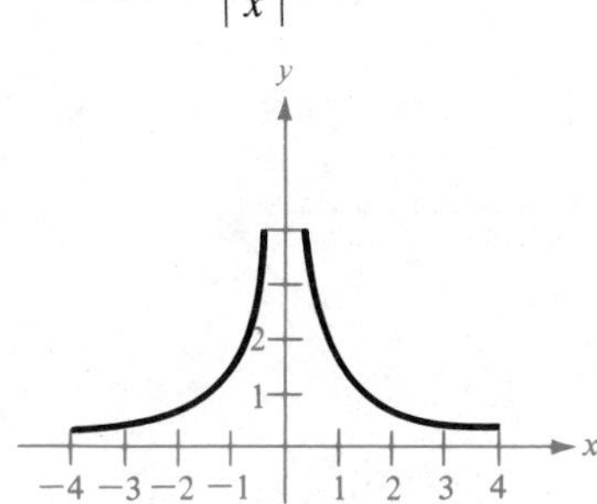

28. $f(x) = |x - 2|$

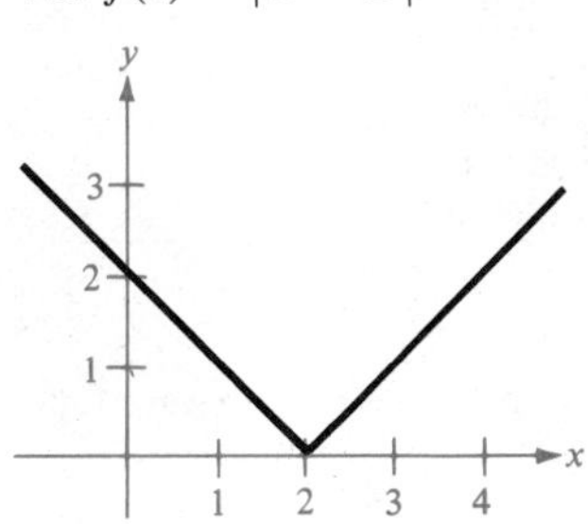

29. $f(x) = \dfrac{|x|}{x}$

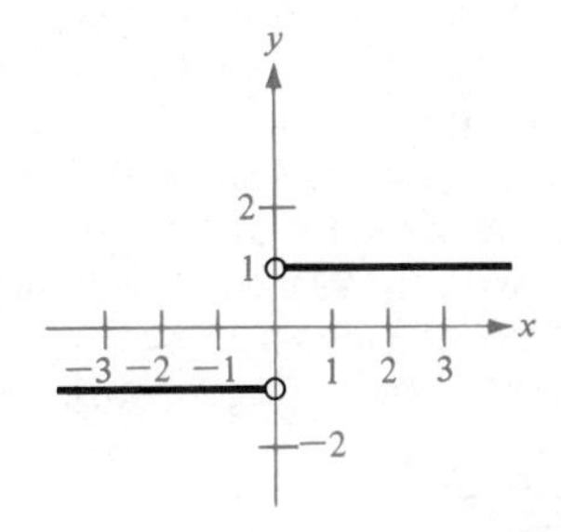

30. $f(x) = \sqrt{x^2 - 4}$

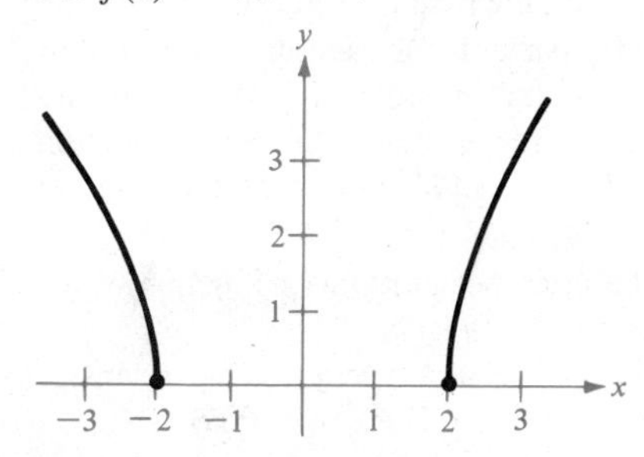

In Exercises 31–40, identify the equations that determine y as a function of x.

31. $x^2 + y^2 = 4$ **32.** $x = y^2$

33. $x^2 + y = 4$ **34.** $x + y^2 = 4$

35. $2x + 3y = 4$

36. $x^2 + y^2 - 2x - 4y + 1 = 0$

37. $y^2 = x^2 - 1$ **38.** $y = \pm\sqrt{x}$

39. $x^2y - x^2 + 4y = 0$

40. $xy - y - x - 2 = 0$

In Exercises 41–48, find the inverse of f, then graph both f and f^{-1}.

41. $f(x) = 2x - 3$ **42.** $f(x) = 3x$

43. $f(x) = x^3$ **44.** $f(x) = x^3 + 1$

45. $f(x) = \sqrt{x}$ **46.** $f(x) = x^2,\ 0 \le x$

47. $f(x) = \dfrac{1}{x}$ **48.** $f(x) = \sqrt[3]{x}$

In Exercises 49–54, determine if the given function is one-to-one and, if so, find its inverse.

49. $f(x) = ax + b$ **50.** $f(x) = \sqrt{x - 2}$

51. $f(x) = x^2$ **52.** $f(x) = x^4$

53. $f(x) = |x - 2|$ **54.** $f(x) = \dfrac{1}{x - 1}$

In Exercises 55–56, find a formula for the given function and give its domain.

55. The value V of a farm having \$500,000 worth of buildings, livestock, and equipment in terms of the number of acres on the farm. (Each acre is valued at \$1750.)

56. The value V of wheat at \$4.45 per bushel as a function of the number of bushels, x.

Ⓒ**57.** A company produces a product for which the variable cost is \$12.30 per unit and the fixed costs are \$98,000. The product sells for \$17.98.
(a) Write the total cost C as a function of the number of units produced x.
(b) Write the revenue R as a function of the number of units produced x.
(c) Write the profit P as a function of the number of units produced x.

Ⓒ**58.** The inventor of a new game believes that the variable cost for producing the game is \$0.95 per unit, and his fixed costs are \$6,000. He plans to wholesale the game for \$1.69.
(a) Write the total cost C as a function of the number of games sold x.
(b) Write the average cost per unit

$$\overline{C} = \frac{C}{x}$$

as a function of x.
(c) How many units must be sold before the average cost per unit falls below the wholesale price per unit?

Ⓒ**59.** The demand function for a particular commodity is given by

$$p = \frac{14.75}{1 + 0.01x}, \qquad 0 \le x$$

where p is the price per unit and x is the number of units sold.
(a) Find the inverse of this function. That is, find x as a function of p.
(b) Use the result of part (a) to find the number of units sold when the price is \$10.00.

Ⓒ**60.** The demand function for a particular product is given by

$$p = 289 - \frac{(x + 5)^2}{10{,}000}, \qquad 0 \le x \le 1{,}695$$

where p is the price per unit and x is the number of units sold.
(a) Find the inverse of this function. That is, find x as a function of p.
(b) Use the result of part (a) to find the number of units sold when the price is \$189.

61. A power station is on one side of a river that is one-half mile wide. A factory is three miles downstream on the other side of the river. It costs \$10 per foot to run the power lines on land and \$15 per foot to run them underwater. Write the cost C of running the line from the power station to the factory as a function of x. (See Figure 1.48.)

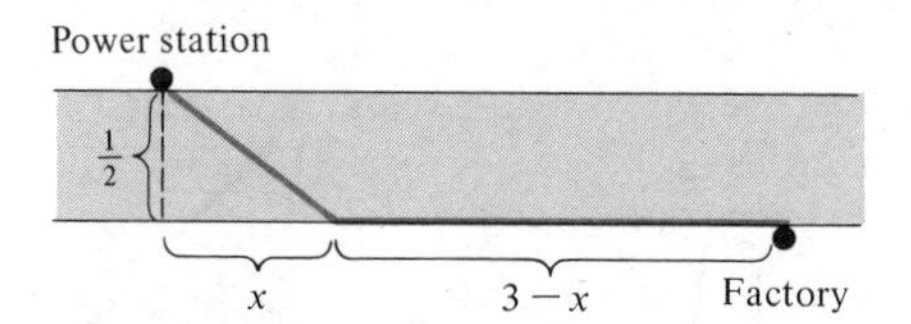

Figure 1.48

Ⓒ**62.** A radio manufacturer charges \$90 per unit for units that cost \$60 to produce. To encourage large orders from distributors, the manufacturer will reduce the price by \$0.01 per unit for each unit in excess of 100. (For example, an order of 200 units would have a price of \$89 per unit.) This price reduction is discontinued when the price per unit drops to \$75.
(a) Write the price per unit p as a function of the order size x.
(b) Write the profit P as a function of the order size x. (Note: $P = R - C = px - 60x$)
(c) Find the total profit for an order of 1000 units.

63. Assume that the amount of money deposited in a bank is proportional to the square of the interest rate the bank pays on the money. That is, $d = kr^2$, where d is the total deposit, r is the interest rate, and k is the proportionality constant. Assuming the bank can reinvest the money for a return of 18%, write the bank's profit P as a function of the interest rate r.

Ⓒ Calculator may be helpful.

Section 1.5

Limits

Introductory Example

Compound Interest

There are many ways of computing the interest earned in a savings account. The most common of these methods involves *compound interest*. Compound interest is interest that is periodically added to the principal. Each time interest is added to the principal, the interest is said to be compounded. The result of compounding interest is that, starting with the second compounding, the account earns interest on interest in addition to earning interest on principal.

Compound interest rates are always given as annual percentages, no matter how many times the interest is compounded per year. As an example of the effect of various compounding periods, consider an account paying 12% (annual percentage rate) on an initial deposit of \$10,000. If $A(t)$ represents the balance after one year and t represents the compounding period, the formula for the balance is

$$A(t) = (10{,}000)(1 + 0.12t)^{1/t}$$

Note in Table 1.9 that no matter how small we allow t to become we cannot seem to obtain a balance that is larger than \$11,274.97. Mathematically, we describe this phenomenon by saying that the **limit** of the balance as t approaches zero is \$11,274.97. (See Figure 1.49.)

Table 1.9 Balance After One Year

Compounding period	Balance (\$)
year: $t = 1$	11,200.00
quarter: $t = 1/4$	11,255.09
month: $t = 1/12$	11,268.25
day: $t = 1/365$	11,274.75
hour: $t = 1/8{,}760$	11,274.96
minute: $t = 1/525{,}600$	11,274.97
second: $t = 1/31{,}536{,}000$	11,274.97

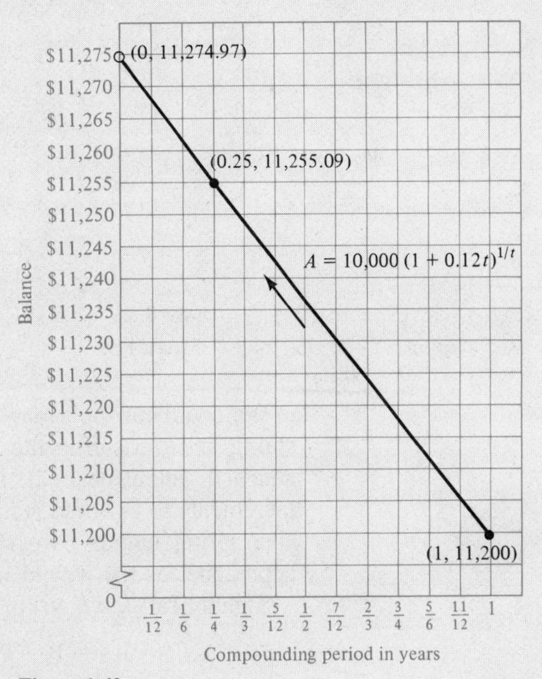

Figure 1.49

Section Objectives: *To provide an intuitive discussion of the limit concept ▪ To identify some properties of limits ▪ To provide some general guidelines for evaluating limits.*

The notion of a limit is fundamental to the study of calculus. Therefore, it is important for you to acquire a good working knowledge of limits before moving on to other calculus topics.

Just what do we mean by the term "limit"? In everyday language we refer to the speed limit, a wrestler's weight limit, the limit of one's endurance, the limits of modern technology, or stretching a spring to its limits. These terms all suggest that a limit is a type of bound, which on some occasions may not be reached but on other occasions may be reached or even exceeded.

Development of Limits

Suppose we are given an ideal spring, which is made so that it will break only if a weight of 10 or more pounds is attached to it. Our task is to determine how far the spring will stretch without breaking (see Figure 1.50).

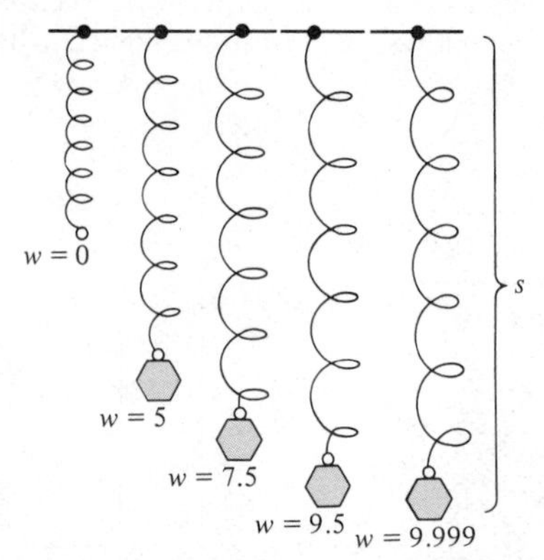

What is the limit of s as w approaches 10 lb?

Figure 1.50

We could carry out our experiment by increasing the weight attached to the spring and measuring the spring length s at each successive weight. As our attached weight nears 10 pounds, we would need to use smaller and smaller increments so as not to reach the 10-pound maximum. By recording the successive spring lengths, we should be able to determine the value L, which s approaches as the weight w approaches 10 pounds.

Symbolically, we write

$$\text{as} \quad w \to 10, \quad s \to L \quad (w < 10)$$

and we say that L is the **limit** of the length s.

Mathematically, our notion of a limit is much like the limit of a spring. For instance, consider the function f given by

$$f(x) = \frac{x^3 - 1}{x - 1}$$

where the domain includes all real numbers other than $x = 1$. Suppose our objective is to determine the limit (if it exists) of $f(x)$ as x approaches 1. Note that we are not interested in determining the value of $f(1)$, because f is not defined at $x = 1$. What we seek, however, is the value (if any) that $f(x)$ approaches as x approaches 1.

To perform this experiment, we could program the function into a calculator and evaluate $f(x)$ for several values of x near 1. In approaching 1 from the left, we use the values

$$x = 0.5, 0.75, 0.9, 0.99, 0.999$$

and from the right, we use

$$x = 1.5, 1.25, 1.1, 1.01, 1.001$$

Table 1.10 gives the values of $f(x)$ that correspond to these 10 different values of x.

Table 1.10

x	0.5	0.75	0.9	0.99	0.999 →	1	← 1.001	1.01	1.1	1.25	1.5
$f(x)$	1.750	2.313	2.710	2.970	2.997 →	?	← 3.003	3.030	3.310	3.813	4.750

From Table 1.10, it seems reasonable to guess that the limit of $f(x)$, as x approaches 1, is 3. We denote this limit by

$$f(x) \to 3 \qquad \text{as} \qquad x \to 1$$

or by the equation

$$\lim_{x \to 1} f(x) = 3$$

Even though $f(x)$ is undefined when $x = 1$, it appears from our table that we can force $f(x)$ to be arbitrarily close to the value 3 by choosing values of x closer and closer to 1. This suggests the following definition.

Definition of Limit

If $f(x)$ becomes arbitrarily close to a single number L as x approaches c from either side, then we write

$$\lim_{x \to c} f(x) = L$$

and say that the limit of $f(x)$, as x approaches c, is L.

Remark

The phrase "x approaches c" means that no matter how close x comes to the value c, there is always another value of x (different from c) in the domain of f that is even closer to c.

Inherent in the definition of a limit is the assumption that a function cannot approach two different limits at the same time. This means that *if the limit of a function exists, it is unique*.

It is important to realize that the limit L of $f(x)$ as $x \to c$ *does not* depend on the value of $f(x)$ *at* $x = c$. Rather, this limit is determined solely from the values of $f(x)$ when x is *near* c. Since the value of L does not depend on the value of $f(x)$ at $x = c$, any one of the following cases can occur:

1. f is undefined when $x = c$, hence $f(c)$ cannot equal L.
2. $f(c)$ exists, but $f(c) \neq L$.
3. $f(c)$ exists, and $f(c) = L$.

The following example illustrates these three cases.

Example 1

Evaluating Limits

Sketch the graphs of

$$f(x) = \frac{x^3 - 1}{x - 1}, \qquad g(x) = \begin{cases} \dfrac{x^3 - 1}{x - 1}, & x \neq 1 \\ 2, & x = 1 \end{cases}, \qquad h(x) = x^2 + x + 1$$

For each function, find the limit as x approaches 1.

Solution The graphs of these three functions are shown in Figure 1.51.

By factoring $x^3 - 1$ we have

$$\frac{x^3 - 1}{x - 1} = \frac{(x - 1)(x^2 + x + 1)}{(x - 1)}$$

$$= x^2 + x + 1, \text{ for all } x \neq 1$$

From this equation and the graphs in Figure 1.51, we can see that the three functions f, g, and h are equal for all x other than $x = 1$. Thus, the limit of each function, as $x \to c$, is the same:

$$\lim_{x \to 1} f(x) = \lim_{x \to 1} g(x) = \lim_{x \to 1} h(x) = 3$$

(Keep in mind that the limit as x approaches 1 does not depend on the value of the function *at* $x = 1$.)

Finally, we note the three cases:

1. $f(1)$ is undefined and cannot equal 3.
2. $g(1)$ exists, but $g(1) = 2 \neq 3$.
3. $h(1)$ exists, and $h(1) = 3$.

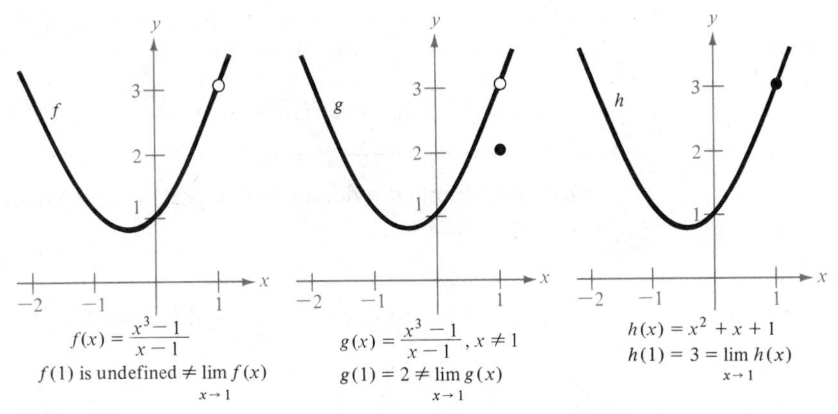

Figure 1.51

□

Of the three functions in Example 1, only the polynomial function h possesses the property that its limit as $x \to 1$ corresponds to the value of the function at $x = 1$. That is,

$$\lim_{x \to 1} h(x) = h(1) = 3$$

All polynomial functions share this special property, that the limit as $x \to c$ can be determined by substituting c for x in the polynomial. Thus, if $p(x)$ is any polynomial, then

$$\lim_{x \to c} p(x) = p(c)$$

We will see later that the limit (as $x \to c$) of some other functions can also be determined by substituting c for x.

In our original example of a spring, we could test for the limit of the length of the spring only by using weights that were less than 10 pounds. However, our definition of the limit,

$$\lim_{x \to c} f(x) = L$$

requires that we let x approach c from the right as well as from the left. To denote these two possible approaches to c, we use the symbols

$$x \to c^-$$

read "x approaches c from the left," and

$$x \to c^+$$

read "x approaches c from the right." Because of the importance of the "limits" obtained from these one-sided approaches, we give each a special name.

One-Sided Limits

$\lim_{x \to c^-} f(x)$ is called the **limit from the left.**

$\lim_{x \to c^+} f(x)$ is called the **limit from the right.**

One-sided limits provide us with a practical way to determine whether or not $\lim_{x \to c} f(x)$ exists.

Theorem 1.8

If f is a function and c and L are numbers, then

$$\lim_{x \to c} f(x) = L$$

if and only if

$$\lim_{x \to c^-} f(x) = L \qquad \text{and} \qquad \lim_{x \to c^+} f(x) = L$$

Example 2

Evaluating One-sided Limits

For the function

$$f(x) = \begin{cases} 4 - x, \text{ for } x < 1 \\ 4x - x^2, \text{ for } x > 1 \end{cases}$$

find $\lim_{x \to 1} f(x)$, if it exists.

Solution Remember, we are concerned about the value of f near $x = 1$ rather than at $x = 1$. Thus, for $x < 1$

$$\lim_{x \to 1^-} f(x) = \lim_{x \to 1^-} (4 - x) = 4 - 1 = 3$$

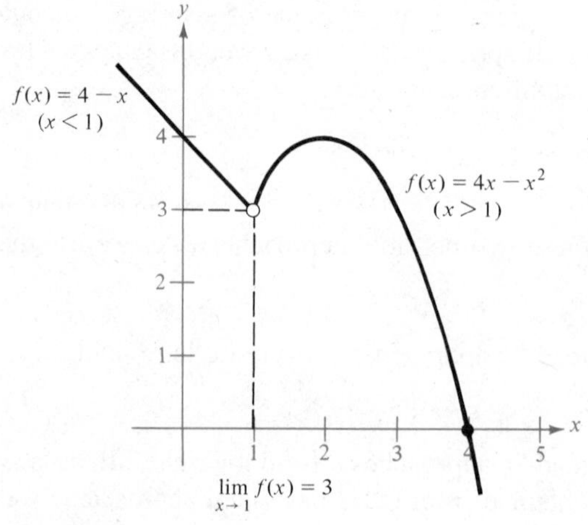

Figure 1.52

and for $x > 1$

$$\lim_{x \to 1^+} f(x) = \lim_{x \to 1^+} (4x - x^2) = 4 - 1 = 3$$

Since the one-sided limits both exist and are equal to 3, we have

$$\lim_{x \to 1} f(x) = 3$$

A sketch will further illustrate our result (see Figure 1.52). □

Another interesting situation involving one-sided limits arises with the "post office" function given in the next example.

Example 3

Limits That Differ from the Right and the Left

Suppose the cost of sending first-class mail is 20¢ for the first ounce and 17¢ for each additional ounce. Letting x represent the weight of a letter and $f(x)$ the cost of mailing the letter first class, we have

$$f(x) = 20 + 17n, \qquad n < x \leq n + 1 \ (n = 0, 1, 2, 3, \ldots)$$

Show that the limit of $f(x)$, as $x \to 2$, does not exist.

Solution The graph of f is shown in Figure 1.53.

Observe that

$$f(x) = \begin{cases} 37, & 1 < x \leq 2 \\ 54, & 2 < x \leq 3 \end{cases}$$

Thus

$$\lim_{x \to 2^-} f(x) = 37$$

whereas,

$$\lim_{x \to 2^+} f(x) = 54$$

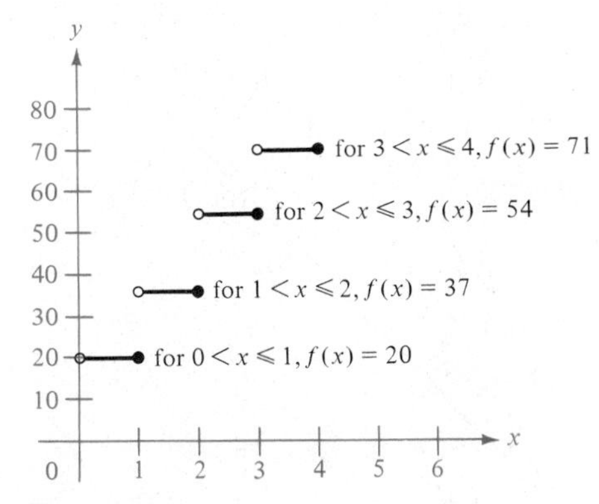

Figure 1.53

Since these one-sided limits are not equal, the limit of $f(x)$, as $x \to 2$, *does not exist*. □

Example 4

An Example of the Failure of Direct Substitution

If

$$f(x) = \frac{x^2 + x - 6}{x + 3}$$

determine $\lim_{x \to -3} f(x)$.

Solution Direct substitution of -3 for x yields the *meaningless result*

$$\lim_{x \to -3} f(x) = \frac{9 - 3 - 6}{-3 + 3} = \frac{0}{0}$$

By factoring the numerator, we obtain

$$f(x) = \frac{x^2 + x - 6}{x + 3} = \frac{(x + 3)(x - 2)}{x + 3}$$

Now, for all $x \neq -3$, and therefore for all x near -3, we can cancel like factors and obtain

$$\frac{x^2 + x - 6}{x + 3} = x - 2$$

It follows that

$$\lim_{x \to -3} \frac{x^2 + x - 6}{x + 3} = \lim_{x \to -3} (x - 2) = -5$$

See Figure 1.54.

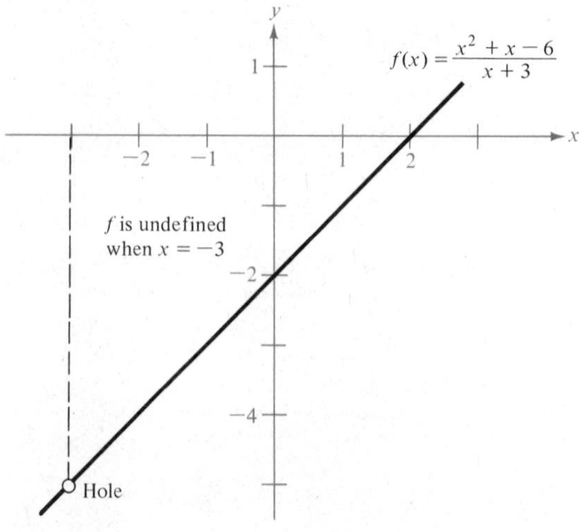

Figure 1.54

□

Remark

Note that we can run into trouble using direct substitution to evaluate limits. For instance, in Example 4, direct substitution produces a fraction with a denominator of zero. This fraction is meaningless since division by zero is undefined. *In this text, we will follow the convention of denoting the failure of direct substitution by printing the meaningless result in color.* Thus, our use of color in the first equation of the solution means that we attempted to use direct substitution but the attempt failed.

In our informal development of the limit concept, we have used, without identifying them, several basic properties of limits — properties which, for the most part, are ones we would expect the limit to possess. We provide here a list of some of these basic properties of limits.

Properties of Limits

1. $\lim\limits_{x \to c} b = b$
2. $\lim\limits_{x \to c} x = c$
3. $\lim\limits_{x \to c} b \cdot f(x) = b\left[\lim\limits_{x \to c} f(x)\right]$
4. $\lim\limits_{x \to c} [f(x) \pm g(x)] = \lim\limits_{x \to c} f(x) \pm \lim\limits_{x \to c} g(x)$
5. $\lim\limits_{x \to c} [f(x)g(x)] = \left[\lim\limits_{x \to c} f(x)\right]\left[\lim\limits_{x \to c} g(x)\right]$
6. $\lim\limits_{x \to c} \dfrac{f(x)}{g(x)} = \dfrac{\lim\limits_{x \to c} f(x)}{\lim\limits_{x \to c} g(x)}$, provided $\lim\limits_{x \to c} g(x) \neq 0$
7. $\lim\limits_{x \to c} x^n = c^n$

As an illustration let us identify the use of these properties in evaluating $\lim_{x \to 2} (x^2 + 2x - 3)$. By Property 7,

$$\lim_{x \to 2} x^2 = 2^2 = 4$$

By Properties 2 and 3,

$$\lim_{x \to 2} 2x = 2 \lim_{x \to 2} x = 2(2) = 4$$

By Property 1,

$$\lim_{x \to 2} 3 = 3$$

Finally, by Property 4,

$$\begin{aligned} \lim_{x \to 2} (x^2 + 2x - 3) &= \lim_{x \to 2} x^2 + \lim_{x \to 2} 2x - \lim_{x \to 2} 3 \\ &= 4 + 4 - 3 = 5 \end{aligned}$$

Before concluding this section, we provide two general guidelines for evaluating the types of limits encountered thus far.

Guidelines for Evaluating $\lim_{x \to c} f(x)$

1. Check to see if f is defined differently for $x \to c^-$ than for $x \to c^+$. If so, evaluate each one-sided limit and draw appropriate conclusions about the limit as $x \to c$.
2. Use direct substitution. If f involves a quotient and $0/0$ is obtained, try to change the fraction algebraically to remove the zero from the denominator. Then take the limit as $x \to c$.

Example 5

Evaluating Limits

Evaluate the following limits, if they exist.

(a) $\lim_{x \to -2} \dfrac{x^2 + x - 2}{x - 2}$

(b) $\lim_{x \to 1} f(x)$, where $f(x) = \begin{cases} 2x - x^3, & x < 1 \\ 2x^2 - 2, & x \geq 1 \end{cases}$

(c) $\lim_{x \to -1} f(x)$, using $f(x)$ in part (b)

(d) $\lim_{x \to 0} \dfrac{(x + 2)^2 - 4}{x}$

Solution

(a) By Property 6 and direct substitution, we obtain

$$\lim_{x \to -2} \frac{x^2 + x - 2}{x - 2} = \frac{4 - 2 - 2}{-2 - 2} = \frac{0}{-4} = 0$$

(b) Since f is defined differently for $x < 1$ than for $x \geq 1$, we consider the one-sided limits

$$\lim_{x \to 1^-} f(x) = \lim_{x \to 1^-} (2x - x^3) = 2 - 1 = 1$$

and $$\lim_{x \to 1^+} f(x) = \lim_{x \to 1^+} (2x^2 - 2) = 2 - 2 = 0$$

Because these one-sided limits are not equal, we conclude that $\lim_{x \to 1} f(x)$ does not exist.

(c) By direct substitution and Property 4, we have, for $x < 1$,

$$\lim_{x \to -1} f(x) = \lim_{x \to -1} (2x - x^3) = -2 + 1 = -1$$

(d) By direct substitution and Property 6, we obtain

$$\lim_{x \to 0} \frac{(x + 2)^2 - 4}{x} = \frac{4 - 4}{0} = \frac{0}{0}$$

which is meaningless. Factoring and reducing, we have

$$\lim_{x \to 0} \frac{x^2 + 4x + 4 - 4}{x} = \lim_{x \to 0} \frac{x(x + 4)}{x} = \lim_{x \to 0} (x + 4) = 4$$

□

Table 1.11

x	-0.5	-0.25	-0.1	-0.01	-0.001	0	0.001	0.01	0.1	0.25	0.5
$\frac{\sqrt{x+1}-1}{x}$	0.5858	0.5359	0.5132	0.5013	0.5001	$\longrightarrow ? \longleftarrow$	0.4999	0.4988	0.4881	0.4721	0.4495

Example 6

Evaluating a Limit by Rationalizing the Numerator

Determine

$$\lim_{x\to 0} \frac{\sqrt{x+1}-1}{x}$$

Solution By direct substitution, we get the meaningless result

$$\lim_{x\to 0} \frac{\sqrt{x+1}-1}{x} = \frac{0}{0}$$

In this case, we change the form of the fraction by rationalizing the numerator:

$$\frac{\sqrt{x+1}-1}{x} = \left(\frac{\sqrt{x+1}-1}{x}\right)\left(\frac{\sqrt{x+1}+1}{\sqrt{x+1}+1}\right) = \frac{(x+1)-1}{x(\sqrt{x+1}+1)}$$

$$= \frac{x}{x(\sqrt{x+1}+1)} = \frac{1}{\sqrt{x+1}+1}$$

Therefore,

$$\lim_{x\to 0} \frac{\sqrt{x+1}-1}{x} = \lim_{x\to 0} \frac{1}{\sqrt{x+1}+1} = \frac{1}{1+1} = \frac{1}{2}$$

Table 1.11 reinforces our conclusion that this limit is $\frac{1}{2}$. □

In several of the examples in this section, we encountered the meaningless expression $\frac{0}{0}$. We call such an expression an **indeterminate form** since we cannot (from the form alone) determine the limit. When you are trying to evaluate a limit and encounter this form, remember that you must change the fraction algebraically so that the new denominator does not have zero as its limit.

In Examples 3 and 5, we looked at limits that fail to exist at a point because the limits from the right and left differ. We close this section by looking at another important way in which a limit may fail to exist. It may happen that, as $x \to c$, $f(x)$ will increase (or decrease) without bound. Our last example describes such a situation.

Example 7

An Unbounded Function

Evaluate (if possible) the limit

$$\lim_{x\to 2} \frac{3}{x-2}$$

Solution From Figure 1.55 and Table 1.12, we can see that $f(x)$ *decreases without bound* as x gets closer and closer to 2 from the left and $f(x)$ *increases without bound* as x approaches 2 from the right. Symbolically, we can write this as

$$\lim_{x \to 2^-} \frac{3}{x - 2} = -\infty$$

and

$$\lim_{x \to 2^+} \frac{3}{x - 2} = \infty$$

Since f is unbounded as x approaches 2, we conclude that the limit *does not exist*.

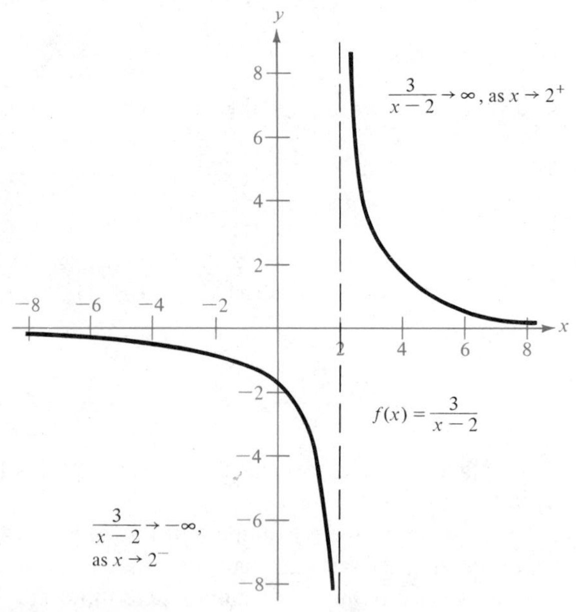

Figure 1.55

Table 1.12

x	0	1	1.5	1.9	1.999 $\longrightarrow$	2	$\longleftarrow$ 2.001	2.1	2.5	3	5
$f(x)$	$-\frac{3}{2}$	-3	-6	-30	$-3000 \longrightarrow$	$-\infty$ | ∞	$\longleftarrow$ 3000	30	6	3	1

□

Remark The equals sign in the statement

$$\lim_{x \to c} f(x) = \infty$$

does not mean that the limit exists! On the contrary, it tells us how the limit *fails to exist* by denoting the unbounded behavior of $f(x)$ as x approaches c. We will follow the convention of printing in color this unusual use of an equals sign.

Section Exercises (1.5)

In Exercises 1–30, determine if the limit exists, and, if it does, determine the limit.

1. $\lim_{x\to 2} x^2$

2. $\lim_{x\to -3} (3x+2)$

3. $\lim_{x\to 0} (5x+4)$

4. $\lim_{x\to 0} \dfrac{x^2-1}{x+1}$

5. $\lim_{x\to -1^-} \dfrac{x^2-1}{x+1}$

6. $\lim_{x\to -1} \dfrac{2x^2-x-3}{x+1}$

7. $\lim_{x\to 3} \dfrac{x-3}{x^2-9}$

8. $\lim_{x\to 1} f(x)$, where $f(x) = \begin{cases} x, & x \le 1 \\ 1-x, & x > 1 \end{cases}$

9. $\lim_{x\to 3} f(x)$, where $f(x) = \begin{cases} \dfrac{x+2}{2}, & x \le 3 \\ \dfrac{12-2x}{3}, & x > 3 \end{cases}$

10. $\lim_{x\to 2} f(x)$, where $f(x) = \begin{cases} x^2-4x+6, & x < 2 \\ -x^2+4x-2, & x \ge 2 \end{cases}$

11. $\lim_{x\to 1} f(x)$, where $f(x) = \begin{cases} x^3+1, & x < 1 \\ x+1, & x \ge 1 \end{cases}$

12. $\lim_{x\to -1} \dfrac{x^3+1}{x+1}$

13. $\lim_{x\to -2} \dfrac{x^3+8}{x+2}$

14. $\lim_{x\to 0} \dfrac{|x|}{x}$

15. $\lim_{x\to 2} \dfrac{|x-2|}{x-2}$

16. $\lim_{\Delta x\to 0} \dfrac{(x+\Delta x)^2-x^2}{\Delta x}$

17. $\lim_{\Delta x\to 0^+} \dfrac{2(x+\Delta x)-2x}{\Delta x}$

18. $\lim_{\Delta x\to 0} \dfrac{(x+\Delta x)^3-x^3}{\Delta x}$

19. $\lim_{\Delta x\to 0} \dfrac{(x+\Delta x)^2-2(x+\Delta x)+1-(x^2-2x+1)}{\Delta x}$

20. $\lim_{\Delta x\to 0} \dfrac{(1+\Delta x)^3-1}{\Delta x}$

21. $\lim_{x\to 5^+} \dfrac{x-5}{x^2-25}$

22. $\lim_{x\to 2^+} \dfrac{2-x}{x^2-4}$

23. $\lim_{x\to 1} \dfrac{x^2+x-2}{x^2-1}$

24. $\lim_{x\to 5} \dfrac{x^2-25}{x+5}$

25. $\lim_{x\to -2^-} \dfrac{x}{\sqrt{x^2-4}}$

26. $\lim_{x\to 0} \dfrac{\sqrt{2+x}-\sqrt{2}}{x}$

27. $\lim_{x\to 0} \dfrac{\sqrt{3+x}-\sqrt{3}}{x}$

28. $\lim_{x\to 0} \dfrac{[1/(x+4)]-\frac{1}{4}}{x}$

29. $\lim_{x\to 0} \dfrac{[1/(2+x)]-\frac{1}{2}}{x}$

30. $\lim_{x\to 4^-} \dfrac{\sqrt{x}-2}{x-4}$

31. If $\lim_{x\to c} f(x) = 2$ and $\lim_{x\to c} g(x) = 3$, find:

(a) $\lim_{x\to c} [f(x)-g(x)]$ (b) $\lim_{x\to c} [f(x)g(x)]$

(c) $\lim_{x\to c} \dfrac{f(x)}{g(x)}$

32. If $\lim_{x\to c} f(x) = \frac{3}{2}$ and $\lim_{x\to c} g(x) = \frac{1}{2}$, find:

(a) $\lim_{x\to c} [f(x)+g(x)]$ (b) $\lim_{x\to c} [f(x)g(x)]$

(c) $\lim_{x\to c} \dfrac{f(x)}{g(x)}$

In Exercises 33–40, compile a table to evaluate the limit of the given function.

Ⓒ**33.** $\lim_{x\to 2} (5x+4)$

Ⓒ**34.** $\lim_{x\to 2} \dfrac{x-2}{x^2-x-2}$

Ⓒ**35.** $\lim_{x\to 2} \dfrac{x-2}{x^2-4}$

Ⓒ**36.** $\lim_{x\to 0} \dfrac{\sqrt{x+3}-\sqrt{3}}{x}$

Ⓒ**37.** $\lim_{x\to 0} \dfrac{\sqrt{x+2}-\sqrt{2}}{x}$

Ⓒ**38.** $\lim_{x\to 2^-} \dfrac{2-x}{\sqrt{4-x^2}}$

Ⓒ**39.** $\lim_{x\to 0} \dfrac{[1/(2+x)]-\frac{1}{2}}{x}$

Ⓒ**40.** $\lim_{x\to 2} \dfrac{x^5-32}{x-2}$

In Exercises 41–46, use the graph to visually determine:

(a) $\lim_{x\to c^+} f(x)$ (b) $\lim_{x\to c^-} f(x)$ (c) $\lim_{x\to c} f(x)$

Ⓒ Calculator may be helpful.

41.

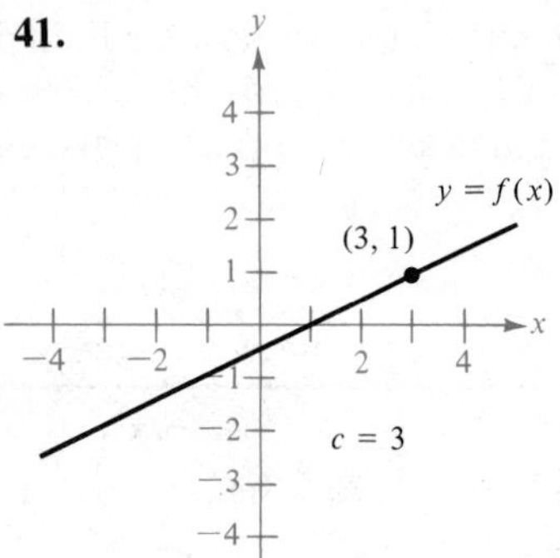

42.

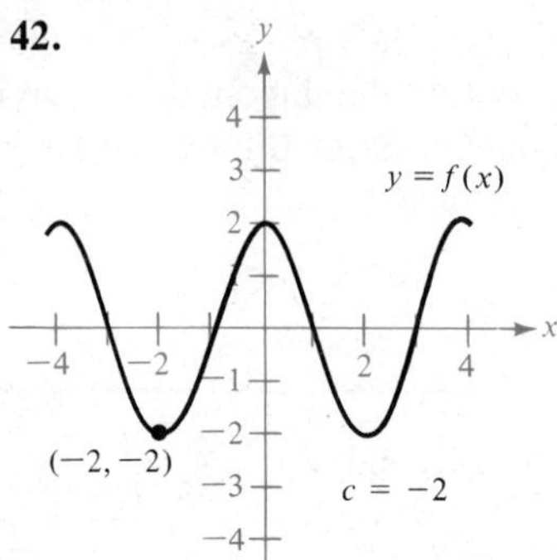

43.

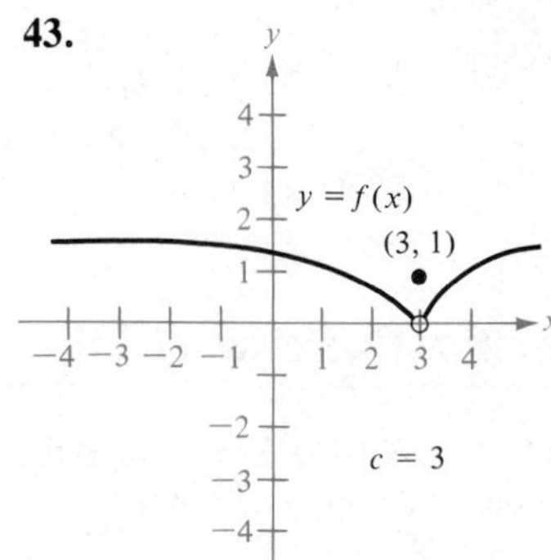

44.

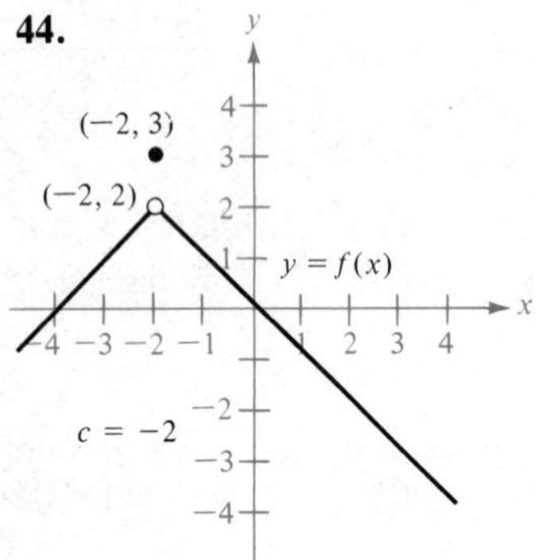

45.

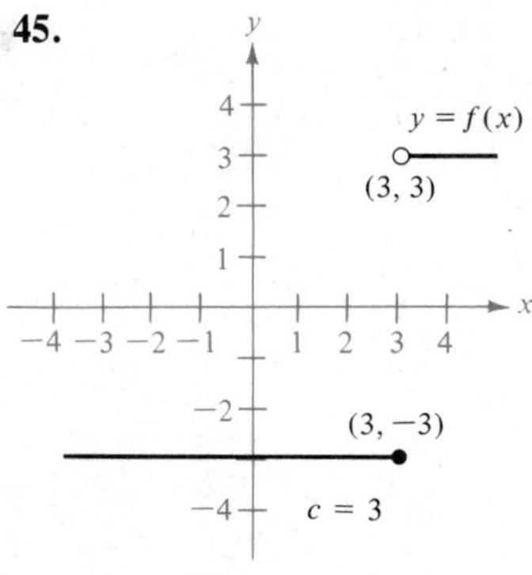

46.

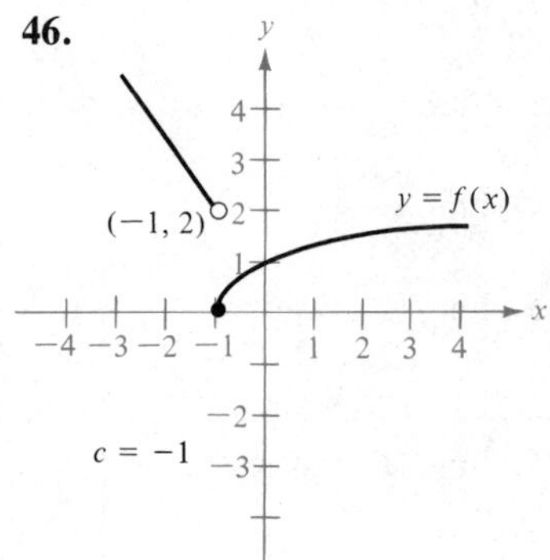

Section 1.6

Continuity

Introductory Example

Continuously Compounded Interest

In the Introductory Example to Section 1.5, we saw that the balance in a savings account tends to increase as the number of compoundings per year increases. For instance, monthly compounding produces a higher balance than does quarterly compounding. Figure 1.56 shows the balance in an account paying 12% compounded quarterly on an initial deposit of $10,000. Note that a full quarter must pass before interest is added. We say that the graph in Figure 1.56 is **discontinuous.** The discontinuities occur at the end of each quarter when interest is added to the account.

In Chapter 5, we will see that the maximum balance that one can obtain by increasing the number of compoundings per year is given by a procedure called *continuous compounding*. Continuous compounding is often compared to compounding every second. However, the distinction between these two types of compounding is comparable to the difference between a digital watch and a watch with a second hand. The digital watch registers abrupt changes from one second to the next, whereas the watch with the second hand continuously registers changes in time. Figure 1.57 shows the balance in an account that is compounded continuously. Note that this graph has no jumps or breaks. In mathematical terms, we say that the graph in Figure 1.57 is **continuous.**

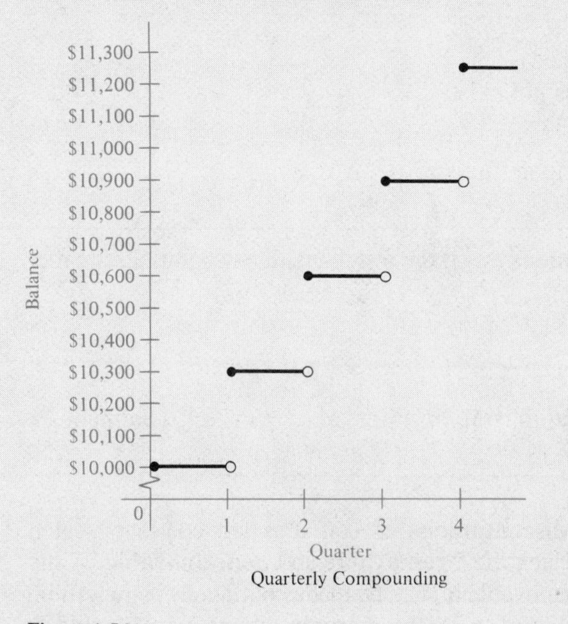

Figure 1.56

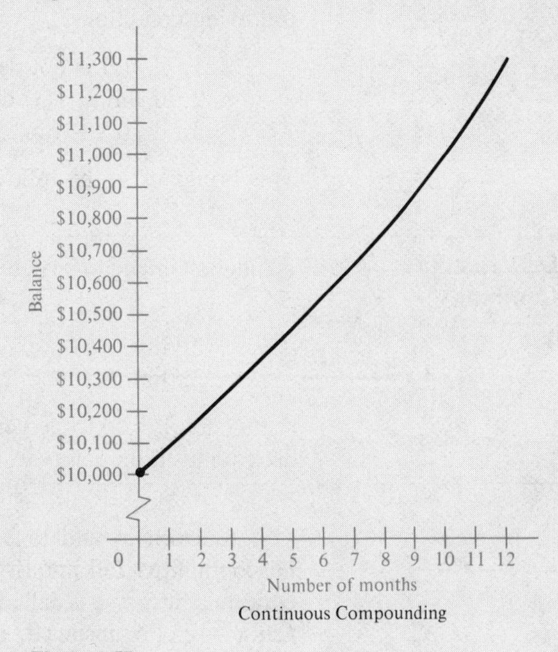

Figure 1.57

Section Objectives: ***To define the term "continuity" ▪ To identify some properties of continuous functions.***

In mathematics the term "continuous" has much the same meaning as it does in our everyday usage. To say that a function is continuous at $x = c$ means that there is no interruption in the graph of f at c. Its graph is unbroken at c, and there are no holes, jumps, or gaps. Roughly speaking, we say that a function is continuous if its graph can be traced without lifting the pencil from the paper. As simple as this concept may seem initially, its precise definition eluded mathematicians for many years. In fact, it was not until the early 1800s that a careful definition was finally developed.

Before looking at this definition, let us consider the function whose graph is shown in Figure 1.58. This figure identifies three values of x at which the graph of f is not continuous. At all other points ($x \neq c_1, c_2, c_3$) of the interval (a, b), the graph of f is uninterrupted, and we say it is continuous at such points. Where the graph is discontinuous, we observe the following:

1. At $x = c_1$, $f(c_1)$ is not defined.
2. At $x = c_2$, $\lim_{x \to c_2} f(x)$ does not exist.
3. At $x = c_3$, $f(c_3) \neq \lim_{x \to c_3} f(x)$.

Thus, it appears that continuity at $x = c$ is destroyed under one or more of the following conditions:

1. if $f(c)$ is not defined
2. if $\lim_{x \to c} f(x)$ does not exist
3. if $f(c) \neq \lim_{x \to c} f(x)$

This brings us to the following definition.

Definition of Continuity

A function is said to be **continuous at** c if the following three conditions are met:

1. $f(c)$ is defined.
2. $\lim_{x \to c} f(x)$ exists.
3. $\lim_{x \to c} f(x) = f(c)$.

A function is said to be **continuous on an interval** (a, b) if it is continuous at each point in the interval.

A function is said to be **discontinuous at** c if it is not continuous at c. Discontinuities fall into two categories: **removable** and **nonremovable.** A discontinuity at $x = c$ is called removable if f can be made continuous by redefining f at $x = c$. For instance, in Figure 1.58 the discontinuity at $x = c_1$ could be removed by defining $f(c_1)$ as $\lim_{x \to c_1} f(x)$. Furthermore, the discontinuity at c_3 can be removed by redefining $f(c_3)$ so as to move the point $(c_3, f(c_3))$ up to plug

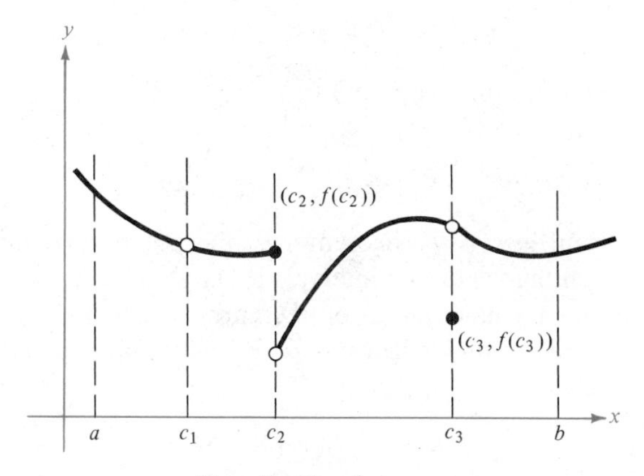

Figure 1.58

the hole. Finally, the discontinuity at $x = c_2$ in Figure 1.58 is not removable since we cannot make the graph continuous by merely redefining $f(c_2)$.

For a function to be continuous on a closed interval $[a, b]$, it is sufficient that it be continuous on the open interval (a, b) and that

$$\lim_{x \to a^+} f(x) = f(a) \qquad \text{and} \qquad \lim_{x \to b^-} f(x) = f(b)$$

In this situation, we say that f is continuous *from the right at a* and continuous *from the left at b*.

In Section 1.5, we noted that for any polynomial function $p(x)$, we have

$$\lim_{x \to c} p(x) = p(c)$$

Now we can see that one consequence of this property is that *every* polynomial function is continuous.

Theorem 1.9 Continuity of a Polynomial

A polynomial function is continuous at every real number.

The sum, difference, or product of two polynomials is a polynomial. The quotient of two polynomials is called a **rational function.** For those points at which the denominator of a rational function is zero, the function is undefined and hence not continuous; for all other points, rational functions are continuous.

Theorem 1.10 Continuity of a Rational Function

A rational function is continuous at every real number in its domain.

Example 1

Checking for Continuity

Discuss the continuity of

$$g(x) = \begin{cases} 5 - x, & -1 \le x \le 2 \\ x^2 - 1, & 2 < x \le 3 \end{cases}$$

Solution By Theorem 1.9, the polynomial functions $5 - x$ and $x^2 - 1$ are continuous on the intervals $[-1, 2)$ and $(2, 3]$, respectively. Thus, to conclude that g is continuous on the entire interval $[-1, 3]$, we need only to check the behavior of g when $x = 2$. By taking the one-sided limits when $x = 2$, we see that

$$\lim_{x\to 2^-} g(x) = \lim_{x\to 2^-} (5 - x) = 3$$

and

$$\lim_{x\to 2^+} g(x) = \lim_{x\to 2^+} (x^2 - 1) = 3$$

Since these two limits are equal, we have

$$\lim_{x\to 2} g(x) = g(2) = 3$$

Thus, g is continuous at $x = 2$, and consequently it is continuous on the entire interval $[-1, 3]$. The graph of g is shown in Figure 1.59.

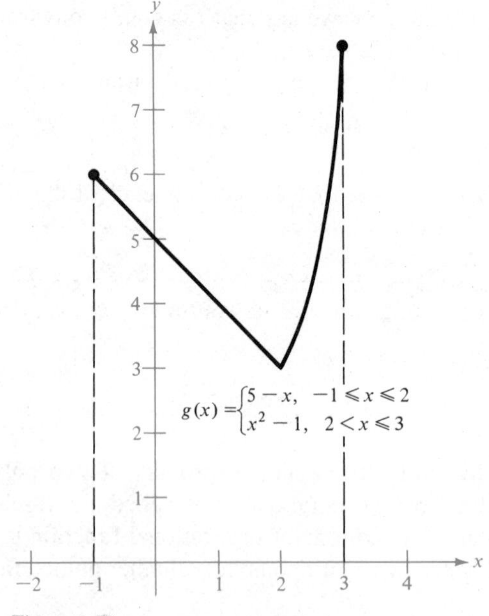

Figure 1.59

□

Example 2

A Removable Discontinuity

Discuss the continuity of

$$f(x) = \begin{cases} \dfrac{x^2 - 2x - 3}{x - 3}, & x \neq 3 \\ 3, & x = 3 \end{cases}$$

on the entire real line.

Solution For $x \neq 3$ the rational function

$$f(x) = \frac{x^2 - 2x - 3}{x - 3}$$

is defined, and by Theorem 1.10 it is continuous for all $x \neq 3$. Thus, we need only test for continuity at $x = 3$. Note that for $x \neq 3$

$$\frac{x^2 - 2x - 3}{x - 3} = \frac{(x - 3)(x + 1)}{x - 3} = x + 1$$

and so

$$\lim_{x \to 3} f(x) = \lim_{x \to 3} (x + 1) = 4 \neq f(3)$$

Therefore, we conclude that f is not continuous at $x = 3$ (see Figure 1.60). Note that this is an example of a removable discontinuity.

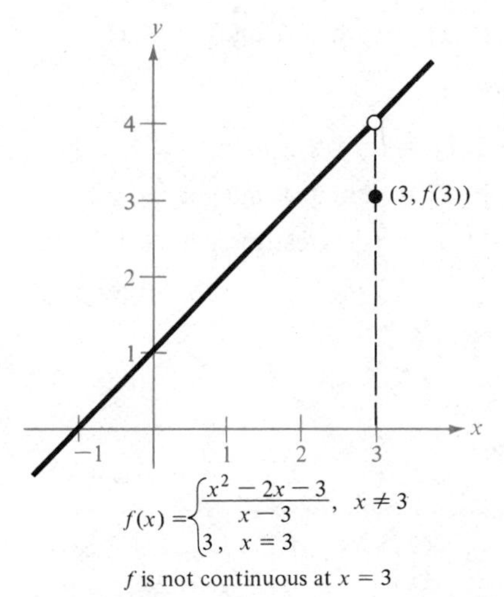

Figure 1.60 □

Example 3

Testing for One-sided Continuity

Discuss the continuity of $f(x) = \sqrt{3 - x}$.

Solution First, f is defined for all $x \leq 3$. Furthermore, f is continuous from the left at $x = 3$, because

$$\lim_{x \to 3^-} f(x) = \lim_{x \to 3^-} \sqrt{3 - x} = 0 = f(3)$$

For all $x \leq 3$ the function f obviously satisfies the three requirements for continuity, and we conclude that f is continuous on the interval $(-\infty, 3]$. (See Figure 1.61.)

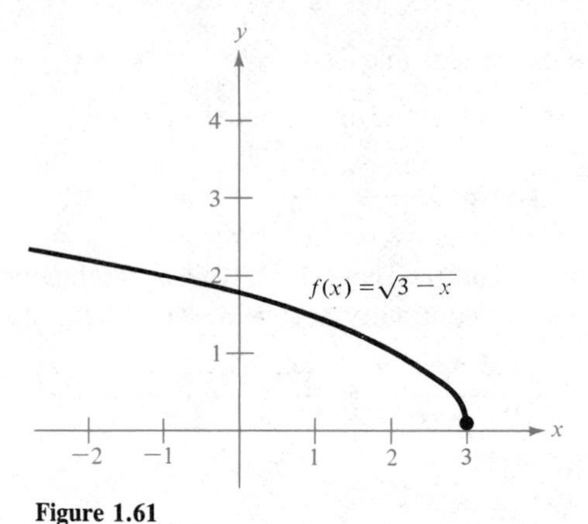

Figure 1.61

□

In the context of one-sided limits and one-sided continuity, it is interesting to look at the **greatest integer function.** This function is denoted by

$$[x] = (\text{greatest integer} \leq x)$$

For instance,

$$[-2.1] = (\text{greatest integer} \leq -2.1) = -3$$
$$[-2] = (\text{greatest integer} \leq -2) = -2$$
$$[1.5] = (\text{greatest integer} \leq 1.5) = 1$$

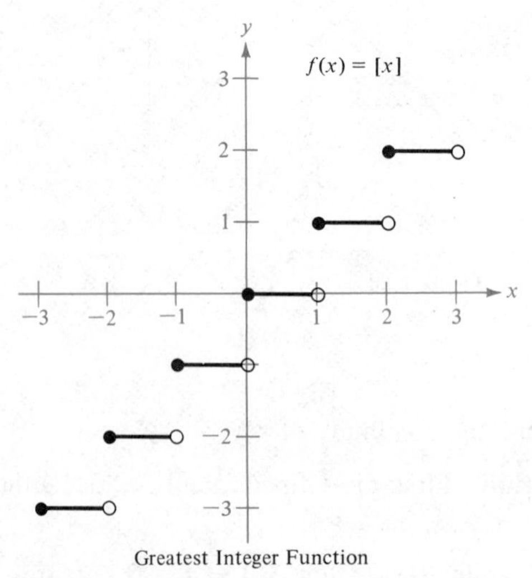

Figure 1.62

From the graph of the greatest integer function (Figure 1.62), note how the function jumps up one unit at each integer and thus is discontinuous at each point.

The most common use of the greatest integer function is with positive values of x. In such cases, this function corresponds to **truncating** the decimal portion of x. For example, $[1.345]$ is truncated to 1, and $[3.57]$ is truncated to 3.

Example 4

An Application of the Greatest Integer Function

A bookbinding company produces 10,000 books in an eight-hour shift. The fixed costs *per shift* amount to \$5,000, while the unit cost *per book* is \$3. Using the greatest integer function, we can write the cost of producing x books as

$$C(x) = 5{,}000\left(1 + \left[\frac{x-1}{10{,}000}\right]\right) + 3x$$

Sketch the graph of this cost function and discuss its continuity.

Solution Note that during the first eight-hour shift

$$1 \le x \le 10{,}000, \qquad \left[\frac{x-1}{10{,}000}\right] = 0$$

and we have

$$C(x) = 5{,}000\left(1 + \left[\frac{x-1}{10{,}000}\right]\right) + 3x = 5{,}000 + 3x$$

During the second eight-hour shift,

$$10{,}001 \le x \le 20{,}000, \qquad \left[\frac{x-1}{10{,}000}\right] = 1$$

and we have

$$C(x) = 5{,}000\left(1 + \left[\frac{x-1}{10{,}000}\right]\right) + 3x = 10{,}000 + 3x$$

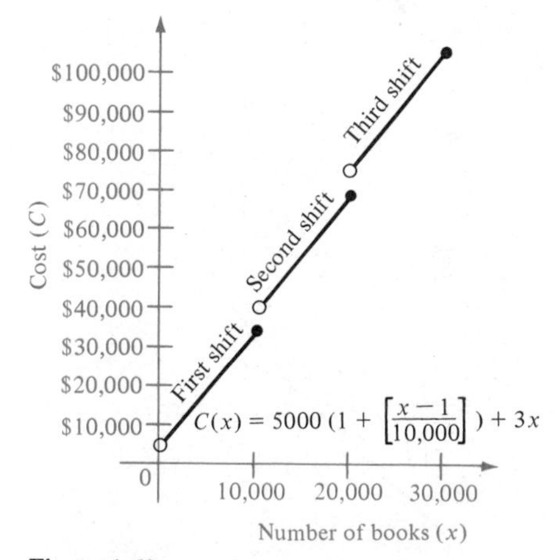

Figure 1.63

The graph of C is given in Figure 1.63. Note that the graph has discontinuities at $x = 10{,}000$, $20{,}000$, and so forth. □

Because continuity is defined in terms of limits, it is not surprising to learn that continuous functions and limits possess many of the same properties.

Properties of Continuous Functions

If the functions f and g are continuous at c, then the following functions are also continuous at c:

1. $f \pm g$
2. af, where a is any constant
3. $f \cdot g$
4. f/g, if $g(c) \neq 0$
5. $f(g(x))$, provided f is continuous at $g(c)$

Section Exercises (1.6)

In Exercises 1–6, find the points of discontinuity (if any).

1. $f(x) = -\dfrac{x^3}{2}$

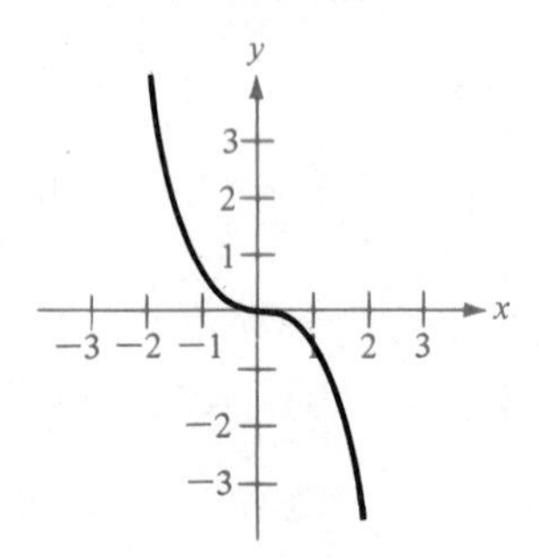

2. $f(x) = \dfrac{x^2 - 1}{x}$

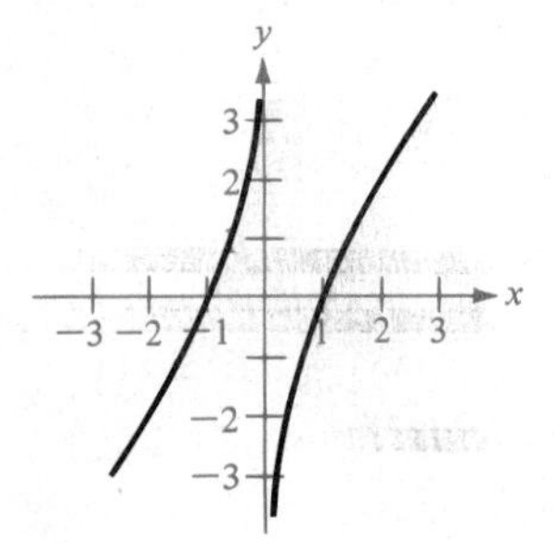

3. $f(x) = \dfrac{x^2 - 1}{x + 1}$

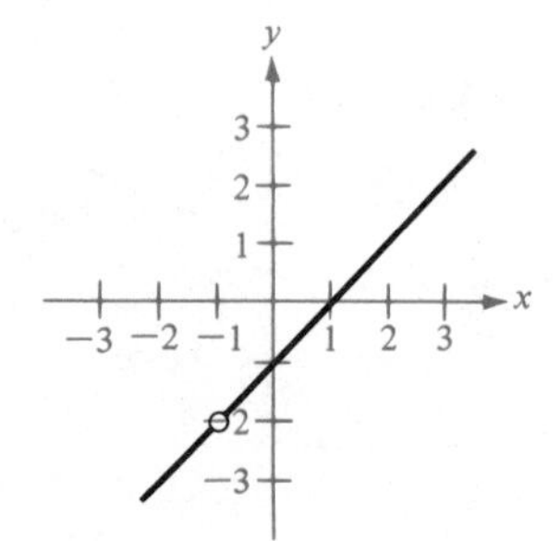

4. $f(x) = \dfrac{1}{x^2 - 4}$

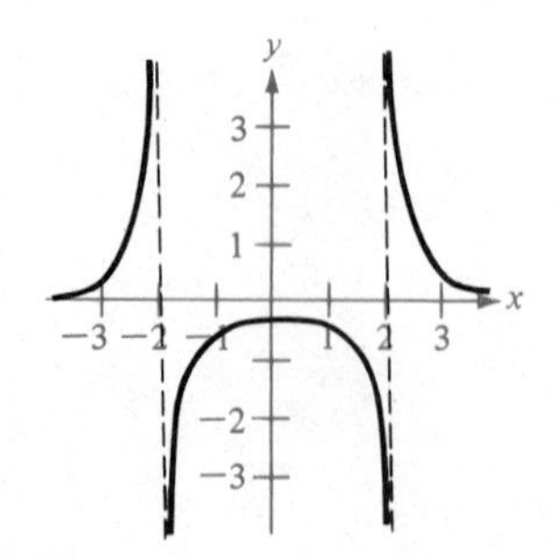

5. $f(x) = \begin{cases} x, & x < 1 \\ 2, & x = 1 \\ 2x - 1, & 1 < x \end{cases}$

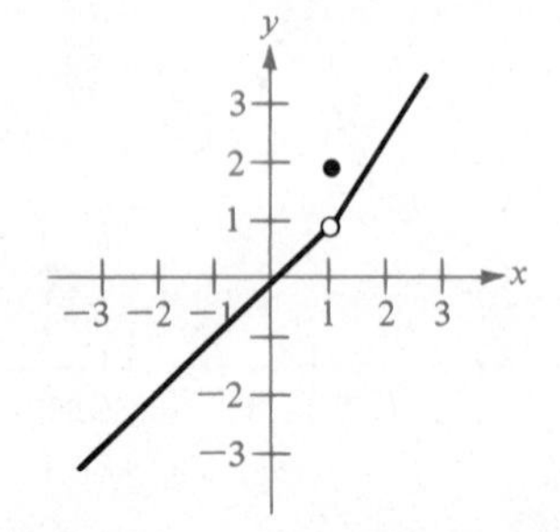

6. 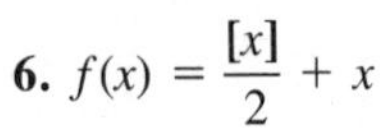 $f(x) = \dfrac{[x]}{2} + x$

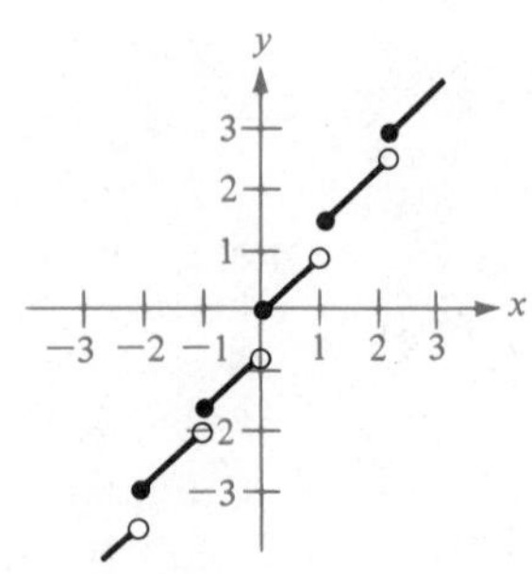

In Exercises 7–26, find the discontinuities (if any) for each function. Which of the discontinuities are removable?

7. $f(x) = x^2 - 2x + 1$

8. $f(x) = \dfrac{1}{x^2 + 1}$

9. $f(x) = \dfrac{1}{x - 1}$

10. $f(x) = \dfrac{x}{x^2 - 1}$

11. $f(x) = \dfrac{x}{x^2 + 1}$

12. $f(x) = \dfrac{x - 3}{x^2 - 9}$

13. $f(x) = \dfrac{x + 2}{x^2 - 3x - 10}$

14. $f(x) = \dfrac{x - 1}{x^2 + x - 2}$

15. $f(x) = \begin{cases} x, & x \leq 1 \\ x^2, & x > 1 \end{cases}$

16. $f(x) = \begin{cases} -2x + 3, & x < 1 \\ x^2, & x \geq 1 \end{cases}$

17. $f(x) = \begin{cases} \dfrac{x}{2} + 1, & x \leq 2 \\ 3 - x, & x > 2 \end{cases}$

18. $f(x) = \begin{cases} -2x, & x \leq 2 \\ x^2 - 4x + 1, & x > 2 \end{cases}$

19. $f(x) = \dfrac{|x + 2|}{x + 2}$

20. $f(x) = \begin{cases} |x - 2| + 3, & x < 0 \\ x + 5, & x \geq 0 \end{cases}$

21. $f(x) = \begin{cases} 3 + x, & x \leq 2 \\ x^2 + 1, & x > 2 \end{cases}$

22. $f(x) = [x - 1]$

23. $f(x) = x - [x]$

24. $f(g(x)), f(x) = x^2, g(x) = x - 1$

25. $f(g(x)), f(x) = \dfrac{1}{\sqrt{x}}, g(x) = x - 1$

26. $f(g(x)), f(x) = \dfrac{1}{x - 1}, g(x) = x^2 + 5$

Ⓒ 27. Mike Jenkin's salary in 1982 is \$28,500, and his contract guarantees a 9% increase for each of the next five years. Thus, his salary S over the next five years is given by

$$S(t) = 28{,}500(1.09)^{[t]}$$

where t is the time in years, with $t = 0$ corresponding to 1982. Sketch a graph of Mike's salary between 1982 and 1987 and discuss the continuity of this function.

Ⓒ 28. A dial-direct long distance call between two particular cities costs \$0.52 for the first two minutes plus \$0.36 for each additional minute or fraction thereof. Use the greatest integer function to write the cost of the call C in terms of the time in minutes t. Sketch a graph of this function and discuss its continuity.

29. The number of units in inventory in a small company is given by

$$N(t) = 25\left(2\left[\frac{t + 2}{2}\right] - t\right)$$

where t is the time in months. Sketch a graph of this function and discuss its continuity. How often does this company replenish its inventory?

Ⓒ 30. \$5,000 is deposited in a savings plan that pays 12% compounded semiannually. The amount in the account after t years is given by the function.

$$A(t) = 5000(1.06)^{[2t]}$$

Sketch a graph of this function and discuss its continuity.

Ⓒ 31. The cost (in millions of dollars) of removing x percent of the pollutants from the smokestack of a certain factory is given by

$$C(x) = \frac{2x}{100 - x}$$

Sketch the graph of this function and discuss its continuity. What is the appropriate domain of this function?

Ⓒ Calculator may be helpful.

Chapter 2

Differentiation

Section 2.1

The Derivative and the Slope of a Curve

Introductory Example

Social Security Contributions

The U.S. Social Security system was created in 1937 by the Federal Insurance Contributions Act (F.I.C.A.). This act requires that most employees (together with their employers) contribute to the federal system for retirement, survivor's, disability, and hospital insurances. F.I.C.A. taxes are levied on an employee's income up to a given amount per year. The maximum amount of tax for which an employee and his or her employer are liable (see Table 2.1) has increased from $60.00 in 1937 to $4542.60 in 1983. The graph in Figure 2.1 approximates this increase in F.I.C.A. tax. Notice that this graph is becoming steeper as time passes. In mathematical terms, we describe this phenomenon by saying that the **slope** of this graph increases with time.

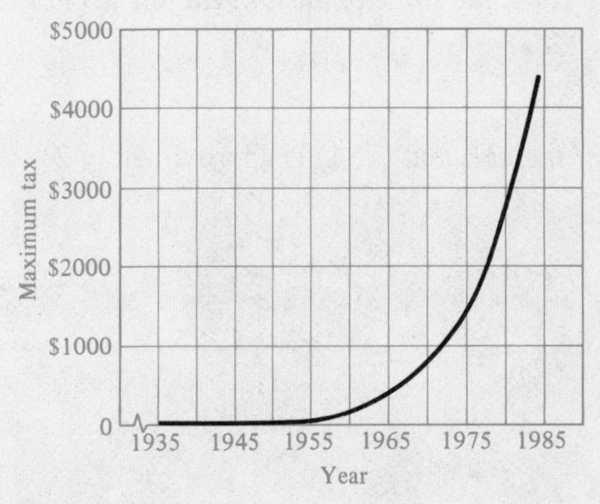

Figure 2.1

Table 2.1 Rate Paid by Employer and Employee

Year	Rate (%)	Maximum Taxable Earnings $	Maximum Tax $
1937–49	1.0	3,000	60.00
1950	1.5	3,000	90.00
1951–53	1.5	3,600	108.00
1954	2.0	3,600	144.00
1955–56	2.0	4,200	168.00
1957–58	2.25	4,200	189.00
1959	2.5	4,800	240.00
1960–61	3.0	4,800	288.00
1962	3.125	4,800	300.00
1963–65	3.625	4,800	348.00
1966	4.2	6,600	554.40
1967	4.4	6,600	580.80
1968	4.4	7,800	686.40
1969–70	4.8	7,800	748.80
1971	5.2	7,800	811.20
1972	5.2	9,000	936.00
1973	5.85	10,800	1,263.60
1974	5.85	13,200	1,544.40
1975	5.85	14,100	1,649.70
1976	5.85	15,300	1,790.10
1977	5.85	16,500	1,930.50
1978	6.05	17,700	2,141.70
1979	6.13	22,900	2,807.54
1980	6.13	25,900	3,175.34
1981	6.65	29,700	3,950.10
1982	6.7	31,800	4,261.20
1983	6.7	33,900	4,542.60

Section Objectives: ▪ *To define the slope of a curve at a point* ▪ *To define the derivative and use it to find the slope of a curve* ▪ *To calculate the derivative by its limit definition.*

To determine the slope of a curve at some point, we make use of the **tangent line** to the curve at the point. For instance, in Figure 2.2 the tangent line to the graph of f at P is the line that best approximates the graph of f at that point. Our problem of finding the slope of a curve at a point thus becomes one of finding the slope of the line tangent to the curve at that point.

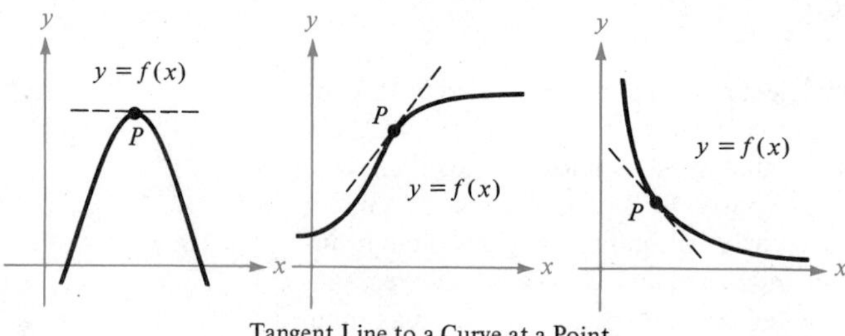

Tangent Line to a Curve at a Point

Figure 2.2

Example 1

Approximating the Slope of a Curve at a Point

Use the graph in Figure 2.3 to approximate the slope of $f(x) = x^2$ at the point $(1, 1)$.

Solution From the graph of $f(x) = x^2$, we see that the tangent line at $(1, 1)$ rises 2 units for each unit change in x. Thus, the *slope of the tangent line* at $(1, 1)$ is

$$\text{slope} \approx \tfrac{2}{1} = 2$$

Finally, we conclude that the *slope of the curve* at $(1, 1)$ is approximately 2.

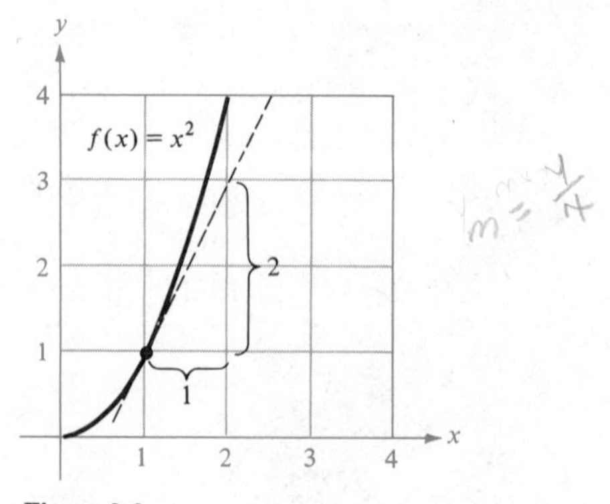

Figure 2.3

□

Remark

When graphically approximating the slope of a curve, you should realize that the scales on the horizontal and vertical axes may differ. When this happens (as it frequently does in applications), the slope of the tangent line is distorted, and you must be careful to account for the scale differences.

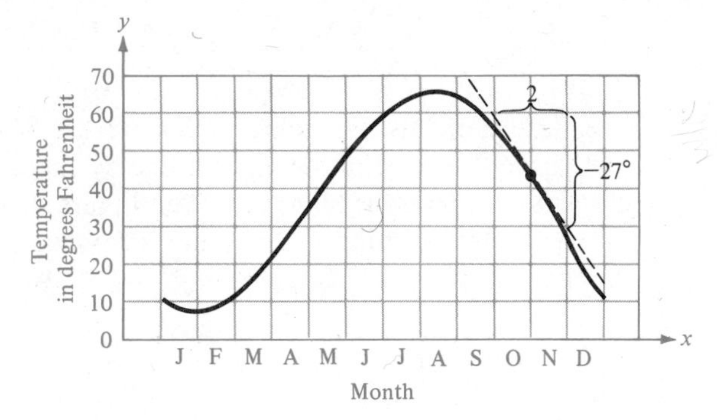

Figure 2.4

Example 2

Approximating the Slope of a Curve at a Point

Figure 2.4 graphically depicts the average daily temperature (in degrees Fahrenheit) in Duluth, Minnesota. Estimate the slope of this curve at the indicated point.

Solution From the graph in Figure 2.4, we see that the tangent line at the given point falls approximately 27 units for each two-unit change in x. Thus, we estimate the slope at the given point to be

$$\text{slope} \approx \tfrac{-27}{2} = -13.5 \text{ (degrees per month)}$$

This means that we can expect the average November temperature to be roughly 13.5 degrees lower than the average October temperature. □

From Examples 1 and 2, we see that we can approximate the slope of a curve at a point by making a careful graph and then "eyeballing" the tangent line at the point of tangency. A more precise method of approximating tangent lines

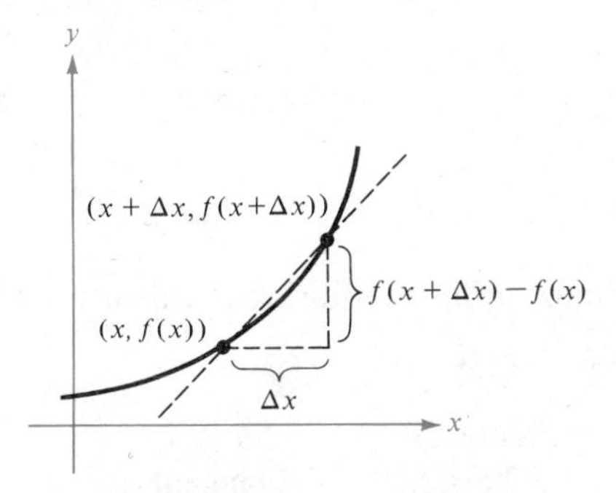

Figure 2.5 The Secant Line through $(x, f(x))$ and $(x + \Delta x, f(x + \Delta x))$

makes use of a **secant line** through the point of tangency and a second point on the curve. (See Figure 2.5.) If $(x, f(x))$ is the point of tangency and $(x + \Delta x, f(x + \Delta x))$ is a second point on the graph of f, then the slope of the secant line through these two points is given by

$$m_{\text{sec}} = \frac{f(x + \Delta x) - f(x)}{x + \Delta x - x} = \frac{f(x + \Delta x) - f(x)}{\Delta x}$$

The beauty of this procedure is that we can obtain better and better approximations to the slope of the tangent line by choosing the second point closer and closer to the point of tangency. (See Figure 2.6.)

Finally, by resorting to the limit process we can determine the *exact* slope of the tangent line at $(x, f(x))$ to be

$$m = \lim_{\Delta x \to 0} \frac{f(x + \Delta x) - f(x)}{\Delta x}$$

We use this limit as our formal definition of the slope of a curve at a point.

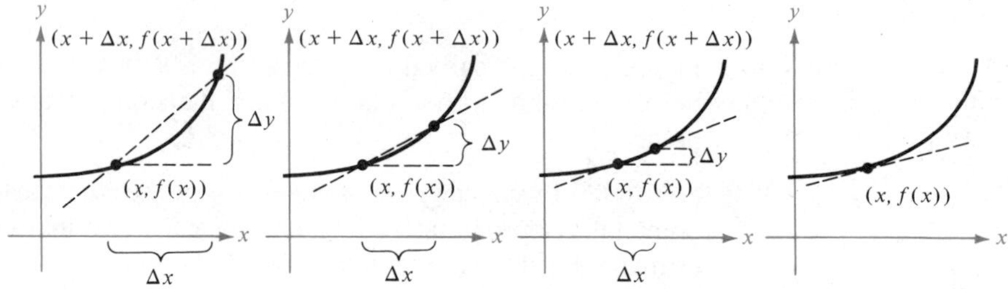

As Δx approaches 0, the secant lines approach the tangent line.

Figure 2.6

Definition of the Slope of a Curve

At $(x, f(x))$ the slope m of the graph of $y = f(x)$ is equal to the slope of its tangent line at $(x, f(x))$, and it is determined by the formula

$$m = \lim_{\Delta x \to 0} \frac{f(x + \Delta x) - f(x)}{\Delta x}$$

provided this limit exists.

Remark

The numerator in this definition is called the **change in *y*** and is denoted by

$$\Delta y = f(x + \Delta x) - f(x)$$

as shown in Figure 2.6.

To calculate the slope of the tangent line to a curve by its limit definition, we follow a general four-step process.

Four-Step Process Given $y = f(x)$:

1. Determine $f(x + \Delta x)$.
2. Calculate $f(x + \Delta x) - f(x)$.
3. Divide by Δx to obtain

$$\frac{f(x + \Delta x) - f(x)}{\Delta x}$$

4. Let $\Delta x \to 0$ to obtain

$$\lim_{\Delta x \to 0} \frac{f(x + \Delta x) - f(x)}{\Delta x} = m$$

Example 3

Finding the Slope by the Four-Step Process

Determine the slope of the line tangent to $f(x) = 2x - 3$ at any point (x, y).

Solution By the four-step process,

1. $$f(x + \Delta x) = 2(x + \Delta x) - 3$$

2. $$\begin{aligned} f(x + \Delta x) - f(x) &= (2x + 2\Delta x - 3) - (2x - 3) \\ &= 2\Delta x \end{aligned}$$

3. $$\frac{f(x + \Delta x) - f(x)}{\Delta x} = \frac{2\Delta x}{\Delta x} = 2$$

4. $$\lim_{\Delta x \to 0} \frac{f(x + \Delta x) - f(x)}{\Delta x} = \lim_{\Delta x \to 0} 2 = 2$$

Therefore $$m = 2$$
(See Figure 2.7.)

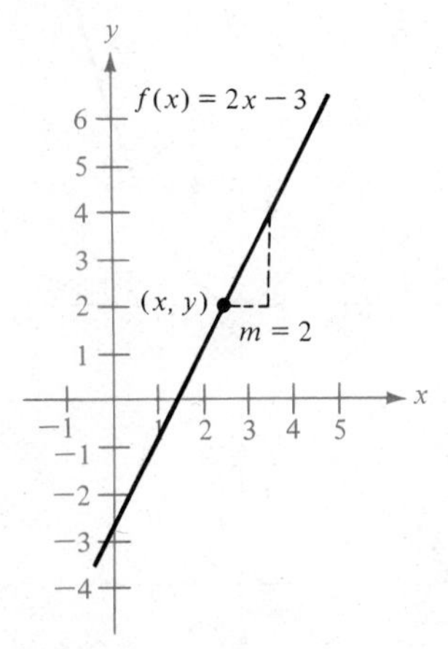

Figure 2.7

□

Note in Example 3 that the slope of f is constant. This is not surprising since we know from Chapter 1 that the graph of f is a line. Of course, not all graphs have constant slope, as the next example illustrates.

Example 4

Finding the Slope by the Four-Step Process

Determine the formula for the slope of the graph of $y = x^2 + 3$ (see Figure 2.8).

(a) What is the slope at (0, 3)?

(b) What is the slope at (−2, 7)?

Solution By the four-step process,

1. $$f(x + \Delta x) = (x + \Delta x)^2 + 3$$

2. $$f(x + \Delta x) - f(x) = x^2 + 2x(\Delta x) + (\Delta x)^2 + 3 - x^2 - 3$$
$$= 2x(\Delta x) + (\Delta x)^2$$

3. $$\frac{f(x + \Delta x) - f(x)}{\Delta x} = 2x + (\Delta x)$$

4. $$\lim_{\Delta x \to 0} \frac{f(x + \Delta x) - f(x)}{\Delta x} = \lim_{\Delta x \to 0} (2x + \Delta x) = 2x$$

Therefore, $m = 2x$

(a) At (0, 3), $m = 2(0) = 0$

(b) At (−2, 7), $m = 2(-2) = -4$

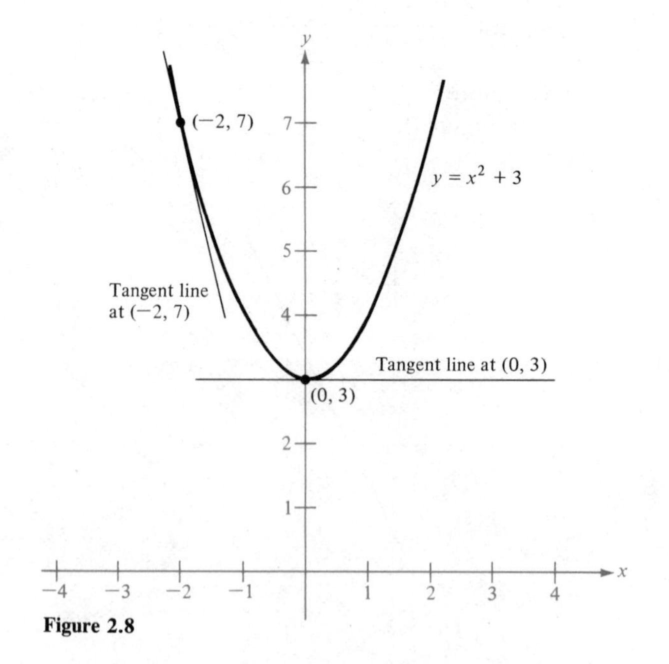

Figure 2.8

□

Example 5

Finding the Slope by the Four-Step Process

Write the equation of the line tangent to $f(x) = \sqrt{x}$ at the point (4, 2). (See Figure 2.9.)

Solution

1. $$f(x + \Delta x) = \sqrt{x + \Delta x}$$

2. $$f(x + \Delta x) - f(x) = \sqrt{x + \Delta x} - \sqrt{x}$$

Rationalizing the numerator, we obtain

$$f(x + \Delta x) - f(x) = (\sqrt{x + \Delta x} - \sqrt{x})\left(\frac{\sqrt{x + \Delta x} + \sqrt{x}}{\sqrt{x + \Delta x} + \sqrt{x}}\right)$$

$$= \frac{(x + \Delta x) - x}{\sqrt{x + \Delta x} + \sqrt{x}} = \frac{\Delta x}{\sqrt{x + \Delta x} + \sqrt{x}}$$

3. $$\frac{f(x + \Delta x) - f(x)}{\Delta x} = \frac{1}{\sqrt{x + \Delta x} + \sqrt{x}}$$

4. $$\lim_{\Delta x \to 0} \frac{f(x + \Delta x) - f(x)}{\Delta x} = \lim_{\Delta x \to 0}\left(\frac{1}{\sqrt{x + \Delta x} + \sqrt{x}}\right) = \frac{1}{2\sqrt{x}}$$

Therefore, at (4, 2) the slope of the tangent line is

$$m = \frac{1}{2\sqrt{4}} = \frac{1}{4}$$

and the equation of this tangent line is

$$y - 2 = \tfrac{1}{4}(x - 4)$$

$$y = \tfrac{1}{4}x + 1$$

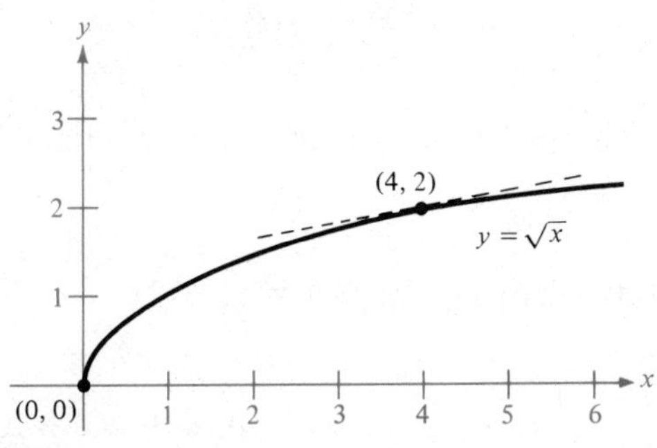

Figure 2.9

□

We have now arrived at a crucial point in our study of calculus, for the limit

$$\lim_{\Delta x \to 0} \frac{f(x + \Delta x) - f(x)}{\Delta x}$$

is used to define one of the two fundamental quantities of calculus — namely, the **derivative.**

Definition of the Derivative

The limit

$$\lim_{\Delta x \to 0} \frac{f(x + \Delta x) - f(x)}{\Delta x}$$

is called the **derivative** of f at x (provided the limit exists), and it is denoted by $f'(x)$.

A function is said to be **differentiable** at x if its derivative exists at x, and the process of finding the derivative is called **differentiation.**

In addition to $f'(x)$ (read "f prime of x"), various other notations are used to denote the derivative. The most commonly used notations are

$$\frac{dy}{dx}, \qquad y', \qquad \frac{d}{dx}[f(x)], \qquad D_x(y)$$

We will frequently use dy/dx to denote the derivative. We read dy/dx as the "derivative of y with respect to x," and, using limit notation, we write

$$\frac{dy}{dx} = \lim_{\Delta x \to 0} \frac{\Delta y}{\Delta x}$$

Thus, we have

$$\frac{dy}{dx} = \lim_{\Delta x \to 0} \frac{\Delta y}{\Delta x} = \lim_{\Delta x \to 0} \frac{f(x + \Delta x) - f(x)}{\Delta x} = f'(x)$$

Since the derivative $f'(x)$ and the slope of the tangent line to the graph of f are both defined by the limit

$$\lim_{\Delta x \to 0} \frac{f(x + \Delta x) - f(x)}{\Delta x}$$

we can use the four-step process to determine $f'(x)$.

Example 6

Finding the Derivative by the Four-Step Process

Find the derivative of $f(x) = x^3 + 2x$.

Solution

1. $$f(x + \Delta x) = (x + \Delta x)^3 + 2(x + \Delta x)$$

2. $$\begin{aligned} f(x + \Delta x) - f(x) &= x^3 + 3x^2(\Delta x) + 3x(\Delta x)^2 + (\Delta x)^3 \\ &\quad + 2x + 2(\Delta x) - (x^3 + 2x) \\ &= 3x^2(\Delta x) + 3x(\Delta x)^2 + (\Delta x)^3 + 2(\Delta x) \end{aligned}$$

3. $$\frac{f(x + \Delta x) - f(x)}{\Delta x} = 3x^2 + 3x(\Delta x) + (\Delta x)^2 + 2$$

4. $$\lim_{\Delta x \to 0} \frac{f(x + \Delta x) - f(x)}{\Delta x} = \lim_{\Delta x \to 0} [3x^2 + 3x(\Delta x) + (\Delta x)^2 + 2]$$
$$= 3x^2 + 2$$

Therefore, $$f'(x) = 3x^2 + 2$$ □

Example 7

Finding the Derivative by the Four-Step Process

Given $y = 2/t$, determine the derivative of y with respect to t.

Solution Considering $y = f(t)$, we have

1. $$f(t + \Delta t) = \frac{2}{t + \Delta t}$$

2. $$f(t + \Delta t) - f(t) = \frac{2}{t + \Delta t} - \frac{2}{t} = \frac{2t - 2t - 2(\Delta t)}{t(t + \Delta t)}$$
$$= \frac{-2(\Delta t)}{t(t + \Delta t)}$$

3. $$\frac{f(t + \Delta t) - f(t)}{\Delta t} = \frac{-2}{t(t + \Delta t)}$$

4. $$\lim_{\Delta t \to 0} \frac{f(t + \Delta t) - f(t)}{\Delta t} = \lim_{\Delta t \to 0} \frac{-2}{t(t + \Delta t)} = \frac{-2}{t^2}$$

Therefore, the derivative of y with respect to t is

$$\frac{dy}{dt} = \frac{-2}{t^2}$$ □

Remark

Before concluding this section, we must point out that a function may have a tangent line at each point on its graph and yet not be differentiable at each point. In particular, the derivative does not exist at points with vertical tangents since the slope at such points is undefined. This is the case in Example 5, where

$$\lim_{\Delta x \to 0} \frac{f(x + \Delta x) - f(x)}{\Delta x} = \frac{1}{2\sqrt{x}}$$

does not exist at the point (0, 0). Figure 2.10 shows another instance of a vertical tangent.

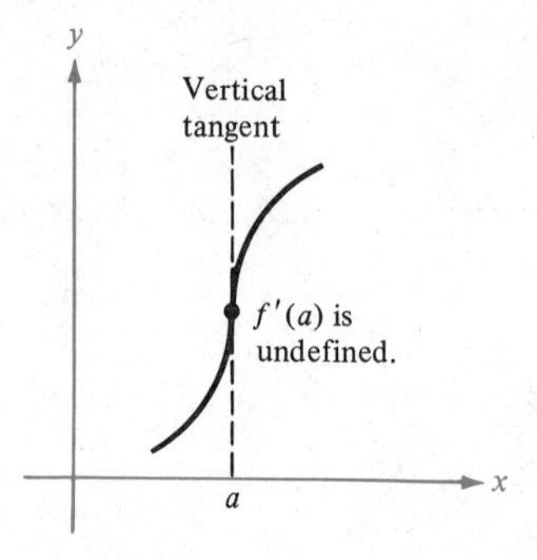

Figure 2.10

Section Exercises (2.1)

In Exercises 1–6, trace the curves on another piece of paper and sketch the tangent line at the given point (x, y).

1.

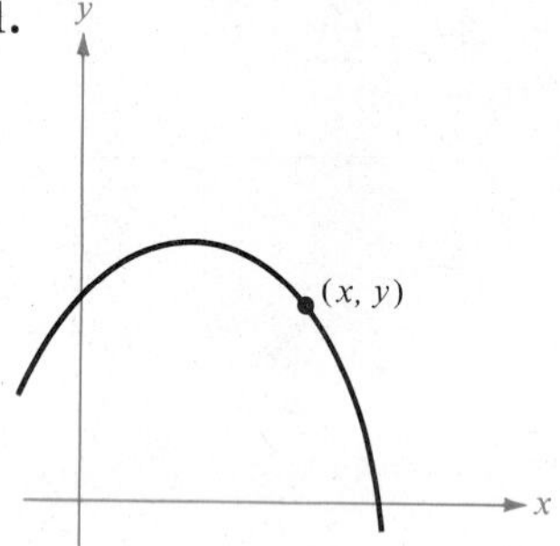

2.

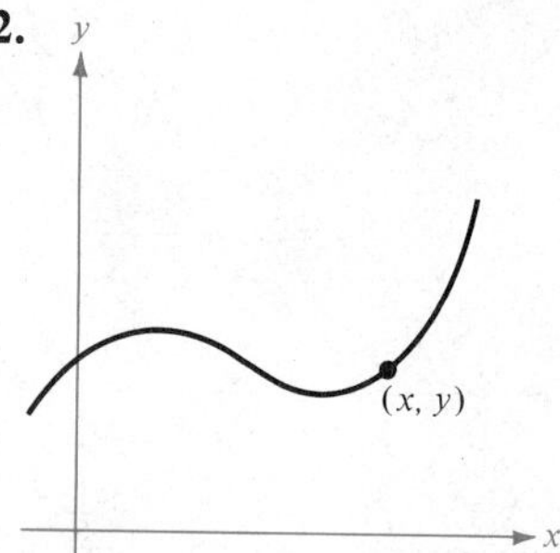

3.

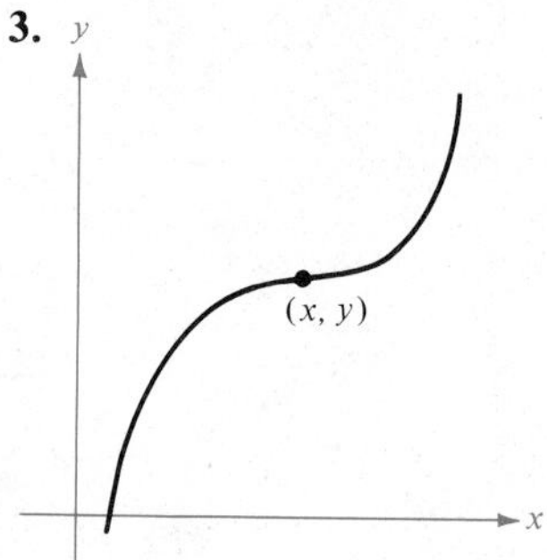

4.

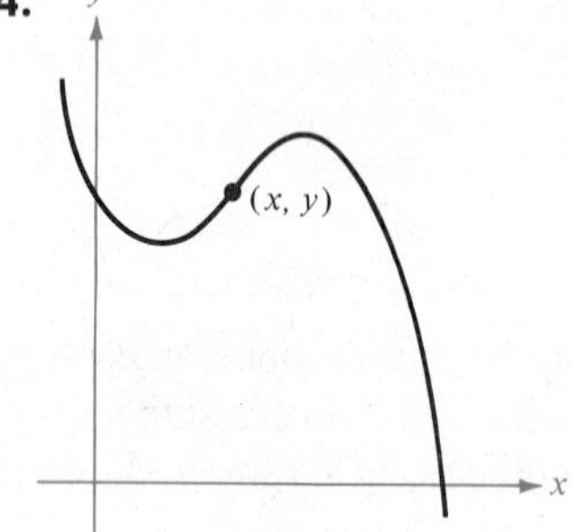

5.

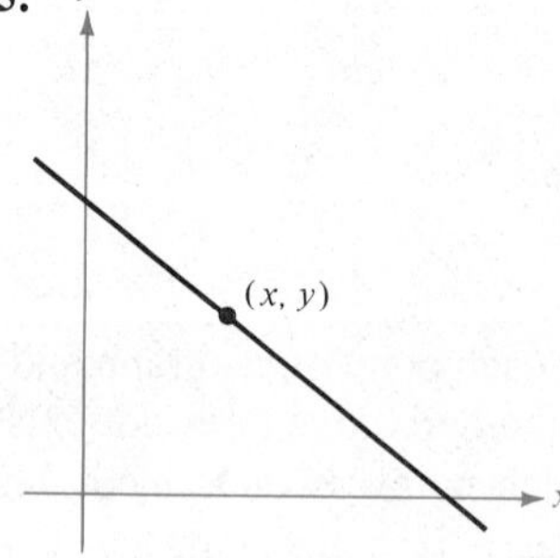

6.

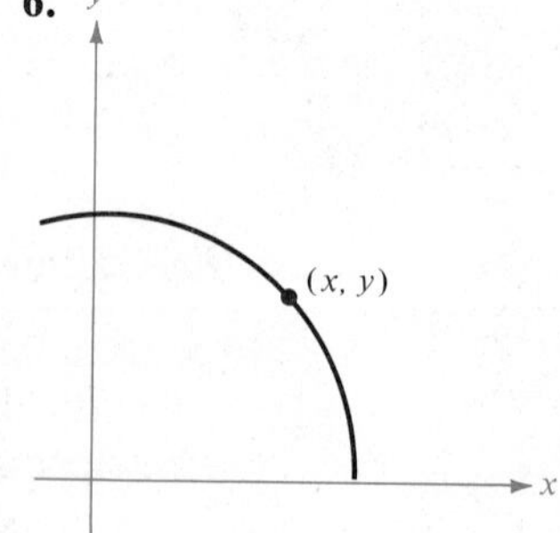

In Exercises 7–12, estimate the slope of the curve at the given point (x, y).

7.

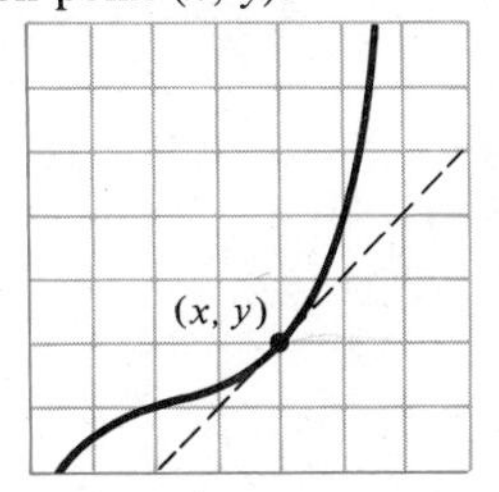

8.

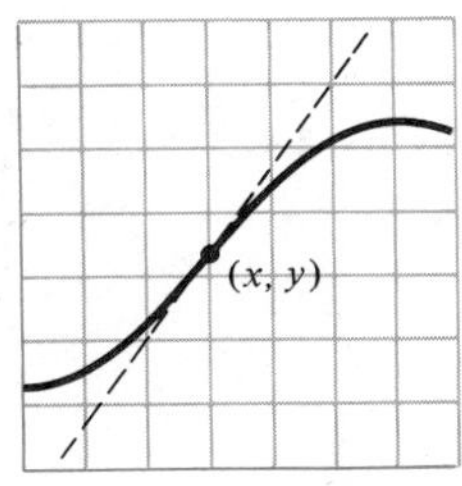

9.

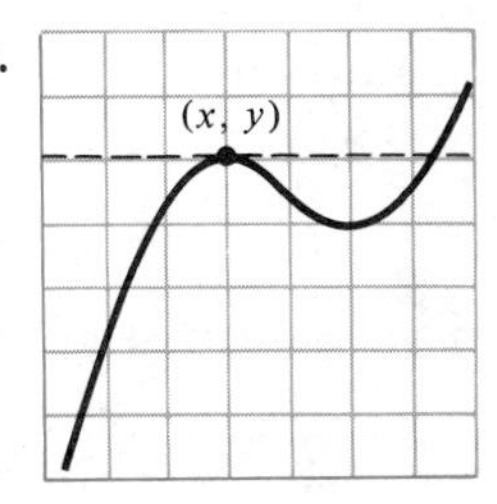

10.

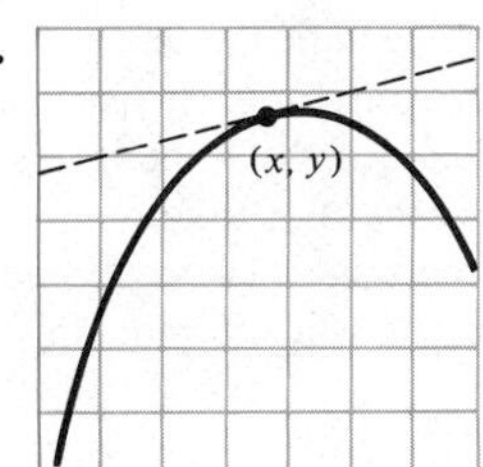

11.

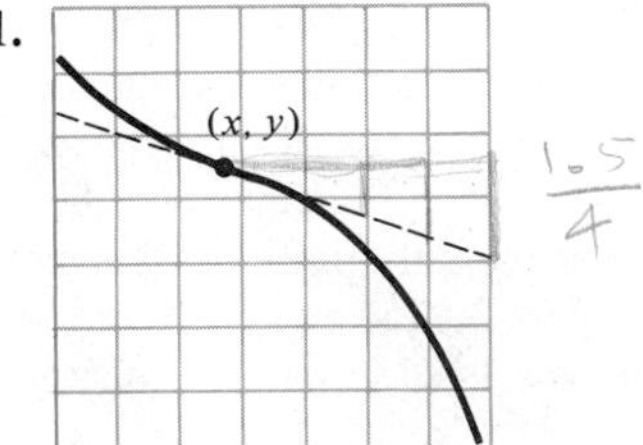

12.

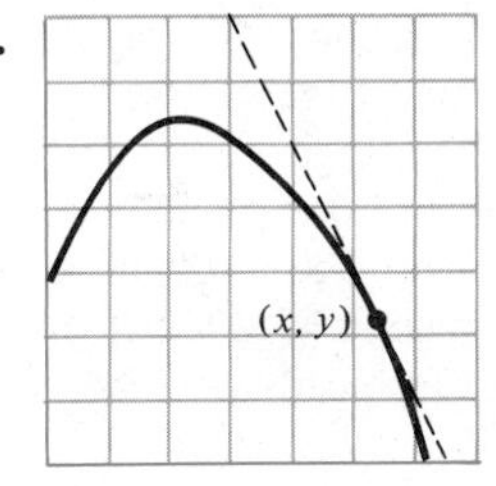

In Exercises 13–22, find the derivative by the four-step process.

13. $f(x) = 3$ **14.** $f(x) = 3x + 2$

15. $f(x) = -5x$ **16.** $f(x) = 1 - x^2$

17. $f(x) = 2x^2 + x - 1$ **18.** $f(x) = \sqrt{x - 4}$

19. $f(x) = \dfrac{1}{x - 1}$ **20.** $f(x) = \dfrac{1}{x^2}$

21. $f(t) = t^3 - 12t$ **22.** $f(t) = t^3 + t^2$

In Exercises 23–30, use the four-step process to find the derivative of each function. Sketch the graph of each function and find the equation of the tangent line at the given point.

23. $f(x) = x^2 + 1$; $(2, 5)$

24. $f(x) = x^2 + 2x + 1$; $(-3, 4)$

25. $f(x) = x^3$; $(2, 8)$ **26.** $f(x) = x^3$; $(-2, -8)$

27. $f(x) = \sqrt{x + 1}$; $(3, 2)$

28. $f(x) = \dfrac{1}{\sqrt{x}}$; $(4, \frac{1}{2})$

29. $f(x) = \dfrac{1}{x}$; $(1, 1)$ **30.** $f(x) = \dfrac{1}{x + 1}$; $(0, 1)$

31. Find the equation of a line that is tangent to the curve $y = x^3$ and is parallel to the line $3x - y + 1 = 0$.

32. Find the equation of a line that is tangent to the curve $y = 1/\sqrt{x}$ and is parallel to the line $x + 2y - 6 = 0$.

33. There are two tangent lines to the curve $y = 4x - x^2$ that pass through the point $(2, 5)$. Find the equations of those two lines and make a sketch to verify your results.

34. Two lines through the point $(1, 3)$ are tangent to the curve $y = x^2$. Find the equations of these two lines and make a sketch to verify your results.

Section 2.2

Some Rules for Differentiation

Introductory Example

Home Mortgage Payments

When a loan is repaid in monthly installments, the size of the monthly payment is a function of the total number of years in the loan period. For example, if a home mortgage of $75,000.00 at 16% is taken out for 20 years, the monthly payment is $1,043.44. If the same mortgage is taken out for 30 years, the monthly payment is $1,008.57. (See Table 2.2.) Figure 2.11 shows graphically how the monthly payment is related to the number of years in the loan period. For this particular graph, we can interpret the slope to be the rate at which the monthly payment is decreasing for each year that is added to the loan period. Note that the slope levels off as the number of years increases. For the homeowner this means that extending a mortgage period beyond 20 years is a relatively inefficient way of reducing the monthly payment.

Table 2.2

Number of years	Monthly payment (dollars)	Slope	Total interest (dollars)
5	1,823.85	−238.83	34,431.00
10	1,256.35	−51.19	75,762.00
15	1,101.53	−17.78	123,275.40
20	1,043.44	−7.20	175,425.60
25	1,019.17	−3.10	230,751.00
30	1,008.57	−1.37	288,085.20
35	1,003.85	−0.61	346,617.00
40	1,001.74	−0.28	405,835.20
45	1,000.78	−0.12	465,421.20
50	1,000.35	−0.06	525,210.00
55	1,000.16	−0.03	585,105.60
60	1,000.07	−0.01	645,050.40

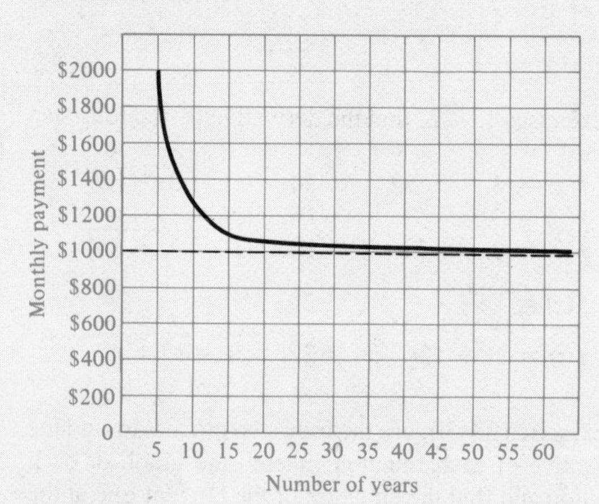

Figure 2.11

Section Objective: ***To prove and demonstrate these differentiation rules: the Constant Rule, the (Simple) Power Rule, the Constant Multiple Rule, and the Sum Rule.***

In Section 2.1, we found derivatives by the limit definition of the derivative. This procedure is rather tedious even for simple functions, and fortunately there are rules that greatly simplify the differentiation process. These rules permit us to calculate the derivative without the direct use of limits. We now derive some of these rules, and in each case we assume that the derivative of the given function exists.

Constant Rule

The derivative of a constant is zero.

$$\frac{d}{dx}[c] = 0, \qquad \text{where } c \text{ is a constant}$$

Proof Let $f(x) = c$; then by the limit definition

$$\frac{d}{dx}[f(x)] = f'(x) = \lim_{\Delta x \to 0} \frac{f(x + \Delta x) - f(x)}{\Delta x}$$

$$= \lim_{\Delta x \to 0} \frac{c - c}{\Delta x} = \lim_{\Delta x \to 0} 0 = 0$$

Therefore,

$$\frac{d}{dx}[c] = 0$$

(See Figure 2.12.)

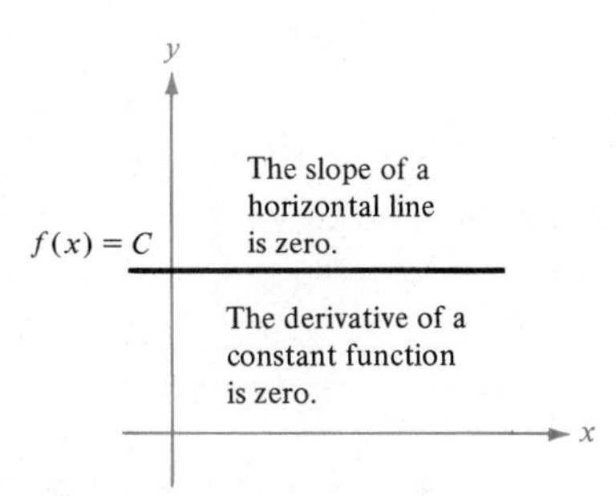

Figure 2.12

□

Example 1

Finding the Derivative of a Constant

(a) $\frac{d}{dx}[7] = 0$

(b) If $f(x) = 0$, then $f'(x) = 0$.

(c) If $y = 2$, then $dy/dx = 0$.

(d) If $g(t) = \frac{-3}{2}$, then $g'(t) = 0$. □

Before deriving the next rule, we review the procedure for expanding a binomial. Recall that

$$(x + \Delta x)^2 = x^2 + 2x\,\Delta x + (\Delta x)^2$$

and $$(x + \Delta x)^3 = x^3 + 3x^2(\Delta x) + 3x(\Delta x)^2 + (\Delta x)^3$$

and, in general, the binomial expansion is

$$(x + \Delta x)^n = x^n + nx^{n-1}(\Delta x) + \frac{n(n-1)x^{n-2}}{2}(\Delta x)^2 + \frac{n(n-1)(n-2)x^{n-3}}{2(3)}(\Delta x)^3 + \cdots + (\Delta x)^n$$

where n is a positive integer. We use this binomial expansion in proving a special case of the following rule.

(Simple) Power Rule

$$\frac{d}{dx}[x^n] = nx^{n-1}, \qquad \text{where } n \text{ is any real number}$$

Proof Although the Power Rule is true for any real number n, the binomial expansion applies only when n is a positive integer. Thus, for the time being, we give a proof for the case when n is a positive integer. Sections 2.4 and 2.7 contain proofs for negative integers and arbitrary rationals, respectively.

$$\frac{d}{dx}[f(x)] = \lim_{\Delta x \to 0} \frac{f(x + \Delta x) - f(x)}{\Delta x} = \lim_{\Delta x \to 0} \frac{(x + \Delta x)^n - x^n}{\Delta x}$$

Now applying the binomial expansion rule, we obtain

$$\frac{d}{dx}[x^n] = \lim_{\Delta x \to 0} \frac{x^n + nx^{n-1}(\Delta x) + [n(n-1)x^{n-2}/2](\Delta x)^2 + \cdots + (\Delta x)^n - x^n}{\Delta x}$$

$$= \lim_{\Delta x \to 0} \left[nx^{n-1} + \frac{n(n-1)x^{n-2}}{2}(\Delta x) + \cdots + (\Delta x)^{n-1} \right]$$

$$= nx^{n-1} + 0 + \cdots + 0 = nx^{n-1}$$ □

Before showing some examples of the use of the Power Rule, we establish yet a third differentiation rule.

Constant Multiple Rule

$$\frac{d}{dx}[cf(x)] = cf'(x), \qquad c \text{ is a constant}$$

Proof Applying the definition of the derivative, we have

$$\frac{d}{dx}[cf(x)] = \lim_{\Delta x \to 0} \frac{cf(x + \Delta x) - cf(x)}{\Delta x}$$

$$= \lim_{\Delta x \to 0} c\left[\frac{f(x + \Delta x) - f(x)}{\Delta x}\right]$$

$$= c\left[\lim_{\Delta x \to 0} \frac{f(x + \Delta x) - f(x)}{\Delta x}\right] = cf'(x)$$

□

Informally, the Constant Multiple Rule states that constants can be "factored" out of the differentiation process:

$$\frac{d}{dx}[cf(x)] = c\frac{d}{dx}[\quad f(x)] = cf'(x)$$

The usefulness and versatility of this rule are often overlooked, especially when the constant appears in the denominator. Note that

$$\frac{d}{dx}\left[\frac{f(x)}{c}\right] = \frac{d}{dx}\left[\frac{1}{c}f(x)\right] = \frac{1}{c}\frac{d}{dx}[\quad f(x)] = \frac{1}{c}f'(x)$$

To use the Constant Multiple Rule to best advantage, be on the lookout for constants that can be factored out *before* differentiating.

Example 2

Applying the Power Rule and the Constant Multiple Rule

(a) If $f(x) = x^3$, then by the Power Rule $f'(x) = 3x^2$.

(b) If $y = 2x^{1/2}$, then by the Constant Multiple Rule and the Power Rule, we have

$$\frac{dy}{dx} = \frac{d}{dx}[2x^{1/2}] = 2\frac{d}{dx}[x^{1/2}] = 2\left(\frac{1}{2}x^{-1/2}\right) = x^{-1/2}$$

(c) If

$$f(t) = \frac{4t^2}{5}$$

we rewrite $f(t)$ as $f(t) = \frac{4}{5}(t^2)$. Then by the Constant Multiple and Power Rules, we have

$$f'(t) = \frac{d}{dt}\left[\frac{4}{5}t^2\right] = \frac{4}{5}\frac{d}{dt}[t^2] = \frac{4}{5}(2t) = \frac{8}{5}t$$

(d) If $y = x^{-2}$, then by the Power Rule

$$\frac{dy}{dx} = \frac{d}{dx}[x^{-2}] = -2x^{-3}$$

□

It is of benefit to see that the Constant Multiple and Power Rules can be combined into one rule. The combination rule is

$$\frac{d}{dx}[cx^n] = cnx^{n-1}, \qquad \text{where } n \text{ is any real number}$$

For instance, in part (c) of Example 2, we can apply this combination rule to obtain

$$\frac{d}{dx}\left[\frac{4}{5}t^2\right] = \left(\frac{4}{5}\right)(2)t = \frac{8}{5}t$$

Remark Two special cases of the Power Rule occur when $n = 0$ and $n = 1$. If $n = 0$, we actually have the Constant Rule. If $n = 1$, we have

$$\frac{d}{dx}[cx^1] = c\frac{d}{dx}[x^1] = c(1)(x^0) = c$$

Example 3

Applying the Constant Multiple Rule

(a) $\dfrac{d}{dx}\left[-\dfrac{3x}{2}\right] = -\dfrac{3}{2}$

(b) $\dfrac{d}{dx}[3\pi x] = 3\pi$

(c) If $y = -\dfrac{x}{2}$, then $y' = -\dfrac{1}{2}$.

(d) If $f(x) = -\dfrac{5x}{6}$, then $f'(x) = -\dfrac{5}{6}$. □

Remark The four problems in Example 3 are very simple, yet errors are frequently made in differentiating a constant multiple of the first power of x. Keep in mind that

$$\frac{d}{dx}[cx] = c, \qquad \text{where } c \text{ is any constant}$$

The next rule is one that you might expect to be true and it is often used without thinking about it. For instance, if you were to find the derivative of $y = 3x + 2x^3$, you would probably write

$$y' = 3 + 6x^2$$

without questioning your answer. The validity of differentiating a sum "term by term" is given in the following rule.

Sum Rule The derivative of the sum of two functions is the sum of their derivatives:

$$\frac{d}{dx}[f(x) + g(x)] = f'(x) + g'(x)$$

Proof Let $S(x) = f(x) + g(x)$. Then the derivative of S is

$$\begin{aligned} S'(x) &= \lim_{\Delta x \to 0} \frac{S(x + \Delta x) - S(x)}{\Delta x} \\ &= \lim_{\Delta x \to 0} \frac{f(x + \Delta x) + g(x + \Delta x) - f(x) - g(x)}{\Delta x} \\ &= \lim_{\Delta x \to 0} \frac{f(x + \Delta x) - f(x) + g(x + \Delta x) - g(x)}{\Delta x} \\ &= \lim_{\Delta x \to 0} \left[\frac{f(x + \Delta x) - f(x)}{\Delta x} + \frac{g(x + \Delta x) - g(x)}{\Delta x}\right] \\ &= \lim_{\Delta x \to 0} \frac{f(x + \Delta x) - f(x)}{\Delta x} + \lim_{\Delta x \to 0} \frac{g(x + \Delta x) - g(x)}{\Delta x} \\ &= f'(x) + g'(x) \end{aligned}$$

Thus,

$$\frac{d}{dx}[f(x) + g(x)] = f'(x) + g'(x)$$ □

By a similar procedure, it can be shown that

$$\frac{d}{dx}[f(x) - g(x)] = f'(x) - g'(x)$$

Furthermore, either rule can be extended to the derivative of the sum or difference of any finite number of functions. For instance, if

$$S(x) = f(x) + g(x) - h(x) - k(x)$$

then $$S'(x) = f'(x) + g'(x) - h'(x) - k'(x)$$

Example 4

Applying the Sum Rule

If $f(x) = x^3 - 4x + 5$, find the value of $f'(2)$.

Solution $f'(x) = 3x^2 - 4$. Therefore,

$$f'(2) = 3(2)^2 - 4 = 12 - 4 = 8$$ □

Example 5

Applying the Sum Rule

If

$$g(x) = -\frac{x^4}{2} + 3x^3 - 2x$$

find the value of $g'(-1)$.

Solution

$$g'(x) = -\tfrac{4}{2}x^3 + 9x^2 - 2 = -2x^3 + 9x^2 - 2$$
$$g'(-1) = -2(-1)^3 + 9(-1)^2 - 2 = 2 + 9 - 2 = 9$$
□

Example 6

Evaluating the Derivative at a Point

If $y = 3x^{-2}$, find the value of dy/dx at the point $(2, \tfrac{3}{4})$.

Solution By the Power Rule with $n = -2$, we obtain

$$\frac{dy}{dx} = 3(-2)x^{-3} = -6x^{-3} = -\frac{6}{x^3}$$

Therefore, when $x = 2$

$$\frac{dy}{dx} = -\frac{6}{2^3} = -\frac{6}{8} = -\frac{3}{4}$$
□

Example 7

Rewriting before Differentiating

If $g(t) = 5 - (1/2t^3)$, find $g'(2)$.

Solution We can rewrite $g(t)$ as

$$g(t) = 5 - \frac{1}{2}t^{-3}$$

then $$g'(t) = 0 - \frac{1}{2}(-3)t^{-4} = \frac{3}{2t^4}$$

Therefore,

$$g'(2) = \frac{3}{2(2)^4} = \frac{3}{32}$$
□

Note in Example 7 that the term

$$\frac{1}{2t^3} \quad \text{was rewritten as} \quad \frac{1}{2}t^{-3}$$

Errors are frequently made in rewriting fractions in negative exponent form. Study the next examples carefully.

Example 8

Parentheses and the Power Rule

Differentiate

$$y = \frac{5}{2x^3}$$

Solution First write

$$y = \frac{5}{2}(x^{-3})$$

then $$y' = \frac{5}{2}(-3x^{-4}) = -\frac{15}{2}x^{-4} = -\frac{15}{2x^4}$$ □

Example 9

Parentheses and the Power Rule

Differentiate

$$y = \frac{5}{(2x)^3}$$

Solution First, write

$$y = \frac{5}{2^3x^3} = \frac{5}{2^3}(x^{-3}) = \frac{5}{8}x^{-3}$$

Then $$y' = \frac{5}{8}(-3x^{-4}) = -\frac{15}{8}x^{-4} = -\frac{15}{8x^4}$$ □

Remark Note the effect of the parentheses in the denominator in Example 9. The exponent applies to the factor 2, whereas in Example 8 it does *not* apply.

Example 10

Parentheses and the Power Rule

Differentiate

$$y = \frac{7}{3x^{-2}}$$

Solution First, write

$$y = \frac{7}{3}x^2$$

then $$y' = \frac{7}{3}(2x) = \frac{14x}{3}$$ □

Example 11

Parentheses and the Power Rule

Differentiate

$$y = \frac{7}{(3x)^{-2}}$$

Solution First, write

$$y = 7(3x)^2 = 7(3^2x^2) = 63x^2$$

then $$y' = 63(2x) = 126x$$ □

Example 12

Rewriting Radicals as Rational Exponents

Differentiate $y = \sqrt{x}$.

Solution First, write

$$y = x^{1/2}$$

Then we apply the Power Rule to obtain

$$\frac{dy}{dx} = \frac{1}{2}x^{-1/2} = \frac{1}{2\sqrt{x}}$$

□

Example 13

Rewriting Radicals as Rational Exponents

Differentiate

$$y = \frac{1}{2\sqrt[3]{x^2}}$$

Solution Rewriting with a rational exponent, we have

$$y = \left(\frac{1}{2}\right)\left(\frac{1}{x^{2/3}}\right) = \left(\frac{1}{2}\right)x^{-2/3}$$

By the Power and Constant Multiple Rules, we have

$$\frac{dy}{dx} = \left(\frac{1}{2}\right)\left(-\frac{2}{3}\right)x^{-5/3} = -\frac{1}{3x^{5/3}}$$

□

Example 14

Finding the Equation of the Tangent Line at a Point

Find the equation of the tangent line to $y = \sqrt{3x}$ at the point (3, 3).

Solution Rewriting we have

$$y = (\sqrt{3})(x^{1/2})$$

By the Power and Constant Multiple Rules, we have

$$y' = \sqrt{3}\left(\frac{1}{2\sqrt{x}}\right) = \frac{\sqrt{3}}{2\sqrt{x}}$$

Now, at the point (3, 3), the derivative is

$$y' = \frac{\sqrt{3}}{2\sqrt{3}} = \frac{1}{2}$$

and since the derivative is the slope of the tangent line, the *equation* of the tangent line is

$$y - y_1 = m(x - x_1)$$

$$y - 3 = \frac{1}{2}(x - 3)$$

$$y = \frac{x}{2} + \frac{3}{2}$$

(See Figure 2.13.)

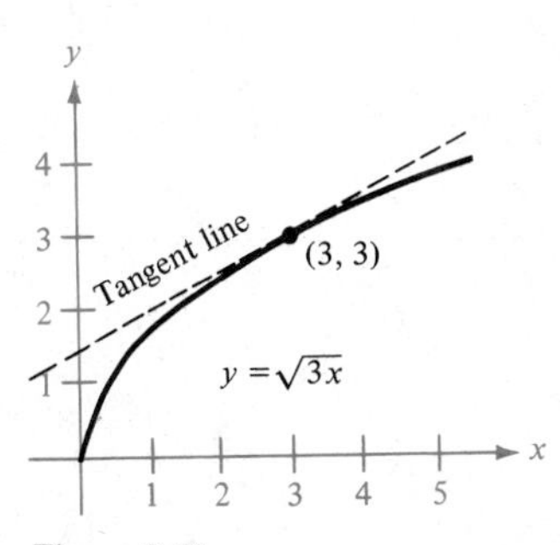

Figure 2.13

□

Remark When differentiating functions involving radical signs, you should rewrite the function in terms of rational exponents.

Section Exercises (2.2)

In Exercises 1–10, differentiate the functions.

1. $y = 3$

2. $f(x) = -2$

3. $f(x) = x + 1$

4. $g(x) = 3x - 1$

5. $g(x) = x^2 + 4$

6. $y = t^2 + 2t - 3$

7. $f(t) = -2t^2 + 3t - 6$

8. $y = x^3 - 9$

9. $s(t) = t^3 - 2t + 4$

10. $f(x) = 2x^3 - x^2 + 3x - 1$

In Exercises 11–18, differentiate the functions and evaluate each derivative at the indicated point.

11. $f(x) = \dfrac{1}{x}$; $(1, 1)$

12. $f(x) = -\frac{1}{2} + \frac{7}{5}x^3$; $(0, -\frac{1}{2})$

13. $f(t) = 3 - \dfrac{3t}{5t^2}$; $(\frac{3}{5}, 2)$

14. $y = \dfrac{1}{(3x)^3}$; $(1, \frac{1}{27})$

15. $y = \dfrac{1}{3x^3}$; $(1, \frac{1}{3})$

16. $y = 3x\left(x^2 - \dfrac{2}{x}\right)$; $(2, 18)$

17. $y = (2x + 1)^2$; $(0, 1)$

18. $f(x) = 3(5 - x)^2$; $(5, 0)$

In Exercises 19–36, find $f'(x)$.

19. $f(x) = -2(1 - 4x^2)^2$

20. $f(x) = [(x - 2)(x + 4)]^2$

21. $f(x) = x^2 - \dfrac{4}{x}$

22. $f(x) = x^2 - 3x - 3x^{-2} + 5x^{-3}$

23. $f(x) = x^3 - 3x - \dfrac{2}{x^4}$

24. $f(x) = \dfrac{2x^2 - 3x + 1}{x}$

25. $f(x) = \dfrac{x^3 - 3x^2 + 4}{x^2}$

26. $f(x) = \dfrac{2}{3x^2}$

27. $f(x) = \dfrac{\pi}{(3x)^2}$

28. $f(x) = (x^2 + 2x)(x + 1)$

29. $f(x) = x(x^2 - 1)$

30. $f(x) = x + \dfrac{1}{x^2}$

31. $f(x) = x^{4/5}$

32. $f(x) = x^{1/3} - 1$

33. $f(x) = \sqrt[3]{x} + \sqrt[5]{x}$

34. $f(x) = \dfrac{1}{\sqrt[3]{x^2}}$

35. $f(x) = \dfrac{1}{x^{1/2}} + \dfrac{1}{x^2} + \dfrac{1}{x^4}$

36. $f(x) = 5x^{3/2} - 3x^{1/2} - x^{-1/2}$

In Exercises 37–45, find the slope of the tangent line to $y = x^n$ at the point (1, 1).

37.

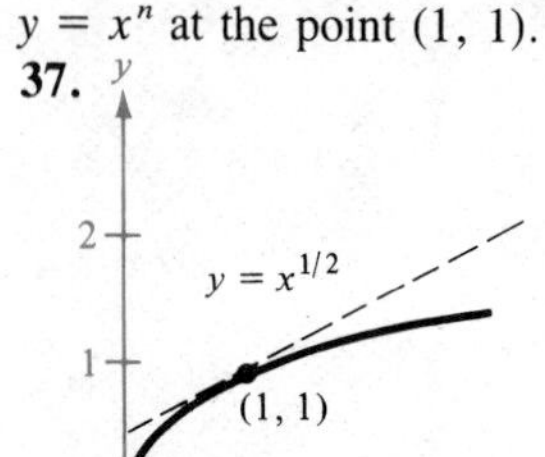

38.

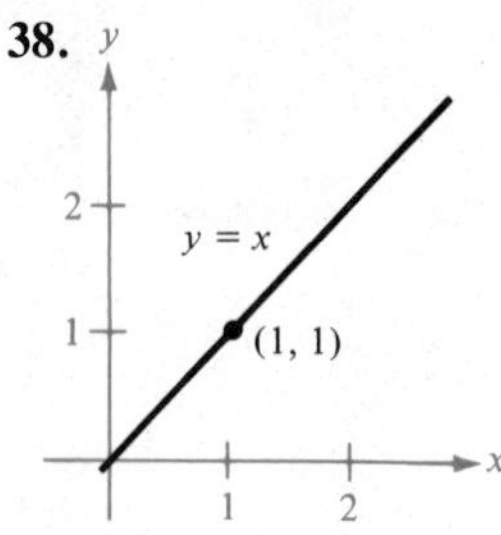

39.

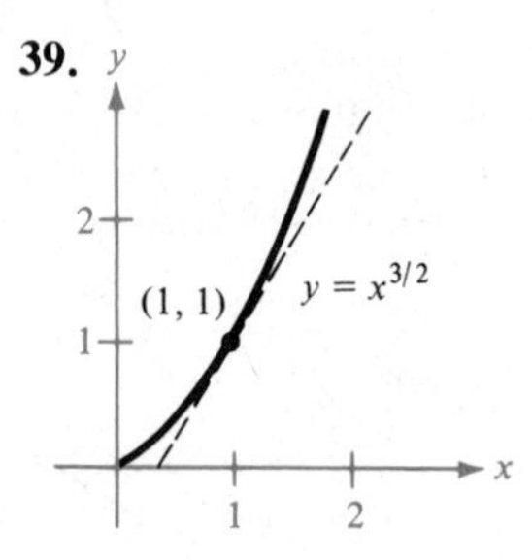

40.

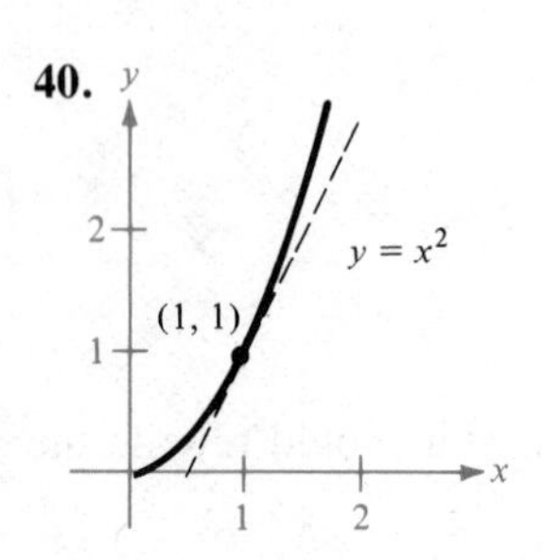

41.

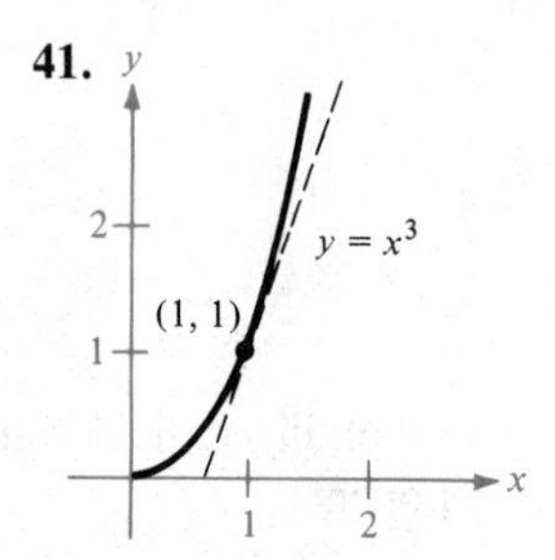

42.

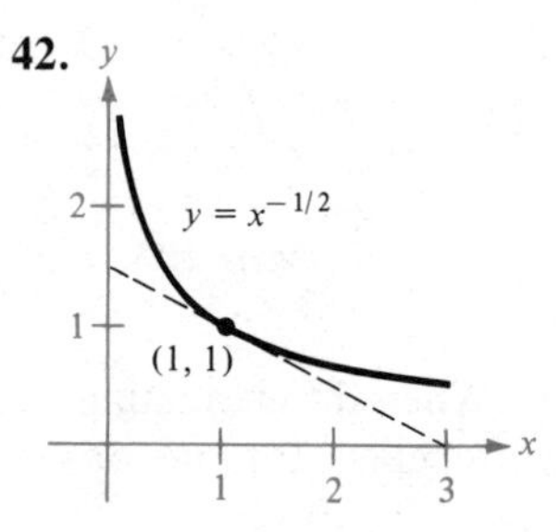

43.

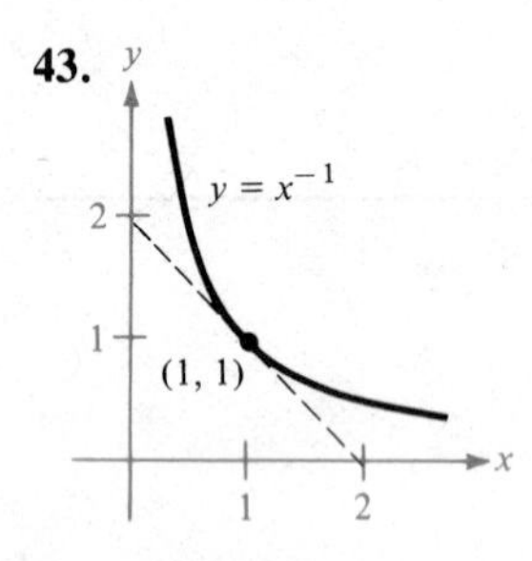

44.

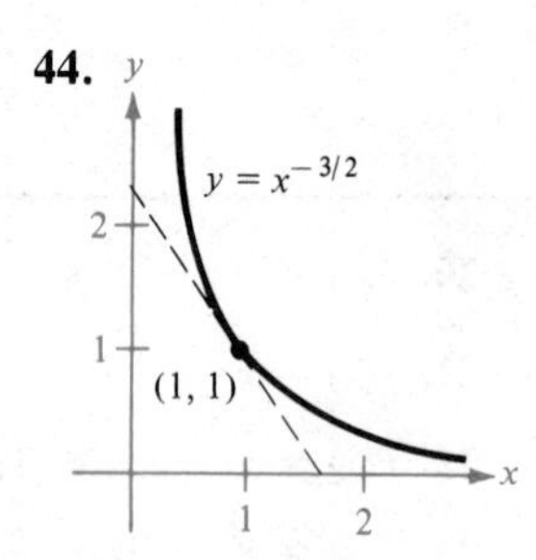

45.

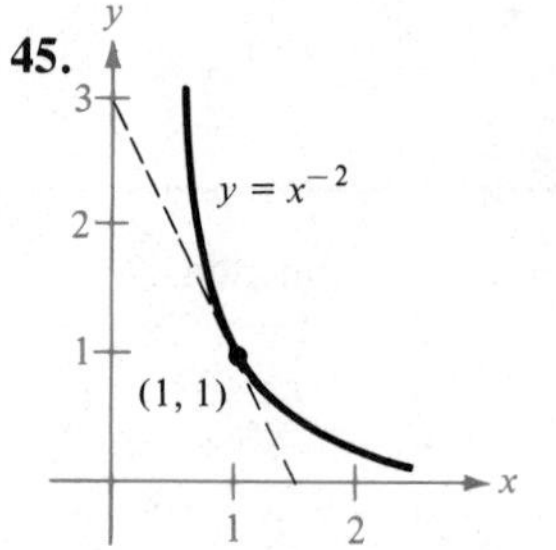

46. Find the equation of the line tangent to $y = x^4 - 3x^2 + 2$ at the point (1, 0).

47. Find the equation of the line tangent to $y = x^3 + x$ at the point $(-1, -2)$.

48. At what points, if any, does $y = x^4 - 3x^2 + 2$ have horizontal tangents?

49. At what points, if any, does $y = x^3 + x$ have horizontal tangents?

50. At what points, if any, does $y = 1/x^2$ have horizontal tangents?

Section 2.3

Rates of Change: Velocity and Marginals

Introductory Example

Inflation and the Consumer Price Index

The "official" annual rate of inflation in the United States is the **average rate of change** in the Consumer Price Index from January 1 of one year to January 1 of the next year. This index is published by the Bureau of Labor Statistics and is a statistical measure of the change in price of a fixed amount of selected goods and services relative to the base year 1967. That is, each number in the Consumer Price Index represents that year's price for goods and services that cost $100.00 in 1967. Table 2.3 lists the Consumer Price Index from 1955 to 1980, and Figure 2.14 graphically depicts the index from 1950 to 1980.

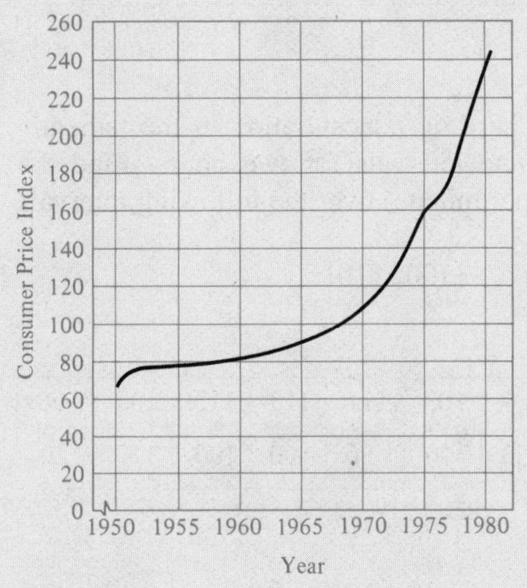

Figure 2.14

Table 2.3

Year	Consumer Price Index	Annual Inflation Rate (%)
1955	80.2	−0.4
1956	81.4	1.5
1957	84.3	3.6
1958	86.6	2.7
1959	87.3	0.8
1960	88.7	1.6
1961	89.6	1.0
1962	90.6	1.1
1963	91.7	1.2
1964	92.9	1.3
1965	94.5	1.7
1966	97.2	2.9
1967	100.0	2.9
1968	104.2	4.2
1969	109.8	5.4
1970	116.3	5.9
1971	121.3	4.3
1972	125.3	3.3
1973	133.1	6.2
1974	147.7	11.0
1975	161.2	9.1
1976	170.5	5.8
1977	181.5	6.5
1978	195.4	7.7
1979	217.4	11.3
1980	242.5	11.5

Section Objectives: ***To distinguish between average and instantaneous rates of change ▪ To use the derivative to calculate instantaneous rates of change ▪ To interpret the economic term "marginal."***

We have already seen how the derivative is used to determine the slope of a curve. We now consider another interpretation—namely, as a way of determining the rate of change in one variable with respect to another. There are numerous applications of the notion of rate of change. A few examples are population growth rates, unemployment rates, production rates, and rate of water flow. Although rates of change often involve change with respect to time, we can actually investigate the rate of change with respect to any related variable.

When determining the rate of change of one variable with respect to another, we must be careful to distinguish between *average* and *instantaneous* rates of change. The distinction between these two rates of change is comparable to the distinction between the slope of the secant line through two points on a curve and the slope of the tangent line at one point on a curve.

Definition of Average Rate of Change

If $y = f(x)$, then the **average rate of change** of y on the interval $[x, x + \Delta x]$ is given by

$$\text{average rate of change} = \frac{\Delta y}{\Delta x}$$

Example 1

Finding the Average Rate of Change over an Interval

A drug is administered to a patient. The drug concentration in the patient's bloodstream is monitored over 10-minute intervals for two hours. Find the average rates of change (in milligrams per minute) over the following intervals for the concentrations in Table 2.4.

(a) [0, 10] (b) [0, 20] (c) [100, 110]

Table 2.4

C (mg)	0	2	17	37	55	73	89	103	111	113	113	103	68
t (min)	0	10	20	30	40	50	60	70	80	90	100	110	120

Solution

(a) For the interval [0, 10] the average rate of change is

$$\frac{\Delta C}{\Delta t} = \frac{2 - 0}{10 - 0} = \frac{2}{10} = 0.2 \text{ mg/min}$$

(b) For the interval [0, 20] the average rate of change is

$$\frac{\Delta C}{\Delta t} = \frac{17 - 0}{20 - 0} = \frac{17}{20} = 0.85 \text{ mg/min}$$

(c) For the interval [100, 110] the average rate of change is

$$\frac{\Delta C}{\Delta t} = \frac{103 - 113}{110 - 100} = \frac{-10}{10} = -1 \text{ mg/min}$$

□

Note in Example 1 that the average rate of change is positive when the concentration increases and negative when the concentration decreases. (See Figure 2.15.)

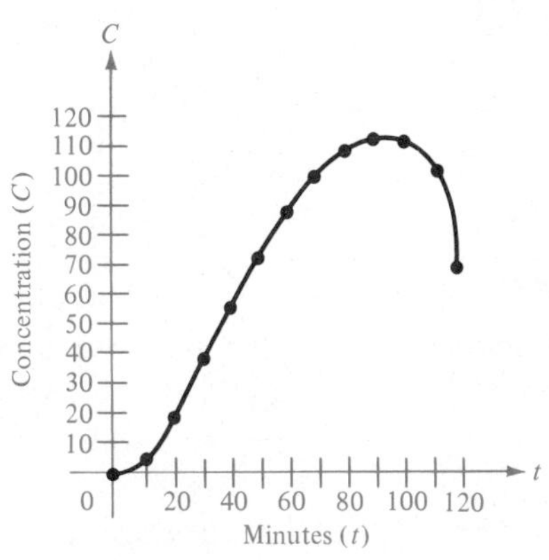

Figure 2.15

Example 2

Finding the Average Rate of Change

If a free-falling object is dropped from a height of 100 feet, its height h at time t is given by

$$h = -16t^2 + 100$$

where h is measured in feet and t is measured in seconds. Find the average rate of change of the height over the following intervals. (See Figure 2.16.)

(a) [1, 2] (b) [1, 1.5] (c) [1, 1.1]

Solution Using the equation $h = -16t^2 + 100$, we determine the heights at $t = 1, 1.1, 1.5$, and 2, as shown in the following table.

h (ft)	84	80.64	64	36
t (sec)	1	1.1	1.5	2

(a) For the interval [1, 2], the object falls from a height of 84 feet to a height of 36 feet. Thus, the average rate of change is

$$\frac{\Delta h}{\Delta t} = \frac{84 - 36}{1 - 2} = \frac{48}{-1} = -48 \text{ ft/sec}$$

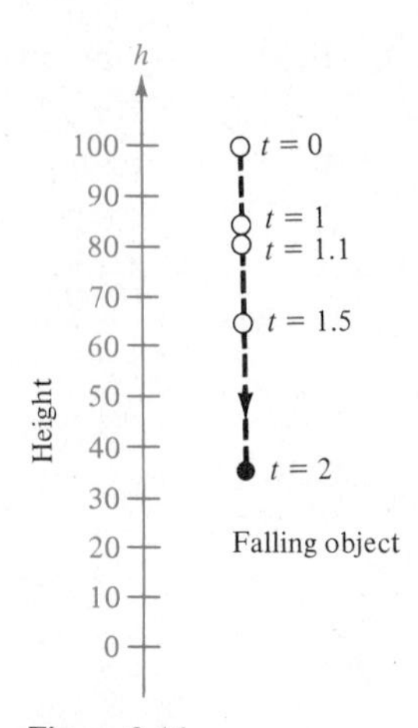

Figure 2.16

(b) For the interval [1, 1.5] the average rate of change is

$$\frac{\Delta h}{\Delta t} = \frac{84 - 64}{1 - 1.5} = \frac{20}{-.5} = -40 \text{ ft/sec}$$

(c) For the interval [1, 1.1] the average rate of change is

$$\frac{\Delta h}{\Delta t} = \frac{84 - 80.64}{1 - 1.1} = \frac{3.36}{-.1} = -33.6 \text{ ft/sec}$$

□

Suppose that in Example 2 we wanted to find the rate of change in h at the instant, $t = 1$ second. We call this the **instantaneous rate of change** of the height when $t = 1$. Just as we approximated the slope of the tangent line by the slope of the secant line, we can approximate the instantaneous rate of change at $t = 1$ by calculating the average rate of change over a small interval $[1, 1 + \Delta t]$, as shown in Table 2.5.

Table 2.5

Δt	1	0.5	0.1	0.01	0.001	0.0001
$\frac{\Delta h}{\Delta t}$	−48	−40	−33.6	−32.16	−32.016	−32.0016

From Table 2.5, it seems reasonable to conclude that the instantaneous rate of change of the height when $t = 1$ is -32 feet per second. We will verify this conclusion after presenting the following definition.

Definition of Instantaneous Rate of Change

The **instantaneous rate of change** of y at x is the limit of the average rate of change on the interval $[x, x + \Delta x]$

$$\lim_{\Delta x \to 0} \frac{\Delta y}{\Delta x} = \lim_{\Delta x \to 0} \frac{f(x + \Delta x) - f(x)}{\Delta x}$$

Remark

The limit in the preceding definition is the same as the limit in the definition of the derivative of f at x. This is an important observation, for we have arrived at a second major interpretation of the derivative—as an *instantaneous rate of change in one variable with respect to another.*

We make the following summary statement concerning the derivative and its interpretations.

Interpretations of the Derivative

Let the function given by $y = f(x)$ be differentiable at x; then its derivative

$$\frac{dy}{dx} = f'(x) = \lim_{\Delta x \to 0} \frac{f(x + \Delta x) - f(x)}{\Delta x}$$

denotes both:

1. the *slope* of the graph of f at x, and
2. the *instantaneous rate of change* in y with respect to x.

Remark

In future work with the derivative, we will use "rate of change" to mean "instantaneous rate of change." Also, it is important to bear in mind that the three quantities—the derivative, the slope of a graph, and the rate of change—*are all equivalent.*

Now, let us return to the problem of finding the rate of change when $t = 1$ of the height of the falling object in Example 2. Considering the derivative as a rate of change, we see that if

$$h = -16t^2 + 100$$

then $$\frac{dh}{dt} = -32t$$

and we conclude that the rate of change in h when $t = 1$ is

$$\frac{dh}{dt} = -32 \text{ ft/sec}$$

We call this particular rate of change the **velocity** of the falling body when $t = 1$. Note that negative velocity indicates that the object's height is decreasing. The use of the derivative to compute velocity is important in a wide variety of problems in physics and engineering.

Another important use of rates of change is in the field of economics. Economists refer to **marginal profit, marginal revenue,** and **marginal cost** as the rates of change of the profit, revenue, and cost with respect to the number of units produced or sold.* An equation that relates these three quantities is

$$P = R - C$$

*A summary of these and other business terms can be found on pages 237 and 238.

where

$$P = \text{total profit}, \qquad R = \text{total revenue} \qquad C = \text{total cost}$$

Differentiating each of these gives the *marginals,* a term used in economics to denote derivatives:

$$\frac{dP}{dx} = \text{marginal profit}$$

$$\frac{dR}{dx} = \text{marginal revenue}$$

$$\frac{dC}{dx} = \text{marginal cost}$$

Example 3

Finding the Marginal Profit

A manufacturer determines that the profit derived from selling x units of a certain item is given by

$$P = 0.0002x^3 + 10x$$

(a) Find the marginal profit for a production level of 50 units.
(b) Compare this to the actual gain in profit obtained by increasing the production level from 50 to 51 units.

Solution

(a) Since

$$P = 0.0002x^3 + 10x$$

the marginal profit is given by

$$\frac{dP}{dx} = 0.0006x^2 + 10$$

When $x = 50$, the marginal profit is

$$\frac{dP}{dx} = (0.0006)(50)^2 + 10 = 1.5 + 10 = \$11.50 \text{ per unit}$$

(b) For $x = 50$, the actual profit is

$$P = (0.0002)(50)^3 + 10(50) = 25 + 500 = \$525.00$$

and for $x = 51$, the actual profit is

$$P = (0.0002)(51)^3 + 10(51) = 26.53 + 510 = \$536.53$$

Thus, the additional profit obtained by increasing the production level from 50 to 51 units is

$$536.53 - 525.00 = \$11.53$$

Note that the actual profit increase of \$11.53 (when x increases from 50 to 51) can be approximated by the marginal profit of \$11.50 (when $x = 50$). (See Figure 2.17.)

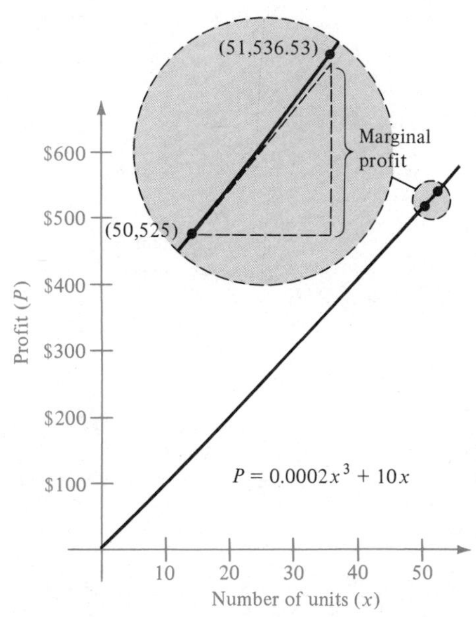

Figure 2.17

Remark

The profit function in Example 3 is quite unusual in that the profit continues to increase as long as the number of units sold increases. In practice, it is more common to encounter situations in which sales can be increased only by lowering the price per item, and such reductions in price will ultimately cause the profit to decline.

We define the number of units, x, consumers are willing to purchase at a given price per unit, p, by the **demand function**

$$p = f(x)$$

The total revenue, R, is then related to the price per unit and the quantity demanded (or sold) by the equation

$$R = xp$$

Example 4

Finding the Marginal Revenue

A fast-food restaurant has determined that the monthly demand for their hamburgers is given by

$$p = \frac{60{,}000 - x}{20{,}000}$$

Table 2.6 shows the demand for hamburgers at various prices. (See Figure 2.18.)

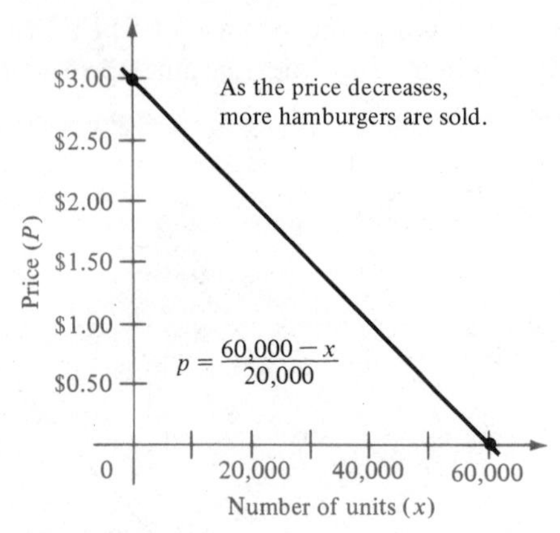

Figure 2.18

Table 2.6

p	0	\$0.50	\$1.00	\$1.50	\$2.00	\$2.50	\$3.00
x	60,000	50,000	40,000	30,000	20,000	10,000	0

Find the increase in revenue per hamburger for monthly sales of 20,000 hamburgers. In other words, find the marginal revenue when $x = 20{,}000$.

Solution Since

$$p = \frac{60{,}000 - x}{20{,}000}$$

we have

$$R = xp = x\left(\frac{60{,}000 - x}{20{,}000}\right) = \frac{1}{20{,}000}(60{,}000x - x^2)$$

and the marginal revenue is

$$\frac{dR}{dx} = \frac{1}{20{,}000}(60{,}000 - 2x)$$

Finally, when $x = 20{,}000$, the marginal revenue is

$$\frac{dR}{dx} = \frac{1}{20{,}000}[60{,}000 - 2(20{,}000)] = \frac{20{,}000}{20{,}000} = \$1/\text{unit}$$

□

Example 5

Finding the Marginal Profit

Suppose that in Example 4 the cost, C, of producing x hamburgers is

$$C = 5{,}000 + 0.56x$$

Find the profit *and* the marginal profit when (a) $x = 20{,}000$, (b) $x = 24{,}400$, and (c) $x = 30{,}000$. (See Figure 2.19.)

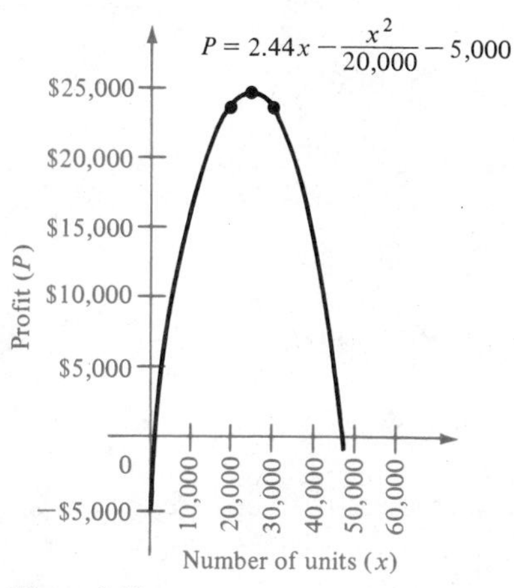

Figure 2.19

Solution Since $P = R - C$ and from Example 4 we know that

$$R = \frac{1}{20{,}000}(60{,}000x - x^2)$$

we have

$$\begin{aligned} P &= \frac{1}{20{,}000}(60{,}000x - x^2) - 5{,}000 - 0.56x \\ &= 3x - \frac{x^2}{20{,}000} - 5{,}000 - 0.56x \\ &= 2.44x - \frac{x^2}{20{,}000} - 5{,}000 \end{aligned}$$

Thus, the marginal profit is

$$\frac{dP}{dx} = 2.44 - \frac{x}{10{,}000}$$

(a) When $x = 20{,}000$ the profit is

$$P = (2.44)(20{,}000) - \frac{(20{,}000)^2}{20{,}000} - 5{,}000 = \$23{,}800.00$$

and the marginal profit is

$$\frac{dP}{dx} = 2.44 - \frac{20{,}000}{10{,}000} = \$0.44 \text{ per unit}$$

(b) When $x = 24{,}400$ the profit is

$$P = (2.44)(24{,}400) - \frac{(24{,}400)^2}{20{,}000} - 5{,}000 = \$24{,}768.00$$

and the marginal profit is

$$\frac{dP}{dx} = 2.44 - \frac{24{,}400}{10{,}000} = 0$$

(c) When $x = 30{,}000$ the profit is

$$P = (2.44)(30{,}000) - \frac{(30{,}000)^2}{20{,}000} - 5{,}000 = \$23{,}200.00$$

and the marginal profit is

$$\frac{dP}{dx} = 2.44 - \frac{30{,}000}{10{,}000} = -\$0.56 \text{ per unit}$$

□

Remark From Example 5, we can see that when more than 24,400 hamburgers are sold the marginal profit is negative. This means that increasing production beyond this point will *reduce* rather than increase profit.

Section Exercises (2.3)

Ⓒ **1.** The average hourly wage paid to U.S. workers from 1970 to 1977 is shown in the following table.

Year	1970	1971	1972	1973
Wage	\$3.22	\$3.44	\$3.67	\$3.92
Year	1974	1975	1976	1977
Wage	\$4.22	\$4.54	\$4.87	\$5.25

Find the average rate of change in hourly wages for the following periods of time.
(a) 1970 to 1971 (b) 1970 to 1975
(c) 1972 to 1977 (d) 1974 to 1976

Ⓒ **2.** Since 1790 the center of population of the United States has been gradually moving westward. Use Figure 2.20 to estimate the rate (in miles per year) at which the center of population was moving *westward* during the period:
(a) from 1790 to 1900
(b) from 1900 to 1970

3. For the function $y = 2x^3 + 3x - 4$ find:
(a) the average rate of change of y with respect to x over the interval $[1, 2]$
(b) the instantaneous rate of change of y with respect to x when $x = 1$

Ⓒ **4.** For the function

$$y = \frac{2}{x}$$

find:
(a) the average rate of change of y with respect to x over the interval $[2, 2.5]$
(b) the instantaneous rate of change of y with respect to x when $x = 2$

5. Suppose a guitar string is plucked and vibrates with a frequency of

$$F = 200\sqrt{T}$$

Ⓒ Calculator may be helpful.

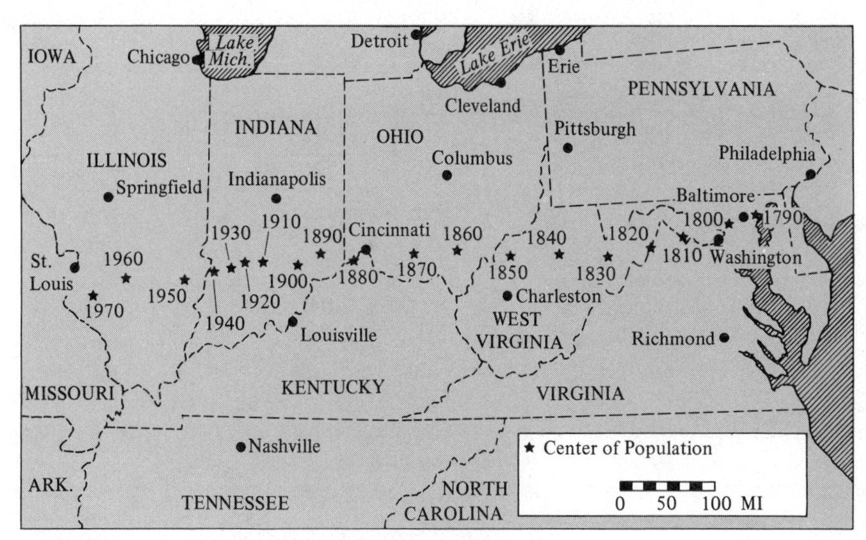

Figure 2.20

where F is measured in vibrations per second and the tension T is measured in pounds. As the string is tightened, find the rate at which the frequency is increasing when the tension is
(a) 4 lb (b) 9 lb

Ⓒ **6.** At 0° Celsius the wind-chill corrected temperature is given by

$$T = 33 - (1.43)(10\sqrt{w} + 10.45 - w)$$

where T is measured in degrees Celsius and the wind speed w is measured in meters per second. As the wind speed increases, find the rate at which T is decreasing when the wind speed is
(a) 4 m/sec (b) 9 m/sec

7. Suppose that the effectiveness E of a pain-killing drug t hours after entering the bloodstream is given by

$$E = \frac{1}{27}(9t + 3t^2 - t^3), \qquad 0 \le t \le 4.5$$

Find the rate of change of E with respect to t when
(a) $t = 1$ (b) $t = 2$
(c) $t = 3$ (d) $t = 4$

8. Suppose that a certain company finds that by charging p dollars per unit its monthly revenue R will be

$$R = 12{,}000p - 1{,}000p^2, \qquad 0 \le p \le 12$$

(Note that the revenue is zero when $p = 12$ since no one is willing to pay that much.) Find the rate of change of R with respect to p when
(a) $p = 1$ (b) $p = 4$
(c) $p = 6$ (d) $p = 10$

9. Suppose that the position of an accelerating car is given by

$$s(t) = 10t^{3/2}, \qquad 0 \le t \le 10$$

where s is measured in feet and t is measured in seconds. Find the velocity of the car when
(a) $t = 0$ (b) $t = 1$
(c) $t = 4$ (d) $t = 9$

Ⓒ **10.** The height of a silver dollar dropped from the top of the World Trade Center is given by

$$s(t) = -16t^2 + 1350$$

where s is measured in feet and t is measured in seconds.
(a) How long will it take for the silver dollar to hit the ground?
(b) How fast will the silver dollar be traveling when it hits the ground?

Ⓒ Calculator may be helpful.

11. Suppose that the height of an object fired straight up from ground level is given by

$$h(t) = 200t - 16t^2$$

where h is measured in feet and t is measured in seconds.
(a) How fast is the object moving as it leaves the ground? (when $t = 0$)
(b) How fast is the object moving after 3 seconds?

12. Derive the equation for the velocity of an object that moves according to the law

$$s(t) = t^2 + 2t$$

where s is measured in meters and t is measured in seconds.

Ⓒ**13.** The revenue from producing x units of a product is

$$R = 12x - 0.001x^2$$

(a) Find the additional revenue if production is increased from 5000 to 5001 units.
(b) Find the marginal revenue when 5000 units are produced.

Ⓒ**14.** The revenue from renting x apartments is

$$R = 25(900 + 32x - x^2)$$

(a) Find the additional revenue if the number of rentals is increased from 14 to 15.
(b) Find the marginal revenue when 14 apartments are rented.

Ⓒ**15.** The profit for producing x units of a product is given by

$$P = 2400 - 403.4x + 32x^2 - 0.664x^3$$

where $10 \le x \le 25$.
(a) Complete the following table and use it to sketch the graph of the profit function on the given interval.

x	10	15	20	23	25
P					

(b) Complete the following table for marginal profit.

x	10	15	20	23	25
$\frac{dP}{dx}$					

Ⓒ**16.** The profit for producing x units of a product is given by

$$P = 36{,}000 + 2{,}048\sqrt{x} - \frac{1}{8x^2}$$

where $150 \le x \le 275$.
(a) Complete the following table and use it to sketch the graph of the profit function on the given interval.

x	150	175	200	225	250	275
P						

(b) Complete the following table for marginal profit.

x	150	175	200	225	250	275
$\frac{dP}{dx}$						

Ⓒ**17.** The Smiths drive an average of 15,000 miles per year in a car that gets x miles per gallon. Assume that the average fuel cost is \$1.75 per gallon.
(a) Show that the Smiths' annual fuel cost is approximately

$$C = \frac{26{,}250}{x}$$

(b) Complete the following table and use it to sketch the graph of the cost function.

x	10	15	20	25	30	35	40
C							

Ⓒ Calculator may be helpful.

(c) Complete the following table for the marginal cost.

x	10	15	20	25	30	35	40
$\frac{dC}{dx}$							

(d) Find the decrease in annual fuel cost when x increases from 10 to 11 and compare this with the marginal cost when $x = 10$.

(e) Find the decrease in annual fuel cost when x increases from 30 to 31 and compare this with the marginal cost when $x = 30$.

Ⓒ**18.** If Q is the order size when the inventory is replenished, then the annual inventory cost for a certain manufacturer is given by

$$C = \frac{1{,}008{,}000}{Q} + 6.3Q$$

(a) Complete the following table for the marginal cost.

Q	300	350	400	450	500
$\frac{dC}{dQ}$					

(b) Find the change in annual cost when Q is increased from 350 to 351 and compare this with the marginal cost when $Q = 350$.

Ⓒ**19.** The monthly demand (in quarts) for motor oil at a local service station is given by

$$p = \frac{1100 - x}{400}$$

(a) Find the monthly revenue as a function of x.

(b) Complete the following table for marginal revenue and price.

x	300	400	500	550	600	700
$\frac{dR}{dx}$						
p						

Ⓒ**20.** The demand function for a store's product is

$$p = 25(20 - \sqrt{x})$$

(a) Find the monthly revenue as a function of x.

(b) Complete the following table for marginal revenue and price.

x	100	150	200	250
$\frac{dR}{dx}$				
p				

Ⓒ**21.** Suppose that the cost (in Exercise 19) to the service station owner for x quarts is

$$C = 65 + 1.25x$$

(a) Find the monthly profit as a function of x.

(b) Find the profit and marginal profit when $x = 250$ and when $x = 300$.

Ⓒ**22.** Suppose that the cost to the store (in Exercise 20) is

$$C = 1.25x + 42.50$$

(a) Find the profit as a function of x.

(b) Find the marginal profit when $x = 100$ and when $x = 125$.

Ⓒ**23.** Repeat Exercise 21 if the cost of a quart of oil is increased to $1.50 so that

$$C = 65 + 1.50x$$

Ⓒ Calculator may be helpful.

Section 2.4

The Product and Quotient Rules

Introductory Example

The Solubility of Oxygen in Water

When air is in contact with undersaturated water, some of the oxygen in the air is absorbed in the water. The amount of oxygen absorbed depends on temperature, salinity, and pressure. Cold water absorbs more oxygen than warm water; fresh water absorbs more oxygen than salt water; and water at low altitudes absorbs more oxygen than water at high altitudes. In 1964, Montgomery, Thom, and Cockburn* published the results of their experiments measuring the solubility of oxygen in pure water at sea level. From their measurements they constructed a mathematical model (see Figure 2.21) for solubility

$$S = \frac{468}{31.6 + t}$$

where S is measured in milligrams of oxygen per liter of water and t is the temperature in degrees Celsius. In this section, we will learn how to differentiate such quotients and then see that the rate of change of the solubility of oxygen in water is given by

$$\frac{dS}{dt} = \frac{-468}{(31.6 + t)^2}$$

Note that the fact that this rate of change is always negative $(0 \leq t)$ coincides with our assertion that cold water absorbs more oxygen than warm water.

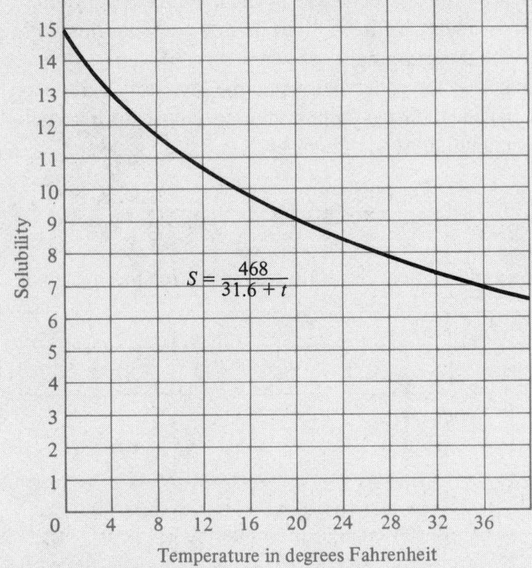

Solubility of Oxygen in Pure Water at Sea Level

Figure 2.21

* Data from H. A. C. Montgomery, N. S. Thom, and A. Cockburn, *Determination of dissolved oxygen by the Winkler method and the solubility of oxygen in pure water and sea water,* J. Appl. Chem. **14** (1964) 180–196.

Section Objectives: ***To derive and use rules for the differentiation of products and quotients ▪ To point out some pitfalls in the use of these differentiation rules.***

In Section 2.2, we noted that the derivative of a sum or difference of two functions is simply the sum or difference of their derivatives. The rules for the derivative of a product or quotient of two functions are not so simple, and you may find the results surprising. As we derive each of the rules, we will assume that the derivatives of the given functions exist. Furthermore, we strongly recommend that you memorize each rule, especially the *verbal* statements of the product and quotient rules.

Product Rule

The derivative of the product of two functions is equal to the first function times the derivative of the second, plus the second times the derivative of the first.

$$\frac{d}{dx}[f(x)g(x)] = f(x)g'(x) + g(x)f'(x)$$

Proof Let $F(x) = f(x)g(x)$; then

$$F'(x) = \lim_{\Delta x \to 0} \frac{F(x + \Delta x) - F(x)}{\Delta x}$$

$$= \lim_{\Delta x \to 0} \frac{f(x + \Delta x)g(x + \Delta x) - f(x)g(x)}{\Delta x}$$

Now because it is both legal and useful, let us subtract and add $f(x + \Delta x)\,g(x)$ in the numerator, giving us

$$F'(x)$$

$$= \lim_{\Delta x \to 0} \frac{f(x + \Delta x)g(x + \Delta x) - f(x + \Delta x)g(x) + f(x + \Delta x)g(x) - f(x)g(x)}{\Delta x}$$

$$= \lim_{\Delta x \to 0} \left[f(x + \Delta x)\frac{g(x + \Delta x) - g(x)}{\Delta x} + g(x)\frac{f(x + \Delta x) - f(x)}{\Delta x}\right]$$

$$= \lim_{\Delta x \to 0} f(x + \Delta x)\frac{g(x + \Delta x) - g(x)}{\Delta x} + \lim_{\Delta x \to 0} g(x)\frac{f(x + \Delta x) - f(x)}{\Delta x}$$

$$= \left[\lim_{\Delta x \to 0} f(x + \Delta x)\right]\left[\lim_{\Delta x \to 0} \frac{g(x + \Delta x) - g(x)}{\Delta x}\right]$$

$$+ \left[\lim_{\Delta x \to 0} g(x)\right]\left[\lim_{\Delta x \to 0} \frac{f(x + \Delta x) - f(x)}{\Delta x}\right]$$

From the differentiability of f and g we obtain the following limits:

$$\lim_{\Delta x \to 0} f(x + \Delta x) = f(x) \qquad \lim_{\Delta x \to 0} \frac{g(x + \Delta x) - g(x)}{\Delta x} = g'(x)$$

$$\lim_{\Delta x \to 0} g(x) = g(x) \qquad \lim_{\Delta x \to 0} \frac{f(x + \Delta x) - f(x)}{\Delta x} = f'(x)$$

Therefore, we have

$$F'(x) = f(x)g'(x) + g(x)f'(x)$$

and

$$\frac{d}{dx}[f(x)g(x)] = f(x)g'(x) + g(x)f'(x)$$

□

Example 1

Finding the Derivative of a Product

Find the derivative of $F(x) = (3x - 2x^2)(5 + 4x)$.

Solution By the Product Rule,

$$\begin{aligned} F'(x) &= (3x - 2x^2)\frac{d}{dx}[5 + 4x] + (5 + 4x)\frac{d}{dx}[3x - 2x^2] \\ &= (3x - 2x^2)(4) + (5 + 4x)(3 - 4x) \\ &= (12x - 8x^2) + (15 - 8x - 16x^2) = 15 + 4x - 24x^2 \end{aligned}$$

Of course, we could have bypassed the Product Rule by first multiplying $(3x - 2x^2)$ by $(5 + 4x)$ to get $15x + 2x^2 - 8x^3$ and then differentiating this result. Although in this case this alternative procedure would have been just as simple, we cannot always avoid the use of the Product Rule by first multiplying the given factors. □

Remark

It is helpful to realize that, in general,

$$\begin{pmatrix}\text{the derivative of the} \\ \text{product of two functions}\end{pmatrix} \neq \begin{pmatrix}\text{the product of the derivatives} \\ \text{of the two functions}\end{pmatrix}$$

This is evident in Example 1, since

$$\frac{d}{dx}[(3x - 2x^2)(5 + 4x)] = 15 + 4x - 24x^2$$

whereas

$$\left(\frac{d}{dx}[3x - 2x^2]\right)\left(\frac{d}{dx}[5 + 4x]\right) = (3 - 4x)(4) = 12 - 16x$$

Example 2

Finding the Derivative of a Product

Differentiate $f(x) = (1 + x^{-1})(x - 1)$.

Solution By the Product Rule,

$$f'(x) = (1 + x^{-1})(1) + (x - 1)(-x^{-2})$$

$$= 1 + \frac{1}{x} - \frac{x - 1}{x^2} = \frac{x^2 + x - x + 1}{x^2} = \frac{x^2 + 1}{x^2}$$ □

Quotient Rule

The derivative of the quotient of two functions is equal to the denominator times the derivative of the numerator minus the numerator times the derivative of the denominator, all divided by the square of the denominator.

$$\frac{d}{dx}\left[\frac{f(x)}{g(x)}\right] = \frac{g(x)f'(x) - f(x)g'(x)}{[g(x)]^2}, \quad \text{where} \quad g(x) \neq 0$$

Proof Let

$$F(x) = \frac{f(x)}{g(x)}$$

Then

$$F'(x) = \lim_{\Delta x \to 0} \frac{F(x + \Delta x) - F(x)}{\Delta x}$$

$$= \lim_{\Delta x \to 0} \frac{\dfrac{f(x + \Delta x)}{g(x + \Delta x)} - \dfrac{f(x)}{g(x)}}{\Delta x}$$

$$= \lim_{\Delta x \to 0} \frac{g(x)f(x + \Delta x) - f(x)g(x + \Delta x)}{\Delta x\, g(x)g(x + \Delta x)}$$

Now by adding and subtracting $f(x)g(x)$ in the numerator, we have

$$F'(x) = \lim_{\Delta x \to 0} \frac{g(x)f(x + \Delta x) - f(x)g(x) + f(x)g(x) - f(x)g(x + \Delta x)}{\Delta x\, g(x)g(x + \Delta x)}$$

$$= \frac{\displaystyle\lim_{\Delta x \to 0} \frac{g(x)[f(x + \Delta x) - f(x)]}{\Delta x} - \lim_{\Delta x \to 0} \frac{f(x)[g(x + \Delta x) - g(x)]}{\Delta x}}{\displaystyle\lim_{\Delta x \to 0} [g(x)g(x + \Delta x)]}$$

$$= \frac{g(x)\left[\displaystyle\lim_{\Delta x \to 0} \frac{f(x + \Delta x) - f(x)}{\Delta x}\right] - f(x)\left[\displaystyle\lim_{\Delta x \to 0} \frac{g(x + \Delta x) - g(x)}{\Delta x}\right]}{\displaystyle\lim_{\Delta x \to 0} [g(x)g(x + \Delta x)]}$$

$$= \frac{g(x)f'(x) - f(x)g'(x)}{[g(x)]^2}$$ □

As suggested previously, you should memorize the verbal statement of the Quotient Rule. The following form may assist you in memorizing the rule:

$$\frac{d}{dx}[\text{quotient}] = \frac{(\text{denominator})\frac{d}{dx}[\text{numerator}] - (\text{numerator})\frac{d}{dx}[\text{denominator}]}{(\text{denominator})^2}$$

This form certainly points out that

$$\begin{pmatrix}\text{the derivative of the} \\ \text{quotient of two functions}\end{pmatrix} \neq \begin{pmatrix}\text{the quotient of the derivatives} \\ \text{of the two functions}\end{pmatrix}$$

Example 3

Finding the Derivative of a Quotient

Differentiate

$$y = \frac{2x^2 - 4x + 3}{2 - 3x}$$

Solution By the Quotient Rule,

$$\begin{aligned} y' &= \frac{(2 - 3x)\frac{d}{dx}[2x^2 - 4x + 3] - (2x^2 - 4x + 3)\frac{d}{dx}[2 - 3x]}{(2 - 3x)^2} \\ &= \frac{(2 - 3x)(4x - 4) - (2x^2 - 4x + 3)(-3)}{(2 - 3x)^2} \\ &= \frac{-12x^2 + 20x - 8 + (6x^2 - 12x + 9)}{(2 - 3x)^2} \\ &= \frac{-6x^2 + 8x + 1}{(2 - 3x)^2} \end{aligned}$$

□

Example 4

Finding the Derivative of a Quotient

In the introductory example to this section, we saw that the solubility of oxygen in pure water at sea level is given by

$$S = \frac{468}{31.6 + t}$$

where S is measured in milligrams of oxygen per liter of water and t is the temperature in degrees Celsius. Use the Quotient Rule to find the rate of change of S with respect to t.

Solution By the Quotient Rule,

$$\frac{dS}{dt} = \frac{(31.6 + t)(0) - (468)(1)}{(31.6 + t)^2} = \frac{-468}{(31.6 + t)^2}$$

□

Remark

You can avoid many algebraic errors when differentiating quotients if you use parentheses liberally. It is a good idea to enclose all factors and derivatives in parentheses and to pay special attention to the subtraction required in the numerator of the Quotient Rule. In fact, liberal use of parentheses is a sound guideline in *all* types of differentiation problems.

Remark

Another sound guideline for differentiation is to consider *rewriting before differentiating*. Study the next two examples carefully to see how rewriting a function can simplify differentiation.

Example 5

Rewriting before Differentiating

Find the derivative of

$$y = \frac{3 - (1/x)}{x + 5}$$

Solution First, let us rewrite y as

$$y = \frac{(3x - 1)/x}{x + 5} = \frac{3x - 1}{x(x + 5)} = \frac{3x - 1}{x^2 + 5x}$$

Now by the Quotient Rule we have

$$\frac{dy}{dx} = \frac{(x^2 + 5x)(3) - (3x - 1)(2x + 5)}{(x^2 + 5x)^2}$$

$$= \frac{(3x^2 + 15x) - (6x^2 + 13x - 5)}{(x^2 + 5x)^2} = \frac{-3x^2 + 2x + 5}{(x^2 + 5x)^2}$$

(Notice the use of parentheses.) □

Example 6

Rewriting before Differentiating

Not every quotient needs to be differentiated by the Quotient Rule. The quotients in the table on page 158 can be considered as products of a constant times a function of x. In such cases the Constant Multiple Rule (Section 2.2) is more convenient and is used in place of the Quotient Rule.

When we introduced the Power Rule,

$$\frac{d}{dx}[x^n] = nx^{n-1}$$

in Section 2.2, we stated that the rule is valid for any real number n. However, we proved only the case where n is a positive integer. We now prove this rule for the case when n is a negative integer.

Given	Write	Then
$y = \dfrac{x^2 + 3x}{-6}$	$y = \dfrac{-1}{6}(x^2 + 3x)$	$y' = \dfrac{-1}{6}(2x + 3)$
$y = \dfrac{5x^4}{8}$	$y = \dfrac{5}{8}x^4$	$y' = \dfrac{5}{8}(4x^3) = \dfrac{5}{2}x^3$
$y = \dfrac{-3(3x - 2x^2)}{7x}$	$y = \dfrac{-3}{7}(3 - 2x)$	$y' = \dfrac{-3}{7}(-2) = \dfrac{6}{7}$
$y = \dfrac{9}{-5x^2}$	$y = \dfrac{9}{-5}(x^{-2})$	$y' = \dfrac{9}{-5}(-2x^{-3}) = \dfrac{18}{5x^3}$

□

Example 7

Using the Quotient Rule to Prove a Special Case of the Power Rule

Use the Quotient Rule to prove the Power Rule for the case when n is a negative integer.

Solution Suppose n is a negative integer; then there exists a positive integer k such that $n = -k$. Let

$$y = x^n = x^{-k} = \frac{1}{x^k}$$

By the Quotient Rule,

$$\frac{dy}{dx} = \frac{x^k(0) - (1)(kx^{k-1})}{(x^k)^2}$$

$$= \frac{0 - kx^{k-1}}{x^{2k}} = -kx^{-k-1} = nx^{n-1}$$

Therefore, if n is a negative integer, we also have

$$\frac{d}{dx}[x^n] = nx^{n-1}$$

□

Example 8

Combining the Product and Quotient Rules

Differentiate

$$y = \frac{(1 - 2x)(3x + 2)}{5x - 4}$$

Solution Here we have a product within a quotient, and we could first multiply the factors in the numerator, then apply the Quotient Rule. However, to show how the Product Rule can be used within the Quotient Rule, we proceed with the equation as written.

$$y' = \frac{(5x-4)\frac{d}{dx}[(1-2x)(3x+2)] - (1-2x)(3x+2)\frac{d}{dx}[5x-4]}{(5x-4)^2}$$

$$= \frac{(5x-4)[(1-2x)(3) + (3x+2)(-2)] - (1-2x)(3x+2)(5)}{(5x-4)^2}$$

$$= \frac{(5x-4)(-12x-1) - (1-2x)(15x+10)}{(5x-4)^2}$$

$$= \frac{(-60x^2+43x+4) - (-30x^2-5x+10)}{(5x-4)^2}$$

$$= \frac{-30x^2+48x-6}{(5x-4)^2}$$

□

Example 9

Rate of Change of Blood Pressure

As blood moves from the heart through the major arteries out to the capillaries and back through the veins, the systolic pressure continuously drops. Suppose that this pressure is given by

$$P = \frac{25t^2+125}{t^2+1}, \qquad 0 \le t \le 10$$

where P is measured in millimeters of mercury and t is measured in seconds. At what rate is the pressure dropping 5 seconds after leaving the heart?

Solution By the Quotient Rule, we have

$$\frac{dP}{dt} = \frac{(t^2+1)(50t) - (25t^2+125)(2t)}{(t^2+1)^2}$$

$$= \frac{50t^3+50t-50t^3-250t}{(t^2+1)^2}$$

$$= -\frac{200t}{(t^2+1)^2}$$

When $t = 5$,

$$\frac{dP}{dt} = -\frac{200(5)}{26^2} = -1.48 \text{ mm/s}$$

Therefore, the pressure is *dropping* at the rate of 1.48 millimeters per second when $t = 5$ seconds.

□

Remark

In the examples in this section much of the work in obtaining the final form of the answer occurs *after* the differentiation is complete. This is often the case; direct application of differentiation rules yields answers that are not in simplest form, as reviewed in Table 2.7. As seen in this table, two characteristics of the simplest form of an algebraic expression are (1) the absence of negative exponents and (2) the combining of like terms.

Table 2.7

	$f'(x)$ after differentiating	$f'(x)$ after simplifying
Example 1	$(3x - 2x^2)(4) + (5 + 4x)(3 - 4x)$	$15 + 4x - 24x^2$
Example 2	$(1 + x^{-1})(1) + (x - 1)(-x^{-2})$	$\dfrac{x^2 + 1}{x^2}$
Example 3	$\dfrac{(2 - 3x)(4x - 4) - (2x^2 - 4x + 3)(-3)}{(2 - 3x)^2}$	$\dfrac{-6x^2 + 8x + 1}{(2 - 3x)^2}$
Example 7	$\dfrac{(5x - 4)[(1 - 2x)(3) + (3x + 2)(-2)] - (1 - 2x)(3x + 2)(5)}{(5x - 4)^2}$	$\dfrac{-30x^2 + 48x - 6}{(5x - 4)^2}$

Section Exercises (2.4)

In Exercises 1–10, differentiate and find $f'(x)$ at the given value of x.

1. $f(x) = \frac{1}{3}(2x^3 - 4); x = 0$

2. $f(x) = \dfrac{5 - 6x^2}{7}; x = 1$

3. $f(x) = \dfrac{7}{3x^3}; x = 1$

4. $f(x) = 5x^{-2}(x + 3); x = 0$

5. $f(x) = (x^2 - 2x + 1)(x^3 - 1); x = 1$

6. $f(x) = (x^3 - 3x)(2x^2 + 3x + 5); x = 0$

7. $f(x) = (x - 1)(x^2 - 3x + 2); x = 0$

8. $f(x) = \left(x^2 - \dfrac{1}{x}\right)(x^2 + 1); x = 1$

9. $f(x) = (x^5 - 3x)\left(\dfrac{1}{x^2} + x\right); x = -1$

10. $f(x) = \dfrac{x + 1}{x - 1}; x = 2$

In Exercises 11–28, differentiate each function

11. $f(x) = \dfrac{3x - 2}{2x - 3}$

12. $f(x) = \dfrac{x^3 + 3x + 2}{x^2 - 1}$

13. $f(x) = \dfrac{3 - 2x - x^2}{x^2 - 1}$

14. $f(x) = \dfrac{9}{3x^2 - 2x}$

15. $f(x) = \dfrac{1}{4 - 3x^2}$

16. $f(x) = x^4\left(1 - \dfrac{2}{x + 1}\right)$

17. $f(x) = \sqrt[3]{x}(\sqrt{x} + 3)$

18. $f(x) = \dfrac{(x + 1)}{\sqrt{x}}$

19. $h(t) = \dfrac{t + 1}{t^2 + 2t + 2}$

20. $h(x) = (x^2 - 1)^2$

21. $h(s) = (s^3 - 2)^2$

22. $f(x) = \left(\dfrac{x^2 - x - 3}{x^2 + 1}\right)(x^2 + x + 1)$

23. $g(x) = \left(\dfrac{x + 1}{x + 2}\right)(2x - 5)$

24. $f(x) = (x^2 - x)(x^2 + 1)(x^2 + x + 1)$

25. $f(x) = (3x^3 + 4x)(x - 5)(x + 1)$

26. $f(x) = \dfrac{x^2 + c^2}{x^2 - c^2}$ (c is a constant)

27. $f(x) = \dfrac{c^2 - x^2}{c^2 + x^2}$

28. $f(x) = \dfrac{x(x^2 - 1)}{x + 3}$

29. Find an equation of the line tangent to the graph of $f(x) = x/(x - 1)$ at the point $(2, 2)$.

30. Find an equation of the line tangent to the graph of $f(x) = (x - 1)(x^2 - 2)$ at the point $(0, 2)$.

31. Find an equation of the tangent line to $f(x) = (x^3 - 3x + 1)(x + 2)$ at $(1, -3)$.

32. Find an equation of the tangent line to $f(x) = (x - 1)/(x + 1)$ at the point $(2, \frac{1}{3})$.

33. At what point(s) does $f(x) = x^2/(x - 1)$ have a horizontal tangent?

ⓒ **34.** A certain automobile depreciates according to the formula

$$V = \frac{7500}{1 + 0.4t + 0.1t^2}$$

where $t = 0$ represents the time of purchase (in years). Find the rate at which the car is depreciating

(a) 1 year after purchase

(b) 2 years after purchase

ⓒ **35.** A population of 500 bacteria is introduced into a culture and grows in number according to the formula

$$P(t) = 500\left(1 + \frac{4t}{50 + t^2}\right)$$

where t is measured in hours. Find the rate at which the population is growing when $t = 2$.

ⓒ **36.** In Section 1.7, Example 7, the function

$$f(t) = \frac{t^2 - t + 1}{t^2 + 1}$$

measured the percentage of the normal level of oxygen in a pond where t is the time in weeks after organic waste is dumped into the pond. Find the rate of change of f with respect to t when

(a) $t = 0.5$ (b) $t = 2$ (c) $t = 8$

ⓒ **37.** Suppose that the temperature T of food placed in a freezer drops according to the equation

$$T = \frac{700}{t^2 + 4t + 10}$$

where t is the time in hours. Find the rate of change of T with respect to t when

(a) $t = 1$ (b) $t = 3$

(c) $t = 5$ (d) $t = 10$

ⓒ **38.** Suppose that the temperature T of food placed in a refrigerator drops according to the equation

$$T = 10\left(\frac{4t^2 + 16t + 75}{t^2 + 4t + 10}\right)$$

where t is the time in hours. What is the initial temperature of the food? What is the limit of T as t approaches infinity. Find the rate of change of T with respect to t when

(a) $t = 1$ (b) $t = 3$

(c) $t = 5$ (d) $t = 10$

ⓒ Calculator may be helpful.

Section 2.5

The Chain Rule and the General Power Rule

Introductory Example

The Flow of Blood Through a Blood Vessel

When fluid flows through a pipe, the walls of the pipe cause a drag that slows the fluid near the walls. This phenomenon is described by Poiseuille's Law, which states that the velocity v of the fluid at a distance of r units from the center of a circular pipe of radius R is proportional to $R^2 - r^2$. That is,

$$v = k(R^2 - r^2)$$

This explains why the flow of blood near the wall of a blood vessel is slower than the flow at the center of the vessel. (See Figure 2.22). Since blood moves more slowly near the wall of a blood vessel, we can see why cholesterol has a tendency to build up on the vessel walls. Such deposits are dangerous because they gradually reduce the radius of the blood vessel, which in turn reduces the flow of blood to the part of the body being fed by the vessel.

To increase the flow of blood through a blood vessel, it is possible to prescribe drugs that dilate the blood vessel (R increases). Suppose that a patient took a dilating drug and that the radius of a certain blood vessel began increasing at the rate of

$$\frac{dR}{dt} = 10^{-4} \text{ cm/min}$$

At what rate would the velocity of the blood flow be changing? In this section, we will see that this can be answered by using the **Chain Rule,** which states that

$$\frac{dv}{dt} = \frac{dv}{dR}\frac{dR}{dt}$$

Since for a fixed r and k we have

$$\frac{dv}{dR} = 2kR$$

we conclude that the velocity is increasing at the rate of

$$\frac{dv}{dt} = 2kR(10^{-4}) \text{ cm/min}^2$$

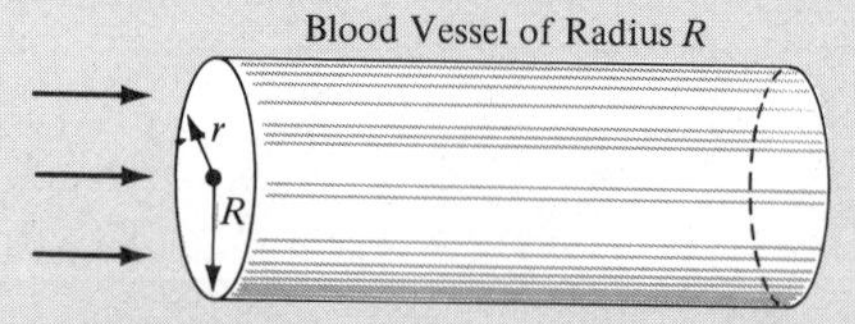

Figure 2.22

Section Objectives: ***To derive and use the Chain and Power Rules for differentiating ▪ To summarize the rules for differentiation and illustrate their use in combination with one another.***

In our development of differentiation rules, we have yet to discuss one of the most important rules in differential calculus, the **Chain Rule.** The term "chain" refers to functions that are composed as a chain, or composite, of other functions. For example, if

$$y = f(u) = 3u^{15}$$

where

$$u = g(x) = 2x - 1$$

then we can express y as a composite function of x as follows:

$$y = f(u) = f(g(x)) = 3(2x - 1)^{15}$$

Now, suppose you were asked to find the derivative of *y with respect to x*. One way to do this would be to expand y as a 15th-degree polynomial and then differentiate as usual. Of course, such an expansion would be very cumbersome, and you should be happy to hear that the Chain Rule provides us with a simple way to find this derivative.

Development of the Chain Rule

Consider two differentiable functions

$$y = f(u) \qquad \text{and} \qquad u = g(x)$$

with the composite function $y = f(g(x))$. Since a derivative indicates a "rate of change," we can say that

$$y \text{ changes } \frac{dy}{du} \text{ times as fast as } u$$

$$u \text{ changes } \frac{du}{dx} \text{ times as fast as } x$$

From this it seems reasonable to conclude that

$$y \text{ changes } \frac{dy}{du}\frac{du}{dx} \text{ times as fast as } x$$

which is equivalent to saying

$$\frac{dy}{dx} = \frac{dy}{du}\frac{du}{dx}$$

In functional notation this can be written as

$$\frac{d}{dx}[f(g(x))] = f'(u)g'(x) = f'(g(x))g'(x)$$

Chain Rule

If $y = f(u)$ is a differentiable function of u, and $u = g(x)$ is a differentiable function of x, then $y = f(g(x))$ is a differentiable function of x and

$$\frac{dy}{dx} = \frac{dy}{du}\frac{du}{dx}$$

or, equivalently,

$$\frac{d}{dx}[f(g(x))] = f'(u)g'(x) = f'(g(x))g'(x)$$

Remark

As an aid to memorizing the Chain Rule, it is helpful to think of dy/du and du/dx as two "fractions." When the two fractions are multiplied together, we can imagine that the two du's cancel each other to produce

$$\frac{dy}{\cancel{du}}\frac{\cancel{du}}{dx} = \frac{dy}{dx}$$

Example 1

Applying the Chain Rule

If $y = 3u^{15}$ and $u = 2x - 1$, find dy/dx.

Solution By the Chain Rule, we have

$$\frac{dy}{dx} = \frac{dy}{du}\frac{du}{dx} = 45(u)^{14}(2) = 90(u)^{14}$$

Substituting for u, we get

$$\frac{d}{dx}[3(2x - 1)^{15}] = 90(2x - 1)^{14}$$

□

The function in Example 1 is an instance of perhaps the most common type of composite function, that is, functions of the form

$$y = [u(x)]^n$$

The rule for differentiating such "power functions" is called the **General Power Rule,** and it is a special case of the Chain Rule.

General Power Rule

If $y = [u(x)]^n$, where u is a differentiable function of x and n is a real number, then

$$\frac{dy}{dx} = n[u(x)]^{n-1}\frac{du}{dx}$$

or, equivalently,

$$\frac{d}{dx}[u^n] = n[u]^{n-1}u'$$

Proof Let $y = u^n$, where u is a differentiable function of x and n is any real number. Then by the Chain Rule, we have

$$\frac{dy}{dx} = \frac{dy}{du}\frac{du}{dx} = \frac{d}{du}[u^n]\frac{du}{dx}$$

But by the simple Power Rule of Section 2.2 (replacing x with u), we have

$$\frac{d}{du}[u^n] = nu^{n-1}$$

Therefore,

$$\frac{dy}{dx} = nu^{n-1}\frac{du}{dx}$$

□

Example 2

Applying the General Power Rule

Differentiate $f(x) = (3x - 2x^2)^3$.

Solution If we let $u = 3x - 2x^2$, then we have

$$f(x) = (3x - 2x^2)^3 = u^n$$

Now, by the Power Rule, the derivative is

$$\begin{aligned} f'(x) &= \overset{n}{3}\overbrace{(3x - 2x^2)^2}^{u^{n-1}}\overbrace{\frac{d}{dx}[3x - 2x^2]}^{u'} \\ &= 3(3x - 2x^2)^2(3 - 4x) \\ &= (9 - 12x)(3x - 2x^2)^2 \end{aligned}$$

□

Keep in mind that the Power Rule is applicable to fractional powers, as demonstrated in the following example.

Example 3

Applying the General Power Rule

Find the derivative of $y = \sqrt[3]{(x^2 + 2)^2}$.

Solution If we rewrite the equation as $y = (x^2 + 2)^{2/3}$, then by the Power Rule, letting $u = x^2 + 2$, we have

$$y' = \overset{n}{\frac{2}{3}}\overbrace{(x^2 + 2)^{-1/3}}^{u^{n-1}}\overbrace{(2x)}^{u'} = \frac{4x}{3\sqrt[3]{x^2 + 2}}$$

□

Remark

The derivative of a quotient may sometimes be found more readily by using the Power Rule than by using the Quotient Rule, especially when the numerator is a constant. Our next example is a case in point.

Example 4

Applying the General Power Rule to Quotients

Differentiate

$$g(t) = \frac{-7}{(2t - 3)^2}$$

Solution If we rewrite the equation as

$$g(t) = -7(2t - 3)^{-2}$$

then by the Power Rule, we obtain

$$g'(t) = (-7)(-2)(2t - 3)^{-3}(2) = 28(2t - 3)^{-3} = \frac{28}{(2t - 3)^3}$$

Alternative Solution By the Quotient and Power Rules,

$$g'(t) = \frac{(2t - 3)^2(0) - (-7)(2)(2t - 3)(2)}{[(2t - 3)^2]^2}$$

$$= \frac{28(2t - 3)}{(2t - 3)^4}$$

$$= \frac{28}{(2t - 3)^3}$$

□

Example 5

Simplifying Derivatives

Differentiate $f(x) = 3x^2\sqrt[3]{9 - 4x^2}$.

Solution Write $f(x) = 3x^2(9 - 4x^2)^{1/3}$. Then by the Product and Power Rules, we obtain

$$f'(x) = 3x^2 \frac{d}{dx}[(9 - 4x^2)^{1/3}] + (9 - 4x^2)^{1/3} \frac{d}{dx}[3x^2]$$

$$= 3x^2[\tfrac{1}{3}(9 - 4x^2)^{-2/3}(-8x)] + (9 - 4x^2)^{1/3}(6x)$$

$$= -8x^3(9 - 4x^2)^{-2/3} + 6x(9 - 4x^2)^{1/3}$$

(Note the use of the Power Rule in the first term of the derivative.) Factoring out the least powers of x and $(9 - 4x^2)$, we have

$$f'(x) = x(9 - 4x^2)^{-2/3}[-8x^2(1) + 6(9 - 4x^2)]$$

$$= \frac{x(-8x^2 + 54 - 24x^2)}{(9 - 4x^2)^{2/3}}$$

$$= \frac{x(54 - 32x^2)}{(9 - 4x^2)^{2/3}}$$

Alternative Method of Simplifying Rather than factoring, you may prefer to remove negative exponents by rationalizing as follows:

$$f'(x) = -8x^3(9 - 4x^2)^{-2/3} + 6x(9 - 4x^2)^{1/3}$$
$$= [-8x^3(9 - 4x^2)^{-2/3} + 6x(9 - 4x^2)^{1/3}]\frac{(9 - 4x^2)^{2/3}}{(9 - 4x^2)^{2/3}}$$
$$= \frac{-8x^3(1) + 6x(9 - 4x^2)}{(9 - 4x^2)^{2/3}} = \frac{54x - 32x^3}{(9 - 4x^2)^{2/3}}$$ □

In Example 5, note that we subtract exponents when factoring. Thus, when $(9 - 4x^2)^{-2/3}$ is factored out of $(9 - 4x^2)^{1/3}$, the *remaining* factor has an exponent of $(\frac{1}{3}) - (-\frac{2}{3}) = 1$. The next example further demonstrates this principle.

Example 6

Combining Differentiation Rules

Differentiate

$$y = \frac{(2x - 3)^3}{\sqrt{4x - 9}}$$

Solution Using the Quotient and Power Rules, we have

$$\frac{dy}{dx} = \frac{(4x - 9)^{1/2}(3)(2x - 3)^2(2) - (2x - 3)^3(\frac{1}{2})(4x - 9)^{-1/2}(4)}{(4x - 9)}$$

Factoring out the least powers of $(4x - 9)$ and $(2x - 3)$, we have

$$\frac{dy}{dx} = \frac{(4x - 9)^{-1/2}(2x - 3)^2[6(4x - 9)^1(2x - 3)^0 - 2(2x - 3)^1(4x - 9)^0]}{(4x - 9)}$$
$$= \frac{(4x - 9)^{-1/2}(2x - 3)^2[6(4x - 9) - 2(2x - 3)]}{(4x - 9)}$$
$$= \frac{(2x - 3)^2(24x - 54 - 4x + 6)}{(4x - 9)^{3/2}} = \frac{(2x - 3)^2(20x - 48)}{(4x - 9)^{3/2}}$$ □

Remark

After studying this chapter, you may be asking the question "Do I have to simplify my derivatives?" The answer is "Yes, if you expect to use them." As you will see in the next chapter, most applications require the derivative to be written in simplified form. Furthermore, if you are asking the question because you have discovered that your algebraic muscles are weak, we encourage you to strengthen them through exercise.

Example 7

Combining Differentiation Rules

Find the derivative of

$$y = \left(\frac{3x - 1}{x^2 + 3}\right)^2$$

First Solution By using the General Power Rule, we have

$$\frac{dy}{dx} = 2\left(\frac{3x - 1}{x^2 + 3}\right)\frac{d}{dx}\left[\frac{3x - 1}{x^2 + 3}\right]$$

$$= \left[\frac{2(3x - 1)}{x^2 + 3}\right]\left[\frac{(x^2 + 3)(3) - (3x - 1)(2x)}{(x^2 + 3)^2}\right]$$

$$= \frac{2(3x - 1)(3x^2 + 9 - 6x^2 + 2x)}{(x^2 + 3)^3}$$

$$= \frac{2(3x - 1)(-3x^2 + 2x + 9)}{(x^2 + 3)^3}$$

Second Solution By rewriting and using the Quotient Rule, we have

$$y = \frac{(3x - 1)^2}{(x^2 + 3)^2}$$

$$\frac{dy}{dx} = \frac{(x^2 + 3)^2(2)(3x - 1)(3) - (3x - 1)^2(2)(x^2 + 3)(2x)}{(x^2 + 3)^4}$$

$$= \frac{(x^2 + 3)(3x - 1)[(x^2 + 3)(2)(3) - (3x - 1)(2)(2x)]}{(x^2 + 3)^4}$$

$$= \frac{(3x - 1)[6x^2 + 18 - 12x^2 + 4x]}{(x^2 + 3)^3}$$

$$= \frac{2(3x - 1)(-3x^2 + 2x + 9)}{(x^2 + 3)^3}$$

□

Example 8

Applying the Chain Rule

The size s of an antelope herd is a function of the annual production x (measured in tons) of edible grass within its grazing territory. If it requires a minimum of 2 tons of grass to sustain an antelope for 1 year, the size of the antelope herd can be estimated by

$$s = \sqrt{x - 2}$$

where $x \geq 2$. We further assume that the grass produced in the herd's territory is a function of the annual rainfall r in the territory

$$x = 50r - r^2$$

where r is measured in inches of rainfall per year. Find the rate of change in the population of the herd with respect to the annual rainfall.

Solution Since

$$s = \sqrt{x - 2} = \sqrt{50r - r^2 - 2}$$

we have

$$\frac{ds}{dr} = \frac{1}{2}(50r - r^2 - 2)^{-1/2}(50 - 2r)$$

$$= \frac{25 - r}{\sqrt{50r - r^2 - 2}}$$

□

We have now discussed all the rules needed to differentiate any algebraic function. For convenience we list these rules here, and we urge you to memorize them, as this will greatly increase your efficiency in differentiating.

Differentiation Rules

(u and v are differentiable functions of x)

Constant Rule:

$$\frac{d}{dx}[c] = 0$$

Constant Multiple Rule:

$$\frac{d}{dx}[cu] = c\frac{du}{dx}$$

Sum Rule:

$$\frac{d}{dx}[u \pm v] = \frac{du}{dx} \pm \frac{dv}{dx}$$

Product Rule:

$$\frac{d}{dx}[uv] = u\frac{dv}{dx} + v\frac{du}{dx}$$

Quotient Rule:

$$\frac{d}{dx}\left[\frac{u}{v}\right] = \frac{v\frac{du}{dx} - u\frac{dv}{dx}}{v^2}$$

Chain Rule (y is a differentiable function of u):

$$\frac{d}{dx}[y] = \frac{dy}{du}\frac{du}{dx}$$

Power Rule:

$$\frac{d}{dx}[u^n] = nu^{n-1}\frac{du}{dx}$$

$$\frac{d}{dx}[x^n] = nx^{n-1}$$

Section Exercises (2.5)

In Exercises 1–52, find the derivative and simplify.

1. $y = (2x - 7)^3$
2. $y = (3x^2 + 1)^4$
3. $f(x) = 2(x^2 - 1)^3$
4. $g(x) = 3(9x - 4)^4$
5. $y = \dfrac{1}{x - 2}$
6. $s(t) = \dfrac{1}{t^2 + 3t - 1}$
7. $f(t) = \left(\dfrac{1}{t - 3}\right)^2$
8. $y = \dfrac{-4}{(t + 2)^2}$
9. $f(x) = \dfrac{3}{x^3 - 4}$
10. $f(x) = \dfrac{1}{(x^2 - 3x)^2}$
11. $f(x) = x^2(x - 2)^4$
12. $f(x) = x(3x - 9)^3$
13. $y = (x - 2)(x + 3)^3$
14. $g(t) = (3t^2 - 2)^2(t + 1)^3$
15. $f(t) = \sqrt{t + 1}$
16. $g(x) = \sqrt{2x + 3}$
17. $s(t) = \sqrt{t^2 + 2t - 1}$
18. $y = \sqrt[3]{3x^3 + 4x}$
19. $y = \sqrt[3]{9x^2 + 4}$
20. $g(x) = \sqrt{x^2 - 2x + 1}$
21. $y = 2\sqrt{x^2 + 4}$
22. $f(x) = -3\sqrt[4]{9x + 2}$
23. $f(x) = (x^2 - 9)^{2/3}$
24. $f(t) = (9t + 2)^{2/3}$
25. $y = \dfrac{1}{\sqrt{x + 2}}$
26. $g(t) = \sqrt{\dfrac{1}{t^2 - 2}}$
27. $g(x) = \dfrac{3}{\sqrt[3]{x^3 - 1}}$
28. $s(x) = \dfrac{1}{\sqrt{x^2 - 3x + 4}}$
29. $y = \dfrac{-1}{\sqrt{x} + 1}$
30. $y = \dfrac{1}{2\sqrt{t} - 3}$
31. $f(x) = 2(x^2 - 1)^3$
32. $f(x) = \sqrt{2x}(x + 2)^2$
33. $g(x) = \sqrt[3]{9x^2 + 4}$
34. $g(t) = (t^2 - 2t - 3)^{10}$
35. $y = \dfrac{x}{\sqrt{x^2 + 1}}$
36. $y = \dfrac{\sqrt{x^2 + 1}}{x}$
37. $g(x) = \dfrac{2x}{\sqrt{x + 1}}$
38. $f(x) = \dfrac{-2x^2}{x - 1}$
39. $y = \dfrac{\sqrt{x + 1}}{x^2 + 1}$
40. $f(x) = \dfrac{x + 1}{2x - 3}$
41. $f(t) = \dfrac{3t + 2}{t - 1}$
42. $y = \sqrt{\dfrac{2x}{x + 1}}$
43. $g(t) = \dfrac{3t^2}{\sqrt{t^2 + 2t - 1}}$
44. $f(x) = \sqrt{x}(x - 2)^2$
45. $f(t) = \sqrt{t + 1}\,\sqrt[3]{t + 1}$
46. $y = \sqrt{3x}(x + 2)^3$
47. $y = \sqrt{\dfrac{x + 1}{x}}$
48. $y = (t^2 - 9)\sqrt{t + 2}$
49. $s(t) = \dfrac{-2(2 - t)\sqrt{1 + t}}{3}$
50. $g(x) = \sqrt{x - 1} + \sqrt{x + 1}$
51. $s(t) = \sqrt{\dfrac{t^2 - 4}{4t}}$
52. $y = \dfrac{\sqrt{x^2 + 1}}{x}$

In Exercises 53–56, find the equation of the tangent line to the curve $y = f(x)$ at the given point.

53. $f(x) = \sqrt{3x^2 - 2}$; $(3, 5)$
54. $f(x) = x\sqrt{x^2 + 5}$; $(2, 6)$
55. $f(x) = \left(x - \dfrac{1}{x}\right)^{3/2}$; $(1, 0)$
56. $f(x) = \dfrac{\sqrt[3]{6x - 4}}{x}$; $(2, 1)$

Ⓒ57. An assembly plant purchases small electric motors to install in one of their products. The plant management has determined that the cost per motor over the next few years is

$$C = 4(1.52t + 10)^{3/2}$$

where t is the time in years. Complete the following table for the marginal cost for each of the next five years.

t	1	2	3	4	5
$\dfrac{dC}{dt}$					

Ⓒ58. A small manufacturing company is concerned that the increase in profit per unit produced will not keep pace with the rate of inflation. Based on present performance, the company predicts that the profit per unit during the next five years will be given by

$$P = \sqrt{0.05t^3 + 40.75t}$$

Ⓒ Calculator may be helpful.

Complete the following table for the marginal profit for each of the next five years.

t	1	2	3	4	5
$\frac{dP}{dt}$					

Ⓒ**59.** The *Doyle Log Rule* is a mathematical model for estimating the volume (in board feet) of a log of length L feet and diameter D inches at the small end. According to this model the volume is

$$V = \left(\frac{D - 4}{4}\right)^2 L$$

Find the rate at which the volume is increasing for a 12-foot log whose diameter is
(a) 8 in. (b) 16 in.
(c) 24 in. (d) 36 in.

Ⓒ**60.** A forester has studied the growth curve for a certain species of tree and has found that the expected height after t years is

$$h = 62.9 + 27.6\left(\frac{t - 120}{40}\right) - 0.4\left(\frac{t - 120}{40}\right)^2 - 0.8\left(\frac{t - 120}{40}\right)^3$$

where $0 \le t \le 240$. Complete the following table showing the height and the rate of increase of the height.

t	40	120	200
h			
$\frac{dh}{dt}$			

Ⓒ**61.** The number N of bacteria in a certain culture after t days is given by

$$N = 400\left[1 - \frac{3}{(t^2 + 2)^2}\right]$$

(a) Complete the following table showing the rate of increase in the number of bacteria.

t	0	1	2	3	4
$\frac{dN}{dt}$					

(b) What is the limiting size of this population?

Ⓒ Calculator may be helpful.

Section 2.6

Higher Order Derivatives

Introductory Example

The Acceleration of a Protein Molecule During Gel Electrophoresis

A standard problem in biochemistry is to determine the chemical makeup of a solution by separating the solution into its component parts. A technique for separating proteins in a solution is called *gel electrophoresis*. This technique consists of placing the solution in the top of a glass cylinder containing a gelatin-type substance and passing an electrical current down through the solution and gel (see Figure 2.23). When the current (of voltage v) is turned on, the different proteins in the solution begin moving down into the gel at differing rates, which depend upon their molecular size m and their negative charge c. In general, smaller or more highly charged protein molecules move faster than do larger or less highly charged ones. If the procedure is successful, then after a certain time t the different proteins in the solution will have separated into bands in the gel and the current can be turned off.

One model for determining the distance moved by a protein molecule is

$$s = (k_1 cv)t + \left(\frac{g}{m} - k_2\right)t^2$$

where the constants k_1 and k_2 depend upon the type of gel and the constant g depends upon gravity.

By differentiating s with respect to t, we obtain the velocity of the molecule at time t:

$$\frac{ds}{dt} = k_1 cv + 2\left(\frac{g}{m} - k_2\right)t$$

In this section, we will see that by differentiating a second time we obtain the **acceleration** of the molecule at time t:

$$\frac{d^2s}{dt^2} = 2\left(\frac{g}{m} - k_2\right)$$

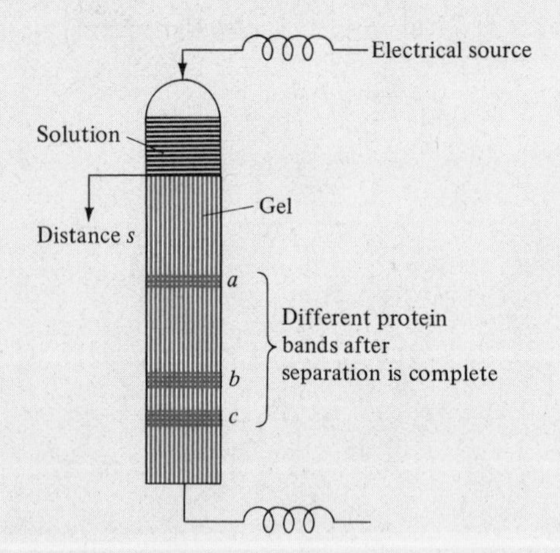

Figure 2.23

Section Objectives: ***To define the second-, third-, and higher-order derivatives of a function ▪ To use derivatives to determine the acceleration of objects traveling along linear paths.***

Since the derivative of a function is itself a function, we may consider finding its derivative. If this is done, the result is again a function that may be differentiated. If we continue in this manner, we have what are called **higher-order derivatives.** For instance, the derivative of f' is called the **second derivative of** f and is denoted by f''. Similarly, we define the **third derivative of** f as the derivative of f'' and we denote it by f'''. So long as each successive derivative is differentiable, we can continue in this manner to obtain derivatives of higher orders.

Example 1

Finding Higher-Order Derivatives

If $f(x) = 2x^4 - 3x^2$, then

$$
\begin{aligned}
f'(x) &= 8x^3 - 6x && \text{(first derivative)} \\
f''(x) &= 24x^2 - 6 && \text{(second derivative)} \\
f'''(x) &= 48x && \text{(third derivative)} \\
f^{(4)}(x) &= 48 && \text{(fourth derivative)} \\
f^{(5)}(x) &= 0 && \text{(fifth derivative)} \\
&\vdots \\
f^{(n)}(x) &= 0 && (n\text{th derivative})
\end{aligned}
$$

□

Remark

We drop the prime notation after the third derivative and use parentheses around the order of the derivative. Note that in this particular example the "fourth derivative of $f(x)$," denoted by $f^{(4)}(x)$, is constant, and therefore each succeeding derivative is zero by our Constant Rule for differentiation.

Example 2

Finding Higher-Order Derivatives

If $g(t) = -t^4 + 2t^3 + t + 4$, find the value of $g'''(2)$.

Solution

$$
\begin{aligned}
g'(t) &= -4t^3 + 6t^2 + 1 \\
g''(t) &= -12t^2 + 12t \\
g'''(t) &= -24t + 12
\end{aligned}
$$

Therefore,

$$g'''(2) = -24(2) + 12 = -36$$

□

Notation for Higher-Order Derivatives

First derivative:	$y', f'(x), \frac{dy}{dx}, \frac{d}{dx}[f(x)], D_x(y)$
Second derivative:	$y'', f''(x), \frac{d^2y}{dx^2}, \frac{d^2}{dx^2}[f(x)], D_x^2(y)$
Third derivative:	$y''', f'''(x), \frac{d^3y}{dx^3}, \frac{d^3}{dx^3}[f(x)], D_x^3(y)$
Fourth derivative:	$y^{(4)}, f^{(4)}(x), \frac{d^4y}{dx^4}, \frac{d^4}{dx^4}[f(x)], D_x^4(y)$
⋮	
nth derivative:	$y^{(n)}, f^{(n)}(x), \frac{d^ny}{dx^n}, \frac{d^n}{dx^n}[f(x)], D_x^n(y)$

Example 3

Finding Higher-Order Derivatives

Find the first four derivatives of

$$y = \frac{3}{x}$$

Solution Since $y = 3x^{-1}$, by the Power rule we obtain

$$y' = -3x^{-2} = -\frac{3}{x^2}$$

$$y'' = 6x^{-3} = \frac{6}{x^3}$$

$$y''' = -18x^{-4} = -\frac{18}{x^4}$$

$$y^{(4)} = 72x^{-5} = \frac{72}{x^5}$$

□

Throughout the remainder of this text, we will encounter various applications of higher-order derivatives. For example, in Chapter 3, we will see that the second derivative can be used to determine the concavity of the graph of a function. Later, in Chapter 9, we will use higher-order derivatives to determine the Taylor Polynomial and Taylor Series for a function. For the time being, we will concentrate on what is perhaps the most common application of a higher-order derivative — that of finding the acceleration of an object moving in a straight line.

In Section 2.3, we saw that the velocity of an object whose position is given by

$$\text{position} = s = f(t)$$

is

$$\text{velocity} = \frac{ds}{dt} = f'(t)$$

The rate of change of the velocity of the object is called its **acceleration** and is given by the second derivative

$$\text{acceleration} = \frac{d^2s}{dt^2} = f''(t)$$

Example 4

Finding the Acceleration of a Moving Object

Suppose that the equation

$$s = -16t^2 + 48t + 160$$

represents the position (in feet after t seconds) above ground of a ball thrown into the air from the top of a cliff. (See Figure 2.24.) Find the acceleration of the ball.

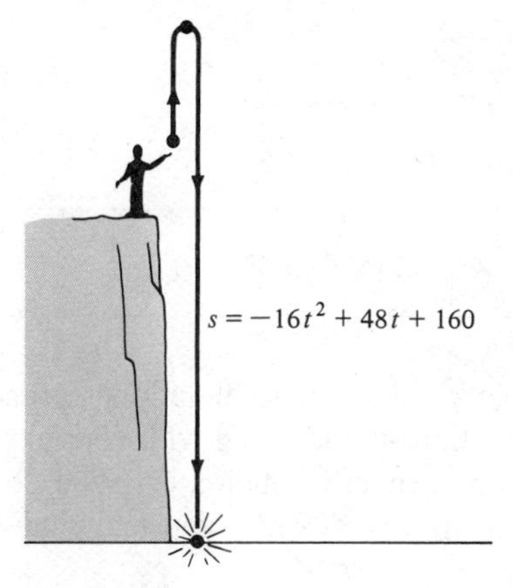

Figure 2.24

Solution Differentiating twice, we obtain

$$s = -16t^2 + 48t + 160 \text{ (ft)}$$

$$\frac{ds}{dt} = -32t + 48 \text{ (ft/sec)}$$

$$\frac{d^2s}{dt^2} = -32 \text{ (ft/sec}^2\text{)}$$

Thus, the acceleration is -32 ft/sec^2. (We call this particular acceleration the **force of gravity.**) □

Note that the acceleration of the ball in Example 4 is negative, which indicates that the ball is being pulled down toward the earth by the force of gravity.

Example 5

Finding the Acceleration of a Moving Object

Suppose that the *velocity* of an automobile starting from rest is given by

$$\frac{ds}{dt} = \frac{80t}{t + 5} \text{ (ft/sec)}$$

Find the velocity and acceleration of the automobile when $t = 0, 5, 10, \ldots,$ 60 seconds.

Solution The acceleration at time t is given by

$$\frac{d^2s}{dt^2} = \frac{(t + 5)(80) - (80t)(1)}{(t + 5)^2} = \frac{400}{(t + 5)^2} \text{ (ft/sec}^2\text{)}$$

Table 2.8 compares the velocity and acceleration of the automobile at five second intervals during its first minute of travel.

Table 2.8

t	0	5	10	15	20	25	30	35	40	45	50	55	60
$\frac{ds}{dt}$	0	40	53.3	60.0	64.0	66.7	68.6	70.0	71.1	72.0	72.7	73.3	73.8
$\frac{d^2s}{dt^2}$	16.0	4.00	1.78	1.00	0.64	0.44	0.33	0.25	0.20	0.16	0.13	0.11	0.09

Note from Table 2.8 that the acceleration approaches zero as the velocity levels off. This observation should agree with your experience of riding in an accelerating car. As a passenger of such a car, you do not feel the velocity; you feel the acceleration. In other words, you feel changes in velocity. □

Section Exercises (2.6)

In Exercises 1–22, find the indicated derivative of the given function.

1. $f(x) = x^2 - 2x + 1; f''(x)$
2. $f(x) = 3x - 1; f''(x)$
3. $f(x) = 5 - 4x; f''(x)$
4. $f(x) = x^5 - 3x^4; f'''(x)$
5. $f(x) = x^4 - 2x^3; f'''(x)$
6. $f(x) = \sqrt{x}; f'''(x)$
7. $f(x) = x^2 - \dfrac{1}{x} + 1; f^{(4)}(x)$
8. $f(t) = 1 - 2t + \dfrac{1}{t^2}; f^{(4)}(t)$
9. $f(t) = t^{-1/3}; f''(t)$
10. $f(t) = \dfrac{3}{4t^2}; f''(t)$
11. $f(x) = \dfrac{3}{(4x)^2}; f'''(x)$
12. $f(x) = (x - 1)^2; f'''(x)$
13. $f(x) = 2x(x - 1)^2; f'''(x)$
14. $f(x) = x\sqrt[3]{x}; f''(x)$

15. $f(x) = 4(x^2 - 1)^3; f''(x)$

16. $g(t) = \dfrac{-4}{(t+2)^2}; g''(t)$

17. $f(x) = \dfrac{x+1}{x-1}; f''(x)$

18. $f(x) = (x-1)(x^2 - 3x + 2); f''(x)$

19. $h(s) = (s^2 - 2s + 1)(s^3 - 1); h''(s)$

20. $s(t) = \sqrt{2t+3}; s'''(t)$

21. $g(x) = \sqrt{4-x}; g'''(x)$

22. $f(x) = \sqrt{9-x^2}; f''(x)$

In Exercises 23–30, find the second derivative and solve the equation $f''(x) = 0$.

23. $f(x) = x^3 - 9x^2 + 27x - 27$

24. $f(x) = 3x^3 - 9x + 1$

25. $f(x) = (x+1)(x-2)(x-5)$

26. $f(x) = 3x^4 - 18x^2$

27. $f(x) = x^4 - 8x^3 + 18x^2 - 16x + 2$

28. $f(x) = \dfrac{x}{x^2+1}$

29. $f(x) = \dfrac{x}{x^2+3}$

30. $f(x) = x\sqrt{4-x^2}$

31. A ball is thrown upward from ground level, and its height above ground is given by

$$s = -16t^2 + 48t$$

(a) Find expressions for the velocity and acceleration of the ball.
(b) Find the time when the ball is at its highest point by finding the time when the velocity is zero.
(c) Find the height at the time given in part (b).

Ⓒ**32.** A coin is dropped from the top of the Washington Monument, and its height above ground is given by

$$s = -16t^2 + 550$$

(a) Find expressions for the velocity and acceleration of the coin.
(b) Find the time when the ball reaches ground level ($s = 0$).
(c) Use the time found in part (b) to find the velocity of the coin when it hits the ground.

Ⓒ**33.** The velocity of an automobile starting from rest is given by

$$\frac{ds}{dt} = \frac{90t}{t+10} \text{ (ft/sec)}$$

Complete the following table showing the velocity and acceleration at 10-second intervals during the first minute of travel.

t	0	10	20	30	40	50	60
$\frac{ds}{dt}$							
$\frac{d^2s}{dt^2}$							

Ⓒ Calculator may be helpful.

Section 2.7

Implicit Differentiation

Introductory Example

The Whispering Gallery

One interesting phenomenon in acoustics is the *whispering gallery*. A whispering gallery contains two special points (see Figure 2.25) that are acoustically close together, even though they may be physically many feet apart. If the gallery is constructed perfectly, all of the sound originating at one of the points bounces off the gallery wall and ceiling in such a way that it is directed toward the other point. Thus, even a low whisper made by a person at one of the points can be heard by a person at the other point.

Mathematically, the two points in the gallery are focus points for each of the gallery's major elliptical cross sections. As sound leaves either one of the focus points, it is reflected off the gallery wall and ceiling in such a way that it makes equal angles (coming and going) with the tangent line to the elliptical cross section at the point of impact (see Figure 2.26). To prove that this reflective property results in the sound from one focus point ending up at the other point we need to determine the slope of the tangent line at a point on an ellipse.

In this section, we will study a procedure called **implicit differentiation** that can be used to show that for the ellipse whose equation is

$$\frac{x^2}{a^2} + \frac{y^2}{b^2} = 1$$

the slope of the tangent line at the point (x, y) is

$$\frac{dy}{dx} = -\frac{xb^2}{ya^2}$$

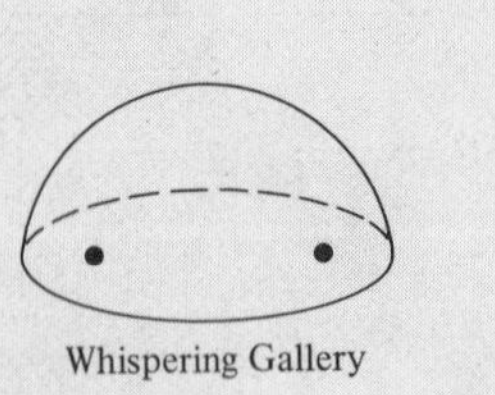

Figure 2.25

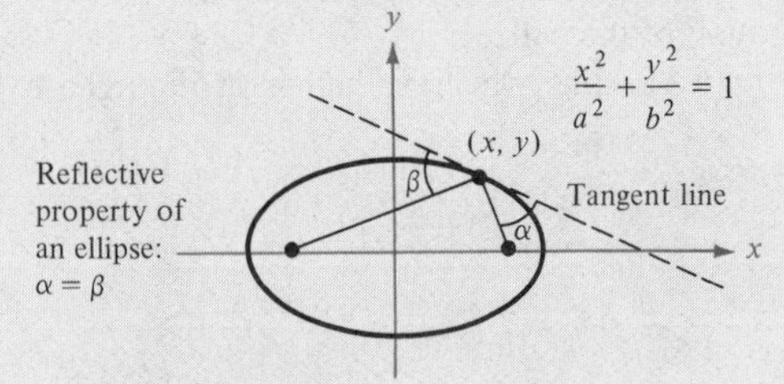

Figure 2.26

Section Objectives: ***To distinguish between explicit and implicit forms of an equation ▪ To demonstrate the technique of implicit differentiation and apply it to equations in implicit form.***

Up to this point, our equations involving two variables were generally expressed in the **explicit form**

$$y = f(x)$$

That is, one of the two variables was explicitly given in terms of the other. For example, the equations

$$y = 3x - 5, \qquad s = -16t^2 + 20t, \qquad u = 3w - w^2$$

are all written in explicit form, and we say that y, s, and u are functions of x, t, and w, respectively.

However, many relationships are not given explicitly and are only implied by a given equation. For instance, we say the equation

$$xy = 1$$

is given in **implicit form.** Suppose that you were asked to find dy/dx in this equation. As it turns out, this is a relatively simple task, and you would probably begin by solving the equation for y.

Implicit form		*Explicit form*
$xy = 1$	Solve for y ⟩	$y = \frac{1}{x} = x^{-1}$

Now, using the explicit form of this equation, you could differentiate as usual to obtain

$$y = x^{-1}$$

$$\frac{dy}{dx} = -x^{-2} = -\frac{1}{x^2}$$

This procedure works well whenever we can easily change an equation from implicit to explicit form. However, we cannot use this procedure in cases in which we are unable to solve for y as a function of x.

For instance, how would you find dy/dx in the following equation?

$$x^2 - 2y^3 + 4y = 2$$

To do this, we use a procedure called **implicit differentiation.**

This procedure is demonstrated in Example 1.

Example 1

Implicit Differentiation

Find dy/dx given that $x^2 - 2y^3 + 4y = 2$.

Solution

1. By differentiating each term with respect to x, we have

$$\frac{d}{dx}[x^2 - 2y^3 + 4y] = \frac{d}{dx}[2]$$

$$2x - 6y^2\left(\frac{dy}{dx}\right) + 4\frac{dy}{dx} = 0$$

(Note the use of the Chain Rule on the $2y^3$ term.)

2. Collecting the dy/dx terms on the left, we have

$$-6y^2\left(\frac{dy}{dx}\right) + 4\frac{dy}{dx} = -2x$$

3. By factoring dy/dx out of the left side, we have

$$\frac{dy}{dx}(-6y^2 + 4) = -2x$$

4. Finally, we divide by $(-6y^2 + 4)$ to obtain

$$\frac{dy}{dx} = \frac{-2x}{-6y^2 + 4}$$

□

Implicit Differentiation

Given an equation involving x and y, we can find dy/dx as follows:

1. Differentiate both sides of the equation *with respect to x*.
2. Collect all terms involving dy/dx on the left side of the equation and move all other terms to the right side of the equation.
3. Factor dy/dx out of the left side of the equation.
4. Solve for dy/dx by dividing both sides of the equation by the left-hand factor that does not contain dy/dx.

Remark

The key to understanding how to find dy/dx implicitly lies in the realization that the differentiation is taking place *with respect to x*. This means that when we differentiate terms involving x alone, we can differentiate as usual. *But* when we differentiate terms involving y, we must apply the Chain Rule. Study the next example carefully. Be sure you understand how the Chain Rule is used to introduce the dy/dx terms.

Example 2

Applying the Chain Rule

Differentiate the following with respect to x.

(a) $3x^2$ (b) $3y^2$

(c) $x + 3y$ (d) xy^2

Solution

(a) $\frac{d}{dx}[3x^2] = 6x$

(b) $\dfrac{d}{dx}[3y^2] = 6y\dfrac{dy}{dx}$

(c) $\dfrac{d}{dx}[x + 3y] = 1 + 3\dfrac{dy}{dx}$

(d)
$$\begin{aligned}\frac{d}{dx}[xy^2] &= x\frac{d}{dx}[y^2] + y^2\frac{d}{dx}[x] \\ &= x\left(2y\frac{dy}{dx}\right) + y^2(1) \\ &= 2xy\frac{dy}{dx} + y^2\end{aligned}$$

(Note the use of the Product Rule.) □

The fact that the derivative in Example 1 involves the two variables x and y is generally of no disadvantage. For instance, in using the derivative to determine the slope of a curve at a point, we are often given both coordinates, and, if not, the remaining one can probably be calculated with a reasonable amount of effort.

Example 3

Using Implicit Differentiation to Find the Slope of a Tangent Line

Determine the slope of the tangent line to the graph of $x^2 + 4y^2 = 4$ at the point $(\sqrt{2}, -1/\sqrt{2})$. (See Figure 2.27.)

Solution Implicitly differentiating the equation $x^2 + 4y^2 = 4$ with respect to x yields

$$2x + 8y\left(\frac{dy}{dx}\right) = 0$$

and $$\frac{dy}{dx} = \frac{-2x}{8y} = \frac{-x}{4y}$$

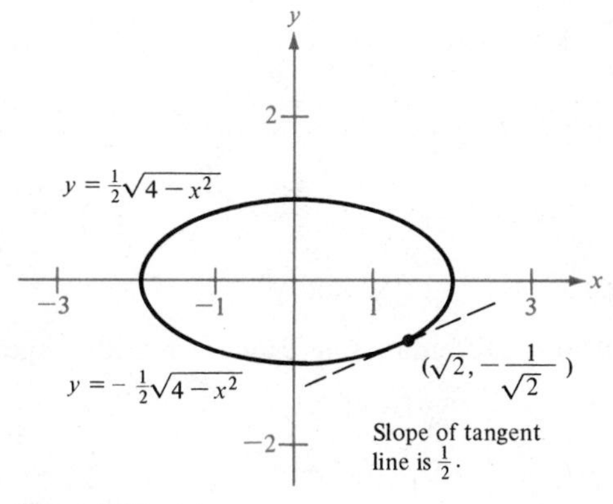

Figure 2.27

Therefore, at $(\sqrt{2}, -1/\sqrt{2})$ the slope is

$$\frac{dy}{dx} = \frac{-\sqrt{2}}{-4/\sqrt{2}} = \frac{1}{2}$$

Alternative Solution Explicitly solving for y in terms of x, we obtain the equations

$$y = \tfrac{1}{2}\sqrt{4 - x^2} \qquad \text{or} \qquad y = -\tfrac{1}{2}\sqrt{4 - x^2}$$

whose graphs are, respectively, the top and bottom halves of the ellipse in Figure 2.27. The point $(\sqrt{2}, -1/\sqrt{2})$ satisfies the equation

$$y = -\tfrac{1}{2}\sqrt{4 - x^2}$$

Therefore, the slope is determined by

$$\frac{dy}{dx} = \frac{-1}{4}(4 - x^2)^{-1/2}(-2x) = \frac{x}{2\sqrt{4 - x^2}}$$

At the point $(\sqrt{2}, -1/\sqrt{2})$, we have

$$\frac{dy}{dx} = \frac{\sqrt{2}}{2\sqrt{4 - 2}} = \frac{1}{2}$$

as was obtained by implicit differentiation. □

Example 4

Using Implicit Differentiation

Find the slope of the curve $3x^2 - 2xy + xy^3 = 7$ at $(1, 2)$.

Solution Differentiating implicitly and using y' for dy/dx, we obtain

$$6x - [2xy' + y(2)] + [x(3y^2)y' + y^3] = 0$$

$$(-2x + 3xy^2)y' = 2y - 6x - y^3$$

$$y' = \frac{2y - 6x - y^3}{3xy^2 - 2x}$$

At the point $(1, 2)$,

$$y' = \frac{4 - 6 - 8}{12 - 2} = \frac{-10}{10} = -1$$

□

Example 5

Using Implicit Differentiation to Find the Second Derivative

Given $x^2 + y^2 = 25$, find y''.

Solution Differentiating each term with respect to x, we obtain

$$2x + 2yy' = 0$$

$$2yy' = -2x$$

$$y' = \frac{-2x}{2y} = -\frac{x}{y}$$

Differentiating a second time with respect to x yields

$$y'' = -\frac{(y)(1) - (x)(y')}{y^2}$$

Finally, substituting $-x/y$ for y', we have

$$y'' = -\frac{y - (x)(-x/y)}{y^2} = -\frac{y^2 + x^2}{y^3} = -\frac{25}{y^3}$$

□

In Section 2.2, we listed the Power Rule as

$$\frac{d}{dx}[x^n] = nx^{n-1}$$

Although we claimed that this rule holds when n is any real number, we have given proofs only for the cases when n is a positive integer (Section 2.2) or a negative integer (Section 2.4). Using implicit differentiation, we can now demonstrate the validity of the Power Rule when n is any rational number.

Example 6

A Proof of the Power Rule for Rational Exponents

Let $y = x^n$, where n is the rational number p/q (p and q are integers). By raising both sides of this equation to the qth power, we have

$$y = x^{p/q}$$
$$y^q = x^p$$

Since p and q are both integers, we implicitly differentiate both sides of the equation with respect to x to obtain

$$qy^{q-1}\frac{dy}{dx} = px^{p-1}$$

Solving for dy/dx yields

$$\frac{dy}{dx} = \frac{px^{p-1}}{qy^{q-1}}$$

Finally, substituting $x^{p/q}$ for y, we have

$$\begin{aligned}\frac{dy}{dx} &= \frac{px^{p-1}}{q(x^{p/q})^{q-1}} = \left(\frac{p}{q}\right)\left[\frac{x^{p-1}}{x^{p(q-1)/q}}\right]\\ &= \left(\frac{p}{q}\right)x^{(p-1)-[p(q-1)/q]} = \left(\frac{p}{q}\right)x^{(pq-q-pq+p)/q}\\ &= \left(\frac{p}{q}\right)x^{(p-q)/q} = \left(\frac{p}{q}\right)x^{(p/q)-1}\\ &= nx^{n-1}\end{aligned}$$

□

Section Exercises (2.7)

In Exercises 1–16, find dy/dx by implicit differentiation and evaluate the derivative at the indicated point.

1. $x^2 + y^2 = 16$
2. $x^2 - y^2 = 16$
3. $xy = 4$
4. $x^2 - y^3 = 0$
5. $x^{1/2} + y^{1/2} = 9$
6. $x^3 + y^3 = 8$
7. $x^3 - xy + y^2 = 4$
8. $x^2y + y^2x = -2$
9. $y^2 = \dfrac{x^2 - 9}{x^2 + 9}$; $(3, 0)$
10. $(x + y)^3 = x^3 + y^3$; $(-1, 1)$
11. $x^3y^3 - y = x$; $(0, 0)$
12. $\sqrt{xy} = x - 2y$; $(4, 1)$
13. $x^{2/3} + y^{2/3} = 5$; $(8, 1)$
14. $(x - y^2)(x + xy) = 4$; $(2, 1)$
15. $x^3 - 2x^2y + 3xy^2 = 38$; $(2, 3)$
16. $x^3 + y^3 = 2xy$; $(1, 1)$

In Exercises 17–20, (a) find two explicit functions defined by the given equation and state their domains; (b) find the derivatives of the functions obtained in part (a); (c) differentiate the given equation implicitly and verify that the result is the same as the result of part (b); (d) sketch the graph of the equation and label the parts of the graph given by the functions of part (a).

17. $x^2 + y^2 = 16$
18. $x^2 + y^2 = 4$
19. $9x^2 + 16y^2 = 144$
20. $4y^2 - x^2 = 4$

In Exercises 21–26, find d^2y/dx^2 in terms of x and y.

21. $x^2 + xy = 5$
22. $x^2y^2 - 2x = 3$
23. $x^2 - y^2 = 16$
24. $1 - xy = x - y$
25. $y^2 = x^3$
26. $y^2 = 4x$

Chapter 3

Applications of the Derivative

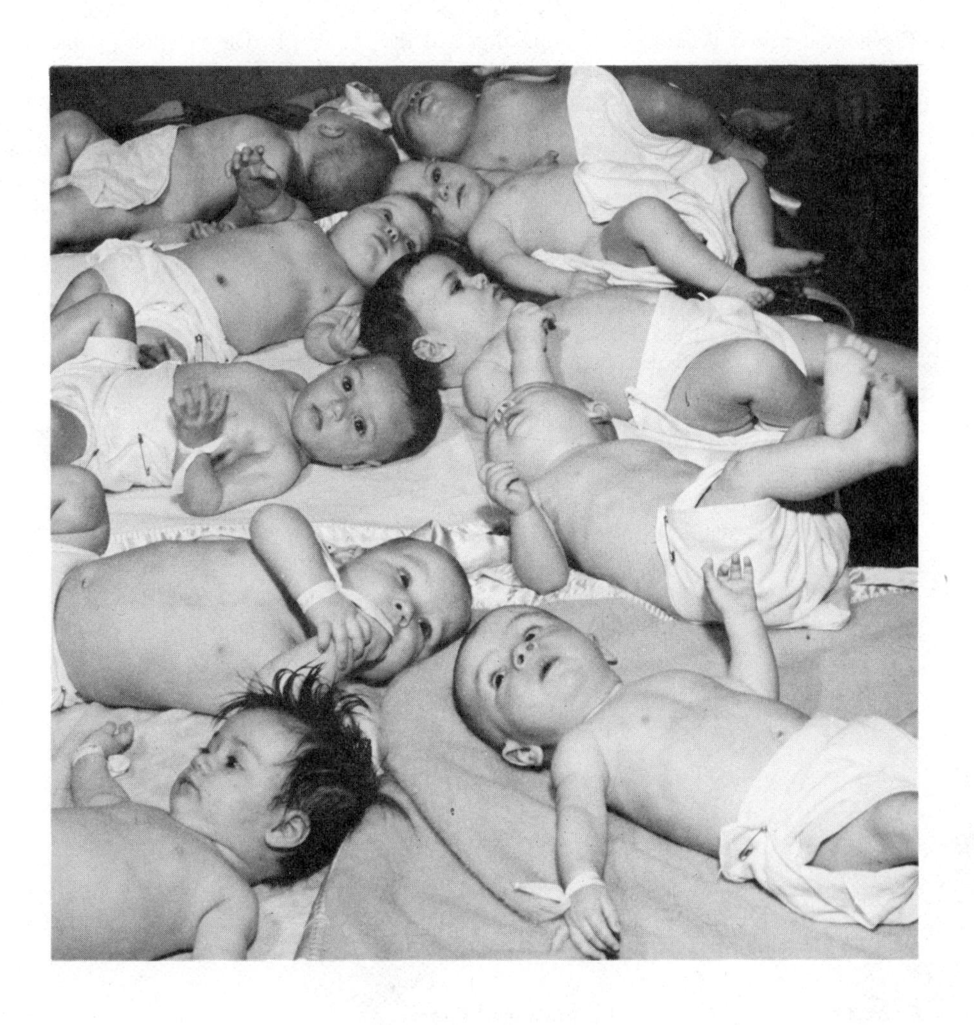

Section 3.1

Increasing and Decreasing Functions

Introductory Example

Egg Yolk Growth Curve

The normal period between the fertilization and the laying of a chicken egg is about 19 days. Although the egg yolk grows continuously during this time, the rate of growth varies significantly. Figure 3.1 shows that almost all of the yolk's growth takes place in the final seven days with the maximum growth rate occurring during the third and fourth days before laying.

We say that this growth function is **increasing** up until the egg is laid, since for each increase in time there is a corresponding *increase* in weight. If we let $f(t)$ represent the weight of the yolk at time t, we can observe that the following three conditions are equivalent:

1. The function f is increasing.
2. The graph of f has a positive slope.
3. The derivative f' is positive.

After the egg is laid and the chick begins to develop, the amount of egg yolk decreases. We say that f is **decreasing** during this period of time, since for each increase in time there is a corresponding *decrease* in weight. Paralleling our observation for increasing functions, we note that the following three conditions are equivalent:

1. The function f is decreasing.
2. The graph of f has a negative slope.
3. The derivative f' is negative.

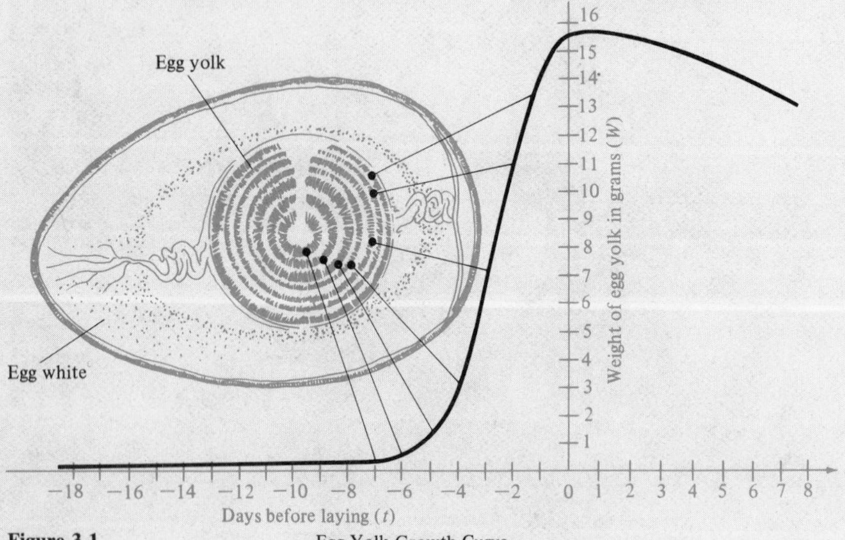

Figure 3.1 Egg Yolk Growth Curve

Section Objectives: ***To define increasing and decreasing functions ▪ To use the derivative to determine when a function is increasing or decreasing.***

Given the graph of a continuous function, it is a simple matter to decide where (on what intervals) the function is increasing, is constant, or is decreasing. For instance, in Figure 3.2, we have identified the behavior of a function on the intervals (a, c_1), (c_1, c_2), (c_2, c_3), (c_3, c_4), and (c_4, b). However, without the graph of a function, it is much more difficult to determine when the function is increasing (or decreasing). We can get some help from the following definition.

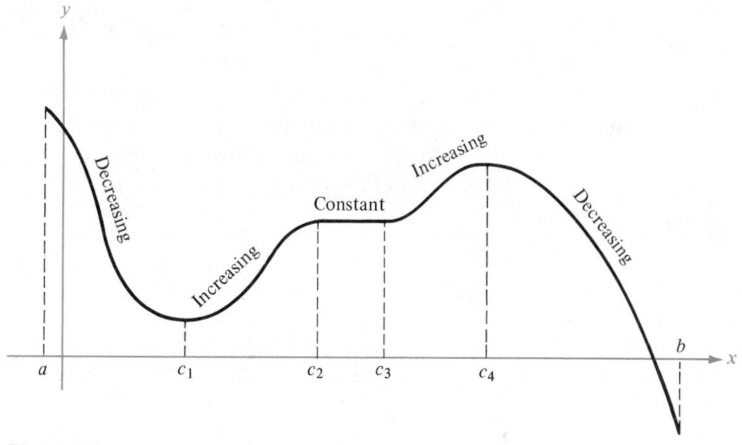

Figure 3.2

Definition of Increasing and Decreasing Functions

A function f is said to be **increasing** on an interval if for any two numbers x_1 and x_2 in the interval

$$x_1 < x_2 \quad \text{implies} \quad f(x_1) < f(x_2)$$

A function f is said to be **decreasing** on an interval if for any two numbers x_1 and x_2 in the interval

$$x_1 < x_2 \quad \text{implies} \quad f(x_1) > f(x_2)$$

The derivative is useful in determining if a function is increasing or decreasing on a specified interval. We know that if the derivative of f is positive at $x = c$, then its graph slopes upward there, and the function is increasing at that point. Similarly, a negative derivative (or downward slope) implies the function is decreasing at that point. The following theorem indicates how the first derivative may be used to determine intervals on which a function is increasing or decreasing.

Theorem 3.1
Test for Increasing or Decreasing Functions

Let f be a function that is differentiable on the interval (a, b).

i. If $f'(x) > 0$ for all x in (a, b), then f is increasing on (a, b).
ii. If $f'(x) < 0$ for all x in (a, b), then f is decreasing on (a, b).
iii. If $f'(x) = 0$ for all x in (a, b), then f is constant on (a, b).

To apply Theorem 3.1, we note in Figure 3.3 that $f'(x)$ can change sign at points where $f'(x) = 0$, as well as at points where $f'(x)$ is undefined. This suggests the following steps for finding intervals on which f is increasing or decreasing.

1. Locate, on the real line, those points where $f'(x) = 0$ and those where f' is undefined.
2. Test the sign of $f'(x)$ at an arbitrary point in each of the intervals determined by the points in the first step.
3. Use Theorem 3.1 to decide whether f is increasing or decreasing on the intervals in question.

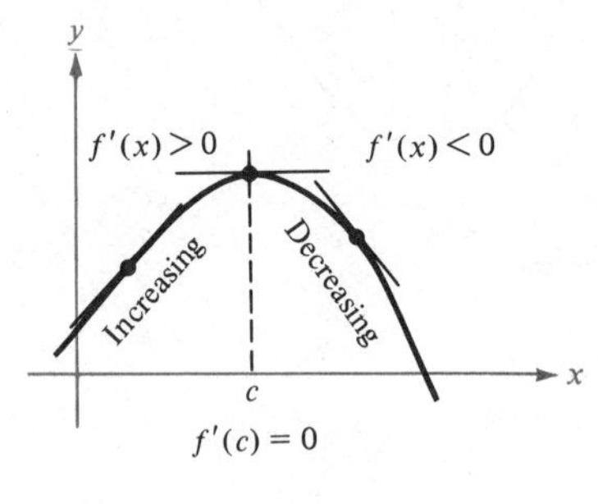

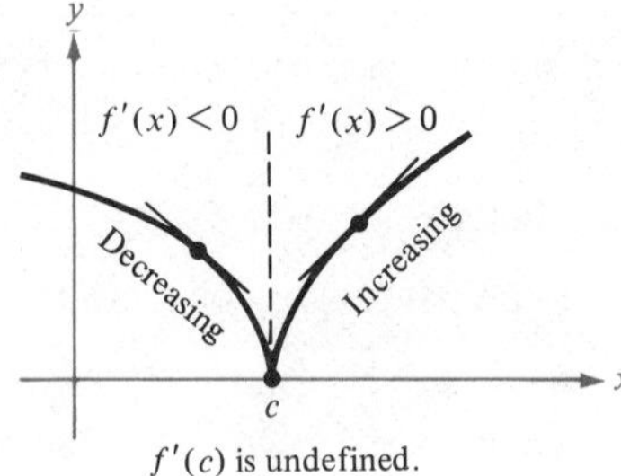

Figure 3.3

Example 1

Determining Intervals on Which f is Increasing or Decreasing

Find the intervals on which $f(x) = x^3 - \frac{3}{2}x^2$ is increasing or decreasing.

Solution Since

$$f'(x) = 3x^2 - 3x = 3(x)(x - 1)$$

we know $f'(x)$ is zero for $x = 0$ and $x = 1$. Thus, the intervals to be tested are

$$(-\infty, 0), \qquad (0, 1), \qquad (1, \infty)$$

Table 3.1 summarizes our testing of these three intervals. (Note that the points in the intervals were chosen for convenience and other points could have been used.)

Table 3.1

Interval	$(-\infty, 0)$	$(0, 1)$	$(1, \infty)$
Point in interval	$x = -1$	$x = \frac{1}{2}$	$x = 2$
Test	$f'(-1) = 6 > 0$	$f'(\frac{1}{2}) = -\frac{3}{4} < 0$	$f'(2) = 6 > 0$
Conclusion	f is increasing	f is decreasing	f is increasing

(See Figure 3.4.)

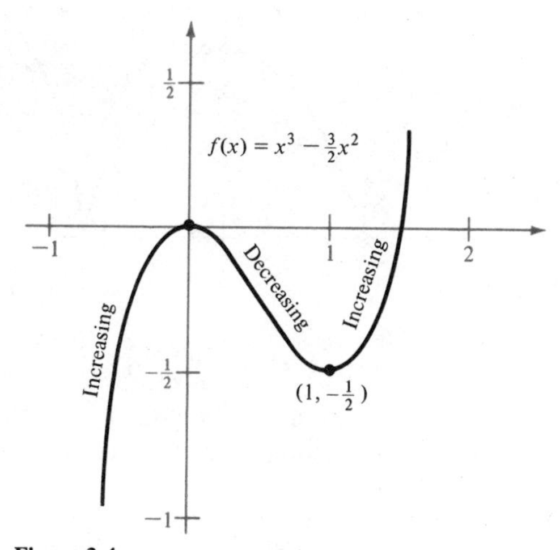

Figure 3.4

□

Example 2

Determining Intervals on Which f is Increasing or Decreasing

Find the intervals on which $f(x) = (x^2 - 4)^{2/3}$ is increasing or decreasing.

Solution Since

$$f'(x) = \frac{2}{3}(x^2 - 4)^{-1/3}(2x) = \frac{4x}{3(x^2 - 4)^{1/3}}$$

we see that $f'(x)$ is zero at $x = 0$ and, furthermore, f' is undefined at $x = \pm 2$. Thus the intervals to be tested are

$$(-\infty, -2), \quad (-2, 0), \quad (0, 2), \quad (2, \infty)$$

Table 3.2 summarizes the testing of these four intervals, and the graph of the function is shown in Figure 3.5.

Table 3.2

Interval	$(-\infty, -2)$	$(-2, 0)$	$(0, 2)$	$(2, \infty)$
Point in interval	$x = -3$	$x = -1$	$x = 1$	$x = 3$
Test	$f'(-3) = \dfrac{-12}{3\sqrt[3]{5}} < 0$	$f'(-1) = \dfrac{-4}{3\sqrt[3]{-3}} > 0$	$f'(1) = \dfrac{4}{3\sqrt[3]{-3}} < 0$	$f'(3) = \dfrac{12}{3\sqrt[3]{5}} > 0$
Conclusion	f is decreasing	f is increasing	f is decreasing	f is increasing

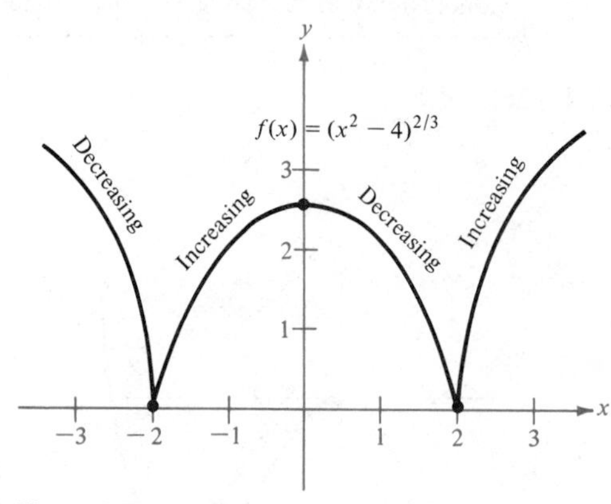

Figure 3.5

□

Since the zeros and undefined points of f' play an important role in studying the behavior of a function f, we give these values of x a special name.

Definition of a Critical Number

If a number c is in the domain of a function f, then c is called a **critical number** of f provided $f'(c) = 0$ or $f'(c)$ does not exist.

Example 3

Finding Critical Numbers

Find the critical numbers for the function

$$f(x) = \frac{x^4 + 1}{x^2}$$

Solution Since

$$f'(x) = \frac{x^2(4x^3) - (x^4 + 1)(2x)}{x^4} = \frac{4x^4 - 2x^4 - 2}{x^3}$$

$$= \frac{2(x^4 - 1)}{x^3} = \frac{2(x^2 + 1)(x - 1)(x + 1)}{x^3}$$

then $f'(x)$ is zero at $x = 1$ and $x = -1$, and f' is undefined at $x = 0$. However, $x = 0$ is not in the domain of f, and hence the only critical numbers of f are $x = 1$ and $x = -1$. (See Figure 3.6.)

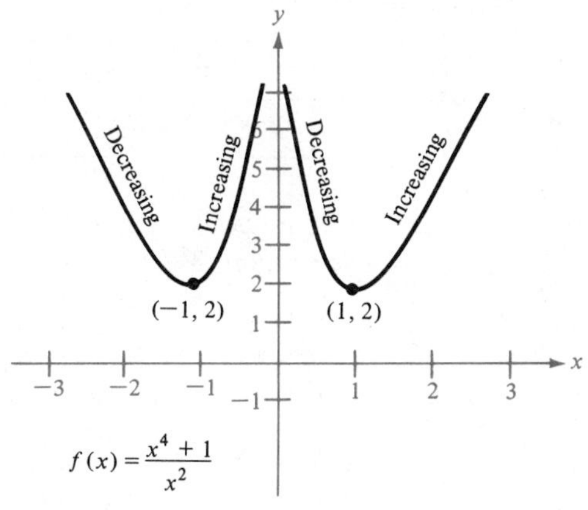

Figure 3.6

□

Remark

Note in Figure 3.6 that the graph of f changes from increasing to decreasing at $x = 0$ even though $x = 0$ is not a critical number of f. This occurs because f is discontinuous at $x = 0$, and you should remember to include such points when you test a function for intervals on which it is increasing or decreasing.

Remark

The converse of Theorem 3.1 is not valid. This means that it is possible for a function to be increasing on an interval even though the derivative is not *positive* at every point in the interval. The next example illustrates such a case.

Example 4

An Increasing Function Whose Derivative is Zero at a Point

Show that the function $f(x) = x^3 - 3x^2 + 3x$ is increasing on the entire real line.

Solution Since

$$f'(x) = 3x^2 - 6x + 3 = 0$$
$$3(x^2 - 2x + 1) = 0$$
$$3(x - 1)^2 = 0$$

f' is defined everywhere and has only one critical number, $x = 1$. Thus the intervals to be tested are $(-\infty, 1)$ and $(1, \infty)$. Table 3.3 summarizes the testing of these two intervals.

Table 3.3

Interval	$(-\infty, 1)$	$(1, \infty)$
Point in interval	$x = 0$	$x = 2$
Test	$f'(0) = 3 > 0$	$f'(2) = 3 > 0$
Conclusion	f is increasing	f is increasing

□

Finally, since f is continuous, it is increasing on the entire real line even though $f'(1) = 0$. (See Figure 3.7.)

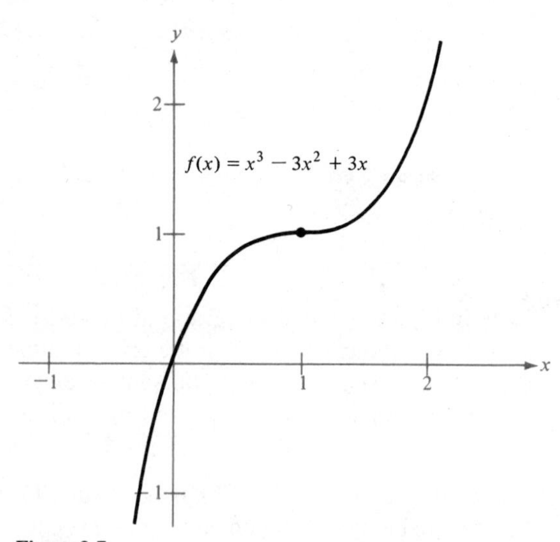

Figure 3.7

Example 5

A Business Application

A national toy distributor sells a certain game with the following cost and revenue functions:

$$C(x) = 2.4x - 0.0002x^2, \qquad 0 \le x \le 6000$$
$$R(x) = 7.2x - 0.001x^2, \qquad 0 \le x \le 6000$$

Determine the interval for which the profit

$$P = R - C$$

is increasing.

Solution Since

$$\begin{aligned} P(x) &= R(x) - C(x) \\ &= (7.2x - 0.001x^2) - (2.4x - 0.0002x^2) \\ &= 4.8x - 0.0008x^2 \end{aligned}$$

we find the value of x for which the marginal profit is zero. (Recall that the marginal profit P' is the derivative of the profit function.)

$$P'(x) = 4.8 - 0.0016x$$

$P'(x) = 0$ when

$$x = \frac{4.8}{0.0016} = 3000 \text{ units}$$

In the interval (0, 3000) P' is positive and the profit is *increasing*, and in the interval (3000, 6000) P' is negative and the profit is *decreasing*. (See Figure 3.8.)

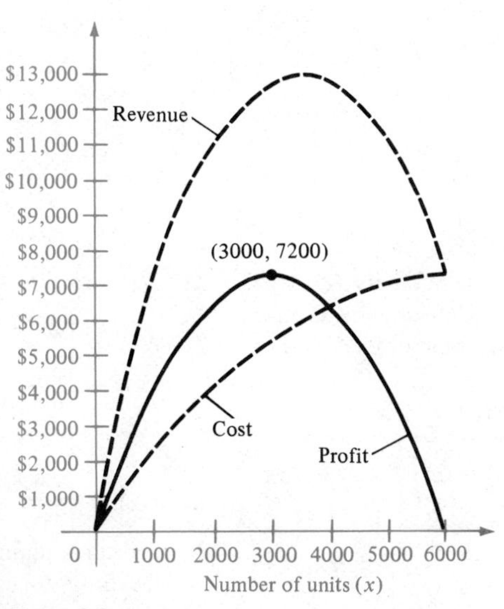

Figure 3.8 □

Note in Figure 3.8 that the profit is increasing when the marginal revenue (the slope of the revenue curve) exceeds the marginal cost (the slope of the cost curve).

Section Exercises (3.1)

In Exercises 1–6, identify the intervals on which the function is increasing or decreasing.

1. $f(x) = x^2 - 6x + 8$

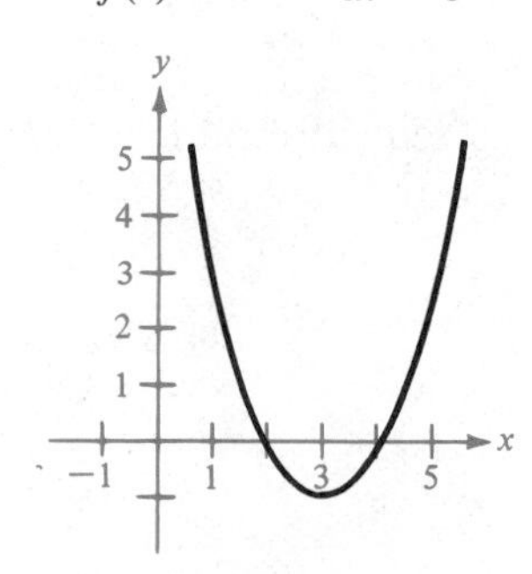

2. $y = -(x + 1)^2$

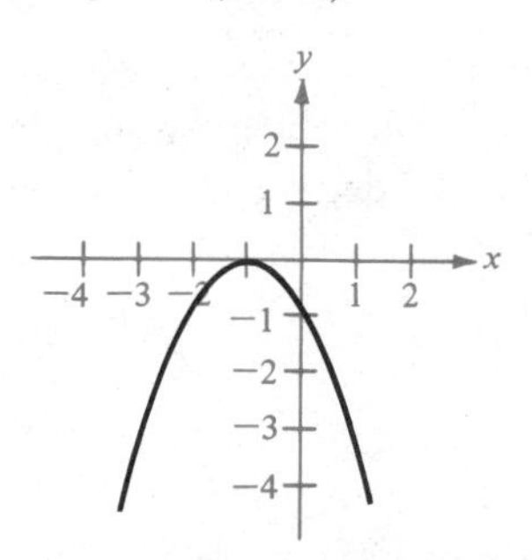

3. $y = \frac{x^3}{4} - 3x$

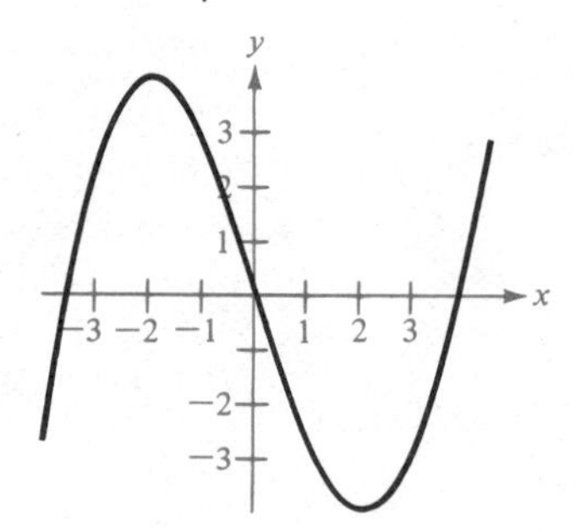

4. $f(x) = x^4 - 2x^2$

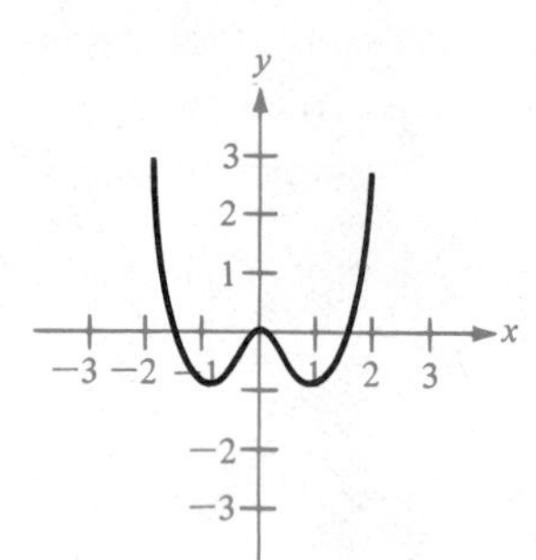

5. $f(x) = \frac{1}{x^2}$

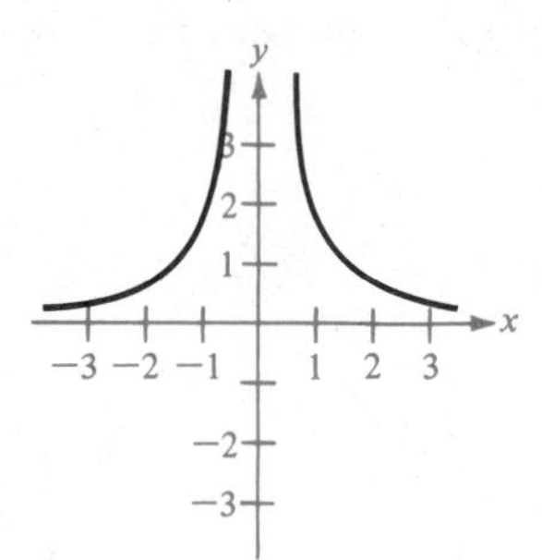

6. $y = \frac{x^2}{x + 1}$

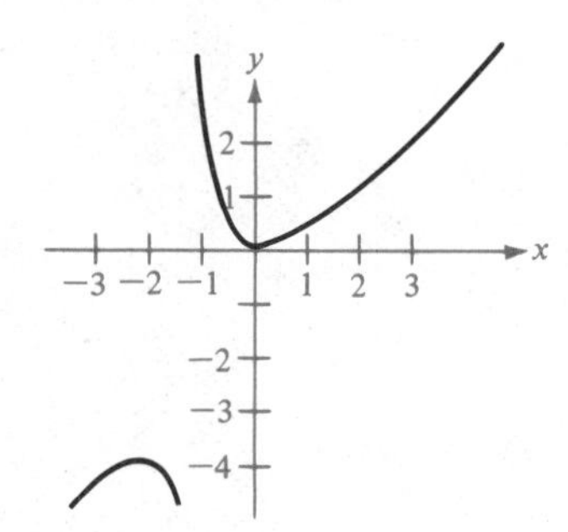

In Exercises 7–24, find the critical numbers (if any exist) and the intervals on which the function is increasing or decreasing. Sketch the graph of each function.

7. $f(x) = 2x - 3$ **8.** $f(x) = 5 - 3x$

9. $f(x) = x^2 - 2x$ **10.** $f(x) = -(x^2 - 2x)$

11. $f(x) = x^3 - 3x^2$ **12.** $f(x) = x^4 - 4x^3 + 15$

13. $y = \dfrac{x}{x^2 + 4}$ **14.** $y = \dfrac{1}{x^2 + 1}$

15. $y = \dfrac{x^2}{x^2 + 4}$ **16.** $f(x) = \sqrt{4 - x^2}$

17. $f(x) = x + \sqrt{x^2 - 1}$

18. $f(x) = x\sqrt{x + 1}$

19. $f(x) = x^{2/3}$ **20.** $f(x) = \sqrt[3]{x - 1}$

21. $f(x) = x\sqrt{6 - x}$

22. $f(x) = \begin{cases} 2x + 1, x \leq -1 \\ x^2 - 2, x > -1 \end{cases}$

23. $f(x) = \begin{cases} 4 - x^2, x \leq 0 \\ -2x + 2, x > 0 \end{cases}$

24. $f(x) = \begin{cases} -x^3 + 1, x \leq 0 \\ -x^2 + 2x + 1, x > 0 \end{cases}$

Ⓒ**25.** A drug is administered to a patient. A model giving the drug concentration C in the patient's bloodstream over a two-hour period is given by

$$C = 0.29483t + 0.04253t^2 - 0.00035t^3, \quad 0 \leq t \leq 120$$

Ⓒ Calculator may be helpful.

where C is measured in milligrams and t is the time in minutes. Find the intervals on which C is increasing or decreasing.

ⓒ**26.** A fast-food restaurant makes a profit P in selling x hamburgers, described by

$$P = 2.44x - \frac{x^2}{20{,}000} - 5000, \quad 0 \le x \le 35{,}000$$

Find the intervals on which P is increasing or decreasing.

ⓒ**27.** After birth, an infant will lose weight for a few days. A model for the average weight W of infants over the first two weeks following birth is

$$W = 0.033t^2 - 0.3974t + 7.3032, \quad 0 \le t \le 14$$

Find the intervals on which W is increasing or decreasing.

28. The ordering and transportation cost C of components used in a manufacturing firm is given by

$$C = 10\left(\frac{1}{x} + \frac{x}{x+3}\right), \quad 1 \le x$$

where C is measured in thousands of dollars and x is the order size in hundreds. Find the intervals on which C is increasing or decreasing.

ⓒ Calculator may be helpful.

Section 3.2

Extrema and the First-Derivative Test

Introductory Example

U.S. Birth Rate from 1900 to 1978

The birth rate in the United States is prepared by the National Center for Health Statistics and is based on vital records received from all states and the District of Columbia. The birth rate is listed in terms of the number of live births per thousand of population. In 1800, the U.S. birth rate was 55.0 per thousand. During the 1800s, this figure gradually dropped to 32.3 in 1900. The rate since 1900 is shown graphically in Figure 3.9.

Note that although the birth rate has decreased overall since 1900 it has done so through several rises and falls rather than by a steady decrease. When the rate changes from decreasing to increasing, we say it has a **relative minimum.** Similarly, when the rate changes from increasing to decreasing, we say it has a **relative maximum.** Moreover, the lowest birth rate between 1900 and 1978 occurred during 1975 when the rate fell to 14.8. We call this value the **absolute minimum** for the interval [1900, 1978].

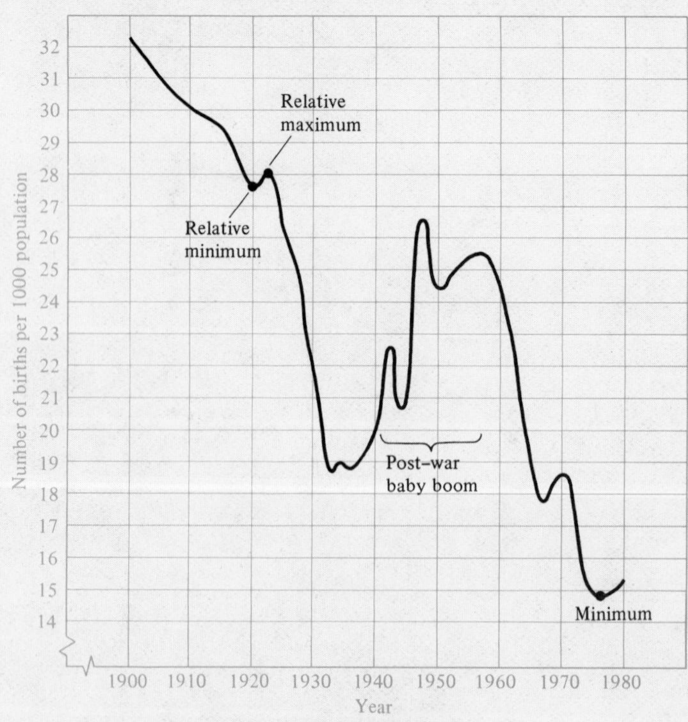

Figure 3.9 United States Birth Rate

Section Objectives: *To use the derivative of a function to locate its relative minimum and relative maximum values* ▪ *To find the maximum and minimum values of a function.*

In the preceding section, we used the derivative of a function to determine the intervals in which the function was increasing or decreasing. In this section, we examine the points at which a function changes from increasing to decreasing, or vice versa. It is often the case that at such points the function achieves a local maximum or a local minimum. We use the terms *relative maximum* and *relative minimum* to mean the local extrema of f.

Definition of Relative Extrema

Let f be a function defined at c.

1. $f(c)$ is called a **relative maximum** of f if there exists an interval (a, b) containing c such that $f(x) \leq f(c)$ for all x in (a, b).
2. $f(c)$ is called a **relative minimum** of f if there exists an interval (a, b) containing c such that $f(x) \geq f(c)$ for all x in (a, b).

(The plurals of maximum and minimum are *maxima* and *minima*.)

Just as we use the terms relative maximum and relative minimum to describe the local behavior of a function, we use the terms *maximum* and *minimum* to describe a function's global or overall behavior.

Definition of Extrema

Let c be a number in the interval $[a, b]$; then we have the following:

1. $f(c)$ is the **minimum** (value) of f on $[a, b]$ if $f(c) \leq f(x)$ for every x in $[a, b]$.
2. $f(c)$ is the **maximum** (value) of f on $[a, b]$ if $f(c) \geq f(x)$ for every x in $[a, b]$.

The minimum and maximum values of a function on an interval are sometimes referred to as the *absolute minimum* and *absolute maximum* of f on $[a, b]$.

In Figure 3.10, the maximum value of f on the interval $[a, b]$ occurs at $x = b$, whereas a relative maximum occurs at $x = c$. The minimum value of f, which occurs at $x = d$, also happens to be a relative minimum.

By observing the locations of the relative maxima and minima shown in Figure 3.10, we might conclude that they always occur at a critical number of the function. This is indeed the case, as the following theorem indicates.

Theorem 3.2 Locating Relative Extrema

If f has a relative minimum or relative maximum when $x = c$, then either (i) $f'(c) = 0$ or (ii) $f'(c)$ is undefined. That is, c is a critical number of f.

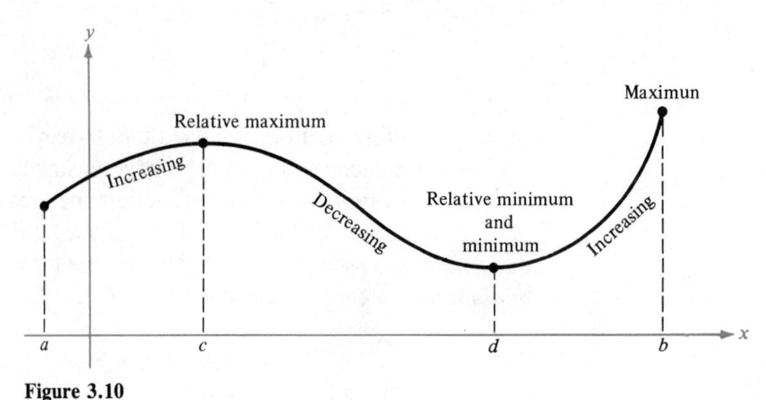

Figure 3.10

Remark

Theorem 3.2 states that, if a function has a relative minimum or a relative maximum when $x = c$, then c must be a critical number of f. Unfortunately, the converse is not true. Indeed, it is possible for a function to have a critical number that does not yield a relative minimum or a relative maximum. This can be seen in the next example.

Example 1

A Function with No Relative Extrema

Show that the function $f(x) = 2 - x^3$ has a critical number but no relative maxima or minima.

Solution We have

$$f(x) = 2 - x^3 \qquad \text{and} \qquad f'(x) = -3x^2$$

Now $f'(x)$ has only one critical number, $x = 0$. However, since

$$f'(x) < 0 \qquad \text{for} \qquad x < 0$$

and $\quad f'(x) < 0 \qquad \text{for} \qquad x > 0$

we conclude that f is *decreasing* for all x and hence has no relative extrema. (See Figure 3.11.)

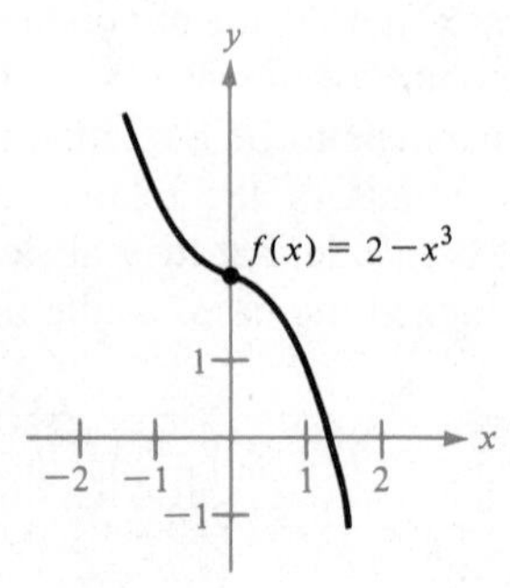

Figure 3.11

□

Remark

Even though Example 1 shows that a critical number may not yield a relative extrema, all relative extrema *must* occur at critical numbers. Thus, we must test each critical number to determine if it yields a relative maximum, a relative minimum, or possibly neither. The following theorem provides such a test.

Theorem 3.3 First-Derivative Test for Relative Extrema

If f is differentiable on (a, b) and c is the only critical number of f in the interval (a, b), then $f(c)$ can be classified as shown in Table 3.4.

Table 3.4

$f(c)$	Sign of f' in (a, c)	Sign of f' in (c, b)	Graphically
Relative maximum	+	−	(+) (−) c
Relative minimum	−	+	(−) (+) c
Neither	+	+	(+) (+) c
Neither	−	−	(−) (−) c

Example 2

Locating Relative Extrema

Locate all relative extrema for the function $f(x) = 2x^3 - 3x^2 - 36x + 14$.

Solution By setting the derivative of f equal to zero, we have

$$
\begin{aligned}
f'(x) = 6x^2 - 6x - 36 &= 0 \\
6(x^2 - x - 6) &= 0 \\
6(x - 3)(x + 2) &= 0
\end{aligned}
$$

Since f' is defined for all real numbers, the only critical numbers of f are -2 and 3. By the method used to test for increasing and decreasing functions, we test the sign of f' in each of the intervals $(-\infty, -2)$, $(-2, 3)$, and $(3, \infty)$. See Table 3.5.

Using Theorem 3.3, we conclude that the critical number -2 yields a relative maximum (f' changes sign from $+$ to $-$) and 3 yields a relative minimum (f' changes sign from $-$ to $+$). (See Figure 3.12.)

Table 3.5

Interval	$(-\infty, -2)$	$(-2, 3)$	$(3, \infty)$
Point in interval	-3	0	4
Sign of $f'(x)$	$+$	$-$	$+$

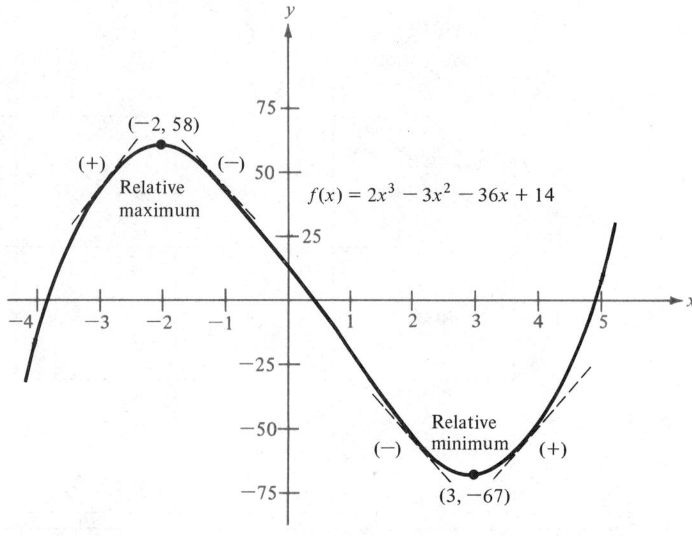

Figure 3.12

□

Example 3

Locating Relative Extrema

Locate all relative extrema for the function $f(x) = x^4 - x^3$.

Solution

$$f'(x) = 4x^3 - 3x^2 = 0$$

$$x^2(4x - 3) = 0$$

Since f' is defined everywhere, the only critical numbers of f are 0 and $\frac{3}{4}$. See Table 3.6.

Using Theorem 3.3, we conclude that the critical number $\frac{3}{4}$ yields a relative minimum, whereas 0 yields neither a relative minimum nor a relative maximum (see Figure 3.13).

Table 3.6

Interval	$(-\infty, 0)$	$(0, \frac{3}{4})$	$(\frac{3}{4}, \infty)$
Point in interval	-1	$\frac{1}{2}$	1
Sign of $f'(x)$	$-$	$-$	$+$

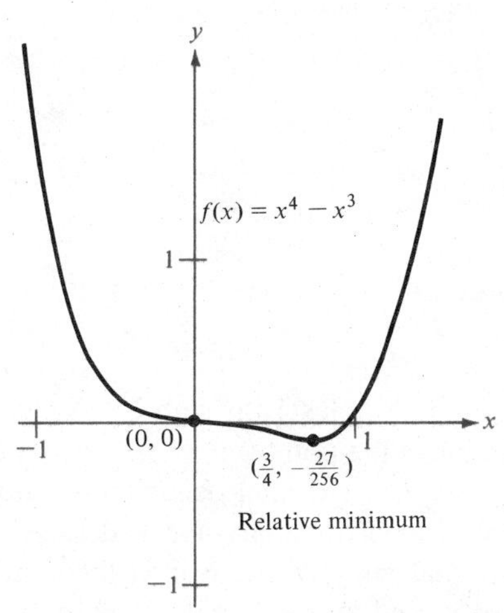

Figure 3.13

□

Example 4

Locating Relative Extrema

Locate all relative extrema for the function $f(x) = 2x - 3x^{2/3}$.

Solution Differentiating we have

$$f'(x) = 2 - \frac{2}{x^{1/3}}$$

Since $f'(x) = 0$ if $x = 1$ and f' is undefined when $x = 0$, the critical numbers of f are 0 and 1. See Table 3.7. Therefore, 0 yields a relative maximum, and 1 yields a relative minimum, as shown in Figure 3.14.

Table 3.7

Interval	$(-\infty, 0)$	$(0, 1)$	$(1, \infty)$
Point in interval	-1	$\frac{1}{2}$	2
Sign of $f'(x)$	$+$	$-$	$+$

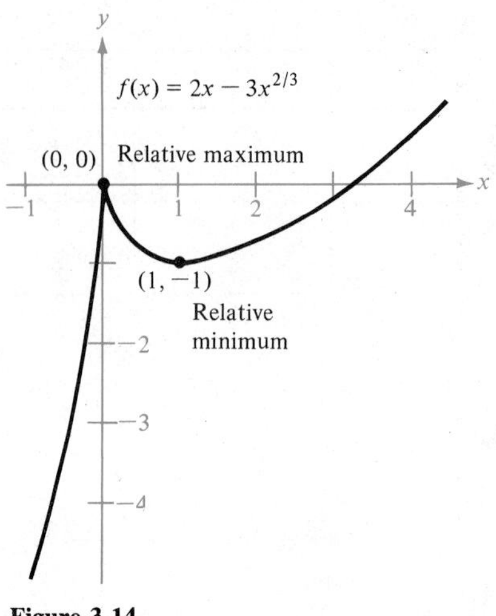

Figure 3.14

□

By reviewing Examples 1 through 4, we can see that the only instance of an *absolute* extrema is the point $(-\frac{3}{4}, -\frac{27}{256})$ encountered in Example 3. In practical applications, absolute extrema occur more often than this. The reason is that practical applications often involve restricted domains that consist of only a portion of the real numbers. Our next theorem points out that if a continuous function has a closed interval as its domain then it *must* have both a minimum and a maximum on the interval.

Theorem 3.4
Extreme Value Theorem

If f is continuous on $[a, b]$, then f takes on both a minimum value and a maximum value on $[a, b]$.

For a continuous function on a closed interval, the existence of the maximum and minimum are illustrated in Figure 3.15. Notice in Figure 3.16 that on the *open* interval $(-1, 2)$, $f(x) = x^2 + 1$ has a minimum value of 1 but does not take on a maximum value. This is the case because, as x approaches 2 (but never takes on the value 2) from the left, $f(x)$ increases toward (but never takes on) the value 5.

If we drop the continuity requirement from the hypothesis of Theorem 3.4, we may destroy the existence of a maximum or a minimum. On the interval $[-1, 2]$ consider the function

$$g(x) = \begin{cases} x^2 + 1, & x \neq 0 \\ 2, & x = 0 \end{cases}$$

This function is discontinuous at $x = 0$ and has a maximum but no minimum on the interval $[-1, 2]$, as shown in Figure 3.17.

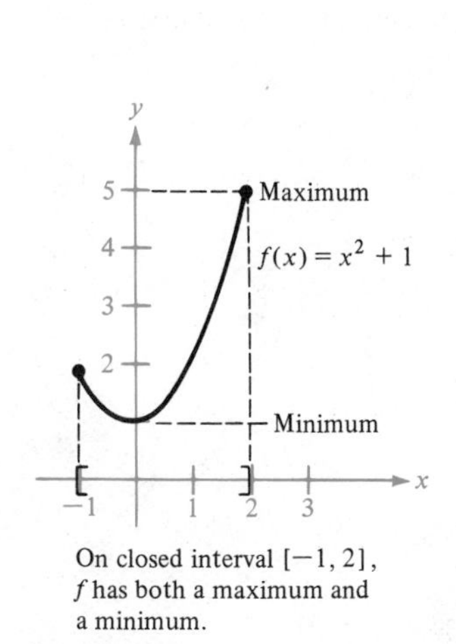

On closed interval $[-1, 2]$, f has both a maximum and a minimum.

Figure 3.15

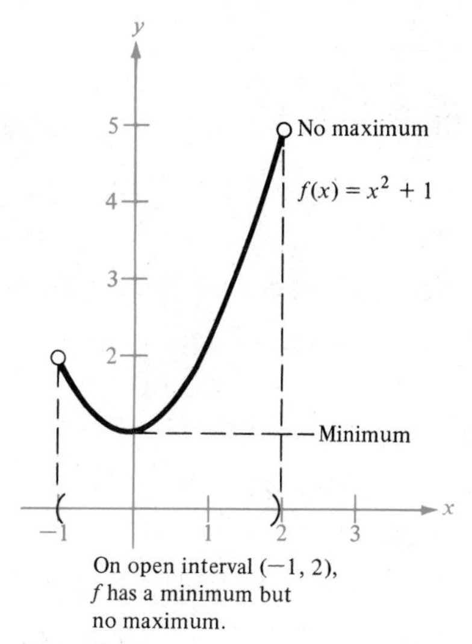

On open interval $(-1, 2)$, f has a minimum but no maximum.

Figure 3.16

This discussion suggests that an *extremum* (maximum or minimum) of a function on an interval $[a, b]$ can occur:

1. at an *interior point* of $[a, b]$, and thus it is also a relative extremum, or
2. at an *endpoint* of $[a, b]$, in which case we call it an **endpoint extremum.**

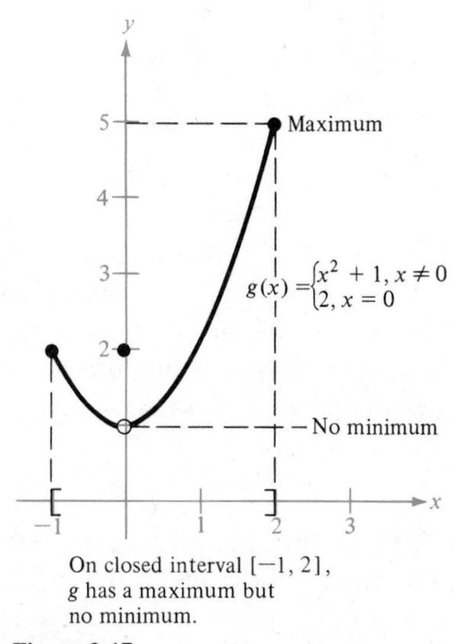

On closed interval $[-1, 2]$, g has a maximum but no minimum.

Figure 3.17

Example 5

Finding the Maximum Profit

In Example 5 of Section 2.3, we considered a fast-food restaurant whose profit function for hamburgers is

$$P(x) = 2.44x - \frac{x^2}{20,000} - 5000$$

Find the production level that produces a maximum profit.

Solution By setting the marginal profit equal to zero, we have

$$P'(x) = 2.44 - \frac{x}{10,000} = 0$$

$$-\frac{x}{10,000} = -2.44$$

$$x = 24,400 \text{ units}$$

Figure 3.18 verifies the fact that this production level corresponds to the maximum profit.

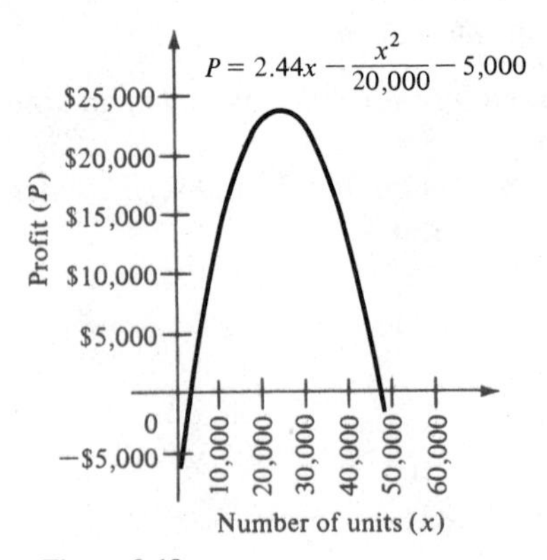

Figure 3.18

□

Remark

In Example 5, we concluded that the maximum profit occurred at $x = 24,400$ without actually verifying this maximum by the First-Derivative Test. It is often the case that the nature of an application clearly indicates whether a critical number yields a maximum or minimum without a formal test. This is further demonstrated in the next example.

Example 6

Finding the Minimum and Maximum Oxygen Levels

Suppose that $f(t)$ measures the level of oxygen in a pond, where $f(t) = 1$ is the normal level and the time t is measured in weeks. When $t = 0$, some organic waste is dumped into the pond, and as the waste material oxidizes, the amount of oxygen in the pond is given by

$$f(t) = \frac{t^2 - t + 1}{t^2 + 1}, \qquad 0 \le t < \infty$$

(a) When is the oxygen level lowest?
(b) When is the oxygen level highest?

Solution
(a) Since

$$f'(t) = \frac{(t^2 + 1)(2t - 1) - (t^2 - t + 1)(2t)}{(t^2 + 1)^2}$$
$$= \frac{2t^3 - t^2 + 2t - 1 - 2t^3 + 2t^2 - 2t}{(t^2 + 1)^2}$$
$$= \frac{t^2 - 1}{(t^2 + 1)^2}$$

is zero when $t^2 - 1 = 0$, and $t = -1$ is not in the domain, the only critical number is $t = 1$. The nature of the problem indicates that this value ($t = 1$ week) yields the minimum oxygen level. (See Figure 3.19.)

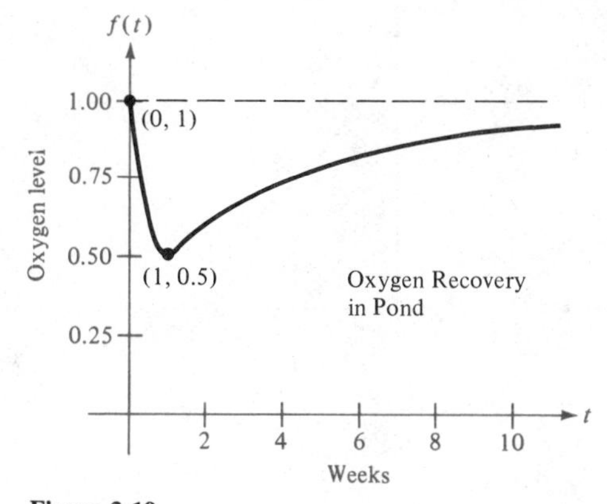

Figure 3.19

(b) Since $y = 1$ is an upper bound for $f(t)$, it is clear that the maximum level of oxygen occurs when $t = 0$ and is an example of an *endpoint extremum*. □

Section Exercises (3.2)

In Exercises 1–20, find all relative extrema.

1. $f(x) = -2x^2 + 4x + 3$
2. $f(x) = x^2 + 8x + 10$
3. $f(x) = x^2 - 6x$
4. $f(x) = (x - 1)^2(x + 2)$
5. $f(x) = 2x^3 + 3x^2 - 12x$
6. $f(x) = (x - 3)^3$

7. $f(x) = x^3 - 6x^2 + 15$
8. $f(x) = x^4 - 1$
9. $f(x) = x^4 - 2x^3$
10. $f(x) = (x - 1)^{2/3}$
11. $f(x) = x^{1/3} + 1$
12. $f(x) = x^{2/3}(x - 5)$
13. $f(x) = x + \dfrac{1}{x}$
14. $f(x) = \dfrac{x}{x + 1}$
15. $f(x) = \dfrac{x^2}{x^2 - 9}$
16. $f(x) = \dfrac{x + 3}{x^2}$
17. $f(x) = x^4 - 32x + 4$
18. $f(x) = \dfrac{x^5 - 5x}{5}$
19. $f(x) = \dfrac{x^2 - 2x + 1}{x + 1}$
20. $f(x) = \dfrac{x^2 - 3x - 4}{x - 2}$

In Exercises 21–30, locate the absolute extrema of the function on the indicated interval.

21. $f(x) = 2(3 - x)$; $[-1, 2]$
22. $f(x) = \dfrac{2x + 5}{3}$; $[0, 5]$
23. $f(x) = -x^2 + 4x$; $[0, 3]$
24. $f(x) = x^2 + 2x - 4$; $[-1, 1]$
25. $f(x) = x^3 - 3x^2$; $[-1, 3]$
26. $f(x) = x^3 - 12x$; $[0, 4]$
27. $f(x) = 3x^{2/3} - 2x$; $[-1, 1]$
28. $g(t) = \dfrac{t^2}{t^2 + 3}$; $[-1, 1]$
29. $h(s) = \dfrac{1}{s - 2}$; $[0, 1]$
30. $h(t) = \dfrac{t}{t - 2}$; $[3, 5]$

In Exercises 31–38, determine from the graph of f if f possesses a relative minimum in the interval (a, b).

31.
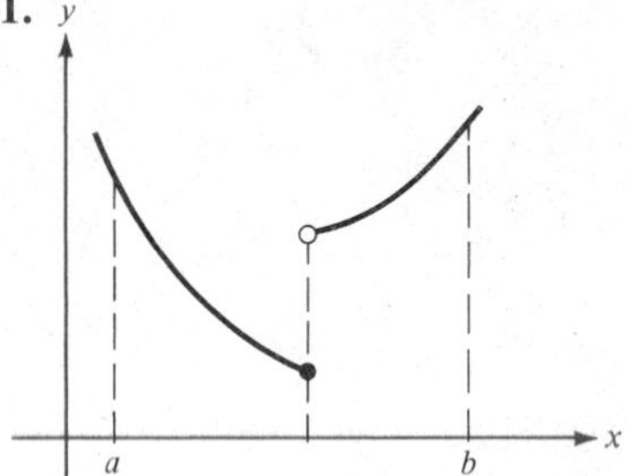

32.
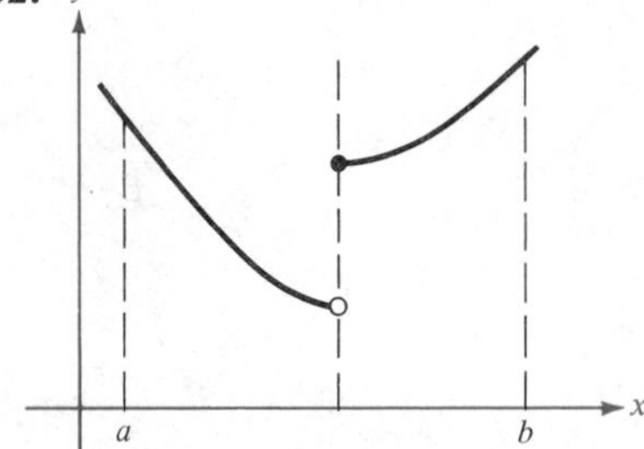

33.
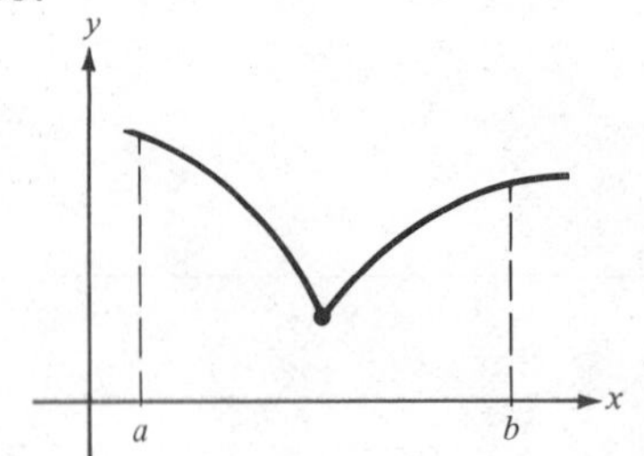

34.
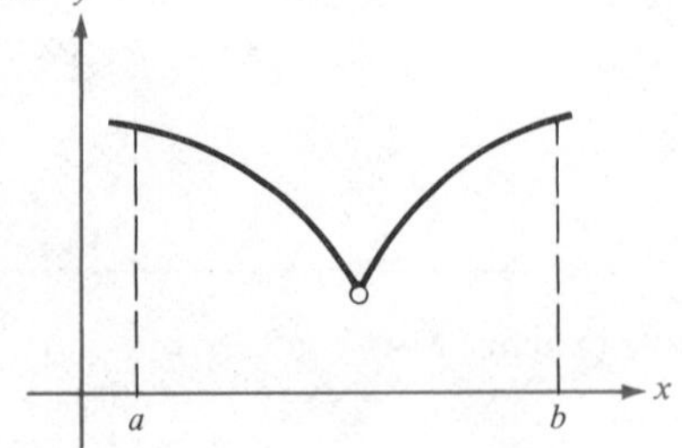

35.

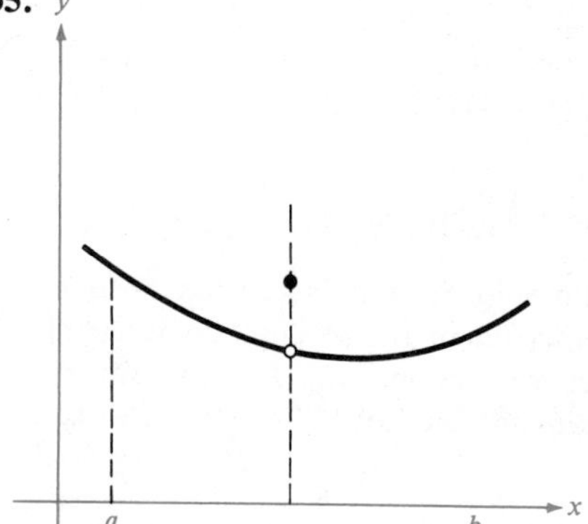

36.

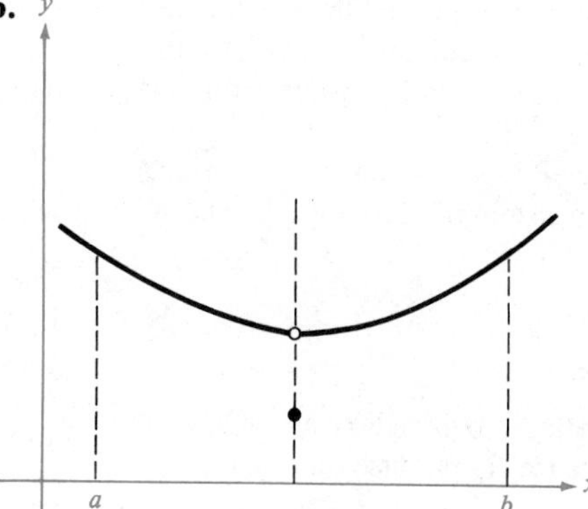

37.

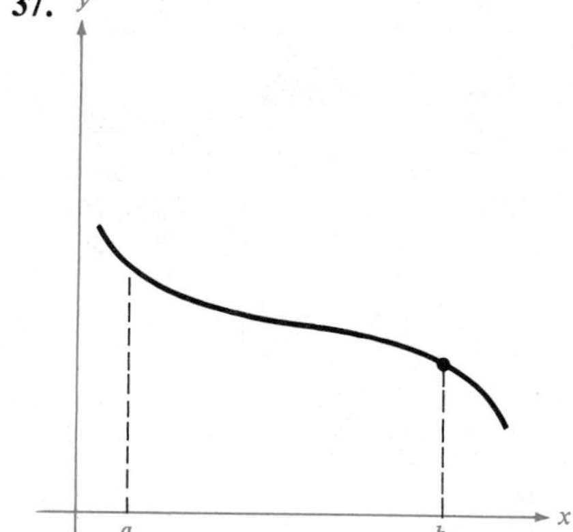

38.

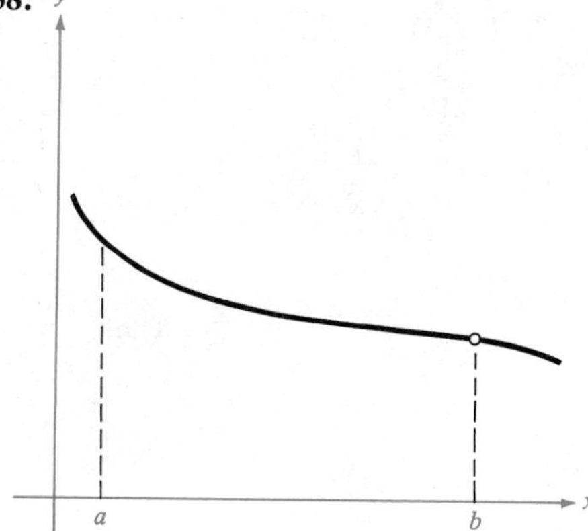

39. The height of a ball at time t is given by the equation $h(t) = 96t - 16t^2$.
(a) What was the initial velocity of the ball?
(b) How high did the ball go?
(c) What direction was the ball moving at time $t = 4$?
(d) Find the height of the ball at time $t = 1$.

40. Repeat Exercise 39 using $h(t) = 64t - 16t^2$ as the position function.

41. Coughing forces the trachea (windpipe) of a person to contract, which in turn affects the velocity v of the air through the trachea. Suppose the velocity of the air during coughing is

$$v = k(R - r)r^2$$

where k is a constant, R is the normal radius of the trachea, and r is the radius during coughing. What radius r will produce the maximum air velocity?

42. The concentration C of a certain chemical in the bloodstream t hours after injection into muscle tissue is given by

$$C = \frac{3t}{27 + t^3}$$

When is the concentration greatest?

43. The resistance R of a certain type of resistor is given by

$$R = \sqrt{0.001T^4 - 4T + 100}$$

where R is measured in ohms and the temperature T is measured in degrees Celsius. What temperature produces a minimum resistance for this type of resistor?

44. The electric power P in watts in a (direct current) circuit with two resistors of resistance R_1 and R_2 connected in series is

$$P = \frac{vR_1R_2}{(R_1 + R_2)^2}$$

where v is the voltage. If v and R_1 are held constant, what resistance R_2 produces the maximum power?

The error estimate for the Trapezoidal Rule (see Section 6.5) involves the maximum of the absolute value of the second derivative in an interval. In Exercises 45–48, find the maximum value of $|f''(x)|$ in the indicated interval.

45. $f(x) = \dfrac{1}{x^2 + 1}$; $[0, 3]$

46. $f(x) = \dfrac{1}{x^2 + 1}$; $[\frac{1}{2}, 3]$

Ⓒ**47.** $f(x) = \sqrt{1 + x^3}$; $[0, 2]$

48. $f(x) = x^3(3x^2 - 10)$; $[0, 1]$

The error estimate for Simpson's Rule (see Section 6.5) involves the maximum of the absolute value of the fourth derivative in an interval. In Exercises 49–52, find the maximum value of $|f^{(4)}(x)|$ in the indicated interval.

49. $f(x) = 15x^4 - \left(\dfrac{2x - 1}{2}\right)^6$; $[0, 1]$

50. $f(x) = x^5 - 5x^4 + 20x^3 + 600$; $[0, \frac{3}{2}]$

51. $f(x) = (x + 1)^{2/3}$; $[0, 2]$

52. $f(x) = \dfrac{1}{x^2}$; $[1, 2]$

Section 3.3

Concavity and the Second-Derivative Test

Introductory Example

Annual Divorce Rate from 1920 to 1978

The divorce rate in the United States is based on the annual number of divorces per 1000 marriages. Figure 3.20 graphically depicts the changes in this rate between 1920 and 1978. Note that the divorce rate increased during the entire period from the beginning of the Depression until the end of World War II (1932–1946). There are no relative minimum or relative maximum points between these two dates, and yet there are definite turning points in the graph. Note in Figure 3.20 that there are some intervals in which the slope of the graph is increasing and other intervals in which the slope of the graph is decreasing. During a period in which the slope is increasing, the graph is said to be **concave upward** (see 1937–1946), and during a period in which the slope is decreasing, the graph is said to be **concave downward** (see 1933–1937). The turning point separating these two types of concavity is called a **point of inflection.**

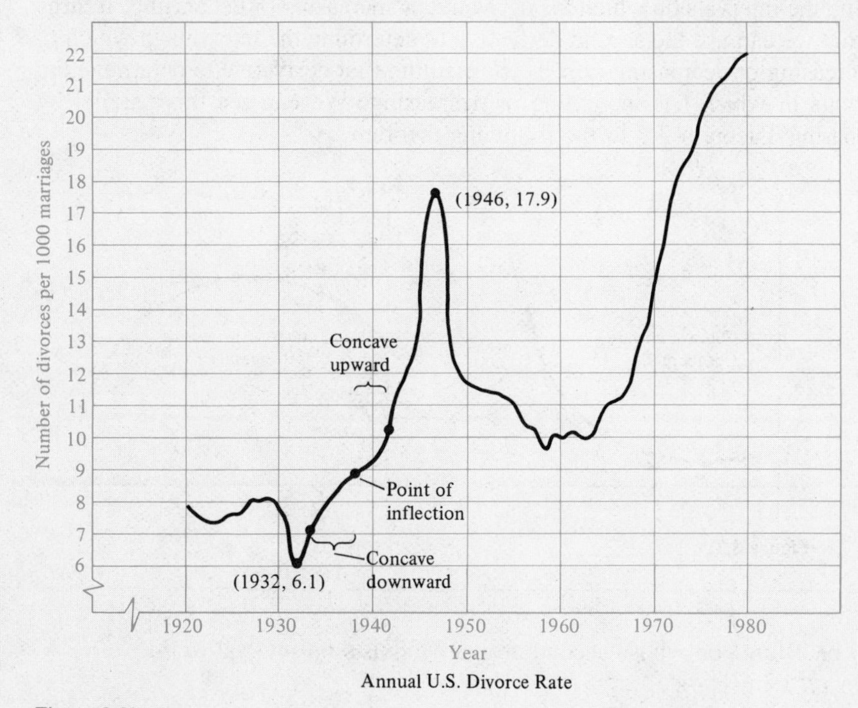

Figure 3.20

Section Objectives: ***To use the first and second derivatives of a function to determine concavity and points of inflection ▪ To introduce the Second-Derivative Test for relative extrema.***

We have already seen that locating the intervals in which a function f increases or decreases is helpful in determining its graph. In this section, we extend this idea and show that, by locating the intervals in which f' increases or decreases, we can determine where the graph of f is curving upward or curving downward. We define this notion of curving upward or downward as **concavity.**

Definition of Concavity

Let f be differentiable on (a, b). We say that the graph of f is:

1. **concave upward** on (a, b) if f' is increasing on (a, b), or
2. **concave downward** on (a, b) if f' is decreasing on (a, b).

Note in Figure 3.21 that:

1. If a curve lies *above* its tangent lines, then it is concave upward.
2. If a curve lies *below* its tangent lines, then it is concave downward.

This visual test for concavity is useful when the graph of a function is given. To determine concavity without seeing a graph, we need an analytic test for finding the intervals on which the derivative is increasing or decreasing. It turns out that we can use the second derivative to determine the intervals in which f' is increasing or decreasing, just as we used the first derivative to determine the intervals in which f is increasing or decreasing. We can see this parallel by comparing Theorem 3.3 to the following theorem.

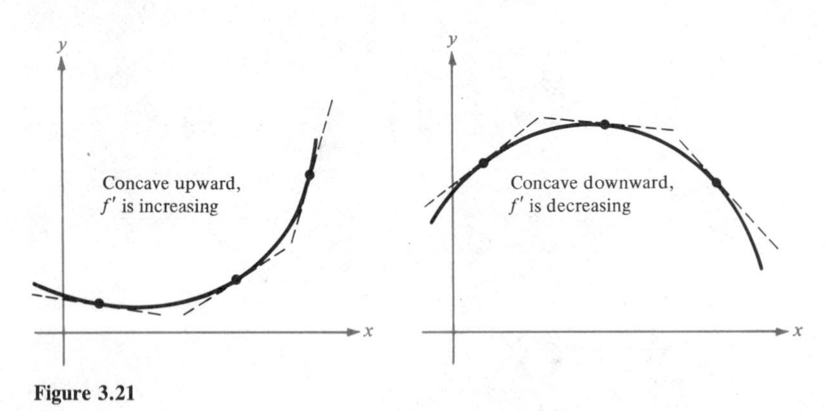

Figure 3.21

Theorem 3.5
Test for Concavity

Let f be a function whose second derivative exists on interval (a, b).
For *all* x in (a, b):

i. If $f''(x) > 0$, then the graph of f is concave upward on (a, b).
ii. If $f''(x) < 0$, then the graph of f is concave downward on (a, b).

Example 1

Determining Concavity

Determine the intervals in which the graph of $f(x) = 6/(x^2 + 3)$ is concave upward or downward.

Solution Since $f(x) = 6(x^2 + 3)^{-1}$, then

$$f'(x) = (-6)(2x)(x^2 + 3)^{-2} = \frac{-12x}{(x^2 + 3)^2}$$

and
$$f''(x) = \frac{(x^2 + 3)^2(-12) - (-12x)(2)(2x)(x^2 + 3)}{(x^2 + 3)^4}$$

$$= \frac{-12(x^2 + 3) + (48x^2)}{(x^2 + 3)^3} = \frac{36(x^2 - 1)}{(x^2 + 3)^3}$$

Since $f''(x) = 0$ when $x = \pm 1$, we test f'' in the intervals $(-\infty, -1)$, $(-1, 1)$, and $(1, \infty)$. The results are shown in Table 3.8 and Figure 3.22.

Table 3.8

Interval	$(-\infty, -1)$	$(-1, 1)$	$(1, \infty)$
Sign of f''	$+$	$-$	$+$
f'	Increasing	Decreasing	Increasing
Graph of f	Concave upward	Concave downward	Concave upward

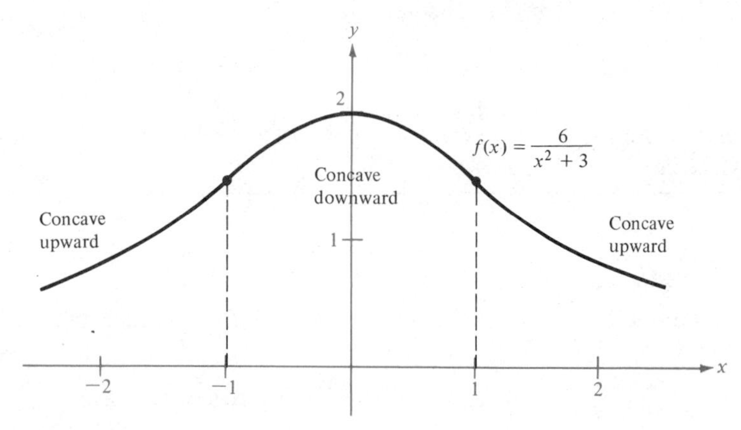

Figure 3.22 □

Remark

In Example 1, f' is increasing on the interval $(1, \infty)$ even though f is decreasing there. Be sure you see that the increasing or decreasing of f' does not necessarily correspond to the respective increasing or decreasing of f.

The graph in Figure 3.22 has two points at which the concavity changes. If at such a point the tangent line to the graph exists, we call the point a **point of inflection.**

Definition of a Point of Inflection

If the graph of a continuous function possesses a tangent line at a point where its concavity changes from upward to downward (or vice versa), we call the point a **point of inflection.** (See Figure 3.23.)

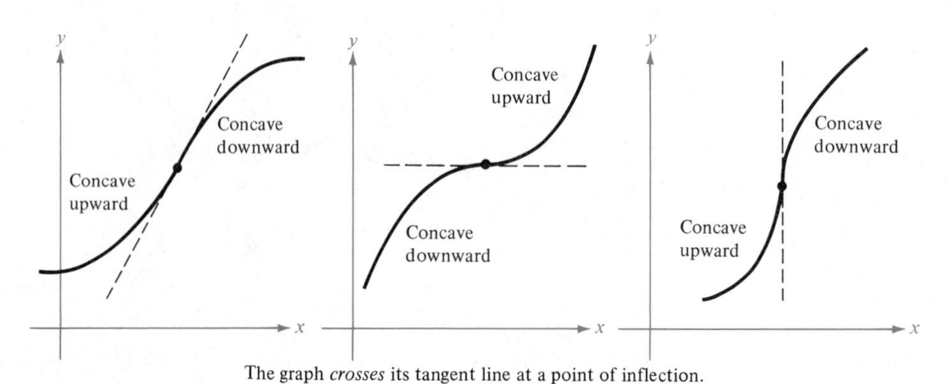

The graph *crosses* its tangent line at a point of inflection.

Figure 3.23

Since a point of inflection occurs where the concavity of a graph changes, it must be true (see Theorem 3.5) that the sign of f'' changes at such points. Thus, to locate possible points of inflection, we need only determine the values of x for which $f''(x) = 0$ or for which $f''(x)$ does not exist. This parallels the procedure for locating relative extrema of f by determining the critical numbers of f (see Theorem 3.2).

Property of Points of Inflection

If $(c, f(c))$ is a point of inflection of f, then either $f''(c) = 0$ or $f''(c)$ does not exist.

Example 2

Finding Points of Inflection

Determine the points of inflection and discuss the concavity of the graph of $f(x) = x^4 + x^3 - 3x^2 + 1$.

Solution Differentiating twice, we have

$$f'(x) = 4x^3 + 3x^2 - 6x$$

$$f''(x) = 12x^2 + 6x - 6 = 6(2x^2 + x - 1) = 6(2x - 1)(x + 1)$$

Possible points of inflection occur at $x = -1$ and $x = \frac{1}{2}$. By testing the intervals $(-\infty, -1)$, $(-1, \frac{1}{2})$, and $(\frac{1}{2}, \infty)$, we determine that f is concave upward in $(-\infty, -1)$, concave downward in $(-1, \frac{1}{2})$, and concave upward in $(\frac{1}{2}, \infty)$. Thus, the numbers -1 and $\frac{1}{2}$ both yield points of inflection. (See Figure 3.24.)

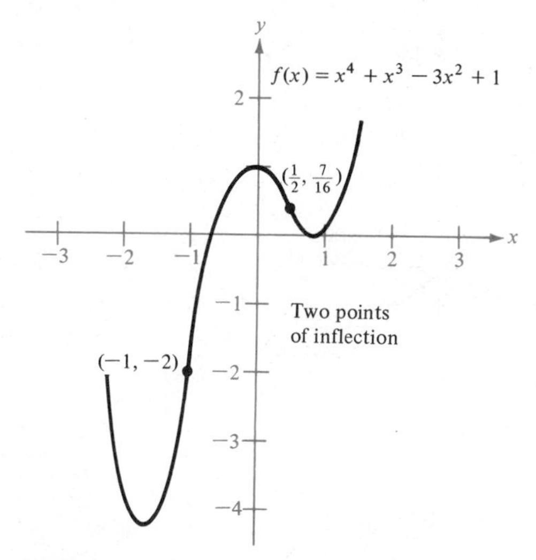

Figure 3.24

□

We should point out that it is possible for the second derivative to be zero at a point which is *not* a point of inflection. For example, compare the graphs of $f(x) = x^3$ and $g(x) = x^4$ (see Figure 3.25). Both second derivatives are zero when $x = 0$, but only the graph of f has a point of inflection at $x = 0$. This shows that, before concluding that a point of inflection exists at a value of x for which $f''(x) = 0$, we should test to be certain that the concavity actually changes there.

If the second derivative exists, we can often use it as a simple test for relative minima and relative maxima. The test is based on the idea that if $f(c)$ is a relative maximum of a differentiable function f, then its graph is concave downward in some interval containing c. Similarly, if $f(c)$ is a relative minimum, its graph is concave upward in some interval containing c (see Figure 3.26). This test is referred to as the Second-Derivative Test for Relative Extrema.

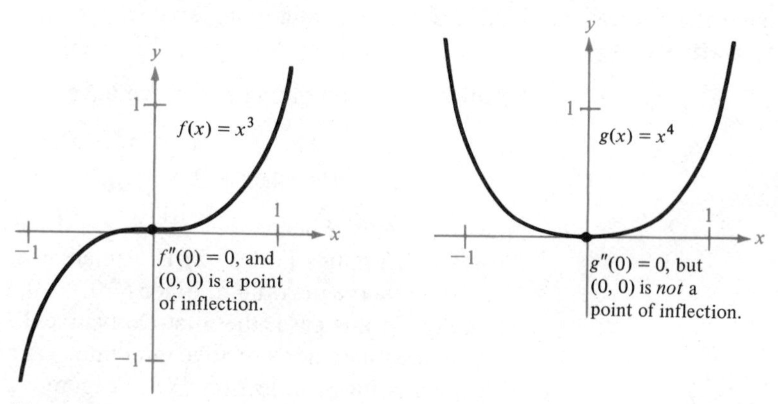

Figure 3.25

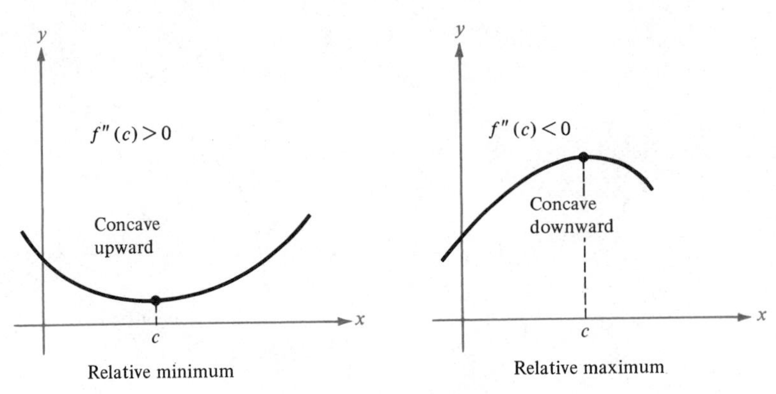

Figure 3.26

Theorem 3.6 Second-Derivative Test for Relative Extrema

Let f'' exist on some open interval containing c and $f'(c) = 0$.

i. If $f''(c) > 0$, then $f(c)$ is a relative minimum.
ii. If $f''(c) < 0$, then $f(c)$ is a relative maximum.

Remark If $f''(c) = 0$, the Second-Derivative Test does not apply. In such cases, we can use the First-Derivative Test.

Example 3

Using the Second-Derivative Test

Find the relative minimum and relative maximum points for the function $f(x) = -3x^5 + 5x^3$.

Solution Differentiating twice, we have

$$f'(x) = -15x^4 + 15x^2 = 15(-x^4 + x^2) = 15x^2(1 - x^2)$$
$$f''(x) = 15(-4x^3 + 2x)$$

Now $f'(x) = 0$ when x is -1, 0, or 1. By the Second-Derivative Test, $f''(-1) > 0$ implies $(-1, -2)$ is a relative minimum, and $f''(1) < 0$ implies $(1, 2)$ is a relative maximum. Since $f''(0) = 0$, the Second-Derivative Test does not apply. In this case, the First-Derivative Test shows that $(0, 0)$ is neither a relative maximum nor a relative minimum. [A test of concavity would show that $(0, 0)$ is a point of inflection.] (See Figure 3.27.)

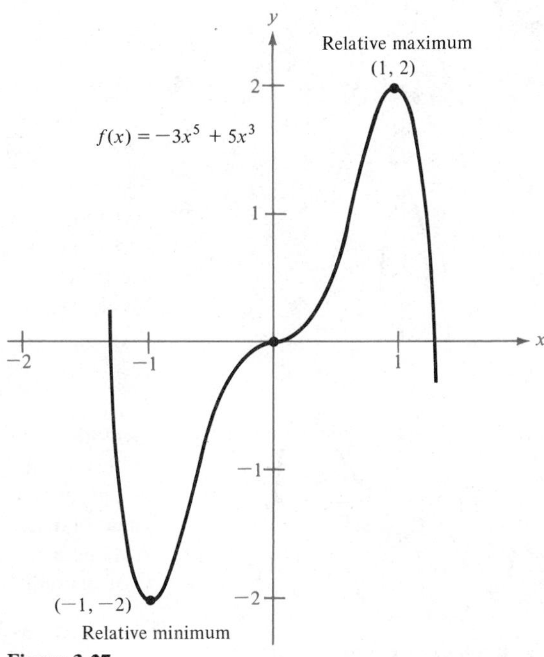

Figure 3.27

□

Section Exercises (3.3)

In Exercises 1–6, find the intervals on which the given function is concave upward and those on which it is concave downward.

1. $y = x^2 - x - 2$

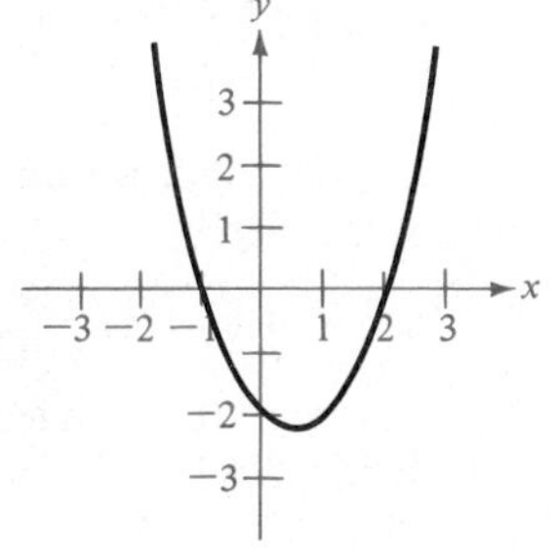

2. $y = -x^3 + 3x^2 - 2$

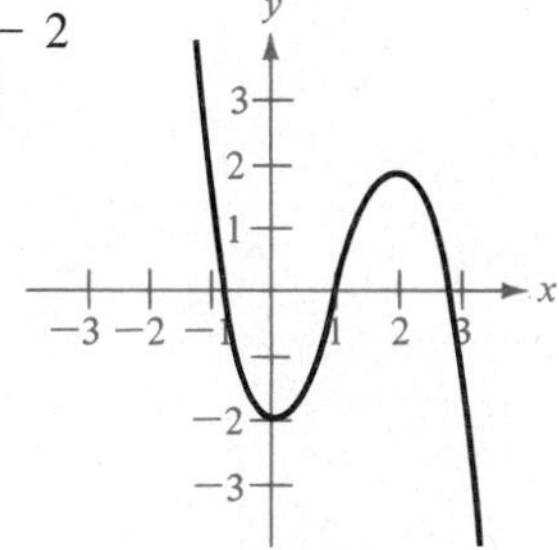

3. $f(x) = \dfrac{24}{x^2 + 12}$

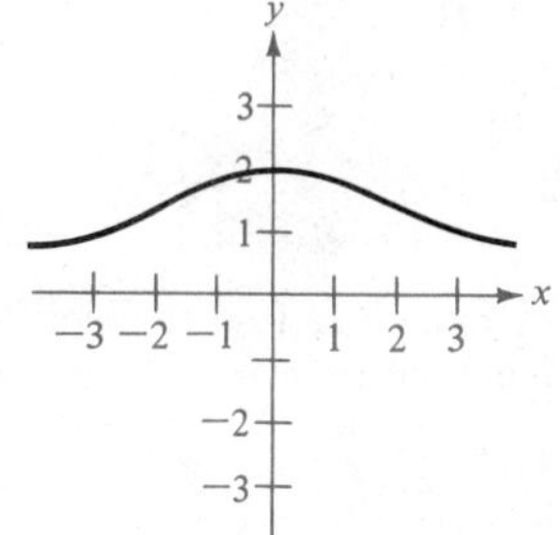

4. $f(x) = \dfrac{x^2 - 1}{2x + 1}$

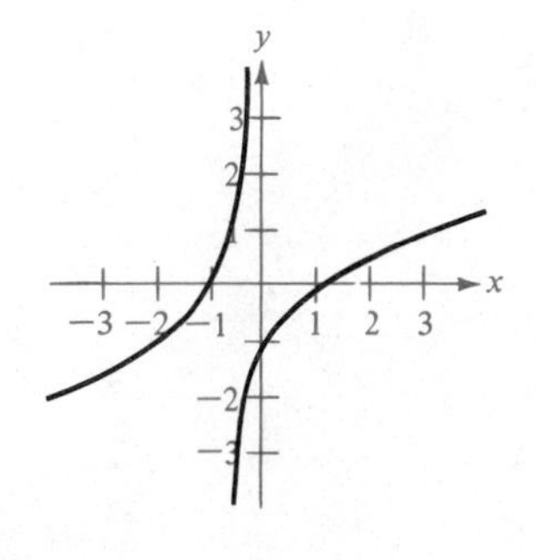

5. $y = \dfrac{x^2 + 1}{x^2 - 1}$

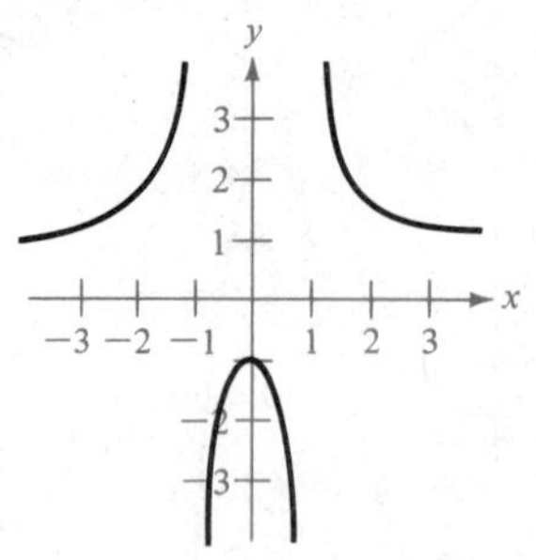

6. $f(x) = \dfrac{1}{270}(-3x^5 + 40x^3 + 135x)$

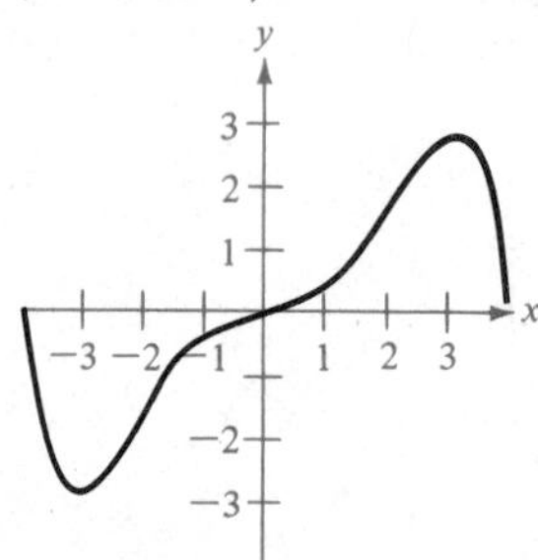

In Exercises 7–18, identify all relative extrema. Use the Second-Derivative Test when applicable.

7. $f(x) = 6x - x^2$
8. $f(x) = x^2 + 3x - 8$
9. $f(x) = (x - 5)^2$
10. $f(x) = -(x - 5)^2$
11. $f(x) = x^3 - 3x^2 + 3$
12. $f(x) = 5 + 3x^2 - x^3$
13. $f(x) = x^4 - 4x^3 + 2$
14. $f(x) = x^3 - 9x^2 + 27x - 26$
15. $f(x) = x^{2/3} - 3$
16. $f(x) = \sqrt{x^2 + 1}$
17. $f(x) = x + \dfrac{4}{x}$
18. $f(x) = \dfrac{x}{x - 1}$

In Exercises 19–31, identify all relative extrema and points of inflection. Sketch the graphs.

19. $f(x) = x^3 - 6x^2 + 12x - 8$
20. $f(x) = x^3 + 1$
21. $f(x) = x^3 - 12x$
22. $f(x) = 2x^3 - 3x^2 - 12x + 8$
23. $f(x) = \dfrac{x^4}{4} - 2x^2$
24. $f(x) = 2x^4 - 8x + 3$
25. $f(x) = \dfrac{4}{1 + x^2}$
26. $f(x) = \dfrac{x^2}{1 + x^4}$
27. $f(x) = x\sqrt{x + 3}$
28. $f(x) = x^3(x + 1)$
29. $f(x) = (x - 3)(x + 2)^3$
30. $f(x) = (x - 1)^3(x - 3)^2$
31. $f(x) = x^{1/3}(x + 3)^{2/3}$
32. The equation

$$E = \frac{T}{(x^2 + a^2)^{3/2}}$$

gives the electric field intensity on the axis of a uniformly charged ring, where T is the total charge on the ring and a is the radius of the ring. At what value of x is E maximum?

33. A manufacturer has determined that the total cost C of operating a certain facility is given by

$$C = 0.5x^2 + 15x + 5000$$

where x is the number of units produced. At what level of production will the average cost per unit be minimum? (The average cost per unit is given by C/x.)

34. Find the optimal order size if the total cost C for ordering and storing x units is

$$C = 2x + \left(\frac{300{,}000}{x}\right)$$

Ⓒ35. The deflection D of a particular type of beam of length L is given by

$$D = 2x^4 - 5Lx^3 + 3L^2x^2$$

where x is the distance (in feet) from one end of the beam. Find the value of x that yields the maximum deflection.

Ⓒ Calculator may be helpful.

Section 3.4

Optimization Problems

Introductory Example

Federal Tax Rates

The federal government's income is obtained primarily by taxing individuals and businesses within the country. For example, in 1980, the federal government had receipts totaling 527 billion dollars (421 billion from individuals and 106 billion from businesses). These 1980 receipts corresponded to roughly 20% of the gross national product for 1980. It seems clear that in order to increase its revenue the government could simply increase its percentage take of the gross national product. The fallacy in this reasoning is that increases in tax rates tend to discourage productivity. Thus, by greatly increasing its tax rate, the government might actually end up getting less in dollars.

What tax rate will produce the maximum income for the government? We call this a problem of **optimization.** Optimization problems are typically characterized by two conflicting phenomena. For instance, the two conflicting phenomena in the government's tax problem are (1) increased rates produce more revenue, *but* (2) increased rates discourage production, which in turn decreases revenue. In an effort to encourage production and at the same time increase revenues, the federal government has gradually shifted the tax burden from business and corporate taxes to personal income and social insurance taxes. (See Figure 3.28.) Table 3.9 compares these two sources of federal income from 1965 to 1980.

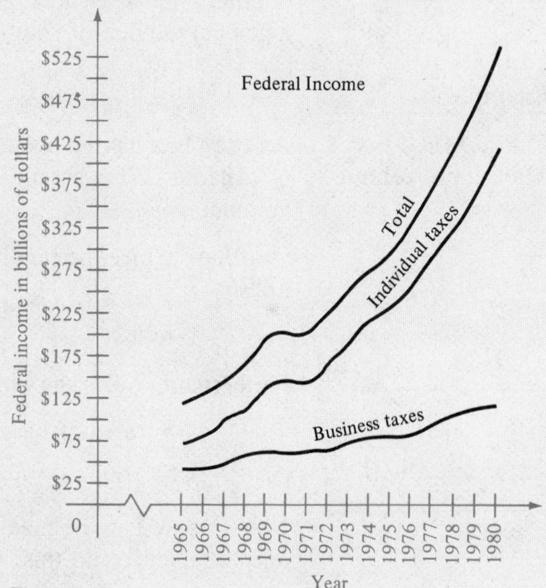

Figure 3.28

Table 3.9

Year	Total receipts (billions of dollars)	Individual: Amount (billions of dollars)	Individual: Percentage of total	Business: Amount (billions of dollars)	Business: Percentage of total
1965	120.0	76.9	64.1	43.1	35.9
1970	194.8	142.8	73.3	52.0	26.7
1975	283.4	219.4	77.4	64.0	22.6
1980	527.3	421.0	79.8	106.3	20.2

Section Objectives: ***To apply extrema of functions to the solution of problems from physics, engineering, and economics ▪ To outline a general approach to solving optimization problems.***

Up to this point in our study of calculus, most of the problems have been given in the form of mathematical equations. Our primary task has been to choose and then apply an appropriate calculus procedure to arrive at a solution. In this section, we focus on application problems that are not originally stated in terms of mathematical equations; hence an additional *problem formulation stage* is required in the solution process. In this and the next section, we provide some guidelines to assist you in constructing mathematical formulations for several common applications of calculus.

One of the most common applications of calculus is the determination of minimum or maximum values. Consider how frequently we hear or read terms like greatest profit, least cost, cheapest product, least time, greatest voltage, optimum size, least area, greatest strength, or greatest distance. Before outlining a general method of solution for such problems, we present an example.

Example 1

Finding the Maximum Volume

An open box having a square base is to be constructed from 108 square inches of material. What should be the dimensions of the box to obtain a maximum volume?

Solution Since the box has a square base, its volume is given by (see Figure 3.29)

$$V = x^2h$$

Furthermore, since the box is open at the top, its surface area is given by

$$S = (\text{area of base}) + (\text{area of four sides})$$

$$S = 108 = x^2 + 4xh$$

Now since V is to be maximized, it is helpful to express V as a function of just one variable. To do this, we solve for h in terms of x to obtain

$$4xh = 108 - x^2$$

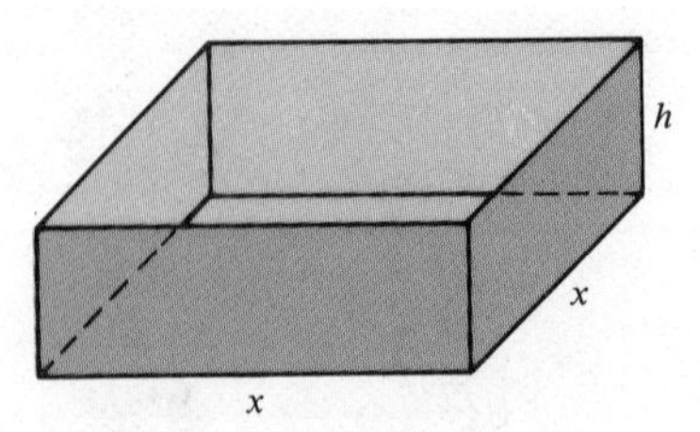

Open Box with Square Base

Figure 3.29

$$h = \frac{108 - x^2}{4x}, \qquad 0 < x \le \sqrt{108}$$

Substituting for h we get

$$V = x^2\left(\frac{108 - x^2}{4x}\right) = 27x - \frac{x^3}{4}$$

On the interval $0 < x \le \sqrt{108}$, the critical numbers for V are the solutions to

$$\frac{dV}{dx} = 27 - \frac{3x^2}{4} = 0$$

$$3x^2 = 108$$

$$x = 6$$

The Second-Derivative Test will verify that V has a relative maximum when $x = 6$. Furthermore, this relative maximum is the maximum of V on $(0, \sqrt{108}]$. Finally, when $x = 6$,

$$h = \frac{108 - (6)^2}{4(6)} = \frac{72}{24} = 3$$

and we conclude that the maximum volume of the box occurs when $x = 6$ inches and $h = 3$ inches. □

Before taking a closer look at the actual steps involved in Example 1, be sure that you understand the basic question it asks. Many students have trouble with word problems because they are too eager to start solving the problem by using a pat formula. For instance, in Example 1, you should realize that there are infinitely many open boxes having 108 square inches of surface area. You might begin this problem by asking yourself which basic shape would seem to yield a maximum volume. Should the box be tall, squatty, or more cubical? You might even try calculating a few volumes to see if you can get a better feeling for what the optimum dimensions should be. (See Figure 3.30.) Remember that you are not ready to begin solving a problem until you have clearly identified what the problem is.

After you are sure you understand what is being asked in an optimization problem, you are ready to begin considering a method for solving the problem.

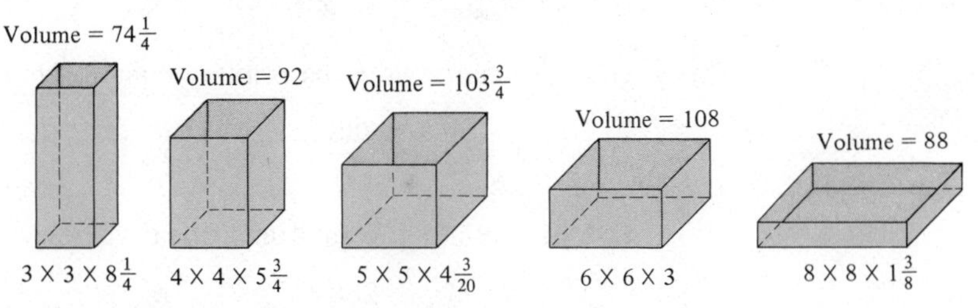

Figure 3.30 Which size box has the maximum volume?

There are several obvious stages in the solution of Example 1. We first made a sketch and assigned symbols to all *known* quantities and quantities *to be determined*. Second, we identified an equation for the quantity to be maximized. Then, we reduced this equation to obtain a function of one independent variable. And, finally, we applied the techniques of calculus to find the value of x that yielded the desired maximum.

Solving for Minimum or Maximum Values

1. Assign symbols to all given quantities and quantities to be determined. When feasible, make a sketch.
2. Write a "primary" equation for the quantity to be maximized or minimized.
3. Reduce this "primary" equation to one having a single independent variable — this may involve the use of "secondary" equations (restrictions) relating the independent variables of the "primary" equation.
4. Determine the desired maximum or minimum value by the techniques of calculus.

Remark When performing step 4, recall that to determine the maximum or minimum value of a continuous function f on a closed interval, we compare the values of f at its relative extrema to the values of f at the endpoints of the interval. The largest (smallest) of these values is the desired maximum (minimum).

Example 2

Finding the Minimum Sum

Find two positive numbers that minimize the sum of twice the first number plus the second if the product of the two numbers is 288.

Solution

1. Let x be the first number, y the second, and S the sum to be minimized.
2. Since we wish to minimize S, the *primary* equation is

$$S = 2x + y$$

3. Since the product of the two numbers is 288, we have the *secondary* equation

$$xy = 288 \qquad \text{or} \qquad y = \frac{288}{x}$$

Thus, we can rewrite the primary equation in terms of x alone as

$$S = 2x + \frac{288}{x}, \qquad 0 < x$$

4. Differentiating to find the critical values yields

$$\frac{dS}{dx} = 2 - \frac{288}{x^2} = 0$$

$$x^2 = 144$$
$$x = \pm 12$$

Choosing the positive x-value, we find that the two numbers are

$$x = 12 \qquad \text{and} \qquad y = \frac{288}{12} = 24$$

The Second-Derivative Test readily verifies that $x = 12$ yields a minimum for S. □

Example 3

Finding the Minimum Distance

Find the points on the graph of $y = 4 - x^2$ that are closest to the point (0, 2).

Solution

1. Figure 3.31 indicates that there are two points at a minimum distance from (0, 2).

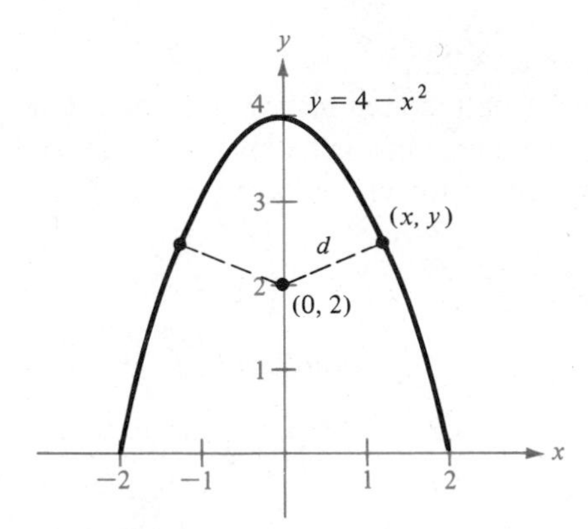

Figure 3.31

2. We are asked to minimize the distance d; hence we use the Distance Formula to obtain the primary equation

$$d = \sqrt{(x - 0)^2 + (y - 2)^2}$$

3. Using the secondary equation $y = 4 - x^2$, we can rewrite the primary equation as

$$d = \sqrt{(x)^2 + (4 - x^2 - 2)^2} = \sqrt{x^2 + (2 - x^2)^2}$$
$$= \sqrt{x^4 - 3x^2 + 4}$$

Since d is smallest when the expression under the radical is smallest, we need only find the critical values of

$$f(x) = x^4 - 3x^2 + 4$$

4. Differentiation yields

$$f'(x) = 4x^3 - 6x = 0$$
$$2x(2x^2 - 3) = 0$$
$$x = 0, \sqrt{\tfrac{3}{2}}, -\sqrt{\tfrac{3}{2}}$$

The Second-Derivative Test verifies that $x = 0$ yields a relative maximum, while both $x = \sqrt{\frac{3}{2}}$ and $x = -\sqrt{\frac{3}{2}}$ yield a minimum distance; hence on the graph of $y = 4 - x^2$, the closest points to $(0, 2)$ are $(\sqrt{\frac{3}{2}}, \frac{5}{2})$ and $(-\sqrt{\frac{3}{2}}, \frac{5}{2})$. □

Example 4

Finding the Minimum Area

A rectangular page is to contain 24 square inches of print. The margins at the top and bottom of the page are each $1\frac{1}{2}$ inches wide. The margins on each side are 1 inch. What should the dimensions of the page be so that the least amount of paper is used?

Solution

1. See Figure 3.32.
2. Letting A be the area to be minimized, our primary equation is $A = xy$.
3. The printed area inside the margins is given by $24 = (x - 3)(y - 2)$. Solving this equation for y, we have

$$y = \frac{24}{x - 3} + 2 = \frac{2x + 18}{x - 3}$$

Thus, the primary equation becomes

$$A = (x)\left(\frac{2x + 18}{x - 3}\right) = 2\left(\frac{x^2 + 9x}{x - 3}\right)$$

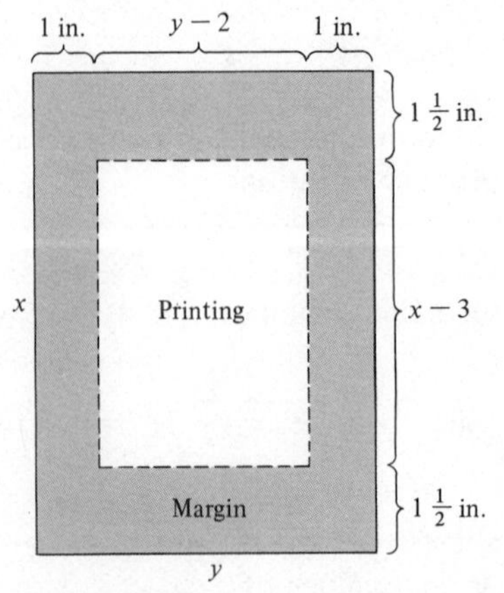

Figure 3.32

Since the margins at the top and the bottom of the page must add up to 3, we are only interested in values of A when $x > 3$.

4. Now, to find the minimum of

$$A = 2\left(\frac{x^2 + 9x}{x - 3}\right), \qquad 3 < x < \infty$$

we differentiate with respect to x and obtain

$$\frac{dA}{dx} = 2\left[\frac{(x - 3)(2x + 9) - (x^2 + 9x)}{(x - 3)^2}\right] = 2\left[\frac{x^2 - 6x - 27}{(x - 3)^2}\right]$$

Setting $dA/dx = 0$, we have

$$x^2 - 6x - 27 = 0$$
$$(x - 9)(x + 3) = 0$$
$$x = 9 \qquad \text{or} \qquad x = -3$$

Since $x = -3 < 0$ has no physical meaning, we choose $x = 9$, and the First-Derivative Test will confirm that A is a relative minimum at this point. Therefore, the dimensions of the page should be $x = 9$ inches and $y = [24/(9 - 3)] + 2 = 6$ inches. □

Remark

In Example 4, we eliminated the negative critical number because it was not a feasible solution. In many practical applications, it is not necessary to test *all* the critical numbers and the endpoints of the given interval to see if they yield the desired maximum or minimum because common sense often permits us to eliminate all but one of them.

Example 5

Finding the Minimum Length

Two posts, one 20 feet high and the other 28 feet high, stand 30 feet apart. They are to be stayed by wires attached to a single stake, running from ground level to the tops of the posts. Where should the stake be placed to use the least wire?

Solution

1. See Figure 3.33.
2. Let W be the length of the wire (to be minimized); then our primary equation is $W = y + z$.
3. In this problem, rather than solving for y in terms of z (or vice versa), we solve for both y and z in terms of a third variable x (see Figure 3.33).

$$x^2 + 20^2 = y^2, \qquad y = \sqrt{x^2 + 400}$$
$$(30 - x)^2 + 28^2 = z^2, \qquad z = \sqrt{x^2 - 60x + 1684}$$

Thus, we have

$$W = y + z = \sqrt{x^2 + 400} + \sqrt{x^2 - 60x + 1684}, \qquad 0 \le x \le 30$$

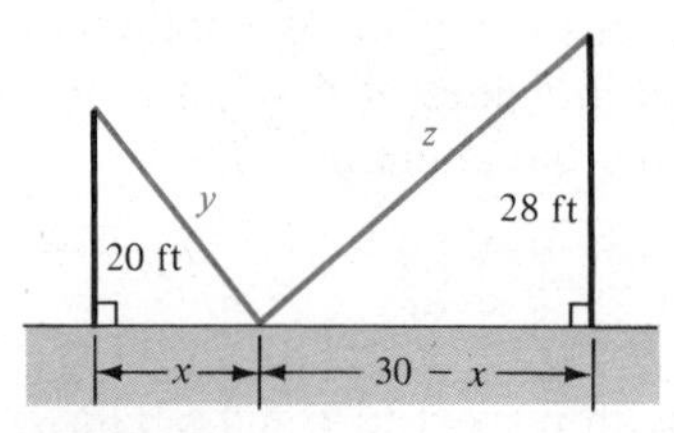

Figure 3.33

4. Differentiating W yields

$$\frac{dW}{dx} = \frac{x}{\sqrt{x^2 + 400}} + \frac{x - 30}{\sqrt{x^2 - 60x + 1684}}$$

Setting $dW/dx = 0$, we obtain

$$x\sqrt{x^2 - 60x + 1{,}684} = (30 - x)\sqrt{x^2 + 400}$$

Squaring both sides, we have

$$x^4 - 60x^3 + 1684x^2 = x^4 - 60x^3 + 1300x^2 - 24{,}000x + 360{,}000$$

$$384x^2 + 24{,}000x - 360{,}000 = 0$$

$$2x^2 + 125x - 1875 = 0$$

$$(x + 75)(2x - 25) = 0$$

$$x = -75 \qquad \text{or} \qquad x = 12.5$$

Since $x = -75$ is not in the interval $[0, 30]$ and since both $x = 0$ and $x = 30$ are not feasible solutions (see Figure 3.33), we conclude that the wire should be stayed at $x = 12.5$ feet from the 20-foot pole. □

Let's review the primary equations developed in the five examples in this section.

Example	Primary equation	Domain
1	$V = 27x - \frac{x^3}{4}$	$0 < x \le \sqrt{108}$
2	$S = 2x + \frac{288}{x}$	$0 < x$
3	$d = \sqrt{x^4 - 3x^2 + 4}$	all real numbers
4	$A = 2\left(\frac{x^2 + 9x}{x - 3}\right)$	$3 < x$
5	$W = \sqrt{x^2 + 400} + \sqrt{x^2 - 60x + 1684}$	$0 \le x \le 30$

As applications go, these five examples are fairly simple, and yet the resulting primary equations are quite complicated. The point is that real applications often involve equations that are at least as messy as these five, and you should learn to expect that. Remember that one of the main goals of this course is that you learn to use the power of calculus to analyze equations that at first glance seem formidable. (See Figure 3.34 for the graphs of these five equations.)

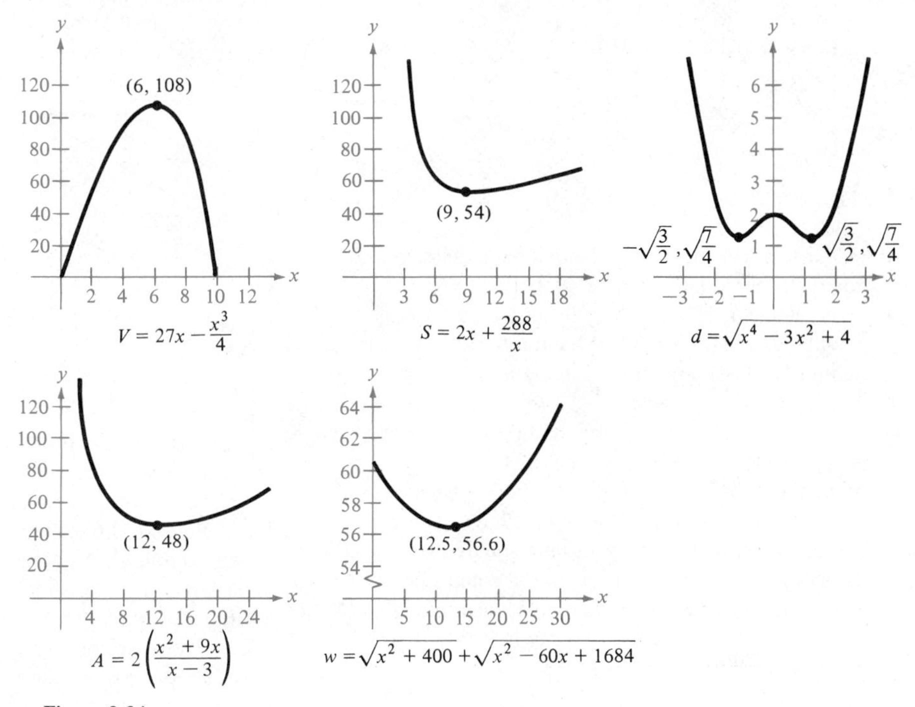

Figure 3.34

Section Exercises (3.4)

1. A rectangle has a perimeter of 100 ft. What length and width should it have so that its area is maximum?
2. What positive number x minimizes the sum of x and its reciprocal?
3. The sum of one number and two times a second number is 24. What numbers should be selected so that their product is as large as possible?
4. The difference of two numbers is 50. Select the two numbers so that their product is as small as possible.
5. Find two positive numbers whose sum is 110 and whose product is a maximum.
6. Find two positive numbers such that the sum of the first and twice the second is 100 and whose product is a maximum.
7. Find two positive numbers whose product is 192 and whose sum is a minimum.
8. The product of two positive numbers is 192. What numbers should be chosen so that the sum of the first plus three times the second is a minimum?

9. A rancher has 200 ft of fencing to enclose two adjacent rectangular corrals. (See Figure 3.35.) What dimensions should be used so that the enclosed area will be a maximum?

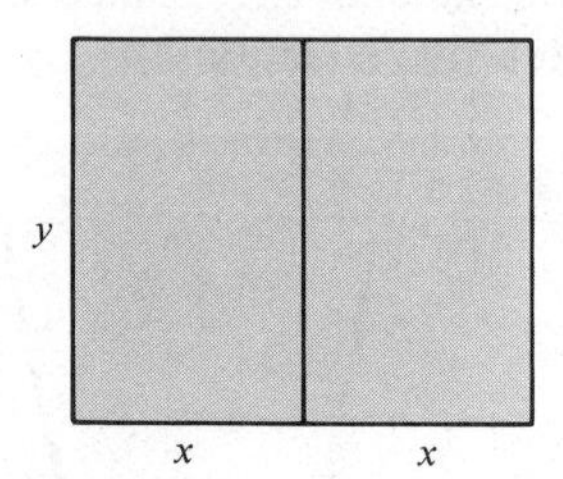

Figure 3.35

10. A dairy farmer plans to fence in a rectangular pasture adjacent to a river. He figures that the pasture must contain 180,000 m^2 in order to provide enough grass for his herd. What dimensions would require the least amount of fencing if no fencing is needed along the river?

11. Find the coordinates of the point on the curve $y = \sqrt{x}$ closest to the point (4, 0).

12. Find the coordinates of the point on the curve $y = x^2$ that is closest to the point $(2, \frac{1}{2})$.

13. An open box is to be made from a square piece of material, 12 in. on a side, by cutting equal squares from each corner and turning up the sides. Find the volume of the largest box that can be made in this manner. (See Figure 3.36.)

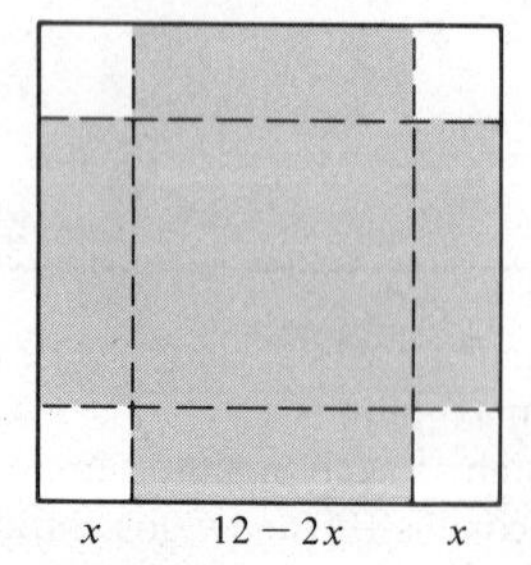

Figure 3.36

14. (a) Solve Exercise 13 if the square piece of material is s inches on a side.
(b) If the dimensions of the square piece of material are doubled, how does the volume change?

Ⓒ**15.** An open box is to be made from a rectangular piece of material by cutting equal squares from each corner and turning up the sides. Find the dimensions of the box of maximum volume if the material has dimensions 2 ft by 3 ft.

16. A net enclosure for golf practice is open at one end, as shown in Figure 3.37. Find the dimensions that require the least amount of netting if the volume of the enclosure is to be $83\frac{1}{3}$ m^3.

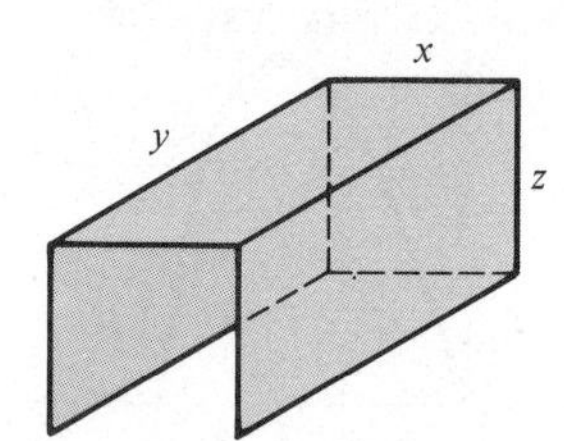

Figure 3.37

17. An indoor physical fitness room consists of a rectangular region with a semicircle on each end. If the perimeter of the room is to be a 200-m running track, find the dimensions that will make the area of the *rectangular* region as large as possible.

18. A page is to contain 30 in^2 of print. The margins at the top and the bottom of the page are each 2 in. wide. The margins on each side are only 1 in. wide. Find the dimensions of the page so that the least paper is used.

Ⓒ**19.** A right circular cylinder is to be designed to hold 12 fluid ounces of a soft drink and to use the minimal amount of material in its construction. Find the required dimensions, assuming that 1 fluid ounce requires 1.80469 in^3.

20. Work Exercise 19 if the right circular cylinder has a volume of V_0 cubic inches.

21. A rectangle is bounded by the x- and y-axes and the graph of

$$y = \frac{(6 - x)}{2}$$

as shown in Figure 3.38. What length and width should the rectangle have so that its area is a maximum?

Ⓒ Calculator may be helpful.

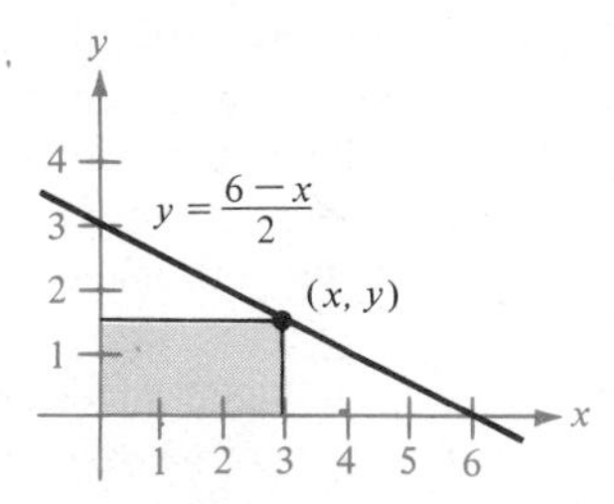

Figure 3.38

22. Find the dimensions of the largest isosceles triangle that can be inscribed in a circle of radius r.
23. A rectangle is bounded by the x-axis and the semicircle $y = \sqrt{25 - x^2}$, as shown in Figure 3.39. What length and width should the rectangle have so that its area is a maximum?

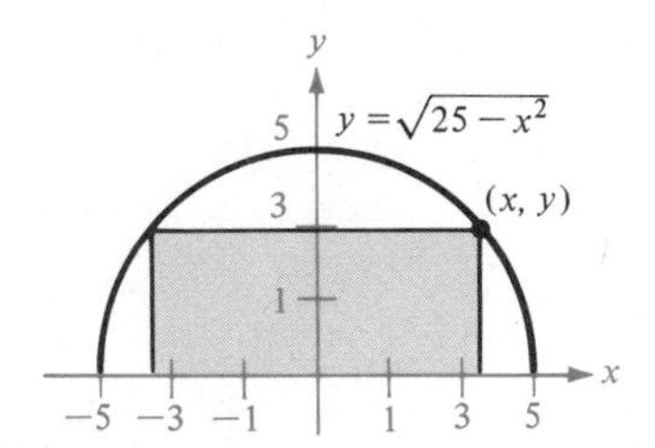

Figure 3.39

24. Find the dimensions of the largest rectangle that can be inscribed in a semicircle of radius r. (See Exercise 23.)
25. A rectangular package to be sent by a postal service can have a maximum combined length and girth (perimeter of a cross section) of 100 in. Find the dimensions of the package of maximum volume that can be sent. (See Figure 3.40.)

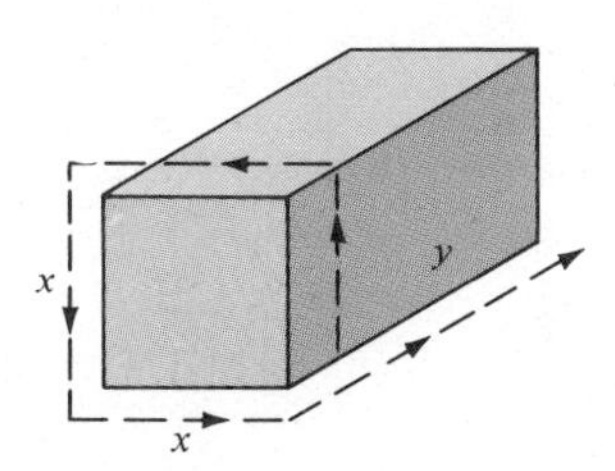
Figure 3.40

26. The combined perimeter of an equilateral triangle and a square is 10. Find the dimensions of the triangle and square that produce a minimum total area.
27. The combined perimeter of a circle and a square is 16. Find the dimensions of the circle and square that produce a minimum total area.
28. A figure is formed by adjoining two hemispheres to each end of a right circular cylinder. The total volume of the figure is 12 in^3. Find the radius of the cylinder that produces the minimum surface area.
29. A man is in a boat 2 mi from the nearest point on the coast. He is to go to a point Q, 3 mi down the coast and 1 mi inland. If he can row at 2 mi/h and walk at 4 mi/h, toward what point on the coast should he row in order to reach point Q in the least time? (See Figure 3.41.)

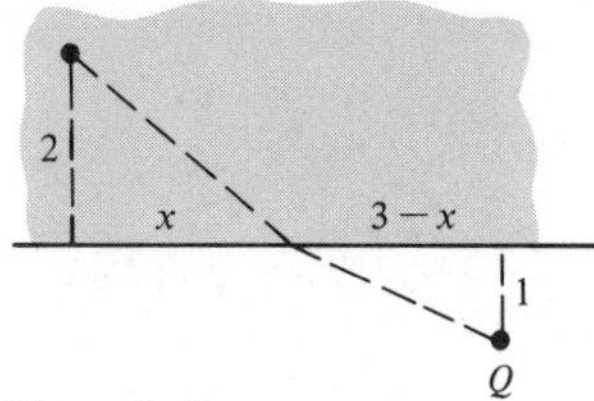

Figure 3.41

30. The conditions are the same as in Exercise 29 except that the man can row at 4 mi/h.
31. A right triangle is formed in the first quadrant by the x- and y-axes and a line through the point (2, 3). Find the vertices of the triangle so that its area is minimum.
32. [c] A right triangle is formed in the first quadrant by the x- and y-axes and a line through the point (1, 2). Find the vertices of the triangle so that the length of the hypotenuse is minimum. (See Figure 3.42.)

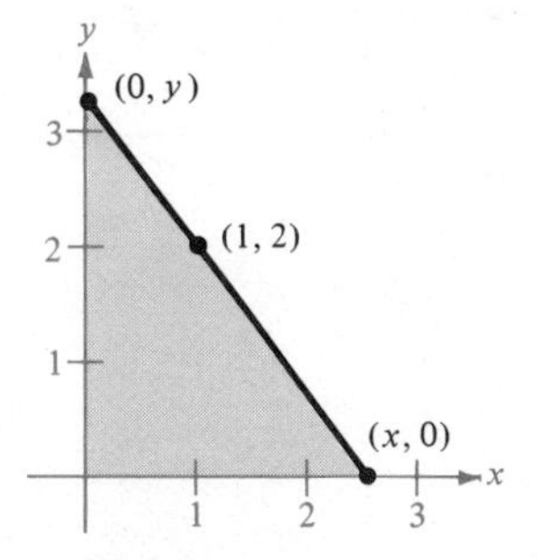

Figure 3.42

[c] Calculator may be helpful.

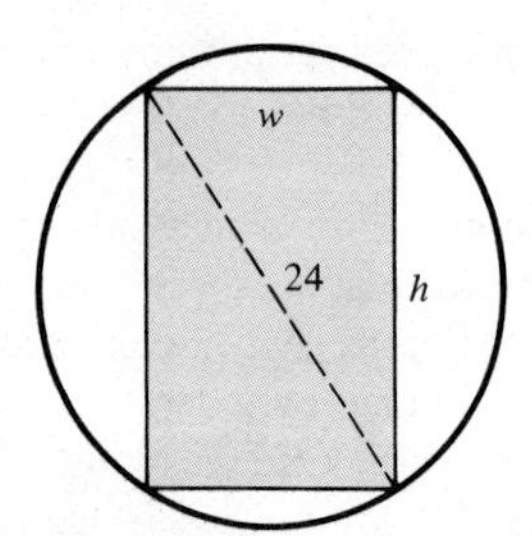

Figure 3.43

33. A wooden beam has a rectangular cross section of height h and width w, as shown in Figure 3.43. The strength S of the beam is directly proportional to the width and the square of the height. What are the dimensions of the strongest beam that can be cut from a round log of diameter 24 in.? (Hint: $S = kh^2w$, where k is the proportionality constant.)

34. Show that among all positive numbers x and y with $x^2 + y^2 = r^2$, the sum $x + y$ is largest when $x = y$.

35. Find the volume of the largest right circular cone that can be inscribed in a sphere of radius r.

36. Find the volume of the largest right circular cylinder that can be inscribed in a sphere of radius r.

Section 3.5

Business and Economics Applications

Introductory Example

Pricing Airline Tickets for Maximum Profit

An airline company offers flights daily between Chicago and Denver. The total monthly revenue and cost for these flights are given by

$$R = \frac{240x}{400 + x^2} \quad \text{and} \quad C = \sqrt{0.2x + 1}$$

where R and C are measured in millions of dollars and x is measured in thousands of passengers. The price per ticket is given by

$$p = \frac{(1{,}000{,}000)R}{1000x} = \frac{1000R}{x}$$

What price will yield a maximum profit for the airline?

One way to find the maximum profit is to examine the marginal revenue and marginal cost functions:

$$\frac{dR}{dx} = \frac{240(400 - x^2)}{(400 + x^2)^2} \quad \text{and} \quad \frac{dC}{dx} = \frac{0.1}{\sqrt{0.2x + 1}}$$

Recall that these marginals measure the *rate of change* of R and C with respect to x. As long as the rate of change of R exceeds that of C, greater sales will yield greater profits. However, when the rate of change of R falls below that of C, greater sales will lessen profits. Thus, the maximum profit must occur at the point where these two marginals are equal (see Figure 3.44). By solving the equation $dR/dx = dC/dx$, we can determine that the maximum profit occurs at $x = 17.4$, or roughly 17,400 passengers. This corresponds to a ticket price of about \$341.50. (See Table 3.10.)

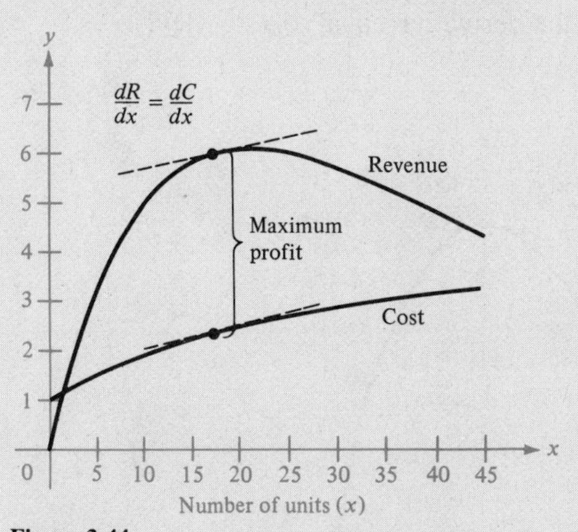

Figure 3.44

Table 3.10

Price (dollars)	Number of passengers	Monthly profit (dollars)
100.00	44,721	1,318,682
150.00	34,641	2,380,446
200.00	28,284	3,076,766
250.00	23,664	3,521,780
300.00	20,000	3,763,932
341.50	17,401	3,825,684
350.00	16,903	3,823,087
400.00	14,142	3,700,218
450.00	11,547	3,376,976

Section Objectives: ***To use calculus to solve optimization problems in business and economics*** ▪ ***To compute the elasticity of a demand function.***

Recall from Section 2.3 that many business applications involve the equation

$$P = R - C$$

where

$$P = \text{total profit}, \qquad R = \text{total revenue}, \qquad C = \text{total cost}$$

It is worthwhile noting that the problems in this section are primarily minimum and maximum problems; hence the four-step procedure used in Section 3.4 is an appropriate model to follow.

Example 1

Finding the Maximum Revenue

Suppose a company has determined that its total revenue R for a given product is given by $R = -x^3 + 450x^2 + 52{,}500x$, where R is measured in dollars and x is the number of units produced. What production level will yield a maximum revenue?

Solution

1. A sketch is given in Figure 3.45.
2. The *primary* equation is the given revenue equation:

$$R = -x^3 + 450x^2 + 52{,}500x$$

3. Since R is already given as a function of one variable, we do not need a *secondary* equation.
4. Differentiating and setting the derivative equal to zero yield

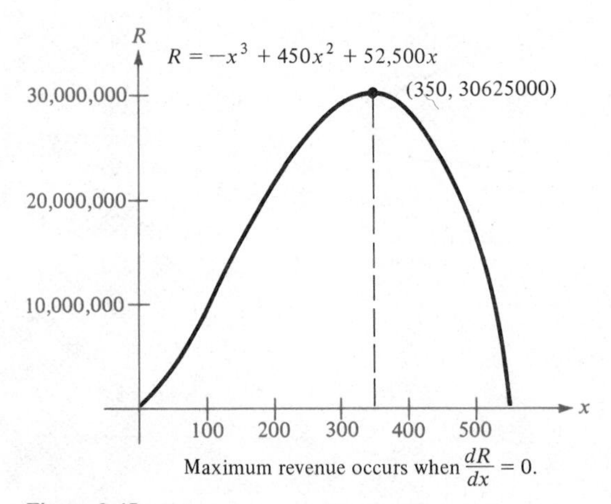

Figure 3.45

$$\frac{dR}{dx} = -3x^2 + 900x + 52{,}500 = 0$$

$$-x^2 + 300x + 17{,}500 = 0$$

$$(-x + 350)(x + 50) = 0$$

The critical values are $x = 350$ and $x = -50$. Choosing the positive value of x, we conclude that the maximum revenue is obtained when 350 units are produced. □

To study the effect of production levels on cost, economists use the **average cost function** $\overline{C}$ defined as

$$\overline{C} = \frac{C}{x}$$

where $C = f(x)$ is the total cost function.

Example 2

Finding the Minimum Average Cost

A company estimates that the cost (in dollars) of producing x units of a certain product is given by

$$C = 800 + 0.04x + 0.0002x^2$$

Find the production level that minimizes the average cost per unit. Compare this minimal average cost to the average cost when 400 units are produced.

Solution

1. We let $\overline{C}$ be the average cost and x be the number of units produced. (See Figure 3.46.)

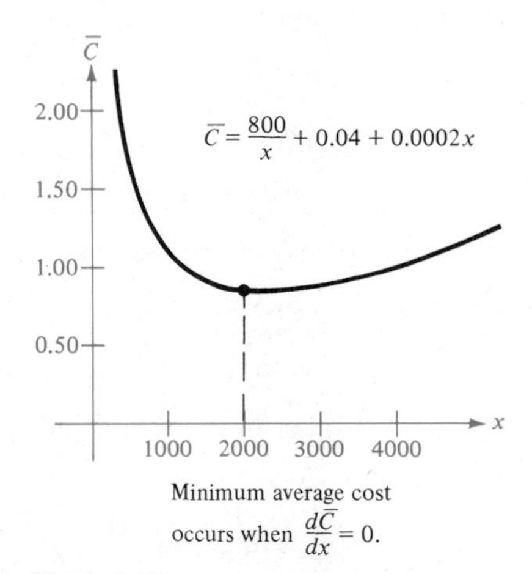

Figure 3.46

2. The primary equation representing the quantity to be minimized is

$$\overline{C} = \frac{C}{x}$$

3. Substituting from the given equation for C, we have

$$\overline{C} = \frac{800 + 0.04x + 0.0002x^2}{x} = \frac{800}{x} + 0.04 + 0.0002x$$

4. Setting the derivative equal to zero yields

$$\frac{d\overline{C}}{dx} = -\frac{800}{x^2} + 0.0002 = 0$$

$$x^2 = \frac{800}{0.0002} = 4{,}000{,}000$$

$$x = 2000$$

For 2000 units the average cost is

$$\overline{C} = \frac{800}{2000} + 0.04 + (0.0002)(2000) = \$0.84$$

For 400 units the average cost is

$$\overline{C} = \frac{800}{400} + 0.04 + (0.0002)(400) = \$2.12$$

□

Example 3

Finding the Maximum Revenue

A certain business sells 2000 items per month at a price of \$10 each. If it can sell 250 more items per month for each \$0.25 reduction in price, what price per item will maximize its monthly revenue?

Solution

1. Let

x = number of items sold per month

p = price of each item (in dollars)

R = monthly revenue (in dollars)

2. Since the monthly revenue is to be maximized, the primary equation is

revenue = (number of units sold)(price per unit)

$$R = xp$$

3. The number of items x is related to p in that x increases 250 units every time p drops \$0.25 from the original cost of \$10. This is described by the equation

$$x = 2000 + 250\left(\frac{10 - p}{0.25}\right) = 12{,}000 - 1000p$$

or

$$p = 12 - \frac{x}{1000}$$

Substituting this result into the revenue equation, we have

$$R = x\left(12 - \frac{x}{1000}\right) = 12x - \frac{x^2}{1000}$$

(See Figure 3.47.)

4. Setting the derivative equal to zero, we obtain

$$\frac{dR}{dx} = 12 - \frac{x}{500} = 0$$

Therefore, $dR/dx = 0$ if $x = 6000$, and we conclude that the price that maximizes the revenue will be

$$p = 12 - \frac{6000}{1000} = \$6$$

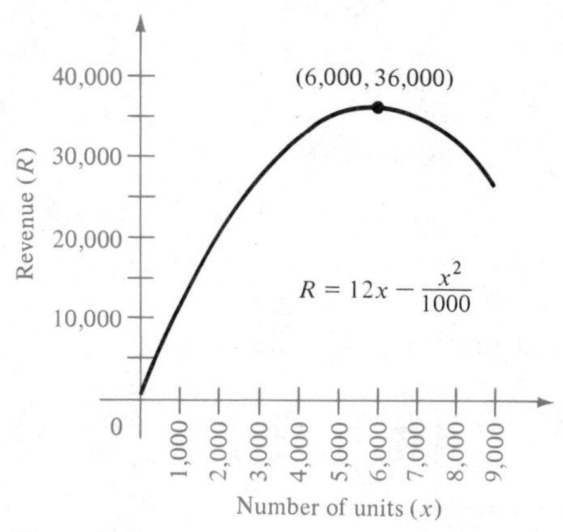

Figure 3.47

□

From the equation

$$p = 12 - \frac{x}{1000}$$

in Example 3 we can determine the number of units that can be sold — that is, the *demand* — at a given price. We call the function given by $p = f(x)$ the **demand function.** Note that as the price increases, the demand decreases, and vice versa.

Example 4

Finding the Maximum Profit

A business, in marketing a certain item, has discovered that the demand for the item is represented by the equation

$$x = \frac{2500}{p^2}$$

Assuming that the total revenue R is given by $R = xp$ and the cost for producing x items is given by $C = 0.5x + 500$, find the price per unit that yields a maximum profit.

Solution

1. Let P represent the profit.
2. Since we are seeking a maximum profit, the primary equation is

$$P = R - C$$

3. Using the equation for the profit P, we have

$$P = R - C = xp - (0.5x + 500)$$

Solving for p in terms of x in the equation

$$x = \frac{2500}{p^2}$$

we have

$$p = \frac{50}{\sqrt{x}}$$

Thus, our profit equation becomes

$$P = x\left(\frac{50}{\sqrt{x}}\right) - (0.5x + 500) = 50\sqrt{x} - 0.5x - 500$$

(See Figure 3.48.)
4. Setting the derivative equal to zero, we obtain

$$\frac{dP}{dx} = \frac{25}{\sqrt{x}} - 0.5 = 0$$

$$\sqrt{x} = \frac{25}{0.5} = 50$$

$$x = 2500$$

Finally, we conclude that the maximum profit occurs when the price is

$$p = \frac{50}{\sqrt{2500}} = \frac{50}{50} = \$1.00$$

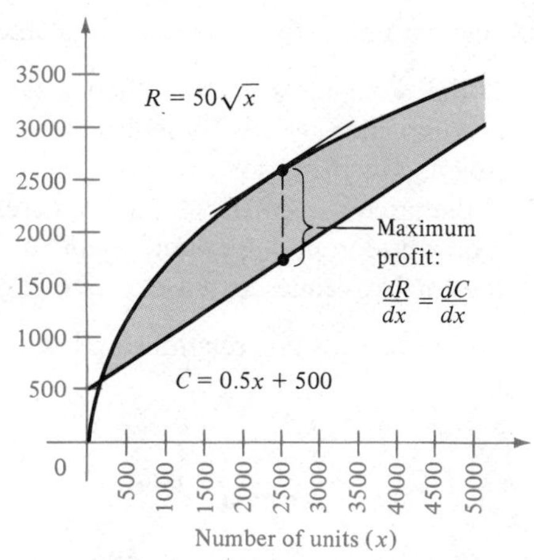

Figure 3.48

□

Remark

To find the maximum profit in Example 4, we differentiated the equation $P = R - C$ and set dP/dx equal to zero. From the equation

$$\frac{dP}{dx} = \frac{dR}{dx} - \frac{dC}{dx} = 0$$

it follows that the maximum profit occurs when the marginal revenue is equal to the marginal cost. (See Figure 3.48.)

One way economists describe the behavior of a demand function is by a term called the **price elasticity of demand.** It describes the relative responsiveness of consumers to a change in the price of an item. If $p = f(x)$ is a differentiable demand function, then the price elasticity of demand is given by

$$\eta = \frac{p/x}{dp/dx}$$

(where η is the lower-case Greek letter eta). For a given price, if $|\eta| < 1$, the demand is said to be **inelastic;** if $|\eta| > 1$, the demand is said to be **elastic.**

Example 5

Test for Elasticity

Show that the demand function in Example 4, $p = 50x^{-1/2}$, is elastic.

Solution

$$\eta = \frac{p/x}{dp/dx} = \frac{(50x^{-1/2})/x}{-25x^{-3/2}}$$

$$= \frac{50x^{-3/2}}{-25x^{-3/2}} = -2$$

Since $|\eta| = 2$, we conclude that the demand is elastic for any price. □

Elasticity has an interesting relationship to the total revenue:

1. If the demand is *elastic,* then a decrease in the price per unit is accompanied by a sufficient increase in unit sales to cause the total revenue to increase.
2. If the price is *inelastic,* then the increase in unit sales accompanying a decrease in price per unit is *not* sufficient to increase revenue, and the total revenue decreases. (See Figure 3.49.)

The next example verifies this relationship between elasticity and total revenue.

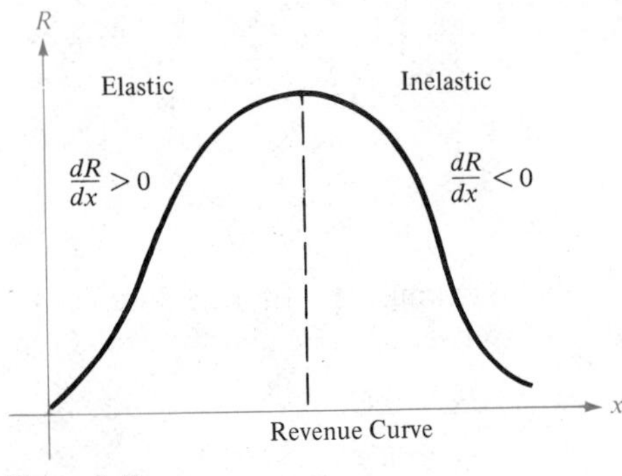

Figure 3.49

Example 6

Elasticity and Marginal Revenue

Show that for a differentiable demand function, the marginal revenue is positive when the demand is elastic and negative when the demand is inelastic. (Assume that the quantity demanded increases as the price decreases, and thus dp/dx is negative.)

Solution Since the revenue is given by $R = xp$, we calculate the marginal revenue to be

$$\frac{dR}{dx} = x\left(\frac{dp}{dx}\right) + p = p\left[\left(\frac{x}{p}\right)\left(\frac{dp}{dx}\right) + 1\right]$$
$$= p\left[\frac{dp/dx}{p/x} + 1\right] = p\left(\frac{1}{\eta} + 1\right)$$

If the demand is inelastic, then $|\eta| < 1$ implies that $1/\eta < -1$, since dp/dx is negative. (We assume that x and p are positive.) Therefore,

$$\frac{dR}{dx} = p\left(\frac{1}{\eta} + 1\right)$$

is negative.

Similarly, if the demand is elastic, then $|\eta| > 1$ and $-1 < 1/\eta$, which implies that dR/dx is positive. □

Summary of Business Terms and Formulas

Basic Terms:

x = number of units produced (or sold)

p = price per unit

R = total revenue from selling x units

$$R = xp$$

C = total cost of producing x units

$\overline{C}$ = average cost per unit

$$\overline{C} = \frac{C}{x}$$

P = total profit from selling x units

$$P = R - C$$

Demand Function:

$p = f(x)$ = price required to sell x units

η = price elasticity of demand

$$\eta = \frac{p/x}{dp/dx}$$

If $|\eta| < 1$, the price is inelastic.
If $|\eta| > 1$, the price is elastic.

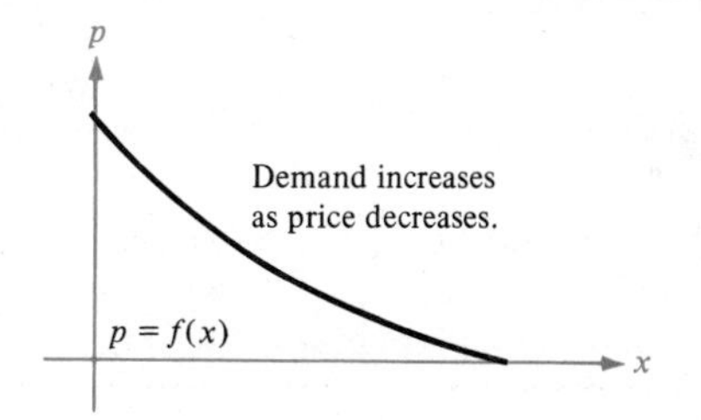

Typical Graphs of Revenue, Cost, and Profit Functions

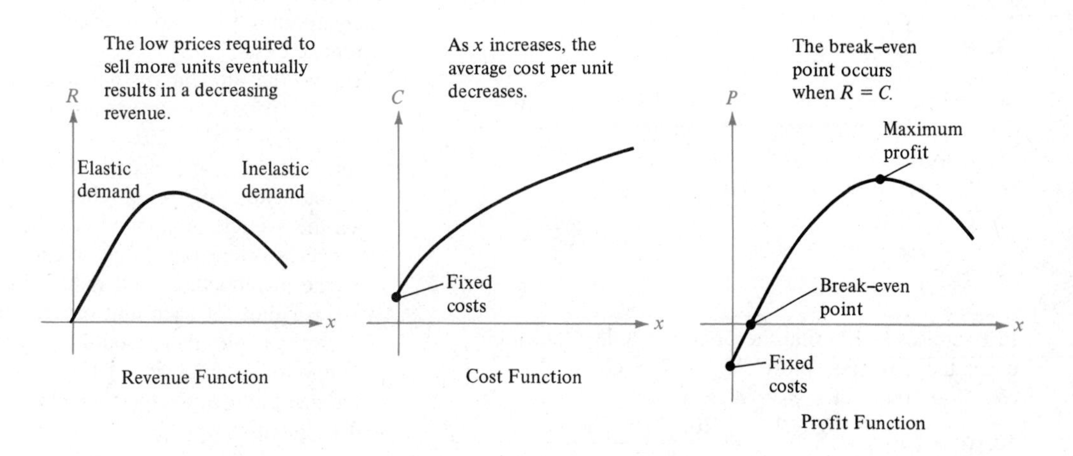

Revenue Function

Cost Function

Profit Function

Marginals

$\dfrac{dR}{dx}$ = marginal revenue = the rate of change of the revenue

$\approx$ the *extra* revenue for selling one additional unit

$\dfrac{dC}{dx}$ = marginal cost = the rate of change of the cost

$\approx$ the *extra* cost of producing one additional unit

$\dfrac{dP}{dx}$ = marginal profit = the rate of change of the profit

$\approx$ the *extra* profit for selling one additional unit

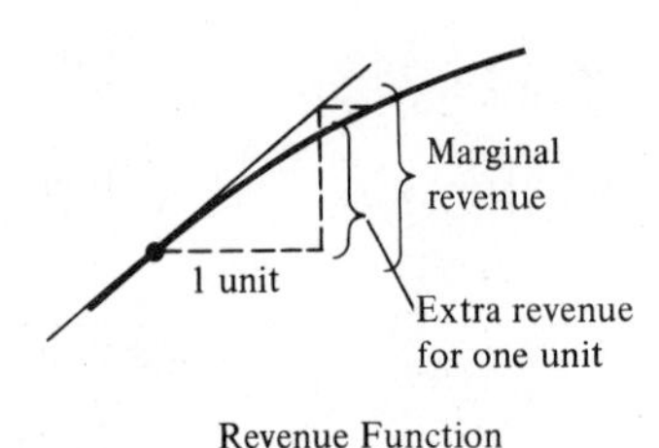

Revenue Function

Section Exercises (3.5)

In Exercises 1–4, find the number of units x that produce a maximum revenue R.

1. $R = 900x - 0.1x^2$
2. $R = 600x^2 - 0.02x^3$
3. $R = \dfrac{1{,}000{,}000x}{0.02x^2 + 1800}$
4. $R = 30x^{2/3} - 2x$

In Exercises 5–8, find the number of units x that produce the minimum average cost per unit $\overline{C}$ ($\overline{C} = C/x$).

5. $C = 0.125x^2 + 20x + 5000$
6. $C = 0.001x^3 - 5x + 250$
7. $C = 3000x - x^2\sqrt{300 - x}$
8. $C = \dfrac{2x^3 - x^2 + 5000x}{x^2 + 2500}$

In Exercises 9–12, find the price per unit p that produces the maximum profit P ($P = R - C$).

9. $C = 100 + 30x$; $p = 90 - x$
10. $C = 2400x + 5200$; $p = \dfrac{30{,}000 - 2x^2}{5}$
11. $C = 4000 - 40x + 0.02x^2$; $p = \dfrac{5000 - x}{100}$
12. $C = 35x + 2\sqrt{x - 1}$; $p = 40 - \sqrt{x - 1}$
13. A manufacturer of lighting fixtures has daily production costs of $C = 800 - 10x + \frac{1}{4}x^2$. How many fixtures x should be produced each day to minimize costs?
14. Let x be the amount (in hundreds of dollars) a company spends on advertising and let P be the profit. If $P = 230 + 20x - \frac{1}{2}x^2$, what amount of advertising gives the maximum profit?
15. A manufacturer of radios charges \$90 per unit when the average production cost per unit is \$60. However, to encourage large orders from distributors, the manufacturer will reduce the charge by \$0.10 per unit for each unit ordered in excess of 100 (for example, there would be a charge of \$88 per radio for an order size of 120). Find the largest order size the manufacturer should allow so as to realize maximum profit.

16. A real estate office handles 50 apartment units. When the rent is \$180 per month, all units are occupied. However, on the average, for each \$10 increase in rent, one unit becomes vacant. Each occupied unit requires an average of \$12 per month for service and repairs. What rent should be charged to realize the most profit?

17. A power station is on one side of a river that is $\frac{1}{2}$ mi wide, and a factory is 6 mi downstream on the other side. It costs \$6/ft to run power lines overland and \$8/ft to run them underwater. Find the most economical path for the transmission line from the power station to the factory.

18. An offshore oil well is 1 mi off the coast. The refinery is 2 mi down the coast. If the cost of laying pipe in the ocean is twice as expensive as on land, in what path should the pipe be constructed in order to minimize the cost?

Ⓒ**19.** Assume that the amount of money deposited in a bank is proportional to the square of the interest rate the bank pays on this money. Furthermore, the bank can reinvest this money at 12%. Find the interest rate the bank should pay to maximize profit. (Use the simple interest formula.)

20. Show that the average cost is minimum at the value of x where the average cost equals the marginal cost.

21. Given the cost function $C = 2x^2 + 5x + 18$,
(a) Find the value of x where the average cost is minimum.
(b) For the value of x found in part (a), show that the marginal cost and average cost are equal (see Exercise 20).

22. Given the cost function $C = x^3 - 6x^2 + 13x$.
(a) Find the value of x where the average cost function is minimum.
(b) For the value of x found in part (a), show that the marginal cost and average cost are equal.

Ⓒ**23.** The demand function for a certain product is given by $x = 20 - 2p^2$.
(a) Consider the point (2, 12). If the price decreases by 5%, determine the corresponding percentage increase in quantity demanded.
(b) **Average elasticity of demand** is defined to be the percentage change in quantity divided by the percentage change in price. Use the percentage of part (a) to find the average elasticity at (2, 12).
(c) Find the exact elasticity at (2, 12) by using the formula in this section. Compare the result with that of part (b).
(d) Find an expression for total revenue ($R = xp$), and find the values of x and p that maximize R.
(e) For the value of x found in part (d), show that $|\eta| = 1$.

24. Assume that the demand equation is $p^3 + x^3 = 9$.
(a) Find η when $x = 2$.
(b) Find the values of x and p that maximize the total revenue ($R = xp$).
(c) Show that $|\eta| = 1$ for the value of x found in part (b).

25. The demand function for a particular commodity is given by $p = (16 - x)^{1/2}$, $0 \le x \le 16$. Determine the price and quantity for which revenue is maximum.

26. The demand function is given by the equation

$$x = \frac{a}{p^m}, \qquad m > 1$$

Show that $\eta = -m$ (i.e., in terms of approximate price changes, a 1% increase in price results in a m% decrease in quantity demanded).

Ⓒ**27.** A given commodity has a demand function given by $p = 100 - \frac{1}{2}x^2$ and the total cost function of $C = 40x + 375$.
(a) What price gives the maximum profit?
(b) What is the average cost per unit if production is set to give maximum profit?

Ⓒ**28.** Rework Exercise 27, using the cost function $C = 50x + 375$.

Ⓒ**29.** When a wholesaler sold a certain product at \$25 per unit, sales were 800 units each week. However, after a price raise of \$5, the average number of units sold dropped to 775 per week. Assume that the demand function is linear, and find the price that will maximize the total revenue.

Ⓒ Calculator may be helpful.

Section 3.6

Asymptotes

Introductory Example

A Cost-Benefit Model for Smokestack Emission

During the 1960s and 1970s, the federal government became increasingly sensitive to environmental concerns. One of the major forms of pollution that came under legislative attack was the pollution of the air. Regulations governing auto exhaust, smokestack emissions, aerial pesticides, and aerosol sprays became the topic of heated controversy. The industries affected by the regulations tended to argue for moderate regulations building up over a period of several years, while environmental groups tended to push for strict regulatory standards realized over a short period of time. The basic objection from industry centered around the cost of meeting the pollution regulations.

As a case in point, consider the air pollutants in the stack emission of a utility company that burns coal to generate electricity. The cost of removing a certain percentage of the existing pollution is typically *not* a linear function. That is, if it costs C dollars to remove 25% of the pollution, it costs more than $2C$ dollars to remove 50% of the pollution. Ultimately, as the percentage removed approaches 100%, the cost tends to become increasingly prohibitive. Suppose that the cost C of removing $p\%$ of the smokestack pollutants is

$$C = \frac{80{,}000p}{100 - p}$$

Observe in Figure 3.50 the significant difference in cost between 85% and 90% removal, as contrasted to the minimal difference between 10% and 15%. Also, note that the cost increases without bound as p approaches 100%. We call the line $p = 100$ a **vertical asymptote** of this graph.

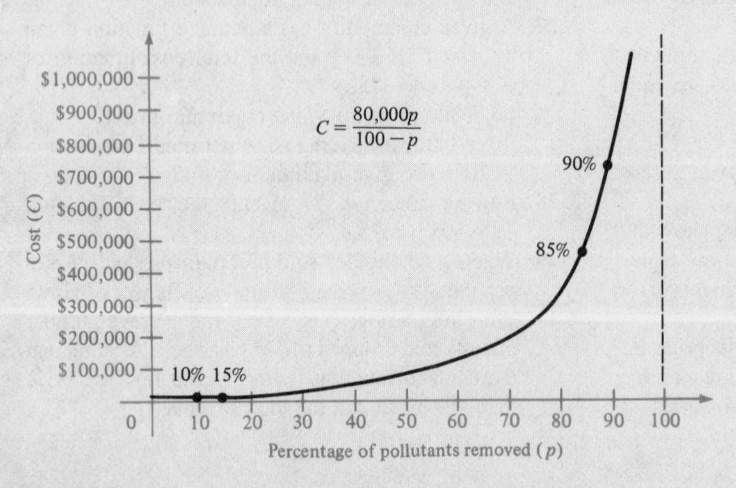

Figure 3.50

Section Objective: ***To introduce vertical and horizontal aysmptotes as an aid to curve sketching.***

In the first three sections of this chapter, we looked at different ways we can use calculus as an aid to sketching the graph of a function. In this section, we look at another valuable aid to curve sketching — the determination of vertical and horizontal asymptotes.

Recall from Section 1.5, Example 7 that the function

$$f(x) = \frac{3}{x - 2}$$

is unbounded as x approaches 2. (See Figure 3.51.) We call the line $x = 2$ a **vertical asymptote** for the function f. Note that the graph of f moves closer and closer to the vertical asymptote as x approaches 2.

In general, the types of limits where $f(x)$ approaches infinity as x approaches c from the left or right are referred to as **infinite limits.**

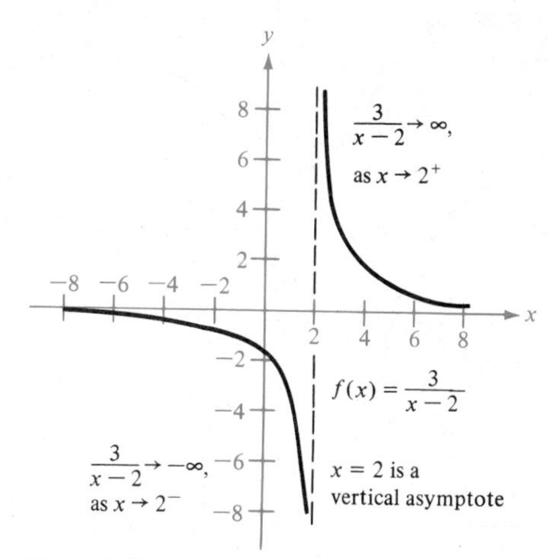

Figure 3.51

Definition of Vertical Asymptote	If $f(x)$ approaches infinity as x approaches c from the right or left, then we call the line $x = c$ a **vertical asymptote** of f.

If both one-sided limits of f at c are infinite and of like sign, then the line $x = c$ is called an **even** vertical asymptote (see Figure 3.52.) If both one-sided limits of f at c are infinite and have unlike signs, then the line $x = c$ is called an **odd** vertical asymptote (see Figure 3.53).

In general, a vertical asymptote occurs when a ratio has a zero denominator and a nonzero numerator. This is demonstrated in the next two examples.

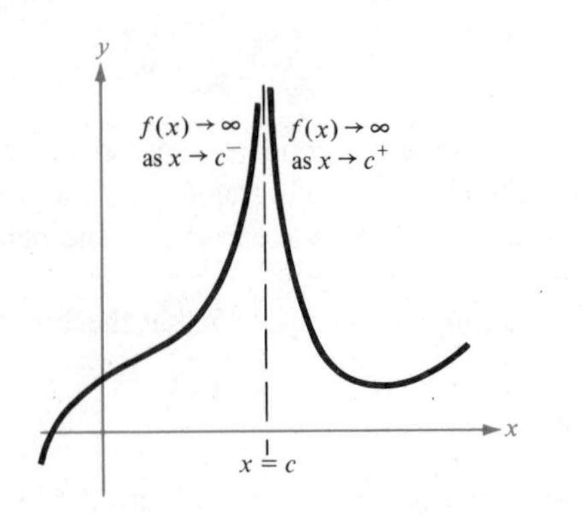

Figure 3.52 Even Asymptotes

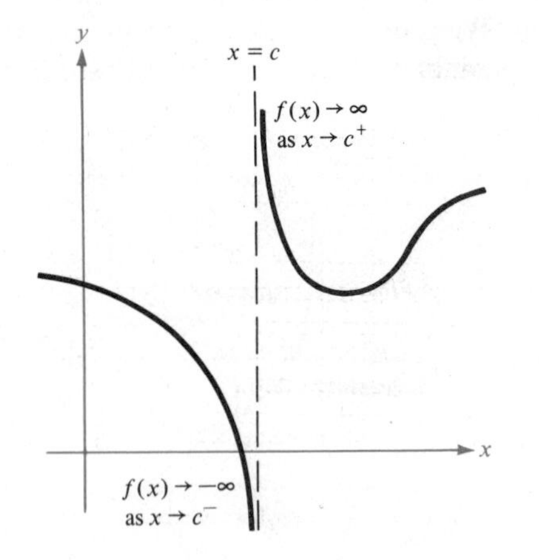

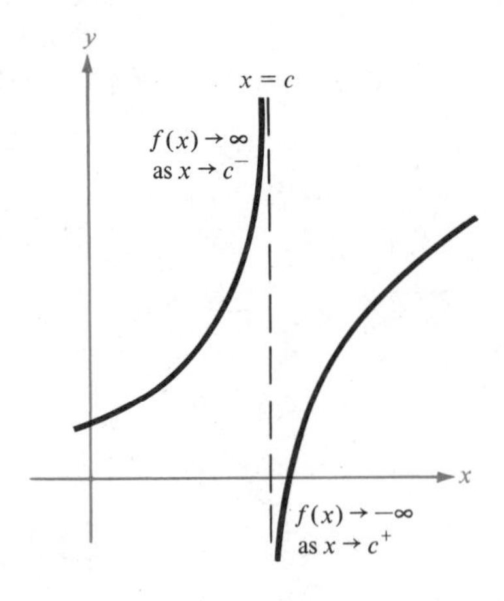

Figure 3.53 Odd Asymptotes

Example 1

Finding Vertical Asymptotes

Determine and classify as odd or even any vertical asymptote for the function

$$f(x) = \frac{x + 2}{x^2 - 2x}$$

Solution The possible vertical asymptotes occur where the denominator of $f(x)$ is zero. Thus we consider the one-sided limits at $x = 2$ and at $x = 0$. We have

$$\lim_{x \to 2^+} \frac{x + 2}{x^2 - 2x} = \frac{4}{0^+} = \infty$$

Note that 0^+ means that $x^2 - 2x$ approaches zero through positive values. We also have

$$\lim_{x \to 2^-} \frac{x + 2}{x^2 - 2x} = \frac{4}{0^-} = -\infty$$

where 0^- indicates that the denominator approaches zero through negative values. Similarly, at $x = 0$ we have

$$\lim_{x\to0^+} \frac{x+2}{x^2-2x} = \frac{2}{0^-} = -\infty$$

$$\lim_{x\to0^-} \frac{x+2}{x^2-2x} = \frac{2}{0^+} = \infty$$

Therefore, the lines $x = 2$ and $x = 0$ are both *odd* vertical asymptotes (see Figure 3.54).

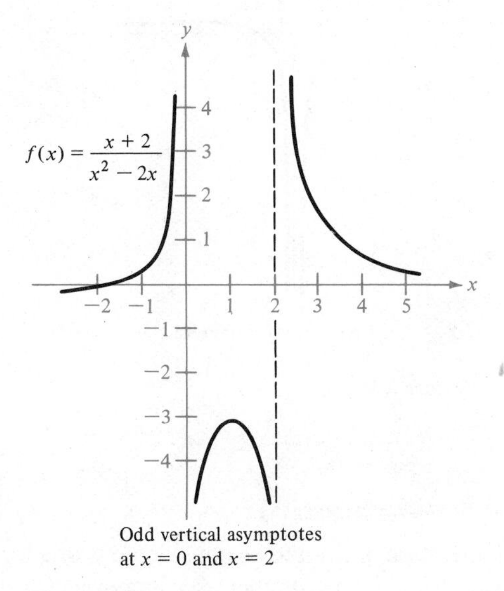

Odd vertical asymptotes at $x = 0$ and $x = 2$

Figure 3.54 □

Remark Remember that the symbol

$$\lim_{x\to c^+} f(x) = \frac{4}{0^+} = \infty$$

in Example 1 does not mean that we are dividing by zero or that the limit exists. We simply use this notation as a way of indicating that the limit is $+\infty$ rather than $-\infty$.

Example 2

Finding Vertical Asymptotes

Determine and classify any vertical asymptotes of

$$g(x) = \frac{2x}{(x+3)^2}$$

Solution The only possible vertical asymptote occurs when $x = -3$ (denominator is zero). Therefore, the one-sided limits are

$$\lim_{x \to -3^+} \frac{2x}{(x + 3)^2} = \frac{-6}{0^+} = -\infty$$

and
$$\lim_{x \to -3^-} \frac{2x}{(x + 3)^2} = \frac{-6}{0^+} = -\infty$$

We conclude that the line $x = -3$ is an *even* vertical asymptote (compare Figure 3.55 with Figure 3.52).

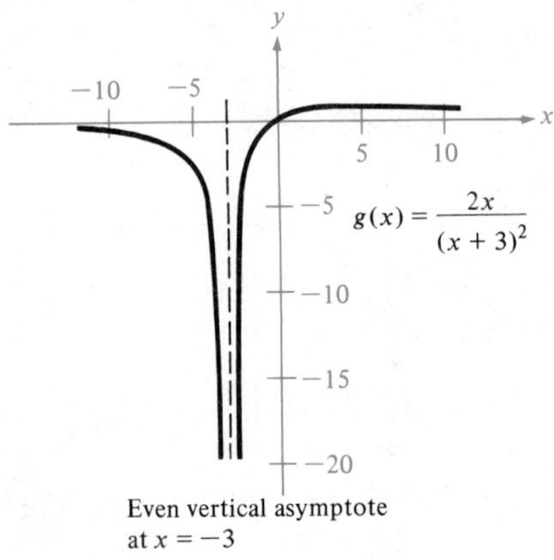

Figure 3.55

□

Another type of limit to be considered is one for which the values of a function approach some finite number as x increases (or decreases) without bound. The types of limits where $f(x)$ approaches some finite value as x becomes infinite are called **limits at infinity.**

Definition of Horizontal Asymptote

If f is a function and L is some number, then the statements:

1. $\lim_{x \to \infty} f(x) = L$
2. $\lim_{x \to -\infty} f(x) = L$

denote **limits at infinity.** In either case, the line $y = L$ is called a **horizontal asymptote** of f.

Figure 3.56 suggests some ways in which the graph of a function f may approach one or more horizontal asymptotes. Observe in Figure 3.56 that it is possible for a function to cross its horizontal asymptote.

When evaluating limits at infinity, the following theorem is frequently used.

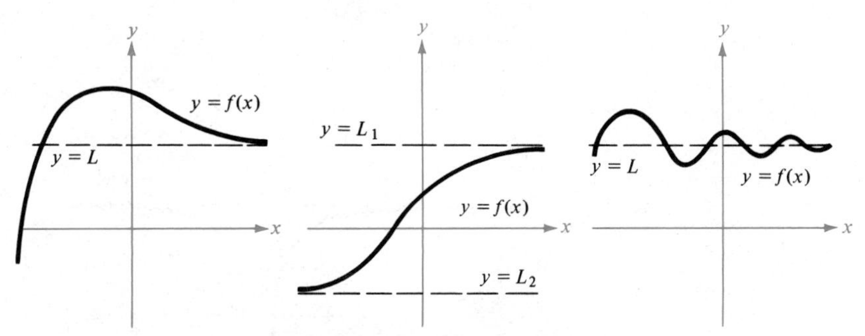

The graph of a function may approach a horizontal asymptote in various ways.

Figure 3.56

Theorem 3.7
Limits at Infinity

If r is positive and c is any real number, then

$$\lim_{x\to\infty} \frac{c}{x^r} = 0$$

Furthermore,

$$\lim_{x\to-\infty} \frac{c}{x^r} = 0$$

if x^r is defined when $x < 0$.

Example 3

Evaluating Limits at Infinity

Evaluate

$$\lim_{x\to\infty}\left(5 - \frac{2}{x^2}\right)$$

Solution Since

$$\lim_{x\to\infty}\left(5 - \frac{2}{x^2}\right) = \lim_{x\to\infty} 5 - \lim_{x\to\infty}\frac{2}{x^2}$$

it follows that

$$\lim_{x\to\infty}\left(5 - \frac{2}{x^2}\right) = 5 - 0 = 5$$

and we conclude that the function

$$y = 5 - \frac{2}{x^2}$$

has a horizontal asymptote of $y = 5$. (See Figure 3.57.)

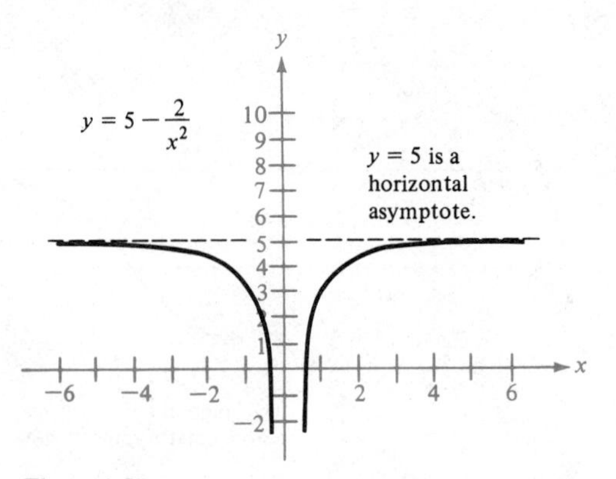

Figure 3.57

□

Example 4

Evaluating Limits at Infinity

Evaluate

$$\lim_{x\to\infty} \frac{2x - 1}{x + 1}$$

Solution Since

$$\lim_{x\to\infty} \frac{2x - 1}{x + 1} = \frac{\lim_{x\to\infty} (2x - 1)}{\lim_{x\to\infty} (x + 1)}$$

we obtain the meaningless result

$$\lim_{x\to\infty} \frac{2x - 1}{x + 1} = \frac{\infty}{\infty}$$

To apply Theorem 3.7 to each term of the fraction, we divide both numerator and denominator by x. Thus

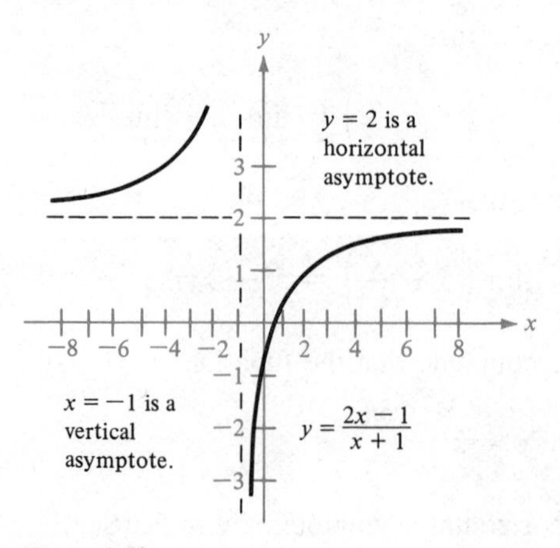

Figure 3.58

$$\frac{2x - 1}{x + 1} = \frac{2 - (1/x)}{1 + (1/x)}$$

Now, we see that

$$\lim_{x \to \infty} \frac{2x - 1}{x + 1} = \lim_{x \to \infty} \frac{2 - (1/x)}{1 + (1/x)} = \frac{2 - 0}{1 + 0} = 2$$

(You may wish to verify this result by substituting large positive values for x.)

By taking the limit as $x \to -\infty$, we can see that $y = 2$ is a horizontal asymptote to the left as well as to the right. Moreover, $x = -1$ is a vertical asymptote, and the graph of this function is shown in Figure 3.58. □

Previously (Section 1.5), we encountered the meaningless expression $\frac{0}{0}$. In Example 4, we obtained another meaningless expression, $\frac{\infty}{\infty}$. In both cases, we were able to resolve the difficulty by rewriting the given *indeterminate form* in an equivalent *determinate form*. To do this it is, of course, necessary to know which forms are determinate. For ready reference, we list six of the most common determinate forms.

Determinate Forms for $\lim \frac{f(x)}{g(x)}$

1. Limit is zero: $\frac{0}{\pm\infty}$, $\frac{L}{\pm\infty}$, $\frac{0}{L}$

2. Limit is infinite: $\frac{\pm\infty}{0}$, $\frac{\pm\infty}{L}$, $\frac{L}{0}$

Example 5

Evaluating Limits at Infinity

Evaluate the following limits:

(a) $\lim_{x \to \infty} \frac{-2x + 3}{3x^2 + 1}$

(b) $\lim_{x \to \infty} \frac{-2x^2 + 3}{3x^2 + 1}$

(c) $\lim_{x \to \infty} \frac{-2x^3 + 3}{3x^2 + 1}$

Solution

(a) To apply Theorem 3.7 to each term of the fraction, divide both numerator and denominator by x^2. Thus, it follows that

$$\lim_{x \to \infty} \frac{-2x + 3}{3x^2 + 1} = \lim_{x \to \infty} \frac{(-2/x) + (3/x^2)}{3 + (1/x^2)}$$

$$= \frac{-0 + 0}{3 + 0} = \frac{0}{3} = 0$$

(b) In this case, we also divide by x^2 in order to apply Theorem 3.7 to each term of the fraction. Thus, we obtain

$$\lim_{x\to\infty} \frac{-2x^2 + 3}{3x^2 + 1} = \lim_{x\to\infty} \frac{-2 + (3/x^2)}{3 + (1/x^2)}$$

$$= \frac{-2 + 0}{3 + 0} = \frac{-2}{3}$$

(c) In this case, we divide each term of the fraction by x^3, and by Theorem 3.7 it follows that

$$\lim_{x\to\infty} \frac{-2x^3 + 3}{3x^2 + 1} = \lim_{x\to\infty} \frac{-2 + (3/x^3)}{(3/x) + (1/x^3)}$$

$$= \frac{-2 + 0}{0 + 0} = \frac{-2}{0^+} = -\infty$$

□

Observe that, in part (a) of Example 5, the degree of the numerator was *less* than the degree of the denominator and the limit of the ratio was zero. In part (b), the degrees of the numerator and denominator were *equal* and the limit was merely the ratio of the coefficients of the highest-powered terms. Finally, in part (c), the degree of the numerator was *greater* than that of the denominator and the limit was infinite. These results suggest an informal alternative way of determining the limits at infinity for functions expressed as the ratio of two polynomials.

Limits at Infinity for Rational Functions

For the rational function $f(x)/g(x)$ where

$$f(x) = a_n x^n + a_{n-1}x^{n-1} + \cdots + a_0$$

and $$g(x) = b_m x^m + b_{m-1}x^{m-1} + \cdots + b_0$$

we have

$$\lim_{x\to\pm\infty} \frac{f(x)}{g(x)} = \begin{cases} 0, \text{ if } n < m \\ \dfrac{a_n}{b_m}, \text{ if } n = m \\ \pm\infty, \text{ if } n > m \end{cases}$$

This result seems reasonable if we realize that for large values of x the highest-powered term of a polynomial is the most "influential" term. That is, a polynomial tends to behave like its highest-powered term as x becomes sufficiently large in either the positive or negative sense.

There are many examples of asymptotic behavior in business and biology. For instance, the following example describes the asymptotic behavior of an average cost function.

Example 6

An Application Involving a Horizontal Asymptote

A business has a cost function of $C = 0.5x + 5000$, where C is measured in dollars and x is the number of units produced. The *average cost per unit* is given by

$$\overline{C} = \frac{C}{x}$$

Find the average cost per unit when $x = 1000$, 10,000, and 100,000. What is the limit of $\overline{C}$ when x approaches infinity?

Solution When $x = 1000$, the average cost per unit is

$$\overline{C} = \frac{0.5(1000) + 5000}{1000} = \$5.50$$

When $x = 10{,}000$, the average cost per unit is

$$\overline{C} = \frac{0.5(10{,}000) + 5000}{10{,}000} = \$1.00$$

When $x = 100{,}000$, the average cost per unit is

$$\overline{C} = \frac{0.5(100{,}000) + 5000}{100{,}000} = \$0.55$$

As x approaches infinity, the limiting average cost is

$$\lim_{x\to\infty} \frac{0.5x + 5000}{x} = \$0.50$$

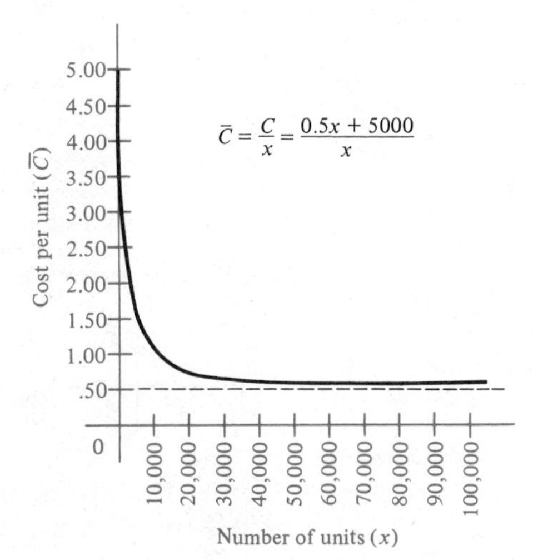

As $x \to \infty$ the average cost per unit approaches \$0.50

Figure 3.59

As shown in Figure 3.59, this example points out one of the major problems of a small business. That is, it is difficult to have competitively low prices when the production level is low. □

Section Exercises (3.6)

In Exercises 1–6, find the vertical and horizontal asymptotes. Classify vertical asymptotes as even or odd.

1. $f(x) = \dfrac{1}{x^2}$

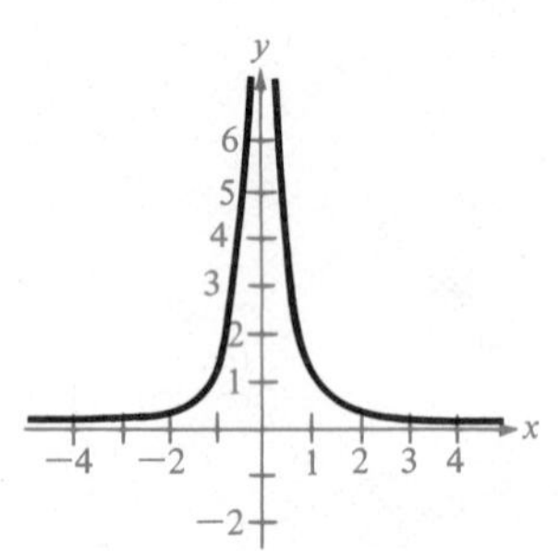

2. $f(x) = \dfrac{4}{(x-2)^3}$

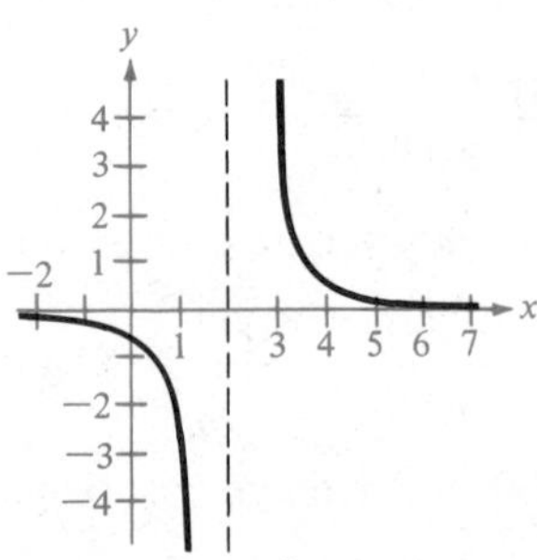

3. $f(x) = \dfrac{x^2 - 2}{x^2 - x - 2}$

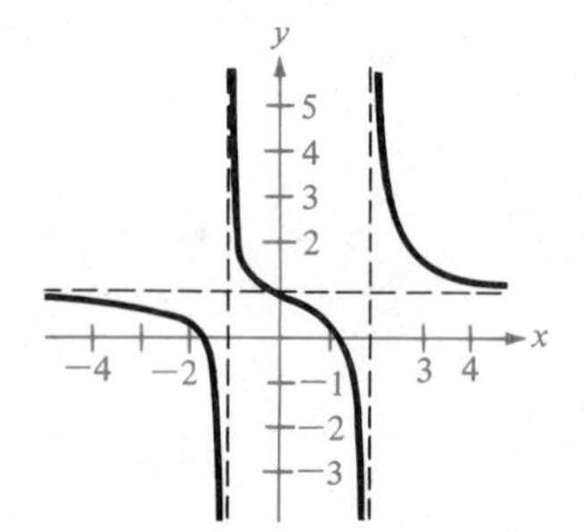

4. $f(x) = \dfrac{2 + x}{1 - x}$

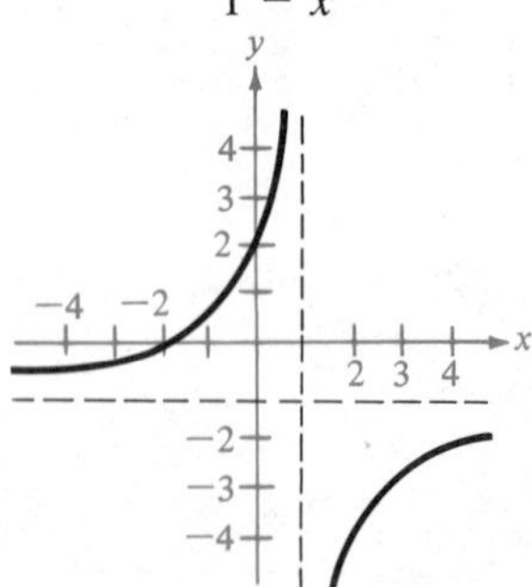

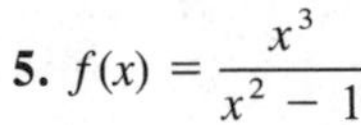

5. $f(x) = \dfrac{x^3}{x^2 - 1}$

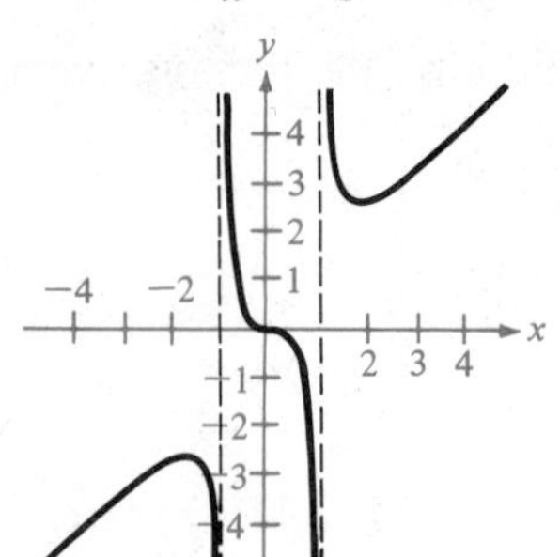

6. $f(x) = \dfrac{-4x}{x^2 + 4}$

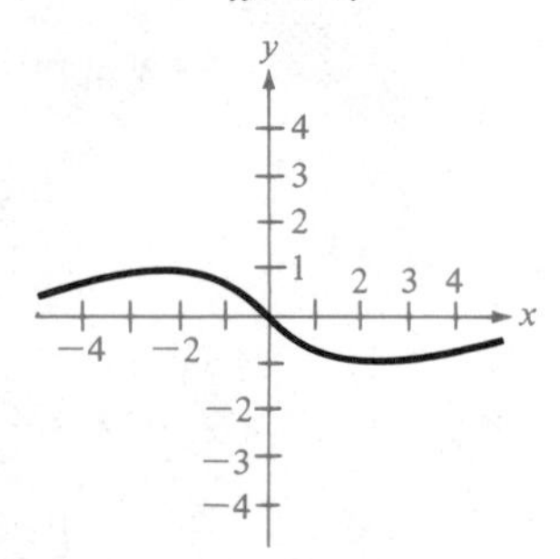

In Exercises 7–16, find the indicated limit.

7. $\lim\limits_{x\to\infty} \dfrac{2x - 1}{3x + 2}$

8. $\lim\limits_{x\to\infty} \dfrac{5x^3 + 1}{10x^3 - 3x^2 + 7}$

9. $\lim\limits_{x\to\infty} \dfrac{x}{x^2 - 1}$

10. $\lim\limits_{x\to\infty} \dfrac{2x^{10} - 1}{10x^{11} - 3}$

11. $\lim\limits_{x\to-\infty} \dfrac{5x^2}{x + 3}$

12. $\lim\limits_{x\to\infty} \dfrac{x^3 - 2x^2 + 3x + 1}{x^2 - 3x + 2}$

13. $\lim\limits_{x\to\infty} \left(2x - \dfrac{1}{x^2}\right)$

14. $\lim\limits_{x\to\infty} (x + 3)^{-2}$

15. $\lim\limits_{x\to-\infty} \left(\dfrac{2x}{x - 1} + \dfrac{3x}{x + 1}\right)$

16. $\lim\limits_{x\to\infty} \left(\dfrac{2x^2}{x - 1} + \dfrac{3x}{x + 1}\right)$

In Exercises 17–19, complete the table for each function and estimate $\lim_{x\to\infty} f(x)$.

ⓒ**17.** $f(x) = \dfrac{x + 1}{x\sqrt{x}}$

x	1	10	10^2	10^4	10^6
$f(x)$					

ⓒ**18.** $f(x) = x - \sqrt{x(x - 1)}$

x	1	10	10^2	10^4	10^6
$f(x)$					

ⓒ Calculator may be helpful.

Ⓒ**19.** $f(x) = x^2 - x\sqrt{x(x-1)}$

x	1	10	10^2	10^4	10^6
$f(x)$					

In Exercises 20–36, sketch the graph of each equation. As a sketching aid, examine each equation for intercepts, relative extrema, and asymptotes.

20. $y = \dfrac{x-3}{x-2}$

21. $y = \dfrac{2+x}{1-x}$

22. $y = \dfrac{x^2}{x^2-9}$

23. $y = \dfrac{x^2}{x^2+9}$

24. $x^2y = 4$

25. $xy^2 = 4$

26. $y = \dfrac{2x}{1-x^2}$

27. $y = \dfrac{2x}{1-x}$

28. $y = 1 + \dfrac{1}{x}$

29. $y = 2 - \dfrac{3}{x^2}$

30. $y = \dfrac{x}{\sqrt{x^2-4}}$

31. $y = \dfrac{x^3}{\sqrt{x^2-4}}$

32. $f(x) = x^2 + \dfrac{1}{x^2}$

33. $f(x) = \dfrac{x^2}{x^2-1}$

34. $f(x) = \dfrac{x}{x^2-4}$

35. $f(x) = \dfrac{1}{x^2-x-2}$

36. $f(x) = \dfrac{x-2}{x^2-4x+3}$

Ⓒ**37.** The cost in millions of dollars to the federal government to intercept and seize $x\%$ of a certain illegal drug as it enters the country is given by

$$C = \frac{528x}{100-x}, \qquad 0 \le x < 100$$

(a) Find the cost of intercepting and seizing
(i) 25% (ii) 50% (iii) 75%
of the illegal drug.

(b) Consider the limit

$$\lim_{x\to 100^-} C$$

According to this model, is it possible to intercept and seize 100% of this drug?

(c) Sketch a graph of this function.

Ⓒ**38.** The game commission in a certain state introduces 50 deer into newly acquired state game lands. It is believed that the size of the herd will increase according to the model

$$N = \frac{10(5+3t)}{1+0.04t}$$

where t is the time in years.

(a) Find the number in the herd when t is
(i) 5 years (ii) 10 years (iii) 25 years

(b) According to this model, what is the limiting size of this herd as time increases?

39. Psychologists have developed mathematical models to predict performance as a function of the number of trials n for a certain task. One such model is

$$P = \frac{b + \theta a(n-1)}{1 + \theta(n-1)}$$

where P is the percentage of correct responses after n trials and a, b, and θ are constants depending upon the actual learning situation. Find the limit of P as n approaches infinity.

Ⓒ**40.** Consider the learning curve given by

$$P = \frac{0.5 + 0.9(n-1)}{1 + 0.9(n-1)}, \qquad 0 < n$$

(a) Complete the following table for this model.

P										
n	1	2	3	4	5	6	7	8	9	10

(b) Find the limit as n approaches infinity.

(c) Sketch a graph of this learning curve.

Ⓒ**41.** The cost function for a certain product is given by

$$C = 1.35x + 4570$$

where C is measured in dollars and x is the number of units produced.

(a) Find the average cost per unit when
(i) $x = 100$ (ii) $x = 1000$

(b) What is the limit of the average cost as x approaches infinity?

Ⓒ**42.** The cost and revenue functions for a particular item are given by

$$C = 34.5x + 15{,}000 \qquad \text{and} \qquad R = 69.9x$$

(a) Find the average profit function

$$\overline{P} = \frac{R - C}{x}$$

(b) Find the average profit per unit when x is
(i) 1000 (ii) 10,000 (iii) 100,000

(c) What is the limit of the average profit function as x approaches infinity?

Ⓒ Calculator may be helpful.

Section 3.7

Curve Sketching: A Summary

Introductory Example

United States Population Predictions

The U.S. Bureau of the Census conducts a nationwide census every ten years. In addition to tallying the population of the United States, the Census Bureau compiles numerous population characteristics regarding age, sex, ethnic background, marital status, immigration patterns, and so on. With this data, the Census Bureau predicts various population growth patterns. Four of the current predictions are shown in Figure 3.60. Note that since 1950 the population curve seems to have changed its rate of growth slightly, and the three lower predictions reflect this change.

All four of these "series" predictions assume a slightly increased life expectancy and (except for Series III) an annual immigration of 400,000. (Series III assumes no immigration.) The difference in the predictions stems from different assumptions regarding the birth rate: Series I uses a birth rate of 2.7 children per "female lifetime"; Series II and III, 2.1; and Series IV, 1.7.

Table 3.11 lists the census results, 1900–1980.

Table 3.11

U.S. Population	75,994,575	91,972,266	105,710,620
Year	1900	1910	1920

U.S. Population	122,775,048	131,669,275	150,697,361
Year	1930	1940	1950

U.S. Population	179,323,175	203,235,298	223,889,000
Year	1960	1970	1980

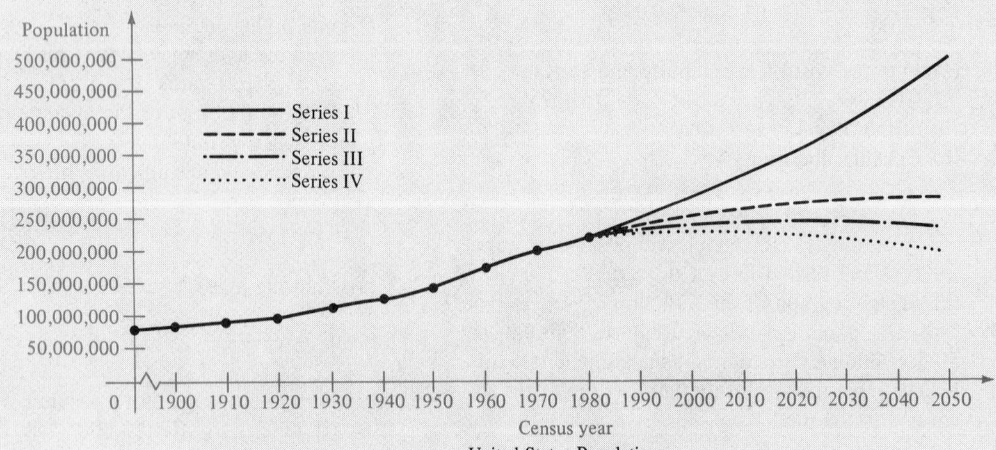

Figure 3.60 United States Population

Section Objective: ***To summarize curve-sketching techniques presented in this and previous chapters.***

The cliché "a picture is worth a thousand words" properly identifies the importance of curve sketching in mathematics. Descartes' introduction of this fruitful concept not only preceded but to a great extent was responsible for the rapid advances in mathematics during the last half of the seventeenth century. Today government, science, industry, business, education, and the social and health sciences all make widespread use of graphs to describe and predict relationships between variables within their domain of interest.

Although we can gain some idea of the relationship between two variables directly from the mathematical equation relating the variables, the true relationship is often best seen from a graph of the equation. In many instances, the graph of a functional relationship is easily made by merely plotting a collection of points. Yet we have seen how some functions behave rather oddly, and to sketch their graphs requires considerable ingenuity.

We have previously discussed several concepts that are useful in sketching the graph of a function. For instance, the following properties of a function f and its graph have been discussed at some length: domain and range of f; continuity; x- and y-intercepts; horizontal and vertical asymptotes; relative minima and maxima; concavity and points of inflection. In this section, we incorporate these concepts into an effective procedure for sketching curves.

Suggestions for Sketching the Graph of a Function

1. Make a rough preliminary sketch, using any easily determined intercepts, vertical asymptotes, or other points of discontinuity.
2. Locate the x-values where $f'(x)$ and $f''(x)$ are either zero or undefined.
3. Test the behavior of f *at* each of these x-values as well as *within* each interval determined by them. (See Table 3.12 in Example 1.)
4. Sharpen the accuracy of the final sketch by plotting the relative extrema points, the points of inflection, and a few points between.

Example 1

Analyzing a Graph

Sketch a graph of

$$f(x) = \frac{x^2 - 2x + 4}{x - 2}$$

Solution f has a vertical asymptote at $x = 2$ and a y-intercept at $(0, -2)$.

$$f'(x) = \frac{x(x - 4)}{(x - 2)^2} \quad \text{and} \quad f''(x) = \frac{8}{(x - 2)^3}$$

The critical numbers are $x = 0$ and $x = 4$. There are no possible points of inflection since $f''(x)$ is never zero. Testing, we have the results shown in Table 3.12. The graph is shown in Figure 3.61.

Table 3.12

	$f(x)$	$f'(x)$	$f''(x)$	Shape of graph
x in $(-\infty, 0)$		+	−	increasing, concave down
$x = 0$	−2	0	−	relative maximum
x in $(0, 2)$		−		decreasing, concave down
$x = 2$	does not exist	does not exist	does not exist	odd vertical asymptote
x in $(2, 4)$		−	+	decreasing, concave up
$x = 4$	6	0	+	relative minimum
x in $(4, \infty)$		+	+	increasing, concave up

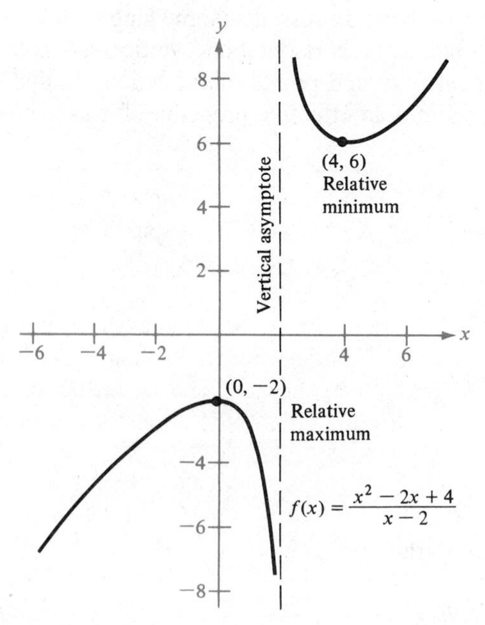

Figure 3.61

□

Example 2

Analyzing a Graph Sketch a graph of $f(x) = \sqrt{x^2 - 3}$.

Solution The function is not defined for $-\sqrt{3} < x < \sqrt{3}$; hence there is no y-intercept.

$$f'(x) = \frac{x}{\sqrt{x^2 - 3}} \quad \text{and} \quad f''(x) = \frac{-3}{(x^2 - 3)^{3/2}}$$

The critical numbers are $x = \pm\sqrt{3}$, and there are no possible points of inflection. The test results are given in Table 3.13; the graph is shown in Figure 3.62.

Table 3.13

	$f(x)$	$f'(x)$	$f''(x)$	Shape of graph
x in $(-\infty, -\sqrt{3})$		$-$	$-$	decreasing, concave down
$x = -\sqrt{3}$	0	does not exist	does not exist	minimum, vertical tangent
x in $(-\sqrt{3}, \sqrt{3})$	does not exist	does not exist	does not exist	undefined
$x = \sqrt{3}$	0	does not exist	does not exist	minimum, vertical tangent
x in $(\sqrt{3}, \infty)$		$+$	$-$	increasing, concave down

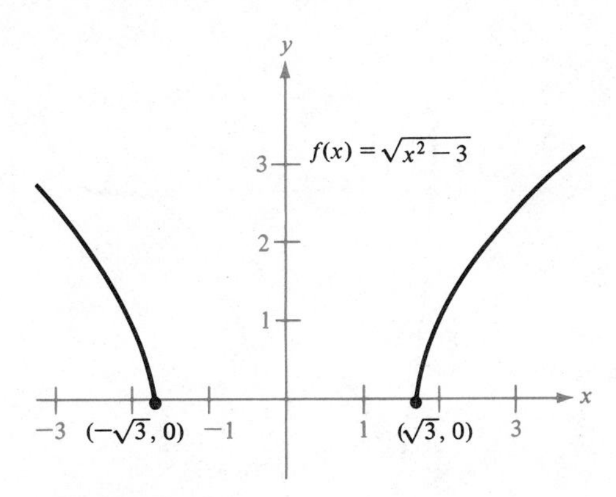

Figure 3.62

□

Example 3

Analyzing a Graph

Sketch the graph of $f(x) = x^4 - 12x^3 + 48x^2 - 64x$.

Solution By factoring, we see that

$$f(x) = x^4 - 12x^3 + 48x^2 - 64x = x(x-4)^3$$

The first derivative is given by

$$f'(x) = 4x^3 - 36x^2 + 96x - 64 = 4(x-1)(x-4)^2$$

which implies that the critical numbers are $x = 1$ and $x = 4$.

The second derivative is

$$f''(x) = 12x^2 - 72x + 96 = 12(x-4)(x-2)$$

which implies that there are possible points of inflection at $x = 2$ and $x = 4$. Testing these values, we have the results shown in Table 3.14. The graph of this function is shown in Figure 3.63.

Table 3.14

	$f(x)$	$f'(x)$	$f''(x)$	Shape of graph
x in $(-\infty, 1)$		$-$	$+0$	decreasing, concave up
$x = 1$	-27		$+$	minimum
x in $(1, 2)$		$+$	$+$	increasing, concave up
$x = 2$	-16	$+$	0	point of inflection
x in $(2, 4)$		$+$	$-$	increasing, concave down
$x = 4$	0	0	0	point of inflection
x in $(4, \infty)$		$+$	$+$	increasing, concave up

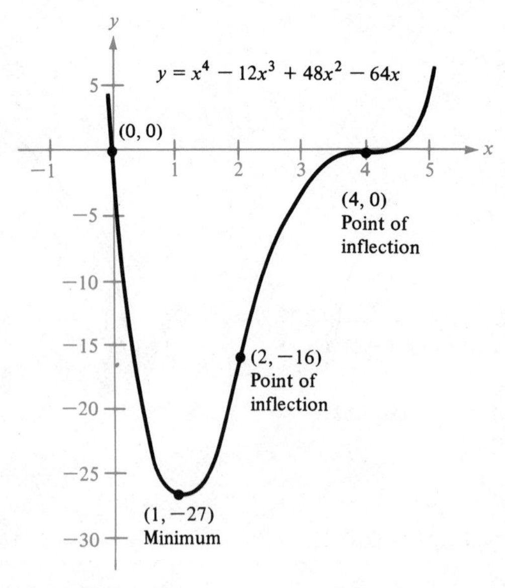

Figure 3.63

□

Example 4

Analyzing a Graph Sketch a graph of $f(x) = 2x^{5/3} - 5x^{4/3}$.

Solution Since

$$f(x) = x^{4/3}(2x^{1/3} - 5)$$

the intercepts are (0, 0) and $(\frac{125}{8}, 0)$. The first and second derivatives are

$$f'(x) = \frac{10x^{1/3}(x^{1/3} - 2)}{3} \quad \text{and} \quad f''(x) = \frac{20(x^{1/3} - 1)}{9x^{2/3}}$$

Testing, we have the results given in Table 3.15. The graph is shown in Figure 3.64.

Table 3.15

	$f(x)$	$f'(x)$	$f''(x)$	Shape of graph
x in $(-\infty, 0)$		+	−	increasing, concave down
$x = 0$	0	0	does not exist	relative maximum
x in (0, 1)		−	−	decreasing, concave down
$x = 1$	−3	−	0	point of inflection
x in (1, 8)		−	+	decreasing, concave up
$x = 8$	−16	0	+	relative minimum
x in $(8, \infty)$		+	+	increasing, concave up

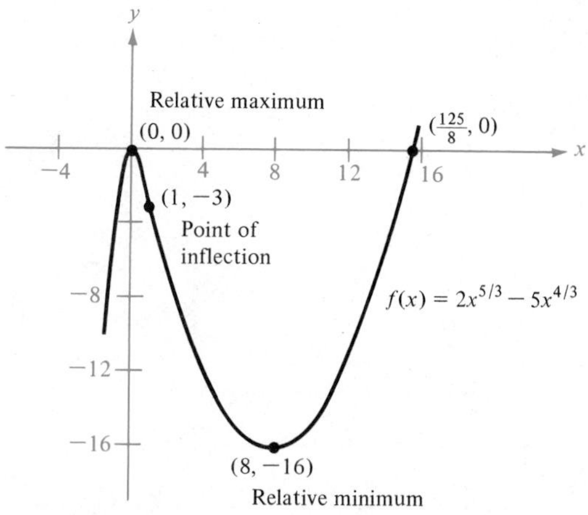

Figure 3.64

□

We conclude this section with a summary of the graphs of polynomial functions of degrees 0, 1, 2, and 3. Because of their simplicity, lower-degree polynomial functions are commonly used as mathematical models, and you will do well to study the following summary carefully.

Graphs of Polynomial Functions

Constant Function (Degree 0):

$$y = a$$

Horizontal Line:

Linear Function (Degree 1):

$$y = ax + b$$

Line (of slope a):

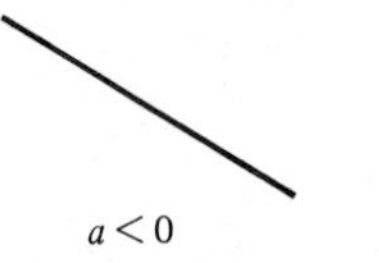

$a<0$

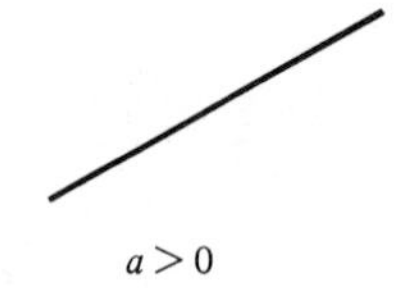

$a>0$

Quadratic Function (Degree 2):

$$y = ax^2 + bx + c$$

Parabola:

$a<0$

$a>0$

Cubic Function (Degree 3):

$$y = ax^3 + bx^2 + cx + d$$

Cubic Curve:

$a<0$

$a>0$

Section Exercises (3.7)

In Exercises 1–22, sketch the graph of each function, choosing a scale that allows all relative extrema and points of inflection to be identified on the sketch.

1. $y = x^3 - 3x^2 + 3$ **2.** $3y = -x^3 + 3x - 2$

3. $y = 2 - x - x^3$

4. $y = x^3 + 3x^2 + 3x + 2$

5. $y = 3x^3 - 9x + 1$

6. $y = (x + 1)(x - 2)(x - 5)$

7. $y = -x^3 + 3x^2 + 9x - 2$

8. $3y = (x - 1)^3 + 6$ **9.** $y = 3x^4 + 4x^3$

10. $y = 3x^4 - 6x^2$ **11.** $y = x^4 - 4x^3 + 16x$

12. $y = x^4 - 8x^3 + 18x^2 - 16x + 5$

13. $y = x^4 - 4x^3 + 16x - 16$

14. $y = x^5 + 1$ **15.** $y = x^5 - 5x$

16. $y = |2x - 3|$ **17.** $y = |x^2 - 6x + 5|$

18. $y = \dfrac{x}{x^2 + 1}$ **19.** $y = \dfrac{x^2}{x^2 + 3}$

20. $y = 3x^{2/3} - x^2$ **21.** $y = 3x^{2/3} - 2x$

22. $y = \dfrac{x}{\sqrt{x^2 + 7}}$

In Exercises 23–30, sketch the graph of each function. In each case label the intercepts, relative extrema, points of inflection, and the domain.

23. $y = \dfrac{1}{x - 2} - 3$

24. $y = \dfrac{x^2 + 1}{x^2 - 2}$

25. $y = \dfrac{2x}{x^2 - 1}$

26. $y = \dfrac{x^2 - 6x + 12}{x - 4}$

27. $y = x\sqrt{4 - x}$

28. $y = x\sqrt{4 - x^2}$

29. $y = \dfrac{x + 2}{x}$

30. $y = x + \dfrac{32}{x^2}$

Chapter 4

Integration

Section 4.1

Antiderivatives and the Indefinite Integral

Introductory Example

Average Weekly Salary for U.S. Workers

The average weekly earnings for full-time (nonsupervisory) workers in the United States increased from \$85.91 in 1962 to \$235.10 in 1980. If we let S be the weekly earnings, then the rate of change of S can be approximated by the mathematical model

$$\frac{dS}{dt} = 0.0273t^2 + 0.2708t + 2.1336$$

where S is measured in dollars and t in years, with $t = 1$ corresponding to 1962. Thus, at the beginning of 1962, weekly salaries were increasing at a rate of \$2.43 per year. To estimate the value of S for 1963, we can add the mid-year ($t = 1.5$) estimated increase to the weekly earnings for 1962. Table 4.1 shows the results of continuing this process through 1980. Note how well this model fits the actual average salaries during this time period.

Suppose that we want to use this model to predict the average weekly salary in the year 2000. We could continue the process shown in Table 4.1, but that would be tedious. A simpler way would be to find the weekly salary function S from its derivative (rate of change) dS/dt. We call the function S an **antiderivative** of dS/dt. Using techniques that will be presented in this section, we find the weekly salary function to be

$$S = 0.0091t^3 + 0.1354t^2 + 2.1336t + 83.2653$$

Finally, by letting $t = 39$ (for the year 2000), we approximate the weekly salary in the year 2000 as \$912.22. (See Figure 4.1 on page 262.)

Table 4.1

Year	Actual salary (\$)	Estimated salary (\$)	Estimated increase (\$)
1962	85.91	85.91	
1963	88.46	88.51	2.60
1964	91.33	91.49	2.98
1965	95.45	94.91	3.42
1966	98.82	98.82	3.91
1967	101.84	103.27	4.45
1968	107.73	108.32	5.05
1969	114.61	114.02	5.70
1970	119.83	120.43	6.41
1971	127.31	127.60	7.17
1972	136.90	135.59	7.99
1973	145.39	144.45	8.86
1974	154.76	154.23	9.78
1975	163.53	164.99	10.76
1976	175.45	176.79	11.80
1977	189.00	189.68	12.89
1978	203.70	203.71	14.03
1979	219.30	218.94	15.23
1980	235.10	235.43	16.49

Section Objectives: ***To define integration as the inverse operation of differentiation*** ▪ ***To derive some rules for evaluating indefinite integrals.***

Figure 4.1

Up to this point in our study of calculus, we have been concerned primarily with this problem:

Given a function, find its derivative.

Many important applications of calculus involve the inverse problem:

Given the derivative of a function, find the original function.

For example, suppose we are given the following derivatives:

$$f'(x) = 2, \qquad g'(x) = 3x^2, \qquad s'(t) = 4t$$

Our problem is to determine the functions f, g, and s that have these respective derivatives. If we make some educated guesses, we might come up with the following functions:

$$f(x) = 2x \qquad \text{because} \qquad \frac{d}{dx}[2x] = 2$$

$$g(x) = x^3 \qquad \text{because} \qquad \frac{d}{dx}[x^3] = 3x^2$$

$$s(t) = 2t^2 \qquad \text{because} \qquad \frac{d}{dt}[2t^2] = 4t$$

This operation of determining the original function from its derivative is the inverse operation of differentiation, and we call it **antidifferentiation.**

Definition of an Antiderivative

A function F is called an **antiderivative** of a function f if $F'(x) = f(x)$ for every x in the domain of f.

We will use the phrase "$F(x)$ is an antiderivative of $f(x)$" synonymously with "F is an antiderivative of f."

It should be emphasized that if $F(x)$ is an antiderivative of $f(x)$, then $F(x) + C$ (where C is any constant) is also an antiderivative of $f(x)$. For instance,

$$F(x) = x^3, \qquad G(x) = x^3 - 5, \qquad H(x) = x^3 + 0.3$$

are all antiderivatives of $3x^2$, because

$$\frac{d}{dx}[x^3] = \frac{d}{dx}[x^3 - 5] = \frac{d}{dx}[x^3 + 0.3] = 3x^2$$

As it turns out, *all* the antiderivatives of $3x^2$ are of the form $x^3 + C$. The point is that the process of antidifferentiation does not determine a unique function but rather a *family* of functions, each differing from the other by a constant. The antidifferentiation process is commonly referred to as **integration** and is denoted by the symbol

$$\int$$

called an **integral sign.** The symbol

$$\int f(x)\,dx$$

is called the **indefinite integral** of $f(x)$, and it denotes the family of antiderivatives of $f(x)$. More specifically, if $F'(x) = f(x)$ for all x, then

$$\int f(x)\,dx = F(x) + C$$

where $f(x)$ is called the **integrand** and C the **constant of integration.**

For the present we simply note that the dx in the indefinite integral identifies the variable of integration. That is, the symbol $\int f(x)\,dx$ denotes the "antiderivative of f *with respect to x*" just as the symbol dy/dx denotes the "derivative of y *with respect to x*." Additional uses of this notation will become apparent later.

The inverse nature of the operations of integration and differentiation can be shown symbolically as follows.

Differentiation and Integration Are Inverse Operations

Differentiation is the inverse of integration:

$$\frac{d}{dx}\left[\int f(x)\,dx\right] = f(x)$$

Integration is the inverse of differentiation:

$$\int f'(x)\,dx = f(x) + C$$

We have not yet provided any rules for determining antiderivatives. Fortunately, since integration (antidifferentiation) is the inverse operation of differentiation, we can readily obtain integration rules from differentiation rules. The following rules are easily verified by differentiation.

Basic Integration Rules

1. Constant Rule: $\int dx = x + C$
2. Constant Multiple Rule: $\int kf(x)\,dx = k\int f(x)\,dx$
3. Sum Rule: $\int [f(x) \pm g(x)]\,dx = \int f(x)\,dx \pm \int g(x)\,dx$
4. Simple Power Rule: $\int x^n\,dx = \frac{x^{n+1}}{n+1} + C$, if $n \neq -1$

Remark

Be sure you see that the Simple Power Rule has the restriction that n cannot be -1. This means that we *cannot* use the Simple Power Rule to evaluate the integral

$$\int \frac{1}{x}\,dx$$

To be able to evaluate this integral, we must wait until the natural logarithmic function is introduced in Section 5.4.

Applications of these rules are demonstrated in the following examples.

Example 1

Evaluating Indefinite Integrals

Evaluate the indefinite integral $\int (3x - 7)\,dx$.

Solution

$$\int (3x - 7)\,dx = \int 3x\,dx - \int 7\,dx \qquad \text{(Rule 3)}$$

$$= 3\int x\,dx - 7\int dx \qquad \text{(Rule 2)}$$

$$= 3\left(\frac{x^2}{2} + C_1\right) - 7(x + C_2) \qquad \text{(Rules 4 and 1)}$$

$$= \tfrac{3}{2}x^2 - 7x + 3C_1 - 7C_2$$

Because $3C_1 - 7C_2$ is just another constant, we write our answer as

$$\int (3x - 7)\,dx = \tfrac{3}{2}x^2 - 7x + C$$

□

We frequently use the properties of integrals without specifically identifying them, and many times we use more than one rule in a given step so as to reduce the total number of steps in solving a problem.

Example 2

Evaluating Indefinite Integrals

Evaluate the indefinite integral $\int \sqrt[3]{y}\,dy$.

Solution

$$\int \sqrt[3]{y}\,dy = \int y^{1/3}\,dy = \frac{y^{(1/3)+1}}{\frac{1}{3} + 1} + C$$

$$= \frac{y^{4/3}}{\frac{4}{3}} + C = \tfrac{3}{4}y^{4/3} + C$$

□

Example 3

Evaluating Indefinite Integrals

Evaluate

$$\int \left(\frac{3}{x^2} - \frac{1}{\sqrt{x^3}}\right) dx$$

Solution

$$\int \left(\frac{3}{x^2} - \frac{1}{\sqrt{x^3}}\right) dx = \int (3x^{-2} - x^{-3/2})\,dx$$

$$= 3\int x^{-2}\,dx - \int x^{-3/2}\,dx$$

$$= 3\left(\frac{x^{-1}}{-1}\right) - \frac{x^{-1/2}}{-\frac{1}{2}} + C = \frac{-3}{x} + \frac{2}{\sqrt{x}} + C$$

□

In the preceding examples we found a family of functions $F(x) + C$, each having the derivative $F'(x)$. Sometimes additional information is given that allows us to determine the unique member of the family we want.

Example 4

Finding a Particular Solution

Given $f'(x) = 6 - x^{1/2}$ and $f(1) = \frac{4}{3}$, find $f(x)$.

Solution

$$f(x) = \int (6 - x^{1/2})\,dx = 6x - \frac{x^{3/2}}{\frac{3}{2}} + C$$

$$= 6x - \tfrac{2}{3}x^{3/2} + C$$

Thus, we know that $f(x)$ is a member of the family given by

$$f(x) = 6x - \tfrac{2}{3}x^{3/2} + C$$

Now, to find the particular antiderivative we want, we must find which value of C will satisfy the given condition

$$f(1) = \tfrac{4}{3}$$

To do this, we substitute 1 into the general form for $f(x)$ and set the result equal to $\frac{4}{3}$ as follows:

$$f(1) = 6(1) - \tfrac{2}{3}(1)^{3/2} + C = \tfrac{4}{3}$$

$$6 - \tfrac{2}{3} + C = \tfrac{4}{3}$$

$$C = \tfrac{4}{3} - 6 + \tfrac{2}{3}$$

$$C = -4$$

Finally, we conclude that the particular solution is

$$f(x) = 6x - \tfrac{2}{3}x^{3/2} - 4$$

□

Example 5

A Business Application

Suppose the marginal cost for producing x units of a commodity is given by

$$\frac{dC}{dx} = 32 - 0.04x$$

If it costs \$50 to make 1 unit, find the total cost of making 200 units.

Solution Since dC/dx is the derivative of the total cost function, we have

$$C = \int (32 - 0.04x)\,dx = 32x - 0.04\left(\frac{x^2}{2}\right) + K$$

$$= 32x - 0.02x^2 + K$$

When $x = 1$, $C = 50$; thus,

$$50 = 32(1) - 0.02(1)^2 + K$$

$$18.02 = K$$

Therefore, the total cost function is

$$C = 32x - 0.02x^2 + 18.02$$

The cost of making 200 units is, therefore,

$$C = 32(200) - 0.02(40{,}000) + 18.02 = \$5618.02$$ □

Example 6

An Application Involving Gravity

A ball is thrown upward with an initial velocity of 64 ft/s from a height of 80 ft. Use antidifferentiation to verify that the equation for the position function is

$$s(t) = -16t^2 + 64t + 80$$

Solution To begin, we let $t = 0$ represent the initial time. Then the two stated conditions in the problem can be written as follows:

Initial height is 80 ft. ➡ $s(0) = 80$

Initial velocity is 64 ft/s. ➡ $s'(0) = 64$

Furthermore, the implied condition (regarding the force of gravity) gives us the following equation for the second derivative of s.

Acceleration due to gravity is -32 ft/s^2. ➡ $s''(t) = -32$

Now, integrating $s''(t)$, we have

$$s'(t) = \int s''(t)\,dt = \int -32\,dt = -32t + C_1$$

where $C_1 = s'(0) = 64$. Similarly, by integrating $s'(t)$, we have

$$s(t) = \int s'(t)\,dt = \int (-32t + 64)\,dt = -16t^2 + 64t + C_2$$

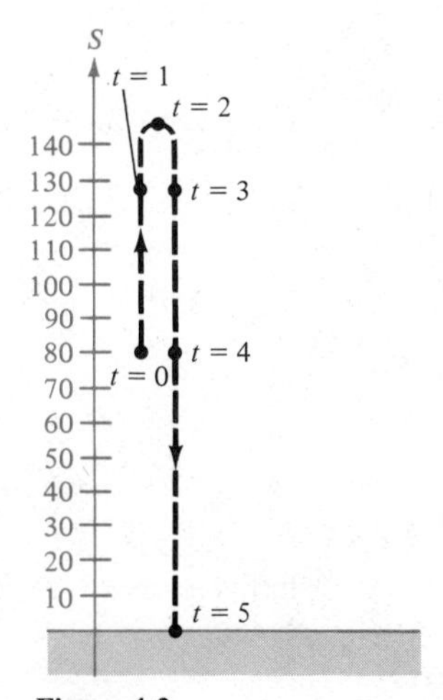

Figure 4.2

where $C_2 = s(0) = 80$. Therefore, we have

$$s(t) = -16t^2 + 64t + 80$$

(See Figure 4.2.) □

Before beginning the exercise set for this section, be sure you realize that one of the most important steps in finding antiderivatives is *rewriting the integrand* in a form that fits the basic integration rules. To further illustrate this point, we list several additional examples in Table 4.2.

Table 4.2

Given	Rewrite	Integrate	Simplify
$\int \sqrt{x}\, dx$	$\int x^{1/2}\, dx$	$\dfrac{x^{3/2}}{\frac{3}{2}} + C$	$\frac{2}{3}x^{3/2} + C$
$\int \dfrac{2}{\sqrt{x}}\, dx$	$2\int x^{-1/2}\, dx$	$2\left(\dfrac{x^{1/2}}{\frac{1}{2}}\right) + C$	$4x^{1/2} + C$
$\int (x^2 + 1)^2\, dx$	$\int (x^4 + 2x^2 + 1)\, dx$	$\dfrac{x^5}{5} + 2\left(\dfrac{x^3}{3}\right) + x + C$	$\dfrac{x^5}{5} + \dfrac{2x^3}{3} + x + C$
$\int \dfrac{x^3 + 3}{x^2}\, dx$	$\int (x + 3x^{-2})\, dx$	$\dfrac{x^2}{2} + 3\left(\dfrac{x^{-1}}{-1}\right) + C$	$\dfrac{x^2}{2} - \dfrac{3}{x} + C$
$\int \sqrt[3]{x}(x - 4)\, dx$	$\int (x^{4/3} - 4x^{1/3})\, dx$	$\dfrac{x^{7/3}}{\frac{7}{3}} - 4\left(\dfrac{x^{4/3}}{\frac{4}{3}}\right) + C$	$\dfrac{3x^{4/3}}{7}(x - 7) + C$

Section Exercises (4.1)

In Exercises 1–24, evaluate the indefinite integrals and check your results by differentiation.

1. $\int (x^3 + 2)\, dx$

2. $\int (x^2 - 2x + 3)\, dx$

3. $\int (x^{3/2} + 2x + 1)\, dx$

4. $\int \left(\sqrt{x} + \dfrac{1}{2\sqrt{x}}\right) dx$

5. $\int \sqrt[3]{x^2}\, dx$

6. $\int (\sqrt[4]{x^3} + 1)\, dx$

7. $\int \dfrac{1}{x^3}\, dx$

8. $\int \dfrac{1}{x^2}\, dx$

9. $\int \dfrac{1}{4x^2}\, dx$

10. $\int (2x + x^{-1/2})\, dx$

11. $\int \dfrac{x^2 + x + 1}{\sqrt{x}}\, dx$

12. $\int \dfrac{x^2 + 1}{x^2}\, dx$

13. $\int (x + 1)(3x - 2)\, dx$

14. $\int (2t^2 - 1)^2\, dt$

15. $\int \dfrac{t^2 + 2}{t^2}\, dt$

16. $\int (1 - 2y + 3y^2)\, dy$

17. $\int y^2\sqrt{y}\, dy$

18. $\int (1 + 3t)t^2\, dt$

19. $\int dx$

20. $\int 3\, dt$

21. $\int t^2\left(t - \dfrac{2}{t}\right)^2 dt$

22. $\int \left(\dfrac{t^3}{3} + \dfrac{1}{4t^2}\right) dt$

23. $\int (9 - y)\sqrt{y}\, dy$

24. $\int 2y(8 - y^{3/2})\, dy$

In Exercises 25–28, find $y = f(x)$ satisfying the given conditions.

25. $f'(x) = 2;\ f(2) = 5$

26. $f'(x) = x^2; f(0) = 6$

27. $f''(x) = x^{-3/2}; f'(4) = 2; f(0) = 0$

28. $f''(x) = x^{-3/2}; f'(1) = 2; f(9) = -4$

29. A company produces a product for which the marginal cost of producing x units is

$$\frac{dC}{dx} = 2x - 12$$

and the fixed costs amount to \$125.
(a) Find the total cost function.
(b) Find the average cost function.
(c) Find the total cost of producing 50 units.

30. A company produces a product for which the marginal cost of producing x units is

$$\frac{dC}{dx} = k, \qquad k \text{ is constant}$$

(In other words, for each unit increase in output, the increase in cost is always k.) Describe the cost function for this product.

31. When a particular marketing firm sells x units, its marginal revenue is

$$\frac{dR}{dx} = 100 - 5x$$

(a) Find the revenue function. (Remember that $R = 0$ when $x = 0$.)
(b) Find the revenue when 18 units are sold.
(c) Find the demand function.

32. A department store has determined its marginal revenue from x units of a particular product to be

$$\frac{dR}{dx} = 100 - 6x - 2x^2$$

(a) Find the revenue function. (Remember that $R = 0$ when $x = 0$.)
(b) Find the revenue when 6 units are sold.
(c) Find the demand function.

33. An evergreen nursery usually sells a certain type of shrub after 6 years of growth and shaping. The growth rate after t years is given by

$$\frac{dh}{dt} = 0.5t + 2$$

where $t = 0$ represents the time the shrubs are 5-in. seedlings ($h = 5$ when $t = 0$).
(a) Find the height h after t years.
(b) How tall are these shrubs when they are sold?

In Exercises 34–38, use $a(t) = -32$ ft/s² as the acceleration due to gravity. (Neglect air resistance.)

34. An object is dropped from a balloon, which is stationary at 1600 ft. Express its height above the ground as a function of t. How long does it take the object to reach the ground?

35. A ball is thrown vertically upward with an initial velocity of 60 ft/s. How high will the ball go?

36. With what initial velocity must an object be thrown upward from the ground to reach a maximum height of 550 ft? (Approximate height of the Washington Monument.)

37. If an object is thrown upward from a point s_0 feet above the ground with an initial velocity of v_0 feet per second, show that its height above the ground is given by the function

$$f(t) = -16t^2 + v_0t + s_0$$

Ⓒ**38.** A balloon, rising vertically with a velocity of 16 ft/s, releases a sandbag at an instant when the balloon is 64 ft above the ground.
(a) How many seconds after its release will the bag strike the ground?
(b) With what velocity will it reach the ground?

Ⓒ Calculator may be helpful.

Section 4.2

The General Power Rule

Introductory Example

Propensity to Consume

The U.S. poverty level for a family of four in 1980 was around \$10,000 (after taxes). Families at or below this income level had to use all their income to purchase the necessities of life, such as food, clothing, and shelter. Economists say that such families consume 100% of their income. As income level increases the average consumption begins to drop below 100%. For instance, a family earning \$11,000 (after taxes) may be willing to save \$185 and thus consume only \$10,815 (98.3%) of their income. The *extra* amount (measured as a percentage) a family is willing to consume when income increases is the *marginal propensity to consume*.

Let us assume that the marginal propensity to consume an *after tax* income x (greater than \$10,000) is

$$\frac{dQ}{dx} = \frac{0.97}{(x - 9999)^{0.03}}, \qquad 10{,}000 \le x$$

Note that as the income x increases, dQ/dx decreases, showing that families with larger incomes tend to be less willing to spend their entire income. How can we use this marginal function to predict the amount a family with an after tax income of \$20,000 is willing to consume? To do this, we observe that the propensity to consume Q, is an antiderivative of dQ/dx. Furthermore, since $Q = \$10{,}000$ when $x = \$10{,}000$, we can apply the **General Power Rule** introduced in this section:

$$Q = 9999 + (x - 9999)^{0.97}, \qquad 10{,}000 \le x$$

Thus, from this model we predict that a family with an after tax income of \$20,000 would be willing to consume \$17,585.51 and save the remaining \$2,414.49. Figure 4.3 shows this consumption function for after tax incomes between \$10,000 and \$150,000.

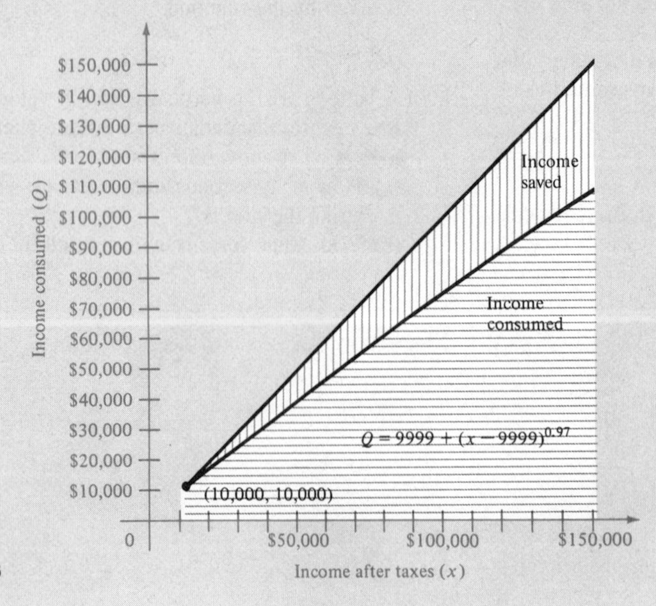

Figure 4.3

Section Objective: *To introduce the General Power Rule for integration.*

In Section 4.1, we used the Simple Power Rule

$$\int x^n\,dx = \frac{x^{n+1}}{n+1} + C, \qquad n \neq -1$$

to find antiderivatives of functions expressed as powers of x alone. In this section, we look at a technique for finding antiderivatives for more general functions.

Development of the General Power Rule

To begin, consider how you might evaluate the following integral:

$$\int 2x(x^2+1)^3\,dx$$

You could expand the integrand into polynomial form and then use the Simple Power Rule on the individual powers of x. *But,* there is a simpler way to evaluate this integral! To see this, you need to remember that what we are hunting for is a function f such that

$$f'(x) = 2x(x^2+1)^3$$

After some experimentation, you might observe that

$$\frac{d}{dx}[(x^2+1)^4] = 4(x^2+1)^3(2x)$$

and, by dividing by 4, you could obtain

$$\frac{d}{dx}\left[\frac{(x^2+1)^4}{4}\right] = (x^2+1)^3(2x)$$

Now, having found a function with the desired derivative, you could conclude that

$$\int 2x(x^2+1)^3\,dx = \frac{(x^2+1)^4}{4} + C$$

Be sure you see that the key to this solution is the presence of the factor $2x$ in the integrand. The importance of $2x$ as a factor of the integrand lies in the fact that it is precisely the derivative of x^2+1. That is, if

$$u = x^2+1$$

then $\quad u' = 2x$

Thus, written in terms of u, our original integral becomes

$$\int \underbrace{2x}_{u'}\underbrace{(x^2+1)^3}_{u^3}\,dx = \int u^3u'\,dx = \frac{u^4}{4} + C$$

This is just one example of an important integration rule called the **General Power Rule.**

General Power Rule for Integration

If u is a differentiable function of x, then

$$\int u^n u' \, dx = \frac{u^{n+1}}{n+1} + C, \qquad \text{where } n \neq -1$$

An important consideration that is often overlooked when using the General Power Rule is the existence of u' as a factor of the integrand. We must first determine u by identifying, within the integrand, a function u that is raised to some power. Then, secondly, we must show that its derivative u' is also a factor of the integrand. Simply stated, the clues for using the General Power Rule for integration are as follows:

1. Identify the function u that is raised to a power.
2. Check to see if u' is also a factor of the integrand.

We demonstrate these considerations in the next example.

Example 1

Applying the General Power Rule

Use the General Power Rule to evaluate each of the following integrals.

(a) $\int 3(3x - 1)^4 \, dx$ (b) $\int (2x + 1)(x^2 + x) \, dx$

(c) $\int 3x^2\sqrt{x^3 - 2} \, dx$ (d) $\int \frac{-4x}{(1 - 2x^2)^2} \, dx$

Solution

(a) By letting $u = 3x - 1$, we have $u' = 3$.

$$\int 3(3x - 1)^4 \, dx = \int \overbrace{(3x - 1)^4}^{u^n} \overbrace{(3)}^{u'} \, dx = \frac{(3x - 1)^5}{5} + C$$

(b) By letting $u = x^2 + x$, we have $u' = 2x + 1$.

$$\int (2x + 1)(x^2 + x) \, dx = \int \overbrace{(x^2 + x)^1}^{u^n} \overbrace{(2x + 1)}^{u'} \, dx = \frac{(x^2 + 1)^2}{2} + C$$

(c) By letting $u = x^3 - 2$, we have $u' = 3x^2$.

$$\int 3x^2\sqrt{x^3 - 2} \, dx = \int \overbrace{(x^3 - 2)^{1/2}}^{u^n} \overbrace{(3x^2)}^{u'} \, dx = \frac{(x^3 - 2)^{3/2}}{\frac{3}{2}} + C$$

(d) By letting $u = 1 - 2x^2$, we have $u' = -4x$.

$$\int \frac{-4x}{(1-2x^2)^2}\,dx = \int \overbrace{(1-2x^2)^{-2}}^{u^n}\overbrace{(-4x)}^{u'}\,dx = \frac{(1-2x^2)^{-1}}{-1} + C \qquad \square$$

Many times, part of u' is missing from the integrand, and in *some* such cases we can make the necessary adjustments in order to apply the General Power Rule. In other instances we cannot adjust appropriately and, therefore, cannot apply the General Power Rule.

Example 2

Applying the General Power Rule

Evaluate $\int x(3-4x^2)^2\,dx$.

Solution Let $u = 3 - 4x^2$; then $u' = -8x$. Now we see that the factor -8 is not a part of the integrand. However, we can adjust the integrand in the following way by multiplying by -8 and its reciprocal.

$$\int x(3-4x^2)^2\,dx = \int \frac{1}{-8}(3-4x^2)^2(-8x)\,dx$$

Now, because $-\frac{1}{8}$ is a constant, we can factor it out of the right-hand integral to obtain

$$-\frac{1}{8}\int \overbrace{(3-4x^2)^2}^{u^2}\overbrace{(-8x)}^{u'}\,dx$$

to which we apply the General Power Rule to get

$$\left(-\frac{1}{8}\right)\frac{(3-4x^2)^3}{3} + C = \frac{-(3-4x^2)^3}{24} + C \qquad \square$$

Example 3

An Example for Which the General Power Rule Fails

Evaluate $\int -8(3-4x^2)^2\,dx$.

Solution If we let $u = 3 - 4x^2$, then $u' = -8x$, and again part of u' is missing from the integrand. Since the missing part of u' is a *variable* rather than a constant, the adjustment would require that we move the variable quantity $1/x$ outside the integral sign. However, this is not possible. That is,

$$\int -8(3-4x^2)^2\,dx = \int \left(\frac{1}{x}\right)(3-4x^2)^2(-8x)\,dx$$

$$\neq \frac{1}{x}\int (3-4x^2)^2(-8x)\,dx$$

(Note: *If* we were permitted to move variable quantities outside the integral sign, we would be able to move the entire integrand out and eliminate the problem entirely.) In this example we cannot apply the General Power Rule since we cannot make the necessary adjustments for u'. However, we can (in this particular case) expand the integrand and write

$$\int -8(3 - 4x^2)^2\,dx = \int -8(9 - 24x^2 + 16x^4)\,dx$$
$$= \int (-72 + 192x^2 - 128x^4)\,dx$$
$$= -72x + 64x^3 - \tfrac{128}{5}x^5 + C \qquad \square$$

Sometimes an integrand contains an extra constant factor that is *not* needed as part of u'. In such cases, we simply move this factor outside the integral sign, insert the necessary factor, and adjust accordingly. The next example illustrates this situation.

Example 4

Applying the General Power Rule to Fractional Powers

Evaluate

$$\int \frac{7x^2\,dx}{\sqrt{4x^3 - 5}}$$

Solution Write

$$\int \frac{7x^2\,dx}{\sqrt{4x^3 - 5}} = \int 7x^2(4x^3 - 5)^{-1/2}\,dx$$

and let $u = 4x^3 - 5$; then $u' = 12x^2$. We need the factor 12, rather than 7, so we write

$$\int 7x^2(4x^3 - 5)^{-1/2}\,dx = 7\int \frac{1}{12}(4x^3 - 5)^{-1/2}(12x^2)\,dx$$
$$= \frac{7}{12}\int (4x^3 - 5)^{-1/2}(12x^2)\,dx$$
$$= \left(\frac{7}{12}\right)\frac{(4x^3 - 5)^{1/2}}{\frac{1}{2}} + C$$
$$= \frac{7}{6}\sqrt{4x^3 - 5} + C \qquad \square$$

Remark Be sure you see the distinction between the two types of constant factors moved outside the integral sign in Example 4. In the equation

$$\int 7x^2(4x^3 - 5)^{-1/2}\,dx = (7)\left(\frac{1}{12}\right)\int (4x^3 - 5)^{-1/2}(12x^2)\,dx$$

the 7 is an unnecessary factor that is moved out *as is*, whereas the $\frac{1}{12}$ is the *reciprocal* adjustment for the factor 12 used to create $u' = 12x^2$.

Section Exercises (4.2)

In Exercises 1–24, evaluate the indefinite integrals and check your results by differentiation.

1. $\int (1 + 2x)^4 \, dx$
2. $\int (x^2 - 1)^3 2x \, dx$
3. $\int x^2(x^3 - 1)^4 \, dx$
4. $\int x(1 - 2x^2)^3 \, dx$
5. $\int x(x^2 - 1)^7 \, dx$
6. $\int \frac{x^2}{(x^3 - 1)^2} \, dx$
7. $\int \frac{4x}{\sqrt{1 + x^2}} \, dx$
8. $\int \frac{6x}{(1 + x^2)^3} \, dx$
9. $\int 5x\sqrt[3]{1 + x^2} \, dx$
10. $\int 3(x - 3)^{5/2} \, dx$
11. $\int \frac{-3}{\sqrt{2x + 3}} \, dx$
12. $\int \frac{4x + 6}{(x^2 + 3x + 7)^3} \, dx$
13. $\int \frac{x + 1}{(x^2 + 2x - 3)^2} \, dx$
14. $\int u^3\sqrt{u^4 + 2} \, du$
15. $\int \frac{1}{\sqrt{x}(1 + \sqrt{x})^2} \, dx$
16. $\int \left(1 + \frac{1}{t}\right)^3 \left(\frac{1}{t^2}\right) dt$
17. $\int \frac{x^2}{(1 + x^3)^2} \, dx$
18. $\int \frac{x^2}{\sqrt{1 + x^3}} \, dx$
19. $\int \frac{x^3}{\sqrt{1 + x^4}} \, dx$
20. $\int \frac{t + 2t^2}{\sqrt{t}} \, dt$
21. $\int \frac{1}{2\sqrt{x}} \, dx$
22. $\int \frac{1}{(3x)^2} \, dx$
23. $\int \frac{1}{\sqrt{2x}} \, dx$
24. $\int \frac{1}{3x^2} \, dx$

Ⓒ25. A lumber company is seeking a model that yields the average weight loss W per (ponderosa pine) log as a function of the number of days of drying time t. The model is to be reliable up to 100 days after the log is cut. Based on the weight loss during the first 30 days, it was determined that

$$\frac{dW}{dt} = \frac{12}{\sqrt{16t + 9}}$$

(a) Find W as a function of t. Note that no weight loss occurs until the tree is cut.
(b) Find the total weight loss after 100 days.

Ⓒ26. A particular company has determined their marginal cost to be

$$\frac{dC}{dx} = \frac{4}{\sqrt{x + 1}}$$

(a) Find the cost function if $C = 50$ when $x = 15$.
(b) Graph the marginal cost function and the cost function on the same set of axes.

Ⓒ27. The marginal cost for a certain commodity has been determined to be

$$\frac{dC}{dx} = \frac{12}{\sqrt[3]{12x + 1}}$$

(a) Find the cost function if $C = 100$ when $x = 13$.
(b) Graph the marginal cost function and the cost function on the same set of axes.

Ⓒ Calculator may be helpful.

Section 4.3

Area and the Fundamental Theorem of Calculus

Introductory Example

The Average Cost of Fuel

A large trucking firm estimates that the price of a gallon of diesel fuel during the next 5 years will rise according to the model

$$p = 1.21 + 0.11t + 0.02t^2$$

where p is measured in dollars per gallon and t is measured in years. To construct its 5-year budget plan, the firm needs to estimate the average cost of fuel over this 5-year period. The problem in estimating this average is that the cost of the fuel is not increasing linearly. (If the increase were linear, we could use the midpoint, $t = 2.5$ years and $p = \$1.61$, to estimate the average.)

The solution to this problem is given by the **definite integral**

$$\text{average price} = \frac{1}{5}\int_0^5 [1.21 + 0.11t + 0.02t^2]\,dt = \$1.65$$

Graphically, this average corresponds to one-fifth of the **area** of the region under the price curve shown in Figure 4.4.

Now, since the trucking firm purchases 100,000 gal of fuel per year, it should plan to spend

$$(1.65)(100{,}000)(5) = \$825{,}000$$

on fuel during the next 5 years.

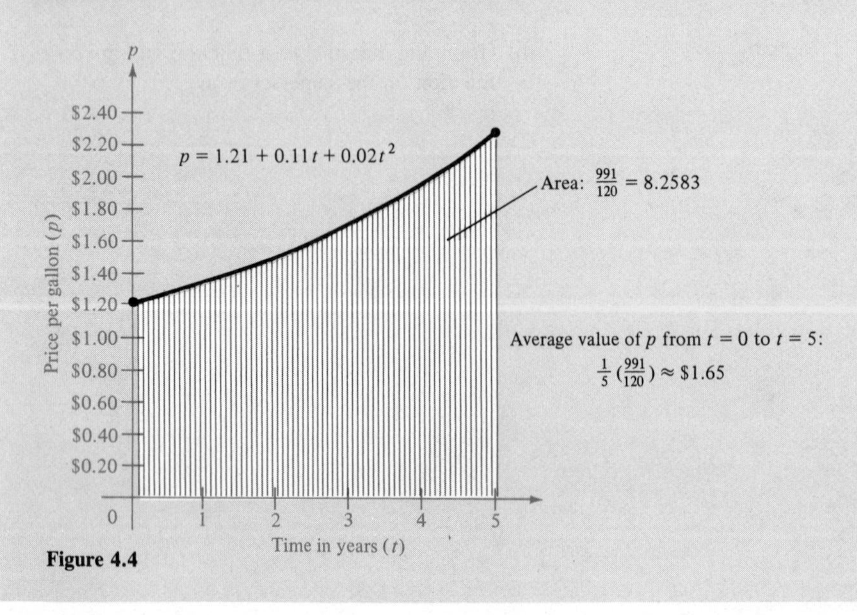

Figure 4.4

Section Objectives: ***To provide an informal development of the Fundamental Theorem of Calculus ▪ To use the Fundamental Theorem to calculate areas under a curve ▪ To derive some properties of the definite integral ▪ To use a definite integral to find the average of a function on an interval.***

Area is a concept familiar to all of us through our study of various geometric figures, such as the rectangle, square, triangle, and circle. We generally think of area as a number that in some way suggests the size of a bounded region. Of course, for simple geometric figures we have specific formulas for calculating their areas.

Our problem here is to develop a way to calculate the area of any plane region R, bounded by the x-axis, the lines $x = a$ and $x = b$, and the graph of a nonnegative continuous function f. (See the shaded region in Figure 4.5.)

The solution of this problem is contained in the Fundamental Theorem of Calculus and represents one of the most famous discoveries in the history of mathematics. From this theorem we will see that, just as the derivative can be used to find slope, the antiderivative can be used to find area. In anticipation of the connection between antiderivatives and area, we denote the area of the region shown in Figure 4.5 by

$$\text{area} = \int_a^b f(x)\,dx$$

The symbol $\int_a^b f(x)\,dx$ is called the **definite integral from *a* to *b*,** where a is the **lower limit of integration** and b is the **upper limit of integration.**

Development of the Fundamental Theorem of Calculus

In developing the Fundamental Theorem of Calculus, we temporarily introduce the area function shown in Figure 4.6. Specifically, if f is continuous and nonnegative on $[a, b]$, we denote the area of the region under the graph of f from a to x by $A(x)$.

Now, if we let x increase by an amount Δx, then the area of the region under the graph of f increases by ΔA. Furthermore, if $f(m)$ and $f(M)$ denote the

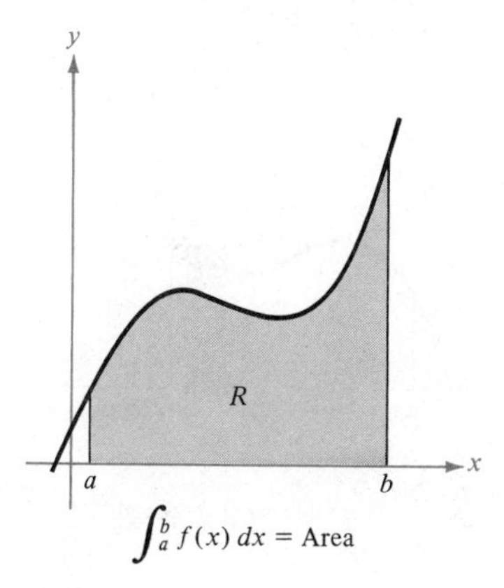

Figure 4.5

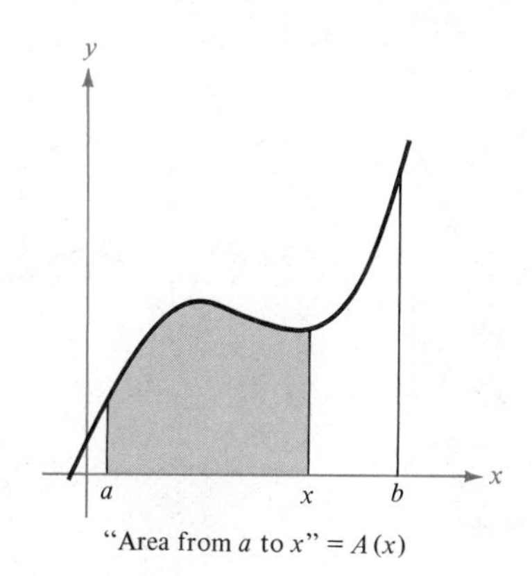

Figure 4.6

minimum and maximum values of f on the interval $[x, x + \Delta x]$, then we have the relationship

$$\text{area of inscribed rectangle} = f(m)\,\Delta x \leq \Delta A \leq f(M)\,\Delta x$$
$$= \text{area of circumscribed rectangle}$$

(See Figure 4.7.)

Dividing each term in

$$f(m)\,\Delta x \leq \Delta A \leq f(M)\,\Delta x$$

by Δx, we have

$$f(m) \leq \frac{\Delta A}{\Delta x} \leq f(M) \qquad \text{since } \Delta x > 0$$

Since both $f(m)$ and $f(M)$ approach $f(x)$ as Δx approaches zero, and since

$$\lim_{\Delta x \to 0} \frac{\Delta A}{\Delta x} = A'(x)$$

it follows that

$$f(x) \leq A'(x) \leq f(x)$$

which means that

$$f(x) = A'(x)$$

Thus, we have established that the area function $A(x)$ is an antiderivative of f and, consequently, must be of the following form:

$$A(x) = F(x) + C$$

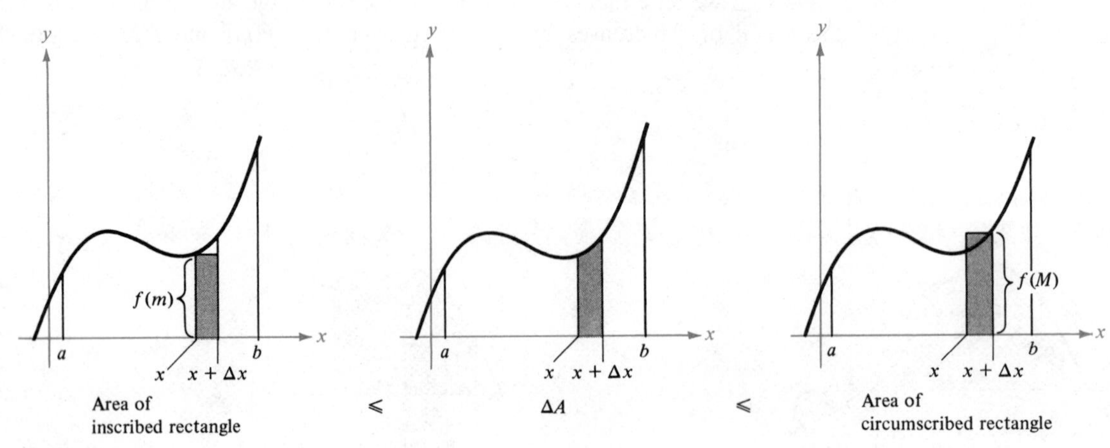

Figure 4.7

where $F(x)$ is any antiderivative of f (Section 4.1). To solve for C, we note that $A(a) = 0$ and thus $C = -F(a)$. Furthermore, by evaluating $A(b)$, we have

$$A(b) = F(b) + C = F(b) - F(a)$$

Finally, by replacing $A(b)$ by its integral form, we have

$$\int_a^b f(x)\,dx = F(b) - F(a)$$

This equation tells us that *if we can find an antiderivative for f*, then we can use the antiderivative to evaluate the definite integral $\int_a^b f(x)\,dx$. We summarize this result in the following theorem.

Theorem 4.1 (Fundamental Theorem of Calculus)

If a function f is continuous on the interval $[a, b]$, then

$$\int_a^b f(x)\,dx = F(b) - F(a)$$

where F is any function such that $F'(x) = f(x)$ for all x in $[a, b]$.

Several comments regarding the Fundamental Theorem of Calculus are in order. First, this theorem describes a means for *evaluating* a definite integral, not a procedure for finding antiderivatives. Second, when applying this theorem, we find it helpful to use the formulation

$$\int_a^b f(x)\,dx = F(x)\Big]_a^b = F(b) - F(a)$$

For instance, we write

$$\int_1^3 x^3\,dx = \frac{x^4}{4}\Big]_1^3 = \frac{(3)^4}{4} - \frac{(1)^4}{4} = \frac{81}{4} - \frac{1}{4} = 20$$

Third, we observe that the constant of integration C can be dropped from the antiderivative, because

$$\begin{aligned}\int_a^b f(x)\,dx &= \Big[F(x) + C\Big]_a^b = [F(b) + C] - [F(a) + C] \\ &= F(b) - F(a) + C - C = F(b) - F(a)\end{aligned}$$

Finally, even though it was convenient to assume f to be nonnegative in our development of the Fundamental Theorem, this is not necessary, and we can apply the Fundamental Theorem in cases where f is negative. We will say more about this possibility later.

We now list some useful properties of the definite integral. In each case we assume the integrability of f and g on $[a, b]$.

Properties of Definite Integrals

1. $\int_a^b kf(x)\,dx = k\int_a^b f(x)\,dx$, where k is a constant

2. $\int_a^b f(x)\,dx = \int_a^c f(x)\,dx + \int_c^b f(x)\,dx$, where $a < c < b$

3. $\int_a^b [f(x) \pm g(x)]\,dx = \int_a^b f(x)\,dx \pm \int_a^b g(x)\,dx$

Example 1

Finding Area by the Fundamental Theorem

Use the Fundamental Theorem to find the area of the region bounded by $f(x) = x^2 - 1$ and the x-axis, $1 \le x \le 2$.

Solution The definite integral representing this area is

$$\text{area} = \int_1^2 (x^2 - 1)\,dx$$

Now, applying the Fundamental Theorem, we have

$$\int_1^2 (x^2 - 1)\,dx = \left[\frac{x^3}{3} - x\right]_1^2 = \left(\frac{8}{3} - 2\right) - \left(\frac{1}{3} - 1\right) = \frac{4}{3}$$

(See Figure 4.8.)

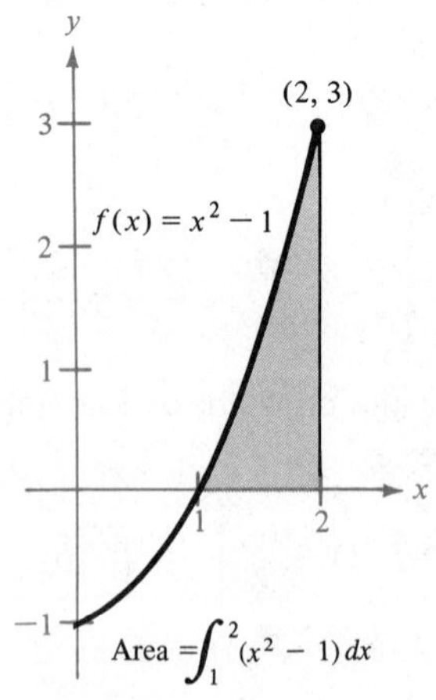

Figure 4.8

□

Example 2

Evaluating Definite Integrals

Evaluate the definite integral $\int_0^1 (4t + 1)^2\,dt$ and sketch the region whose area is represented by this integral.

Solution

$$\int_0^1 (4t+1)^2\,dt = \frac{1}{4}\int_0^1 (4t+1)^2(4)\,dt$$

$$= \frac{1}{4}\left[\frac{(4t+1)^3}{3}\right]_0^1 = \frac{1}{4}\left(\frac{125}{3} - \frac{1}{3}\right) = \frac{31}{3}$$

(Note the use of Property 1 in the first line of the solution.) The region whose area is represented by this integral is shown in Figure 4.9.

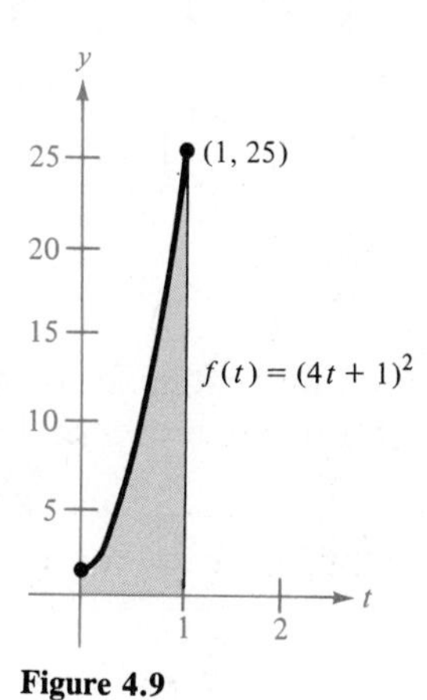

Figure 4.9

□

Example 3

Evaluating a Definite Integral

Evaluate $\int_1^4 3\sqrt{x}\,dx$.

Solution

$$\int_1^4 3\sqrt{x}\,dx = 3\int_1^4 x^{1/2}\,dx = 3\left[\frac{x^{3/2}}{\frac{3}{2}}\right]_1^4 = 2(4)^{3/2} - 2(1)^{3/2} = 14$$

□

Example 4

Integrating Absolute Values

Evaluate $\int_0^2 |2x-1|\,dx$.

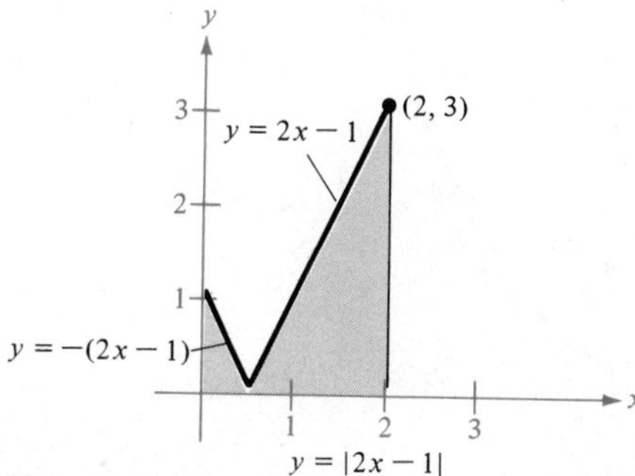

Figure 4.10

Solution From Figure 4.10 and the definition of absolute value, we note that

$$|2x - 1| = \begin{cases} -(2x - 1), & \text{for } x < \frac{1}{2} \\ (2x - 1), & \text{for } x \geq \frac{1}{2} \end{cases}$$

Hence, we rewrite the integral in two parts as follows:

$$\begin{aligned} \int_0^2 |2x - 1|\, dx &= \int_0^{1/2} -(2x - 1)\, dx + \int_{1/2}^2 (2x - 1)\, dx \\ &= \Big[-x^2 + x\Big]_0^{1/2} + \Big[x^2 - x\Big]_{1/2}^2 \\ &= \left(-\frac{1}{4} + \frac{1}{2}\right) - (0 + 0) + (4 - 2) - \left(\frac{1}{4} - \frac{1}{2}\right) \\ &= \frac{5}{2} \end{aligned}$$

□

In addition to being used for finding areas, definite integrals can be used to find the *average value* of a function over an interval.

Definition of the Average Value of a Function

If f is continuous on $[a, b]$, then the **average value** of f on this interval is given by

$$\text{average of } f \text{ on } [a, b] = \frac{\int_a^b f(x)\, dx}{b - a}$$

Example 5

Finding the Average Value of a Function

What is the average value of $f(x) = 3x^2 - 2x$ on $[1, 4]$?

Solution The average value of $f(x) = 3x^2 - 2x$ on $[1, 4]$ is

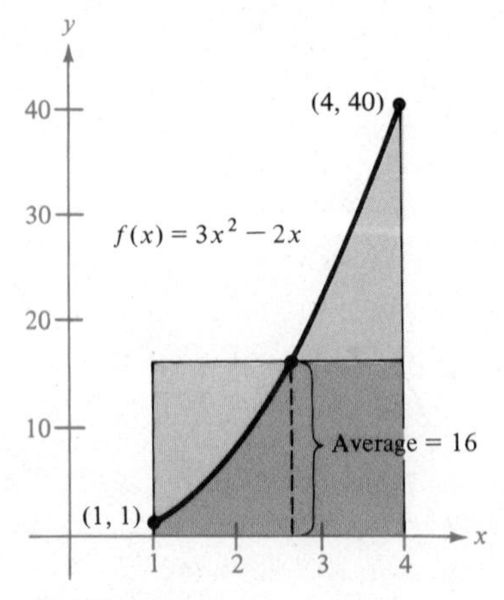

Figure 4.11

$$\frac{\int_1^4 (3x^2 - 2x)\,dx}{4 - 1} = \frac{1}{3}\left[\frac{3x^3}{3} - \frac{2x^2}{2}\right]_1^4$$
$$= \tfrac{1}{3}[64 - 16 - (1 - 1)]$$
$$= \tfrac{48}{3} = 16$$

(See Figure 4.11.) □

Note in Figure 4.11 that the area under the curve is equal to the area of the rectangle whose height is the average value. In fact, we could have defined the average value of f on the interval $[a, b]$ to be that value K such that

$$\int_a^b f(x)\,dx = \int_a^b K\,dx = K(b - a)$$

Example 6

Finding the Average Value of a Function

Find the average value of $f(x) = x^3 - x^2$ on the interval $[-1, 1]$.

Solution Note from Figure 4.12 that $f(x)$ is nonpositive over the entire interval $[-1, 1]$. Therefore, it is not surprising that the average value of f over this interval turns out to be negative. The average of f is

$$\frac{\int_{-1}^1 (x^3 - x^2)\,dx}{1 - (-1)} = \frac{1}{2}\left[\frac{x^4}{4} - \frac{x^3}{3}\right]_{-1}^1 = -\frac{1}{3}$$

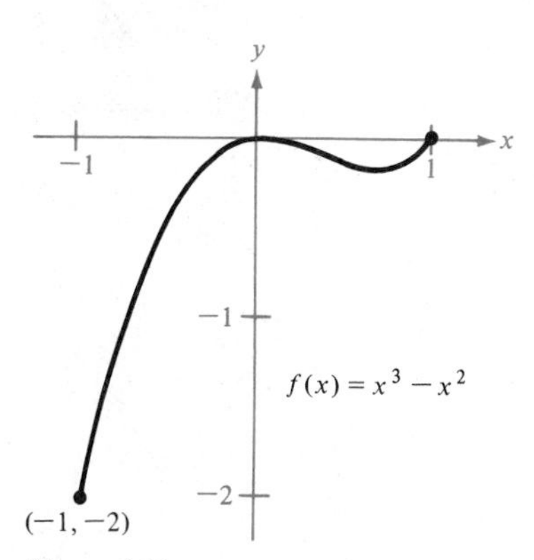

Figure 4.12

□

Before concluding this section, we must make two important observations regarding integrals. First, be sure you realize from Example 6 that a definite integral can have a negative value. Even though one of the main applications of definite integrals is finding area (which must be positive), this is not the only application, and it is common to encounter a definite integral whose value is negative. Second, be sure you recognize the distinction between indefinite and definite integrals. The *indefinite integral* $\int f(x)\,dx$ denotes a family of *functions* (the antiderivatives of $f(x)$), whereas the *definite integral* $\int_a^b f(x)\,dx$ is a *number*.

Section Exercises (4.3)

In Exercises 1–24, evaluate the definite integrals.

1. $\int_0^1 2x\,dx$
2. $\int_2^7 3\,dv$
3. $\int_{-1}^0 (x-2)\,dx$
4. $\int_2^5 (-3v+4)\,dv$
5. $\int_{-1}^1 (t^2-2)\,dt$
6. $\int_0^3 (3x^2+x-2)\,dx$
7. $\int_0^1 (2t-1)^2\,dt$
8. $\int_{-1}^1 (t^3-9t)\,dt$
9. $\int_1^2 \left(\frac{3}{x^2}-1\right)dx$
10. $\int_0^1 (3x^3-9x+7)\,dx$
11. $\int_1^2 (5x^4+5)\,dx$
12. $\int_{-3}^3 v^{1/3}\,dv$
13. $\int_{-1}^1 (\sqrt[3]{t}-2)\,dt$
14. $\int_{-2}^{-1} \sqrt{\frac{-2}{x}}\,dx$
15. $\int_1^4 \frac{u-2}{\sqrt{u}}\,du$
16. $\int_{-2}^{-1} \left(\frac{-1}{u^2}+u\right)du$
17. $\int_0^1 \frac{x-\sqrt{x}}{3}\,dx$
18. $\int_0^2 (2-t)\sqrt{t}\,dt$
19. $\int_{-1}^1 |x|\,dx$
20. $\int_0^3 |2x-3|\,dx$
21. $\int_0^4 \frac{1}{\sqrt{2x+1}}\,dx$
22. $\int_0^1 x\sqrt{1-x^2}\,dx$
23. $\int_{-1}^1 x(x^2+1)^3\,dx$
24. $\int_0^2 \frac{x}{\sqrt{1+2x^2}}\,dx$

In Exercises 25–30, determine the area of each region having the given boundaries.

25. $y = x - x^2;\ y = 0$

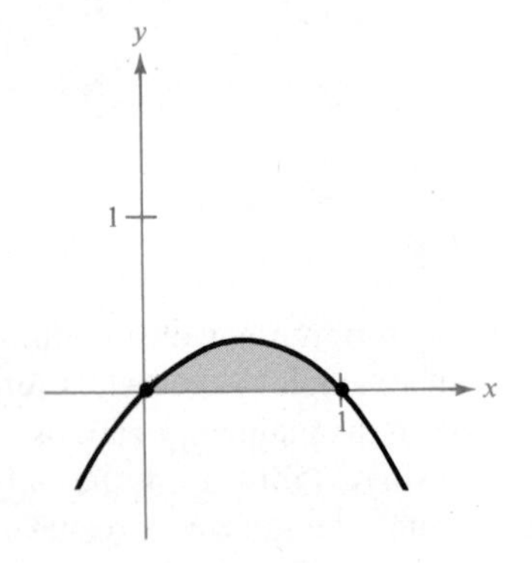

26. $y = -x^2 + 2x + 3;\ y = 0$

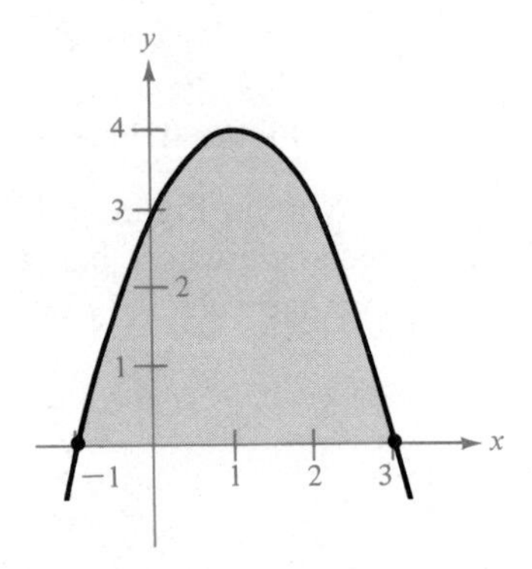

27. $y = 1 - x^4;\ y = 0$

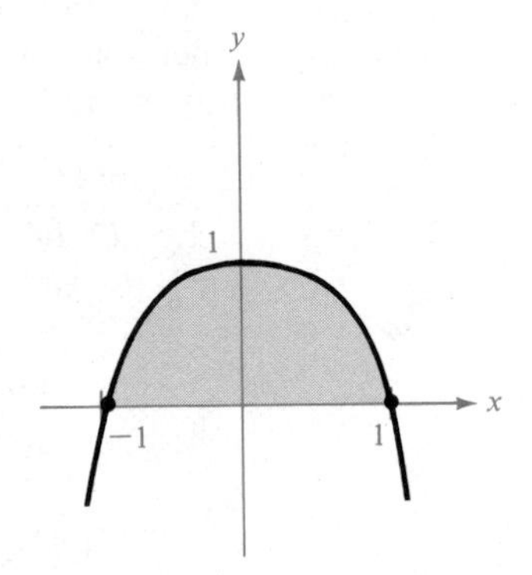

28. $y = \frac{1}{x^2};\ x = 1,\ x = 2,\ y = 0$

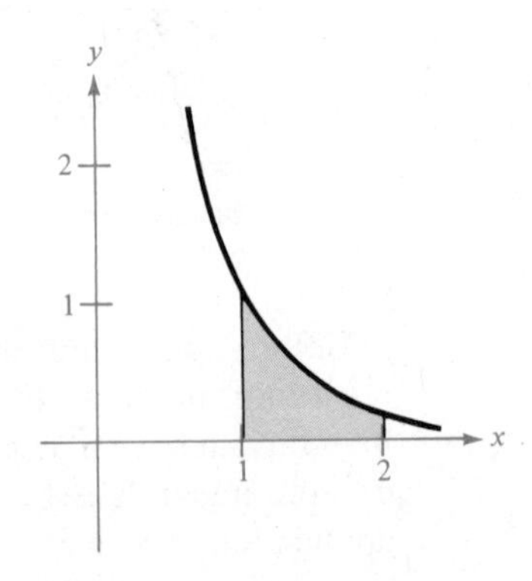

29. $y = \sqrt[3]{2x};\ x = 0,\ x = 4,\ y = 0$

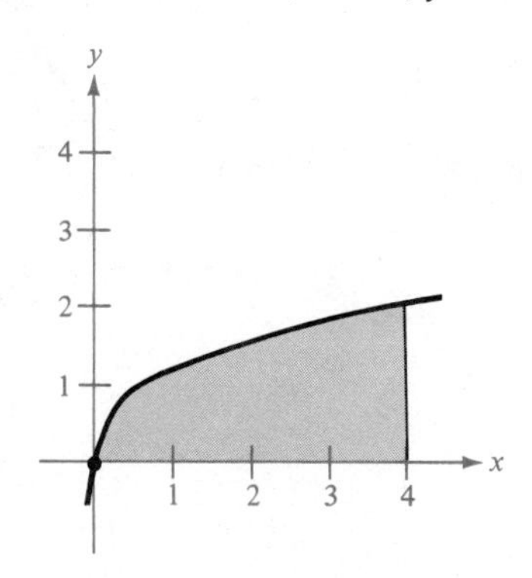

30. $y = (3 - x)\sqrt{x};\ y = 0$

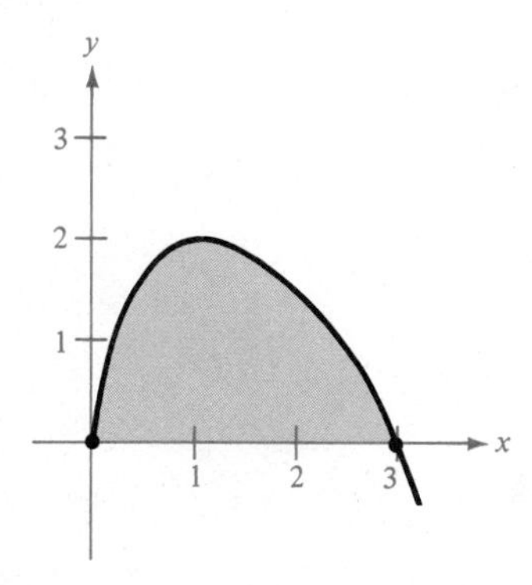

In Exercises 31–36, evaluate the definite integrals and make a sketch of the region whose area is given by the integral.

31. $\int_1^3 (2x - 1)\,dx$ **32.** $\int_0^2 (x + 4)\,dx$

33. $\int_3^4 (x^2 - 9)\,dx$ **34.** $\int_{-1}^2 (-x^2 + x + 2)\,dx$

35. $\int_0^1 (x - x^3)\,dx$ **36.** $\int_0^1 \sqrt{x}(1 - x)\,dx$

In Exercises 37–42, sketch the graph of each function over the given interval. Find the average value of each function over the given interval.

37. $f(x) = 4 - x^2;\ [-2, 2]$

38. $f(x) = x^2 - 2x + 1;\ [0, 1]$

39. $f(x) = x\sqrt{4 - x^2};\ [0, 2]$

40. $f(x) = \dfrac{x^2 + 1}{x^2};\ [\frac{1}{2}, 2]$

41. $f(x) = x - 2\sqrt{x};\ [0, 4]$

42. $f(x) = \dfrac{1}{(x - 3)^2};\ [0, 2]$

Ⓒ**43.** The volume V in liters of air in the lungs during one 5-s respiratory cycle is approximated by the model

$$V = 0.1729t + 0.1522t^2 - 0.0374t^3, \quad 0 \le t \le 5$$

where t is the time in seconds. Approximate the average volume of air in the lungs during one respiratory cycle.

44. The velocity v of the flow of blood at a distance r from the central axis of an artery of radius R is given by

$$v = k(R^2 - r^2)$$

where k is the constant of proportionality. Find the average rate of flow of blood along a radius of the artery. (Use zero and R as the limits of integration.)

Ⓒ**45.** The air temperature during a period of 12 h is given by the model

$$T = 53 + 5t - 0.3t^2, \quad 0 \le t \le 12$$

where t is measured in hours and T in degrees Fahrenheit. (See Figure 4.13.) Find the average temperature during:

(a) the first 6 h of the period
(b) the entire period

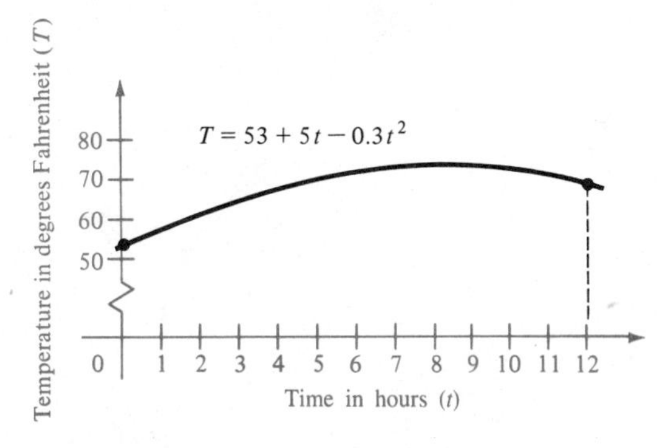

Figure 4.13

Ⓒ Calculator may be helpful.

46. (C) The annual U.S. death rate (per 1000 people x years of age) can be approximated by the model

$$R = 0.036x^2 - 2.8x + 58.14, \qquad 40 \le x \le 60$$

(See Figure 4.14.)

(a) Find the average death rate for people between 40 and 50 years of age.

(b) Find the average death rate for people between 50 and 60 years of age.

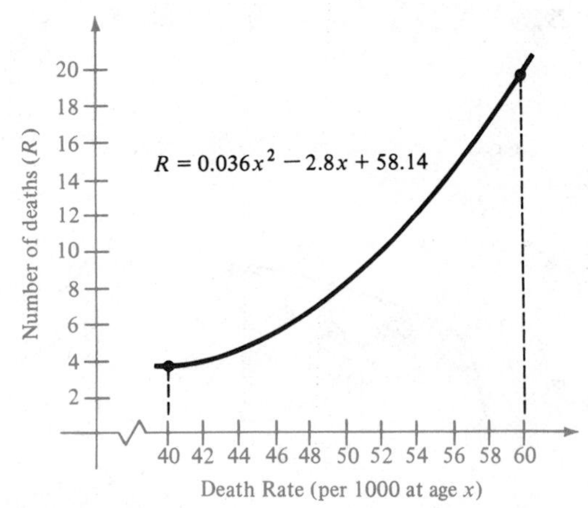

Figure 4.14

(C) Calculator may be helpful.

Section 4.4

The Area of a Region Between Two Curves

Introductory Example

Consumer Surplus

During the 1970s the price of hand-held calculators dropped dramatically, and with this decline the sale of these electronic wonders followed the classic demand pattern. That is, as the price dropped, more and more people were willing to purchase calculators. Suppose that the annual demand function for a particular model of calculator is given by

$$p(x) = 0.3625x^3 - 4.35x^2 - 21.75x + 300, \quad 0 \le x \le 10$$

where the price p is measured in dollars and the number sold x is measured in units of 10,000. From Figure 4.15 we can see that no one would be willing to pay \$300 for this model of calculator, but 100,000 people would be willing to pay \$10.

Suppose that the actual selling price is \$91.20 (this corresponds to sales of 60,000 units). Then the total revenue, $R = xp$, is represented by the area of the rectangle shown in Figure 4.15. The area of the region lying between this rectangle and the demand curve represents a quantity that economists call the *consumer surplus*. The consumer surplus is the extra amount of money consumers would have been willing to spend but in fact did not have to spend because the price was low. The formula for the **area of the region** between the demand curve and the actual price is

$$\text{area} = \int_0^6 [(0.3625x^3 - 4.35x^2 - 21.75x + 300) - (91.2)]\,dx = 665.55$$

Since x is measured in units of 10,000, the consumer surplus for this particular price and demand function is as follows:

$$\text{consumer surplus} = (10{,}000)(665.55) = \$6{,}655{,}500.00$$

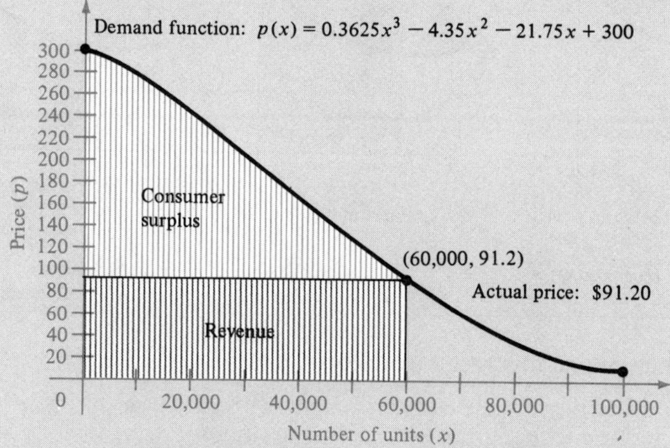

Figure 4.15

Section Objective: ***To demonstrate a procedure for determining the area of a region between two curves.***

With a few modifications we can extend the application of definite integrals from the area of a region *under* a curve to the area of a region *between* two curves. Let us consider the region bounded by $y = f(x)$, $y = g(x)$, $x = a$, and $x = b$.

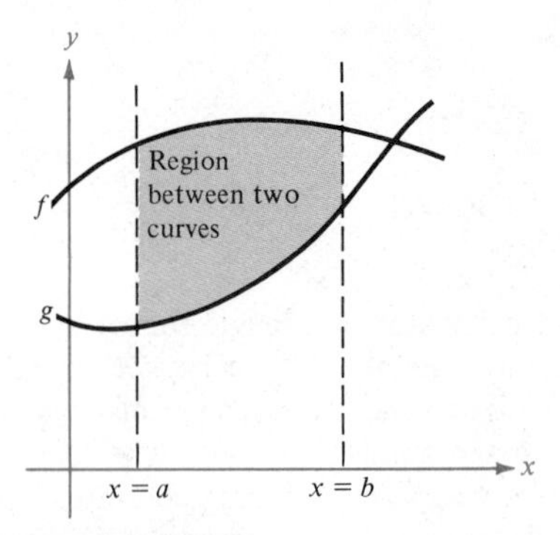

Figure 4.16

(See Figure 4.16.) Assume that both f and g are continuous in $[a, b]$ and that $g(x) \leq f(x)$ for all x in $[a, b]$. If both f and g are above the x-axis, we can geometrically interpret the area of the region between f and g as simply the area of the region under g subtracted from the area of the region under f. (See Figure 4.17.)

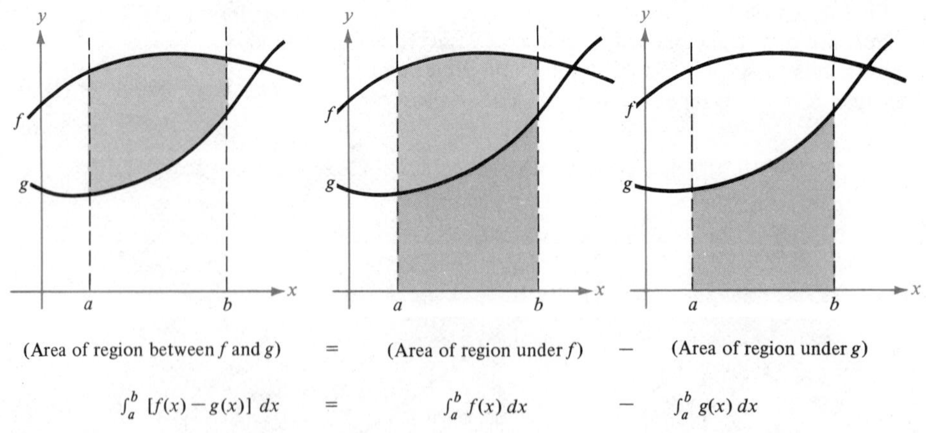

(Area of region between f and g) = (Area of region under f) − (Area of region under g)

$$\int_a^b [f(x) - g(x)]\,dx = \int_a^b f(x)\,dx - \int_a^b g(x)\,dx$$

Figure 4.17

We summarize this result as follows.

Area Between Two Curves

If f and g are continuous on $[a, b]$ and $g(x) \leq f(x)$ for all x in $[a, b]$, then the area of the region bounded by $y = f(x)$, $y = g(x)$, $x = a$, and $x = b$ is given by

$$A = \int_a^b [f(x) - g(x)]\, dx$$

Remark

Even though Figure 4.17 shows f and g above the x-axis, it is important to realize that the formula for the area between two curves only requires that g lie below f. In other words, this same formula can be applied to cases for which f and g are negative on all (or part) of the interval $[a, b]$.

Example 1

Finding the Area Between Two Curves

Find the area of the region bounded by $y = x^2 + 2$, $y = x$, $x = 0$, and $x = 1$.

Solution Since $x \leq x^2 + 2$ for all x in $[0, 1]$, we determine the area:

$$\text{area} = \int_0^1 [(x^2 + 2) - (x)]\, dx = \int_0^1 (x^2 - x + 2)\, dx$$

$$= \left[\frac{x^3}{3} - \frac{x^2}{2} + 2x\right]_0^1 = \frac{1}{3} - \frac{1}{2} + 2 = \frac{11}{6}$$

(See Figure 4.18.)

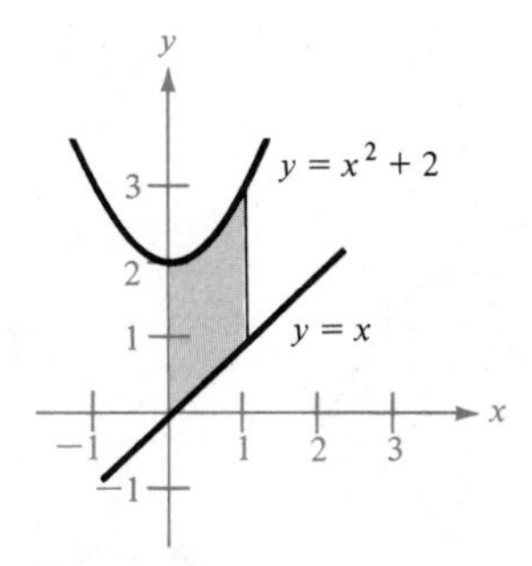

Figure 4.18

□

In Example 1, the two curves $y = x$ and $y = x^2 + 2$ do not intersect, and the values of a and b are explicitly given. A more common type of problem involves the area of the region bounded by two *intersecting* curves. In this type of problem, the values of a and b must be calculated.

Example 2

Finding the Area Between Two Intersecting Curves

Find the area of the region bounded by the curve $y = 2 - x^2$ and the line $y = x$.

Solution In this case a and b are determined by the points of intersection of $y = x$ and $y = 2 - x^2$. In order to find these points, we set these two functions equal to each other and solve for x.

$$\begin{aligned} x &= 2 - x^2 \\ x^2 + x - 2 &= 0 \\ (x + 2)(x - 1) &= 0 \\ x &= -2 \text{ or } 1 \end{aligned}$$

Since $x \leq 2 - x^2$ on the interval $[-2, 1]$, we have

$$\begin{aligned} \text{area} &= \int_{-2}^{1} [(2 - x^2) - (x)]\, dx \\ &= \int_{-2}^{1} (2 - x^2 - x)\, dx \\ &= \left[2x - \frac{x^3}{3} - \frac{x^2}{2}\right]_{-2}^{1} \\ &= \left(2 - \frac{1}{3} - \frac{1}{2}\right) - \left(-4 + \frac{8}{3} - 2\right) \\ &= \frac{9}{2} \end{aligned}$$

(See Figure 4.19.)

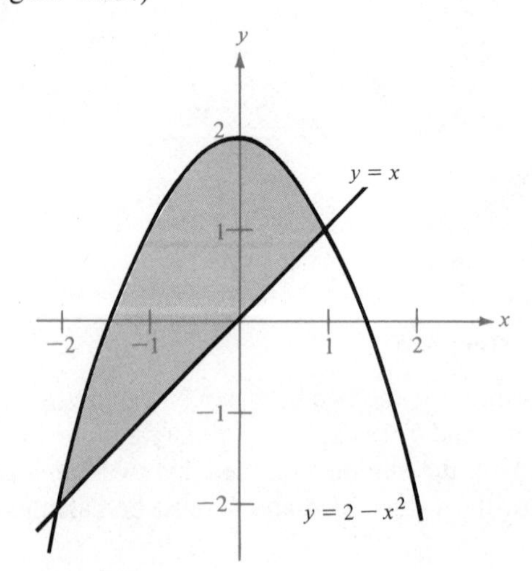

Figure 4.19

□

Example 3

Finding the Area of a Region Below the x-Axis

Find the area of the region bounded by $y = x^2 - 3x - 4$ and the x-axis.

Solution Refer to Figure 4.20. Since $x^2 - 3x - 4$ intersects the x-axis when $x = -1$ and $x = 4$, and since $x^2 - 3x - 4 \leq 0$ for all x in $[-1, 4]$, the desired area is

$$\begin{aligned}
\text{area} &= \int_{-1}^{4} [(0) - (x^2 - 3x - 4)]\, dx \\
&= \int_{-1}^{4} (-x^2 + 3x + 4)\, dx \\
&= \left[\frac{-x^3}{3} + \frac{3x^2}{2} + 4x\right]_{-1}^{4} \\
&= \left(\frac{-64}{3} + 24 + 16\right) - \left(\frac{1}{3} + \frac{3}{2} - 4\right) = \frac{125}{6}
\end{aligned}$$

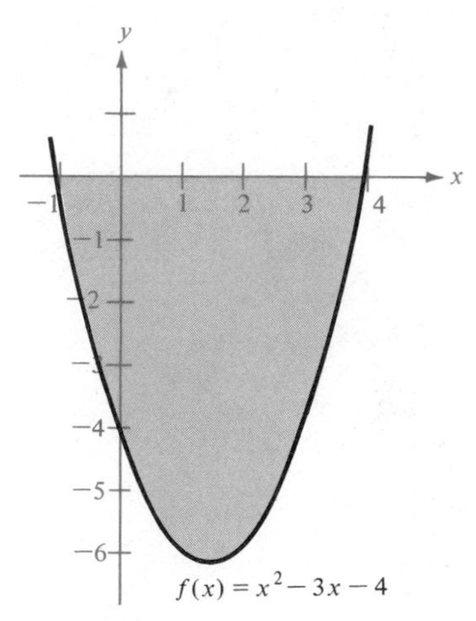

Figure 4.20 □

Occasionally, two curves intersect in more than two points. In determining the area of the region between two such curves, we must find *all* points of intersection and check to see which curve is above the other in each interval determined by these points.

Example 4

Curves Having Multiple Points of Intersection

Find the area of the region between the curves $f(x) = x^3 - 2x^2 + x - 1$ and $g(x) = -x^2 + 3x - 1$.

Solution Solving for x in the equation $f(x) = g(x)$, we have

$$
\begin{aligned}
f(x) - g(x) = x^3 - x^2 - 2x &= 0 \\
x(x^2 - x - 2) &= 0 \\
x(x - 2)(x + 1) &= 0
\end{aligned}
$$

Thus, the zeros are $x = -1$, $x = 0$, and $x = 2$.

By graphing these two functions, we see that $g(x) \le f(x)$ on $[-1, 0]$, but the two curves switch at the point $(0, -1)$, and $f(x) \le g(x)$ on $[0, 2]$. (See Figure 4.21.) Hence, we need to use two integrals to determine the area of the region between f and g, one for the interval $[-1, 0]$ and one for $[0, 2]$.

$$
\begin{aligned}
\text{area} &= \int_{-1}^{0} [f(x) - g(x)]\,dx + \int_{0}^{2} [g(x) - f(x)]\,dx \\
&= \int_{-1}^{0} (x^3 - x^2 - 2x)\,dx + \int_{0}^{2} (-x^3 + x^2 + 2x)\,dx \\
&= \left[\frac{x^4}{4} - \frac{x^3}{3} - x^2\right]_{-1}^{0} + \left[\frac{-x^4}{4} + \frac{x^3}{3} + x^2\right]_{0}^{2} \\
&= -\left(\frac{1}{4} + \frac{1}{3} - 1\right) + \left(-4 + \frac{8}{3} + 4\right) = \frac{37}{12}
\end{aligned}
$$

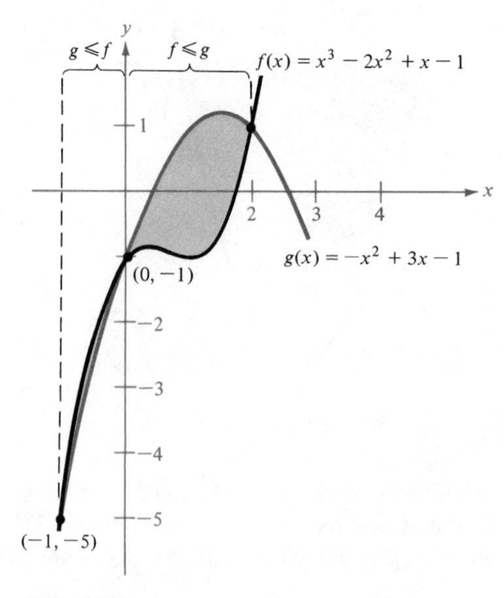

Figure 4.21

□

Example 5

A Business Application

Based on U.S. Department of Energy statistics, the total consumption of gasoline in the United States from 1950 to 1974 followed a growth pattern given by the quadratic model

$$f(t) = 2.158 + 0.082t + 0.001t^2$$

where $f(t)$ is measured in billions of gallons and t is measured in years, with $t = 0$ corresponding to the year 1970. With the onset of dramatic increases in crude oil prices in 1974, the growth pattern for gasoline consumption changed and began following the pattern described by the model

$$g(t) = 1.95 + 0.174t - 0.009t^2$$

Find the total amount of gasoline saved from 1974 to 1982 as a result of consuming gasoline at the post-1974 rate rather than the pre-1974 rate. (See Figure 4.22.)

Solution Since the pre-1974 curve f lies above the post-1974 curve g on the interval [4, 12], the amount of gasoline saved is given by the following integral.

$$\begin{aligned}\int_4^{12} [f(t) - g(t)]\,dt &= \int_4^{12} [(2.158 + 0.082t + 0.001t^2) \\ &\qquad - (1.95 + 0.174t - 0.009t^2)]\,dt \\ &= \int_4^{12} (0.208 - 0.092t + 0.01t^2)\,dt \\ &= \left[0.208t - 0.046t^2 + 0.01\frac{t^3}{3}\right]_4^{12} \\ &= 1.632 - 0.309 \\ &= 1.323\end{aligned}$$

Therefore, 1.323 billion gallons of gasoline were saved.

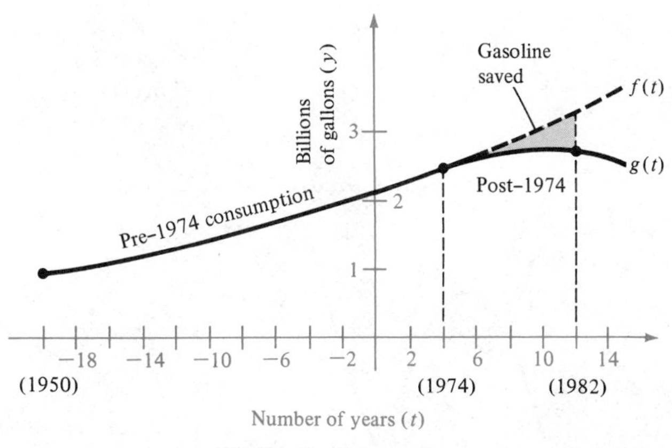

Figure 4.22

□

Section Exercises (4.4)

In Exercises 1–6, find the area of the region bounded by the given curves.

1. $f(x) = x^2 - 6x$
$g(x) = 0$

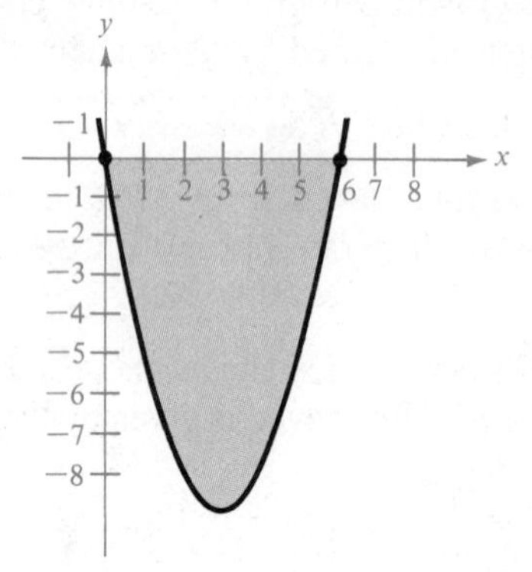

2. $f(x) = x^2 + 2x + 1$
$g(x) = 2x + 5$

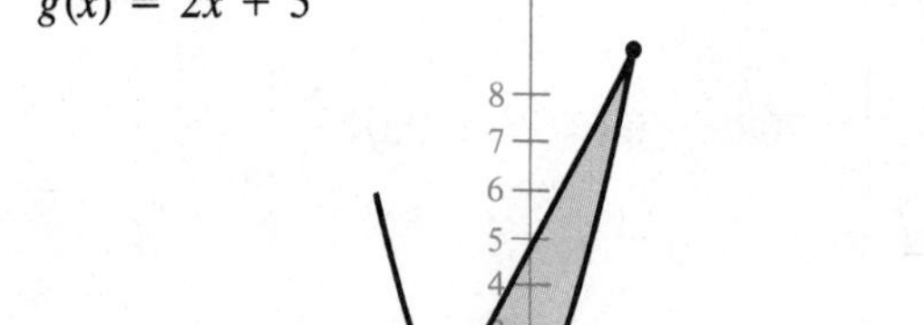

3. $f(x) = x^2 - 4x + 3$
$g(x) = -x^2 + 2x + 3$

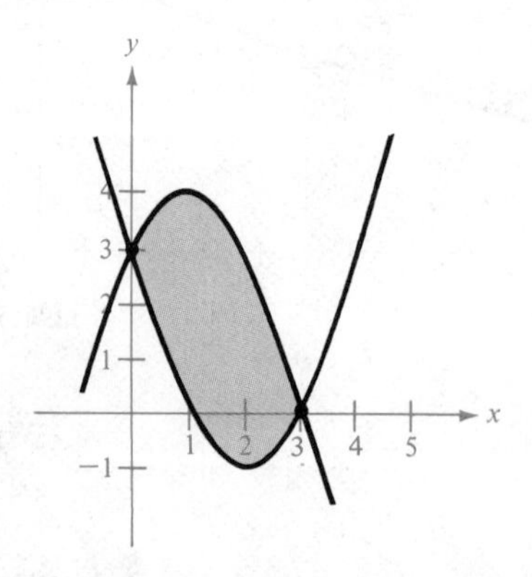

4. $f(x) = x^2$
$g(x) = x^3$

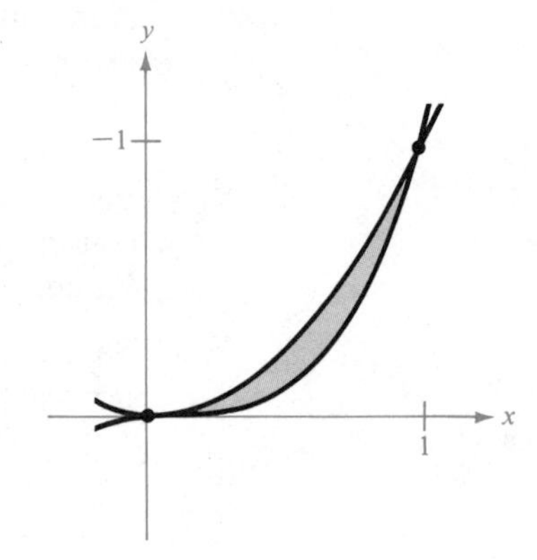

5. $f(x) = 3(x^3 - x)$
$g(x) = 0$

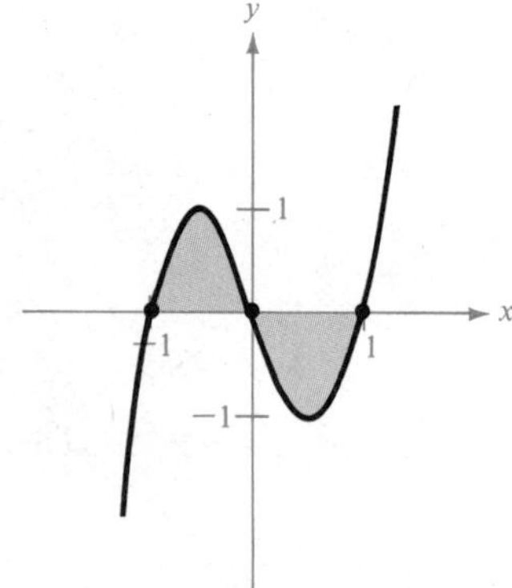

6. $f(x) = (x - 1)^3$
$g(x) = x - 1$

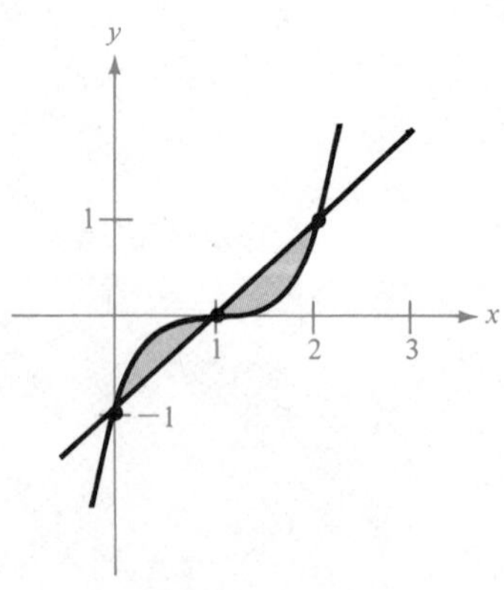

In Exercises 7–20, sketch the region bounded by the graphs of the given equations and find the area of each region by means of definite integrals.

7. $f(x) = x^2 - 4x,\ g(x) = 0$
8. $f(x) = 3 - 2x - x^2,\ g(x) = 0$
9. $f(x) = x^2 + 2x + 1,\ g(x) = 3x + 3$
10. $f(x) = -x^2 + 4x + 2,\ g(x) = x + 2$
11. $y = x,\ y = 2 - x,\ y = 0$
12. $y = \dfrac{1}{x^2},\ y = 0,\ x = 1,\ x = 5$
13. $f(x) = 3x^2 + 2x,\ g(x) = 8$
14. $f(x) = x(x^2 - 3x + 3),\ g(x) = x^2$
15. $f(x) = x^3 - 2x + 1,\ g(x) = -2x,\ x = 1$
16. $f(x) = \sqrt[3]{x},\ g(x) = x$
17. $f(x) = \sqrt{3x} + 1,\ g(x) = x + 1$
18. $f(x) = x^2 + 5x - 6,\ g(x) = 6x - 6$
19. $y = x^2 - 4x + 3,\ y = 3 + 4x - x^2$
20. $y = x^4 - 2x^2,\ y = 2x^2$

Ⓒ**21.** Based on U.S. Department of Agriculture statistics, the total consumption of beef in the United States from 1950 to 1970 followed a growth pattern approximated by

$$f(t) = 23.703 + 1.002t + 0.015t^2$$

where $f(t)$ is measured in billions of pounds and t is measured in years, with $t = 0$ representing 1970. From 1970 to 1980, the growth pattern was more closely approximated by

$$g(t) = 22.93 + 0.678t - 0.037t^2$$

Estimate the total reduction in consumption of beef from 1970 to 1980 due to the change in consumption rate.

Ⓒ**22.** Prior to 1980, the revenue (in billions of dollars) of a large corporation was given by the model

$$R = 7.21 + 0.58t$$

where t is measured in years, with $t = 0$ corresponding to 1980. Due to conditions within the corporation, a more accurate model after 1980 is

$$R = 7.21 + 0.45t$$

Approximate the total reduction in revenue from 1980 to 1985 due to the decrease in growth of corporate sales.

Ⓒ**23.** Repeat Exercise 22 with the two models being

$$R = 7.21 + 0.26t + 0.02t^2$$

and

$$R = 7.21 + 0.1t + 0.01t^2$$

Ⓒ**24.** For the years from 1980 to 1990 the projected fuel cost C (in millions of dollars) for a large corporation was given by

$$C = 568.5 + 7.15t$$

where t is the time in years, with $t = 0$ corresponding to 1980. Due to the installation of fuel-saving equipment, a more accurate model for the fuel cost for the period is

$$C = 525.60 + 6.43t$$

Approximate the total savings for the 10-year period due to the installation of the new equipment.

In economics, the equilibrium price p_0 for a product is defined to be the price at which the supply and demand curves intersect. The *consumer surplus* is defined to be the area between the demand curve and the line $p = p_0$, as shown in Figure 4.23. The *producer surplus* is defined similarly using the supply curve. In Exercises 31–34, find the consumer surplus and producer surplus for each of the given supply and demand curves.

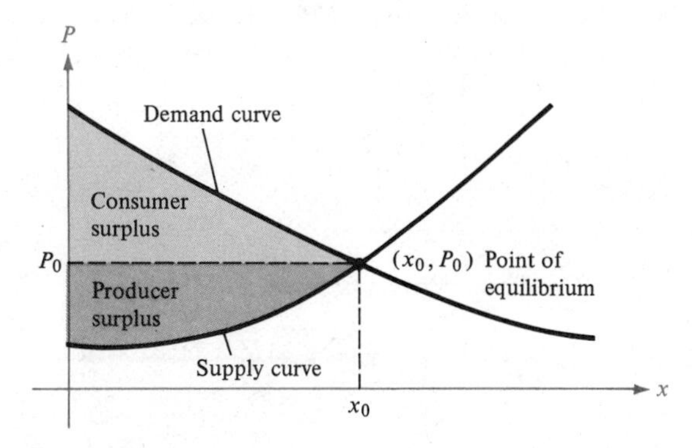

Figure 4.23

Ⓒ**25.** Demand equation: $p(x) = 50 - 0.5x$
Supply equation: $p(x) = 0.125x$

Ⓒ**26.** Demand equation: $p(x) = 1000 - 0.4x^2$
Supply equation: $p(x) = 42x$

Ⓒ**27.** Demand equation: $p(x) = 10{,}000/\sqrt{x + 100}$
Supply equation: $p(x) = 100\sqrt{0.05x + 10}$

Ⓒ**28.** Demand equation: $p(x) = \sqrt{25 - 0.1x}$
Supply equation: $p(x) = \sqrt{9 + 0.1x} - 2$

Ⓒ Calculator may be helpful.

Section 4.5

Volumes of Solids of Revolution

Introductory Example

The Cost of Material in a Wedding Ring

A manufacturer plans to make gold wedding rings out of a gold alloy that costs \$1200 per cubic inch. A typical ring is shown in Figure 4.24. To estimate the cost of material for one ring, we can use the dimensions given in Figure 4.24 to find the volume of the ring. We classify this ring in mathematical terms as a **solid of revolution,** and we imagine that it is formed by revolving a plane region about a line. For this particular solid the plane region is bounded above by the graph of

$$f(x) = \tfrac{7}{16} - 16x^2$$

and below by the graph of

$$g(x) = \tfrac{3}{8}$$

As this region is revolved about the x-axis, it traces the outline of the ring, as indicated in Figure 4.24.

In this section, we will see that the **volume** of this solid of revolution is

$$\begin{aligned}
\text{volume} &= \pi\int_{-1/16}^{1/16} [[f(x)]^2 - [g(x)]^2]\, dx \\
&= \pi\int_{-1/16}^{1/16} \left[\left(\frac{7}{16} - 16x^2\right)^2 - \left(\frac{3}{8}\right)^2\right] dx \\
&= \pi\int_{-1/16}^{1/16} \left(\frac{13}{256} - 14x^2 + 256x^4\right) dx \\
&= \pi\left[\frac{13x}{256} - \frac{14x^3}{3} + \frac{256x^5}{5}\right]_{-1/16}^{1/16} \\
&\approx 0.01309 \text{ in}^3
\end{aligned}$$

Now, since the alloy for the ring costs \$1200 per cubic inch, the total cost of material for the ring is

$$\text{cost} = (0.01309)(1200) = \$15.71$$

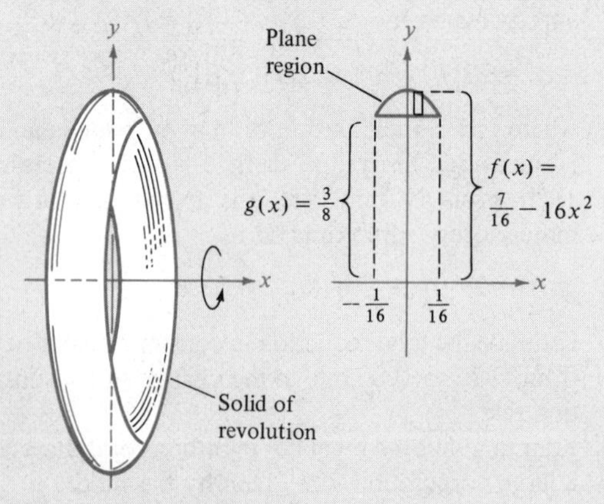

Figure 4.24

Section Objectives: *To define a solid of revolution* ▪ *To introduce the Disc Method for using a definite integral to find the volume of a solid of revolution.*

In Sections 4.3 and 4.4, we saw that a definite integral can be used to find the *area* of a region and the *average value* of a function. These are only two of a wide variety of applications of definite integrals. For example, definite integrals can be used to find the *work* done by a force, the *arc length* of a curve, the *surface area* of a solid, the *center of mass* of a solid, and many other quantities that are useful in the physical sciences.

In this section, we will see how integration can be used to find the volume of a solid. Specifically, we will look at solids that are formed by revolving plane regions about an axis.

Definition of a Solid of Revolution

If a region in the plane is revolved about a line, the resulting solid is called a **solid of revolution** and the line is called the **axis of revolution.** (See Figure 4.25.)

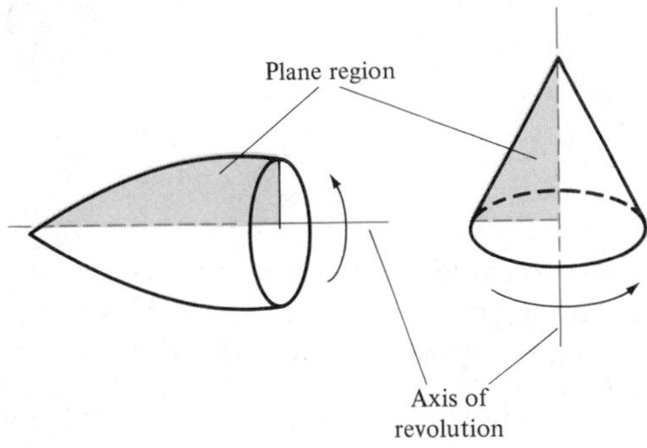

Figure 4.25

Development of a Definite Integral Formula for Volume

To develop a formula for finding the volume of a solid of revolution, we consider a continuous function $y = f(x)$ that is nonnegative on the interval $[a, b]$. Figure 4.26 shows that the area of the region bounded by the graph of f and the x-axis $(a \leq x \leq b)$ can be approximated by n rectangles, each of width

$$\Delta x = \frac{b - a}{n}$$

Using a definite integral, we can write this approximation as

$$\int_a^b f(x)\, dx \approx f(x_1)\,\Delta x + f(x_2)\,\Delta x + \cdots + f(x_n)\,\Delta x$$

where $f(x_i)$ is the height of the ith rectangle. It is important to realize that this approximation improves as n becomes larger. (In fact, taking the limit as n approaches infinity makes the approximation exact.)

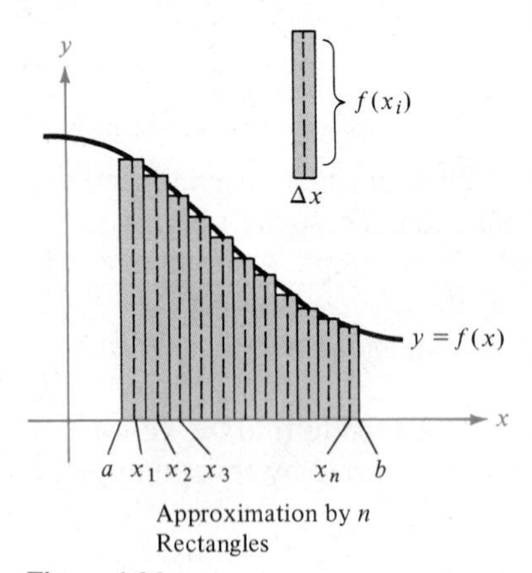

Figure 4.26

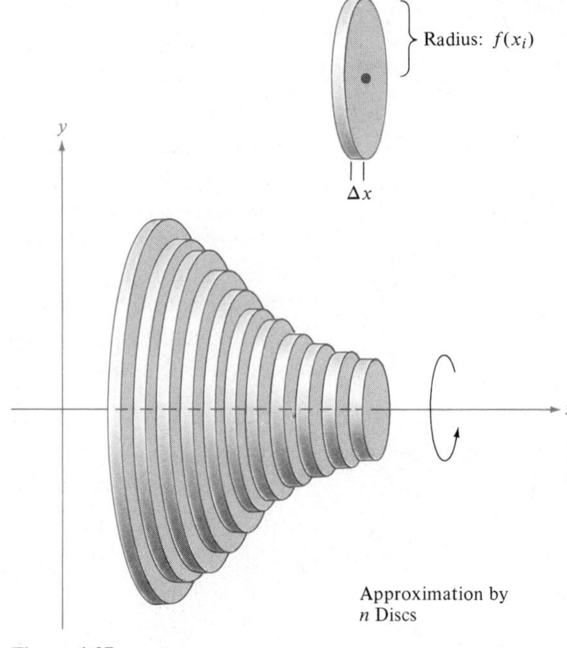

Figure 4.27

Now, to extend this procedure to three dimensions, we consider the circular discs formed by revolving each of the rectangles in Figure 4.26 about the x-axis. The volume of the ith disc (see Figure 4.27) is given by

$$\text{volume of } i\text{th disc} = \pi(\text{radius})^2(\text{width}) = \pi[f(x_i)]^2\,\Delta x$$

and by summing the volumes of all n discs, we have

$$\begin{array}{l}\text{volume of solid}\\ \text{of revolution}\end{array} \approx \pi[f(x_1)]^2\,\Delta x + \pi[f(x_2)]^2\,\Delta x + \cdots + \pi[f(x_n)]^2\,\Delta x$$

as shown in Figure 4.27. Finally, by letting n approach infinity, we can conclude that

$$\text{volume of solid of revolution} = \pi\int_a^b [f(x)]^2\,dx$$

Disc Method for the Volume of a Solid of Revolution

The volume of the solid formed by revolving the region bounded by $y = f(x)$ and the x-axis ($a \leq x \leq b$) about the x-axis is

$$\text{volume} = \pi\int_a^b [f(x)]^2\,dx$$

(See Figure 4.28.)

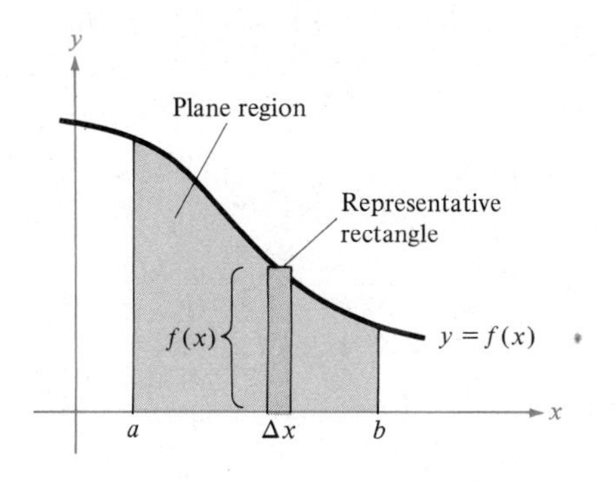

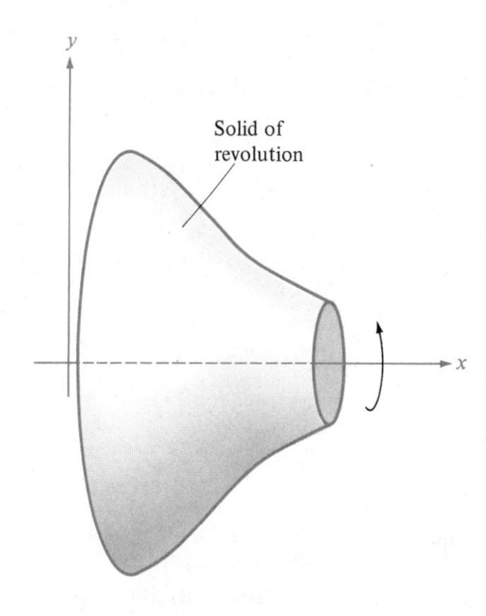

Figure 4.28

Remark

The *representative rectangle* shown in the plane region in Figure 4.28 is useful in visualizing the radius $f(x)$.

Example 1

Finding the Volume of a Solid of Revolution

Find the volume of the solid formed by revolving the region bounded by $y = -x^2 + x$ and $y = 0$ about the x-axis.

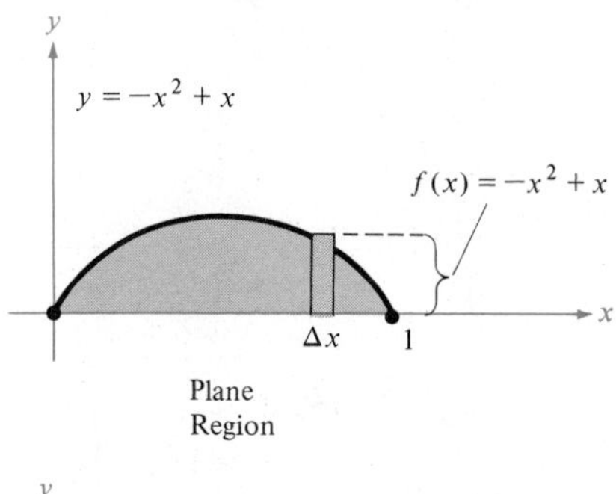

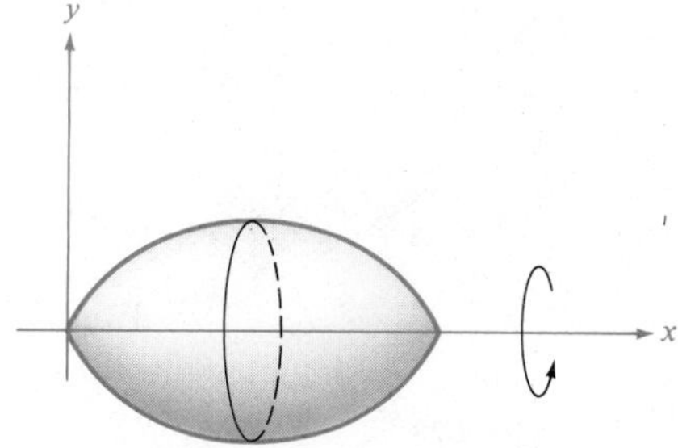

Figure 4.29 Solid of Revolution

Solution From the representative rectangle in Figure 4.29, we see that the radius of this solid is given by

$$\text{radius} = f(x) = -x^2 + x$$

and it follows that its volume is

$$\begin{aligned}\text{volume} &= \pi\int_0^1 [f(x)^2]\, dx = \pi\int_0^1 (-x^2 + x)^2\, dx \\ &= \pi\int_0^1 (x^4 - 2x^3 + x^2)\, dx \\ &= \pi\left[\frac{x^5}{5} - \frac{x^4}{2} + \frac{x^3}{3}\right]_0^1 = \frac{\pi}{30} \approx 0.105\end{aligned}$$

□

Remark Note in Example 1 that the entire problem was worked *without* any reference to the three-dimensional sketch given in Figure 4.29. In general, to set up the integral for calculating the volume of a solid of revolution, a sketch of the plane region is more useful than a sketch of the solid, since the radius is more readily visualized in the plane region.

We can extend the Disc Method to finding the volume of a solid of revolution with a *hole* as follows. Suppose that a region is bounded by $y = f(x)$ and

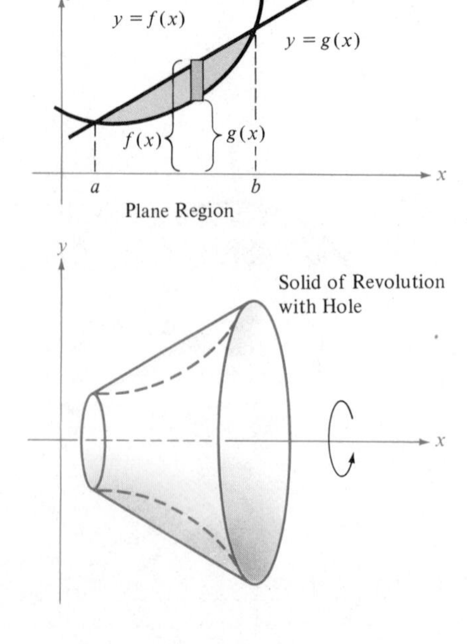

Figure 4.30

$y = g(x)$, as shown in Figure 4.30. If this region is revolved about the x-axis, then the volume of the resulting solid is given by

$$\pi \int_a^b [f(x)]^2\,dx - \pi \int_a^b [g(x)]^2\,dx = \pi \int_a^b [[f(x)]^2 - [g(x)]^2]\,dx$$

We call $f(x)$ the *outer radius* of the solid and $g(x)$ the *inner radius*. Note that the integral involving the inner radius represents the volume of the hole and is *subtracted* from the integral involving the outer radius.

Example 2

Finding the Volume of a Solid of Revolution with a Hole

Find the volume of the solid formed by revolving the region bounded by $f(x) = \sqrt{25 - x^2}$ and $g(x) = 3$ about the x-axis. (See Figure 4.31.)

Solution We begin by finding the points of intersection of f and g as follows:

$$\begin{aligned} f(x) &= g(x) \\ \sqrt{25 - x^2} &= 3 \\ 25 - x^2 &= 9 \\ 16 &= x^2 \\ \pm 4 &= x \end{aligned}$$

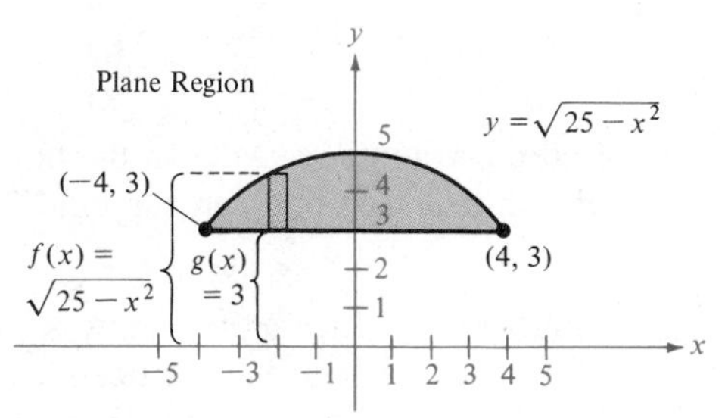

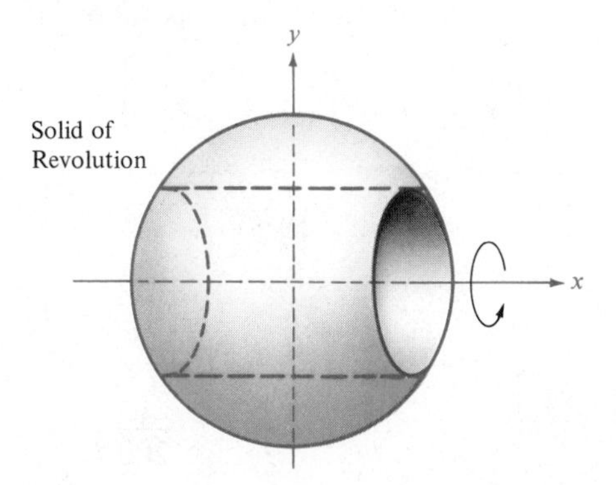

Figure 4.31

Now, using $f(x)$ as the outer radius and $g(x)$ as the inner radius, we integrate from -4 to 4 to obtain the volume.

$$\text{volume} = \pi\int_{-4}^{4} [[f(x)]^2 - [g(x)]^2]\,dx$$

$$= \pi\int_{-4}^{4} ([\sqrt{25 - x^2}]^2 - [3]^2)\,dx$$

$$= \pi\int_{-4}^{4} (16 - x^2)\,dx = \pi\left[16x - \frac{x^3}{3}\right]_{-4}^{4} = \frac{256\pi}{3} \qquad \square$$

Occasionally, an application of the Disc Method involves a vertical axis of revolution. In such cases we must find the radius of the solid as a function of y and integrate along the y-axis, as demonstrated in the next example.

Example 3

An Application to Biology

A pond is to be stocked with a certain species of fish. It has been determined that the food supply in 500 cubic feet of this pond water can adequately support one fish. The pond is (nearly) circular in circumference, is 20 feet deep at its center, and has a radius of 200 feet. Assuming that the bottom of the pond can be approximated by the model

$$y = 20\left[\left(\frac{x}{200}\right)^2 - 1\right]$$

find the volume of water in the pond and then estimate the maximum number of fish that the pond can support.

Solution We see from Figure 4.32 that the radius of the pond is measured *horizontally* rather than vertically. Hence, we must adapt the Disc Method by integrating along the y-axis from -20 to 0. To find the radius, we solve for x in the given equation as follows:

$$20\left[\left(\frac{x}{200}\right)^2 - 1\right] = y$$

$$\left(\frac{x}{200}\right)^2 - 1 = \frac{y}{20}$$

$$\left(\frac{x}{200}\right)^2 = \frac{y}{20} + 1$$

$$\frac{x}{200} = \sqrt{\frac{y}{20} + 1}$$

$$x = 200\sqrt{\frac{y}{20} + 1}$$

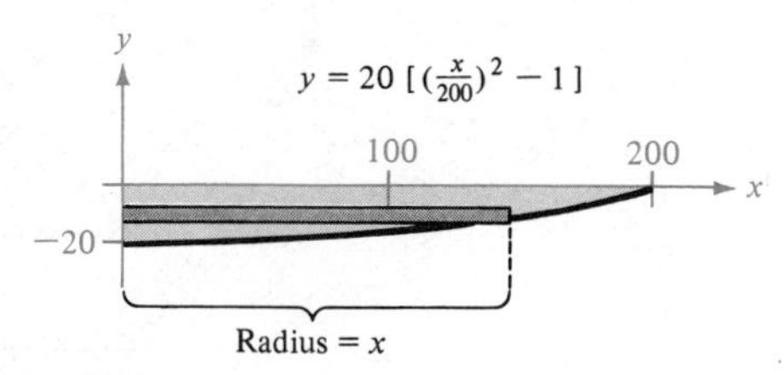

Figure 4.32

Now, using x as the radius, we find the volume of the pond:

$$\begin{aligned}\text{volume} &= \pi\int_{-20}^{0}\left(200\sqrt{\frac{y}{20}+1}\right)^2 dy \\ &= 40{,}000\pi\int_{-20}^{0}\left(\frac{y}{20}+1\right) dy \\ &= 40{,}000\pi\left[\frac{y^2}{40}+y\right]_{-20}^{0} \\ &= 40{,}000\pi\left[0-\left(\frac{400}{40}-20\right)\right] \approx 1{,}256{,}637 \text{ ft}^3\end{aligned}$$

Finally, since each fish requires 500 ft^3 of water, the maximum number of fish that this pond can support is

$$\frac{1{,}256{,}637}{500} \approx 2{,}513 \text{ fish}$$

□

Section Exercises (4.5)

In Exercises 1–12, find the volume of the solid formed by revolving the given region about the x-axis.

1.

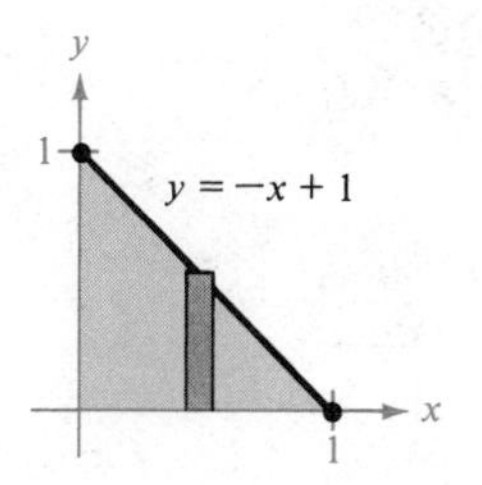

2.

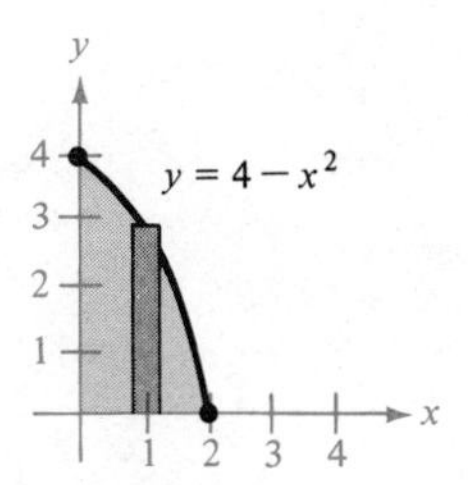

3.

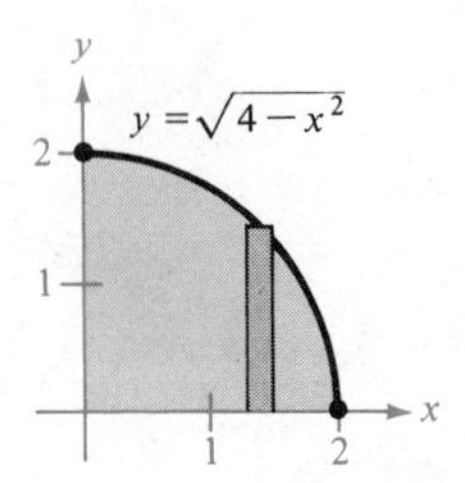

4.

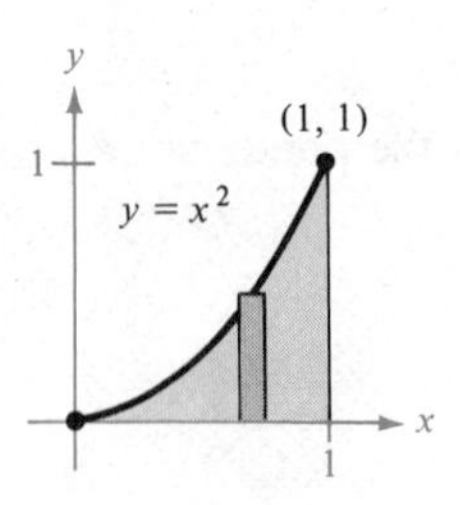

5.

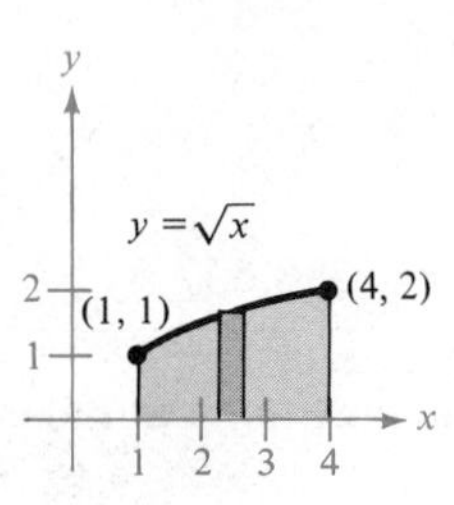

6.

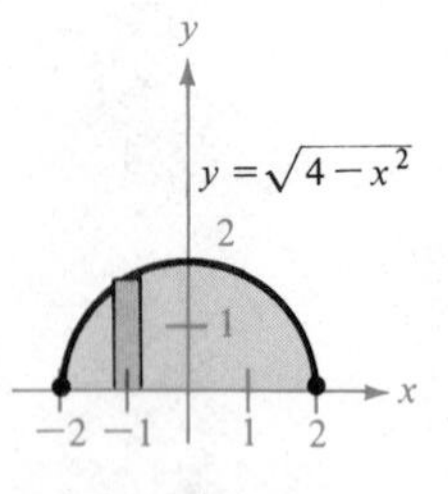

7.

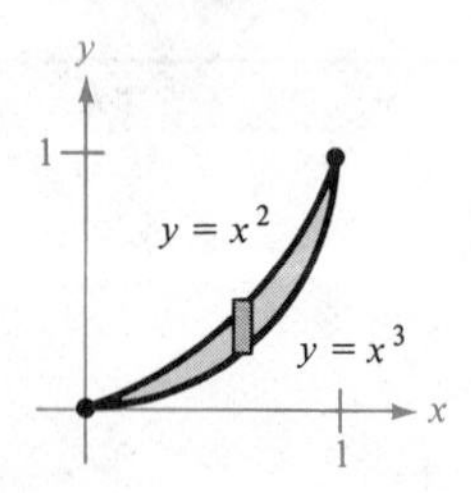

8.

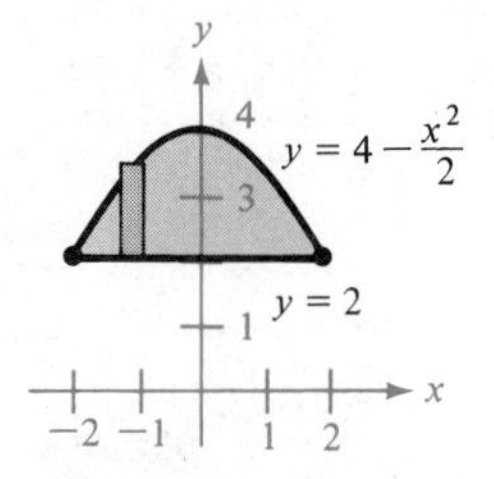

9. The region bounded by $y = 2x^2$, $y = 0$, and $x = 2$.

10. The region bounded by $y = 1 - \dfrac{x^2}{4}$ and $y = 0$.

11. The region bounded by $y = 6 - 2x - x^2$ and $y = x + 6$.

12. The region bounded by $y = x^2$ and $y = 4x - x^2$.

In Exercises 13–16, find the volume of the solid formed by revolving the given region about the y-axis.

13.

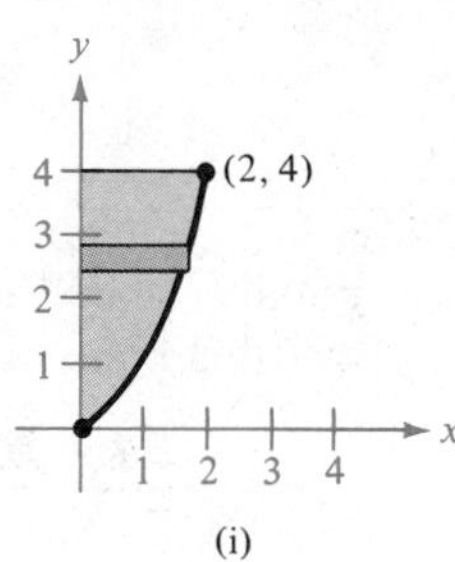

(i)

14.

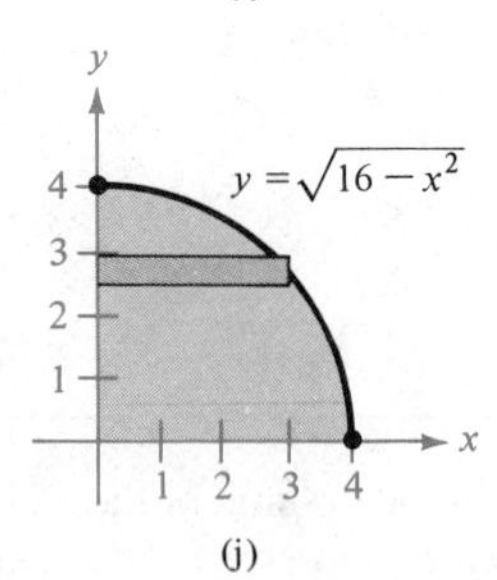

(j)

15.

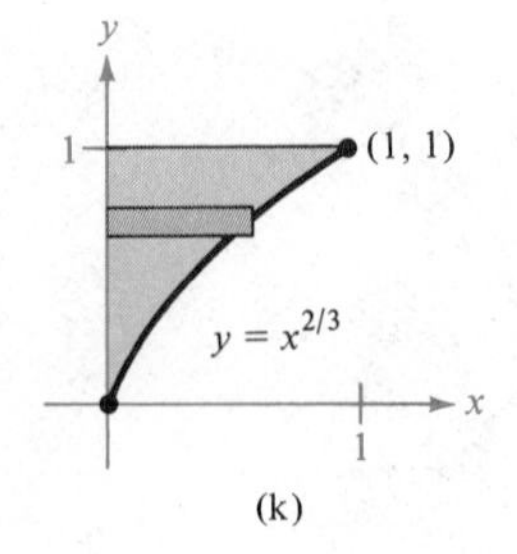

(k)

16.

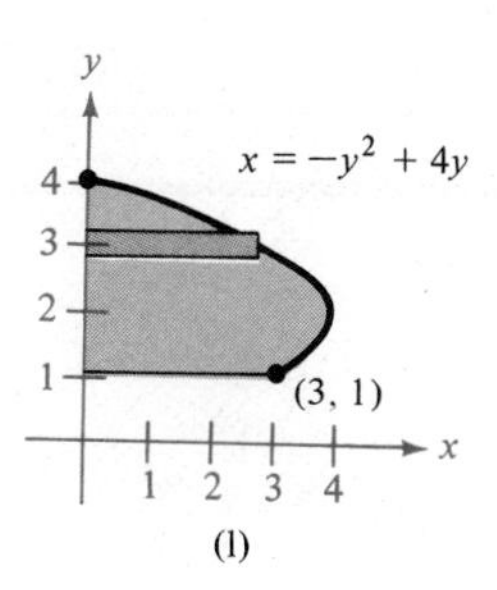

17. Use the Disc Method to verify that the volume of a right circular cone is $\frac{1}{3}\pi r^2 h$, where r is the radius of the base and h is the height.

18. Use the Disc Method to verify that the volume of a sphere of radius r is $\frac{4}{3}\pi r^3$.

19. The upper half of the ellipse $9x^2 + 25y^2 = 225$ is revolved about the x-axis to form a prolate spheroid (shaped like a football). Find the volume of the spheroid.

20. The right half of the ellipse $9x^2 + 25y^2 = 225$ is revolved about the y-axis to form an oblate spheroid (shaped like an M & M candy). Find the volume of this spheroid.

Chapter 5

Exponential and Logarithmic Functions

Section 5.1

Exponential Functions

Introductory Example

Radioactive Carbon Dating

In living organic material, the ratio of radioactive carbon isotopes to the total number of carbon atoms is about 1 to 10^{12}. When organic material dies, its radioactive carbon isotopes begin to decay, with a half-life of about 5700 years. This means that after 5700 years the ratio of isotopes to atoms will have decreased to one-half of the original ratio, after a second 5700 years the ratio will have decreased to one-fourth of the original, and so on. Figure 5.1 shows this declining ratio. The formula for the ratio R of carbon isotopes to carbon atoms is given by the **exponential function**

$$R = 10^{-12}2^{-t/5700}.$$

where t is measured in years.

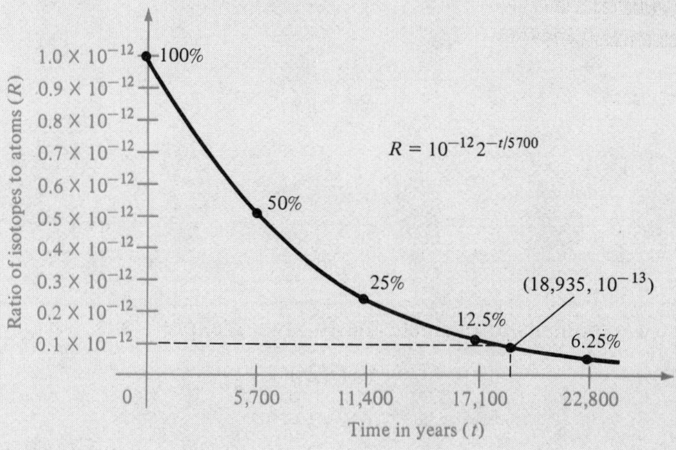

Figure 5.1

Suppose that we wish to estimate the age of a certain fossil. Testing the carbon makeup of the fossil, we determine that the ratio is 1 isotope in 10^{13} atoms. Thus, we have $R = 10^{-13}$, and we can estimate the age of the fossil by solving the equation

$$10^{-13} = 10^{-12}2^{-t/5700}$$

In Figure 5.1, note that $0.1 \times 10^{-12} = (10^{-1})(10^{-12}) = 10^{-13}$.

In Section 5.3, we will see that this equation can be solved by means of *logarithms*. The solution turns out to be

$$t = (5,700)\frac{\ln 10}{\ln 2} \approx 18,935 \text{ years}$$

Section Objectives: *To review the properties of exponents and exponential functions* ▪ *To introduce the number* **e.**

We are quite familiar with the behavior of functions such as

$$f(x) = x^2 \qquad g(x) = \sqrt{x}, \qquad h(x) = x^{-1}$$

that involve a variable raised to a constant power. By interchanging roles and raising a constant to a variable power, we obtain an important class of functions called **exponential functions.** Some simple examples are

$$f(x) = 2^x, \qquad g(x) = \left(\frac{1}{10}\right)^x = \frac{1}{10^x}, \qquad h(x) = 3^{2x} = 9^x$$

In general, we can use any positive base $a \neq 1$ for exponential functions.

Definition of an Exponential Function

If $a > 0$ and $a \neq 1$, then we call the function $y = a^x$ the **exponential function** with base a.

Before discussing the behavior of exponential functions, we list some familiar properties of exponents.

Properties of Exponents ($a, b > 0$)

1. $a^0 = 1$
2. $a^x a^y = a^{x+y}$
3. $\dfrac{a^x}{a^y} = a^{x-y}$
4. $(a^x)^y = a^{xy}$
5. $(ab)^x = a^x b^x$
6. $\left(\dfrac{a}{b}\right)^x = \dfrac{a^x}{b^x}$
7. $a^{-x} = \dfrac{1}{a^x}$

Example 1

Applying the Properties of Exponents

Use the properties of exponents to evaluate the following expressions.

(a) $(2^2)(2^3)$

(b) $(2^2)(2^{-3})$

(c) $(3^2)^3$

(d) $(\frac{1}{3})^{-2}$

(e) $\dfrac{3^2}{3^3}$

(f) $(2^{1/2})(3^{1/2})$

Solution

(a) By Property 2, we have

$$(2^2)(2^3) = 2^{2+3} = 2^5 = 32$$

(b) By Properties 2 and 7, we have

$$(2^2)(2^{-3}) = 2^{2-3} = 2^{-1} = \tfrac{1}{2}$$

(c) By property 4, we have

$$(3^2)^3 = 3^{2(3)} = 3^6 = 729$$

(d) By Property 7, we have

$$(\tfrac{1}{3})^{-2} = \frac{1}{(\frac{1}{3})^2} = 3^2 = 9$$

(e) By Properties 3 and 7, we have

$$\frac{3^2}{3^3} = 3^{2-3} = 3^{-1} = \tfrac{1}{3}$$

(f) By Property 5, we have

$$(2^{1/2})(3^{1/2}) = [(2)(3)]^{1/2} = 6^{1/2} = \sqrt{6}$$ □

Remark

Although we demonstrated the seven properties of exponents with integer and rational values for x and y, it is important to realize that the properties hold for *any* real values for x and y. With a calculator we can readily obtain approximate values for a^x when x is nonintegral or irrational. For example,

$$\frac{1}{2^{0.6}} = 2^{-0.6} \approx 0.660$$

$$2^{3/4} = 2^{0.75} \approx 1.682$$

$$2^{\sqrt{2}} = 2^{1.4142} \approx 2.665$$

Example 2

Graphing Exponential Functions

Sketch the graphs of the exponential functions $f(x) = 2^x$, $g(x) = (\frac{1}{2})^x = 2^{-x}$, and $h(x) = 3^x$.

Solution Table 5.1 lists some values for these functions, and Figure 5.2 shows their graphs.

Table 5.1

x	-3	-2	-1	0	1	2	3	4
2^x	$\frac{1}{8}$	$\frac{1}{4}$	$\frac{1}{2}$	1	2	4	8	16
2^{-x}	8	4	2	1	$\frac{1}{2}$	$\frac{1}{4}$	$\frac{1}{8}$	$\frac{1}{16}$
3^x	$\frac{1}{27}$	$\frac{1}{9}$	$\frac{1}{3}$	1	3	9	27	81

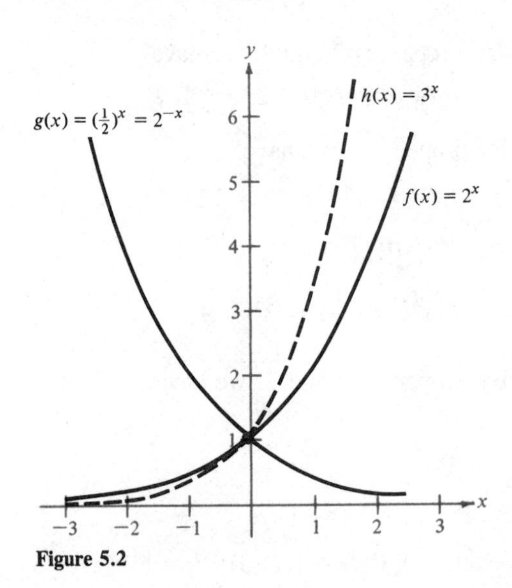

Figure 5.2

□

The forms of the graphs in Figure 5.2 are typical of those of the exponential functions a^x and a^{-x} $(a > 1)$. We summarize the characteristics of these graphs in Figure 5.3.

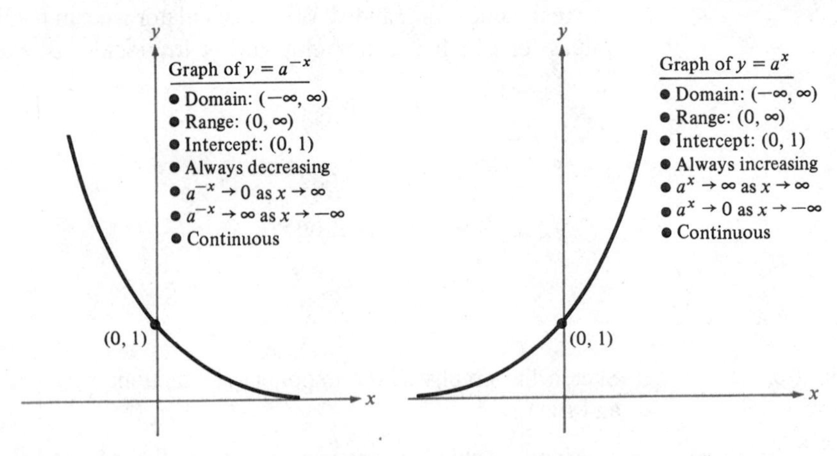

Figure 5.3

Example 3

Graphing an Exponential Function

Sketch the graph of $f(x) = 3^{-x} - 1$.

Solution By plotting the points listed in Table 5.2 and determining that $y = -1$ is a horizontal asymptote to the right,

$$\lim_{x\to\infty} (3^{-x} - 1) = \lim_{x\to\infty} 3^{-x} - \lim_{x\to\infty} 1 = 0 - 1 = -1$$

we obtain the graph shown in Figure 5.4.

Table 5.2

x	$f(x)$
-2	8
-1	2
0	0
1	$-\frac{2}{3}$
2	$-\frac{8}{9}$

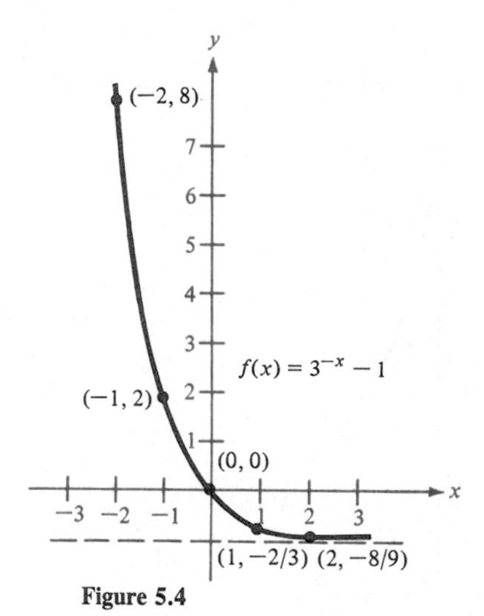

Figure 5.4

□

We have introduced exponential functions using an unspecified base a. It turns out that in calculus the natural (or convenient) choice for a base is the irrational number e, whose decimal approximation is

$$e \approx 2.71828 \ldots$$

This choice may seem anything but natural; however, the convenience of this particular choice will become apparent as we develop the rules for differentiating and integrating exponential and logarithmic functions. In the development of these differentiation rules, we encounter the limit used in the following definition of e.

Definition of e

$$e = \lim_{x \to 0} (1 + x)^{1/x}$$

(To twelve significant digits, $e \approx 2.71828182846$.)

Table 5.3 lists some values of the function $f(x) = (1 + x)^{1/x}$ for x near zero, and Figure 5.5 shows how the graph of f approaches e as x approaches zero.

Table 5.3

x	$(1+x)^{1/x}$	x	$(1+x)^{1/x}$
-0.5	4.0000	1.0	2.0000
-0.1	2.8680	0.5	2.2500
-0.01	2.7320	0.1	2.5937
-0.001	2.7196	0.01	2.7048
-0.0001	2.7184	0.001	2.7169
$\downarrow$	$\downarrow$	0.0001	2.7181
0	$e \approx 2.71828$	$\downarrow$	$\downarrow$
		0	$e \approx 2.71828$

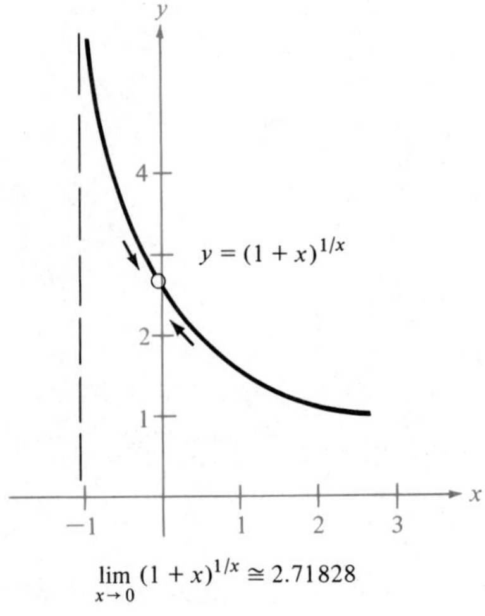

Figure 5.5

Example 4

Graphing an Exponential Function

Sketch the graph of $f(x) = e^x$.

Solution Figure 5.6 shows the graph of $f(x) = e^x$ determined from the values in Table 5.4. (See Appendix for table of exponential values.)

Table 5.4

x	-2	-1	0	1	2
e^x	$\frac{1}{e^2} \approx 0.135$	$\frac{1}{e} \approx 0.368$	1	$e \approx 2.718$	$e^2 \approx 7.389$

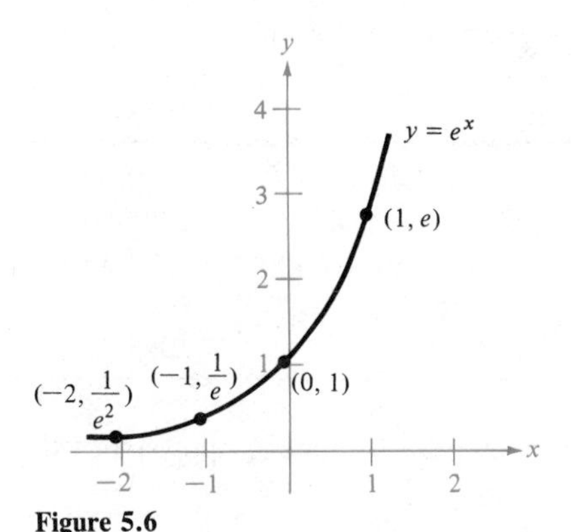

Figure 5.6

□

To give you greater insight into the usefulness of the natural base e, we take another look at the compound interest problem discussed in the Introductory Examples of Sections 1.5 and 1.6.

Compound Interest

Suppose P dollars are deposited in a savings account at an annual interest rate r. If accumulated interest is deposited into the account, what is the balance in the account at the end of one year? Of course, the answer depends on the number of times the interest is compounded, as indicated in Table 5.5. For instance, the results for a deposit of $1000 at 8% interest compounded n times a year are as shown in Table 5.6.

Table 5.5

Number of times compounded	Balance after one year
1 (annually)	$A = P(1 + r)$
2 (semiannually)	$A = P\left(1 + \frac{r}{2}\right)^2$
4 (quarterly)	$A = P\left(1 + \frac{r}{4}\right)^4$
⋮	⋮
n	$A = P\left(1 + \frac{r}{n}\right)^n$

Table 5.6

n	Balance ($)
1 (annually)	1080.00
2 (semiannually)	1081.50
4 (quarterly)	1082.43
12 (monthly)	1082.99
365 (daily)	1083.22

As n increases, the balance A approaches the limit

$$\lim_{n\to\infty} P\left(1 + \frac{r}{n}\right)^n$$

We determine this limit in the following manner:

$$\lim_{n\to\infty} P\left(1 + \frac{r}{n}\right)^n = P \lim_{n\to\infty}\left[\left(1 + \frac{r}{n}\right)^{n/r}\right]^r$$

If we let $x = r/n$, then $x \to 0$ as $n \to \infty$. Thus, we have

$$P\left[\lim_{n\to\infty}\left(1 + \frac{r}{n}\right)^{n/r}\right]^r = P\left[\lim_{x\to 0}(1 + x)^{1/x}\right]^r = P[e]^r$$

The equation

$$A = \lim_{n\to\infty} P\left(1 + \frac{r}{n}\right)^n = Pe^r$$

denotes the balance at the end of one year, during which time the interest is said to be compounded *continuously*. For a deposit of $1000 at 8% interest compounded continuously, the balance at the end of one year would be

$$A = 1000e^{0.08} \approx \$1083.29$$

In general, we have the following two formulas:

1. compounded n times per year, $\quad A = P\left(1 + \frac{r}{n}\right)^{nt}$

2. compounded continuously, $\quad A = Pe^{rt}$

where

P = amount of deposit
r = interest rate
n = number of times compounded per year
t = number of years
A = balance after t years

Example 5

Finding Compound Interest

A deposit of \$2500 is made in a savings account that pays an annual interest rate of 10%. Find the balance in the account at the end of 5 years if the interest is compounded:

(a) quarterly (b) continuously

Solution

(a) The balance after quarterly compounding is

$$\begin{aligned} A &= P\left(1 + \frac{r}{n}\right)^{nt} \\ &= 2500\left(1 + \frac{0.1}{4}\right)^{(4)(5)} \\ &= 2500(1.025)^{20} \\ &= \$4096.54 \end{aligned}$$

(b) The balance after continuous compounding is

$$\begin{aligned} A &= Pe^{rt} \\ &= 2500[e^{(0.1)(5)}] \\ &= 2500e^{0.5} \\ &= \$4121.80 \end{aligned}$$

□

Example 6

Evaluating Exponential Functions

A certain bacterial culture is growing according to the *logistics growth function*

$$y = \frac{1.25}{1 + 0.25e^{-0.4t}}, \qquad 0 \le t$$

where y is the weight of the culture in grams and t is the time in hours. Find the weight of the culture after:

(a) 0 h (b) 1 h (c) 10 h

What is the limit of this function as t approaches ∞?

Solution

(a) When $t = 0$, we have

$$y = \frac{1.25}{1 + 0.25e^{0}} = \frac{1.25}{1.25} = 1 \text{ g}$$

(b) When $t = 1$, we have

$$y = \frac{1.25}{1 + 0.25e^{-0.4}} \approx 1.070 \text{ g}$$

(c) When $t = 10$, we have

$$y = \frac{1.25}{1 + 0.25e^{-4}} \approx 1.244 \text{ g}$$

Finally, taking the limit as t approaches ∞, we have

$$\lim_{t\to\infty} \frac{1.25}{1 + 0.25e^{-0.4t}} = \lim_{t\to\infty} \frac{1.25}{1 + \dfrac{0.25}{e^{0.4t}}} = \frac{1.25}{1 + 0} = 1.25 \text{ g}$$

(See Figure 5.7.)

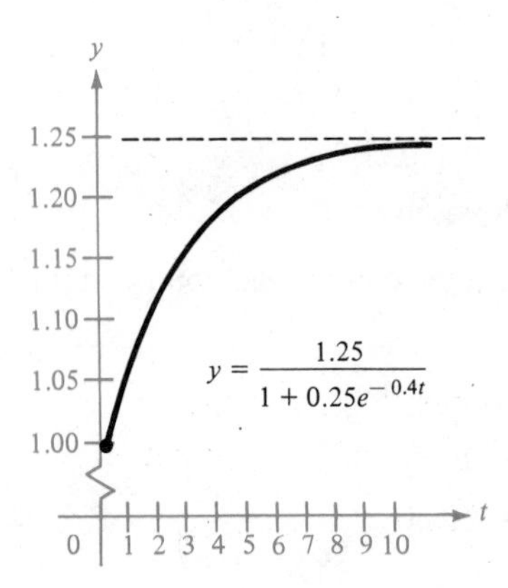

Figure 5.7

□

Section Exercises (5.1)

1. Evaluate each of the following expressions.
(a) $5(5^3)$ (b) $27^{2/3}$
(c) $64^{3/4}$ (d) $81^{1/2}$
(e) $25^{3/2}$

2. Evaluate each of the following expressions.
(a) $(\frac{1}{5})^3$ (b) $(\frac{1}{8})^{1/3}$
(c) $64^{2/3}$ (d) $(\frac{5}{8})^2$
(e) $100^{3/2}$

In Exercises 3–5, use the properties of exponents to simplify each expression.

3. (a) $(5^2)(5^3)$ (b) $(5^2)(5^{-3})$
(c) $(5^2)^2$ (d) $(5)^{-3}$
(e) $\dfrac{5^3}{5^6}$ (f) $(\frac{1}{5})^{-2}$
(g) $(8^{1/2})(2^{1/2})$ (h) $(32^{3/2})(\frac{1}{2})^{3/2}$

4. (a) $(4^3)(4^2)$ (b) $(\frac{1}{4})^2(4^2)$
(c) $(4^6)^{1/2}$ (d) $[(8^{-1})(8^{2/3})]^3$
(e) $[(25^{1/2})(25^2)]^{1/5}$ (f) $(8^2)(4^3)$
(g) $\dfrac{5^3}{25^2}$ (h) $(9^{2/3})(3)(3^{2/3})$

5. (a) $e^2(e^4)$ (b) $(e^3)^4$
(c) $(e^3)^{-2}$ (d) $\dfrac{e^5}{e^3}$
(e) $\left(\dfrac{1}{e}\right)^{-2}$ (f) $\left(\dfrac{e^5}{e^2}\right)^{-1}$

In Exercises 6–22, sketch the graph of the given exponential function.

6. $f(x) = 4^x$ **7.** $g(x) = 5^x$
8. $h(x) = (\frac{1}{4})^x = 4^{-x}$ **9.** $f(x) = (\frac{1}{5})^x = 5^{-x}$
10. $f(x) = (\frac{1}{4})^{-x}$ **11.** $g(x) = (\frac{1}{5})^{-x}$

12. $y = 2^{-x^2}$

13. $y = 3^{-x^2}$

14. $y = 3^{|x|}$

15. $y = 3^{-|x|}$

16. $s(t) = 2^{-t} + 3$

17. $s(t) = \dfrac{3^{-t}}{4}$

Ⓒ**18.** $f(x) = e^{2x}$

Ⓒ**19.** $h(x) = e^{x-2}$

Ⓒ**20.** $A(t) = 500e^{0.15t}$

Ⓒ**21.** $N(t) = 1000e^{-0.2t}$

Ⓒ**22.** $g(x) = \dfrac{10}{1 + e^{-x}}$

Ⓒ**23.** If \$1000 is invested at 10% interest, find the amount after 10 years if the interest is compounded:
(a) annually (b) semiannually
(c) quarterly (d) monthly
(e) daily (f) continuously

Ⓒ**24.** If \$2500 is invested at 12% interest, find the amount after 20 years if the interest is compounded:
(a) annually (b) semiannually
(c) quarterly (d) monthly
(e) daily (f) continuously

Ⓒ**25.** The demand equation for a certain product is given by

$$p = 5000\left(1 - \frac{4}{4 + e^{-0.002x}}\right)$$

(See Figure 5.8.) Find the price of the product if the demand is:
(a) $x = 100$ units (b) $x = 500$ units

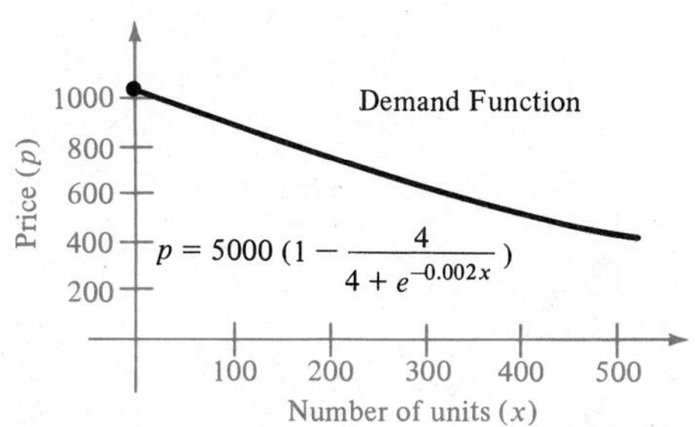

Figure 5.8

Ⓒ**26.** The demand equation for a certain product is given by

$$p = 500 - 0.5e^{0.004x}$$

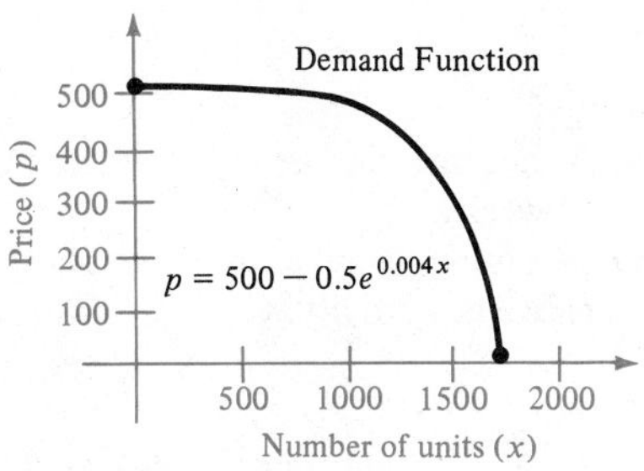

Figure 5.9

(See Figure 5.9) Find the price of the product if the demand is:
(a) $x = 1000$ units (b) $x = 1500$ units

Ⓒ**27.** A bacterial culture is growing according to the logistics growth function

$$y = \frac{850}{1 + 3e^{-0.05t}}$$

(a) Find the limit of this function as t approaches infinity.
(b) Sketch the graph of this function.

Ⓒ**28.** The yield V (in millions of cubic feet per acre) for a forest at age t years is given by

$$V = 6.7e^{-48.1/t}$$

(a) Find the limiting volume of wood per acre as t approaches infinity.
(b) Find the volume per acre after:
(i) 20 years (ii) 50 years
(c) Sketch the graph of this function.

Ⓒ**29.** In a group project in learning theory, a mathematical model for the proportion P of correct responses after n trials was found to be

$$P = \frac{0.83}{1 + e^{-0.2n}}$$

(a) Find the proportion of correct responses after 10 trials.
(b) Find the limiting proportion of correct responses as n approaches infinity.

Ⓒ**30.** In a typing class the average number N of words per minute typed after t weeks of lessons was found to be

Ⓒ Calculator may be helpful.

$$N = \frac{157}{1 + 5.4e^{-0.12t}}$$

(a) Find the average number of words per minute after 10 weeks.
(b) Find the limiting number of words per minute as t approaches infinity.

Ⓒ**31.** Find the amount of money that should be deposited in an account paying 12% compounded continuously to produce a final balance of $100,000 in:
(a) 1 year
(b) 10 years
(c) 20 years
(d) 50 years

Ⓒ Calculator may be helpful.

Section 5.2

Differentiation and Integration of Exponential Functions

Introductory Example

The Heights of 18- to 24-Year-Old Males

In a given population, how many people would you expect to be a certain height? One way to answer this question is to use the **normal probability density function** as a model. This model was developed by a famous mathematician named Carl Friedrich Gauss. It has since been used by statisticians to represent bell-shaped distributions that occur in a wide variety of fields.

As an example of one of these normal distributions, let us consider the heights of American males between 18 and 24 years of age. The average height of this population is estimated to be 70 in. and roughly 68% of the heights lie within $3\frac{1}{3}$ in. of this mean (average). The normal probability density function describing this distribution is

$$f(x) = 0.12e^{-0.045(x-70)^2}$$

To find the percentage of heights between h_1 and h_2, we can use the **definite integral**

$$\text{percentage} = 0.12\int_{h_1}^{h_2} e^{-0.045(x-70)^2}\,dx$$

Figure 5.10 (see page 319) shows the graph of f including an area represented by this integral. Table 5.7 lists the percentages corresponding to heights between 60 and 80 in. Note that 99.7% of the total population have heights between 60 and 80 in.

Table 5.7

Height	Percentage
$60 \le x < 61$	0.22
$61 \le x < 62$	0.47
$62 \le x < 63$	0.97
$63 \le x < 64$	1.80
$64 \le x < 65$	3.09
$65 \le x < 66$	4.83
$66 \le x < 67$	6.90
$67 \le x < 68$	9.02
$68 \le x < 69$	10.78
$69 \le x < 70$	11.79
$70 \le x < 71$	11.79
$71 \le x < 72$	10.78
$72 \le x < 73$	9.02
$73 \le x < 74$	6.90
$74 \le x < 75$	4.83
$75 \le x < 76$	3.09
$76 \le x < 77$	1.80
$77 \le x < 78$	0.97
$78 \le x < 79$	0.47
$79 \le x < 80$	0.22

Section Objectives: ***To develop differentiation formulas for exponential functions ▪ To develop integration formulas for exponential functions.***

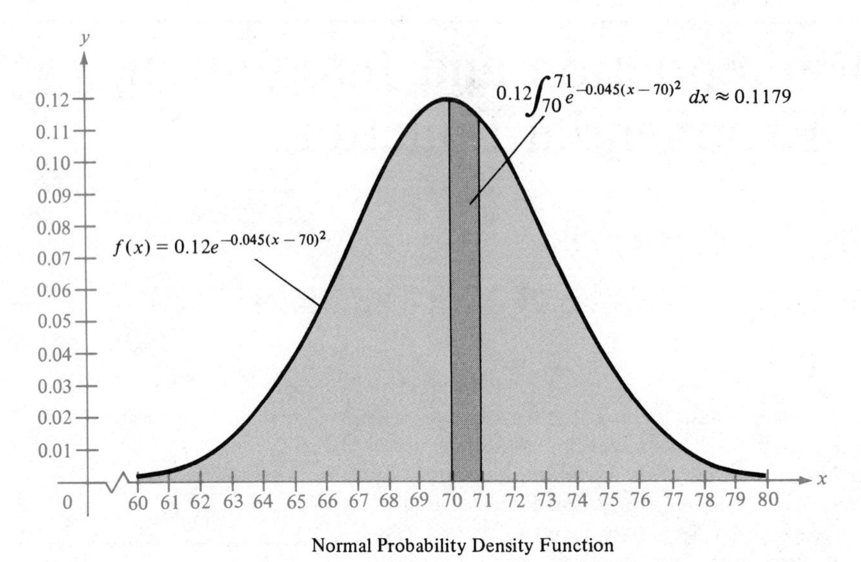

Figure 5.10

In the preceding section, we promised to defend the choice of e as the most convenient base for an exponential function. Perhaps the most convincing argument in this defense is that by choosing e as the exponential base we have the very nice result that the function $y = e^x$ *is its own derivative*. (This is not true of other exponential functions of the form $y = a^x$, $a \neq e$.) To see that this is true, let us return to the limit definition of the derivative as given in Section 2.1.

Recall that

$$f'(x) = \lim_{\Delta x \to 0} \frac{f(x + \Delta x) - f(x)}{\Delta x}$$

Thus,

$$\frac{d}{dx}[e^x] = \lim_{\Delta x \to 0} \frac{e^{x+\Delta x} - e^x}{\Delta x}$$

$$= \lim_{\Delta x \to 0} \frac{e^x(e^{\Delta x} - 1)}{\Delta x}$$

Now, the limit definition of e

$$e = \lim_{\Delta x \to 0} (1 + \Delta x)^{1/\Delta x}$$

tells us that for small values of Δx we have

$$e \approx (1 + \Delta x)^{1/\Delta x} \qquad \text{and} \qquad e^{\Delta x} \approx 1 + \Delta x$$

Finally, replacing $e^{\Delta x}$ by this approximation, we have

$$\frac{d}{dx}[e^x] = \lim_{\Delta x \to 0} \frac{e^x[(1 + \Delta x) - 1]}{\Delta x} = \lim_{\Delta x \to 0} \frac{e^x(\Delta x)}{\Delta x} = e^x$$

Thus, we have the following theorem.

Theorem 5.1 (Derivative of e^x)

$$\frac{d}{dx}[e^x] = e^x$$

Remark We can interpret Theorem 5.1 geometrically to say that the slope of the graph of $f(x) = e^x$ at any point (x, e^x) is numerically equal to the y-coordinate of the point (Figure 5.11).

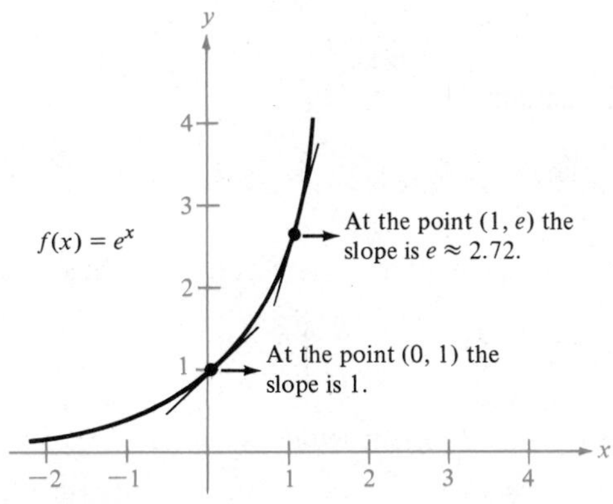

Figure 5.11

By applying the Chain Rule to the formula

$$\frac{d}{dx}[e^x] = e^x$$

we can obtain the differentiation formula for e^u where u is a differentiable function of x.

Derivatives of Exponential Functions

$$\frac{d}{dx}[e^x] = e^x$$

$$\frac{d}{dx}[e^u] = e^u u'$$

Example 1

Differentiating Exponential Functions

Differentiate the following functions.

(a) $y = e^{2x-1}$ (b) $y = e^{-3/x}$

Solution

(a) Considering $u = 2x - 1$, we have $u' = 2$, and thus

$$\frac{dy}{dx} = e^u u' = e^{2x-1}(2) = 2e^{2x-1}$$

(b) Considering $u = -3/x$, we have $u' = 3/x^2$, and thus

$$\frac{dy}{dx} = e^u u' = e^{-3/x}\left(\frac{3}{x^2}\right) = \frac{3e^{-3/x}}{x^2}$$

□

Example 2

Combining Differentiation Rules

Differentiate

$$f(x) = e^2 + \frac{e^{-2x}}{x^3}$$

Solution Note that e^2 is a constant; thus, we have

$$f'(x) = 0 + \frac{d}{dx}\left[\frac{e^{-2x}}{x^3}\right]$$

Now by the Quotient Rule, we obtain

$$f'(x) = \frac{x^3[e^{-2x}(-2)] - e^{-2x}[3x^2]}{(x^3)^2} = \frac{-2x^3e^{-2x} - 3x^2e^{-2x}}{x^6}$$

$$= \frac{-x^2e^{-2x}(2x + 3)}{x^6} = \frac{-e^{-2x}(2x + 3)}{x^4}$$

□

Example 3

The Normal Probability Density Function

Show that the *normal probability density function*

$$f(x) = \frac{1}{\sqrt{2\pi}}e^{-x^2/2}$$

has points of inflection when $x = \pm 1$.

Solution First,

$$f'(x) = \frac{1}{\sqrt{2\pi}}(-x)e^{-x^2/2}$$

and by the Product Rule, we have

$$f''(x) = \frac{1}{\sqrt{2\pi}}[(-x)(-x)e^{-x^2/2} + (-1)e^{-x^2/2}]$$

$$= \frac{1}{\sqrt{2\pi}}(e^{-x^2/2})(x^2 - 1)$$

Therefore, $f''(x) = 0$ when $x = \pm 1$, and we can apply the techniques in Section 3.3 to conclude that these values of x yield the two points of inflection. (See Figure 5.12.)

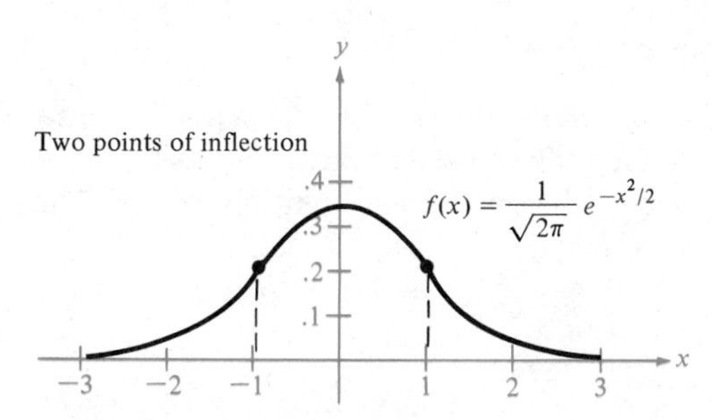

Bell-Shaped Curve Given By Normal Probability Density Function

Figure 5.12

□

Remark

The general form of a normal probability density function is given by

$$f(x) = \frac{1}{\sigma\sqrt{2\pi}}e^{-x^2/2\sigma^2}$$

where σ is the standard deviation (σ is the lower-case Greek letter sigma). By following the procedure of Example 3, we can show that the bell-shaped graph of f has points of inflection when $x = \pm\sigma$.

Each of the formulas for differentiating exponential functions has its corresponding integration formula, as shown next.

Integrals of Exponential Functions

$$\int e^x\,dx = e^x + C$$

$$\int e^u u'\,dx = e^u + C$$

Example 4

Integrating Exponential Functions

Evaluate $\int e^{3x+1}\,dx$.

Solution Considering $u = 3x + 1$, we have $u' = 3$. Introducing the missing factor 3 in the integrand and then multiplying the integral by the reciprocal factor $\frac{1}{3}$, we write

$$\int e^{3x+1}\,dx = \frac{1}{3}\int e^{3x+1}(3)\,dx$$

$$= \frac{1}{3}\int e^{u}u'\,dx = \frac{1}{3}e^{u} + C = \frac{e^{3x+1}}{3} + C$$

□

Example 5

Integrating Exponential Functions

Evaluate $\int 5xe^{-x^2}\,dx$.

Solution If we let $u = -x^2$, then $u' = -2x$. Now, we adjust our integrand by moving the unneeded factor 5 outside the integral sign. Then, we introduce the needed factor -2 and multiply by $\frac{-1}{2}$. Thus, we write

$$\int 5xe^{-x^2}\,dx = 5\int e^{-x^2}(x)\,dx$$

$$= 5\left(\frac{-1}{2}\right)\int e^{-x^2}(-2x)\,dx = -\frac{5}{2}\int e^{u}u'\,dx$$

$$= -\tfrac{5}{2}e^{-x^2} + C$$

□

Remark

Keep in mind that we cannot introduce a missing *variable* factor in the integrand. For instance, we cannot evaluate $\int 5e^{x^2}\,dx$ by introducing the factor $2x$ and then multiplying the integral by $1/(2x)$. That is,

$$\int 5e^{x^2}\,dx \neq \frac{5}{2x}\int e^{x^2}(2x)\,dx$$

Example 6

Integrating Exponential Functions

Evaluate

$$\int \frac{e^{1/x}}{x^2}\,dx$$

Solution By choosing $u = 1/x$, we have

$$u' = -\frac{1}{x^2}$$

Thus,

$$\int \frac{e^{1/x}}{x^2}\,dx = -\int e^{1/x}\left(-\frac{1}{x^2}\right)dx = -e^{1/x} + C$$

□

Section Exercises (5.2)

In Exercises 1–6, find the slope of the tangent line to the given exponential function at the point (0, 1).

1.

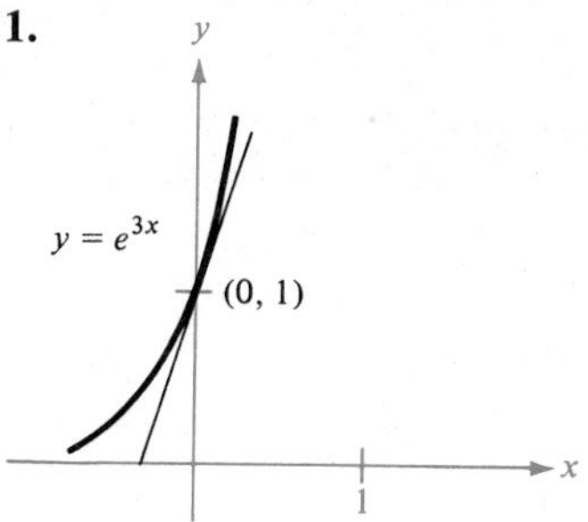

2.

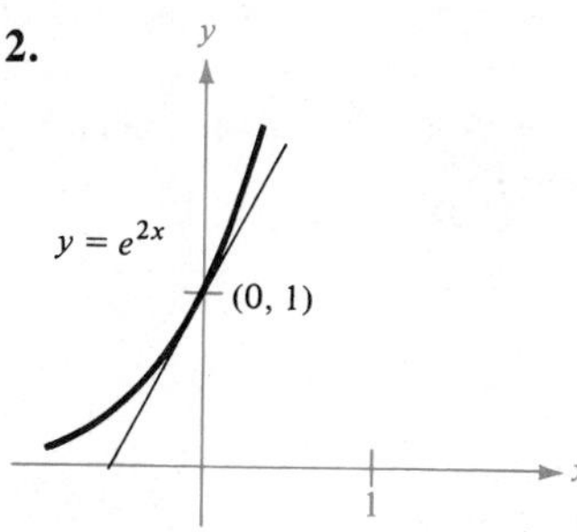

3.

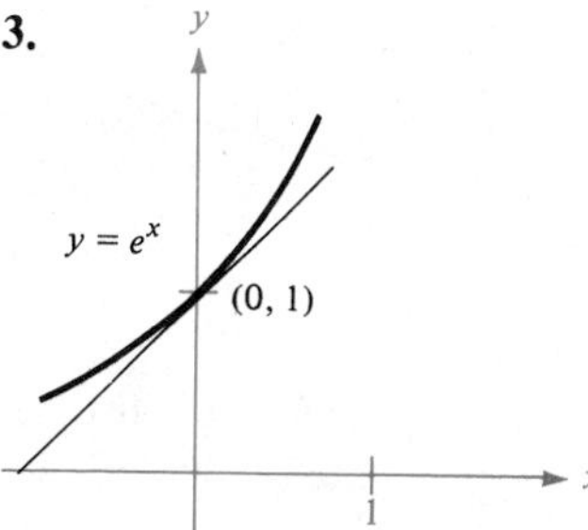

4.

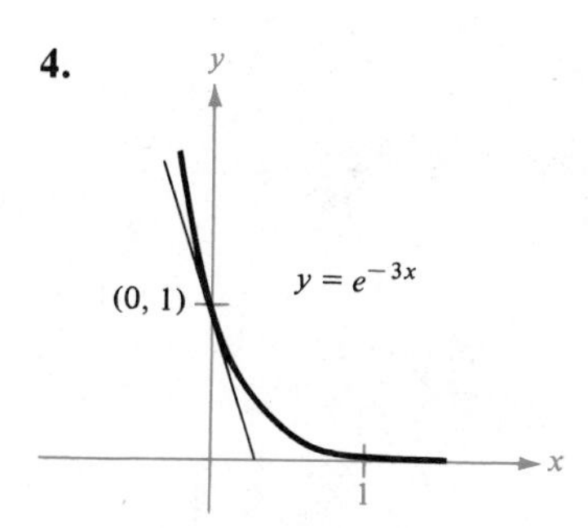

5.

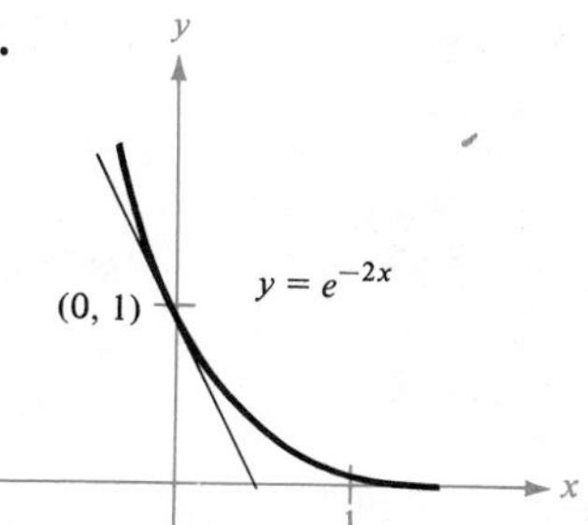

6.

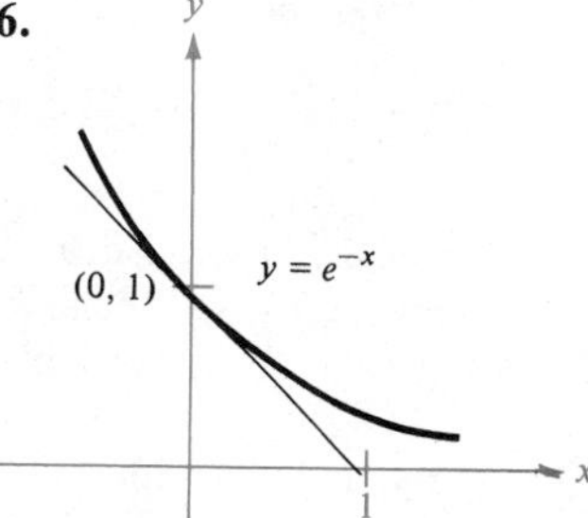

In Exercises 7–18, find dy/dx.

7. $y = e^{2x}$ **8.** $y = e^{1-x}$

9. $y = e^{1-2x+x^2}$ **10.** $y = e^{-x^2}$

11. $y = e^{\sqrt{x}}$ **12.** $y = x^2e^{-x}$

13. $y = (e^{-x} + e^{x})^3$ **14.** $y = e^{-1/x^2}$

15. $y = \dfrac{2}{e^x + e^{-x}}$ **16.** $y = \dfrac{e^x - e^{-x}}{2}$

17. $y = xe^x - e^x$

18. $y = x^2e^x - 2xe^x + 2e^x$

In Exercises 19–30, evaluate the integral.

19. $\displaystyle\int_0^1 e^{-2x}\,dx$ **20.** $\displaystyle\int_1^2 e^{1-x}\,dx$

21. $\displaystyle\int_0^2 (x^2 - 1)e^{x^3-3x+1}\,dx$ **22.** $\displaystyle\int x^2e^{x^3}\,dx$

23. $\displaystyle\int xe^{ax^2}\,dx$ **24.** $\displaystyle\int_0^{\sqrt{2}} xe^{-(x^2/2)}\,dx$

25. $\displaystyle\int_1^3 \frac{e^{3/x}}{x^2}\,dx$ **26.** $\displaystyle\int (e^x - e^{-x})^2\,dx$

27. $\displaystyle\int \frac{e^{2x} + 2e^x + 1}{e^x}\,dx$ **28.** $\displaystyle\int \frac{5 - e^x}{e^{2x}}\,dx$

29. $\displaystyle\int e^x\sqrt{1 - e^x}\,dx$ **30.** $\displaystyle\int \frac{2e^x - 2e^{-x}}{(e^x + e^{-x})^2}\,dx$

In Exercises 31–34, find (if any exist) the extrema and the points of inflection, and sketch the graph of each function.

Ⓒ**31.** $f(x) = \dfrac{2}{1 + e^{-x}}$ Ⓒ**32.** $f(x) = \dfrac{e^x - e^{-x}}{2}$

Ⓒ**33.** $f(x) = x^2e^{-x}$ Ⓒ**34.** $f(x) = xe^{-x}$

Ⓒ**35.** A certain lake is stocked with 500 fish and their population increases according to the *logistics growth curve*

$$p(t) = \frac{10{,}000}{1 + 19e^{-t/5}}$$

where t is measured in months. (See Figure 5.13.)

Ⓒ Calculator may be helpful.

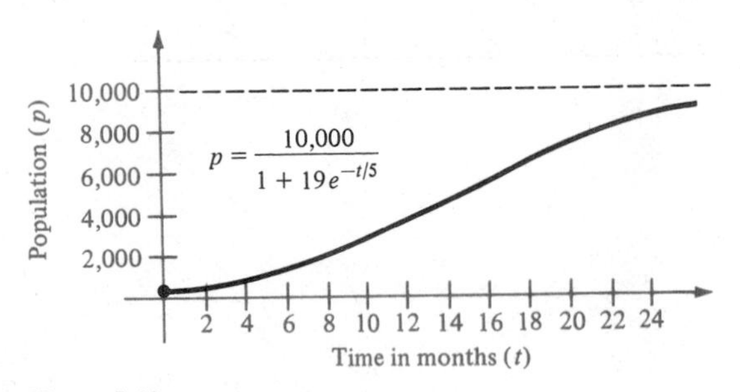

Figure 5.13

At what rate is the fish population of the lake increasing at the end of 1 month? At the end of 10 months? After how many months is the population increasing most rapidly?

Ⓒ**36.** The *Ebbinghaus Model* for human memory is

$$p(t) = (100 - a)e^{-bt} + a$$

where $p(t)$ is the percentage retained after t weeks. (The constants a and b vary from one person to another.) If $a = 20$ and $b = 0.5$, at what rate is information being retained after 1 week? After 3 weeks?

Ⓒ**37.** The yield V (in millions of cubic feet per acre) for a forest stand at age t is given by

$$V = 6.7e^{-48.1/t}$$

where t is measured in years. Find the rate at which the yield is increasing when:
(a) $t = 15$ years
(b) $t = 20$ years
(c) $t = 60$ years

Ⓒ**38.** The average typing speed (in the number of words per minute) after t weeks of lessons is given by

$$N = \frac{157}{1 + 5.4e^{-0.12t}}$$

Find the rate at which the speed is increasing when:
(a) $t = 5$ weeks
(b) $t = 10$ weeks
(c) $t = 30$ weeks

Ⓒ**39.** The balance in a certain savings account is given by

$$A = (5000)e^{0.08t}$$

where A is measured in dollars and t is measured in years. Find the rate at which the balance is increasing when:
(a) $t = 1$ year
(b) $t = 10$ years
(c) $t = 50$ years

Ⓒ**40.** Find the area of the region bounded by $y = 3e^{-x/2}$, $y = 0$, $x = 0$, and $x = 4$. (See Figure 5.14.)

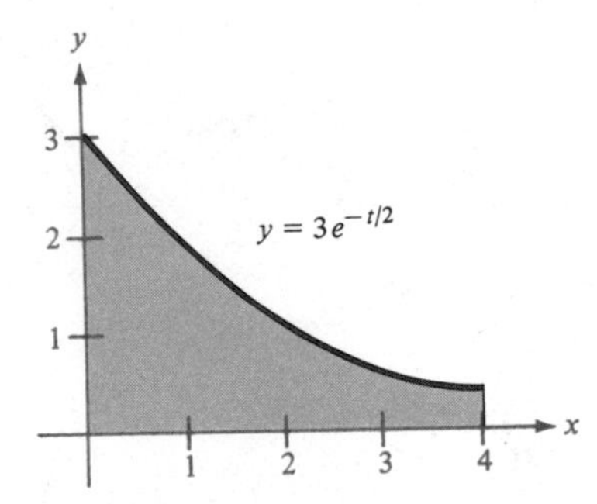

Figure 5.14

Ⓒ**41.** Find the area of the region bounded by $y = xe^{-x^2}$, $y = 0$, $x = 0$, and $x = 4$. (See Figure 5.15.)

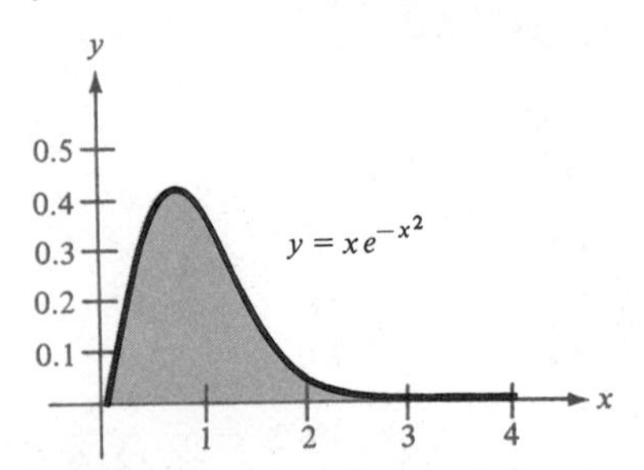

Figure 5.15

Ⓒ**42.** A deposit of $2500 is made in a savings account at an annual percentage rate of 12% compounded continuously. Find the average balance in this account during the first five years.

Ⓒ**43.** In the Ebbinghaus Model for human memory (see Exercise 36), let $a = 20$ and $b = 0.5$. Then the percentage retained in memory after t weeks is

$$p = 80e^{-0.5t} + 20$$

Ⓒ Calculator may be helpful.

Find the *average* percentage retained during:
(a) the first 2 weeks
(b) the second 2 weeks
(See Figure 5.16.)

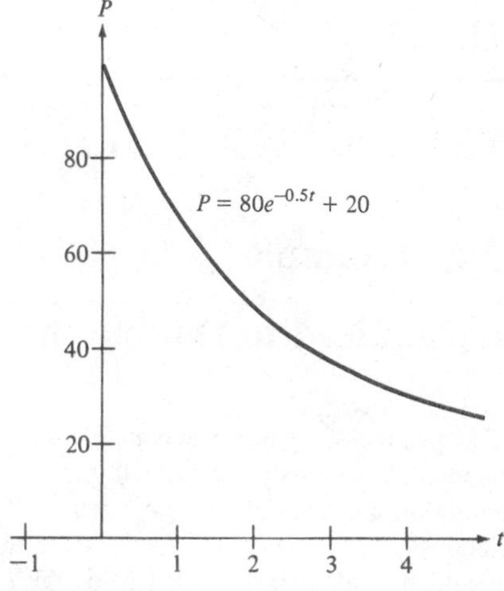

Figure 5.16

Section 5.3

The Natural Logarithmic Function

Introductory Example

The Time Required to Double the Balance in a Savings Account

When money is deposited in a savings account earning interest compounded continuously at r percent, the balance in the account after t years is $A = Pe^{rt}$, where P is the original deposit.

How long would it take such a deposit to double? This depends, of course, upon the value of r. If r is large, we expect that it would not take as long as it would when r is small. To find the doubling time for a given value of r, we could try several different values of t to find a time that doubles the original deposit. For instance, \$1000 is deposited at 8%; the balances at the ends of successive years are shown in Table 5.8. From the table it appears that this balance doubles sometime between eight and nine years after being deposited.

Table 5.8

t	1	2	3
$A = 1000e^{0.08t}$	\$1083.29	\$1173.51	\$1271.25

t	4	5	6
$A = 1000e^{0.08t}$	\$1377.13	\$1491.82	\$1616.07

t	7	8	9
$A = 1000e^{0.08t}$	\$1750.67	\$1896.48	\$2054.43

The problem with this "trial-and-error" procedure is that it is very time-consuming. Furthermore, it did not give us the exact doubling time. In this section, we will see that there is a simple way to solve this problem by using the "inverse" exponential function. We call this inverse function the **natural logarithmic function,** and we can use it to solve our "doubling time" problem in the following way. Let $A = 2P$ to obtain

$$Pe^{rt} = 2P$$

$$e^{rt} = 2$$

Now, by taking the natural logarithm of both sides of this equation, we have

$$\ln (e^{rt}) = \ln 2$$

$$rt = \ln 2$$

$$t = \frac{1}{r}(\ln 2) \approx \frac{0.6931}{r} \text{ years}$$

Table 5.9 shows the time required to double a balance for several different interest rates.

Table 5.9

Rate (%)	Doubling time (years)	Rate (%)	Doubling time (years)
5	13.86	13	5.33
6	11.55	14	4.95
7	9.90	15	4.62
8	8.66	16	4.33
9	7.70	17	4.08
10	6.93	18	3.85
11	6.30	19	3.65
12	5.78	20	3.47

Section Objective: ***To introduce the natural logarithmic function and discuss its properties.***

One of the theorems of advanced calculus states that if a continuous function is always increasing (or decreasing), then it possesses an inverse. From the properties of exponential functions identified in Section 5.1, we saw that the function $f(x) = e^x$ has the characteristics necessary for possessing an inverse. We call this inverse of the exponential function the **natural logarithmic function,** and we define it as follows.

Definition of the Natural Logarithmic Function

$$\ln x = b \quad \text{if and only if} \quad e^b = x$$

($\ln x$ is read "the natural log of x.")

This definition suggests that logarithmic equations can be written in an equivalent exponential form, and vice versa. For example,

Logarithmic Form	*Exponential Form*
$\ln 1 = 0$	$e^0 = 1$
$\ln e = 1$	$e^1 = e$
$\ln e^{-1} = -1$	$e^{-1} = \dfrac{1}{e}$
$\ln 2 \approx 0.693$	$e^{0.693} \approx 2$

Since the functions $f(x) = e^x$ and $f^{-1}(x) = \ln x$ are defined to be inverses of one another, their graphs should be reflections of each other in the line $y = x$. (See Section 1.4.) This reflective property is illustrated in Figure 5.17 along with a summary of some other properties of the natural logarithmic function.

Recall from Section 1.4 that inverse functions possess the property that

$$f(g(x)) = f(f^{-1}(x)) = x \quad \text{and} \quad g(f(x)) = f^{-1}(f(x)) = x$$

From this, we conclude that for $f(x) = e^x$ and $f^{-1}(x) = \ln x$ we have

$$e^{\ln x} = f(\ln x) = f(f^{-1}(x)) = x$$

and $$\ln (e^x) = f^{-1}(e^x) = f^{-1}(f(x)) = x$$

Inverse Properties of $\ln x$ and e^x

$$\ln (e^x) = x \quad \text{and} \quad e^{\ln x} = x$$

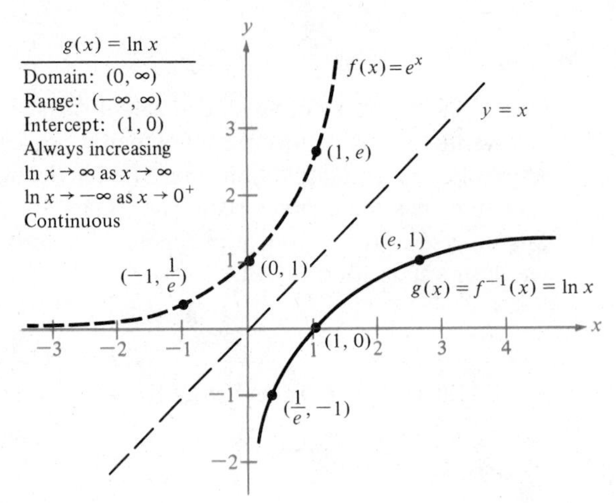

Figure 5.17

Example 1

Applying the Inverse Properties of $\ln x$ and e^x

Simplify the following expressions.

(a) $\ln (e^{\sqrt{2}})$ (b) $e^{\ln 3x}$

Solution

(a) $\ln (e^{\sqrt{2}}) = \sqrt{2}$ (b) $e^{\ln 3x} = 3x$ □

In Section 5.1, we saw that we can multiply two exponential functions by adding exponents:

$$e^x e^y = e^{x+y}$$

The logarithmic version of this rule states that the log of the product of two numbers is equal to the sum of the logs of the numbers, that is,

$$\ln (xy) = \ln x + \ln y$$

This rule and the rules for evaluating the log of a quotient and the log of a power are summarized as follows.

Properties of Logarithms

1. $\ln (xy) = \ln x + \ln y$
2. $\ln \dfrac{x}{y} = \ln x - \ln y$
3. $\ln (x^y) = y \ln x$

Example 2

Expanding Logarithmic Expressions

Use the properties of logarithms to write each of the following expressions as a sum, difference, or multiple of logarithms.

(a) $\ln \frac{10}{9}$ (b) $\ln \sqrt{3x + 2}$

(c) $\ln \frac{xy}{5}$ (d) $\ln \frac{(x - 3)^2}{x\sqrt[3]{x - 1}}$

Solution

(a) By Property 2,

$$\ln \tfrac{10}{9} = \ln 10 - \ln 9$$

(b) By Property 3,

$$\ln \sqrt{3x + 2} = \ln (3x + 2)^{1/2} = \tfrac{1}{2} \ln (3x + 2)$$

(c) By Properties 1 and 2,

$$\ln \frac{xy}{5} = \ln (xy) - \ln 5 = \ln x + \ln y - \ln 5$$

(d) By properties 1, 2, and 3,

$$\begin{aligned} \ln \frac{(x - 3)^2}{x\sqrt[3]{x - 1}} &= \ln (x - 3)^2 - \ln (x\sqrt[3]{x - 1}) \\ &= 2 \ln (x - 3) - [\ln x + \ln (x - 1)^{1/3}] \\ &= 2 \ln (x - 3) - \ln x - \ln (x - 1)^{1/3} \\ &= 2 \ln (x - 3) - \ln x - \tfrac{1}{3} \ln (x - 1) \end{aligned}$$

□

Example 3

Condensing Logarithmic Expressions

Use the properties of logarithms to rewrite the following expressions as the logarithm of a single quantity.

(a) $\ln x + 2 \ln y$ (b) $\ln (x + 1) - \frac{1}{2} \ln x - \ln (x^2 - 1)$

Solution

(a) $\ln x + 2 \ln y = \ln x + \ln y^2 = \ln xy^2$

(b) $\ln (x + 1) - \frac{1}{2} \ln x - \ln (x^2 - 1)$

$$\begin{aligned} &= \ln (x + 1) - \ln x^{1/2} - \ln (x^2 - 1) \\ &= \ln (x + 1) - [\ln \sqrt{x} + \ln (x^2 - 1)] \\ &= \ln (x + 1) - \ln [\sqrt{x}(x^2 - 1)] \\ &= \ln \left[\frac{x + 1}{\sqrt{x}(x^2 - 1)}\right] = \ln \left[\frac{1}{\sqrt{x}(x - 1)}\right] \end{aligned}$$

□

Example 4

Solving Equations

Solve for x in each of the following equations.

(a) $\ln x - \ln (x - 8) = 1$

(b) $y = e^{2x-5}$

Solution

(a) $\ln x - \ln (x - 8) = 1$

$$\ln\left(\frac{x}{x-8}\right) = 1$$

The equivalent exponential form is

$$\frac{x}{x-8} = e^1$$

Thus,

$$x = (x - 8)(e)$$

$$(1 - e)x = -8e$$

$$x = \frac{-8e}{1-e} \approx 12.66$$

(b) We begin by taking the natural logarithm of both sides of the given equation as follows:

$$y = e^{2x-5}$$

$$\ln y = \ln (e^{2x-5})$$

Now, applying the inverse property of the exponential and natural logarithmic function, we have

$$\ln y = 2x - 5$$

$$5 + \ln y = 2x$$

$$\frac{5 + \ln y}{2} = x$$

□

Example 5

An Application to Compound Interest

If P dollars are deposited at 8% interest compounded continuously, how long will it take to double the original deposit? (See the Introductory Example of this section.)

Solution To double the deposit, we write

$$Pe^{0.08t} = 2P$$

Thus,

$$e^{0.08t} = 2$$

$$0.08t = \ln 2$$

$$t = \frac{\ln 2}{0.08} \approx 8.66$$

Therefore, the balance will double by the end of 8 years and 8 months. □

Example 6

An Application to Radioactive Carbon Dating

In the Introductory Example to Section 5.1, we looked at the exponential equation

$$10^{-13} = 10^{-12}2^{-t/5700}$$

where t represented the age of a particular fossil in years. Use the properties of logarithms to solve this equation for t.

Solution Dividing both sides of the given equation by 10^{-12}, we have

$$10^{-1} = 2^{-t/5700}$$

Now, by taking the natural logarithm of both sides, we have

$$\ln (10^{-1}) = \ln (2^{-t/5700})$$

$$(-1)(\ln 10) = \left(\frac{-t}{5700}\right)(\ln 2)$$

$$\frac{\ln 10}{\ln 2}(5700) = t$$

$$t \approx 18{,}935 \text{ years}$$

□

Section Exercises (5.3)

In Exercises 1–8, write each logarithmic equation as an exponential equation, and vice versa.

1. $\ln 2 = 0.6931 \ldots$
2. $\ln 8.4 = 2.128 \ldots$
3. $\ln 0.5 = -0.6931 \ldots$
4. $\ln 0.056 = -2.2882 \ldots$
5. $e^0 = 1$
6. $e^2 = 7.389 \ldots$
7. $e^{-2} = 0.1353 \ldots$
8. $e^{0.25} = 1.284 \ldots$

In Exercises 9–14, sketch the graph of each equation.

Ⓒ **9.** $y = \ln (x - 1)$
Ⓒ **10.** $y = \ln |x|$
Ⓒ **11.** $y = \ln 2x$
Ⓒ **12.** $y = \ln x + 5$
Ⓒ **13.** $y = 3 \ln x$
Ⓒ **14.** $y = \dfrac{\ln x}{4}$

In Exercises 15–17, show that the given functions are inverses of each other by sketching their graphs on the same coordinate axes.

Ⓒ **15.** $f(x) = e^{2x}$; $g(x) = \ln \sqrt{x}$
Ⓒ **16.** $f(x) = e^x - 1$; $g(x) = \ln (x + 1)$
Ⓒ **17.** $f(x) = e^{x-1}$; $g(x) = 1 + \ln x$

In Exercises 18–23, apply the inverse properties of $\ln x$ and e^x to simplify each expression.

18. $\ln e^{2x-1}$
19. $\ln e^{x^2}$
20. $\ln e^{2x} - 1$
21. $e^{\ln (5x+2)}$
22. $e^{\ln x^3} - 8$
23. $e^{\ln \sqrt{x}}$

In Exercises 24–30, use the properties of logarithms to write each expression as a sum, difference, or constant multiple of logarithms.

24. $\ln \dfrac{xy}{z}$
25. $\ln \sqrt{a - 1}$
26. $\ln [z(z - 1)^2]$
27. $\ln \left(\dfrac{x^2 - 1}{x^3}\right)^3$
28. $\ln (3e^2)$
29. $\ln \dfrac{1}{e}$
30. $\ln \dfrac{2x}{\sqrt{x^2 - 1}}$

Ⓒ Calculator may be helpful.

In Exercises 31–34, rewrite each expression as a single logarithm.

31. $3 \ln x + 2 \ln y - 4 \ln z$
32. $\frac{1}{3}[2 \ln (x + 3) + \ln x - \ln (x^2 - 1)]$
33. $2[\ln x - \ln (x + 1) - \ln (x - 1)]$
34. $2 \ln 3 - \frac{1}{2}\ln (x^2 + 1)$

In Exercises 35–50, find the value of the unknown.

35. $e^{\ln x} = 4$
36. $e^{\ln x^2} - 9 = 0$
37. $\ln x = 0$
38. $2 \ln x = 4$
39. $e^{x+1} = 4$
40. $e^{-0.5x} = 0.075$
Ⓒ**41.** $e^{3x/4} = 2.197225$
Ⓒ**42.** $e^{0.09t} = 3$
Ⓒ**43.** $500e^{0.11t} = 600$
Ⓒ**44.** $e^{-0.0174t} = 0.5$
Ⓒ**45.** $3 = 10e^{-0.0315t}$
Ⓒ**46.** $58 = 75 - e^{0.4t}$
Ⓒ**47.** $5^{2x} = 15$
Ⓒ**48.** $2^{1-x} = 6$
Ⓒ**49.** $500(1.07)^t = 1000$
Ⓒ**50.** $1000\left(1 + \dfrac{0.07}{12}\right)^{12t} = 3000$

Ⓒ**51.** A deposit of \$1000 is made into a fund at an annual percentage rate of 11%. Find the time for the investment to double if the interest is compounded:
(a) annually (b) monthly
(c) daily (d) continuously

Ⓒ**52.** A deposit of \$1000 is made into a fund at an annual percentage rate of $10\frac{1}{2}\%$. Find the time for the investment to triple if the interest is compounded:
(a) annually (b) monthly
(c) daily (d) continuously

Ⓒ**53.** Complete the following table for the time t necessary for P dollars to triple if it is compounded continuously at the rate r.

r	2%	4%	6%	8%	10%	12%
t						

Ⓒ**54.** From the Introductory Example of Section 5.1, the formula for the ratio R of carbon isotopes to carbon atoms is given by

$$R = 10^{-12}2^{-t/5700}$$

Estimate the age of a fossil if the ratio of carbon isotopes to atoms is

$$R = 0.32 \times 10^{-12}$$

Ⓒ**55.** The demand equation for a certain product is given by

$$p = 500 - 0.5e^{0.004x}$$

Find the demand if the price charged is
(a) $p = \$350$ (b) $p = \$300$

Ⓒ**56.** The population of a city is given by

$$P = 105{,}300e^{0.015t}$$

where t is the time in years, with $t = 0$ corresponding to 1980. According to this model, in what year will the city have a population of 150,000?

Ⓒ Calculator may be helpful.

Section 5.4

Logarithmic Functions: Differentiation and Integration

Introductory Example

Farm Labor in the United States

During the early development of the United States, its population was almost entirely agricultural. As the country became more industrialized, technological advances made it possible for a smaller and smaller percentage of the population to produce enough food and other farm products to sustain the entire population. This phenomenon has continued right up to the present day, with the result that less than 4% of the total labor force worked on farms in 1980. Table 5.10 lists the total U.S. farm employment (including farm operators and family members doing farm work without wages) over the 30-year period from 1950 to 1980.

A mathematical model for the number of farm workers, p, is given by

$$p = \frac{1{,}000{,}000}{0.0925 + 0.0058t}, \qquad 0 \le t \le 30$$

where t measures the time in years, with 1950 corresponding to $t = 0$.

To estimate the total labor (in worker-years) required to run U.S. farms over a given time period from t_1 to t_2, we can integrate this function between t_1 and t_2. In this section, we will introduce the **log rule for integration.** Using this rule, we have

$$\int_{t_1}^{t_2} \frac{1{,}000{,}000}{0.0925 + 0.0058t}\,dt$$

$$= \frac{1{,}000{,}000}{0.0058}\Big[\ln(0.0925 + 0.0058t)\Big]_{t_1}^{t_2}$$

$$= \frac{1{,}000{,}000}{0.0058}\ln\left(\frac{0.0925 + 0.0058t_2}{0.0925 + 0.0058t_1}\right)$$

Now, using this model to estimate the farm labor over the past three decades, we have

Decade	*Estimated Labor in Worker-Years*
1950–1960	83,923,179
1960–1970	56,202,751
1970–1980	42,316,414

(See Figure 5.18, page 335.)

Table 5.10

Year	Farm employment
1950	9,926,000
1952	9,149,000
1954	8,651,000
1956	7,852,000
1958	7,503,000
1960	7,057,000
1962	6,700,000
1964	6,110,000
1966	5,214,000
1968	4,749,000
1970	4,523,000
1972	4,373,000
1974	4,389,000
1976	4,374,000
1978	3,957,000
1980	3,754,000

Section Objectives: ***To develop differentiation formulas for logarithmic functions* ▪ *To evaluate integrals of the form* ∫ (u′/u) dx.**

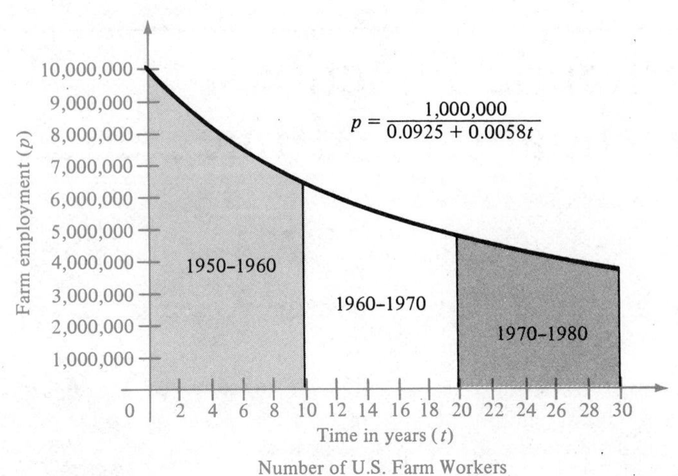

Figure 5.18

To find the derivative of the natural logarithmic function, we could resort to the definition of the derivative and attempt to evaluate the resulting limit:

$$\frac{d}{dx}[\ln x] = \lim_{\Delta x \to 0} \frac{\ln (x + \Delta x) - \ln x}{\Delta x}$$

However, since we already know the derivative of e^x, it is much simpler to use implicit differentiation together with the inverse relationship between e^x and $\ln x$. To do this, we let

$$y = \ln x$$

then

$$e^y = x$$

and

$$\frac{d}{dx}[e^y] = \frac{d}{dx}[x]$$

$$e^y \frac{dy}{dx} = 1$$

$$\frac{dy}{dx} = \frac{1}{e^y} = \frac{1}{x}$$

We summarize this rule, together with its Chain Rule version, as follows. As usual, we assume that u is a differentiable function of x.

Derivative of Natural Logarithmic Function

$$\frac{d}{dx}[\ln x] = \frac{1}{x}$$

$$\frac{d}{dx}[\ln u] = \frac{1}{u}u' = \frac{u'}{u}$$

Example 1

Differentiating a Logarithmic Function

Find the derivative of $f(x) = \ln(2x^2 + 4)$.

Solution By letting $u = 2x^2 + 4$, we have

$$f'(x) = \frac{1}{u}\frac{du}{dx}$$

$$= \frac{1}{2x^2 + 4}(4x) = \frac{2x}{x^2 + 2}$$

□

Example 2

Differentiating a Logarithmic Function

Differentiate $f(x) = \ln \sqrt{x + 1}$.

Solution

$$f'(x) = \frac{1}{\sqrt{x + 1}}\frac{d}{dx}[(x + 1)^{1/2}]$$

$$= \frac{1}{\sqrt{x + 1}}\left(\frac{1}{2}\right)(x + 1)^{-1/2}$$

$$= \frac{1}{2\sqrt{x + 1}\sqrt{x + 1}} = \frac{1}{2(x + 1)}$$

□

Remark

We know from our previous work with differentiation that *it is often helpful to rewrite a function before differentiating*. For instance, in Example 2 if we rewrite $\ln \sqrt{x + 1}$ as $(\frac{1}{2}) \ln(x + 1)$, we can simplify the differentiation process:

$$f'(x) = \frac{d}{dx}[\ln \sqrt{x + 1}] = \frac{d}{dx}\left[\frac{1}{2}\ln(x + 1)\right] = \frac{1}{2}\left(\frac{1}{x + 1}\right)$$

Our next example is an even more dramatic illustration of the benefit of rewriting a function before differentiating.

Example 3

Rewriting before Differentiating

Differentiate

$$f(x) = \ln\left[\frac{x(x^2 + 1)^2}{\sqrt{2x^3 - 1}}\right]$$

Solution Without rewriting the expression for $f(x)$, we have

$$f'(x) = \frac{1}{[x(x^2 + 1)^2/\sqrt{2x^3 - 1}]}\frac{d}{dx}\left[\frac{x(x^2 + 1)^2}{\sqrt{2x^3 - 1}}\right]$$

Of course, we like to avoid differentiating expressions like

$$\frac{x(x^2 + 1)^2}{\sqrt{2x^3 - 1}}$$

if at all possible. In this case, we can do just that by using the properties of logarithms to write

$$f(x) = \ln \left[\frac{x(x^2 + 1)^2}{\sqrt{2x^3 - 1}}\right]$$

as

$$f(x) = (\ln x) + 2 \ln (x^2 + 1) - \tfrac{1}{2} \ln (2x^3 - 1)$$

In this form, our derivative is simply

$$f'(x) = \frac{1}{x} + \frac{4x}{x^2 + 1} - \frac{3x^2}{2x^3 - 1}$$ □

The differentiation formulas

$$\frac{d}{dx}[\ln x] = \frac{1}{x} \quad \text{and} \quad \frac{d}{dx}[\ln u] = \frac{u'}{u}$$

allow us to patch up the hole in our General Power Rule for integration. Recall from Section 4.2 that

$$\int u^n u' \, dx = \frac{u^{n+1}}{n + 1} + C \qquad \text{provided } n \neq -1$$

Having the derivative formulas for logarithmic functions, we are now in a position to evaluate this integral for $n = -1$, as stated in the following theorem.

The Log Rule for Integration

If u is a differentiable function of x, then

$$\int \frac{u'}{u} \, dx = \ln |u| + C$$

In particular,

$$\int \frac{1}{x} \, dx = \ln |x| + C$$

Remark

Be sure to note the use of absolute value in the Log Rule. For those special cases in which u or x is restricted to positive values, we can omit the absolute value sign and write

$$\int \frac{u'}{u} \, dx = \ln u + C, \qquad u > 0$$

$$\int \frac{1}{x} \, dx = \ln x + C, \qquad x > 0$$

Example 4

Integrating with the Log Rule

Evaluate the integral

$$\int \frac{dx}{2x - 1}$$

Solution By letting $u = 2x - 1$, we have $u' = 2$. Thus, to apply the Log Rule, we multiply and divide by 2 and write

$$\int \frac{dx}{2x - 1} = \frac{1}{2} \int \frac{2}{2x - 1}\, dx = \frac{1}{2} \int \frac{u'}{u}\, dx$$

$$= \tfrac{1}{2} \ln |u| + C = \tfrac{1}{2} \ln |2x - 1| + C$$

□

Example 5

Finding Area with the Log Rule

Find the area of the region bounded by

$$y = \frac{x}{x^2 + 1}$$

$y = 0$, and $x = 3$.

Solution From Figure 5.19, we see that the area of the specified region is given by the definite integral

$$\int_0^3 \frac{x}{x^2 + 1}\, dx$$

By letting $u = x^2 + 1$, we have $u' = 2x$. Thus, to apply the Log Rule, we multiply and divide by 2 and write

$$\int_0^3 \frac{x}{x^2 + 1}\, dx = \frac{1}{2} \int_0^3 \frac{2x}{x^2 + 1}\, dx$$

$$= \frac{1}{2} \Big[\ln (x^2 + 1) \Big]_0^3$$

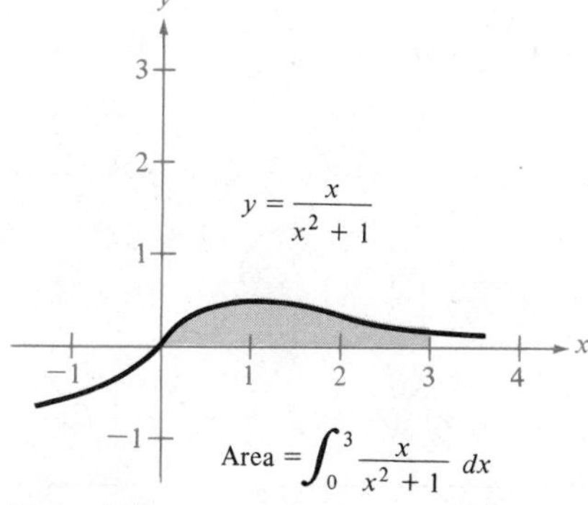

Figure 5.19

(Note that an absolute value sign is unnecessary since $x^2 + 1 > 0$ on the interval under consideration.) Thus,

$$\int_0^3 \frac{x}{x^2+1}\, dx = \frac{1}{2}(\ln 10 - \ln 1) = \frac{1}{2} \ln 10 \approx 1.15$$

□

Integrals to which the Log Rule can be applied are often given in disguised form. For instance, if a rational function has a numerator of degree greater than or equal to that of the denominator, division may reveal a form to which we can apply the Log Rule. Our next example is a case in point.

Example 6

Rewriting before Integrating

Evaluate the integral

$$\int \frac{x^2+x+1}{x^2+1}\, dx$$

Solution By dividing, we first obtain the following form:

$$\frac{x^2+x+1}{x^2+1} = 1 + \frac{x}{x^2+1}$$

Thus,

$$\begin{aligned}\int \frac{x^2+x+1}{x^2+1}\, dx &= \int \left(1 + \frac{x}{x^2+1}\right) dx \\ &= \int (1)\, dx + \frac{1}{2}\int \frac{2x}{x^2+1}\, dx \\ &= x + \tfrac{1}{2} \ln (x^2+1) + C\end{aligned}$$

□

We next summarize some additional situations for which it is helpful to rewrite the integrand in order to recognize the antiderivative. Study this list carefully.

Disguised Uses of the Log Rule

1. $$\int \frac{3x^2+2x-1}{x^2}\, dx = \int \left(\frac{3x^2}{x^2} + \frac{2x}{x^2} - \frac{1}{x^2}\right) dx = \int \left(3 + \frac{2}{x} - \frac{1}{x^2}\right) dx$$
$$= 3x + 2 \ln |x| + \frac{1}{x} + C$$

2. $$\int \frac{1}{1+e^{-x}}\, dx = \int \left(\frac{e^x}{e^x}\right) \frac{1}{1+e^{-x}}\, dx = \int \frac{e^x}{e^x+1}\, dx = \ln (e^x+1) + C$$

3. $$\int \frac{x^3-1}{(x-1)^2}\, dx = \int \left(x + 2 + \frac{3}{x-1}\right) dx = \frac{x^2}{2} + 2x + 3 \ln |x-1| + C$$

Section Exercises (5.4)

In Exercises 1–6 find the slope of the tangent line to the given logarithmic function at the point (1, 0).

1.

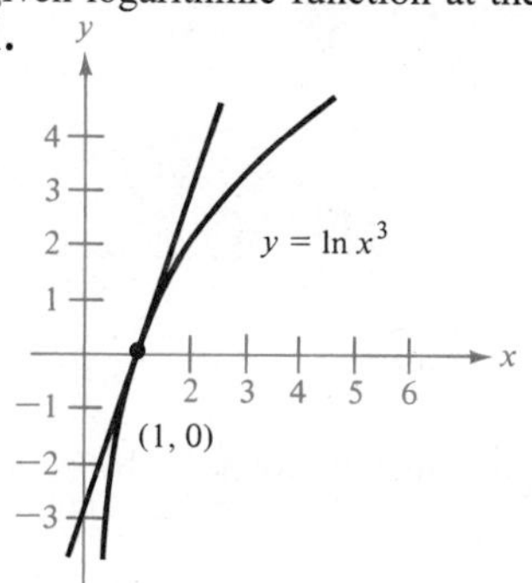

2.

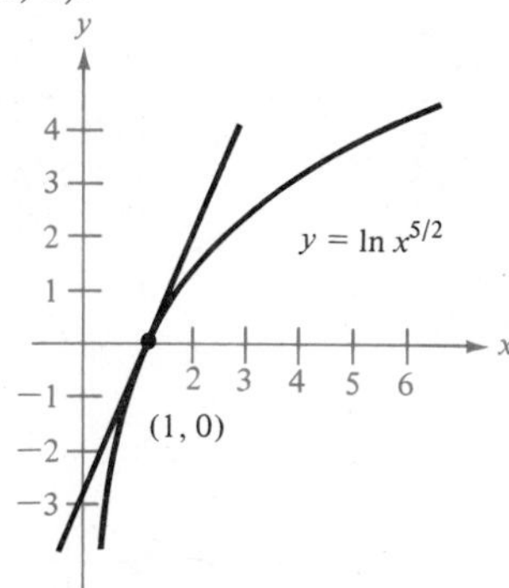

3.

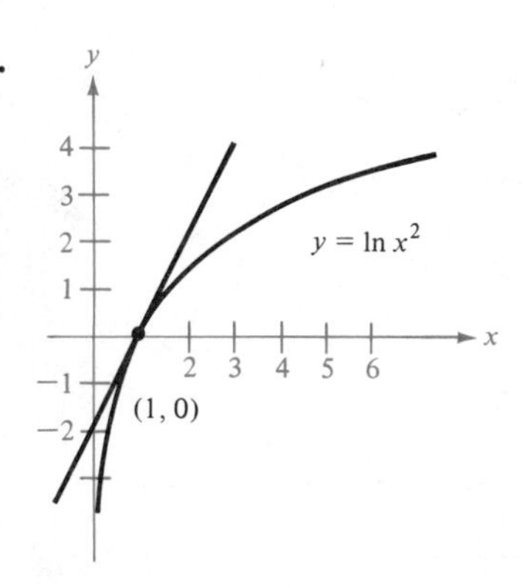

4.

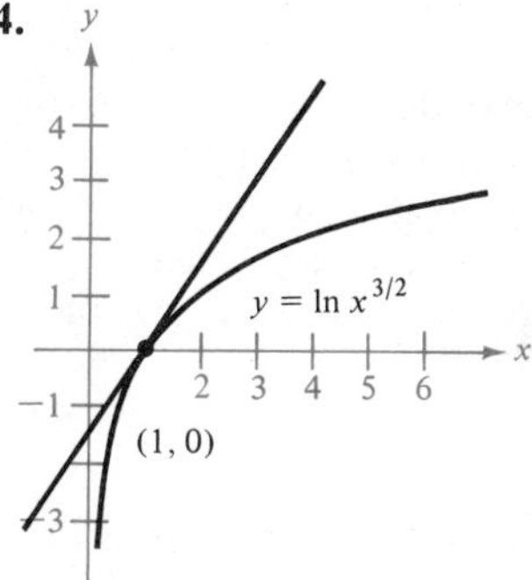

5.

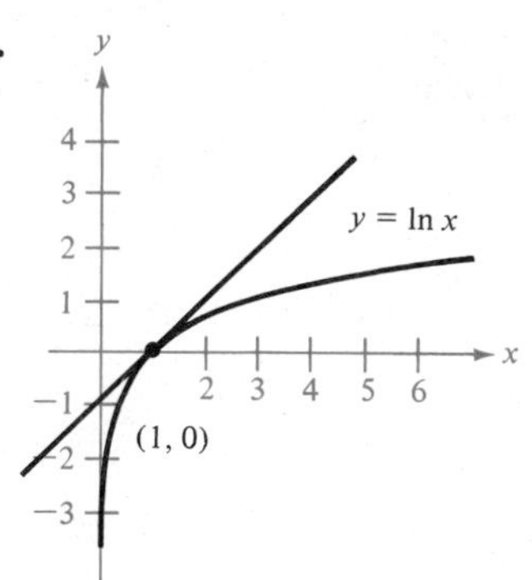

6.

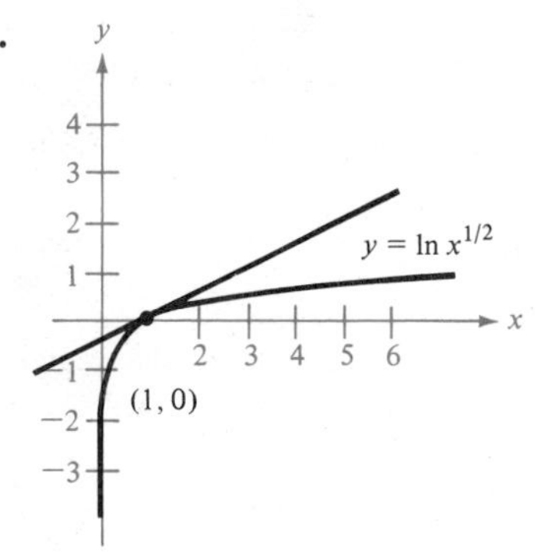

In Exercises 7–26 find dy/dx. (Remember that it may be helpful to use logarithmic properties to rewrite the given function *before* differentiating.)

7. $y = \ln (x^2)$

8. $y = \ln (x^2 + 3)$

9. $y = \ln \sqrt{x^4 - 4x}$

10. $y = \ln (1 - x)^{3/2}$

11. $y = (\ln x)^4$

12. $y = x(\ln x)$

13. $y = \ln (x\sqrt{x^2 - 1})$

14. $y = \ln \left(\dfrac{x}{x + 1}\right)$

15. $y = \ln \left(\dfrac{x}{x^2 + 1}\right)$

16. $y = \ln e^x$

17. $y = \dfrac{\ln x}{x^2}$

18. $y = x^2(\ln x)$

19. $y = \ln \left(\dfrac{\sqrt{4 + x^2}}{x}\right)$

20. $y = \ln \sqrt{\dfrac{x - 1}{x + 1}}$

21. $y = \ln \sqrt{\dfrac{x + 1}{x - 1}}$

22. $y = \ln \sqrt{x^2 - 4}$

23. $y = e^{-x} \ln x$

24. $y = \ln \left(\dfrac{e^x + e^{-x}}{2}\right)$

25. $y = \ln (e^{x^2})$

26. $y = \ln \left(\dfrac{1 + e^x}{1 - e^x}\right)$

In Exercises 27–46 evaluate each integral.

27. $\displaystyle\int \frac{1}{x + 1}\, dx$

28. $\displaystyle\int \frac{1}{x - 5}\, dx$

29. $\displaystyle\int \frac{1}{3 - 2x}\, dx$

30. $\displaystyle\int \frac{1}{6x + 1}\, dx$

31. $\displaystyle\int \frac{x}{x^2 + 1}\, dx$

32. $\displaystyle\int \frac{x^2}{3 - x^3}\, dx$

33. $\int_{-2}^{-1} \frac{x^2 - 4}{x}\, dx$

34. $\int_{-2}^{-1} \frac{x + 5}{x}\, dx$

35. $\int_{0}^{2} \frac{x^2 - 2}{x + 1}\, dx$

36. $\int \frac{1}{(x + 1)^2}\, dx$

37. $\int_{1}^{e} \frac{(1 + \ln x)^2}{x}\, dx$

38. $\int_{0}^{1} \frac{x - 1}{x + 1}\, dx$

39. $\int \frac{x}{1 - x}\, dx$

40. $\int \frac{2x}{(x - 1)^2}\, dx$

41. $\int \frac{1}{\sqrt{x + 1}}\, dx$

42. $\int \frac{x + 3}{x^2 + 6x + 7}\, dx$

43. $\int \frac{e^x + e^{-x}}{e^x - e^{-x}}\, dx$

44. $\int \frac{x^2 + 2x + 3}{x^3 + 3x^2 + 9x + 1}\, dx$

45. $\int \frac{e^{-x}}{1 + e^{-x}}\, dx$

46. $\int \frac{e^{2x}}{1 + e^{2x}}\, dx$

In Exercises 47–51, find any relative extrema and inflection points, and sketch the graph of the function.

Ⓒ**47.** $y = x - \ln x$

Ⓒ**48.** $y = \frac{x^2}{2} - \ln x$

Ⓒ**49.** $y = \frac{\ln x}{x}$

Ⓒ**50.** $y = x(\ln x)$

Ⓒ**51.** $y = x^2(\ln x)$

52. Find the area of the region bounded by $y = (x^2 + 4)/x$, $y = 0$, $x = 1$, and $x = 4$. (See Figure 5.20.)

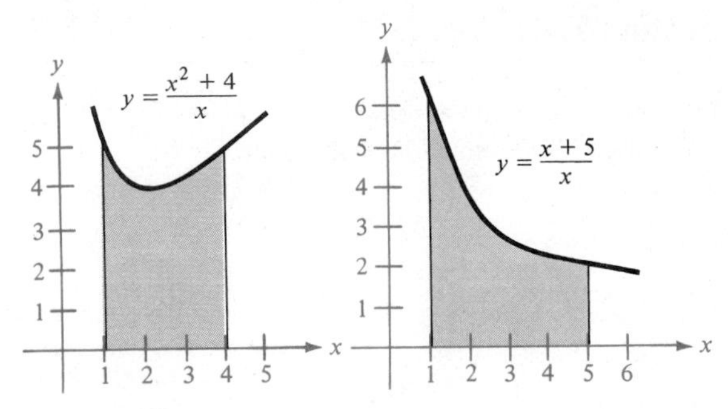

Figure 5.20 **Figure 5.21**

53. Find the area of the region bounded by $y = (x + 5)/x$, $y = 0$, $x = 1$, and $x = 5$. (See Figure 5.21.)

Ⓒ**54.** A population of bacteria is growing at the rate of

$$\frac{dP}{dt} = \frac{3000}{1 + 0.25t}$$

where t is the time in days. Assuming that the initial population (when $t = 0$) is 1000, write an equation that gives the population at any time t, and then find the population when $t = 3$ days.

Ⓒ**55.** The demand equation for a product is given by

$$p = \frac{90{,}000}{400 + 3x}$$

Find the *average* price p on the interval $40 \leq x \leq 50$.

Ⓒ Calculator may be helpful.

Section 5.5

Exponential Growth and Decay

Introductory Example

World Population

It has been stated that the world's population is growing exponentially. Is this true, and if so, what does it mean? Answering the second question first, we say that **exponential growth** implies that the rate of change of the population is proportional to the amount of population at any time. In this section, we will see that the *only* functions having this property are those of the form

$$y = Ce^{kt}$$

Now, to see whether or not the world population is growing exponentially, we use the world population at different times in history as listed in the *Information Please Almanac* for 1981. With the aid of a computer, we found the best fit of an exponential model to these figures to be

$$y = 6164e^{0.00667t}$$

where y is the population and t is the time in years dating from A.D. 0. Table 5.11 lists the estimates for this model, and Figure 5.22 compares the exponential model with the actual growth curve. We can see that the fit is not good. On this basis, we may conclude that world population has *not* followed an exponential growth pattern. It is important to realize that mathematicians use the term *exponential growth* in a technical sense and not simply to describe the growth of any curve whose rate of change is increasing.

Table 5.11

Year	Actual population	Estimated population
1650	470,000,000	371,000,000
1750	695,000,000	723,000,000
1850	1,091,000,000	1,409,000,000
1900	1,571,000,000	1,966,000,000
1930	2,070,000,000	2,402,000,000
1950	2,501,000,000	2,745,000,000
1960	2,986,000,000	2,934,000,000
1970	3,610,000,000	3,137,000,000
1978	4,258,000,000	3,308,000,000

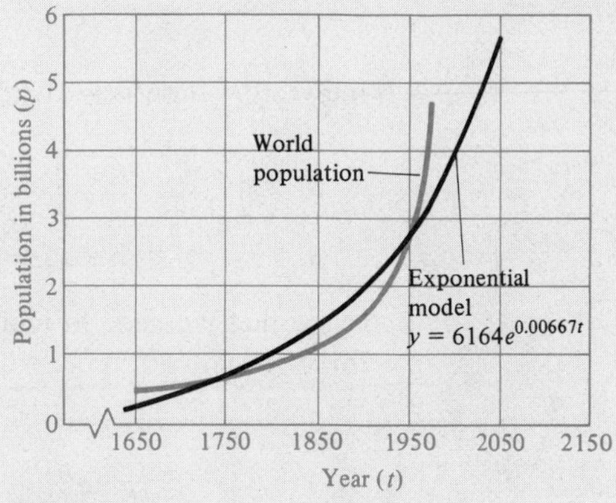

Figure 5.22

Section Objective: ***To introduce the law of exponential growth (or decay) and discuss several of its applications.***

One of the more common applications of integrals of the form

$$\int \frac{u'}{u}\, dx$$

arises in situations involving exponential growth or decay. In this type of application, we deal with a substance whose *rate of growth (or decay) at a particular time is proportional to the amount of the substance present at that time*. For example, the rate of decomposition of a radioactive substance is proportional to the amount of radioactive substance at a given instant. Or, similarly, the rate of growth of a bacteria culture may be proportional to the number of bacteria in the culture at time t. In its simplest form this relationship is described by the differential equation

$$\frac{dy}{dt} = ky$$

where k is a constant and y is a function of t. [A **differential equation** is an equation involving derivatives (of first or higher order) of one variable with respect to another. We discuss such equations in more detail in Chapter 7.]

An obvious solution to the differential equation $dy/dt = ky$ is $y = 0$. To find the other solutions, we assume $y \neq 0$ and divide both sides of the equation by y to obtain

$$\frac{1}{y}\frac{dy}{dt} = k$$

Integrating both sides with respect to t, we have

$$\int \frac{1}{y}\frac{dy}{dt}\, dt = \int k\, dt$$

$$\ln|y| = kt + C_1$$

(Note that we need to include only one constant of integration.) Now, solving for y, we have

$$|y| = e^{kt+C_1} = e^{C_1}e^{kt}$$

$$y = \pm e^{C_1}e^{kt}$$

Finally, since $y = \pm e^{C_1}e^{kt}$ represents all nonzero solutions, and $y = 0$ is already known to be a solution, we can write the general form of the solution as

$$y = Ce^{kt}$$

where C is a real number. This equation is often referred to as the **law of exponential growth (or decay),** and it has wide applications, as we will see in the examples in this section.

Law of Exponential Growth (or Decay) If y is a quantity whose rate of change (with respect to time) is proportional to the quantity present at any time t, then y is of the form

$$y = Ce^{kt}$$

where C is called the **initial value** and k is called the **constant of proportionality.**

Example 1

Radioactive Decay

Let y represent the mass of a particular radioactive element whose half-life is 25 years. (In other words, if we began with 1 g of the element, only $\frac{1}{2}$ g would remain after 25 years, $\frac{1}{4}$ g after 50 years, etc.) How much of 1 g would remain after 15 years? (See Figure 5.23.)

Solution Assuming that the rate of decay is proportional to y, we then have

$$y = Ce^{kt}$$

Since we are given that $y = 1$ when $t = 0$, we know the initial value is

$$1 = Ce^0 = C$$

Furthermore, since $y = \frac{1}{2}$ when $t = 25$, we have

$$\frac{1}{2} = e^{k(25)}$$

$$\ln\left(\tfrac{1}{2}\right) = 25k$$

$$k = \frac{\ln \frac{1}{2}}{25} \approx -0.0277$$

and the equation for y as a function of time is

$$y = Ce^{kt} \approx e^{-0.0277t}$$

Finally, when $t = 15$ years, the amount remaining is

$$y \approx e^{-0.0277(15)} \approx 0.660 \text{ g}$$

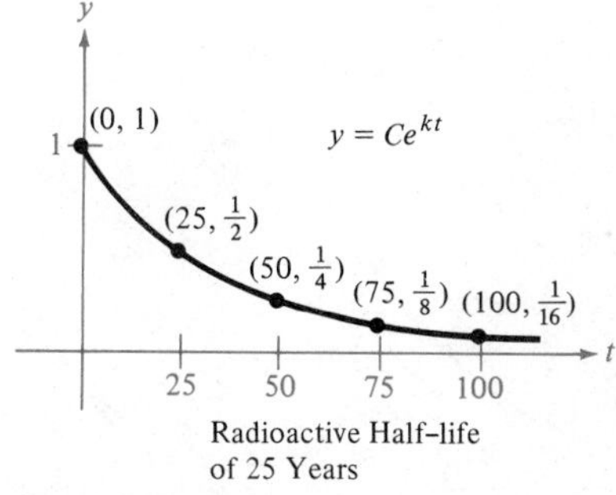

Figure 5.23

□

Remark

Example 1 demonstrates that problems in finding exponential growth functions basically center around solving for the constants C and k. Furthermore, we can see in Example 1 that it is easy to solve for C when we are given the value of y at $t = 0$. In the next example, we demonstrate a procedure for solving for C and k when we do not know the value of y at $t = 0$.

Example 2

Population Growth

In a certain research experiment, a population of fruit flies increases according to the law of exponential growth. If there are 100 flies after the second day of the experiment and 300 flies after the fourth day, how many flies were in the original population? (See Figure 5.24.)

Solution Let y be the number of flies at time t. Since the population increases according to the law of exponential growth, we know y is of the form $y = Ce^{kt}$, where $y = Ce^{k(0)} = C$ is the number of flies in the original population. Since $y = 100$ when $t = 2$ and $y = 300$ when $t = 4$, we have the equations

$$100 = Ce^{2k} \qquad \text{and} \qquad 300 = Ce^{4k}$$

Substituting

$$C = \frac{100}{e^{2k}}$$

into the second equation, we have

$$300 = \left(\frac{100}{e^{2k}}\right)e^{4k} = 100e^{2k}$$

$$\frac{300}{100} = e^{2k}$$

$$\ln 3 = 2k$$

$$k = \frac{\ln 3}{2} \approx 0.5493$$

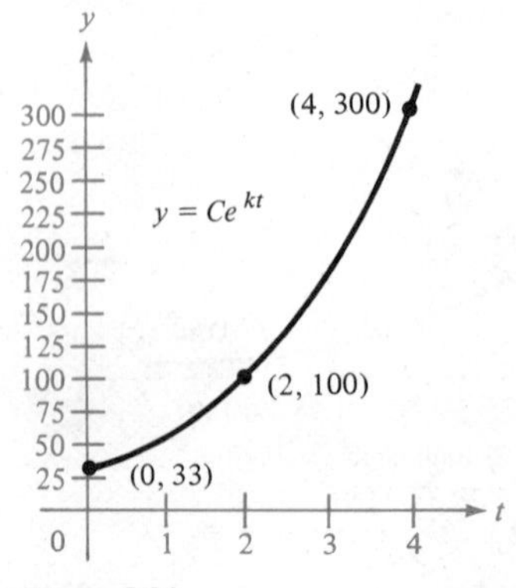

Figure 5.24

Thus, the original population is

$$C = \frac{100}{e^{2(0.5493)}} \approx 33 \text{ flies}$$

□

Example 3

Continuously Compounded Interest

Money is deposited in a savings account for which the interest is compounded continuously. If the balance in the account doubles in 6 years, what is the annual percentage rate? (See Figure 5.25.)

Solution The formula for the balance using continuously compounded interest is given by the exponential growth function

$$A = Pe^{rt}$$

When $t = 6$, $A = 2P$. Thus,

$$2P = Pe^{6r}$$

$$2 = e^{6r}$$

$$\ln 2 = 6r$$

$$r = \frac{\ln 2}{6} \approx 0.1155 = 11.55\%$$

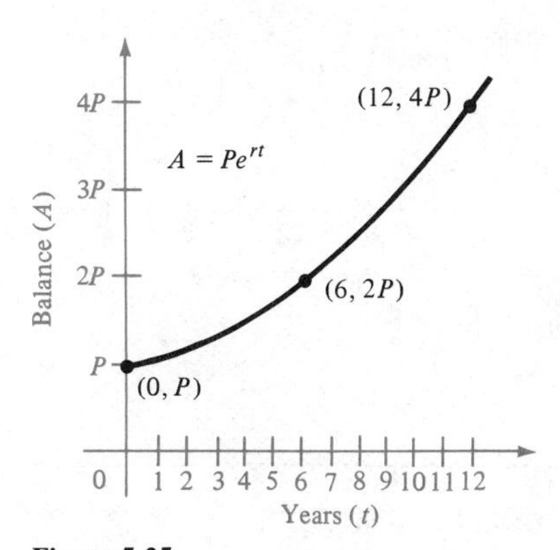

Figure 5.25

□

Example 4

Declining Sales Following Discontinuation of Advertising

Four months after it discontinued its advertising campaign, a toothpaste manufacturer noticed that its sales had dropped from 100,000 units per month to 80,000 units. If the sales follow an exponential pattern of decline, what will they be after another 4 months? (See Figure 5.26.)

Solution Using the exponential model

$$y = Ce^{kt}$$

where t is measured in months, we have

$$100{,}000 = Ce^{0} \qquad \text{and} \qquad 80{,}000 = Ce^{4k}$$

Thus,

$$80{,}000 = 100{,}000e^{4k}$$
$$0.8 = e^{4k}$$
$$\ln 0.8 = 4k$$
$$k = \frac{\ln 0.8}{4} \approx -0.0558$$

Finally, after 4 more months ($t = 8$), we can expect sales to drop to

$$y = 100{,}000e^{-0.0558(8)} \approx 64{,}000 \text{ units}$$

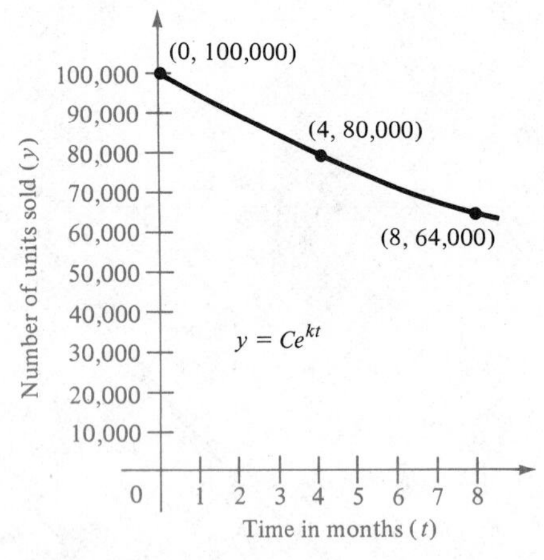

Figure 5.26

□

Section Exercises (5.5)

In Exercises 1–6, find the exponential growth function $y = Ce^{kt}$ that passes through the two given points.

1.

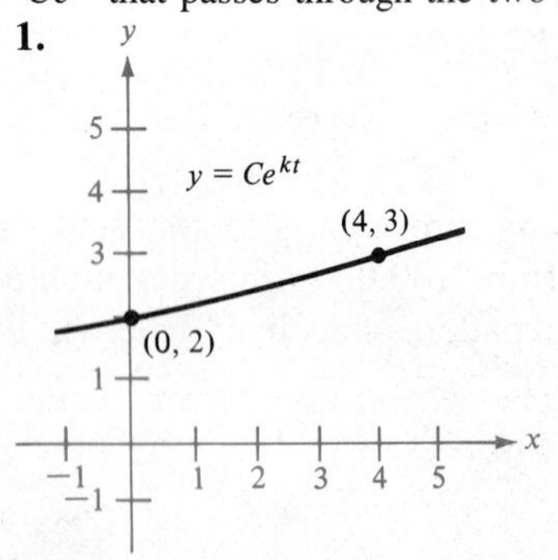

2.

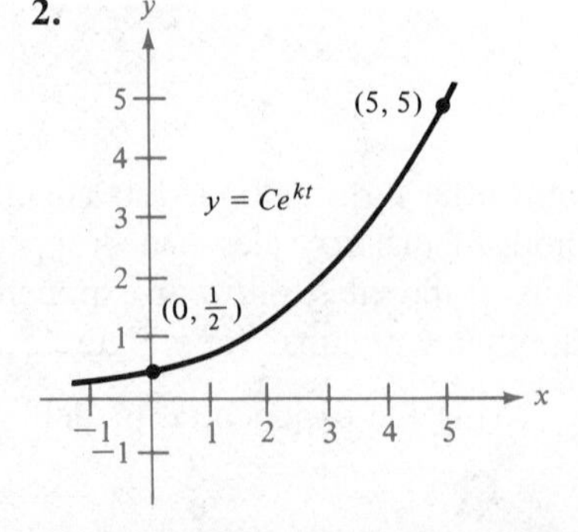

3.

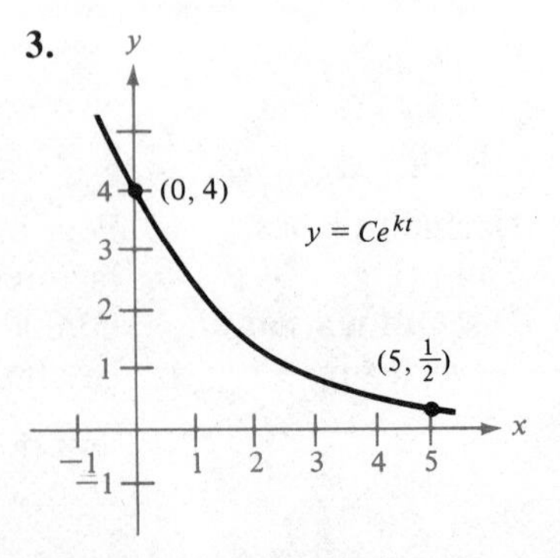

4.

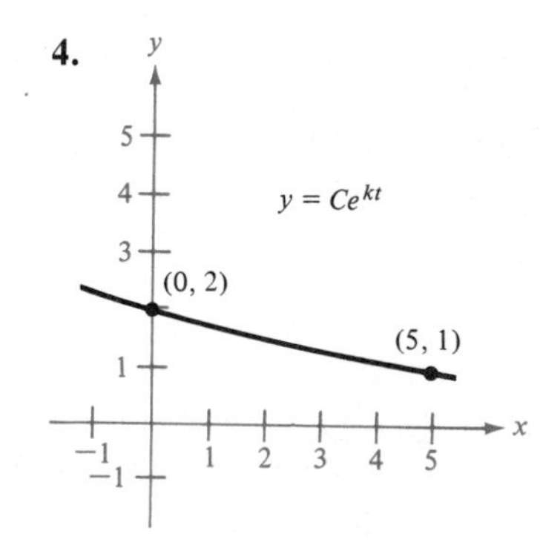

5.

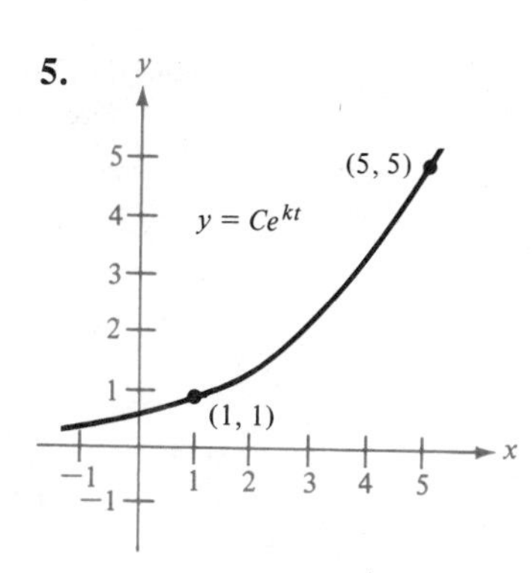

6.

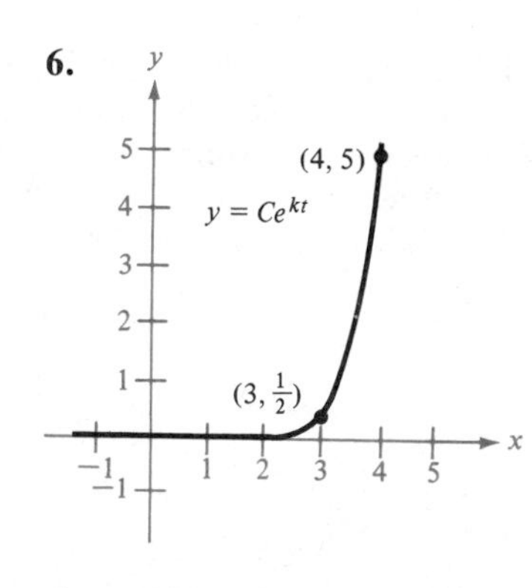

Ⓒ **7.** A deposit of \$750 is made in a savings account for which the interest is compounded continuously. The balance will double in $7\frac{3}{4}$ years.
(a) What is the annual percentage rate for this account?
(b) Find the balance in the account after 10 years.

Ⓒ **8.** A deposit of \$10,000 is made in a savings account for which the interest is compounded continuously. The balance will double in 5 years.
(a) What is the annual percentage rate for this account?
(b) Find the balance after 1 year.

Ⓒ **9.** Due to a slump in the economy, a company finds that its annual revenues have dropped from \$742,000 in 1980 to \$632,000 in 1982. If the revenue is following an exponential pattern of decline, what is the expected revenue for 1983? (Let $t = 0$ represent 1980.)

Ⓒ **10.** Atmospheric pressure P (measured in millimeters of mercury) decreases exponentially with increasing altitude x (measured in meters). If the pressure is 760 mm Hg at sea level ($x = 0$) and 672.71 mm Hg at an altitude of 1000 m, find the pressure at an altitude of 3000 m.

Ⓒ **11.** The management at a certain factory has found that the maximum number of units a worker can produce in a day is 30. The learning curve for the number of units N produced per day after a new employee has worked t days is given by

$$N = 30(1 - e^{kt})$$

After 20 days on the job, a particular worker produced 19 units.
(a) Find the learning curve for this worker.
(b) How many days should pass before this worker is producing 25 units per day?

Ⓒ **12.** If in Exercise 11 the management requires that a new employee be producing at least 20 units per day after 30 days on the job, find:
(a) the learning curve describing this minimum requirement
(b) the number of days before a minimal achiever is producing 25 units per day

Ⓒ **13.** The sales S (in thousands of units) of a new product after it is on the market t years is given by

$$S = Ce^{k/t}$$

(a) Find S as a function of t if 5000 units have been sold after 1 year and the saturation point for the market is 30,000 (i.e., $\lim_{t\to\infty} S = 30$).
(b) How many units have been sold after 5 years?
(c) Sketch a graph of this sales function.

Ⓒ **14.** The sales S (in thousands of units) of a new product after it is on the market t years is given by

$$S = 30(1 - e^{kt})$$

(a) Find S as a function of t if 5000 units have been sold after 1 year.
(b) How many units will saturate this market?
(c) How many units have been sold after 5 years?
(d) Sketch a graph of this sales function.

Ⓒ **15.** A certain type of bacteria increases continuously at a rate proportional to the number present. If there are 100 present at a given time and 300 present 5 h later, how many will there be 10 h after the initial given time?

Ⓒ **16.** Given the conditions of Exercise 15, how long does it take for the number of bacteria to double?

Ⓒ **17.** In 1960 the population of a town was 2500, and in 1970 it was 3350. Assuming the population increases continuously at a constant rate proportional to the existing population, estimate the population in 1990.

Ⓒ**18.** Given the conditions of Exercise 17, how many years are necessary for the population to double?

Ⓒ**19.** Radioactive radium has a half-life of approximately 1600 years. What percentage of a present amount remains after 100 years?

Ⓒ**20.** If radioactive material decays continuously at a rate proportional to the amount present, find the half-life of the material if after 1 year 99.57% of an initial amount still remains.

Ⓒ Calculator may be helpful.

Chapter 6

Techniques of Integration

Section 6.1

Integration by Substitution

Introductory Example

Drug Retention

Most mathematical models used to predict human responses must allow considerable leeway to account for basic human differences. For example, some people have body chemistries that allow them to expel various drugs quickly, while others retain the drugs much longer. In Figure 6.1, we have pictured three different distributions showing various drug retention patterns. Note that each is of the form

$$y = x^r(1 - x)^s, \qquad 0 \le r, s$$

Suppose that a drug such as calcium is given to a patient and that the percentage of the original dosage remaining in the person's system after 24 hours is measured. Assume the retention pattern is given by

$$y = x^2(1 - x)^{1/2}$$

The graph in Figure 6.1(a) shows that most people would retain from 60% to 90% of the dosage after 24 hours, while very few would retain 25% or less. If we want to find the exact portion of the population that is expected to retain between 60% and 90% of the drug, we can compute the ratio of the area under the curve between 0.6 and 0.9 to the total area under the curve. That is,

$$\begin{array}{c}\text{portion of population retaining} \\ \text{between 60\% and 90\%}\end{array} = \frac{\int_{0.6}^{0.9} x^2(1 - x)^{1/2}\,dx}{\int_0^1 x^2(1 - x)^{1/2}\,dx}$$

To evaluate these two integrals, we use a procedure called **substitution.** In this particular case the ratio is 0.529, and we conclude that 52.9% of the population would retain between 60% and 90% of the drug.

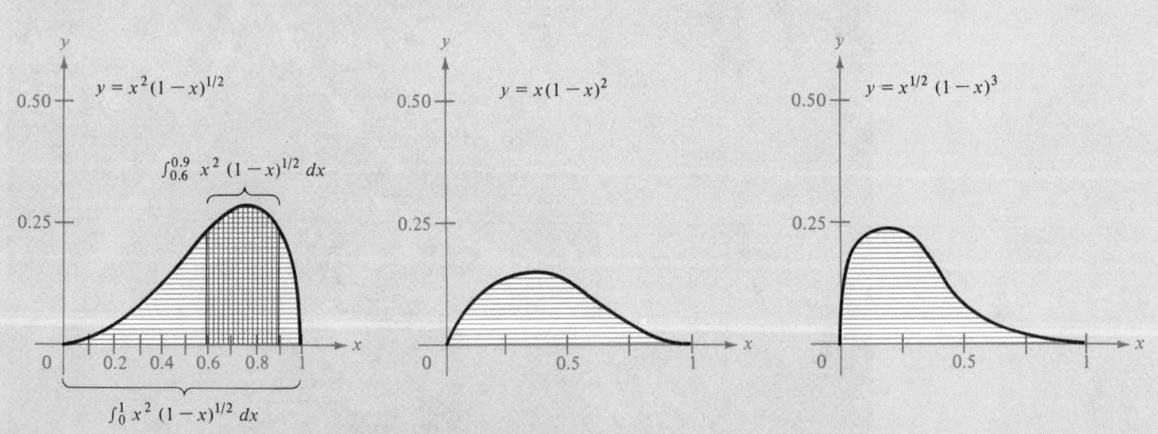

Figure 6.1

Section Objective: ***To expand the use of the basic integration formulas by means of the integration technique known as substitution.***

In Chapters 2, 4, and 5, we presented a number of rules for differentiating and integrating functions. In general, the rules for integrating were derived from corresponding differentiation formulas. It may surprise you to learn that, although we now have all the necessary tools for *differentiating* algebraic, exponential, and logarithmic functions, our set of tools for *integrating* these functions is by no means complete! The primary objective of this chapter is to develop several techniques that greatly expand the set of integrals to which the basic integration formulas can be applied. □

Basic Integration Formulas

1. Constant Rule:

$$\int dx = x + C$$

2. Simple Power Rule:

$$\int x^n \, dx = \frac{x^{n+1}}{n+1} + C, \qquad n \neq -1$$

3. General Power Rule:

$$\int u^n \cdot u' \, dx = \int u^n \, du = \frac{u^{n+1}}{n+1} + C, \qquad n \neq -1$$

4. Simple Log Rule:

$$\int \frac{1}{x} \, dx = \ln|x| + C$$

5. General Log Rule:

$$\int \frac{u'}{u} \, dx = \int \frac{1}{u} \, du = \ln|u| + C$$

6. Simple Exponential Rule:

$$\int e^x \, dx = e^x + C$$

7. General Exponential Rule:

$$\int e^u \cdot u' \, dx = \int e^u \, du = e^u + C$$

We need not work very long with integration problems before we realize that integration is not nearly as straightforward as differentiation is. A major part of any integration problem is the recognition of which basic integration formula to use to solve the problem. Skill in determining which formula to use requires memorization of the basic formulas, familiarity with various procedures for rewriting integrands in the basic forms, and a lot of practice.

For example, consider the ingenuity required in the integration of

$$\int \frac{x}{(x+1)^2}\,dx$$

None of the seven basic formulas listed here can be applied directly to this integral. However, by adding and subtracting 1 in the numerator, we can apply the Log Rule and the Power Rule as follows:

$$\begin{aligned}\int \frac{x}{(x+1)^2}\,dx &= \int \frac{x+1-1}{(x+1)^2}\,dx\\ &= \int \left[\frac{x+1}{(x+1)^2} - \frac{1}{(x+1)^2}\right] dx\\ &= \int \left[\frac{1}{x+1} - (x+1)^{-2}\right] dx\\ &= \ln|x+1| + \frac{1}{x+1} + C\end{aligned}$$

How did we know that adding and subtracting 1 would be the key to evaluating this integral? We've had many years of practice — that's how we knew! Fortunately, (since you have had considerably less time to practice) there is a nice integration technique for solving this integral in a straightforward manner. This technique is called **substitution,** and it involves a change of variable that permits us to rewrite the integrand in a form to which we can apply the basic integration rules.

One of the critical steps in this procedure is rewriting dx in terms of the new variable. In general, if we use the substitution

$$x = f(u)$$

we have

$$\frac{dx}{du} = f'(u)$$

and $\quad dx = f'(u)\,du$

Example 1

Integration by Substitution

Use the substitution $u = x + 1$ to evaluate the integral

$$\int \frac{x}{(x+1)^2}\,dx$$

Solution From the substitution $u = x + 1$, we have

$$x = u - 1, \qquad \frac{dx}{du} = 1, \qquad \text{and} \qquad dx = du$$

Now, replacing *all* instances of x and dx with the appropriate u-variable forms, we have

$$\int \frac{x}{(x+1)^2}\,dx = \int \frac{u-1}{u^2}\,du$$

Then, we write

$$\begin{aligned}\int \frac{u-1}{u^2}\,du &= \int \left(\frac{u}{u^2} - \frac{1}{u^2}\right) du \\ &= \int \left(\frac{1}{u} - \frac{1}{u^2}\right) du \\ &= \ln|u| + \frac{1}{u} + C\end{aligned}$$

Finally, we return to x-variable form by resubstituting $u = x + 1$ to obtain

$$\int \frac{x}{(x+1)^2}\,dx = \ln|x+1| + \frac{1}{x+1} + C$$

□

We outline the basic steps involved in integration by substitution as follows.

Integration by Substitution

1. Let u be some function of x (usually one that appears in the integrand).
2. Solve for x and dx in terms of u and du.
3. Convert the entire integral to u-variable form and try to fit it to one (or more) of the basic integration formulas. If none seems to fit, consider trying a different substitution.
4. After integrating, rewrite the antiderivative as a function of x.

Example 2

Integration by Substitution

Evaluate $\int x\sqrt{x-1}\,dx$.

Solution First, we note that, as it stands, this integral *does not* fit any of the basic integration formulas. Now, consider the substitution $u = \sqrt{x-1}$. Then

$$u^2 = x - 1 \qquad \text{and} \qquad u^2 + 1 = x$$

Solving for dx, we have

$$x = u^2 + 1$$

$$\frac{dx}{du} = 2u$$

$$dx = 2u\,du$$

We now substitute for x, $\sqrt{x-1}$, and dx as follows:

$$\int x\sqrt{x-1}\,dx = \int (u^2+1)(u)\underbrace{2u\,du}$$

Therefore,

$$\int x\sqrt{x-1}\,dx = \int (2u^4 + 2u^2)\,du = \frac{2u^5}{5} + \frac{2u^3}{3} + C$$

Substituting back to variable x, we have

$$\begin{aligned}
\frac{2u^5}{5} + \frac{2u^3}{3} + C &= \frac{2}{5}(\sqrt{x-1})^5 + \frac{2}{3}(\sqrt{x-1})^3 + C \\
&= \frac{6}{15}(x-1)^{5/2} + \frac{10}{15}(x-1)^{3/2} + C \\
&= \frac{2}{15}(x-1)^{3/2}[3(x-1)+5] + C \\
&= \frac{2}{15}(x-1)^{3/2}(3x+2) + C
\end{aligned}$$ □

Example 2 demonstrates one of the characteristics of integration by substitution. That is, the form of the antiderivative as it exists immediately after resubstituting into x-variable form can often be simplified. Thus, when working the exercises in this section, don't be too quick to think your answer is incorrect just because it doesn't look exactly like the answer given at the end of the text. You may be able to reconcile the two answers by algebraic simplification.

A second way to obtain a different form of an antiderivative is demonstrated in the next example.

Example 3

An Extra Constant Term in the Solution

Evaluate

$$\int \frac{1}{\sqrt{x}+1}\,dx$$

Solution Since none of the basic formulas apply, we try $u = \sqrt{x} + 1$. Then,

$$\sqrt{x} = u - 1, \qquad x = (u-1)^2, \qquad \text{and} \qquad dx = 2(u-1)\,du$$

Thus,

$$\int \frac{1}{\sqrt{x}+1}\,dx = \int \frac{1}{u}\,2(u-1)\,du$$
$$= 2\int \frac{u-1}{u}\,du$$
$$= 2\int \left(1-\frac{1}{u}\right) du$$
$$= 2(u - \ln|u|) + C_1$$
$$= 2[\sqrt{x} + 1 - \ln(\sqrt{x}+1)] + C_1$$
$$= 2\sqrt{x} - 2\ln(\sqrt{x}+1) + C$$

Note that the constant term 1 can be incorporated into the constant C_1 to produce C. □

Remark

You should not conclude that there is only one substitution that will work in a given integral. For instance, the substitution $u = \sqrt{x}$ would have worked just as well in Example 3, and we suggest that you try reworking the example with this substitution.

The fourth step outlined above suggests that we convert back to variable x. However, for *definite* integrals it is often more convenient to determine the limits of integration for variable u than to convert back to variable x and evaluate the antiderivative at the original limits. The next example illustrates this procedure.

Example 4

Converting the Limits of Integration for Definite Integrals

Find the area A of the region bounded by

$$y = \frac{x}{\sqrt{2x-1}}$$

and the x-axis, from $x = 1$ to $x = 5$.

Solution From Figure 6.2, we can see that the area of the region is given by

$$A = \int_1^5 \frac{x}{\sqrt{2x-1}}\,dx$$

To evaluate this integral, we let $u = \sqrt{2x-1}$; then,

$$u^2 = 2x - 1$$
$$x = \frac{u^2+1}{2}$$
$$dx = u\,du$$

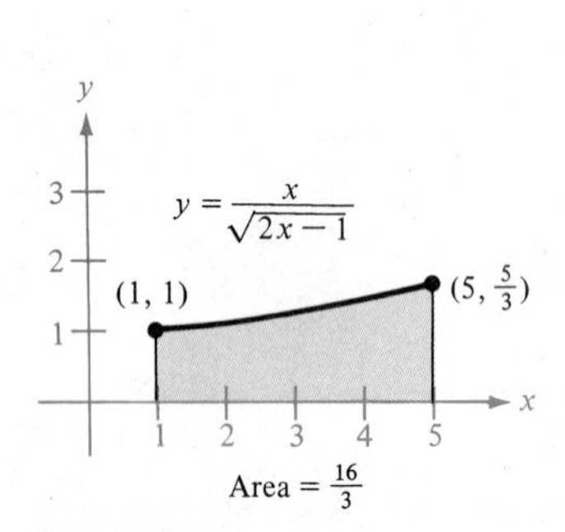

Figure 6.2

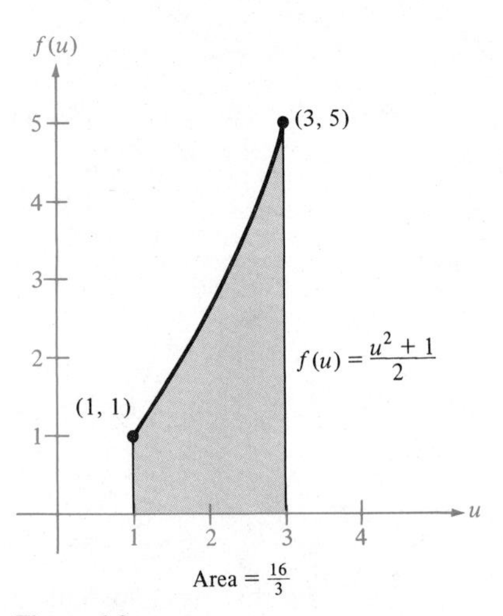

Figure 6.3

Furthermore, when $x = 5$, $u = \sqrt{10 - 1} = 3$; and when $x = 1$, $u = \sqrt{2 - 1} = 1$. (See Figure 6.3.) Thus, it follows that

$$\int_1^5 \frac{x}{\sqrt{2x - 1}}\,dx = \int_1^3 \frac{1}{2}\left(\frac{u^2 + 1}{u}\right)u\,du = \frac{1}{2}\int_1^3 (u^2 + 1)\,du$$

$$= \frac{1}{2}\left[\frac{u^3}{3} + u\right]_1^3 = \frac{1}{2}\left(9 + 3 - \frac{1}{3} - 1\right) = \frac{16}{3}$$

Geometrically, we can interpret the equation

$$\int_1^5 \frac{x}{\sqrt{2x - 1}}\,dx = \int_1^3 \frac{u^2 + 1}{2}\,du$$

to mean that the two *different* regions shown in Figures 6.2 and 6.3 have the *same* area. □

When solving definite integrals by substitution, don't be surprised if the upper limit of integration of the u-variable form is smaller than the lower limit. If this happens, don't rearrange the limits. Simply evaluate as usual. For example, after substituting $u = \sqrt{1 - x}$ in the integral

$$\int_0^1 x^2(1 - x)^{1/2}\,dx$$

we have $u = \sqrt{1 - 1} = 0$ when $x = 1$ and $u = \sqrt{1 - 0} = 1$ when $x = 0$. Thus, the correct u-variable form of this integral is

$$-2\int_1^0 (1 - u^2)^2u^2\,du$$

This situation is illustrated in the next example.

Example 5

An Application to Probability

A psychologist finds that the probability of recall in a certain memory experiment is given by the model

$$p_{a,b} = \int_a^b \frac{28}{9} x\sqrt[3]{1-x}\,dx \qquad 0 \le a \le b \le 1$$

where x represents the percentage of recall. (See Figure 6.4.) For a randomly chosen individual, find the probability that he or she will recall between 0% and $87\frac{1}{2}$% of the material.

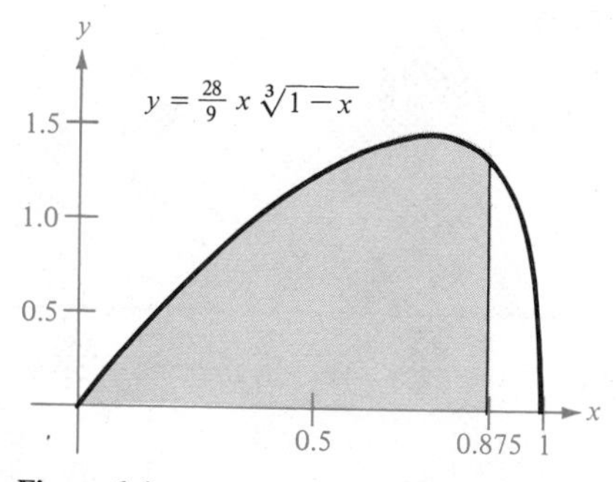

Figure 6.4

Solution Since $a = 0$ and $b = 0.875$, the probability is given by

$$\int_0^{0.875} \frac{28}{9} x\sqrt[3]{1-x}\,dx$$

Now, letting $u = \sqrt[3]{1-x}$, we have

$$u^3 = 1 - x$$
$$x = 1 - u^3$$
$$dx = -3u^2\,du$$

Furthermore, when $x = 0$, we have

$$u = \sqrt[3]{1-0} = 1$$

and when $x = 0.875$,

$$u = \sqrt[3]{1-0.875} = \sqrt[3]{0.125} = 0.5 = \tfrac{1}{2}$$

Thus, the u-variable form of the integral is

$$\int_0^{0.875} \frac{28}{9} x\sqrt[3]{1-x}\, dx = \int_1^{1/2} \left[\frac{28}{9}(1-u^3)(u)(-3u^2)\right] du$$
$$= 3\left(\frac{28}{9}\right)\int_1^{1/2} (u^6 - u^3)\, du$$
$$= \frac{28}{3}\left[\frac{u^7}{7} - \frac{u^4}{4}\right]_1^{1/2}$$
$$= \frac{28}{3}\left[\left(\frac{1}{896} - \frac{1}{64}\right) - \left(\frac{1}{7} - \frac{1}{4}\right)\right]$$
$$= \frac{83}{96} \approx 0.8646$$

Finally, we conclude that the probability of recalling between 0% and $87\frac{1}{2}$% of the material is

$$p_{0,0.875} \approx 86.46\%$$

□

Section Exercises (6.1)

In Exercises 1–20, find the indefinite integral.

1. $\int \frac{x}{(x+1)^3}\, dx$
2. $\int \frac{x^2}{(x+1)^3}\, dx$
3. $\int \frac{x}{(3x-1)^2}\, dx$
4. $\int \frac{5x}{(x-4)^3}\, dx$
5. $\int x(1-x)^4\, dx$
6. $\int x^2(1-x)^3\, dx$
7. $\int \frac{x-1}{x^2-2x}\, dx$
8. $\int \frac{x}{(x^2-1)^2}\, dx$
9. $\int x\sqrt{x-3}\, dx$
10. $\int x\sqrt{2x+1}\, dx$
11. $\int x^2\sqrt{1-x}\, dx$
12. $\int \frac{2x-1}{\sqrt{x+3}}\, dx$
13. $\int \frac{x^2-1}{\sqrt{2x-1}}\, dx$
14. $\int x^3\sqrt{x+2}\, dx$
15. $\int \frac{2\sqrt{t}-\sqrt{3t}}{t}\, dt$
16. $\int t\sqrt[3]{t+1}\, dt$
17. $\int \frac{1}{1+\sqrt{x}}\, dx$
18. $\int \frac{1-\sqrt{x}}{1+\sqrt{x}}\, dx$
19. $\int \frac{\sqrt{2x}}{6x+\sqrt{2x}}\, dx$
20. $\int \frac{1}{\sqrt{x}+\sqrt{2x}}\, dx$

In Exercises 21–30, evaluate the definite integral.

21. $\int_0^5 \frac{x}{(x+5)^2}\, dx$
22. $\int_0^1 x(x+5)^4\, dx$
23. $\int_0^{0.5} x(1-x)^3\, dx$
24. $\int_0^{0.5} x^2(1-x)^3\, dx$
25. $\int_3^7 x\sqrt{x-3}\, dx$
26. $\int_0^4 \frac{x}{\sqrt{2x+1}}\, dx$
27. $\int_0^7 x\sqrt[3]{x+1}\, dx$
28. $\int_1^2 (x-1)\sqrt{2-x}\, dx$
29. $\int_0^2 \frac{1}{1+\sqrt{2x}}\, dx$
30. $\int_0^1 \frac{1}{\sqrt{x}+\sqrt{x+1}}\, dx$
31. Find the area of the region bounded by $y = x\sqrt{x+1}$ and $y = 0$. (See Figure 6.5.)

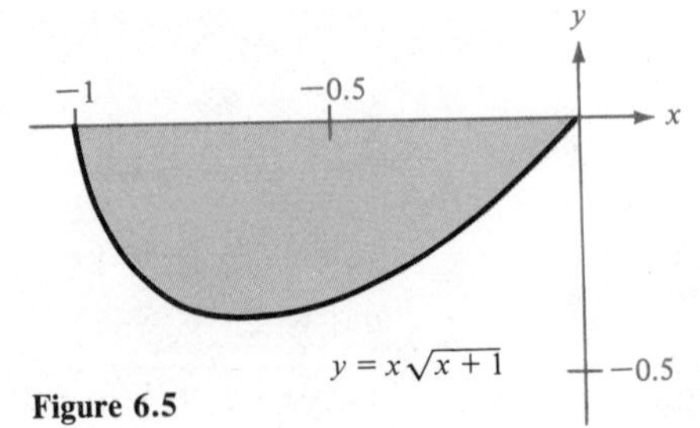

Figure 6.5

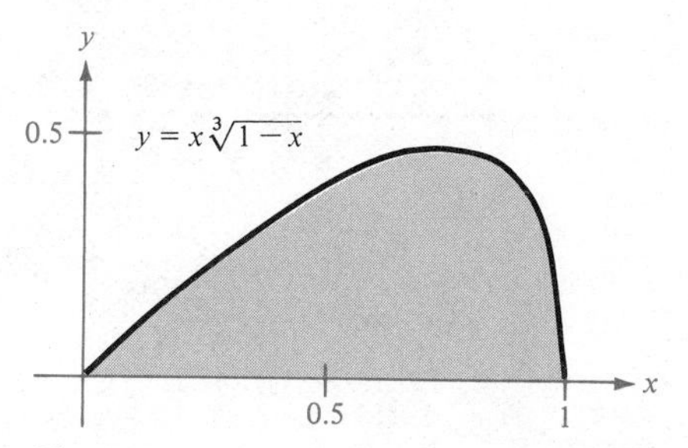

Figure 6.6

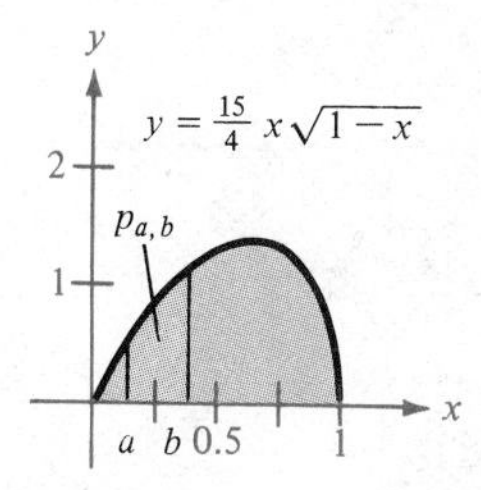

Figure 6.7

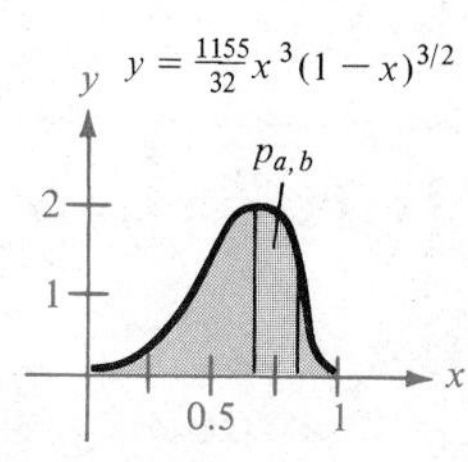

Figure 6.8

32. Find the area of the region bounded by $y = x\sqrt[3]{1-x}$ and $y = 0$. (See Figure 6.6.)

Ⓒ33. In the Introductory Example of this section, the drug retention pattern was given by

$$y = x^2(1-x)^{1/2}$$

Find the portion of the population retaining between 0% and 50% of the drug after 24 hours.

Ⓒ34. Verify the result in the Introductory Example of this section. That is, show that 52.9% of the population retain between 60% and 90% of the drug after 24 hours by showing that

$$\frac{\int_{0.6}^{0.9} x^2(1-x)^{1/2}\,dx}{\int_0^1 x^2(1-x)^{1/2}\,dx} = 0.529$$

Ⓒ35. The probability of recall in a certain experiment is found to be

$$p_{a,b} = \int_a^b \frac{15}{4}x\sqrt{1-x}\,dx$$

where x represents the percentage of recall. (See Figure 6.7.)

(a) For a randomly chosen individual, what is the probability that he or she will recall between 50% and 75% of the material?

(b) What is the median percentage recall? That is, for what value of b is it true that the probability from 0 to b is 0.5?

Ⓒ36. The probability of finding between a and b percent of iron in ore samples taken from a certain region is given by

$$p_{a,b} = \int_a^b \frac{1155}{32}x^3(1-x)^{3/2}\,dx$$

(See Figure 6.8.) Find the probability that a sample will contain between:

(a) 0% and 25% (b) 50% and 100%

Ⓒ37. A company sells a seasonal product that has a daily revenue approximated by

$$R = 0.06t^2(365-t)^{1/2} + 1250, \quad 0 \le t \le 365$$

Find the average daily revenue over the period of one year.

Ⓒ38. Find the *average* amount by which the function

$$f(x) = \frac{1}{x+1}$$

exceeds the function

$$g(x) = \frac{x}{(x+1)^2}$$

on the interval [0, 1].

Ⓒ39. Find the *average* amount by which the function

$$f(x) = x(4x+1)^{1/2}$$

exceeds the function

$$g(x) = 2x^{3/2}$$

on the interval [0, 2].

Ⓒ Calculator may be helpful.

Section 6.2

Integration by Parts

Introductory Example

Diminishing Rate of Return for Poultry Growers

Livestock and poultry growers have long been familiar with the economic principle of *diminishing returns*. This principle relates the total investment (cost) in a certain product to the total return (revenue) from the product. If the ratio of the total revenue to the total cost decreases as the cost increases, we say that the rate of return is diminishing. The reason this principle is important to livestock and poultry growers is that young animals tend to put on more weight relative to food intake than older animals do. Thus, the money spent on feeding young animals brings a higher rate of return than does money spent on feeding older animals.

To illustrate this more clearly, suppose that a certain breed of fryer gains weight according to the model

$$W = 10.85 \ln (1 + t) + 0.1, \qquad 0 \le t \le 1$$

where W is the fryer's weight in pounds and t is the time in years, with $t = 0$ corresponding to the time the chick is hatched. Furthermore, suppose that each fryer's daily food intake is proportional to its present weight. Then, the total cost of feed necessary to raise a bird to age t is proportional to the area under the weight curve between 0 and t. (See Figure 6.9.) Now, let us compare the return ratios for selling a fryer after six months and after one year.

For a six-month-old fryer the return ratio is

$$\frac{\text{revenue}}{\text{cost}} = \frac{k_2[10.85 \ln (1 + 0.5) + 0.1]}{k_1 \int_0^{0.5} [10.85 \ln (1 + t) + 0.1]\, dt} = \frac{4.5k_2}{1.22k_1} = 3.68 \frac{k_2}{k_1}$$

where k_2 is the selling price per pound of animal weight and k_1 is determined from the cost of feed. In this section, you will see that the logarithmic integral representing the cost can be solved using an integration technique called integration by parts. For the one-year-old fryer the return ratio turns out to be

$$\frac{\text{revenue}}{\text{cost}} = \frac{7.62k_2}{4.29k_1} = 1.78 \frac{k_2}{k_1}$$

Thus, even though the older bird can be sold for more, it took more than three times the feed to raise it, and the return ratio is less than half the return ratio for the younger bird. If we include the cost of new chicks and daily overhead, this difference lessens but still favors the younger bird.

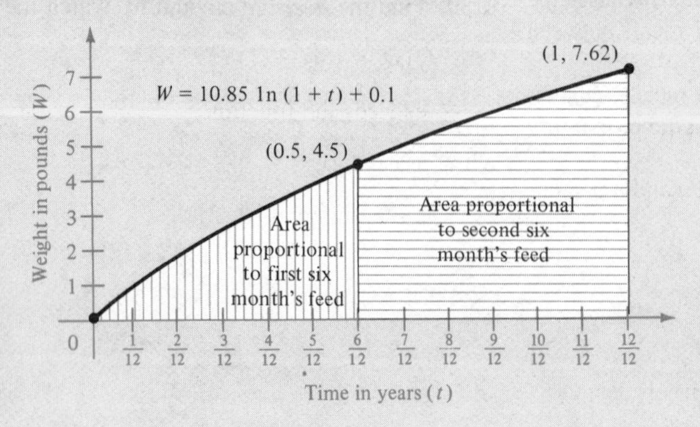

Figure 6.9

Section Objective: *To introduce the technique of integration by parts.*

In Section 6.1, we discussed the technique of integration by substitution. In this section, we introduce a second basic method known as **integration by parts.** This method of integration applies to a wide variety of functions and is particularly useful for integrands involving a product of algebraic and exponential (or logarithmic) functions. For instance, integration by parts works well with products like

$$x \ln x \qquad \text{and} \qquad x^2 e^x$$

Integration by parts is based on the formula for the derivative of a product

$$\frac{d}{dx}[uv] = uv' + vu'$$

where both u and v are differentiable functions of x. Integrating both sides of this equation with respect to x, we have

$$\int \frac{d}{dx}[uv]\,dx = \int uv'\,dx + \int vu'\,dx$$

Now, since integration and differentiation are inverse operations, the left side of the equation simplifies to uv, and we obtain

$$uv = \int uv'\,dx + \int vu'\,dx$$

By rewriting this equation, we have the following integration by parts formula. You should memorize this basic integration formula.

Integration by Parts

$$\int uv'\,dx = uv - \int vu'\,dx$$

Note that the integration by parts formula expresses the original integral in terms of another integral. Depending on the choices for u and v', it may be easier to evaluate the second integral than the original one. Since the choices for u and v' are critical in the integration by parts process, we provide the following general guidelines.

Guidelines for Using Integration by Parts: $\int uv'\,dx$

1. Let v' be the most complicated portion of the integrand that can be "easily" integrated.
2. Let u be that portion of the integrand whose derivative u' is a "simpler" function than u itself.

These are only suggested guidelines, and they should not be followed blindly or without some thought. Furthermore, it is usually best to consider the guidelines above in the stated order, giving greater consideration to the first one.

Example 1

Integration by Parts

Evaluate $\int xe^x\,dx$.

Solution Since e^x is easily integrated and the derivative of x is "simpler" than x itself, we let

$$v' = e^x \qquad \text{and} \qquad u = x$$

Then we have

$$v = e^x + C_1 \qquad \text{and} \qquad u' = 1$$

Therefore,

$$\begin{aligned}\int xe^x\,dx &= uv - \int vu'\,dx \\ &= x(e^x + C_1) - \int (e^x + C_1)(1)\,dx \\ &= xe^x + C_1x - e^x - C_1x + C = xe^x - e^x + C\end{aligned}$$

□

Remark

Note in Example 1 that the first constant of integration C_1 does not appear in the final result. This will always happen, as we can see by replacing v by $v + C_1$ in the general formula:

$$\begin{aligned}\int uv'\,dx &= u(v + C_1) - \int (v + C_1)u'\,dx \\ &= uv + C_1u - \int C_1u'\,dx - \int vu'\,dx \\ &= uv + C_1u - C_1u - \int vu'\,dx = uv - \int vu'\,dx\end{aligned}$$

Thus, we may drop the first constant of integration when integrating by parts.

Example 2

Integration by Parts

Evaluate $\int x^2 \ln x\,dx$.

Solution In this case x^2 is more easily integrated than $\ln x$, and, furthermore, the derivative of $\ln x$ is simpler than $\ln x$. Therefore, we choose

$$\begin{aligned} v' &= x^2 \longrightarrow u = \ln x \\ v &= \frac{x^3}{3} \longrightarrow u' = \frac{1}{x}\end{aligned}$$

Therefore,

$$\begin{aligned}\int x^2 \ln x \, dx &= \frac{x^3}{3} \ln x - \int \frac{x^3}{3}\left(\frac{1}{x}\right) dx \\ &= \frac{x^3}{3} \ln x - \frac{1}{3}\int x^2 \, dx \\ &= \frac{x^3}{3} \ln x - \frac{x^3}{9} + C\end{aligned}$$

□

Remark

The *arrow diagram* in Example 1 is simply a memory device that we have found helpful in remembering the integration by parts formula. After calculating v', u, v, and u', we can follow the arrows to generate the three terms in the formula as follows:

Top Row *Diagonal* *Bottom Row*

$$\int uv' \, dx = \quad uv \quad - \quad \int vu' \, dx$$

It may happen that a particular integral requires repeated application of integration by parts. This is demonstrated in the next example.

Example 3

Repeated Application of Integration by Parts

Evaluate $\int x^2 e^x \, dx$.

Solution We may consider x^2 and e^x to be equally easy to integrate; however, the derivative of x^2 becomes simpler while the derivative of the e^x does not. Therefore, we have $u = x^2$ and

$$\begin{aligned} v' &= e^x \longrightarrow u = x^2 \\ v &= e^x \longleftrightarrow u' = 2x \end{aligned}$$

and it follows that

$$\int x^2 e^x \, dx = x^2 e^x - \int 2xe^x \, dx$$

Now, we apply integration by parts to the new integral, and, with the same considerations in mind, we have

$$\begin{aligned} v' &= e^x \longrightarrow u = 2x \\ v &= e^x \longleftrightarrow u' = 2 \end{aligned}$$

and it follows that

$$\int x^2e^x\,dx = x^2e^x - 2xe^x + \int 2e^x\,dx$$
$$= x^2e^x - 2xe^x + 2e^x + C$$
$$= e^x(x^2 - 2x + 2) + C \qquad \square$$

When making repeated applications of integration by parts, we need to be careful not to interchange the substitutions in successive applications. For instance, in Example 3 our first substitutions were

$$v' = e^x \qquad \text{and} \qquad u = x^2$$

and for the second application of integration by parts, we let

$$v' = e^x \qquad \text{and} \qquad u = 2x$$

If, in the second application, we had switched our substitutions to

$$\begin{aligned} v' &= 2x \longrightarrow u = e^x \\ v &= x^2 \longleftarrow u' = e^x \end{aligned}$$

we would have obtained

$$\int x^2e^x\,dx = x^2e^x - \int 2xe^x\,dx$$
$$= x^2e^x - x^2e^x + \int x^2e^x\,dx$$
$$= \int x^2e^x\,dx$$

which tells us nothing. By switching substitutions, we undid the previous integration and therefore returned to our original integral.

Example 4

Integration by Parts Applied to a Quotient

Evaluate

$$\int \frac{xe^x}{(x+1)^2}\,dx$$

Solution We let

$$v' = \frac{1}{(x+1)^2}$$

since this choice is easily integrated. Thus, we have

$$v' = \frac{1}{(x+1)^2} \longrightarrow u = xe^x$$

$$v = -\frac{1}{x+1} \longrightarrow u' = e^x(x+1)$$

Therefore,

$$\int \frac{xe^x}{(x+1)^2}\,dx = xe^x\left(\frac{-1}{x+1}\right) - \int (x+1)e^x\left(\frac{-1}{x+1}\right)dx$$

$$= -\frac{xe^x}{x+1} + \int e^x\,dx$$

$$= -\frac{xe^x}{x+1} + e^x + C$$

$$= \frac{e^x}{x+1} + C$$

□

Example 5

Integration by Parts Applied to a Single Factor

Evaluate $\int_1^e \ln x\,dx$

Solution At first it may appear that integration by parts does not apply. However, if we let $v' = 1$, we have

$$v' = 1 \longrightarrow u = \ln x$$

$$v = x \longrightarrow u' = \frac{1}{x}$$

Therefore, we have

$$\int \ln x\,dx = x\ln x - \int \frac{1}{x}(x)\,dx$$

$$= x\ln x - \int 1\,dx$$

$$= x\ln x - x + C$$

Now, using this antiderivative to find the definite integral, we have

$$\int_1^e \ln x\,dx = [x\ln x - x]_1^e$$

$$= [e\ln(e) - e] - [1\ln(1) - 1]$$

$$= (e - e) - (0 - 1) = 1$$

(See Figure 6.10.)

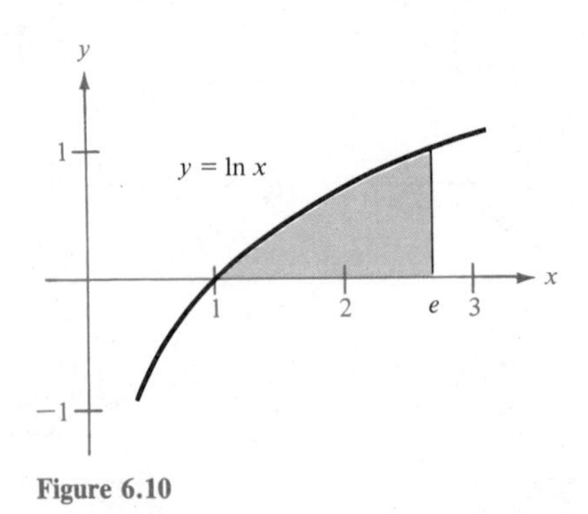

Figure 6.10

□

Before starting the exercises in this section remember that merely knowing *how* to use the various integrating techniques is not enough; you also need to know *when* to use them. Integration is first and foremost a problem of recognition — recognizing which formula or technique to apply to obtain an antiderivative. Frequently, the slightest alteration of an integrand will necessitate the use of a different integration technique. For example, consider the integrals

$$\int x \ln x \, dx, \qquad \int \frac{\ln x}{x} \, dx, \qquad \int \frac{dx}{x \ln x}$$

whose antiderivatives are obtained in the manner shown in Table 6.1.

Table 6.1

Integral	Technique	Antiderivative
$\int x \ln x \, dx$	Integration by parts	$\frac{x^2}{2} \ln x - \frac{x^2}{4} + C$
$\int \frac{\ln x}{x} \, dx$	Power Rule $\int u^n u' \, dx$	$\frac{(\ln x)^2}{2} + C$
$\int \frac{1}{x \ln x} \, dx$	Log Rule $\int \frac{u'}{u} \, dx$	$\ln \lvert \ln x \rvert + C$

Section Exercises (6.2)

In Exercises 1–22, evaluate the given integral. [Note: Solve by the simplest means; not all exercises require integration by parts.]

1. $\int xe^{2x} \, dx$

2. $\int x^2 e^{2x} \, dx$

3. $\int xe^{x^2} \, dx$

4. $\int x^2 e^{x^3} \, dx$

5. $\int xe^{-2x} \, dx$

6. $\int \frac{x}{e^x} \, dx$

7. $\int x^3 e^x \, dx$

8. $\int \frac{e^{1/t}}{t^2} \, dt$

9. $\int x^3 \ln x \, dx$

10. $\int x^2 \ln x \, dx$

11. $\int t \ln (t + 1) \, dt$

12. $\int \frac{1}{x(\ln x)^3} \, dx$

13. $\int (\ln x)^2 \, dx$

14. $\int \ln 3x \, dx$

15. $\int \frac{(\ln x)^2}{x} \, dx$

16. $\int \frac{\ln x}{x^2} \, dx$

17. $\int \frac{xe^{2x}}{(2x + 1)^2} \, dx$

18. $\int \frac{x^3 e^{x^2}}{(x^2 + 1)^2} \, dx$

19. $\int x\sqrt{x - 1} \, dx$

20. $\int x^2\sqrt{x - 1} \, dx$

21. $\int (x^2 - 1)e^x \, dx$

22. $\int \frac{\ln 2x}{x^2} \, dx$

23. Integrate $\int 2x\sqrt{2x - 3} \, dx$ by:
(a) parts, letting $v' = \sqrt{2x - 3}$
(b) substitution, letting $u = \sqrt{2x - 3}$

24. Integrate $\int x\sqrt{4 + x} \, dx$ by:
(a) parts, letting $v' = \sqrt{4 + x}$
(b) substitution, letting $u = \sqrt{4 + x}$

25. Integrate

$$\int \frac{x^3}{\sqrt{4 + x^2}} \, dx$$

by:
(a) parts, letting

$$v' = \frac{x}{\sqrt{4 + x^2}}$$

(b) substitution, letting $u = \sqrt{4 + x^2}$

26. Integrate $\int x\sqrt{4 - x} \, dx$ by:
(a) parts, letting $v' = \sqrt{4 - x}$
(b) substitution, letting $u = \sqrt{4 - x}$

27. Integrate

$$\int \frac{x}{\sqrt{4 + 5x}} \, dx$$

by:
(a) parts, letting $v' = \frac{1}{\sqrt{4 + 5x}}$
(b) substitution, letting $u = \sqrt{4 + 5x}$

28. Use integration by parts to verify that

$$\int x^n e^{ax} \, dx = \frac{x^n e^{ax}}{a} - \frac{n}{a} \int x^{n-1} e^{ax} \, dx$$

29. Use the result of Ex. 28 to integrate $\int x^2 e^{5x} \, dx$.

30. Use the result of Ex. 28 to integrate $\int xe^{-3x} \, dx$.

31. Evaluate the definite integral $\int_0^1 x^2 e^x \, dx$.

32. Evaluate the definite integral $\int_0^1 \ln (1 + x) \, dx$.

33. Find the area of the region bounded by $y = \ln x$, $y = 0$, and $x = 4$.

34. Find the area of the region bounded by $y = xe^{-x}$, $y = 0$, and $x = 4$.

Ⓒ**35.** A model for the ability M, measured on a scale from 0 to 10, of a child to memorize is given by

$$M = 1.6t \ln t + 1, \qquad 0 < t \leq 4$$

where t is the child's age in years. Find the average value of this function for a child:
(a) between his or her 1st and 2nd birthdays
(b) between his or her 3rd and 4th birthdays

Ⓒ**36.** A company sells a seasonal product, and the model for the daily revenue from the product is

$$R = 410.5t^2 e^{-t/30} + 25{,}000, \qquad 0 \leq t \leq 365$$

where t is the time in days.
(a) Sketch the graph of this revenue function.
(b) Find the average daily receipts during the 1st quarter, $0 \leq t \leq 91$.
(c) Find the average daily receipts during the 4th quarter, $274 \leq t \leq 365$.

Ⓒ Calculator may be helpful.

Section 6.3

Partial Fractions

Introductory Example

The Logistics Growth Function

In Section 5.5, the exponential growth function was derived under the assumption that the rate of growth was proportional to the existing quantity. In many situations, there exists some upper limit L past which growth cannot occur. (Population growth is limited by the food supply or space; growth in sales of some new product will stop when the market becomes saturated; and so on.) Under such conditions, we assume that the rate of growth is proportional not only to the existing amount y but *also* to the difference between the existing amount and the limit L. That is,

$$\frac{dy}{dt} = ky(L - y)$$

In integral form, we can express this relationship as follows:

$$\int \frac{1}{y(L - y)}\, dy = \int k\, dt$$

To solve the integral on the left, we can use an integration technique called **partial fractions.** The solution turns out to be the *logistics growth function* discussed in Sections 5.1 and 5.2.

For example, if a bacterial culture weighs 1 g when $t = 0$ and has an upper limit of $L = 10$ g, then its growth function is of the form

$$y = \frac{10}{1 + 9e^{-kt}}$$

where y is the weight (in grams) and t is the time (in hours). Furthermore, if we know the weight at some other point in time, then we can solve for k. For instance, if $y = 2$ when $t = 1$, we can determine that $k = \ln \frac{9}{4} \approx 0.8109$ (see Figure 6.11). On the other hand, if $y = 2$ when $t = 2$, we can determine that $k = \ln \frac{3}{2} \approx 0.4055$ (see Figure 6.12).

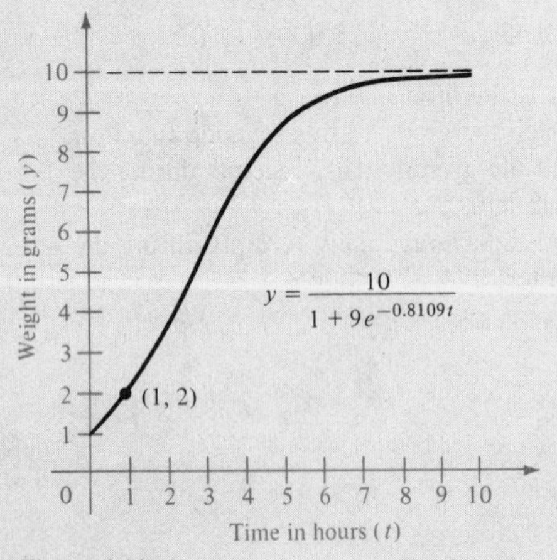

Figure 6.11

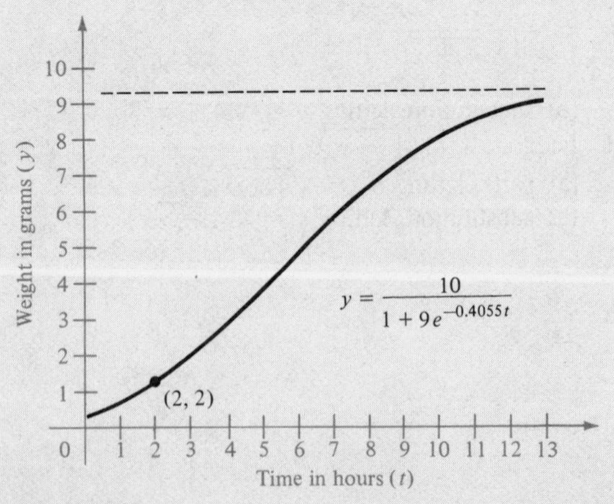

Figure 6.12

Section Objective: ***To introduce the method of partial fractions as a means of integrating rational functions.***

In Sections 6.1 and 6.2, we introduced two techniques that allowed us to apply our fundamental integration rules to a wide variety of functions. In this section, we discuss yet a *third* technique that permits us to rewrite algebraic functions (rational functions, in particular) in a form to which we can apply our fundamental integration rules. This technique involves the decomposition of a rational function into the sum of two or more "simpler" rational functions. We call this procedure the method of **partial fractions.** The need for this procedure can be seen in the discussion of the following problem.

How can we evaluate the following integral?

$$\int \frac{x + 7}{x^2 - x - 6}\,dx$$

Suppose we knew that

$$\frac{x + 7}{x^2 - x - 6} = \frac{2}{x - 3} - \frac{1}{x + 2}$$

Then, we could write

$$\begin{aligned}\int \frac{x + 7}{x^2 - x - 6}\,dx &= \int \left(\frac{2}{x - 3} - \frac{1}{x + 2}\right) dx \\ &= 2\int \frac{dx}{x - 3} - \int \frac{dx}{x + 2} \\ &= 2 \ln |x - 3| - \ln |x + 2| + C\end{aligned}$$

The use of this method depends on our ability to factor the denominator, $(x^2 - x - 6)$, and to find the *partial fractions*

$$\frac{2}{x - 3} \quad \text{and} \quad \frac{-1}{x + 2}$$

Recall that one of the basic concerns of algebra is finding the factors of a polynomial. For instance, the polynomial $x^3 + x^2 - x - 1$ can be written as

$$x^3 + x^2 - x - 1 = (x - 1)(x + 1)^2$$

where $(x - 1)$ is a linear factor and $(x + 1)^2$ is a *repeated* linear factor. We can use this factorization to find the partial fraction decomposition of any rational function having $x^3 + x^2 - x - 1$ as its denominator. Specifically, if $N(x)$ is a polynomial of degree less than three, then the partial fraction decomposition of

$$\frac{N(x)}{x^3 + x^2 - x - 1}$$

has the form

$$\frac{N(x)}{x^3 + x^2 - x - 1} = \frac{N(x)}{(x - 1)(x + 1)^2}$$

$$= \frac{A}{x - 1} + \frac{B}{x + 1} + \frac{C}{(x + 1)^2}$$

Note that the repeated linear factor $(x + 1)^2$ results in *two* fractions. One for $(x + 1)$ and one for $(x + 1)^2$. If $(x + 1)^3$ were a factor, then we would use three fractions: one for $(x + 1)$, one for $(x + 1)^2$, and one for $(x + 1)^3$. In general, the number of fractions resulting from a repeated linear factor is equal to the number of times the factor is repeated.

An algebraic technique for determining the value of the constants in the numerators is demonstrated in the examples that follow.

Example 1

Distinct Linear Factors

Write the partial fraction decomposition for the rational function

$$\frac{x + 7}{x^2 - x - 6}$$

Solution Since $x^2 - x - 6 = (x - 3)(x + 2)$, we include one partial fraction for each factor, and write

$$\frac{x + 7}{x^2 - x - 6} = \frac{A}{x - 3} + \frac{B}{x + 2}$$

Multiplying this equation by the lowest common denominator (LCD), $(x - 3)(x + 2)$, leads to the **basic equation**

$$x + 7 = A(x + 2) + B(x - 3)$$

Since this equation is to be true for all x, we can substitute *convenient* values for x to obtain equations in A and B, which we then solve. If $x = -2$, then

$$-2 + 7 = A(0) + B(-5)$$
$$5 = -5B$$
$$-1 = B$$

If $x = 3$, then

$$3 + 7 = A(5) + B(0)$$
$$10 = 5A$$
$$2 = A$$

The decomposition is, therefore,

$$\frac{x + 7}{x^2 - x - 6} = \frac{2}{x - 3} - \frac{1}{x + 2}$$

as indicated at the beginning of this section. □

Remark

The substitutions for x in Example 1 were chosen for their convenience in determining values for A and B. We chose $x = -2$ so as to eliminate the term $A(x + 2)$, while $x = 3$ was chosen to eliminate the term $B(x - 3)$. The goal is to make *convenient* substitutions, whenever possible.

Example 2

A Substitution Leading to Partial Fractions

Evaluate

$$\int \frac{1}{x\sqrt{x + 1}}\, dx$$

Solution Substituting $u = \sqrt{x + 1}$, we have

$$x = u^2 - 1 \qquad \text{and} \qquad dx = 2u\, du$$

Thus,

$$\int \frac{1}{x\sqrt{x + 1}}\, dx = \int \frac{1}{(u^2 - 1)u} 2u\, du = \int \frac{2}{(u^2 - 1)}\, du$$

Now, since $u^2 - 1 = (u + 1)(u - 1)$, we can apply partial fractions to obtain

$$\frac{2}{(u^2 - 1)} = \frac{A}{u + 1} + \frac{B}{u - 1}$$

which leads to the basic equation

$$2 = A(u - 1) + B(u + 1)$$

Letting $u = -1$, we have

$$2 = A(-2) + B(0)$$
$$-1 = A$$

and letting $u = 1$, we have

$$2 = A(0) + B(2)$$
$$1 = B$$

Thus,

$$\int \frac{2}{(u^2 - 1)}\, du = \int \left(\frac{-1}{u + 1} + \frac{1}{u - 1}\right) du$$
$$= -\ln |u + 1| + \ln |u - 1| + C$$

Finally, by resubstituting $u = \sqrt{x + 1}$, we have

$$\int \frac{1}{x\sqrt{x + 1}}\, dx = -\ln |\sqrt{x + 1} + 1| + \ln|\sqrt{x + 1} - 1| + C$$
$$= \ln \left|\frac{\sqrt{x + 1} - 1}{\sqrt{x + 1} + 1}\right| + C$$

□

Example 3

Repeated Linear Factors

Evaluate

$$\int \frac{5x^2 + 20x + 6}{x^3 + 2x^2 + x}\, dx$$

Solution Since $x^3 + 2x^2 + x = x(x^2 + 2x + 1) = x(x + 1)^2$, we include one fraction for each power of x and $(x + 1)$, and we write

$$\frac{5x^2 + 20x + 6}{x(x + 1)^2} = \frac{A}{x} + \frac{B}{x + 1} + \frac{C}{(x + 1)^2}$$

Multiplying by the LCD, $x(x + 1)^2$, leads to the basic equation

$$5x^2 + 20x + 6 = A(x + 1)^2 + Bx(x + 1) + Cx$$

Substituting $x = -1$ eliminates the A and B terms and yields

$$\begin{aligned} 5 - 20 + 6 &= 0 + 0 - C \\ C &= 9 \end{aligned}$$

If $x = 0$, then

$$\begin{aligned} 6 &= A(1) + 0 + 0 \\ 6 &= A \end{aligned}$$

At this point, we have exhausted the most convenient choices for x and have yet to find the value of B. Under such circumstances, we use any other value for x along with the calculated values of A and C. [Note that it is necessary to make as many substitutions for x as there are unknowns (A, B, C, . . .) to be determined.] Thus, for $x = 1$, $A = 6$, and $C = 9$, we have

$$\begin{aligned} 5 + 20 + 6 &= A(4) + B(2) + C \\ 31 &= 6(4) + 2B + 9 \\ -2 &= 2B \\ -1 &= B \end{aligned}$$

Therefore,

$$\frac{5x^2 + 20x + 6}{x(x + 1)^2} = \frac{6}{x} - \frac{1}{x + 1} + \frac{9}{(x + 1)^2}$$

and it follows that

$$\begin{aligned} \int \frac{5x^2 + 20x + 6}{x^3 + 2x^2 + x}\, dx &= \int \frac{6}{x}\, dx - \int \frac{dx}{x + 1} + \int 9(x + 1)^{-2}\, dx \\ &= 6 \ln |x| - \ln |x + 1| + 9\, \frac{(x + 1)^{-1}}{-1} + C \\ &= \ln \left| \frac{x^6}{x + 1} \right| - \frac{9}{x + 1} + C \end{aligned}$$

□

Remark

We can only apply the partial fraction decomposition outlined in Examples 1, 2, and 3 to a *proper* rational function, that is, to a rational function whose numerator is of lesser degree than its denominator. If the numerator is of equal or greater degree, we must divide first. This is demonstrated in the next example.

Example 4

Dividing to Obtain a Proper Rational Function

Evaluate

$$\int \frac{x^5 + x - 1}{x^4 - x^3}\,dx$$

Solution Since the numerator is fifth degree and the denominator is only fourth degree, we divide to obtain

$$\int \frac{x^5 + x - 1}{x^4 - x^3}\,dx = \int \left(x + 1 + \frac{x^3 + x - 1}{x^4 - x^3}\right) dx$$

Now, applying partial fractions, we have

$$\frac{x^3 + x - 1}{x^3(x - 1)} = \frac{A}{x} + \frac{B}{x^2} + \frac{C}{x^3} + \frac{D}{x - 1}$$

Using techniques similar to those in the first three examples, we find that

$$A = 0, \qquad B = 0, \qquad C = 1, \qquad \text{and} \qquad D = 1$$

Thus, we have

$$\int \frac{x^5 + x - 1}{x^4 - x^3}\,dx = \int \left(x + 1 + \frac{1}{x^3} + \frac{1}{x - 1}\right) dx$$

$$= \frac{x^2}{2} + x - \frac{1}{2x^2} + \ln|x - 1| + C$$

□

Before concluding this section, we must add a few comments. First, it is not necessary to use the partial fractions technique on all integrals of the form

$$\frac{N(x)}{D(x)}$$

For instance, even though

$$\int \frac{x^2 + 2x}{x^3 + 3x^2 - 4}\,dx$$

can be evaluated by partial fractions as follows:

$$\int \frac{x^2 + 2x}{x^3 + 3x^2 - 4}\,dx = \int \frac{x^2 + 2x}{(x - 1)(x + 2)^2}\,dx$$

$$= \int \left(\frac{A}{x - 1} + \frac{B}{x + 2} + \frac{C}{(x + 2)^2}\right) dx$$

it is more easily evaluated by using the Log Rule and writing

$$\int \frac{x^2 + 2x}{x^3 + 3x^2 - 4}\,dx = \frac{1}{3}\int \frac{3x^2 + 6x}{x^3 + 3x^2 - 4}\,dx$$
$$= \frac{1}{3}\ln|x^3 + 3x^2 - 4| + C$$

Second, if the given integral is not in reduced form, reducing it may eliminate the need to use partial fractions. For instance,

$$\int \frac{x^2 - x - 2}{x^3 - 2x - 4}\,dx = \int \frac{(x + 1)(x - 2)}{(x - 2)(x^2 + 2x + 2)}\,dx = \int \frac{x + 1}{x^2 + 2x + 2}\,dx$$

The latter integral is a logarithmic form.

Finally, the partial fractions technique can be used with some quotients involving exponential functions. For instance, the substitution $u = e^x$ allows us to write

$$\int \frac{e^x}{e^x(e^x - 1)}\,dx = \int \frac{du}{u(u - 1)} = \int \left(\frac{A}{u} + \frac{B}{u - 1}\right) du$$

Section Exercises (6.3)

In Exercises 1–20, find the indefinite integral.

1. $\int \frac{1}{x^2 - 1}\,dx$ **2.** $\int \frac{9}{x^2 - 9}\,dx$

3. $\int \frac{1}{x^2 + x}\,dx$ **4.** $\int \frac{-4}{x^2 - 4}\,dx$

5. $\int \frac{1}{2x^2 + x}\,dx$ **6.** $\int \frac{5}{x^2 + x - 6}\,dx$

7. $\int \frac{3}{x^2 + x - 2}\,dx$ **8.** $\int \frac{1}{4x^2 - 9}\,dx$

9. $\int \frac{5 - x}{2x^2 + x - 1}\,dx$ **10.** $\int \frac{x + 1}{x^2 + 4x + 3}\,dx$

11. $\int \frac{x^2 + 12x + 12}{x^3 - 4x}\,dx$ **12.** $\int \frac{3x^2 - 7x - 2}{x^3 - x}\,dx$

13. $\int \frac{2x^3 - 4x^2 - 15x + 5}{x^2 - 2x - 8}\,dx$

14. $\int \frac{x^3 - x + 3}{x^2 + x - 2}\,dx$

15. $\int \frac{x + 2}{x^2 - 4x}\,dx$ **16.** $\int \frac{4x^2 + 2x - 1}{x^3 + x^2}\,dx$

17. $\int \frac{2x - 3}{(x - 1)^2}\,dx$ **18.** $\int \frac{x^4}{(x - 1)^3}\,dx$

19. $\int \frac{4x^2 - 1}{(2x)(x^2 + 2x + 1)}\,dx$ **20.** $\int \frac{3x}{x^2 - 6x + 9}\,dx$

In Exercises 21–24, evaluate the definite integral.

21. $\int_3^4 \frac{1}{x^2 - 4}\,dx$ **22.** $\int_0^1 \frac{3}{2x^2 + 5x + 2}\,dx$

23. $\int_1^5 \frac{x - 1}{x^2(x + 1)}$ **24.** $\int_0^1 \frac{x^2 - x}{x^2 + x + 1}\,dx$

In Exercises 25–30, evaluate the indefinite integral by using the indicated substitution.

25. $\int \frac{e^x}{(e^x - 1)(e^x + 4)}\,dx$; let $u = e^x$

26. $\int \frac{e^x}{(e^{2x} - 1)(e^x + 1)}\,dx$; let $u = e^x$

27. $\int \frac{1}{x\sqrt{4 + x^2}}\,dx$; let $u = \sqrt{4 + x^2}$

28. $\int \frac{1}{x\sqrt{4 + x}}\,dx$; let $u = \sqrt{4 + x}$

29. $\int \frac{1}{\sqrt{x}(\sqrt{x} + 1)^2}\,dx$; let $u = \sqrt{x}$

30. $\int \frac{1}{x[1 - (\ln x)^2]^2}\,dx$; let $u = \ln x$

31. (a) Express

$$\frac{1}{a^2 - x^2}$$

as a sum of partial fractions.

(b) Use the result of part (a) to show that

$$\int \frac{1}{a^2 - x^2}\, dx = \frac{1}{2a} \ln \left| \frac{a + x}{a - x} \right| + C$$

32. Find the area of the region bounded by

$$y = \frac{7}{16 - x^2}$$

and $y = 1$.

33. Referring to the Introductory Example of this section use integration by partial fractions to show that

$$y = \frac{10}{1 + 9e^{-kt}}$$

is a solution to the equation

$$\int \frac{1}{y(10 - y)}\, dy = \int k\, dt$$

(Note: In order to solve for the constant of integration, use the condition that $y = 1$ when $t = 0$.)

Ⓒ**34.** A conservation organization releases 100 animals of an endangered species into a game preserve. The organization believes that the preserve has a carrying capacity of 1000 of these animals and that the growth of the herd will be logistic. That is, the size of the herd will follow the equation

$$\int \frac{1}{y(1000 - y)}\, dy = \int k\, dt$$

Find this logistics curve. (In order to solve for the constant of integration C and the proportionality constant k, assume that $y = 100$ when $t = 0$ and $y = 134$ when $t = 2$.) Sketch the graph of your solution.

35. A single infected individual enters a community of n individuals susceptible to the disease. Let x be the number of newly infected individuals after time t. The common *Epidemic Model* assumes that the disease spreads at a rate proportional to the product of the total number infected and the number of susceptible not yet infected. Thus,

$$\frac{dx}{dt} = k(x + 1)(n - x)$$

and we obtain

$$\int \frac{1}{(x + 1)(n - x)}\, dx = \int k\, dt$$

Solve for x as a function of t.

Ⓒ Calculator may be helpful.

Section 6.4

Integration by Tables

Introductory Example

Total Revenue as a Function of Time

In several applications in this text, we have considered the total revenue function

$$R = xp$$

where x is the number of units sold at a price of p dollars per unit. Another way to look at revenue is as a function of time.

As a case in point, suppose that a company is beginning to market a new type of home care product. Marketing tests indicate that this product should eventually be able to capture 4% of the national market. If the national market is 12,000,000 units per year, this means that the company expects its annual sales to be approaching 480,000 units per year. Of course, since the product is unknown when it is first introduced, initial sales are expected to come in at a lower rate. Suppose the projected revenue rate function is

$$\frac{dR}{dt} = 480{,}000\left(1 - \frac{1}{\sqrt{t^2 + 1}}\right)$$

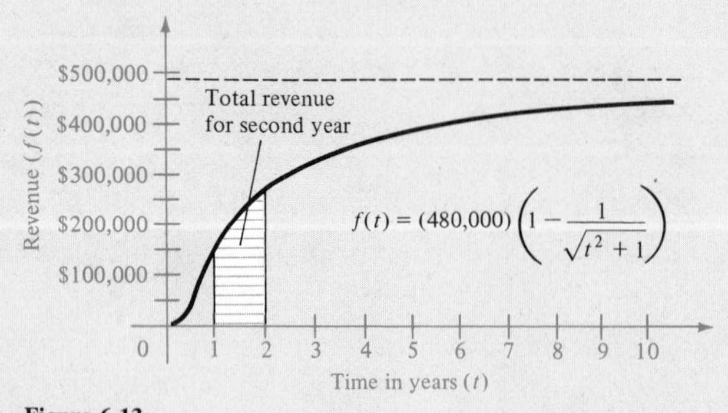

Figure 6.13

where dR/dt is measured in dollars per year and t is the time in years. (See Figure 6.13.) What is the total revenue during the 1st year? During the 2nd year? These questions can be answered by integration. For instance, the total revenue during the 1st year is

$$\begin{aligned} R &= \int_0^1 480{,}000\left(1 - \frac{1}{\sqrt{t^2 + 1}}\right) dt \\ &= 480{,}000\left[t - \ln\left|t + \sqrt{t^2 + 1}\right|\right]_0^1 \\ &\approx \$56{,}900 \end{aligned}$$

Note that the antiderivative of

$$\frac{1}{\sqrt{t^2 + 1}}$$

was obtained from Formula 27 of the **integration tables** listed in this section. Table 6.2 lists the projected total revenue for the first 10 years of sales. Note that the annual sales are approaching $480,000.

Table 6.2

Year	Total Revenue ($)
1	56,900
2	210,100
3	300,100
4	347,400
5	375,500
6	393,900
7	406,900
8	416,900
9	423,900
10	429,700

Section Objectives: *To introduce the technique of integration by tables ▪ To derive and demonstrate the use of reduction formulas ▪ To expand the application of integration by tables by the technique of completing the square.*

So far in this chapter, we have discussed a number of integration techniques to use along with the fundamental integration formulas. Certainly we have not considered every possible method for finding an antiderivative, but we have considered some of the most important ones.

In this section, we expand our list of integration formulas to form a table of integrals. As we add new integration formulas to our basic list, two things occur. On the one hand, it becomes inceasingly more difficult to memorize, or even become familiar with, the entire list of formulas. On the other hand, with a longer list we will likely need fewer techniques for fitting an integral to one of the formulas on the list. We call the procedure of integrating by means of a long unmemorized list of formulas **integration by tables.**

Integration by tables is not to be considered a trivial task. It requires considerable thought and insight, and it often involves a substitution procedure. Many persons find a table of integrals to be a valuable supplement to the integration techniques discussed in previous sections of this chapter. As you continue to improve in the use of the various integrating techniques, we encourage you to gain competence in the use of the table of integrals as well. In doing so you should find that a combination of techniques and tables is the most versatile approach to integration.

To assist you in using the table of integrals, we have grouped the formulas according to the form of the integrand as follows:

Forms involving u^n	Forms involving $a + bu$
Forms involving $\sqrt{a + bu}$	Forms involving $(u^2 - a^2)$
Forms involving $\sqrt{u^2 \pm a^2}$	Forms involving $\sqrt{a^2 - u^2}$
Forms involving e^u	Forms involving $\ln u$

Tables of Integrals

Forms Involving u^n

1. $\displaystyle\int u^n u' \, dx = \frac{u^{n+1}}{n+1} + C, \; n \neq -1$

2. $\displaystyle\int \frac{u'}{u} \, dx = \ln|u| + C$

Forms Involving $a + bu$

3. $$\int \frac{uu'}{a + bu}\,dx = \frac{1}{b^2}[bu - a\ln|a + bu|] + C$$

4. $$\int \frac{uu'}{(a + bu)^2}\,dx = \frac{1}{b^2}\left[\frac{a}{a + bu} + \ln|a + bu|\right] + C$$

5. $$\int \frac{uu'}{(a + bu)^n}\,dx = \frac{1}{b^2}\left[\frac{-1}{(n - 2)(a + bu)^{n-2}} + \frac{a}{(n - 1)(a + bu)^{n-1}}\right] + C,\ n \neq 1, 2$$

6. $$\int \frac{u^2u'}{a + bu}\,dx = \frac{1}{b^3}\left[-\frac{bu}{2}(2a - bu) + a^2\ln|a + bu|\right] + C$$

7. $$\int \frac{u^2u'}{(a + bu)^2}\,dx = \frac{1}{b^3}\left[bu - \frac{a^2}{a + bu} - 2a\ln|a + bu|\right] + C$$

8. $$\int \frac{u^2u'}{(a + bu)^3}\,dx = \frac{1}{b^3}\left[\frac{2a}{a + bu} - \frac{a^2}{2(a + bu)^2} + \ln|a + bu|\right] + C$$

9. $$\int \frac{u^2u'}{(a + bu)^n}\,dx = \frac{1}{b^3}\left[\frac{-1}{(n - 3)(a + bu)^{n-3}} + \frac{2a}{(n - 2)(a + bu)^{n-2}} - \frac{a^2}{(n - 1)(a + bu)^{n-1}}\right] + C,\ n \neq 1, 2, 3$$

10. $$\int \frac{u'}{u(a + bu)}\,dx = \frac{1}{a}\ln\left|\frac{u}{a + bu}\right| + C$$

11. $$\int \frac{u'}{u(a + bu)^2}\,dx = \frac{1}{a}\left[\frac{1}{a + bu} + \frac{1}{a}\ln\left|\frac{u}{a + bu}\right|\right] + C$$

12. $$\int \frac{u'}{u^2(a + bu)}\,dx = -\frac{1}{a}\left[\frac{1}{u} + \frac{b}{a}\ln\left|\frac{u}{a + bu}\right|\right] + C$$

13. $$\int \frac{u'}{u^2(a + bu)^2}\,dx = -\frac{1}{a^2}\left[\frac{a + 2bu}{u(a + bu)} + \frac{2b}{a}\ln\left|\frac{u}{a + bu}\right|\right] + C$$

Forms Involving $\sqrt{a + bu}$

14. $$\int u^n\sqrt{a + bu}\,u'\,dx = \frac{2}{b(2n + 3)}\left[u^n(a + bu)^{3/2} - na\int u^{n-1}\sqrt{a + bu}\,u'\,dx\right]$$

15. $$\int \frac{u'}{u\sqrt{a + bu}}\,dx = \frac{1}{\sqrt{a}}\ln\left|\frac{\sqrt{a + bu} - \sqrt{a}}{\sqrt{a + bu} + \sqrt{a}}\right| + C,\ 0 < a$$

16. $$\int \frac{u'}{u^n\sqrt{a + bu}}\,dx = \frac{-1}{a(n - 1)}\left[\frac{\sqrt{a + bu}}{u^{n-1}} + \frac{(2n - 3)b}{2}\int \frac{u'}{u^{n-1}\sqrt{a + bu}}\,dx\right],\quad n \neq 1$$

17. $\displaystyle\int \frac{\sqrt{a+bu}}{u}\,u'\,dx = 2\sqrt{a+bu} + a\int \frac{u'}{u\sqrt{a+bu}}\,dx$

18. $\displaystyle\int \frac{\sqrt{a+bu}}{u^n}\,u'\,dx = \frac{-1}{a(n-1)}\left[\frac{(a+bu)^{3/2}}{u^{n-1}} + \frac{(2n-5)b}{2}\int \frac{\sqrt{a+bu}}{u^{n-1}}\,u'\,dx\right],\quad n \neq 1$

19. $\displaystyle\int \frac{uu'}{\sqrt{a+bu}}\,dx = \frac{2(2a-bu)}{3b^2}\sqrt{a+bu} + C$

20. $\displaystyle\int \frac{u^n u'}{\sqrt{a+bu}}\,dx = \frac{2}{(2n+1)b}\left[u^n\sqrt{a+bu} - na\int \frac{u^{n-1}}{\sqrt{a+bu}}\,u'\,dx\right]$

Forms Involving $u^2 - a^2$, $0 < a$

21. $\displaystyle\int \frac{u'}{u^2-a^2}\,dx = -\int \frac{u'}{a^2-u^2}\,dx = \frac{1}{2a}\ln\left|\frac{u-a}{u+a}\right| + C$

22. $\displaystyle\int \frac{u'}{(u^2-a^2)^n}\,dx = \frac{-1}{2a^2(n-1)}\left[\frac{u}{(u^2-a^2)^{n-1}} + (2n-3)\int \frac{u'}{(u^2-a^2)^{n-1}}\,dx\right],\ n \neq 1$

Forms Involving $\sqrt{u^2 \pm a^2}$, $0 < a$

23. $\displaystyle\int \sqrt{u^2 \pm a^2}\,u'\,dx = \frac{1}{2}[u\sqrt{u^2 \pm a^2} \pm a^2\ln|u + \sqrt{u^2 \pm a^2}|] + C$

24. $\displaystyle\int u^2\sqrt{u^2 \pm a^2}\,u'\,dx = \frac{1}{8}[u(2u^2 \pm a^2)\sqrt{u^2 \pm a^2} - a^4\ln|u + \sqrt{u^2 \pm a^2}|] + C$

25. $\displaystyle\int \frac{\sqrt{u^2+a^2}}{u}\,u'\,dx = \sqrt{u^2+a^2} - a\ln\left|\frac{a+\sqrt{u^2+a^2}}{u}\right| + C$

26. $\displaystyle\int \frac{\sqrt{u^2 \pm a^2}}{u^2}\,u'\,dx = \frac{-\sqrt{u^2 \pm a^2}}{u} + \ln|u + \sqrt{u^2 \pm a^2}| + C$

27. $\displaystyle\int \frac{u'}{\sqrt{u^2 \pm a^2}}\,dx = \ln|u + \sqrt{u^2 \pm a^2}| + C$

28. $\displaystyle\int \frac{u'}{u\sqrt{u^2+a^2}}\,dx = \frac{-1}{a}\ln\left|\frac{a+\sqrt{u^2+a^2}}{u}\right| + C$

29. $\displaystyle\int \frac{u^2u'}{\sqrt{u^2 \pm a^2}}\,dx = \frac{1}{2}[u\sqrt{u^2 \pm a^2} \mp a^2\ln|u + \sqrt{u^2 \pm a^2}|] + C$

30. $\displaystyle\int \frac{u'}{u^2\sqrt{u^2 \pm a^2}}\,dx = \mp \frac{\sqrt{u^2 \pm a^2}}{a^2 u} + C$

31. $\displaystyle\int \frac{u'}{(u^2 \pm a^2)^{3/2}}\,dx = \frac{\pm u}{a^2\sqrt{u^2 \pm a^2}} + C$

Forms Involving $\sqrt{a^2 - u^2}$, $0 < a$

32. $\displaystyle\int \frac{\sqrt{a^2 - u^2}}{u}\,u'\,dx = \sqrt{a^2 - u^2} - a \ln \left|\frac{a + \sqrt{a^2 - u^2}}{u}\right| + C$

33. $\displaystyle\int \frac{u'}{u\sqrt{a^2 - u^2}}\,dx = \frac{-1}{a} \ln \left|\frac{a + \sqrt{a^2 - u^2}}{u}\right| + C$

34. $\displaystyle\int \frac{u'}{u^2\sqrt{a^2 - u^2}}\,dx = \frac{-\sqrt{a^2 - u^2}}{a^2 u} + C$

35. $\displaystyle\int \frac{u'}{(a^2 - u^2)^{3/2}}\,dx = \frac{u}{a^2\sqrt{a^2 - u^2}} + C$

Forms Involving e^u

36. $\displaystyle\int e^u u'\,dx = e^u + C$

37. $\displaystyle\int u e^u u'\,dx = (u - 1)e^u + C$

38. $\displaystyle\int u^n e^u u'\,dx = u^n e^u - n \int u^{n-1} e^u u'\,dx$

39. $\displaystyle\int \frac{u'}{1 + e^u}\,dx = u - \ln(1 + e^u) + C$

40. $\displaystyle\int \frac{u'}{1 + e^{nu}}\,dx = u - \frac{1}{n} \ln(1 + e^{nu}) + C$

Forms Involving $\ln u$

41. $\displaystyle\int (\ln u)u'\,dx = u[-1 + \ln u] + C$

42. $\displaystyle\int u(\ln u)u'\,dx = \frac{u^2}{4}[-1 + 2\ln u] + C$

43. $\displaystyle\int u^n(\ln u)u'\,dx = \frac{u^{n+1}}{(n+1)^2}[-1 + (n+1)\ln u] + C,\ n \neq -1$

44. $\displaystyle\int (\ln u)^2 u'\,dx = u[2 - 2\ln u + (\ln u)^2] + C$

45. $\displaystyle\int (\ln u)^n u'\,dx = u(\ln u)^n - n\int (\ln u)^{n-1} u'\,dx$

In the remaining examples of this section, we demonstrate the use of this table of integrals to find antiderivatives.

Example 1

Integration by Tables

Use the table of integrals to evaluate $\int x\sqrt{x^4 - 9}\,dx$.

Solution By considering $u = x^2$ and $a = 3$, we can use Formula 23.

$$\int \sqrt{u^2 - a^2}\,u'\,dx = \frac{1}{2}\left[u\sqrt{u^2 - a^2} - a^2 \ln\left|u + \sqrt{u^2 - a^2}\right|\right] + C$$

and write

$$\begin{aligned}\int x\sqrt{x^4 - 9}\,dx &= \frac{1}{2}\int \sqrt{(x^2)^2 - (3)^2}(2x)\,dx \\ &= \frac{1}{4}\left[x^2\sqrt{x^4 - 9} - 9\ln\left|x^2 + \sqrt{x^4 - 9}\right|\right] + C\end{aligned}$$

□

Example 2

Integration by Tables

Use the table of integrals to evaluate

$$\int \frac{dx}{x\sqrt{x+1}}$$

Solution Considering forms involving $\sqrt{a + bu}$, where $a = 1, b = 1$, and $u = x$, we choose Formula 15, which states that

$$\int \frac{u'}{u\sqrt{a + bu}}\,dx = \frac{1}{\sqrt{a}}\ln\left|\frac{\sqrt{a + bu} - \sqrt{a}}{\sqrt{a + bu} + \sqrt{a}}\right| + C, \qquad 0 < a$$

Therefore,

$$\int \frac{dx}{x\sqrt{x+1}} = \ln\left|\frac{\sqrt{x+1} - 1}{\sqrt{x+1} + 1}\right| + C$$

□

Example 3

Integration by Tables

Use the table of integrals to evaluate

$$\int \frac{x}{1 + e^{-x^2}}\, dx$$

Solution Of the forms involving e^u, we use Formula 39,

$$\int \frac{u'}{1 + e^u}\, dx = u - \ln(1 + e^u) + C$$

with $u = -x^2$ and $u' = -2x$. Thus, we obtain

$$\int \frac{x}{1 + e^{-x^2}}\, dx = -\frac{1}{2}\int \frac{-2x}{1 + e^{-x^2}}\, dx$$

$$= -\frac{1}{2}[-x^2 - \ln(1 + e^{-x^2})] + C$$

$$= \frac{1}{2}[x^2 + \ln(1 + e^{-x^2})] + C$$

□

Perhaps you have noticed that a number of the formulas in our table of integrals have the form

$$\int f(x)\, dx = g(x) + \int h(x)\, dx$$

where the right-hand member of the formula contains another integral. Such integration formulas are referred to as **reduction formulas,** since they reduce a given integral to the sum of a function and a simpler integral. We demonstrate the use of reduction formulas in the next two examples.

Example 4

Using Reduction Formulas

Use the table of integrals to evaluate $\int x^2 e^x\, dx$.

Solution We have Formula 38,

$$\int u^n e^u u'\, dx = u^n e^u - n\int u^{n-1} e^u u'\, dx$$

Thus, letting $u = x$ and $n = 2$, we have

$$\int x^2 e^x\, dx = x^2 e^x - 2\int x e^x\, dx$$

Then, from Formula 37,

$$\int u e^u u'\, dx = (u - 1)e^u + C$$

we have

$$\int x^2e^x\,dx = x^2e^x - 2(x-1)e^x + C$$
$$= e^x(x^2 - 2x + 2) + C$$

□

Example 5

Using Reduction Formulas

Use the table of integrals to evaluate

$$\int \frac{\sqrt{3-5x}}{2x}\,dx$$

Solution By Formula 17

$$\int \frac{\sqrt{a+bu}}{u}\,u'\,dx = 2\sqrt{a+bu} + a\int \frac{u'}{u\sqrt{a+bu}}\,dx$$

with $a = 3$, $b = -5$, and $u = x$, it follows that

$$\frac{1}{2}\int \frac{\sqrt{3-5x}}{x}\,dx = \frac{1}{2}\left[2\sqrt{3-5x} + 3\int \frac{dx}{x\sqrt{3-5x}}\right]$$
$$= \sqrt{3-5x} + \frac{3}{2}\int \frac{dx}{x\sqrt{3-5x}}$$

Now, by Formula 15, with $u = x$, $a = 3$, and $b = -5$, it follows that

$$\int \frac{\sqrt{3-5x}}{2x}\,dx = \sqrt{3-5x} + \frac{3}{2}\left[\frac{1}{\sqrt{3}}\ln\left|\frac{\sqrt{3-5x}-\sqrt{3}}{\sqrt{3-5x}+\sqrt{3}}\right|\right] + C$$
$$= \sqrt{3-5x} + \frac{\sqrt{3}}{2}\ln\left|\frac{\sqrt{3-5x}-\sqrt{3}}{\sqrt{3-5x}+\sqrt{3}}\right| + C$$

□

Several of the integration formulas listed in our table involve the sum or difference of two squares. We can *extend* the application of these formulas by the use of an algebraic technique called **completing the square.** This technique provides a means for writing any quadratic polynomial as the sum or difference of two squares. For instance, the polynomial $x^2 + bx + c$ can be written as follows:

$$x^2 + bx + c = x^2 + bx + \left(\frac{b}{2}\right)^2 - \left(\frac{b}{2}\right)^2 + c$$
$$= \left(x + \frac{b}{2}\right)^2 + \left[c - \left(\frac{b}{2}\right)^2\right]$$

Thus, we have written $x^2 + bx + c$ in the form $u^2 \pm a^2$, where

$$u = x + \frac{b}{2}$$

Example 6

Completing the Square

Evaluate

$$\int \frac{dx}{x^2 - 4x + 1}$$

Solution By completing the square, we obtain

$$x^2 - 4x + 1 = (x^2 - 4x + 4) - 4 + 1 = (x - 2)^2 - 3$$

Therefore,

$$\int \frac{dx}{x^2 - 4x + 1} = \int \frac{dx}{(x - 2)^2 - 3}$$

Considering $u = x - 2$ and $a = \sqrt{3}$, we apply Formula 21,

$$\int \frac{u'}{u^2 - a^2}\, dx = \frac{1}{2a} \ln \left| \frac{u - a}{u + a} \right| + C$$

to conclude that

$$\int \frac{dx}{x^2 - 4x + 1} = \frac{1}{2\sqrt{3}} \ln \left| \frac{x - 2 - \sqrt{3}}{x - 2 + \sqrt{3}} \right| + C$$

□

Example 7

Completing the Square

Evaluate

$$\int \frac{dx}{\sqrt{x^2 + 2x}}$$

Solution By completing the square we obtain

$$\int \frac{dx}{\sqrt{x^2 + 2x}} = \int \frac{dx}{\sqrt{(x^2 + 2x + 1) - 1}} = \int \frac{dx}{\sqrt{(x + 1)^2 - 1}}$$

Now, letting $u = x + 1$ and $a = 1$, we apply Formula 27,

$$\int \frac{u'}{\sqrt{u^2 - a^2}}\, dx = \ln \left| u + \sqrt{u^2 - a^2} \right| + C$$

to obtain

$$\int \frac{1}{\sqrt{x^2 + 2x}}\, dx = \ln \left| (x + 1) + \sqrt{(x + 1)^2 - 1} \right| + C = \ln \left| (x + 1) + \sqrt{x^2 + 2x} \right| + C$$

□

Section Exercises (6.4)

In Exercises 1–30, use the table of integrals in this section to find the indefinite integral.

1. $\int \frac{1}{x(1+x)}\,dx$

2. $\int \frac{1}{x(1+x)^2}\,dx$

3. $\int \frac{1}{x\sqrt{x^2+1}}\,dx$

4. $\int \frac{1}{\sqrt{x^2-1}}\,dx$

5. $\int x \ln x\,dx$

6. $\int \frac{1}{1+e^x}\,dx$

7. $\int \frac{1}{x\sqrt{4-x^2}}\,dx$

8. $\int \frac{\sqrt{x^2-9}}{x^2}\,dx$

9. $\int \frac{e^x}{e^{2x}(1+e^x)}\,dx$

10. $\int \frac{x}{x^4-9}\,dx$

11. $\int x\sqrt{x^4-9}\,dx$

12. $\int \frac{(\ln \sqrt{x})^2}{\sqrt{x}}\,dx$

13. $\int \frac{t^2}{(2+3t)^3}\,dt$

14. $\int \frac{\sqrt{3+4t}}{t}\,dt$

15. $\int \frac{s}{s^2\sqrt{3+s}}\,ds$

16. $\int \sqrt{3+x^2}\,dx$

17. $\int \frac{x^2}{1+x}\,dx$

18. $\int \frac{1}{x^2\sqrt{x^2-4}}\,dx$

19. $\int \frac{1}{x^2\sqrt{1-x^2}}\,dx$

20. $\int xe^{x^2}\,dx$

21. $\int x^2 \ln x\,dx$

22. $\int \frac{2x}{(1-3x)^2}\,dx$

23. $\int \frac{x^2}{(3x-5)^2}\,dx$

24. $\int \frac{1}{2x^2(2x-1)^2}\,dx$

25. $\int \frac{1}{1+e^{2x}}\,dx$

26. $\int \frac{1}{\sqrt{x}(1+2\sqrt{x})}\,dx$

27. $\int \frac{\ln x}{x(3+2\ln x)}\,dx$

28. $\int \frac{e^x}{(1-e^{2x})^{3/2}}\,dx$

29. $\int (2x-3)^2\sqrt{(2x-3)^2+4}\,dx$

30. $\int (\ln x)^3\,dx$

31. Express each polynomial as the sum or difference of squares.
(a) x^2+6x (b) x^2-8x+9
(c) x^4+2x^2-5

32. Express each polynomial as the sum or difference of squares.
(a) $2x^2+12x+14$ (b) $3x^2-12x-9$
(c) x^2-2x

In Exercises 33–40, complete the square and then use the table of integrals in this section to find the indefinite integral.

33. $\int \frac{1}{(x-1)^2-4}\,dx$

34. $\int \frac{1}{(x^2+4x-5)^{3/2}}\,dx$

35. $\int \frac{1}{(x-1)\sqrt{x^2-2x+2}}\,dx$

36. $\int \sqrt{x^2-6x}\,dx$

37. $\int \frac{1}{2x^2-4x-6}\,dx$

38. $\int \frac{\sqrt{7-6x-x^2}}{x+3}\,dx$

39. $\int \frac{x}{\sqrt{x^4+2x^2+2}}\,dx$

40. $\int \frac{x\sqrt{x^4+4x^2+5}}{x^2+2}\,dx$

41. Find the area of the region bounded by

$$y = \frac{x}{\sqrt{x+1}}$$

$y = 0$, and $x = 8$.

42. Find the area of the region bounded by

$$y = \frac{x}{1+e^{x^2}}$$

$y = 0$, and $x = 2$.

Ⓒ**43.** The growth in a bacterial colony is given by the logistic curve

$$N(t) = \frac{50}{1+e^{4.8-1.9t}}$$

Ⓒ Calculator may be helpful.

where t is the time in days. Find the average number in the colony during the third day. That is, find the average value of the function for t in the interval [3, 4].

Ⓒ**44.** The growth of a colony of fruit flies is given by the logistic curve

$$N(t) = \frac{375}{1 + e^{4.20 - 0.25t}}$$

where t is the time in days. Find the average number of flies in the colony for t in the interval [21, 28].

Ⓒ**45.** Find the consumer surplus and the producer surplus if for a given product the demand equation is

$$p = \frac{60}{\sqrt{q^2 + 81}}$$

and the supply equation is

$$p = \frac{q}{3}$$

(Note: For a definition of consumer and producer surplus, refer to Exercises 31–34 in Section 4.4.)

Ⓒ Calculator may be helpful.

Section 6.5

Numerical Integration

Introductory Example

Standard Deviation and the Normal Distribution

In Section 5.2, we looked at the normal probability density function

$$f(x) = \frac{1}{\sigma\sqrt{2\pi}} e^{-x^2/2\sigma^2}$$

where $x = 0$ represents the *mean* of the distribution and σ is the *standard deviation* from the mean. (σ is the lowercase Greek letter sigma.) One of the facts regarding this normal distribution is that approximately 68% of the area under this curve lies within one standard deviation from the mean. (See Figure 6.14.) This area is represented by the integral

$$\frac{1}{\sigma\sqrt{2\pi}} \int_{-\sigma}^{\sigma} e^{-x^2/2\sigma^2}\, dx$$

However, at this stage we know of no way to find an antiderivative for $e^{-x^2/2\sigma^2}$. In this section we will see that, when we cannot find an antiderivative for a certain definite integral, we can approximate its value by a procedure called **numerical integration.** Specifically, using **Simpson's Rule** (with $n = 10$), we have

$$\begin{aligned}\frac{1}{\sigma\sqrt{2\pi}} \int_{-\sigma}^{\sigma} e^{-x^2/2\sigma^2}\, dx &\approx \frac{2\sigma}{30}\left(\frac{1}{\sigma\sqrt{2\pi}}\right)[e^{-0.5} \\ &\quad + 4e^{-0.32} + 2e^{-0.18} \\ &\quad + 4e^{-0.08} + 2e^{-0.02} \\ &\quad + 4e^{0} + 2e^{-0.02} + 4e^{-0.08} \\ &\quad + 2e^{-0.18} + 4e^{-0.32} \\ &\quad + e^{-0.5}] \\ &\approx 0.6826 = 68.26\%\end{aligned}$$

Extending this procedure to find the area of the region lying within two, three, and four standard deviations from the mean, we have the results shown in Table 6.3.

Table 6.3

n	$\frac{1}{\sigma\sqrt{2\pi}} \int_{-n\sigma}^{n\sigma} e^{-x^2/2\sigma^2}\, dx$
1	68.26%
2	95.44%
3	99.74%
4	100.00%
Rounded to the nearest 0.01%	

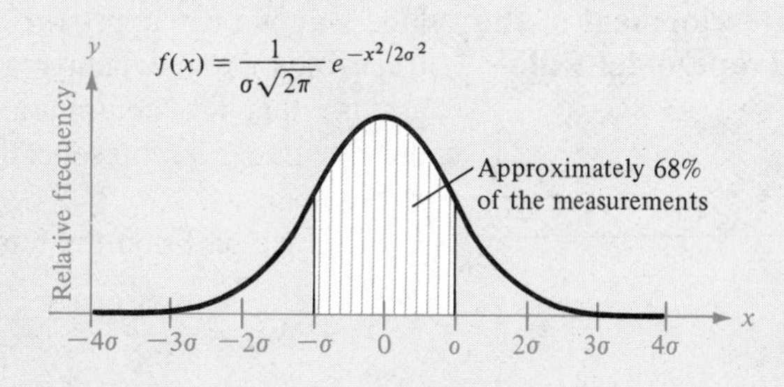

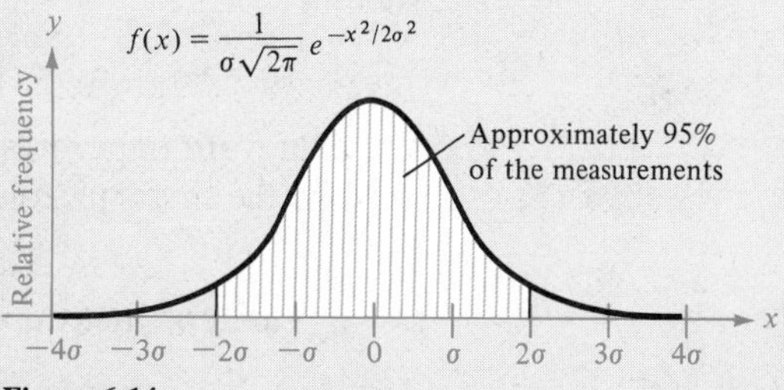

Figure 6.14

Section Objective: ***To introduce two numerical methods for approximating definite integrals: the Trapezoidal Rule and Simpson's Rule.***

When we began our discussion of integration techniques, we mentioned that the search for an antiderivative is not nearly as straightforward a process as is differentiation. Occasionally, we encounter functions for which we cannot find antiderivatives. Of course, inability to find an antiderivative for a particular function may be due to a lack of cleverness on our part. On the other hand, some elementary functions* simply do not possess antiderivatives that are elementary functions. For example, there are no elementary functions that have any one of the following functions as their derivative:

$$\sqrt{x}\,\sqrt{1-x} \qquad \sqrt{1-x^3} \qquad e^{-x^2}$$

$$\frac{1}{\sqrt{x}\,e^x} \qquad \frac{e^x}{x} \qquad \frac{1}{\ln x}$$

Up to this point we have been evaluating definite integrals by means of antiderivatives and the Fundamental Theorem of Calculus. However, if we wish to evaluate a definite integral involving a function whose antiderivative we cannot find, or which does not exist, the Fundamental Theorem cannot be applied, and we must resort to some other technique.

Development of the Trapezoidal Rule

One way we can approximate the definite integral $\int_a^b f(x)\,dx$ is by the use of n trapezoids as shown in Figure 6.15. In the development of this method, we assume that f is continuous and positive on the interval $[a, b]$ and that this definite integral represents the area of a region bounded by f and the x-axis, from $x = a$ to $x = b$.

First, we partition the interval $[a, b]$ into n equal subintervals, each of width

$$\Delta x = \frac{b-a}{n}$$

such that

$$a = x_0 < x_1 < x_2 < \cdots < x_{n-1} < x_n = b$$

We then form trapezoids for each subinterval, as shown in Figure 6.15.

Now, the first trapezoid (standing on end as shown in Figure 6.16) has an area of

$$\begin{aligned} \text{area of first trapezoid} &= (\text{average of two bases})\,(\text{height}) \\ &= \left[\frac{f(x_0) + f(x_1)}{2}\right]\left(\frac{b-a}{n}\right) \end{aligned}$$

*An *elementary function* is one that is formed from algebraic, exponential, logarithmic, trigonometric, or inverse trigonometric functions.

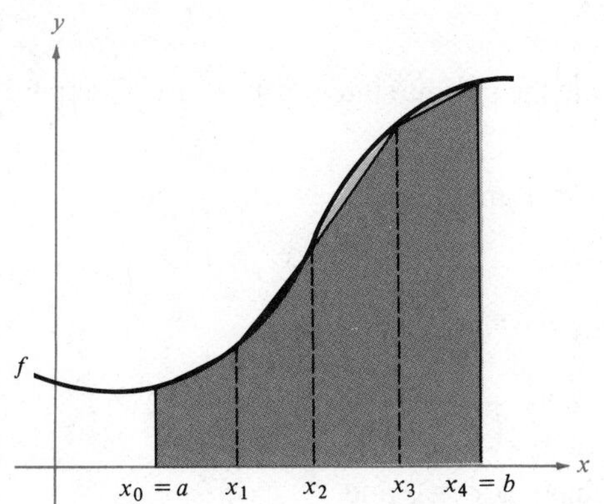

Figure 6.15

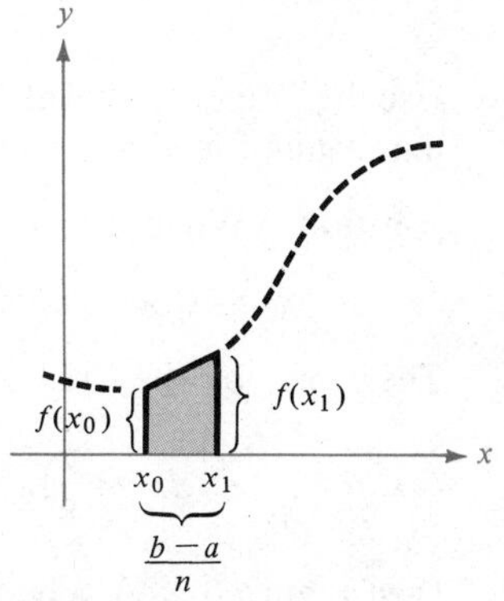

y, x, $f(x_0)$, $f(x_1)$, x_0, x_1, $\frac{b-a}{n}$

Figure 6.16

Extending this procedure to all n trapezoids, we have

$$\text{area of first trapezoid} = \left[\frac{f(x_0) + f(x_1)}{2}\right]\left(\frac{b-a}{n}\right)$$

$$\text{area of second trapezoid} = \left[\frac{f(x_1) + f(x_2)}{2}\right]\left(\frac{b-a}{n}\right)$$

$$\vdots$$

$$\text{area of } n\text{th trapezoid} = \left[\frac{f(x_{n-1}) + f(x_n)}{2}\right]\left(\frac{b-a}{n}\right)$$

Finally, the sum of the areas of the n trapezoids is

$$\left(\frac{b-a}{n}\right)\left[\frac{f(x_0) + f(x_1)}{2} + \frac{f(x_1) + f(x_2)}{2} + \cdots + \frac{f(x_{n-1}) + f(x_n)}{2}\right]$$

$$= \left(\frac{b-a}{2n}\right)[f(x_0) + \underbrace{f(x_1) + f(x_1)}_{2f(x_1)} + \underbrace{f(x_2) + \cdots}_{2f(x_2)} + \underbrace{\cdots + f(x_{n-1})}_{2f(x_{n-1})} + f(x_n)]$$

$$= \left(\frac{b-a}{2n}\right)[f(x_0) + 2f(x_1) + 2f(x_2) + \cdots + 2f(x_{n-1}) + f(x_n)]$$

Thus, we arrive at the following approximation, which we call the Trapezoidal Rule.

Trapezoidal Rule

$$\int_a^b f(x)\,dx \approx \frac{b-a}{2n}[f(x_0) + 2f(x_1) + 2f(x_2) + \cdots + 2f(x_{n-1}) + f(x_n)]$$

Example 1

Approximation with the Trapezoidal Rule

Use the Trapezoidal Rule to approximate the definite integral $\int_0^1 e^x\,dx$. Compare the results for $n = 4$ and $n = 8$.

Solution When $n = 4$, we have $\Delta x = \frac{1}{4}$ and

$$x_0 = 0, \quad x_1 = \tfrac{1}{4}, \quad x_2 = \tfrac{1}{2}, \quad x_3 = \tfrac{3}{4}, \quad x_4 = 1$$

Therefore, by the Trapezoidal Rule we have

$$\int_0^1 e^x\,dx \approx \frac{1}{8}[e^0 + 2e^{0.25} + 2e^{0.5} + 2e^{0.75} + e^1] \approx 1.7272$$

(See Figure 6.17.) When $n = 8$, we have $\Delta x = \frac{1}{8}$ and

$$x_0 = 0, \quad x_1 = \tfrac{1}{8}, \quad x_2 = \tfrac{1}{4}, \quad x_3 = \tfrac{3}{8}, \quad x_4 = \tfrac{1}{2}$$

$$x_5 = \tfrac{5}{8}, \quad x_6 = \tfrac{3}{4}, \quad x_7 = \tfrac{7}{8}, \quad x_8 = 1$$

and it follows that

$$\begin{aligned}\int_0^1 e^x\,dx &\approx \frac{1}{16}[e^0 + 2e^{0.125} + 2e^{0.25} + 2e^{0.375} + 2e^{0.5} + 2e^{0.625} \\ &\qquad + 2e^{0.75} + 2e^{0.875} + e^1] \\ &\approx 1.7205\end{aligned}$$

(See Figure 6.18.) Of course, *for this particular example* we could have found an antiderivative and determined that the exact area of the region is

$$e - 1 \approx 1.7183$$

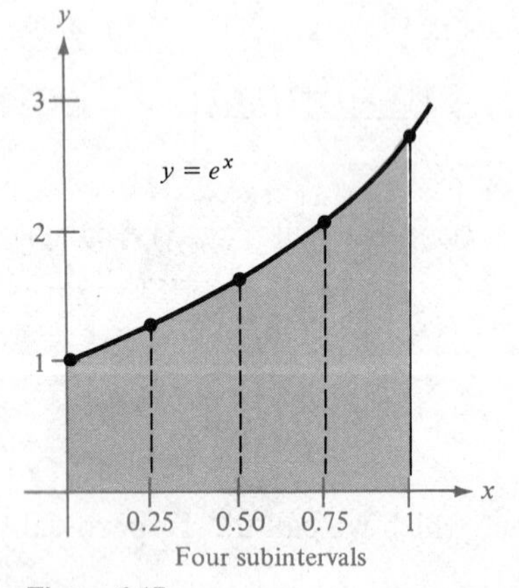

Figure 6.17

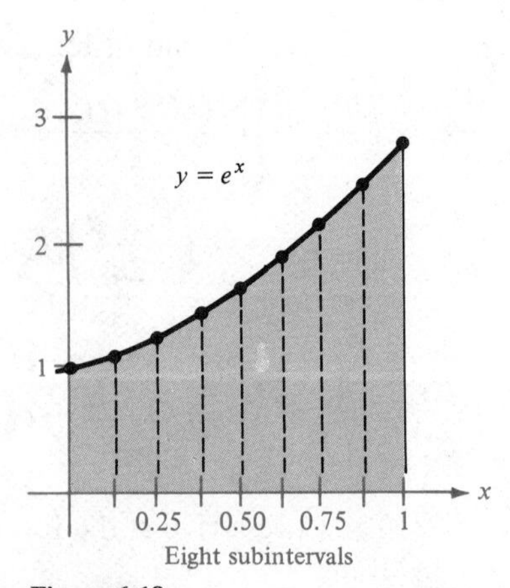

Figure 6.18

□

Remark

Although you may not yet be excited about using a rather lengthy approximation method to evaluate the integral $\int_0^1 e^x\,dx$, two important points must be brought to your attention. First, this approximation method, using trapezoids, becomes more accurate as n increases. (For $n = 16$ in Example 1, the Trapezoidal Rule yields an approximation of 1.7188.) Second, though we could have used the Fundamental Theorem to evaluate the integral in Example 1, this theorem cannot be used to evaluate an integral so simple looking as $\int_0^1 e^{x^2}\,dx$, because e^{x^2} has no elementary antiderivative. Yet the Trapezoidal Rule can be readily applied to this integral.

One way to view this trapezoidal approximation of a definite integral is to say that on each subinterval we approximate f by a *first*-degree polynomial. In Simpson's Rule, which follows, we carry this procedure one step further and approximate f by *second*-degree polynomials.

Before presenting Simpson's Rule, we give the following theorem for evaluating integrals of second-degree polynomials.

Theorem 6.1 If $p(x) = Ax^2 + Bx + C$, then

$$\int_a^b p(x)\,dx = \left(\frac{b-a}{6}\right)\left[p(a) + 4p\left(\frac{a+b}{2}\right) + p(b)\right]$$

Proof Although the proof of this theorem is logically straightforward, it is quite messy in terms of the algebra involved. Thus, for the sake of brevity, we merely give an outline for the proof; we encourage you to supply the missing algebraic steps.

$$\begin{aligned}
&\int_a^b (Ax^2 + Bx + C)\,dx \\
&\quad = \left[\frac{Ax^3}{3} + \frac{Bx^2}{2} + Cx\right]_a^b \\
&\quad = \frac{A(b^3 - a^3)}{3} + \frac{B(b^2 - a^2)}{2} + C(b - a) \\
&\quad = \left(\frac{b-a}{6}\right)[2A(a^2 + ab + b^2) + 3B(b + a) + 6C] \\
&\quad = \frac{b-a}{6}\left[(Aa^2 + Ba + C) + 4\left\{A\left(\frac{b+a}{2}\right)^2 + B\left(\frac{b+a}{2}\right) + C\right\}\right. \\
&\qquad\qquad \left. + (Ab^2 + Bb + C)\right] \\
&\quad = \frac{b-a}{6}\left[p(a) + 4p\left(\frac{a+b}{2}\right) + p(b)\right]
\end{aligned}$$

□

Development of Simpson's Rule

To develop Simpson's Rule for approximating the value of the definite integral $\int_a^b f(x)\,dx$, we again partition the interval $[a, b]$ into n equal parts, each of width

$$\frac{b-a}{n}$$

However, this time we require n to be even and then group the subintervals into pairs such that

$$a = \underbrace{x_0 < x_1 < x_2}_{[x_0, x_2]} < \underbrace{x_3 < x_4}_{[x_2, x_4]} < \ldots < \underbrace{x_{n-2} < x_{n-1} < x_n}_{[x_{n-2}, x_n]} = b$$

Then on the subinterval $[x_0, x_2]$, we approximate $y = f(x)$ by the second-degree polynomial p that passes through the points

$$(x_0, y_0), \qquad (x_1, y_1), \qquad (x_2, y_2)$$

as shown in Figure 6.19. Now, using p as an approximation for f, we have

$$\begin{aligned}
\int_{x_0}^{x_2} f(x)\,dx &\approx \int_{x_0}^{x_2} p(x)\,dx \\
&= \frac{x_2 - x_0}{6}\left[p(x_0) + 4p\left(\frac{x_2 + x_0}{2}\right) + p(x_2)\right] \\
&= \frac{2[(b-a)/n]}{6}[p(x_0) + 4p(x_1) + p(x_2)] \\
&= \frac{b-a}{3n}[f(x_0) + 4f(x_1) + f(x_2)]
\end{aligned}$$

Repeating this procedure on each subinterval $[x_{i-2}, x_i]$ of interval $[a, b]$ results in the formula

$$\int_a^b f(x)\,dx \approx \frac{b-a}{3n}\{[f(x_0) + 4f(x_1) + f(x_2)] + [f(x_2) + 4f(x_3) + f(x_4)] + \cdots + [f(x_{n-2}) + 4f(x_{n-1}) + f(x_n)]\}$$

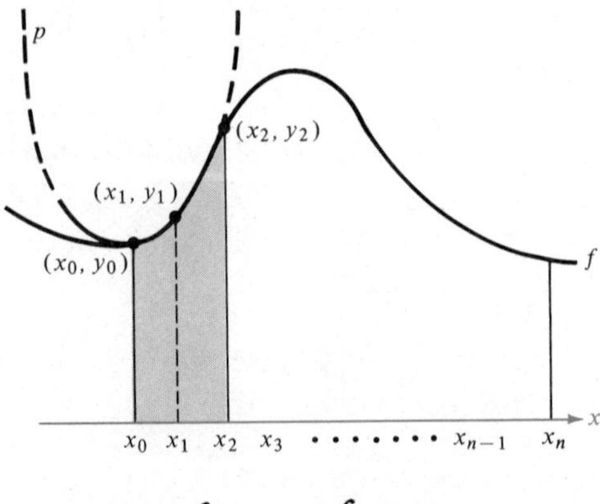

Figure 6.19

By grouping like terms, we obtain the following approximation formula, known as *Simpson's Rule*.

Simpson's Rule (n is even)

$$\int_a^b f(x)\,dx \approx \frac{b-a}{3n}[f(x_0) + 4f(x_1) + 2f(x_2) + 4f(x_3) + \cdots + 4f(x_{n-1}) + f(x_n)]$$

In Example 1, we used the Trapezoidal Rule to estimate $\int_0^1 e^x\,dx$. In the next example, we see how well Simpson's Rule does for the same integral.

Example 2

Approximation with Simpson's Rule

Use Simpson's Rule to approximate the integral $\int_0^1 e^x\,dx$. Compare the results for $n = 4$ and $n = 8$.

Solution When $n = 4$, we have

$$\int_0^1 e^x\,dx \approx \frac{1}{12}[e^0 + 4e^{0.25} + 2e^{0.5} + 4e^{0.75} + e^1] \approx 1.71831884$$

When $n = 8$, we have

$$\int_0^1 e^x\,dx \approx \frac{1}{24}[e^0 + 4e^{0.125} + 2e^{0.25} + 4e^{0.375} + 2e^{0.5} + 4e^{.625} + 2e^{0.75} + 4e^{0.875} + e^1] \approx 1.718284154$$

Recall that the actual value of this integral is

$$e - 1 \approx 1.718281828$$

Thus, with only eight subintervals we have obtained an approximation that is correct to the nearest 0.00001 — an impressive result! □

In Examples 1 and 2, we were able to calculate the exact value of the integral and compare that to our approximations to see how good they were. Of course, in practice we would not bother with an approximation if it were possible to evaluate the integral exactly. However, if it is necessary to use an approximation technique to determine the value of a definite integral, then it is important to know how good we can expect our approximation to be. The following theorem, which we list without proof, gives the formulas for estimating the error involved in the use of the Trapezoidal and Simpson's Rules.

Theorem 6.2 indicates that the errors generated by the Trapezoidal and Simpson's Rules have upper bounds dependent upon the extreme values of $f''(x)$ and $f^{(4)}(x)$ respectively, in the interval $[a, b]$. Furthermore, it is evident from Theorem 6.2 that the bounds for these errors can be made arbitrarily small by *increasing n*, provided, of course, that f'' and $f^{(4)}$ exist and are bounded in

Theorem 6.2
Error in Trapezoidal and Simpson's Rules

The error E in approximating $\int_a^b f(x)\,dx$ is for the Trapezoidal Rule:

$$|E| \le \frac{(b-a)^3}{12n^2}\,[\max|f''(x)|], \qquad a \le x \le b$$

and for Simpson's Rule:

$$|E| \le \frac{(b-a)^5}{180n^4}\,[\max|f^{(4)}(x)|], \qquad a \le x \le b$$

$[a, b]$. The next examples show how to determine a value of n that will bound the error within a predetermined tolerance interval.

Example 3

Approximating the Error in the Trapezoidal Rule

Using the Trapezoidal Rule, estimate the value of $\int_0^1 e^{-x^2}\,dx$. Determine n so that the approximation error is less than 0.01.

Solution If $f(x) = e^{-x^2}$, then

$$f'(x) = -2xe^{-x^2}$$

$$f''(x) = 4x^2e^{-x^2} - 2e^{-x^2} = 2e^{-x^2}(2x^2 - 1)$$

By the methods of differential calculus, we can determine that f'' has only one critical value ($x = 0$) in the interval $[0, 1]$ and that the maximum value of $|f''(x)|$ on this interval is $|f''(0)| = 2$. Thus, by Theorem 6.2, we have

$$|E| \le \frac{(b-a)^3}{12n^2}(2) = \frac{1}{12n^2}(2) = \frac{1}{6n^2}$$

To ensure that our approximation has an error less than 0.01, we must choose n so that

$$\frac{1}{6n^2} \le 0.01 = \frac{1}{100}$$

$$100 \le 6n^2$$

$$\frac{50}{3} \le n^2$$

$$4.08 \approx \sqrt{\frac{50}{3}} = n$$

Therefore, we choose $n = 5$ (since n must be greater than or equal to 4.08) and apply the Trapezoidal Rule to obtain

$$\int_0^1 e^{-x^2}\,dx \approx \frac{1}{10}\left(\frac{1}{e^0} + \frac{2}{e^{0.04}} + \frac{2}{e^{0.16}} + \frac{2}{e^{0.36}} + \frac{2}{e^{0.64}} + \frac{1}{e^1}\right)$$

$$\approx 0.744$$

(See Figure 6.20.) Finally, with an error no larger than 0.01, we know that

$$0.734 \le \int_0^1 e^{-x^2}\,dx \le 0.754$$

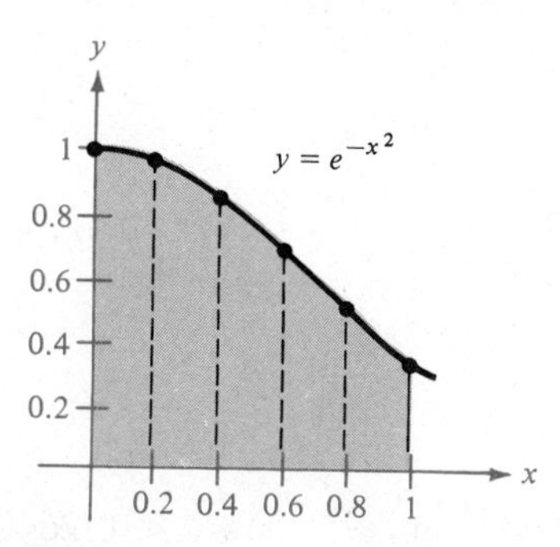

Figure 6.20

□

Example 4

Approximating the Error in Simpson's Rule

Use Simpson's Rule to estimate the definite integral $\int_0^1 \ln(x^2 + 1)\,dx$. Determine n so that the approximation error is less than 0.0001.

Solution According to Theorem 6.2, the error in Simpson's Rule involves the fourth derivative. Hence, by successive differentiations we have

$$f(x) = \ln(x^2 + 1)$$

$$f'(x) = \frac{2x}{x^2 + 1}$$

$$f''(x) = \frac{-2(x^2 - 1)}{(x^2 + 1)^2}$$

$$f'''(x) = \frac{4x(x^2 - 3)}{(x^2 + 1)^3}$$

$$f^{(4)}(x) = \frac{-12(x^4 - 6x^2 + 1)}{(x^2 + 1)^4}$$

On the interval $[0, 1]$, $|f^{(4)}(x)|$ has a maximum value of 12. Thus, we know that

$$|E| \le \frac{(b - a)^5}{180n^4}(12) = \frac{1}{15n^4}$$

Now, by setting this value less than 0.0001, we have

$$\frac{1}{15n^4} < 0.0001$$

and we find $n = 6$ satisfies this inequality. Finally, we calculate the approximation:

$$\int_0^1 \ln(x^2 + 1)\,dx \approx \frac{1}{18}\left[\ln(1) + 4\ln\left(\frac{1}{36} + 1\right) + 2\ln\left(\frac{4}{36} + 1\right) + 4\ln\left(\frac{9}{36} + 1\right) + 2\ln\left(\frac{16}{36} + 1\right) + 4\ln\left(\frac{25}{36} + 1\right) + \ln(2)\right]$$

$$\approx 0.26394$$

and we conclude that

$$0.26384 \leq \int_0^1 \ln(x^2 + 1)\,dx \leq 0.26404$$

(See Figure 6.21.)

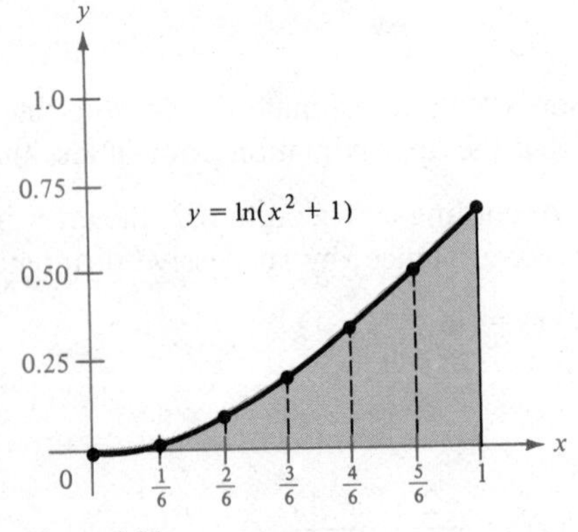

Figure 6.21

□

Remark Looking back at the examples in this section, you may wonder why we introduced the Trapezoidal Rule, since for a fixed n, Simpson's Rule usually gives a better approximation. The main reason for including the Trapezoidal Rule is that its error can be more easily estimated than can the error involved in Simpson's Rule. Certainly an approximation method is of little benefit if we have no idea of the potential error in the approximation. For instance, if $f(x) = \sqrt{x}\ln(x + 1)$, then to estimate the error in Simpson's Rule, we would need to determine the fourth derivative of f, which is a monumental task. Therefore, since the estimation of the error in Simpson's Rule involves the fourth derivative, we sometimes prefer to use the Trapezoidal Rule, even though we may have to use a larger n to obtain the desired accuracy.

Section Exercises (6.5)

In Exercises 1–10, use the Trapezoidal Rule and Simpson's Rule to approximate the value of each definite integral. Compare these results with the exact value of the definite integral. Round your answers to four decimal places.

Ⓒ **1.** $\int_0^2 x^2\,dx,\ n = 4$

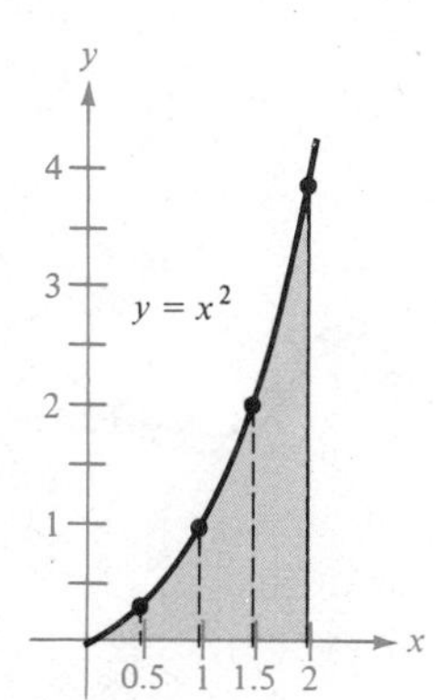

Ⓒ **2.** $\int_0^1 \left(\frac{x^2}{2} + 1\right) dx,\ n = 4$

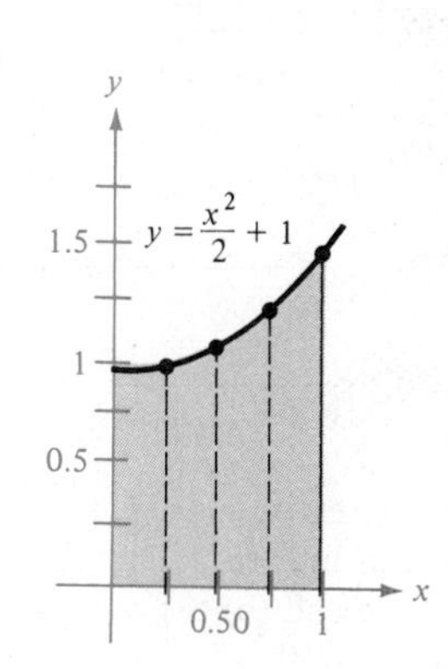

Ⓒ **3.** $\int_0^2 x^3\,dx,\ n = 4$

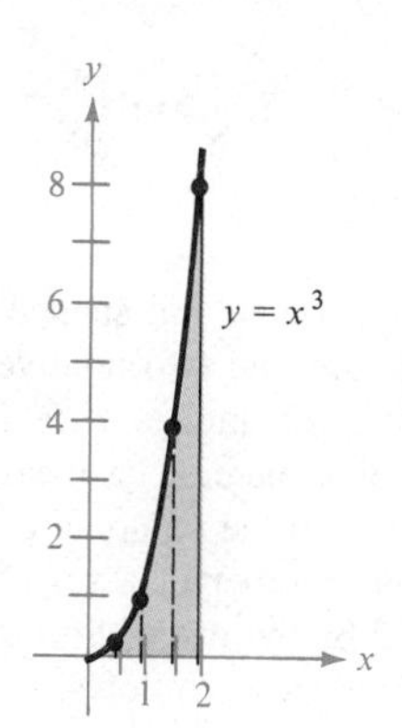

Ⓒ **4.** $\int_1^2 \frac{1}{x}\,dx,\ n = 4$

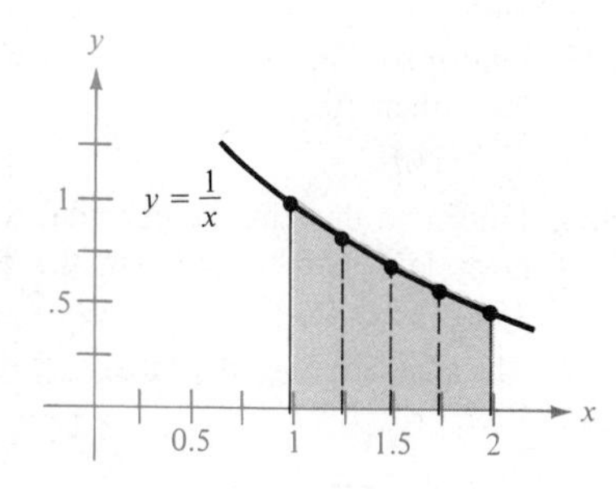

Ⓒ **5.** $\int_0^2 x^3\,dx,\ n = 8$

Ⓒ **6.** $\int_1^2 \frac{1}{x}\,dx,\ n = 8$

Ⓒ **7.** $\int_1^2 \frac{1}{x^2}\,dx,\ n = 4$

Ⓒ **8.** $\int_0^4 \sqrt{x}\,dx,\ n = 8$

Ⓒ **9.** $\int_0^1 \frac{1}{1 + x}\,dx,\ n = 4$

Ⓒ **10.** $\int_0^2 x\sqrt{x^2 + 1}\,dx,\ n = 4$

In Exercises 11–20, approximate each integral using (a) the Trapezoidal Rule and (b) Simpson's Rule.

Ⓒ **11.** $\int_0^2 \sqrt{1 + x^3}\,dx,\ n = 2$

Ⓒ Calculator may be helpful.

Ⓒ**12.** $\int_0^2 \frac{1}{\sqrt{1+x^3}}\,dx,\ n=4$

Ⓒ**13.** $\int_0^1 \sqrt{x}\sqrt{1-x}\,dx,\ n=4$

Ⓒ**14.** $\int_0^1 \frac{1}{1+x^2}\,dx,\ n=4$

Ⓒ**15.** $\int_0^1 \sqrt{1-x^2}\,dx,\ n=4$

Ⓒ**16.** $\int_0^2 e^{-x^2}\,dx,\ n=2$

Ⓒ**17.** $\int_0^1 \sqrt{1-x^2}\,dx,\ n=8$

Ⓒ**18.** $\int_0^2 e^{-x^2}\,dx,\ n=4$

Ⓒ**19.** $\int_1^3 \frac{1}{2+x+x^2}\,dx,\ n=2$

Ⓒ**20.** $\int_0^3 \frac{x}{2+x+x^2}\,dx,\ n=6$

In Exercises 21–24, find the maximum possible error if each integral is approximated by (a) the Trapezoidal Rule and (b) Simpson's Rule.

Ⓒ**21.** $\int_0^2 x^3\,dx,\ n=4$ Ⓒ**22.** $\int_0^1 \frac{1}{x+1}\,dx,\ n=4$

Ⓒ**23.** $\int_0^1 e^{x^3}\,dx,\ n=4$ Ⓒ**24.** $\int_0^1 e^{-x^2}\,dx,\ n=4$

Ⓒ**25.** Find n so that Simpson's Rule will have an error less than 0.00001 in the approximation of $\int_0^1 e^{-x^2}\,dx$.

Ⓒ**26.** Find n so that the Trapezoidal Rule will have an error less than 0.00001 in the approximation of $\int_0^1 e^{-x^2}\,dx$.

Ⓒ**27.** The standard normal probability density function is

$$f(z) = \frac{1}{\sqrt{2\pi}}e^{-z^2/2}$$

The probability that z is in the interval $[a, b]$ is the area of the region defined by $y = f(z)$, $y = 0$, $z = a$, and $z = b$ and is denoted by $\Pr(a \le z \le b)$. Estimate the following probabilities. (Choose n so that the error is less than 0.0001.)
(a) $\Pr(0 \le z \le 1)$ (b) $\Pr(0 \le z \le 2)$

Ⓒ**28.** A family purchases an odd-shaped lot bounded by a stream and two relatively straight roads that meet at right angles. (See Figure 6.22.) At distances of x feet, they measure the distances (y feet) from the one road to the stream. Approximate the number of square feet in the lot by use of Simpson's Rule if the measurements are as follows.

x	0	10	20	30	40	50	60
y	75	81	84	76	67	68	69

x	70	80	90	100	110	120
y	72	68	56	42	23	0

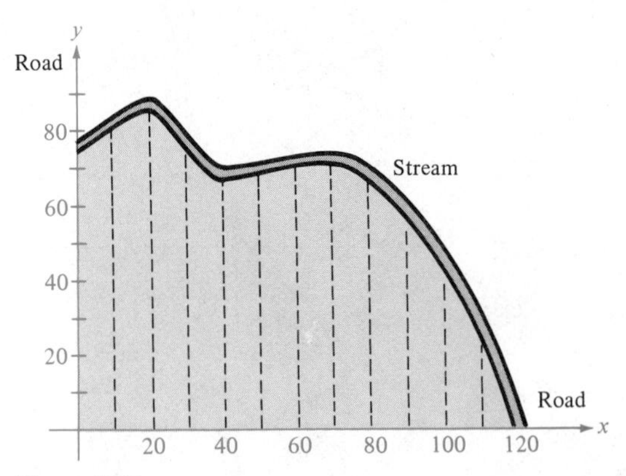

Figure 6.22

Ⓒ**29.** A farmer has an odd-shaped plot of land bounded by a stream and two relatively straight roads that meet at right angles. (See Figure 6.23.) At distances of x meters, he measures the distances (y meters) from the one road to the stream. Approximate the number of acres (1 acre $\approx$ 4047 m^2) in the field by the use of Simpson's Rule if the measurements are as follows:

x	0	100	200	300	400	500
y	125	125	120	112	90	90

x	600	700	800	900	1000
y	95	88	75	35	0

Ⓒ Calculator may be helpful.

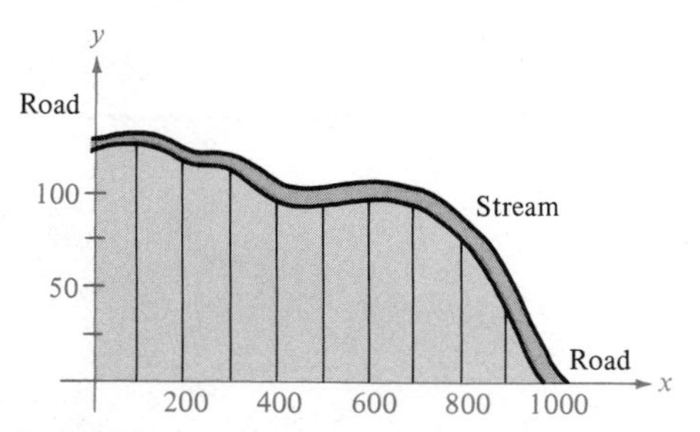

Figure 6.23

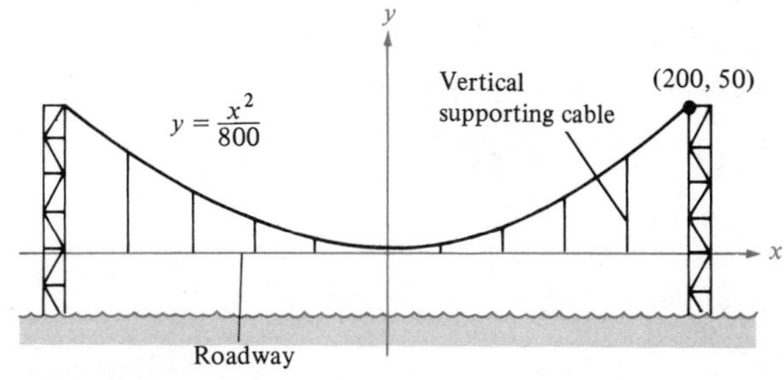

Figure 6.24

Ⓒ **30.** The suspension cable on a 400-ft bridge is in the shape of the parabola given by

$$y = \frac{x^2}{800}$$

(See Figure 6.24.) The length of the cable is given by the integral

$$\int_{-200}^{200} \sqrt{1 + (y')^2}\, dx = 2\int_0^{200} \sqrt{1 + \left(\frac{x}{400}\right)^2}\, dx$$

(a) Use the table of integrals in Section 6.4 to find the length of the cable.
(b) Use Simpson's Rule with $n = 10$ to approximate the length of the cable.

31. Prove that Simpson's Rule is exact when estimating the integral

$$\int_{x_1}^{x_2} (a_0 + a_1x + a_2x^2 + a_3x^3)\, dx.$$

32. Demonstrate the result of Exercise 31 by using Simpson's Rule to evaluate the integral $\int_0^1 x^3\, dx$ with $n = 2$.

Ⓒ Calculator may be helpful.

Section 6.6

Improper Integrals

Introductory Example

More about the Normal Distribution

In the Introductory Example to the preceding section, we discussed the normal probability density function. In that discussion we mentioned that approximately 99.74% of the measurements in a normal distribution fall within three standard deviations of the mean. Furthermore, we saw that nearly 100% (but *not* all) of the measurements fall within four standard deviations of the mean. Since even four standard deviations do not guarantee the inclusion of all measurements, we are led to ask: How many standard deviations away from the mean must we move before we can be sure to capture *all* measurements in a normal distribution? For example, if the mean height of a normally distributed population is 70 in. with a standard deviation of 3 in., can we be sure that no one in the population has a height that differs from the mean by more than five, six, or even seven standard deviations from the mean? The answer (at least theoretically) is that we can never be sure that *all* measurements in a truly normal distribution lie within n standard deviations of the mean, no matter how large we take n to be. Thus, we say that only in the limit does the area under the normal curve equal 1. That is,

$$\lim_{n\to\infty} \frac{1}{\sigma\sqrt{2\pi}} \int_{-n\sigma}^{n\sigma} e^{-x^2/2\sigma^2}\, dx = 1$$

We indicate this limit by the improper integral

$$\frac{1}{\sigma\sqrt{2\pi}} \int_{-\infty}^{\infty} e^{-x^2/2\sigma^2}\, dx$$

which represents the area bounded above by the normal curve and below by the x-axis and extending without bound to the right and left. (See Figure 6.25.)

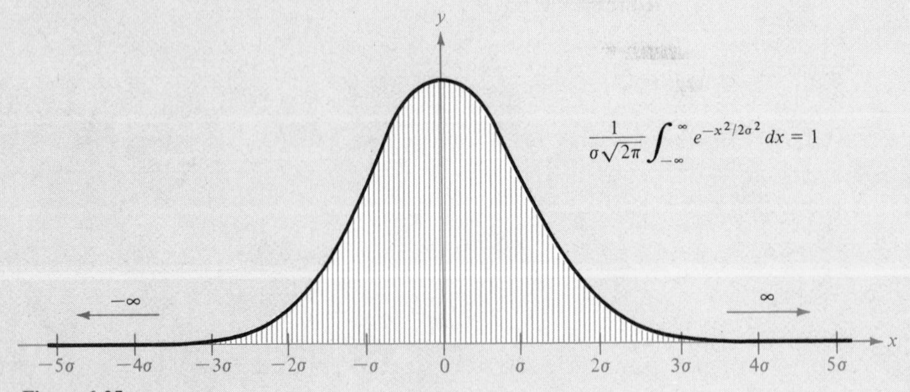

Figure 6.25

Section Objectives: *To introduce the concept of an improper integral ▪ To use limits to determine the convergence or divergence of improper integrals.*

Our definition of the definite integral $\int_a^b f(x)\,dx$ included the requirements that the interval $[a, b]$ is finite and that f is bounded on $[a, b]$. Furthermore, the Fundamental Theorem of Calculus, by which we have been evaluating definite integrals, requires f to be continuous on $[a, b]$. In this section, we discuss a limit procedure for evaluating integrals that do not satisfy these requirements because either:

1. one or both of the limits of integration are infinite, or
2. $f(x)$ has a finite number of infinite discontinuities on the interval $[a, b]$.

Integrals possessing either of properties 1 and 2 are called **improper integrals.**

For instance, the integrals

$$\int_0^{\infty} e^{-x}\,dx \qquad \text{and} \qquad \int_{-\infty}^{\infty} \frac{dx}{x^2+1}$$

are improper because one or both of their limits of integration are infinite. (See Figures 6.26 and 6.27.) On the other hand, the integrals

$$\int_1^5 \frac{dx}{\sqrt{x-1}} \qquad \text{and} \qquad \int_{-2}^{2} \frac{dx}{(x+1)^2}$$

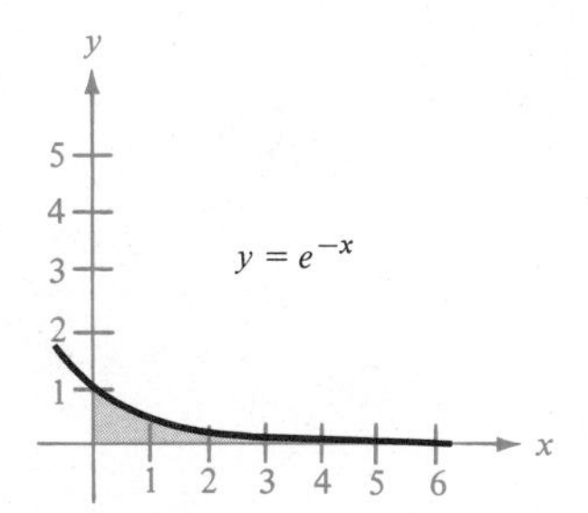

Figure 6.26

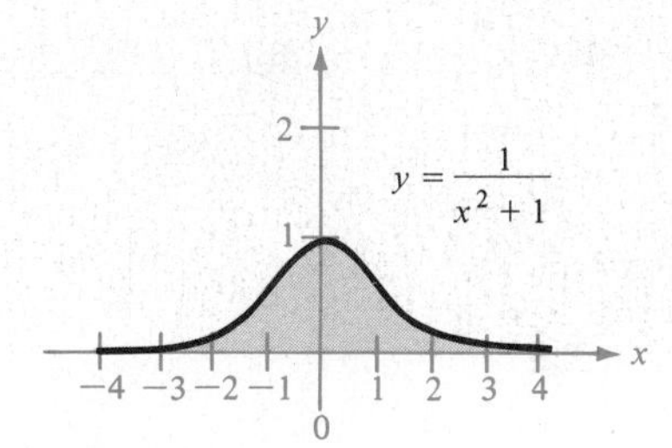

Figure 6.27

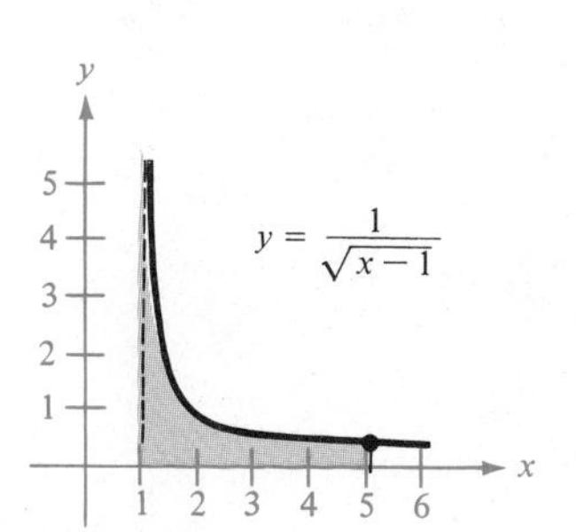

Figure 6.28

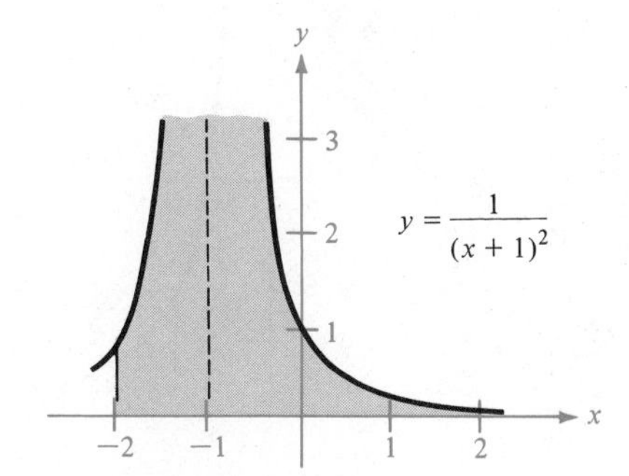

Figure 6.29

are improper because their integrands approach infinity somewhere in the interval of integration. (See Figures 6.28 and 6.29.)

Development of Improper Integrals

To get an idea how we might evaluate an improper integral, consider the integral

$$\int_1^b \frac{1}{x^2}\,dx$$

Remember that as long as b is a real number (no matter how large) this is a *definite* integral (not an *improper* integral), and we can evaluate it as follows:

$$\int_1^b \frac{1}{x^2}\,dx = -\frac{1}{x}\Big]_1^b = -\frac{1}{b} + 1 = 1 - \frac{1}{b}$$

Figure 6.30 shows the region whose area is represented by this integral, and Table 6.4 lists the region's area for several values of b.

Table 6.4

b	2	5	10	100	1000	10,000
$\int_1^b \frac{1}{x^2}\,dx$	0.5	0.8	0.9	0.99	0.999	0.9999

Now, if we let $b \to \infty$, then we have

$$\lim_{b\to\infty} \int_1^b \frac{1}{x^2}\,dx = \lim_{b\to\infty} \left(\frac{-1}{x}\Big]_1^b\right) = \lim_{b\to\infty}\left(1 - \frac{1}{b}\right) = 1$$

We denote this limit by the improper integral

$$\int_1^\infty \frac{1}{x^2}\,dx$$

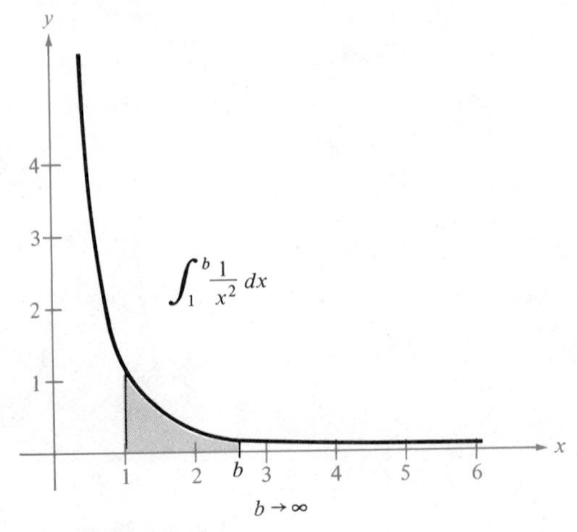

Figure 6.30

and we can interpret it as the area of the unbounded region under $y = 1/x^2$, above the x-axis, and to the right of $x = 1$.

More generally, we define improper integrals having infinite limits of integration as follows.

Improper Integrals (Infinite Limits of Integration)

1. If f is continuous on the interval $[a, \infty)$, then

$$\int_a^{\infty} f(x)\,dx = \lim_{b\to\infty} \int_a^b f(x)\,dx$$

2. If f is continous on the interval $(-\infty, b]$, then

$$\int_{-\infty}^{b} f(x)\,dx = \lim_{a\to-\infty} \int_a^b f(x)\,dx$$

3. If f is continuous on the interval $(-\infty, \infty)$, then

$$\int_{-\infty}^{\infty} f(x)\,dx = \int_{-\infty}^{c} f(x)\,dx + \int_c^{\infty} f(x)\,dx$$

where c is any real number.

In each case above, if the limit exists, then the improper integral is said to **converge;** otherwise the improper integral **diverges.** This means that in the third case the integral will diverge if either one of the integrals on the right diverges.

Example 1

A Divergent Improper Integral

Determine the convergence or divergence of the integral

$$\int_1^{\infty} \frac{1}{x}\,dx$$

Solution

$$\int_1^{\infty} \frac{1}{x}\,dx = \lim_{b\to\infty} \int_1^b \frac{1}{x}\,dx = \lim_{b\to\infty} \left(\ln x\Big]_1^b\right)$$

$$= \lim_{b\to\infty} (\ln b - 0) = \infty$$

Thus, the integral diverges since the limit does not exist. □

Remark

In Example 1, we were careful to use limit notation to determine the divergence of the improper integral. However, in practice we can use the same notation that we use for definite integrals, *provided we keep in mind that a limit is involved.* For instance, the solution to Example 1 could be written as

$$\int_1^{\infty} \frac{1}{x}\,dx = \ln x\Big]_1^{\infty} = \underbrace{\infty}_{\lim_{x\to\infty} \ln x} - \underbrace{0}_{\ln(1)} = \infty$$

Remember that this definite integral notation is a shorthand symbol for an implied limit. Furthermore, the nonexistence of the limit at either endpoint is sufficient to imply divergence.

A comparison of the improper integrals

$$\int_1^{\infty} \frac{1}{x^2}\,dx$$

which converges to 1, and

$$\int_1^{\infty} \frac{1}{x}\,dx$$

which diverges, suggests the somewhat unpredictable nature of improper integrals. The functions being integrated have similar-looking graphs, as shown in Figure 6.31, yet the shaded region under $y = 1/x^2$ to the right of $x = 1$ has a *finite* area, whereas the corresponding region under $y = 1/x$ has an infinite area (see Example 1).

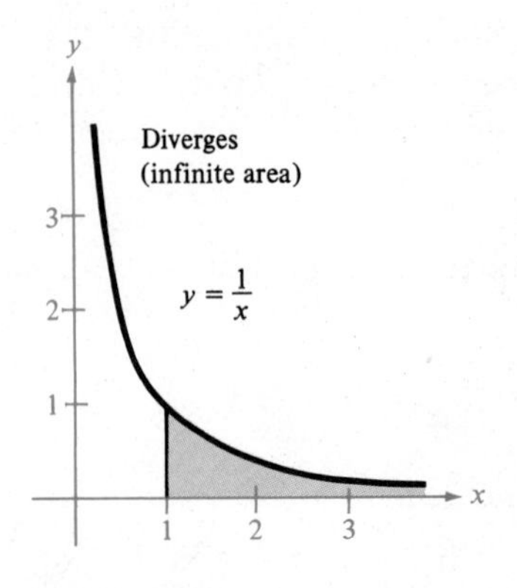

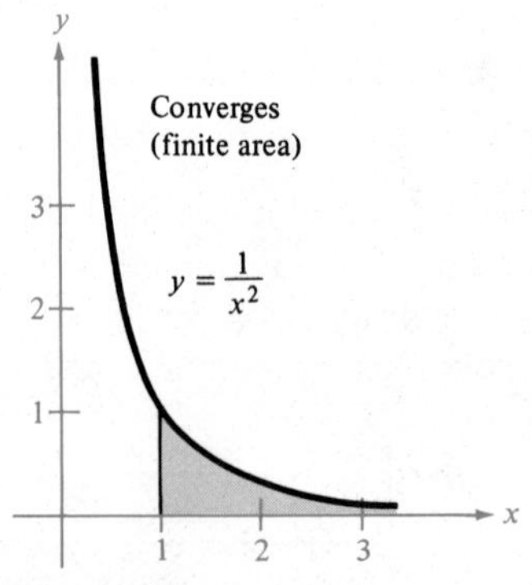

Figure 6.31

Example 2

A Convergent Improper Integral

Evaluate the improper integral

$$\int_{-\infty}^{0} \frac{dx}{(1-2x)^{3/2}}$$

Solution

$$\int_{-\infty}^{0} \frac{dx}{(1-2x)^{3/2}} = \frac{1}{\sqrt{1-2x}}\Bigg]_{-\infty}^{0} = 1 - \underbrace{0}_{\lim\limits_{x\to-\infty} \frac{1}{\sqrt{1-2x}}} = 1$$

Consequently, this improper integral converges. (See Figure 6.32.)

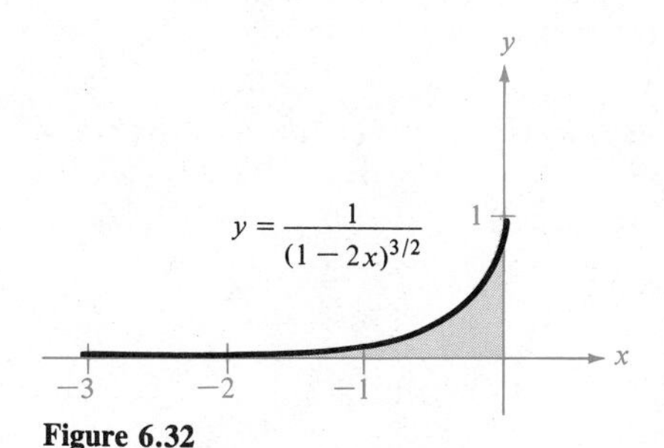

Figure 6.32

□

Example 3

A Convergent Improper Integral

Evaluate the improper integral

$$\int_{0}^{\infty} 2xe^{-x^2}\,dx$$

Solution

$$\int_{0}^{\infty} 2xe^{-x^2}\,dx = -e^{-x^2}\Bigg]_{0}^{\infty} = -(\underbrace{0}_{\lim\limits_{x\to\infty} e^{-x^2}} - 1) = 1$$

(See Figure 6.33.)

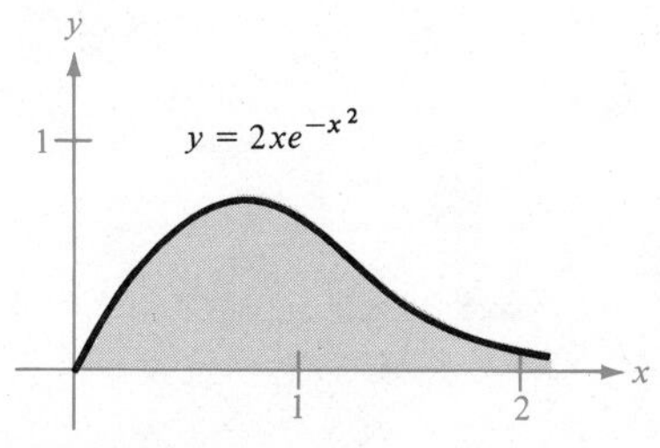

Figure 6.33

□

A second general type of improper integral involves integrands that approach infinity at one of the limits of integration. Such integrals are described in the following definitions.

Improper Integrals (Infinite Integrands)

1. If f is continuous on the intervals $[a, b)$ and approaches infinity at b, then

$$\int_a^b f(x)\,dx = \lim_{c \to b^-} \int_a^c f(x)\,dx$$

2. If f is continuous on the interval $(a, b]$ and approaches infinity at a, then

$$\int_a^b f(x)\,dx = \lim_{c \to a^+} \int_c^b f(x)\,dx$$

3. If f is continuous on the interval $[a, b]$, except for some c in (a, b) at which f approaches infinity, then

$$\int_a^b f(x)\,dx = \int_a^c f(x)\,dx + \int_c^b f(x)\,dx$$

Example 4

A Convergent Improper Integral

Evaluate the improper integral

$$\int_1^2 \frac{dx}{\sqrt[3]{x-1}}$$

Solution Since the integrand approaches infinity at $x = 1$, we write

$$\int_1^2 \frac{dx}{\sqrt[3]{x-1}} = \frac{3}{2}(x-1)^{2/3}\Big]_1^2 = \frac{3}{2}(1 - \underbrace{0}_{\lim\limits_{x \to 1^+} (x-1)^{2/3}}) = \frac{3}{2}$$

and the integral converges. (See Figure 6.34.)

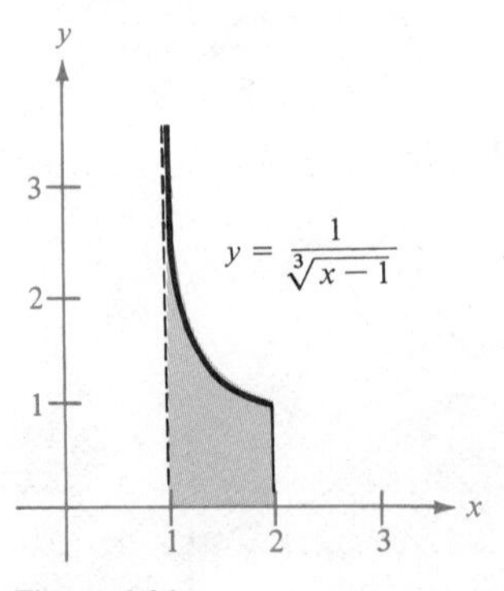

Figure 6.34

□

Example 5

A Divergent Improper Integral

Evaluate the improper integral

$$\int_1^2 \frac{2\,dx}{x^2 - 2x}$$

Solution By separating the integrand into its partial fractions, we obtain

$$\int_1^2 \frac{2dx}{x^2 - 2x} = \int_1^2 \left(\frac{1}{x - 2} - \frac{1}{x}\right) dx$$

$$= \Bigg[\ln|x - 2| - \ln|x|\Bigg]_1^2$$

The integral diverges since

$$\lim_{x\to 2^-} [\ln|x - 2| - \ln|x|] = -\infty - \ln(2) = -\infty$$

(See Figure 6.35.)

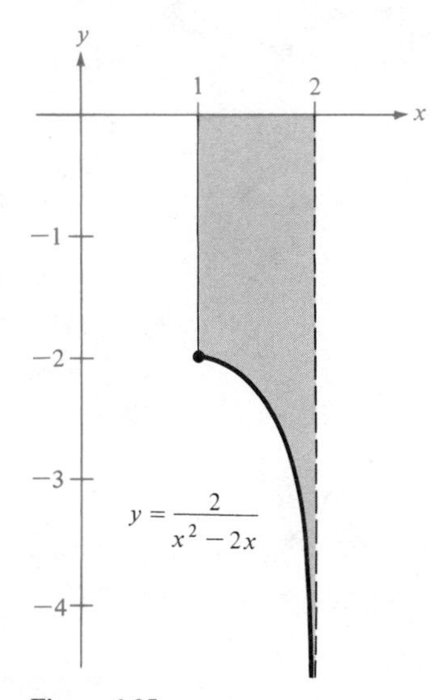

Figure 6.35

□

Example 6

A Divergent Improper Integral

Evaluate the integral $\int_0^3 (x - 1)^{-3}\,dx$.

Solution This integral is improper because the integrand approaches infinity at $x = 1$, which lies between the limits of integration. (See Figure 6.36.) Thus, we must write

$$\int_0^3 (x-1)^{-3}\,dx = \int_0^1 (x-1)^{-3}\,dx + \int_1^3 (x-1)^{-3}\,dx$$

$$= \left.\frac{-1}{2(x-1)^2}\right]_0^1 + \left.\frac{-1}{2(x-1)^2}\right]_1^3$$

The integral diverges since

$$\lim_{x\to 1^-} \frac{-1}{2(x-1)^2} = -\infty$$

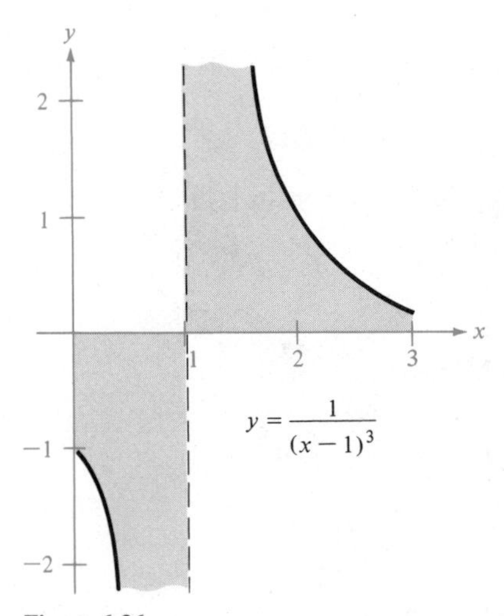

Figure 6.36

Remember that the nonexistence of the limit at *any* endpoint is sufficient to imply divergence. □

Remark

Had we not recognized that the integral in Example 6 was improper, we would have obtained the incorrect result

$$\int_0^3 (x-1)^{-3}\,dx = \left.\frac{-1}{2(x-1)^2}\right]_0^3 = -\frac{1}{8} + \frac{1}{2} = \frac{3}{8}$$

Improper integrals in which the integrand approaches infinity at some point *between* the limits of integration are quite often overlooked, so keep alert for such possibilities.

Section Exercises (6.6)

In Exercises 1–20, determine the divergence or convergence of each improper integral and evaluate those that converge. Use the table of integrals in Section 6.4 when necessary.

1. $\displaystyle\int_0^4 \frac{1}{\sqrt{x}}\,dx$

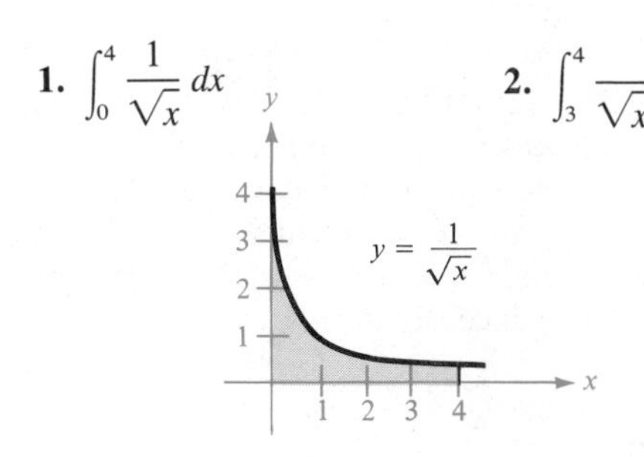

2. $\displaystyle\int_3^4 \frac{1}{\sqrt{x-3}}\,dx$

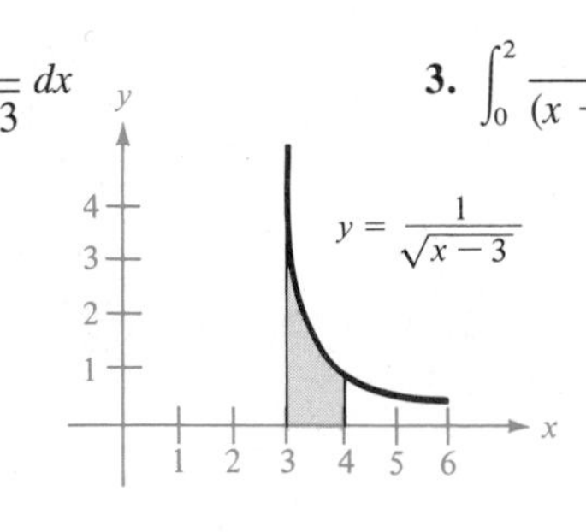

3. $\displaystyle\int_0^2 \frac{1}{(x-1)^{2/3}}\,dx$

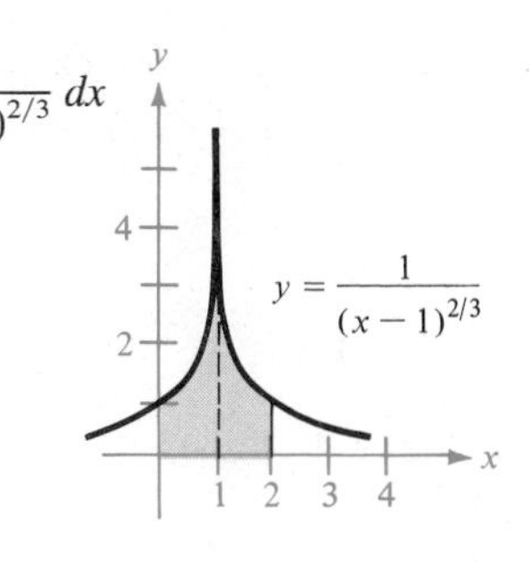

4. $\displaystyle\int_0^2 \frac{1}{(x-1)^2}\,dx$

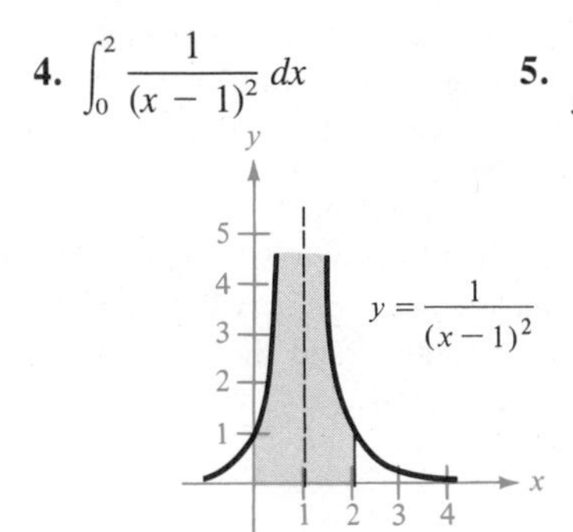

5. $\displaystyle\int_0^\infty e^{-x}\,dx$

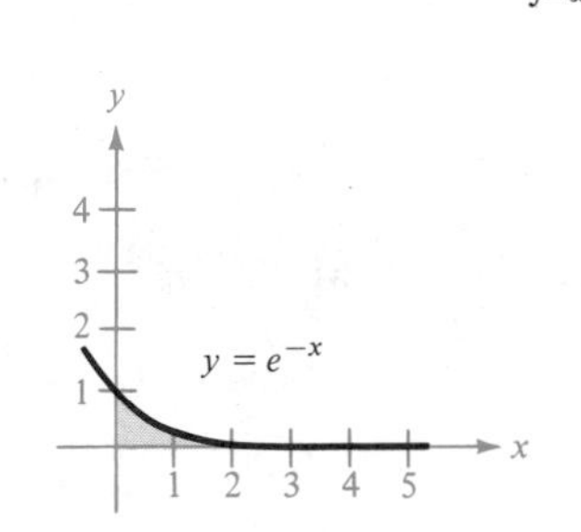

6. $\displaystyle\int_{-\infty}^0 e^{2x}\,dx$

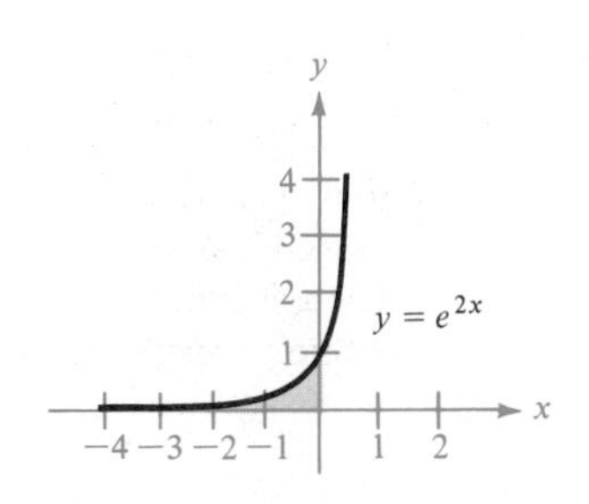

7. $\displaystyle\int_0^8 \frac{1}{\sqrt[3]{8-x}}\,dx$ **8.** $\displaystyle\int_0^1 \frac{1}{x}\,dx$

9. $\displaystyle\int_0^1 \frac{1}{x^2}\,dx$ **10.** $\displaystyle\int_0^2 \frac{1}{\sqrt{4-x^2}}\,dx$

11. $\displaystyle\int_2^4 \frac{1}{\sqrt{x^2-4}}\,dx$ **12.** $\displaystyle\int_0^2 \frac{1}{4-x^2}\,dx$

13. $\displaystyle\int_0^2 \frac{1}{\sqrt[3]{x-1}}\,dx$ **14.** $\displaystyle\int_0^2 \frac{1}{(x-1)^{4/3}}\,dx$

15. $\displaystyle\int_1^\infty \frac{1}{x^2}\,dx$ **16.** $\displaystyle\int_1^\infty \frac{1}{\sqrt{x}}\,dx$

17. $\displaystyle\int_{-\infty}^\infty x^2e^{-x^3}\,dx$ **18.** $\displaystyle\int_{1/2}^\infty \frac{1}{\sqrt{2x-1}}\,dx$

19. $\displaystyle\int_5^\infty \frac{x}{\sqrt{x^2-16}}\,dx$ **20.** $\displaystyle\int_0^\infty \frac{x^3}{(x^2+1)^2}\,dx$

ⓒ**21.** Demonstrate that $\lim_{x\to\infty}[x^ne^{-x}] = 0$ by completing the following table for:

(a) $n = 1$ (b) $n = 2$ (c) $n = 5$

x	1	10	25	50
x^ne^{-x}				

In Exercises 22–24, use the results of Exercise 21 to determine the divergence or convergence of each improper integral, and evaluate those that converge.

22. $\displaystyle\int_0^\infty xe^{-x}\,dx$ **23.** $\displaystyle\int_0^\infty x^2e^{-x}\,dx$

24. $\displaystyle\int_0^\infty (x-1)e^{-x}\,dx$

A nonnegative function $y = f(t)$ is a probability density function for a continuous random variable T if $\int_0^\infty f(t)\,dt = 1$. The probability that T takes a value between a and b is given by $\int_a^b f(t)\,dt$. The *expected value of* T, denoted by $E(T)$, is a measure of the center of the distribution and is given by $E(T) = \int_0^\infty tf(t)\,dt$.

ⓒ**25.** The probability density function for the time, in minutes, between the arrival of patients at a doctor's office is given by

$$f(t) = \begin{cases} \frac{1}{7}e^{-t/7}; & t \geq 0 \\ 0; & t < 0 \end{cases}$$

(a) Show that $\int_0^\infty \frac{1}{7}e^{-t/7}\,dt = 1$.

(b) If a patient has just arrived, what is the probability that it will be more than 5 min until the

ⓒ Calculator may be helpful.

arrival of the next patient? That is, evaluate the integral

$$\int_5^\infty \frac{1}{7} e^{-t/7}\, dt$$

(c) Using the results of Exercise 21, show that

$$E(T) = \int_0^\infty t[\tfrac{1}{7} e^{-t/7}]\, dt = 7$$

Ⓒ**26.** The probability density function for the time, in minutes, between the arrival of calls at a switchboard in a business office is given by

$$f(t) = \begin{cases} \frac{2}{5} e^{-2t/5}; \ t \geq 0 \\ 0; \ t < 0 \end{cases}$$

(a) Show that $\int_0^\infty \frac{2}{5} e^{-2t/5}\, dt = 1$.

(b) If a call just came into the switchboard, what is the probability that it will be at least 3 min before the next call arrives? That is, evaluate the integral

$$\int_3^\infty \frac{2}{5} e^{-2t/5}\, dt$$

(c) Using the results of Exercise 21, show that

$$E(T) = \int_0^\infty t\left[\frac{2}{5} e^{-2t/5}\right] dt = 2.5$$

The capitalized cost C of an asset is the original cost C_0 plus the present value P of all future costs related to its upkeep or replacement. The capitalized cost can be used to estimate the endowment required for the set up and continued operation of a particular enterprise. The present value of future costs for n years at an annual rate r compounded continuously is given by

$$P = \int_0^n c(t)e^{-rt}\, dt$$

where $c(t)$ is the cost per year in dollars at time t.

Ⓒ**27.** A philanthropist wants to provide an endowment to construct and maintain a memorial in the town square. It is estimated that the construction costs will amount to $C_0 = \$650{,}000$ and the annual maintenance cost will be $c(t) = \$25{,}000$. Find the size of the endowment if it will earn 12% compounded continuously and the philanthropist wants it to meet all costs:

(a) for 5 years
(b) for 10 years
(c) forever

Ⓒ**28.** Repeat Exercise 27 assuming that the annual maintenance cost will increase at 8% per year, and therefore, $c(t) = 25{,}000(1 + 0.08t)$. [Use the results of Exercise 21 in evaluating the improper integral of part (c)].

Ⓒ Calculator may be helpful.

Chapter 7

Differential Equations

Section 7.1

Solutions of Differential Equations

Introductory Example

A Model for National Income

Economists refer to the total output of a nation's economy as the *national income*. Moreover, this national income is commonly assumed to be proportional to the total stock of *capital* in the country. Thus, we have

$$y = kx$$

where y is the national income, x is the capital, and k is the proportionality constant.

That portion of the national income that is fed back into the economy in the form of new capital is called *capital investment*. The total capital investment is actually a measure of the rate of change of the capital, and we have

$$\text{investment} = \frac{dx}{dt} = \frac{1}{k}\frac{dy}{dt}$$

That portion of the national income that is consumed is called the *national consumption*. Hence, income, investment, and consumption are related by the equation

$$\text{income} = \text{consumption} + \text{investment}$$

One useful economic model for this relationship assumes that consumption is linearly related to income and that it has two parts: necessities and luxuries. A country must consume a certain amount to survive, and this model assumes that after the survival level is reached people are willing to consume a fixed percentage of their extra income on luxuries. Thus, for an income of y, we have

$$\begin{aligned}\text{income} &= [\text{necessities} + \text{luxuries}] + \text{investment}\\ &= [a + b(y - a)] + \frac{1}{k}\frac{dy}{dt}\end{aligned}$$

which yields the following **differential equation:**

$$y = a(1 - b) + by + \frac{1}{k}\frac{dy}{dt}$$

$$\frac{dy}{dt} = k(1 - b)(y - a)$$

The **solution** to this equation is

$$y = a + Ce^{k(1-b)t}$$

where C represents the income above the necessity level at the time $t = 0$. Figure 7.1 shows a graph for national income in which the factor b is 0.75. In other words, this model assumes people will consume 75% of their "extra" income on luxuries.

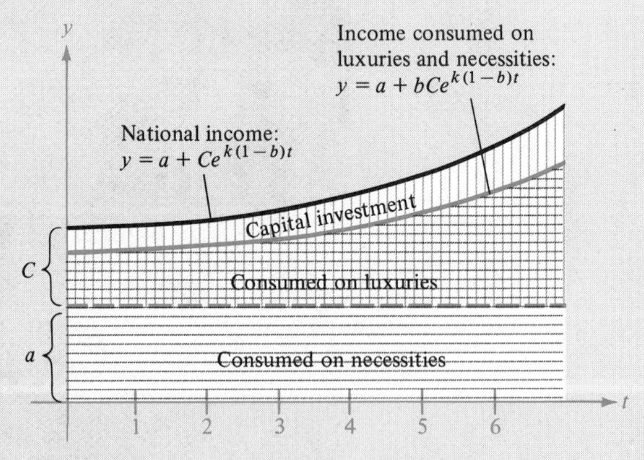

Figure 7.1

Section Objectives: *To introduce the concepts of the general and particular solutions of a differential equation* ▪ *To find the particular solution that satisfies a set of initial conditions.*

A **differential equation** is an equation involving an unknown function (dependent variable) and one or more of its derivatives. For instance,

$$3\frac{dy}{dx} - 2xy = 0$$

is a differential equation in which $y = f(x)$ is a differentiable function of x.

A function $y = f(x)$ is called a **solution** of a given differential equation if the equation is satisfied when y and its derivatives are replaced by $f(x)$ and its derivatives. For example,

$$y = e^{-2x}$$

is a solution to the differential equation

$$y' + 2y = 0$$

because $y' = -2e^{-2x}$ and, by substitution, we have

$$\overbrace{-2e^{-2x}}^{y'} + \overbrace{2e^{-2x}}^{2y} = 0$$

Furthermore, we can readily see that

$$y = 2e^{-2x}, \qquad y = 3e^{-2x}, \qquad \text{and} \qquad y = \tfrac{1}{2}e^{-2x}$$

are also solutions to this same differential equation. In fact, the functions

$$y = Ce^{-2x}$$

where C is any real number, are all solutions. We call this family of solutions the **general solution** of the differential equation $y' + 2y = 0$.

Recall from Example 6 in Section 4.1 that the differential equation

$$s''(t) = -32$$

has the general solution

$$s(t) = -16t^2 + C_1t + C_2$$

Note that this general solution has *two* arbitrary constants. A differential equation usually has a general solution involving as many arbitrary constants as the order of the highest derivative appearing in the equation.

A **particular solution** of a differential equation is any solution that is obtained by assigning specific values to the constants in the general solution. (Occasionally, a differential equation has other solutions, not obtainable from the general solution by assigning values to the arbitrary constants; such solutions are called *singular solutions*. In our brief discussion of differential equations, we will omit any further reference to singular solutions.)

Geometrically, the general solution of a given differential equation represents a family of curves, called **solution curves,** one for each value assigned to the arbitrary constants. For instance, we can easily verify that

$$y = \frac{C}{x}$$

is the general solution to the differential equation

$$xy' + y = 0$$

Figure 7.2 shows some of the solution curves corresponding to different values of C.

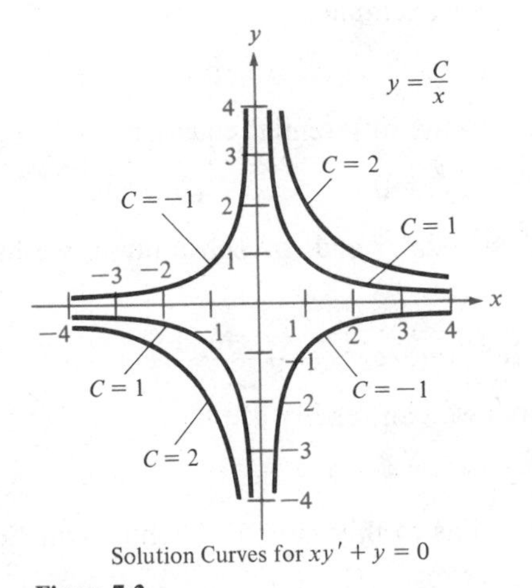

Solution Curves for $xy' + y = 0$

Figure 7.2

In practice, particular solutions of a differential equation are usually obtained from **initial conditions** (or **boundary conditions**) placed on the unknown function and its derivatives. For instance, for Figure 7.2, if we wanted to obtain the particular solution whose curve passes through the point $(-1, 1)$, we have the boundary condition

$$f(-1) = 1 \qquad \text{or} \qquad y = 1 \text{ when } x = -1$$

Substituting this condition into the general solution yields

$$y = \frac{C}{x}$$

$$1 = \frac{C}{-1}$$

$$-1 = C$$

and the particular solution is

$$y = -\frac{1}{x}$$

Example 1

Checking Possible Solutions of a Differential Equation

Determine whether or not the given functions are solutions of the differential equation $y'' - y = 0$.

(a) $y = e^x$ (b) $y = 4e^{-x}$
(c) $y = 0$ (d) $y = e^{2x}$
(e) $y = Ce^x$

Solution

(a) Since $y = e^x$, $y' = e^x$, and $y'' = e^x$, we have

$$y'' - y = e^x - e^x = 0$$

Hence, $y = e^x$ *is* a solution.
(b) Since $y = 4e^{-x}$, $y' = -4e^{-x}$, and $y'' = 4e^{-x}$, we have

$$y'' - y = 4e^{-x} - 4e^{-x} = 0$$

Hence, $y = 4e^{-x}$ *is* a solution.
(c) Since $y = 0$, $y' = 0$, and $y'' = 0$, we have

$$y'' - y = 0 - 0 = 0$$

Hence, $y = 0$ *is* a solution.
(d) Since $y = e^{2x}$, $y' = 2e^{2x}$, and $y'' = 4e^{2x}$, we have

$$y'' - y = 4e^{2x} - e^{2x} = 3e^{2x} \neq 0$$

Hence, $y = e^{2x}$ *is not* a solution.
(e) Since $y = Ce^x$, $y' = Ce^x$, and $y'' = Ce^x$, we have

$$y'' - y = Ce^x - Ce^x = 0$$

Hence, $y = Ce^x$ *is* a solution for any value of C. □

Example 2

Finding a Particular Solution of a Differential Equation

Verify that $y = Cx^3$ is the general solution of the differential equation $xy' - 3y = 0$. Then find the particular solution determined by the boundary condition $y = 2$ when $x = -3$.

Solution Since $y = Cx^3$ and $y' = 3Cx^2$, we have

$$xy' - 3y = x(3Cx^2) - 3(Cx^3) = 0$$

Hence, $y = Cx^3$ is the general solution. Furthermore, from the boundary condition, we determine that

$$y = Cx^3$$
$$2 = C(-3)^3$$
$$-\tfrac{2}{27} = C$$

Therefore, the particular solution is

$$y = -\tfrac{2}{27}x^3$$ □

Example 3

Finding a Particular Solution of a Differential Equation

Given the general solution

$$s(t) = -16t^2 + C_1t + C_2$$

of the differential equation

$$s''(t) = -32$$

find the particular solution satisfying the initial conditions

$$s(0) = 80 \qquad \text{and} \qquad s'(0) = 64$$

Solution From the initial condition $s(0) = 80$, we have

$$80 = -16(0)^2 + C_1(0) + C_2$$
$$80 = C_2$$

Now, by differentiating the general solution, we have

$$s'(t) = -32t + C_1$$

and from the initial condition $s'(0) = 64$, we have

$$64 = -32(0) + C_1$$
$$64 = C_1$$

Finally, we conclude that the particular solution satisfying the given initial conditions is

$$s(t) = -16t^2 + 64t + 80$$ □

Remark

Note from Examples 2 and 3 that we usually refer to conditions for which the independent variable is zero as *initial conditions,* while we use the more general term *boundary conditions* to describe conditions for which the independent variable is not zero.

Section Exercises (7.1)

In Exercises 1–14, verify that y is a solution of the given differential equation.

1. $y = Ce^{4x}$; $\dfrac{dy}{dx} = 4y$
2. $y = Ce^{-2x}$; $\dfrac{dy}{dx} = -2y$
3. $y = Ce^{-t/2} + 5$; $2y' + y - 5 = 0$
4. $y = Ce^{-t} + 10$; $y' + y - 10 = 0$
5. $y = Cx^2 - 3x$; $xy' - 3x - 2y = 0$
6. $y^2 + 2xy - x^2 = C$; $(x + y)y' - x + y = 0$
7. $y = x \ln x + Cx - 2$; $x(y' - 1) - (y + 2) = 0$
8. $y = x \ln x^2 + 2x^{3/2} + Cx$; $y' - \dfrac{y}{x} = 2 + \sqrt{x}$
9. $y = x^2 + 2x + \dfrac{C}{x}$; $\dfrac{dy}{dx} + \dfrac{y}{x} = 3x + 4$
10. $y = C_1 + C_2e^x$; $y'' - y' = 0$
11. $y = C_1e^{x/2} + C_2e^{-2x}$; $2y'' + 3y' - 2y = 0$
12. $y = \dfrac{bx^4}{4 - a} + Cx^a$; $y' - \dfrac{ay}{x} = bx^3$
13. $y = C_1e^{(3+\sqrt{5})x/2} + C_2e^{(3-\sqrt{5})x/2}$;
    $y'' - 3y' + y = 0$
14. $y = \dfrac{x^3}{5} - x + C\sqrt{x}$; $2xy' - y = x^3 - x$

In Exercises 15–20, verify that the general solution satisfies the given differential equation. Then find the particular solution satisfying the given initial or boundary conditions.

15. $y = Ce^{-2x}$; $y' + 2y = 0$; $y = 3$ when $x = 0$
16. $2x^2 + 3y^2 = C$; $2x + 3yy' = 0$;
    $y = 2$ when $x = 1$
17. $y = C_1 + C_2 \ln |x|$; $xy'' + y' = 0$;
    $y = 5$ and $y' = \frac{1}{2}$ when $x = 1$
18. $y = C_1x + C_2x^3$; $x^2y'' - 3xy' + 3y = 0$;
    $y = 0$ and $y' = 4$ when $x = 2$
19. $y = C_1e^{6x} + C_2e^{-5x}$; $y'' - y' - 30y = 0$;
    $y = 0$ and $y' = -4$ when $x = 0$
20. $y = Ce^{x-x^2}$; $y' + (2x - 1)y = 0$;
    $y = 2$ when $x = 1$

In Exercises 21–24, the general solution to the differential equation is given. Sketch the solution curves for the given values of C.

21. $x^2 + y^2 = C$; $yy' + x = 0$;
    $C = 0, C = 1, C = 4$
22. $4y^2 - x^2 = C$; $2yy' - x = 0$;
    $C = 0, C = \pm 1, C = \pm 4$
23. $y = C(x + 2)^2$; $(x + 2)y' - 2y = 0$;
    $C = 0, C = \pm 1, C = \pm 2$
24. $y = Ce^{-x}$; $y' + y = 0$;
    $C = 0, C = \pm 1, C = \pm 2$

Ⓒ 25. The limiting capacity of the habitat for a particular wildlife herd is L. The growth rate dN/dt of the herd is proportional to the unutilized opportunity for growth as described by the differential equation.

$$\frac{dN}{dt} = k(L - N)$$

The solution to this differential equation is

$$N = L - Ce^{-kt}$$

Suppose 100 animals are released into a tract of land that can support up to 750 of these animals. After 2 years the herd has grown to 160 animals.
   (a) Find the population function in terms of the time t in years.
   (b) Sketch the graph of this population function.

Ⓒ 26. The rate of growth of an investment is proportional to the amount in the investment at any time t. That is

$$\frac{dA}{dt} = kA$$

   (a) Show that $A = Ce^{kt}$ is a solution to this differential equation.
   (b) Find the particular solution to this differential equation if the initial investment is \$1000, and 10 years later the amount is \$3320.12.

Ⓒ Calculator may be helpful.

Section 7.2

Separation of Variables

Introductory Example

Newton's Law of Cooling

Newton's Law of Cooling describes the rate at which a small object will change temperature when placed in surroundings of a different temperature. Specifically, if we let T be the temperature (in degrees Fahrenheit) of the object at time t (in minutes) then the rate of change of T is proportional to the difference between T and the temperature of the surroundings.

As an example, consider a (cold-blooded) snake that has been lying in the sun and has reached a body temperature of 120°. Suppose the snake is disturbed by a predator and seeks refuge in a cave where the air temperature is 70°. The (approximate) rate of change in the snake's body temperature is given by the differential equation

$$\frac{dT}{dt} = k(T - 70), \qquad 70 \leq T \leq 120$$

If the snake's body temperature drops to 100° after 10 min, what will its body temperature be after 20 min?

To solve this differential equation, we can use a technique called **separation of variables** to obtain

$$\frac{dT}{T - 70} = k\,dt$$

$$\int \frac{dT}{T - 70} = \int k\,dt$$

which yields

$$T = 70 + Ce^{kt}$$

Using the conditions $T(0) = 120$ and $T(10) = 100$, we solve for the constants C and k to obtain the particular solution

$$T = 70 + 50e^{-0.0511t}$$

Finally, when $t = 20$ min, we have $T \approx 88°$. Figure 7.3 shows the snake's body temperature as it asymptotically approaches 70°.

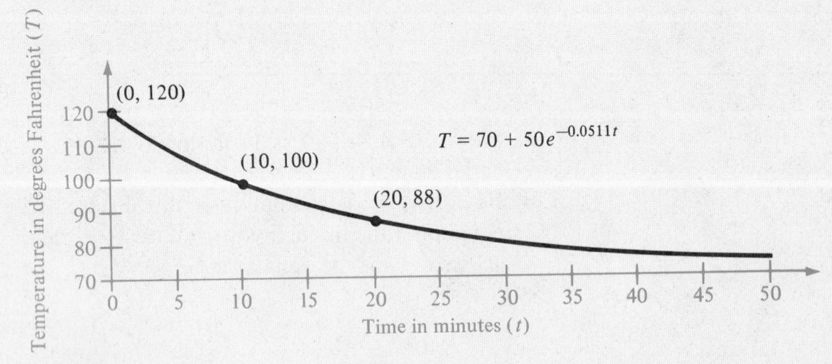

Figure 7.3

Section Objective: ***To introduce the technique of separation of variables to solve differential equations of the form $dy/dx = f(x)g(y)$.***

The simplest types of differential equations are those of the form

$$\frac{dy}{dx} = f(x)$$

In Chapter 4, we saw that we can solve such an equation by means of anti-differentiation (integration) to obtain

$$y = \int f(x)\, dx$$

In this section, we will see that integration can be used to solve another important class of differential equations — those in which the variables can be separated. This method is called **separation of variables** and is outlined as follows.

Separation of Variables

If f and g are continuous functions, then the differential equation

$$\frac{dy}{dx} = f(x)g(y)$$

has a general solution of the form

$$\int \frac{1}{g(y)}\, dy = \int f(x)\, dx + C$$

where C is an arbitrary constant.

Example 1

Separation of Variables

Find the general solution of the equation

$$(x^2 + 4)\frac{dy}{dx} = xy$$

by separation of variables.

Solution The given equation has the form

$$\frac{dy}{dx} = \frac{xy}{x^2 + 4} = \left(\frac{x}{x^2 + 4}\right)(y)$$

Thus, we can separate the variables to obtain

$$\int \frac{1}{y}\, dy = \int \frac{x}{x^2 + 4}\, dx$$

$$\ln |y| = \tfrac{1}{2} \ln (x^2 + 4) + C_1$$

For convenience, we let $C_1 = \ln |C|$ and write

$$\begin{aligned}\ln |y| &= \ln \sqrt{x^2 + 4} + \ln |C| \\ &= \ln |C\sqrt{x^2 + 4}|\end{aligned}$$

Therefore, the general solution is

$$y = C\sqrt{x^2 + 4}$$

□

Remark We can readily verify that the general solution obtained in Example 1 satisfies the given differential equation. We encourage you to verify each solution obtained in this section. For some differential equations it is not feasible to write the solution in the explicit form $y = F(x)$. In such cases implicit differentiation can be used to verify the solution. The next example is a case in point.

Example 2

Implicit Solution Form

Solve the differential equation

$$ye^{x^2} + \frac{y^2 - 1}{x}y' = 0$$

subject to the initial condition $y(0) = 1$.

Solution To separate the variables, we write

$$\begin{aligned}\frac{y^2 - 1}{x}\frac{dy}{dx} &= -ye^{x^2} \\ \frac{y^2 - 1}{y}\,dy &= -xe^{x^2}\,dx \\ \int \frac{y^2 - 1}{y}\,dy &= -\int xe^{x^2}\,dx \\ \int \left(y - \frac{1}{y}\right) dy &= -\frac{1}{2}\int 2xe^{x^2}\,dx \\ \frac{y^2}{2} - \ln |y| &= -\frac{1}{2}e^{x^2} + C_1 \\ y^2 - \ln (y^2) &= -e^{x^2} + C\end{aligned}$$

Since $y = 1$ when $x = 0$, we find C to be

$$\begin{aligned}1 - \ln (1) &= -1 + C \\ 2 &= C\end{aligned}$$

Finally, in implicit form the particular solution satisfying the given initial condition is

$$y^2 - \ln(y^2) = -e^{x^2} + 2$$

□

Remark

In our description of the technique of separation of variables, we assumed that we begin with an equation of the form

$$\frac{dy}{dx} = f(x)g(y)$$

In practice, it often happens that f is a constant function. Don't let this bother you. You can still "separate the variables" as demonstrated in the next example.

Example 3

Separating Variables When f Is a Constant Function

In the Introductory Example of Section 7.1, we discussed the differential equation

$$\frac{dy}{dt} = k(1 - b)(y - a)$$

where y is the national income, t is the time in years, and a, b, and k are constants. Use the method of separation of variables to solve this equation.

Solution Considering $k(1 - b)$ to be a *constant* function of t, we separate variables as follows:

$$\int \frac{1}{y - a}\, dy = \int k(1 - b)\, dt$$

$$\ln|y - a| = k(1 - b)t + C_1$$

$$y - a = Ce^{k(1-b)t}$$

$$y = a + Ce^{k(1-b)t}$$

□

Example 4

Finding a Particular Solution

Find the equation of the curve having the following characteristics.

1. At *each* point (x, y) on the curve, the slope of the curve is y/x^2.
2. The curve passes through the point $(1, 3)$.

Solution Since the slope of the curve is given, we have the differential equation

$$\frac{dy}{dx} = \frac{y}{x^2}$$

with the boundary condition $y(1) = 3$. Separating the variables and integrating, we get

$$\frac{1}{y}\,dy = \frac{1}{x^2}\,dx$$

$$\int \frac{1}{y}\,dy = \int \frac{1}{x^2}\,dx$$

$$\ln|y| = -\frac{1}{x} + C_1$$

$$|y| = e^{-(1/x)+C_1} = e^{-1/x}e^{C_1}$$

$$y = Ce^{-1/x}$$

Now, since $y = 3$ when $x = 1$, it follows that

$$3 = Ce^{-1}$$

$$3e = C$$

Therefore, the equation of the specified curve is

$$y = (3e)e^{-1/x} = 3e^{(x-1)/x}$$

(See Figure 7.4.)

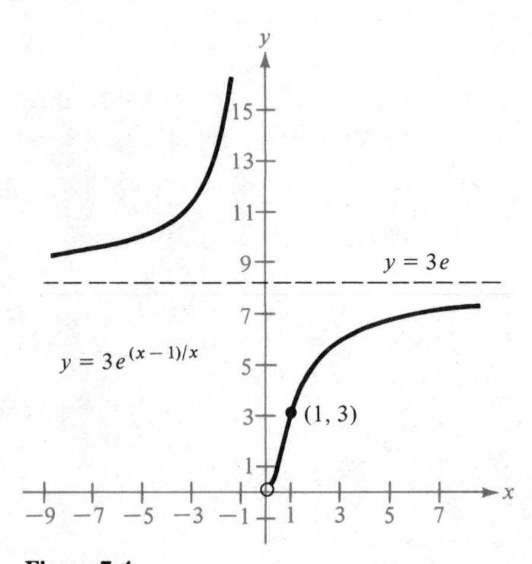

Figure 7.4

□

Section Exercises (7.2)

In Exercises 1–16, use separation of variables to find the general solution of the given differential equation.

1. $\dfrac{dy}{dx} = 2x$

2. $\dfrac{dy}{dx} = \dfrac{1}{x}$

3. $3y^2\dfrac{dy}{dx} = 1$

4. $(y + 1)\dfrac{dy}{dx} = 2x$

5. $y' - xy = 0$

6. $(1 + y)y' - 4x = 0$

7. $(1 + x)y' - 2y = 0$

8. $y' - y = 5$

9. $e^y\dfrac{dy}{dt} = 3t^2 + 1$

10. $\dfrac{dy}{dt} = \sqrt{\dfrac{t}{y}}$

11. $\dfrac{dy}{dx} = \dfrac{x}{y}$

12. $\dfrac{dy}{dx} = \dfrac{x^2 + 2}{3y^2}$

13. $(2 + x)y' = 2y$

14. $xy' = y$

15. $y \ln x - xy' = 0$

16. $\dfrac{dx}{dy} - y(x + 1) = 0$

In Exercises 17–22, use separation of variables to find the particular solution of the given differential equation.

17. $yy' - e^x = 0;\ y(0) = 4$

18. $\sqrt{x} + \sqrt{y}y' = 0;\ y(1) = 4$

19. $y(x + 1) + y' = 0;\ y(-2) = 1$

20. $xyy' - \ln x = 0;\ y(1) = 0$

21. $\dfrac{dP}{dt} - kP = 0;\ P(0) = P_0$

22. $\dfrac{dT}{dt} + k(T - 70) = 0;\ T(0) = 140$

23. Find an equation for the curve that passes through the point (1, 1) and has a slope of $-9x/16y$.

24. Find an equation for the curve that passes through the point (8, 2) and has a slope of $2y/3x$.

Ⓒ**25.** Two young people are tobogganing on a slope with an incline of approximately 12°. After considering the force of gravity, the friction, and the air resistance, the following differential equation was derived to describe the motion of the toboggan

$$12.5 \frac{dv}{dt} = 43.2 - 1.25v$$

where v is the velocity in feet per second. Find the velocity as a function of time if $v = 0$ when $t = 0$. As the value of t increases, what is the value of v that the speed of the toboggan approaches?

Ⓒ**26.** Repeat Exercise 25 with the differential equation

$$12.5 \frac{dv}{dt} = 43.2 - 1.75v$$

(The coefficient of v indicates that the air resistance is 1.75 times the speed of the toboggan.)

Ⓒ**27.** In a chemical reaction a certain compound changes into another compound at a rate proportional to the unchanged amount. If initially there was 20 g of the original compound and 16 g after 1 h, when will 75% of the compound be changed?

Ⓒ**28.** Newton's Law of Cooling states that the rate of change in the temperature of an object is proportional to the difference between its temperature and the temperature of the surrounding air. Suppose a room is kept at a constant temperature of 70° and an object cooled from 350° to 150° in 45 min. At what time will the object cool to a temperature of 80°?

Ⓒ**29.** Use Newton's Law of Cooling in Exercise 28 to find the temperature at $t = 6$ h of a food product that is placed in a freezer whose temperature is 0°. Assume the original temperature of the food is 70° and that after 1 h the temperature of the food has dropped to 48°. At what time will the product cool to a temperature of 10°?

30. The absorption of x rays through a body is defined to be the rate of change of the intensity I of the x ray with respect to the depth of penetration r. The absorption is proportional to the density ρ of the body and to the intensity. This is described by the differential equation

$$\text{absorption} = \frac{dI}{dr} = -k\rho I$$

Solve this differential equation to find the intensity as a function of the depth of penetration. (Assume that the intensity is I_0 when $r = 0$.)

Ⓒ Calculator may be helpful.

Section 7.3

Applications of Differential Equations

Introductory Example

Crop Yield

In this text we have made many references to mathematical models. By now you should be familiar with the two major goals in developing a model: accuracy and simplicity. In practice, we often find a model by making various simplifying assumptions and then checking the accuracy produced by these assumptions.

As a case in point, let us consider three models of differential equations to predict the yield of a crop as a function of time. For our basic model we assume that w is the *current* dry weight of the crop, W is the *maximum* dry weight, and the rate of growth of the crop is

$$\frac{dw}{dt} = f(t)[(W - w)w]$$

where $f(t)$ is the coefficient of growth as a function of time. For simplicity, we measure the time in growing seasons. That is, $t = 0$ represents the time of germination of the crop seeds, and $t = 1$ represents the harvest time.

Now, we consider possibilities for the function $f(t)$:

(a) $f_1(t) = k_1$ (b) $f_2(t) = k_2t$ (c) $f_3(t) = k_3t(1 - t)$

The solutions to the three resulting differential equations are, respectively, as follows:

$$\text{(a)}\quad w = W\left[\frac{1}{1 + C_1e^{-k_1t}}\right]$$

$$\text{(b)}\quad w = W\left[\frac{1}{1 + C_2e^{-k_2t^2/2}}\right]$$

$$\text{(c)}\quad w = W\left[\frac{1}{1 + C_3e^{-k_3t^2(3-2t)/6}}\right]$$

Figure 7.5 compares the different growth patterns of these three models. Of course, to decide which model is best, we would need to compare each model's values with the actual measurements for crop weight.

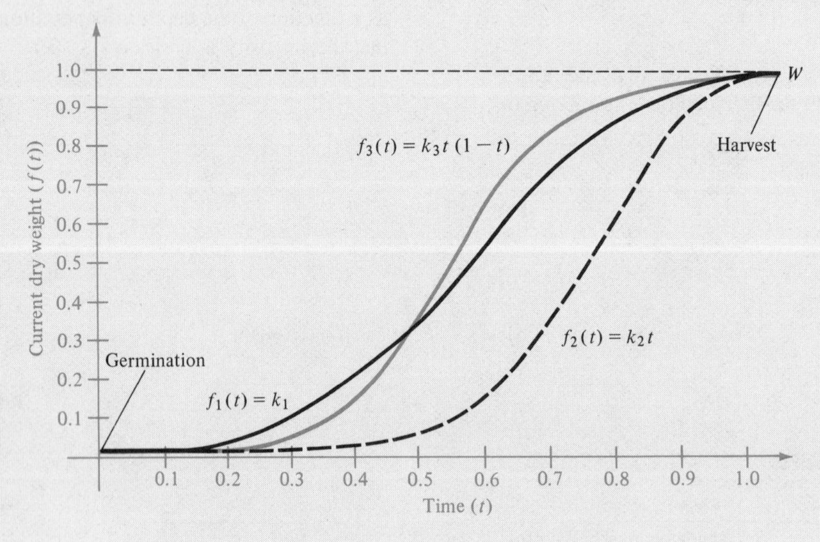

Figure 7.5

Section Objective: ***To introduce several common applications of differential equations.***

In this concluding section on differential equations, we review the general types of applications discussed thus far and introduce some additional types.

Recall from Section 5.5 that the mathematical model for *exponential growth (or decay)* resulted from the differential equation

$$\frac{dy}{dt} = ky$$

The rate of change of y — *is* — *proportional to* — *y*

The *logistics growth curve* was developed from the differential equation

$$\frac{dy}{dt} = ky(L - y)$$

The rate of change of y — *is* — *proportional to* — *y* — *and* — *the difference between L and y*

The model for *Newton's Law of Cooling* was developed from the differential equation

$$\frac{dy}{dt} = k(L - y)$$

The rate of change of y — *is* — *proportional to* — *the difference between L and y*

These three instances suggest that the key terms to look for when constructing a differential equation for a given relationship are "rate of change" and "proportional."

Example 1

A Chemical Reaction Model

During a certain chemical reaction, substance A is converted into substance B at a rate proportional to the square of the amount of A. If 60 g of A are present when $t = 0$ and only 10 g remain unconverted after 1 h ($t = 1$), what is the mathematical model for the amount of A present at any time? How much of A is present after 2 h?

Solution Letting y be the amount of (unconverted) A at any time t, we have the differential equation

$$\frac{dy}{dt} = ky^2$$

The rate of change of y — *is* — *proportional to* — *the square of y*

Separating variables, we have

$$\int \frac{dy}{y^2} = \int k\,dt$$

$$-\frac{1}{y} = kt + C$$

$$y = \frac{-1}{kt + C}$$

Since $y = 60$ when $t = 0$, we have

$$60 = -\frac{1}{C}$$

$$C = -\tfrac{1}{60}$$

Furthermore, since $y = 10$ when $t = 1$, we have

$$10 = \frac{-1}{k - \frac{1}{60}}$$

$$10 = \frac{-60}{60k - 1}$$

$$k = \tfrac{-1}{12}$$

Thus, the model describing this chemical reaction is

$$y = \frac{-1}{\frac{-1}{12}t - \frac{1}{60}} = \frac{60}{5t + 1}$$

Finally, when $t = 2$,

$$y = \tfrac{60}{11} \approx 5.45 \text{ g}$$

Figure 7.6 shows the graph of this funcion.

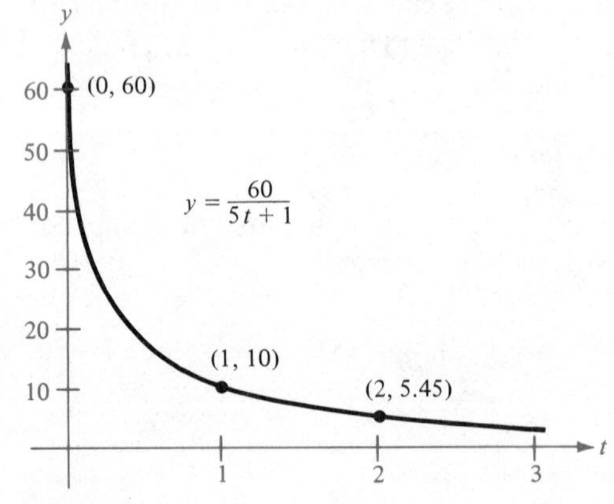

Figure 7.6

□

Example 2

The Gompertz Growth Equation

The *Gompertz growth equation* is a mathematical model describing a limited growth situation in which the rate of growth of y is proportional to y *and* to the natural log of the ratio of L to y, where L is the maximum population size. Set up a differential equation for this model and sketch its graph. Use the conditions $L = 100$, $y = 10$ when $t = 0$, and $y = 50$ when $t = 10$, where t is measured in days.

Solution Since the log of the ratio of L to y is $\ln (L/y)$, we have the differential equation

$$\frac{dy}{dt} = ky \ln \left(\frac{L}{y}\right)$$

The rate of change of y — *is* — *proportional to* — *y* — *and* — *the log of the ratio of L to y*

Separating variables gives us

$$\int \frac{dy}{y \ln (L/y)} = \int k\, dt$$

Letting

$$u = \ln \left(\frac{L}{y}\right) = \ln L - \ln y$$

we have $u' = -1/y$ which yields

$$-\int \frac{-(1/y)}{\ln (L/y)}\, dy = \int k\, dt$$

$$-\ln \left|\ln \left(\frac{L}{y}\right)\right| = kt + C_1$$

$$\ln \left(\frac{L}{y}\right) = Ce^{-kt}$$

$$\frac{L}{y} = e^{Ce^{-kt}}$$

$$y = \frac{L}{e^{Ce^{-kt}}} = Le^{-Ce^{-kt}}$$

Now, since $L = 100$ and $y = 10$ when $t = 0$, we have

$$10 = 100e^{-C}$$

$$C = -\ln \left(\tfrac{1}{10}\right) \approx 2.3026$$

Furthermore, since $y = 50$ when $t = 10$, we have

$$50 = 100e^{-2.3026e^{-10k}}$$

$$\ln\left(\tfrac{1}{2}\right) = -2.3026e^{-10k}$$

$$\frac{-0.6931}{-2.3026} = e^{-10k}$$

$$k = -\tfrac{1}{10}\ln(0.3010) \approx 0.12006$$

Thus, the model for this particular growth pattern is

$$y = 100e^{-2.3026e^{-0.12006t}}$$

Figure 7.7 shows the graph of this equation.

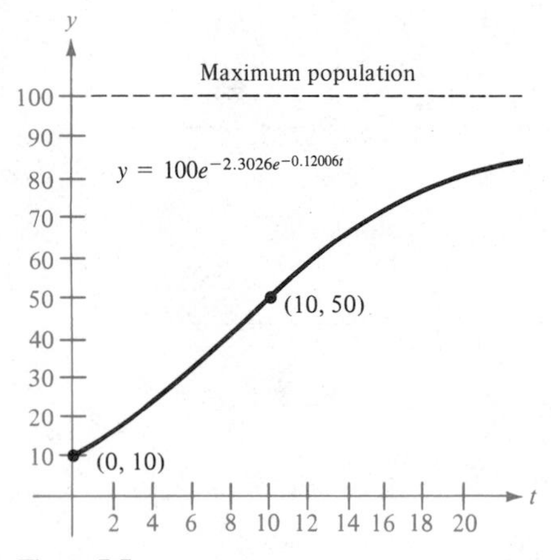

Figure 7.7

□

Example 3

Hybrid Selection Model

In the study of genetics, one encounters a family of differential equations of the form

$$\frac{dy}{dt} = ky(1 - y)(a - by)$$

where y represents the portion of the population having a certain characteristic and t represents the time measured in "generations." If $y = 0.5$ when $t = 0$, $y = 0.8$ when $t = 4$, $a = 2$, and $b = 1$, find the model relating y and t.

Solution Separating variables, we have

$$\int \frac{1}{y(1 - y)(2 - y)}\,dy = \int k\,dt$$

Now, integrating by partial fractions, we obtain

$$\int \left(\frac{\frac{1}{2}}{y} + \frac{1}{1 - y} - \frac{\frac{1}{2}}{2 - y}\right) dy = \int k\,dt$$

$$\frac{1}{2}\ln|y| - \ln|1-y| + \frac{1}{2}\ln|2-y| = kt + C_2$$

$$\ln\left|\frac{\sqrt{y}\sqrt{2-y}}{(1-y)}\right| = kt + C_2$$

$$\left|\frac{\sqrt{y}\sqrt{2-y}}{(1-y)}\right| = C_1 e^{kt}$$

$$\frac{y(2-y)}{(1-y)^2} = Ce^{2kt}$$

Since $y = 0.5$ when $t = 0$, we have

$$\frac{0.5(1.5)}{(0.5)^2} = C$$

$$3 = C$$

Also, since $y = 0.8$ when $t = 4$, we have

$$\frac{0.8(1.2)}{(0.2)^2} = 3e^{8k}$$

$$8 = e^{8k}$$

$$\frac{1}{8}\ln(8) = k$$

$$0.2599 \approx k$$

Thus, the model relating y and t is

$$\frac{y(2-y)}{(1-y)^2} = 3e^{0.2599t}$$

Figure 7.8 shows the graph of this model.

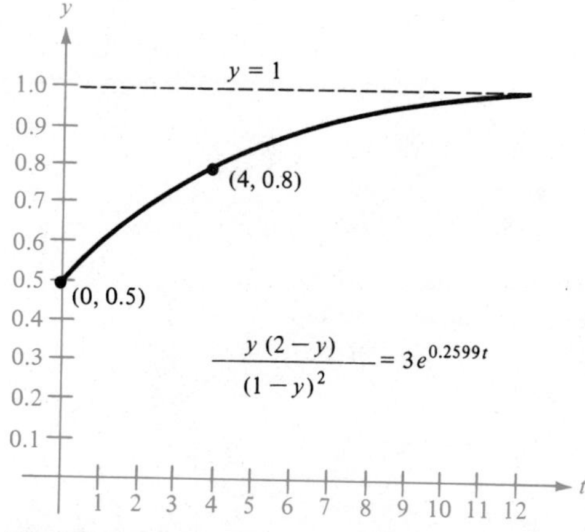

Figure 7.8

□

Remark From the examples in this section, you can see that a great deal of the work in applying differential equations involves solving for the constants k and C. Because of the many opportunities for error, it is a good idea to check to see if the boundary conditions satisfy your final equation.

Section Exercises (7.3)

In Exercises 1 and 2, use the chemical reaction model of Example 1 to find the amount A as a function of t and sketch the graph of the function.

Ⓒ **1.** $A = 100$ g when $t = 0$, and $A = 37$ g when $t = 2$

Ⓒ **2.** $A = 75$ g when $t = 0$ and $A = 12$ g when $t = 1$

In Exercises 3 and 4, use the Gompertz growth model described in Example 2 to find the growth function and sketch its graph.

Ⓒ **3.** $L = 500$; $y = 100$ when $t = 0$, and $y = 150$ when $t = 2$

Ⓒ **4.** $L = 5000$; $y = 500$ when $t = 0$, and $y = 625$ when $t = 1$

In Exercises 5–8, suppose that a medical researcher wants to determine the concentration C (in moles per liter) of a tracer drug injected into a moving fluid. We can start solving this problem by considering a single-compartment dilution model. (See Figure 7.9.) We assume that the fluid in the compartment is continuously mixed and that the volume of fluid in the compartment is constant.

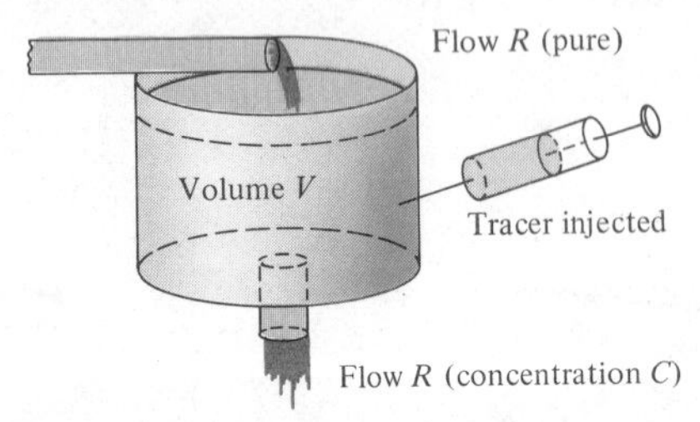

Figure 7.9

5. If the tracer is injected instantaneously at time $t = 0$, then the concentration of the fluid in the compartment begins diluting according to the differential equation

$$\frac{dC}{dt} = \left(-\frac{R}{V}\right)C, \qquad C = C_0 \text{ when } t = 0$$

(a) Solve this differential equation to find the concentration as a function of time.

(b) Find the limit of C as $t \to \infty$.

Ⓒ **6.** Using the solution of the differential equation in Exercise 5, find the concentration as a function of time and sketch its graph if:

(a) $V = 2$ L, $R = 0.5$ L/min, and $C_0 = 0.6$ mole/L

(b) $V = 2$ L, $R = 1.5$ L/min, and $C_0 = 0.6$ mole/L

7. In Exercises 5 and 6, we assumed that there was a single initial injection of the tracer drug into the compartment. Now let us consider the case in which the tracer is continuously injected (beginning at t = 0) at the rate of Q mole/min. By considering Q to be negligible compared to R, we have the differential equation

$$\frac{dC}{dt} = \frac{Q}{V} - \left(\frac{R}{V}\right)C, \qquad C = 0 \text{ when } t = 0$$

(a) Solve this differential equation to find the concentration as a function of time.

(b) Find the limit of C as $t \to \infty$.

Ⓒ **8.** Use the solution of Exercise 7 to find the concentration if $V = 2$ L, $R = 1$ L/min, and $Q = 0.75$ mole/min.

Ⓒ **9.** Assume that the rate of change in the number of miles s of road cleared of snow per hour by a snowplow is inversely proportional to the height h of snow. That is

$$\frac{ds}{dh} = \frac{k}{h}$$

Ⓒ Calculator may be helpful.

Find s as a function of h if $s = 25$ when $h = 2$ and $s = 12$ when $h = 6$ $(2 \le h \le 15)$.

Ⓒ**10.** A wet towel is hung from a clothesline to dry and loses moisture through evaporation at a rate proportional to its moisture content. If after 1 h the towel has lost 40% of its original moisture content, at what time will it have lost 80%?

11. In a certain learning theory model, it is assumed that the rate of increase in the proportion P of correct responses after n trials is proportional to P and to $L - P$ where L is the limiting proportion of correct responses. Write the differential equation for this model and solve it for P as a function of n.

Ⓒ**12.** Use the solution of Exercise 11 to write P as a function of n and then sketch the graph of the solution.

(a) $L = 1.00$; $P = 0.50$ when $n = 0$, and $P = 0.85$ when $n = 4$

(b) $L = 0.80$; $P = 0.25$ when $n = 0$, and $P = 0.60$ when $n = 10$

13. Let x and y be the sizes of two organs of a particular mammal at time t. Empirical data indicate that the *relative* growth rates of these two organs are equal, and hence we have

$$\frac{1}{x}\frac{dx}{dt} = \frac{1}{y}\frac{dy}{dt}$$

Solve this differential equation, writing y as a function of x.

14. When predicting population growth, demographers must consider birth and death rates, as well as the net change caused by the difference between the rates of immigration and emigration. Let P be the population at time t and N be the net increase per unit of time due to the difference between immigration and emigration. Thus, the rate of growth of the population is given by

$$\frac{dP}{dt} = kP + N$$

Solve this differential equation to find P as a function of time.

15. A large corporation starts at time $t = 0$ to invest part of its receipts at a rate of P dollars per year in a fund for future corporate expansion. Assume that the fund earns r percent per year compounded continuously. Thus, the rate of growth of the amount A in the fund is given by

$$\frac{dA}{dt} = rA + P$$

where $A = 0$ when $t = 0$. Solve this differential equation for A as a function of t.

Ⓒ**16.** Use the result of Exercise 15 to find the amount in the fund if:

(a) $P = \$100{,}000$, $r = 12\%$, and $t = 5$ years

(b) $P = \$250{,}000$, $r = 15\%$, and $t = 10$ years

Ⓒ**17.** Use the result of Exercise 15 to find P if the corporation needs $120,000,000 in 8 years and the fund earns $16\frac{1}{4}\%$ compounded continuously.

Ⓒ**18.** Use the result of Exercise 15 to find the time t if the corporation needs $800,000. Assume that it can invest $75,000 per year and the fund earns 13% compounded continuously.

Ⓒ Calculator may be helpful.

Chapter 8

Functions of Several Variables

Section 8.1

Surfaces in Space and the Three-Dimensional Coordinate System

Introductory Example

Total Revenue from Several Products

We have already considered the total revenue derived from the sale of x units of a single product at a price of p dollars per unit,

$$R = xp$$

This formula can be generalized as follows to cover the revenue of several different products:

$$R = x_1p_1 + x_2p_2 + \cdots + x_np_n$$

where x_i represents the number of units and p_i represents the price per unit of the ith product. Although this multiproduct formula is the result of a fairly simple extension of the single-product case, a graphical interpretation of the multiproduct case is not so easy to come by. The problem lies in the fact that *each* of the variables $x_1, x_2, \ldots, x_n$, and R requires its own dimension in the graph. Since we are visually limited to a maximum of three dimensions, our graphical interpretation of revenue functions is limited to one- and two-product cases.

As an example of a two-product situation, consider a manufacturer who sells x units of one product at \$2.00 per unit and y units of another product at \$0.50 per unit. The total revenue for these two products is given by

$$R = (2)(x) + (0.5)(y) = 2x + \frac{y}{2}$$

Graphically, we can represent this equation as a **surface in space.** (In this particular case the surface happens to be a plane.) The values of x and y are represented in the (horizontal) xy-plane, and the revenue R is represented by the height of the surface. (See Figure 8.1.)

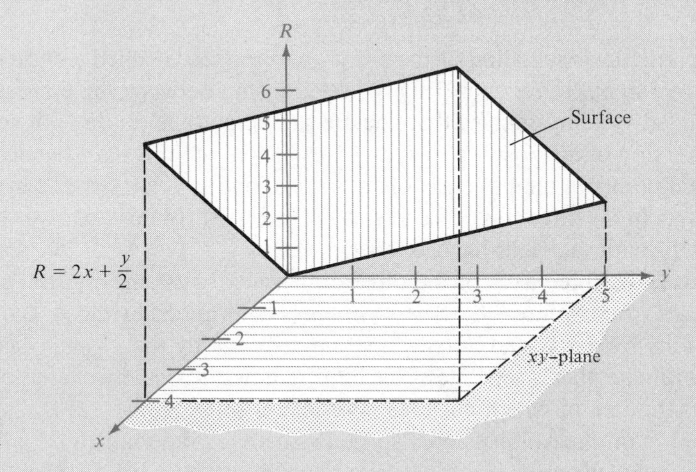

Figure 8.1

Section Objectives: ***To introduce the three-dimensional coordinate system and the formula for the distance between two points in space ▪ To introduce some basic surfaces in space.***

In Chapter 1, we described the Cartesian plane as the plane determined by two mutually perpendicular number lines called the x- and y-axes. These axes together with their point of intersection (the origin) allowed us to develop a two-dimensional coordinate system for identifying points in a plane and for discussing topics in plane analytic geometry. To identify a point in space, we need to introduce a third dimension to our model. The geometry of this three-dimensional model is referred to as **solid analytic geometry.**

To construct a three-dimensional coordinate system, we begin with the two-dimensional system (the xy-plane in a horizontal position) and through the origin, 0, we pass a vertical z-axis that is perpendicular to both the x- and y-axes (Figure 8.2).

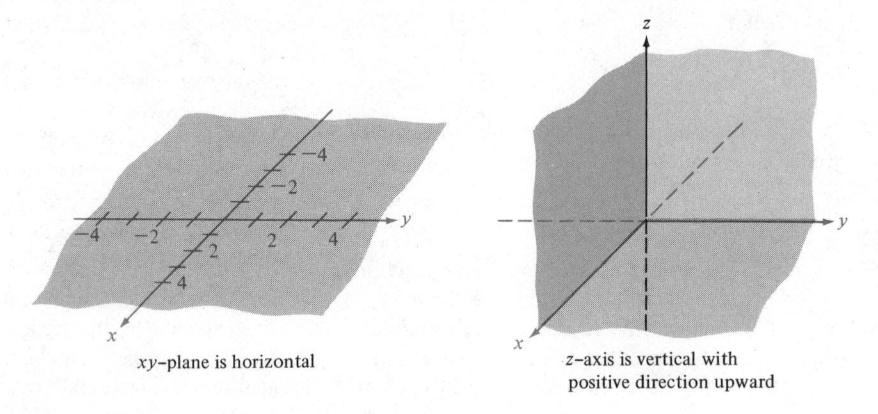

Figure 8.2

This particular orientation of the x-, y-, and z-axes is called a **right-handed system.** As an aid to remembering the relationship between the three axes in a right-handed system, imagine that you are standing at the origin with your arms in the direction of the positive x- and y-axes. The system is right-handed if your *right* hand points in the direction of the x-axis, and it is left-handed if your *left* hand points in the direction of the x-axis (Figure 8.3). In this text, we will work exclusively with the right-handed system.

Pairwise, the three axes of the three-dimensional system form three *coordinate planes*. (See Figure 8.2.) The *xy-plane* is determined by the x- and y-axes, the *xz-plane* by the x- and z-axes, and the *yz-plane* by the y- and z-axes. The three coordinate planes separate space into eight **octants,** and we refer to the *first octant* as the one in which all three coordinates are positive.

A point P in three-dimensional space (3-space) is determined by an ordered triple (x, y, z), where the x-coordinate denotes the directed distance from the yz-plane to P, the y-coordinate the directed distance from the xz-plane to P, and

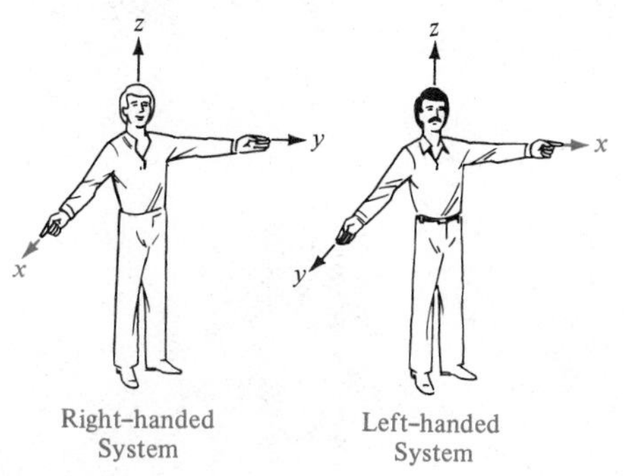

Figure 8.3

the z-coordinate the directed distance from the xy-plane to P. Several points are shown in Figure 8.4.

Many of the geometric properties of 3-space are simple extensions of those established for the plane in Chapter 1. For instance, by using the Pythagorean Theorem twice (see Figure 8.5), we can establish that the *Distance Formula* for the distance d between two points (x_1, y_1, z_1) and (x_2, y_2, z_2) in 3-space is

$$d = \sqrt{(x_2 - x_1)^2 + (y_2 - y_1)^2 + (z_2 - z_1)^2}$$

The *Midpoint Rule* is also an obvious extension of the one for plane geometry. If (x_1, y_1, z_1) and (x_2, y_2, z_2) are two points in space, then the midpoint of the line segment connecting the two points is

$$\left(\frac{x_1 + x_2}{2}, \frac{y_1 + y_2}{2}, \frac{z_1 + z_2}{2}\right)$$

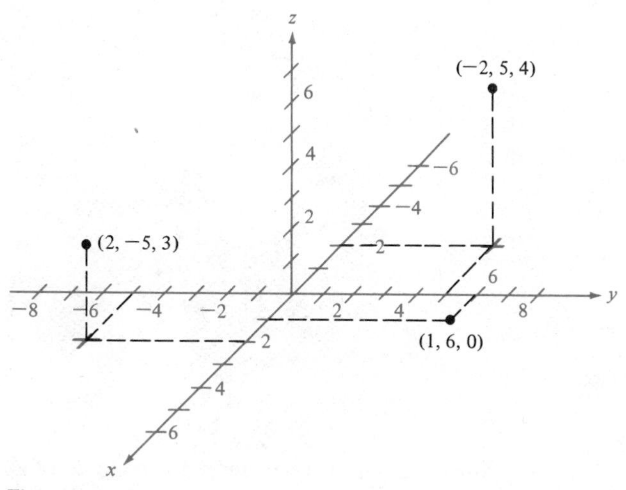

Figure 8.4

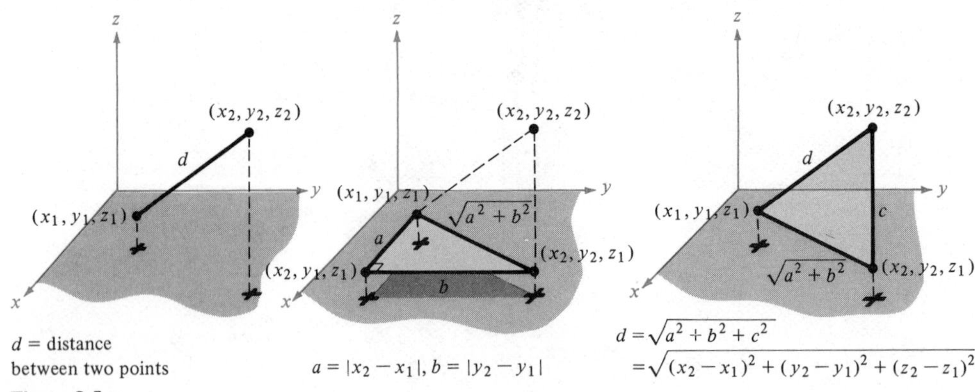

Figure 8.5

Example 1

The Distance and Midpoint Between Two Points

Find the distance between $(5, -2, 3)$ and $(0, 4, -3)$. What is the midpoint of the line segment connecting these two points?

Solution By the Distance Formula,

$$d = \sqrt{(0-5)^2 + (4+2)^2 + (-3-3)^2} = \sqrt{25 + 36 + 36}$$
$$= \sqrt{97}$$

The midpoint is

$$\left(\frac{5+0}{2}, \frac{-2+4}{2}, \frac{3-3}{2}\right) = \left(\frac{5}{2}, 1, 0\right)$$

(See Figure 8.6.)

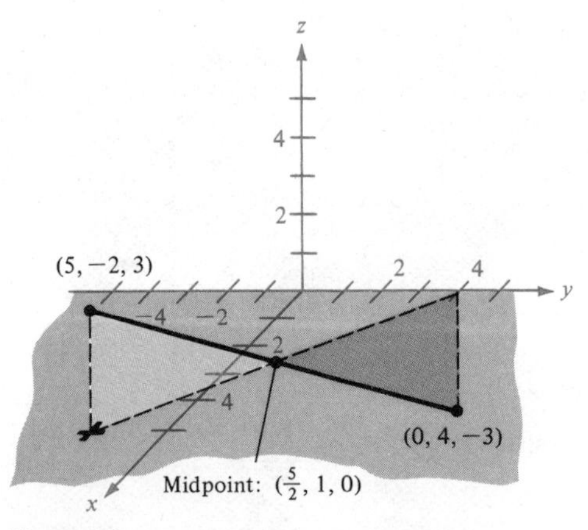

Figure 8.6

□

Since a *sphere* is considered to be the set of all points in space that lie at a fixed distance from a given point, we can use the Distance Formula to obtain the following **standard equation of a sphere** of radius r with center at (h, k, l):

$$(x - h)^2 + (y - k)^2 + (z - l)^2 = r^2$$

Example 2

Finding the Center and Radius of a Sphere

Find the center and radius of the sphere whose equation is

$$x^2 + y^2 + z^2 - 2x + 4y - 6z + 8 = 0$$

Solution We can obtain the standard equation of this sphere by completing the square with each variable, as follows:

$$(x^2 - 2x \qquad) + (y^2 + 4y \qquad) + (z^2 - 6z \qquad) = -8$$

$$(x^2 - 2x + 1) + (y^2 + 4y + 4) + (z^2 - 6z + 9) = -8 + 1 + 4 + 9$$

$$(x - 1)^2 + (y + 2)^2 + (z - 3)^2 = 6$$

Therefore, the center of the sphere is at $(1, -2, 3)$, and its radius is $\sqrt{6}$. (See Figure 8.7.)

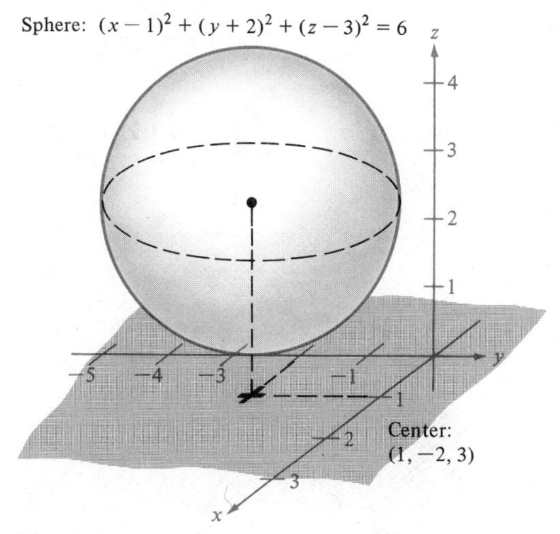

Figure 8.7

□

Note in Example 2 that the points satisfying the equation of the given sphere are the points on the sphere's *surface* (not points inside the sphere). In general, we call the collection of points satisfying an equation involving x, y, and z a **surface in space.** The visualization of a general surface in space is greatly aided by finding the intersection of the surface with one (or more) of the three coordinate planes. We call such intersections **traces.** For example, the xy trace of a surface consists of all points that are common to both the surface *and* the xy-plane. Similarly, the xz- trace of a surface consists of all points that are common to both the surface *and* the xz-plane. The next example demonstrates the use of a trace in analyzing a surface.

Example 3

Finding a Trace of a Surface

Sketch the xy trace of the sphere whose equation is

$$(x - 3)^2 + (y - 2)^2 + (z + 4)^2 = 5^2$$

Solution To find the xy trace of this surface, we use the fact that every point in the xy plane has a z coordinate of zero. This means that if we substitute $z = 0$ into the given equation, the resulting equation (involving the two variables x and y) will represent the intersection of the surface with the xy plane. In other words, by letting $z = 0$, we have

$$(x - 3)^2 + (y - 2)^2 + (0 + 4)^2 = 25$$
$$(x - 3)^2 + (y - 2)^2 + 16 = 25$$
$$(x - 3)^2 + (y - 2)^2 = 9$$
$$(x - 3)^2 + (y - 2)^2 = 3^2$$

Now, since this equation represents a circle (of radius 3) in the xy-plane, we can sketch the xy trace as shown in Figure 8.8.

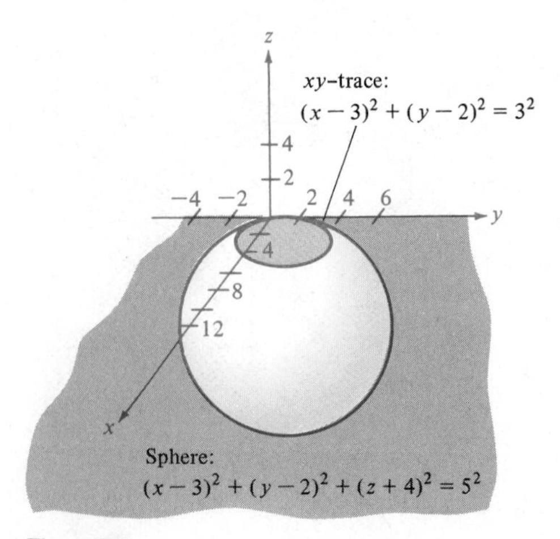

Figure 8.8

□

A second type of surface in space is that of a plane. The **general equation of a plane** is

$$ax + by + cz = d$$

Note the similarity of this equation to the general equation of a line in the plane. As a matter of fact, if we intersect the plane represented by this equation with each of the three coordinate planes, we see that the resulting traces are lines. (See Figure 8.9.) This is what we would expect since we know from geometry that two (nonparallel) planes intersect in a line.

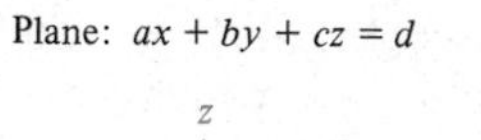

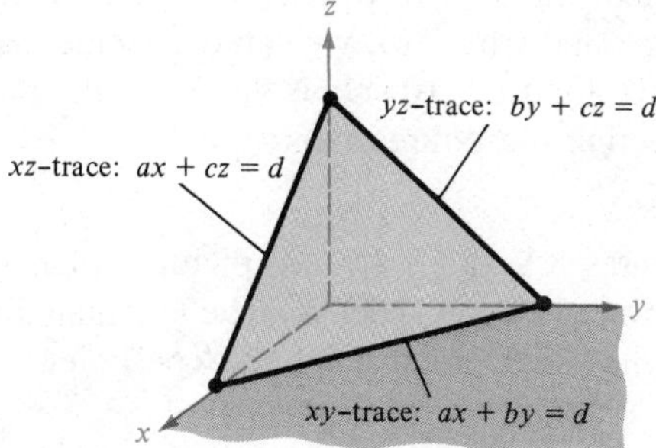

Figure 8.9

In Figure 8.9, the points where the given plane intersects the three coordinate axes are called the x-, y-, and z-intercepts. By connecting these three points, we can form a triangular region, which helps us to visualize the plane in space. Example 4 demonstrates this procedure.

Example 4

Sketching a Plane in Space

Find the x-, y-, and z-intercepts of the plane given by

$$3x + 2y + 4z = 12$$

and then sketch the triangular region formed by connecting these three intercepts.

Solution To find the x-intercept, we let both y and z be zero:

$$3x + 2(0) + 4(0) = 12$$
$$3x = 12$$
$$x = 4$$

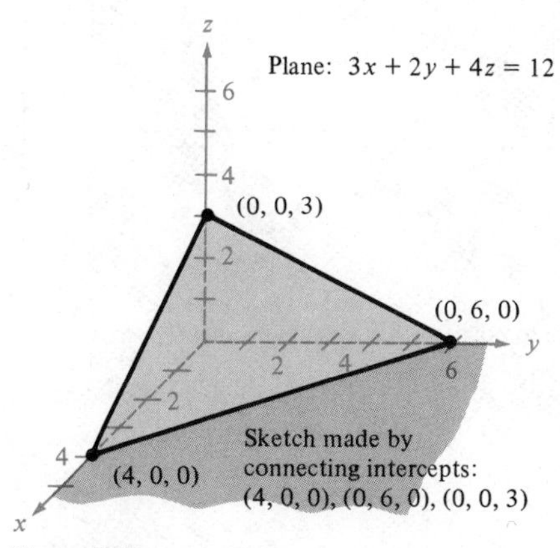

Figure 8.10

Thus, the x-intercept is (4, 0, 0). To find the y-intercept, we let x and z be zero and conclude that $y = 6$, which means that the y-intercept is (0, 6, 0). Similarly, by letting x and y be zero, we can determine that $z = 3$ and that the z-intercept is (0, 0, 3). Figure 8.10 shows the triangular portion of the given plane formed by connecting these three intercepts. □

In Figures 8.9 and 8.10, we pictured planes that had three intercepts. It is possible for a plane in space to have less than three intercepts. In particular, this occurs when one or more of the coefficients in $ax + by + cz = d$ is zero.

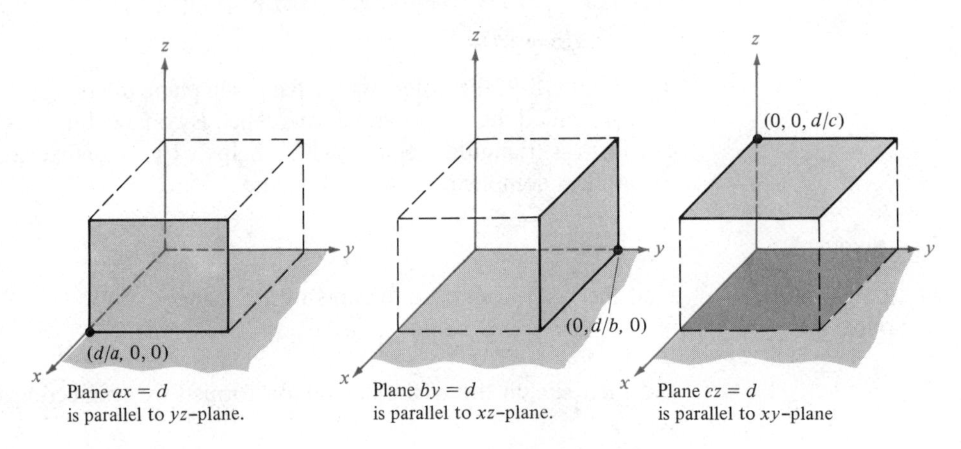

Figure 8.11

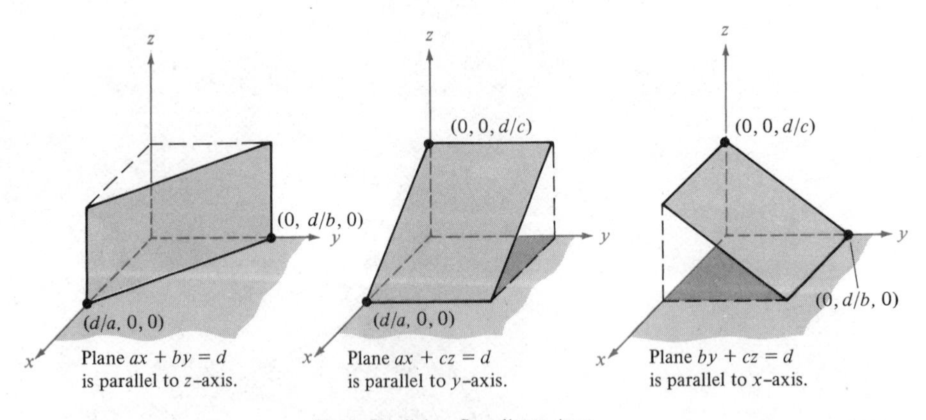

Figure 8.12

Figure 8.11 shows some planes in space that have only one intercept, and Figure 8.12 shows some that have only two intercepts.

So far we have looked at two types of surfaces in space represented by the following equations:

$$\text{sphere: } (x - h)^2 + (y - k)^2 + (z - l)^2 = r^2$$

$$\text{plane: } ax + by + cz = d$$

A third common type of surface in space is the one whose equation is of the form

$$Ax^2 + By^2 + Cz^2 + Dx + Ey + Fz + G = 0$$

We call the graph of such an equation a **quadric surface.** We confine our discussion of quadric surfaces to those having their center at the origin and axes along the coordinate axes.

The six basic quadric surfaces are as follows:

1. the ellipsoid
2. the elliptic paraboloid
3. the hyperbolic paraboloid
4. the hyperboloid of one sheet
5. the elliptic cone
6. the hyperboloid of two sheets

In this text, you will *not* be asked to sketch quadric surfaces in space. However, familiarity with some of the common surfaces will prove to be a great aid in visualizing some of the concepts in this chapter. For this reason, we suggest that you study the equations and quadric surfaces shown in Table 8.1 carefully.

Remark

Table 8.1 includes several computer-generated drawings of quadric surfaces in space. Note that these drawings make use of trace analysis to enhance our three-dimensional perspective. For these particular surfaces, the computer sketched the traces in several evenly spaced planes (all of which were taken parallel to the yz-plane).

When working the exercises in this section, remember that for the purpose of this text, our main goal is to be able to *read* three-dimensional graphs, not to be able to *create* them. In the following exercise set, the only graphing you are asked to do is that involving points or planes in space.

Table 8.1 Quadric Surfaces

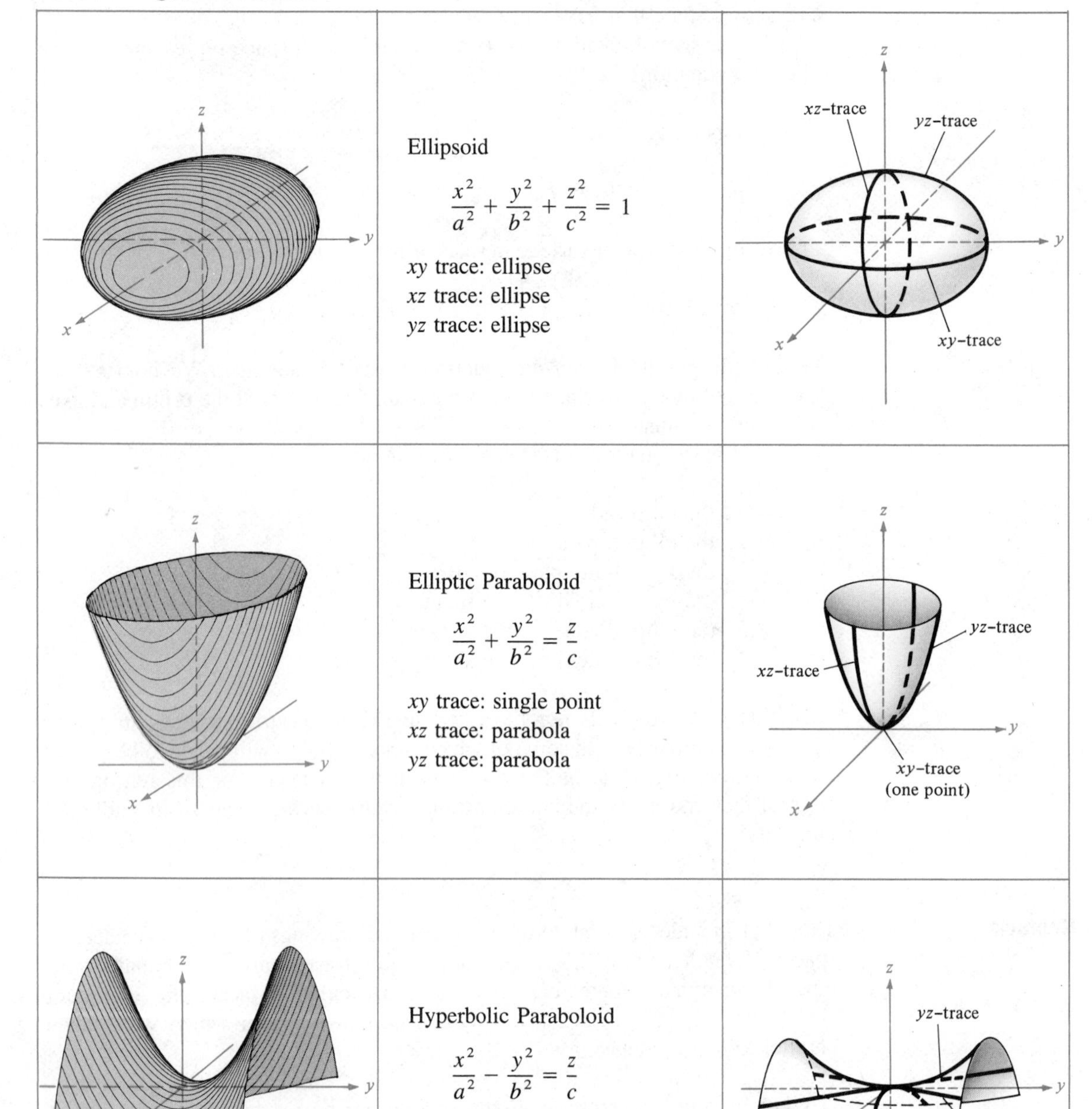

Ellipsoid

$$\frac{x^2}{a^2} + \frac{y^2}{b^2} + \frac{z^2}{c^2} = 1$$

xy trace: ellipse
xz trace: ellipse
yz trace: ellipse

Elliptic Paraboloid

$$\frac{x^2}{a^2} + \frac{y^2}{b^2} = \frac{z}{c}$$

xy trace: single point
xz trace: parabola
yz trace: parabola

Hyperbolic Paraboloid

$$\frac{x^2}{a^2} - \frac{y^2}{b^2} = \frac{z}{c}$$

xy trace: two intersecting lines
xz trace: parabola
yz trace: parabola

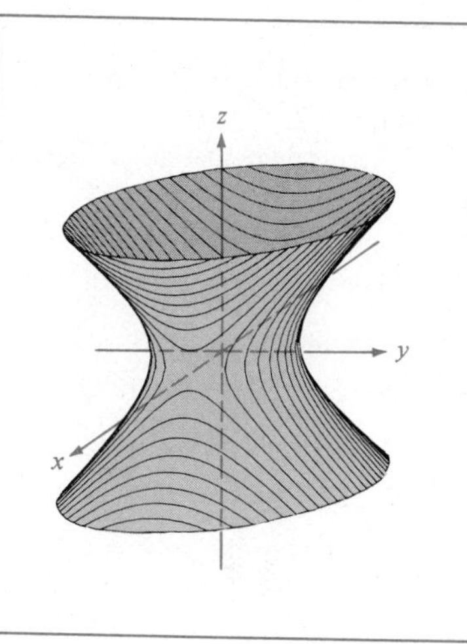

Hyperboloid of One Sheet

$$\frac{x^2}{a^2} + \frac{y^2}{b^2} - \frac{z^2}{c^2} = 1$$

xy trace: ellipse
xz trace: hyperbola
yz trace: hyperbola

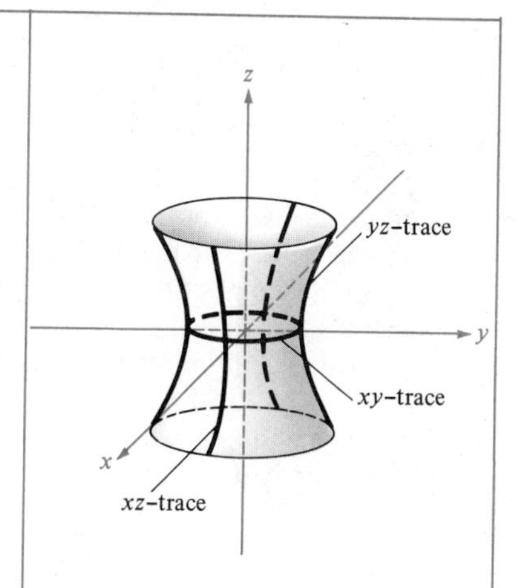

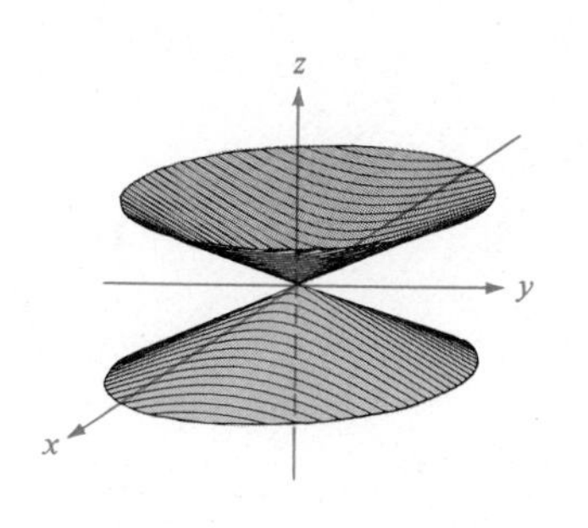

Elliptic Cone

$$\frac{x^2}{a^2} + \frac{y^2}{b^2} - \frac{z^2}{c^2} = 0$$

xy trace: single point
xz trace: two intersecting lines
yz trace: two intersecting lines

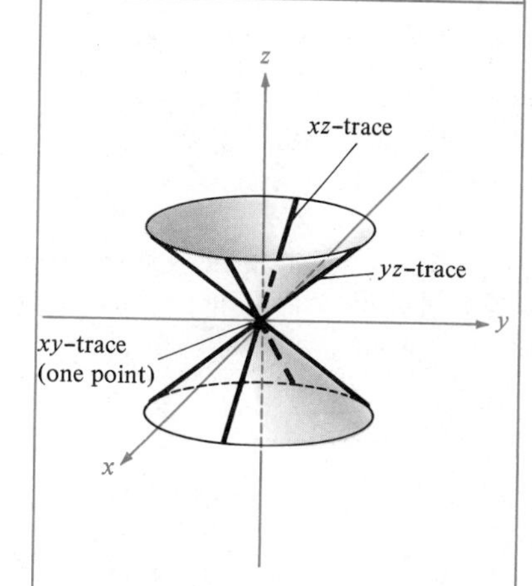

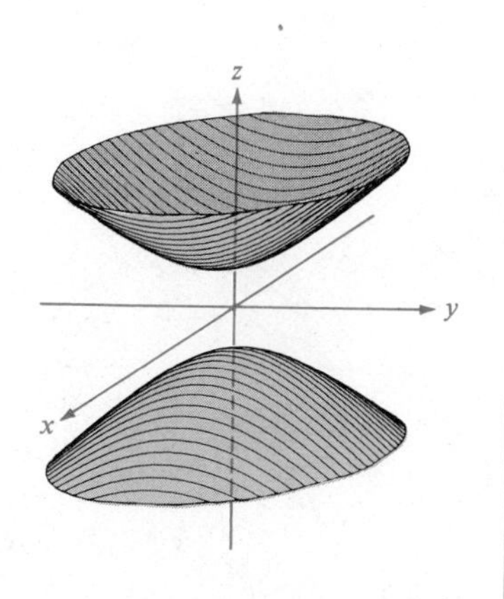

Hyperboloid of Two Sheets

$$\frac{x^2}{a^2} + \frac{y^2}{b^2} - \frac{z^2}{c^2} = -1$$

xy trace: none
xz trace; hyperbola
yz trace: hyperbola

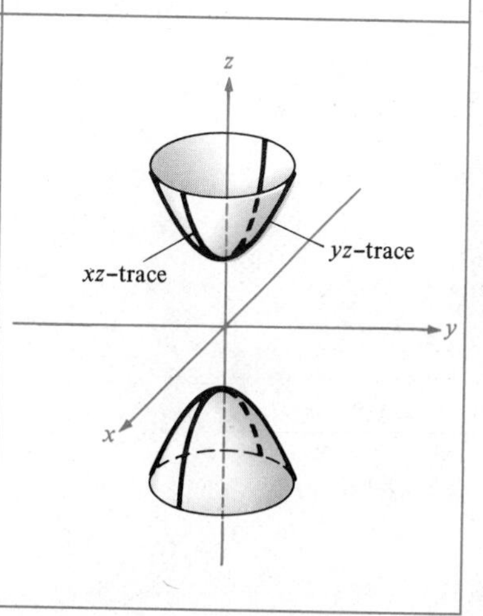

Section Exercises (8.1)

1. Plot the points (2, 1, 3), (−1, 2, 1), and (3, −2, 5).
2. Plot the points ($\frac{3}{2}$, 4, −2), (5, −2, 2), and (5, −2, −2).
3. Find the midpoint of the line segment joining the points (5, −9, 7) and (−2, 3, 3).
4. Find the midpoint of the line segment joining the points (0, −2, 5) and (4, 2, 7).
5. Find the midpoint of the line segment joining the points (−5, −2, 5.5) and (6.3, 4.2, −7.1).
6. The endpoint of a line segment is (−2, 1, 1), and its midpoint is (0, 2, 5). Find the other endpoint.

In Exercises 7–10, find the distance between the two points.

7. (4, 1, 5), (8, 2, 6)
8. (−4, −1, 1), (2, −1, 5)
9. (−1, −5, 7), (−3, 4, −4)
10. (8, −2, 2), (8, −2, 4)

11. Find an equation for the sphere with center at (−2, 1, 1) and radius of 2.
12. Find an equation for the sphere that has a diameter with endpoints (0, 0, 4) and (6, 6, 0).
13. Find the center and radius of the sphere

$$x^2 + y^2 + z^2 - 2x + 6y + 8z + 1 = 0$$

14. Find the center and radius of the sphere

$$4x^2 + 4y^2 + 4z^2 - 4x - 32y + 8z + 33 = 0$$

15. Find the center and radius of the sphere

$$9x^2 + 9y^2 - 6x + 18y + 9z^2 - 26 = 0$$

16. Find the center and radius of the sphere

$$x^2 + y^2 + z^2 - 5x = 0$$

In Exercises 17–22, sketch the graph of each plane.

17. $3x + 3y + 5z = 15$
18. $x + y + z = 3$
19. $z = 3$
20. $x + 2y = 8$
21. $x + y - z = 0$
22. $y = -4$

In Exercises 23–30, match the equation with its graph. (Note that the graphs show only that portion of each surface that lies above the xy-plane.)

23. $\dfrac{x^2}{9} + \dfrac{y^2}{16} + \dfrac{z^2}{9} = 1$
24. $15x^2 - 4y^2 + 15z^2 = -4$
25. $4x^2 - y^2 + 4z^2 = 4$
26. $y^2 = 4x^2 + 9z^2$
27. $4x^2 - 4y + z^2 = 0$
28. $12z = -3y^2 + 4x^2$
29. $4x^2 - y^2 + 4z = 0$
30. $x^2 + y^2 + z^2 = 9$

(a)

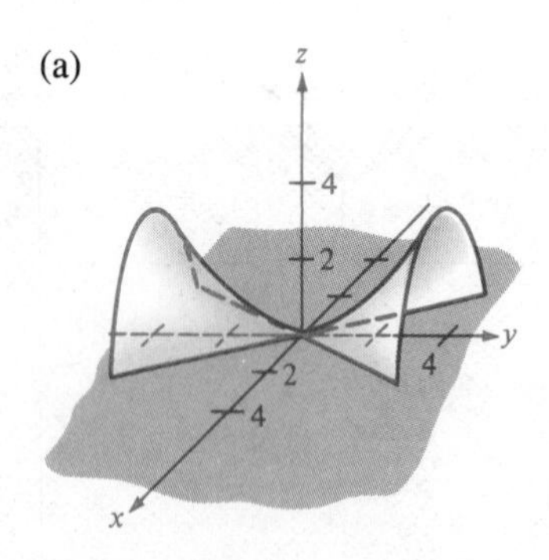

(b)

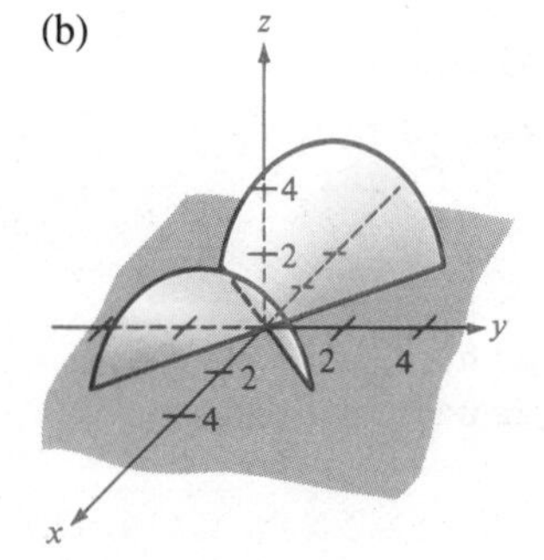

(c)

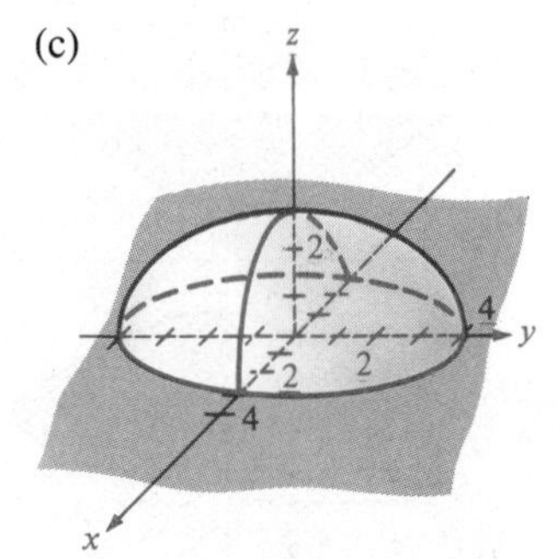

(d)

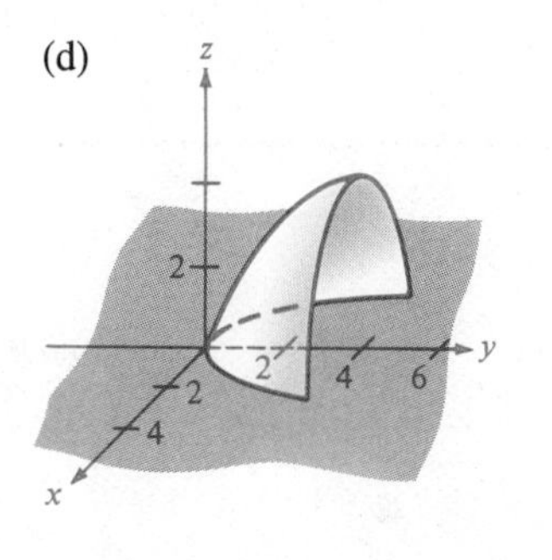

(e)

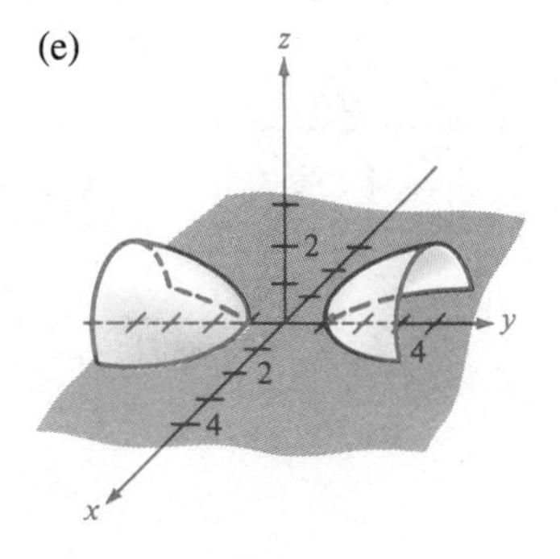

(f)

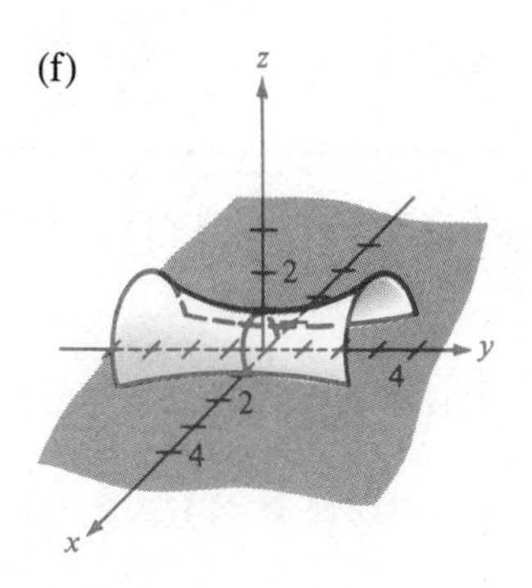

(g)

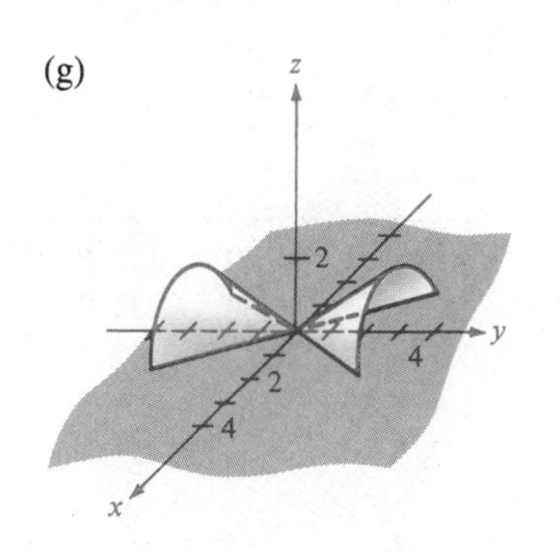

(h)

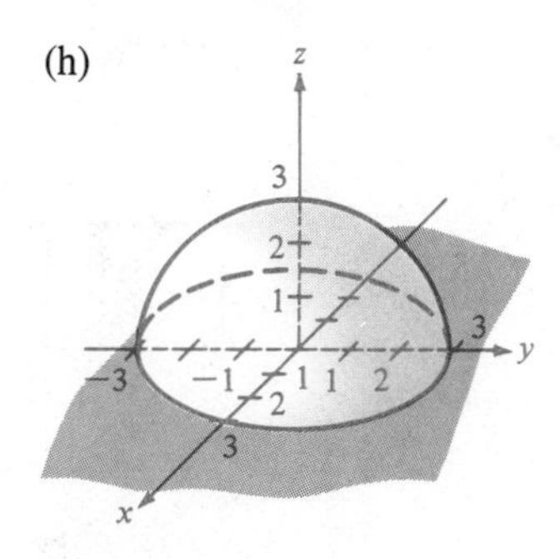

Section 8.2

Functions of Several Variables

Introductory Example

Human Surface Area

The skin surface area A (in square feet) of a person depends on both the weight and the height of the person. One model for estimating skin surface area is given by the following **function of two variables:**

$$A = f(x, y) = 6.579x^{0.425}y^{0.725}$$

where x is the person's weight (in pounds) and y is the person's height (in feet). For example, someone whose weight is 160 lb and whose height is 6 ft has an estimated surface area of

$$A = f(160, 6) = 6.579(160^{0.425})(6^{0.725})$$
$$\approx 208.5 \text{ ft}^2$$

Table 8.2 lists the surface areas (in square feet) given by this model for several different heights and weights.

Figure 8.13 shows the surface in space given by this function. Note that some of the xy-values pictured in Figure 8.13 make no sense for this particular model. For example, a person could not weigh 200 pounds if they were only two feet high! To compensate for this problem, we should restrict this function to an appropriate **domain** in the xy-plane for which the model's estimate is reasonable.

Table 8.2

Weight (lb)	Height 5′	5′4″	5′8″	6′	6′4″
100	149.6	156.8	163.8	170.7	177.6
150	177.7	186.2	194.6	202.8	211.0
200	200.8	210.5	219.9	229.2	238.4
250	220.8	231.4	241.8	252.0	262.1

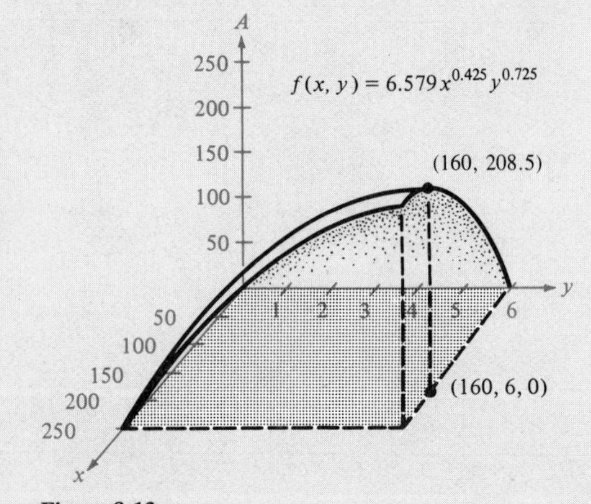

Figure 8.13

Section Objectives: *To introduce the concept of a function of several variables ▪ To discuss the graphs of functions of two variables ▪ To introduce some applications of functions of several variables.*

In the first seven chapters of this book, we have dealt only with functions of a single independent variable. Many familiar quantities are functions not of one, but of two or more variables. For instance, it may be more realistic to consider the demand for a particular product to be a function of the two variables price *and* amount spent on advertising (rather than a function of price alone). Similarly, the growth of a plant may be considered to be a function of the three variables rainfall, hours of sunshine, and amount of fertilizer.

We denote functions of two or more variables by a notation similar to that used for functions of a single variable. For example,

$$f(x, y) = x^2 + xy \qquad \text{and} \qquad g(x, y) = e^{x+y}$$

are functions of two variables, and

$$f(x, y, z) = x + 2y - 3z$$

is a function of three variables.

We give the following definition of a function of *two* variables. Similar definitions can be given for functions of three, four, or n variables.

Definition of a Function of Two Variables

If to each ordered pair (x, y) in some set D there corresponds a unique real number $f(x, y)$, then f is called a **function of x and y**. The set D is the **domain** of f and the corresponding set of values for $f(x, y)$ is the **range** of f.

A function of two variables can be represented geometrically by a surface in space by letting $z = f(x, y)$. Note in Figure 8.14 that, while the surface is three-dimensional, the domain of the function is two-dimensional since it consists of the points in the xy-plane for which the function is defined. In this

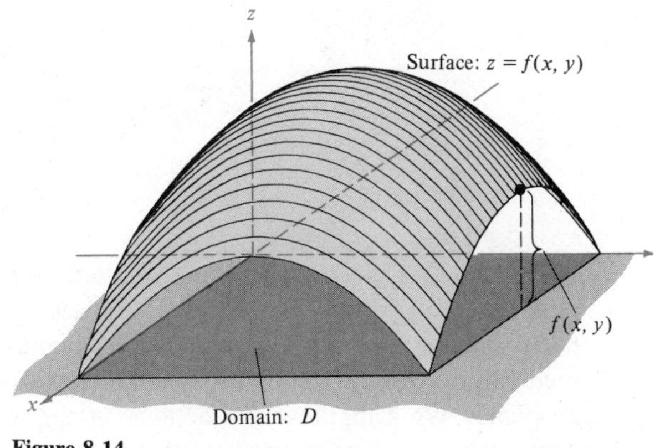

Figure 8.14

chapter, when we refer to the function given by the equation $z = f(x, y)$, we will assume (unless specifically restricted) that the domain is the set of all points (x, y) for which the equation has meaning.

Example 1

Finding the Domain and Range of a Function

Determine the domain and range of the function defined by

$$f(x, y) = \sqrt{64 - x^2 - y^2}$$

Solution Since D is not otherwise specified, we assume the domain of f to be the set of all points (x, y) such that

$$64 - x^2 - y^2 \geq 0 \qquad \text{or} \qquad x^2 + y^2 \leq 64$$

In other words, D is the set of all points lying on or inside the circle

$$x^2 + y^2 = 8^2$$

The range of f is all values $z = f(x, y)$ such that

$$0 \leq z \leq 8$$

The graph of f is a hemisphere, as shown in Figure 8.15.

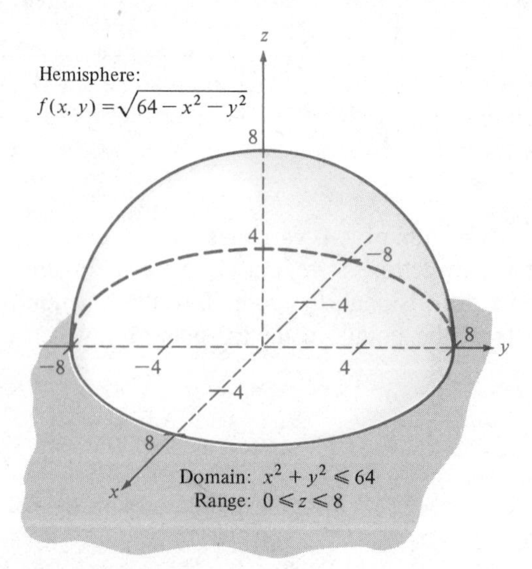

Figure 8.15

□

In Table 8.1 (in the preceding section), we saw that it is often possible to obtain a good picture of a surface in space by sketching traces of the surface taken in several evenly spaced parallel planes. A valuable adaptation of this method involves the use of horizontal planes whose traces are *projected* onto the xy-plane. We call these projections in the xy-plane **level curves.** A graph constructed by sketching several level curves on the same xy-plane is called a

contour map. This process is commonly used in constructing topographical maps, such as the one shown in Figure 8.16. Each of the level curves in this map represents the intersection of the surface $z = f(x, y)$ with the plane $z = c$ ($c = 0, 10, 20, \ldots, 70$). We demonstrate the construction of a contour map in the next example.

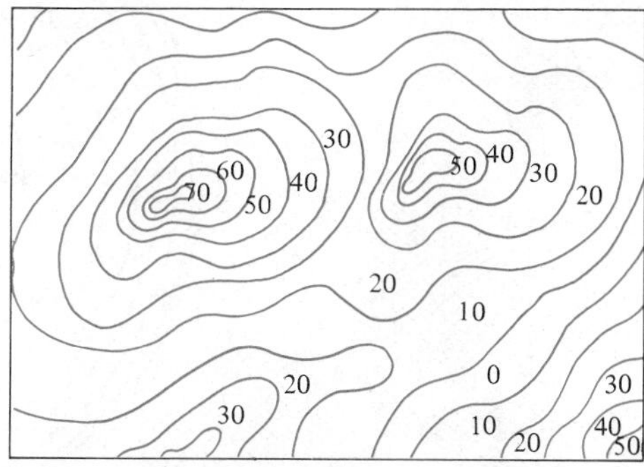

Topographical Map of Two Peaks.
One is over 70 units high and
the other is over 50 units high.

Figure 8.16

Example 2

Sketching a Contour Map

Sketch a contour map for the function $f(x, y) = \sqrt{64 - x^2 - y^2}$ by using $c_1 = 0, c_2 = 1, c_3 = 2, \ldots, c_9 = 8$.

Solution For each of these nine values of c, we obtain a circular level curve.

$$c_1 = 0: \quad 0 = \sqrt{64 - x^2 - y^2};\ x^2 + y^2 = 64 = 8^2$$

$$c_2 = 1: \quad 1 = \sqrt{64 - x^2 - y^2};\ x^2 + y^2 = 63 \approx 7.94^2$$

$$c_3 = 2: \quad 2 = \sqrt{64 - x^2 - y^2};\ x^2 + y^2 = 60 \approx 7.75^2$$

$$c_4 = 3: \quad 3 = \sqrt{64 - x^2 - y^2};\ x^2 + y^2 = 55 \approx 7.42^2$$

$$c_5 = 4: \quad 4 = \sqrt{64 - x^2 - y^2};\ x^2 + y^2 = 48 \approx 6.93^2$$

$$c_6 = 5: \quad 5 = \sqrt{64 - x^2 - y^2};\ x^2 + y^2 = 39 \approx 6.24^2$$

$$c_7 = 6: \quad 6 = \sqrt{64 - x^2 - y^2};\ x^2 + y^2 = 28 \approx 5.29^2$$

$$c_8 = 7: \quad 7 = \sqrt{64 - x^2 - y^2};\ x^2 + y^2 = 15 \approx 3.87^2$$

$$c_9 = 8: \quad 8 = \sqrt{64 - x^2 - y^2};\ x^2 + y^2 = 0 = 0^2$$

The nine level curves representing this hemisphere are shown in Figure 8.17.

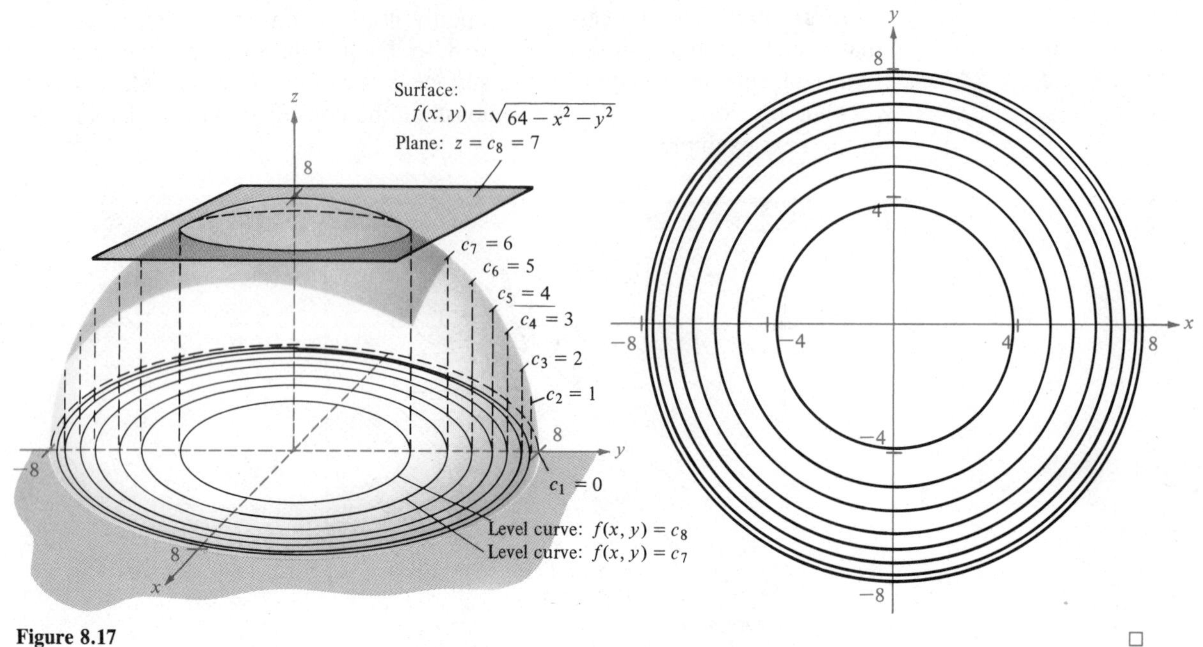

Figure 8.17

□

Remark

Note in Example 2 that a contour map depicts the variation of z with respect to x and y by the change in the spacing of the level curves. Much space between level curves indicates z is changing slowly, while little space between level curves indicates a rapid change in z. Furthermore, remember that in order to give the correct impression in a contour map it is important to choose c-values that are *evenly spaced*.

Example 3

Evaluating a Function of Several Variables

Evaluate the following functions at the indicated point.

(a) Given $f(x, y) = 2x^2 - y^2$, find $f(2, 3)$.

(b) Given $f(x, y, z) = e^x(y + z)$, find $f(0, -1, 4)$.

Solution

(a) Since $f(x, y) = 2x^2 - y^2$, we have

$$\begin{aligned} f(2, 3) &= 2(2^2) - 3^2 \\ &= 2(4) - 9 \\ &= 8 - 9 = -1 \end{aligned}$$

(b) Since $f(x, y, z) = e^x(y + z)$, we have

$$\begin{aligned} f(0, -1, 4) &= e^0(-1 + 4) \\ &= (1)(3) = 3 \end{aligned}$$

□

Example 4

An Application in Economics

The *Cobb-Douglas production function* is used in economics as a model to represent the number of units produced by varying amounts of labor and capital. If x measures the units of labor and y measures the units of capital, then the number of units produced is given by

$$f(x, y) = Cx^a y^{1-a}$$

where C is constant and $0 < a < 1$. Suppose that a manufacturer estimates a particular production function to be

$$f(x, y) = 100x^{0.6}y^{0.4}$$

(a) What is the production level when $x = 1000$ and $y = 500$?
(b) What is the production level when $x = 2000$ and $y = 1000$?

Solution

(a) When $x = 1000$ and $y = 500$, we have a production level of

$$\begin{aligned} f(1000, 500) &= 100(1000^{0.6})(500^{0.4}) \\ &\approx 100(63.10)(12.01) \\ &\approx 75{,}786 \end{aligned}$$

(b) When $x = 2000$ and $y = 1000$, we have a production level of

$$\begin{aligned} f(2000, 1000) &= 100(2000^{0.6})(1000^{0.4}) \\ &\approx 100(95.64)(15.85) \\ &\approx 151{,}572 \end{aligned}$$

Note that by doubling *both* x and y we doubled the production level. In Exercise 30 at the end of this section, you are asked to show that this result is characteristic of the Cobb-Douglas production function. □

Example 5

An Application in Banking

The monthly payment for an installment loan of P dollars taken out over t years at an annual percentage rate of r is given by

$$\text{monthly payment} = f(P, r, t) = \frac{Pr}{12}\left[\frac{1}{1 - \left(\dfrac{1}{1 + \dfrac{r}{12}}\right)^{12t}}\right]$$

(a) Find the monthly payment for a home mortgage of \$75,000 at 16% taken out for 30 years. (See the Introductory Example to Section 2.2.)

(b) Find the monthly payment for a home mortgage of \$75,000 at 8% taken out for 30 years.

Solution

(a) If $P = \$75{,}000$, $r = 0.16$, and $t = 30$, the monthly payment is

$$f(75{,}000,\ 0.16,\ 30) = \frac{(75{,}000)(0.16)}{12}\left[\frac{1}{1-\left(\dfrac{1}{1+\dfrac{0.16}{12}}\right)^{360}}\right]$$

$$= \$1{,}008.57$$

(b) If $P = \$75{,}000$, $r = 0.08$, and $t = 30$, the monthly payment is

$$f(75{,}000,\ 0.08,\ 30) = \frac{(75{,}000)(0.08)}{12}\left[\frac{1}{1-\left(\dfrac{1}{1+\dfrac{0.08}{12}}\right)^{360}}\right]$$

$$= \$550.32$$

□

Section Exercises (8.2)

1. For $f(x, y) = x/y$, find the following:
(a) $f(3, 2)$ (b) $f(-1, 4)$
(c) $f(30, 5)$ (d) $f(5, y)$
(e) $f(x, 2)$ (f) $f(5, t)$

2. For $f(x, y) = 4 - x^2 - 4y^2$, find the following:
(a) $f(0, 0)$ (b) $f(0, 1)$
(c) $f(2, 3)$ (d) $f(1, y)$
(e) $f(x, 0)$ (f) $f(t, 1)$

3. For $f(x, y) = xe^y$, find the following:
(a) $f(5, 0)$ (b) $f(3, 2)$
(c) $f(2, -1)$ (d) $f(5, y)$
(e) $f(x, 2)$ (f) $f(t, t)$

4. For $g(x, y) = \ln|x + y|$, find the following:
(a) $g(2, 3)$ (b) $g(5, 6)$
(c) $g(e, 0)$ (d) $g(0, 1)$
(e) $g(2, -3)$ (f) $g(e, e)$

5. For $h(x, y, z) = xy/z$, find the following:
(a) $h(2, 3, 9)$ (b) $h(1, 0, 0)$

6. For $f(x, y, z) = \sqrt{x + y + z}$, find the following:
(a) $f(0, 5, 4)$ (b) $f(6, 8, -3)$

7. For $V(r, h) = \pi r^2 h$, find the following:
(a) $V(3, 10)$ (b) $V(5, 2)$

Ⓒ **8.** For $F(r, N) = 500[1 + (r/12)]^N$, find the following:
(a) $F(0.09, 60)$ (b) $F(0.14, 240)$

9. For $f(x, y) = \int_x^y (2t - 3)\, dt$, find the following:
(a) $f(0, 4)$ (b) $f(1, 4)$

10. For $g(x, y) = \int_x^y (1/t)\, dt$, find the following:
(a) $g(4, 1)$ (b) $g(6, 3)$

In Exercises 11–20, describe the region R in the xy-plane that corresponds to the domain of the function, and find the range of the function.

11. $f(x, y) = \sqrt{4 - x^2 - y^2}$

12. $f(x, y) = \sqrt{x^2 + y^2 - 1}$

13. $g(x, y) = 4 - x^2 - y^2$

14. $h(x, y) = e^{x/y}$

15. $F(x, y) = \ln(4 - x - y)$

16. $f(x, y) = \ln(4 - xy)$

17. $h(x, y) = \dfrac{1}{xy}$

18. $g(x, y) = \dfrac{1}{x - y}$

19. $f(x, y) = x\sqrt{y}$

20. $f(x, y) = x^2 + y^2$

In Exercises 21–26, sketch a contour map for the given function and the given c-values.

Ⓒ Calculator may be helpful.

21. $f(x, y) = \sqrt{25 - x^2 - y^2}$; $c = 0, 1, 2, 3, 4, 5$

22. $f(x, y) = x^2 + y^2$; $c = 0, 2, 4, 6, 8$

23. $f(x, y) = xy$; $c = 1, -1, 3, -3$

24. $f(x, y) = 6 - 2x - 3y$; $c = 0, 2, 4, 6, 8, 10$

[C] **25.** $f(x, y) = \ln (x - y)$; $c = -2, -1.5, -1, -0.5, 0, 0.5, 1, 1.5, 2$

[C] **26.** $f(x, y) = \dfrac{x + y}{x - y}$; $c = -3, -2, -1, 0, 1, 2, 3$

[C] **27.** The *Doyle Log Rule* is one of several methods used to determine the lumber yield of a log (in board-feet) in terms of its diameter d (in inches) and its length l (in feet). The number of board-feet is given by

$$N(d, l) = \left(\frac{d - 4}{4}\right)^2 (l)$$

(a) Find the number of board-feet of lumber in a log with diameter $d = 22$ in. and length $l = 12$ ft.

(b) Find the number of board-feet of lumber in a log with diameter $d = 30$ in. and length $l = 12$ ft.

[C] **28.** A principal of $1000 is deposited in a savings account earning r percent compounded daily. The amount $A(r, t)$ after t years is given by

$$A(r, t) = 1000\left(1 + \frac{r}{365}\right)^{365t}$$

Use this function of two variables to complete the following table.

	Number of Years			
Rate	5	10	15	20
0.08				
0.10				
0.12				
0.14				

[C] **29.** In a certain queuing (waiting in line) theory model, the average length of time that a customer waits in line for service is given by the function

$$W(x, y) = \frac{1}{x - y}, \qquad y < x$$

where y is the average arrival rate expressed in the number of customers per unit of time and x is the average service rate expressed in the same units. Find the expected waiting time for the following.

(a) $y = 10$ arrivals per hour, $x = 15$ services per hour

(b) $y = 9$ arrivals per hour, $x = 12$ services per hour

(c) $y = 6$ arrivals per hour, $x = 12$ services per hour

30. Use the Cobb-Douglas production function (see Example 4 of this section) to show that, if the number of units of labor and the number of units of capital are *both* doubled, then the production level is also doubled.

[C] Calculator may be helpful.

Section 8.3

Partial Derivatives

Introductory Example

Complementary and Competitive Demand Relationships

Economists often classify related consumer products as *complementary* or *competitive*. If two products have a complementary demand relationship, then an increase in the sale of the one product will be accompanied by an increase in the sale of the other product. For example, video cassette recorders and video cassettes have a complementary demand relationship. Thus, we would expect a drop in the price of video cassette recorders to stimulate the sale of video cassettes *even if the price of video cassettes remains constant*. Similarly, we would expect a drop in the price of video cassettes to stimulate the sale of video cassette recorders *even if the price of the recorders remains constant*.

If two products have a competitive demand relationship, then an increase in the sale of the one product will be accompanied by a decrease in the sale of the other product. For instance, video cassette recorders and video disc recorders both compete for the same home entertainment market, and we would expect a drop in the price of video cassette recorders to serve as a deterrent for the sale of video disc recorders.

Recall from earlier work that the *demand function* for a given product relates the number of units sold to the price per unit. For competitive (or complementary) products, the demand for one of the products is a function not only of its own price but also of the price of the other (one or more) products. Thus, for two products,

$$\text{demand for product } 1 = x_1 = f(p_1, p_2)$$

where p_1 and p_2 are the prices per unit for products 1 and 2, respectively. If these two products are competitive, then an increase in the price of product 2 results in an increase in the demand for product 1. Mathematically, we describe this phenomenon by saying that the **partial derivative** of f with respect to p_2 is positive, and we write

$$\frac{\partial f}{\partial p_2} > 0$$

(∂ is the lower-case Greek letter delta.) Figure 8.18 illustrates a competitive demand situation. For complementary products, this partial derivative is negative, as shown in Figure 8.19.

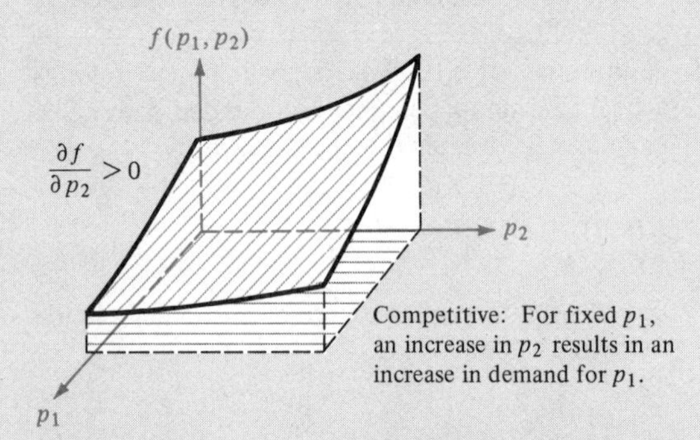

Figure 8.18

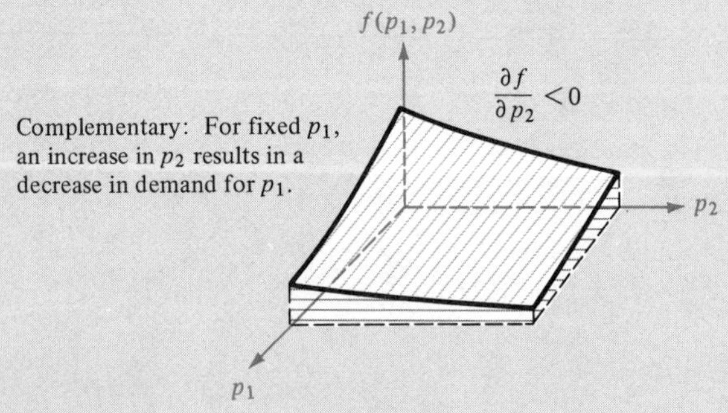

Figure 8.19

Section Objectives: *To define the partial derivatives of a function of two variables ▪ To provide a geometric interpretation of partial derivatives ▪ To demonstrate how to find partial derivatives of order two and higher.*

In the applications of functions of several variables, the question often arises, "How will the function be affected if I change one or some or all of its independent variables?" We can answer this question, at least in part, by considering the independent variables one at a time. For instance, the gross national product is a function of many variables such as tax rates, unemployment, and wars. Thus, an economist who wants to determine the effect of a tax increase holds all other variables constant while raising or lowering taxes. Or to determine the effect of a certain catalyst in an experiment, a chemist conducts the experiment several times using varying amounts of the catalyst, while keeping constant other variables such as temperature and pressure.

Mathematically, we follow a similar procedure to determine the rate of change of a function f with respect to one of its several independent variables. In this procedure, we find the derivative of f with respect to one independent variable at a time, while holding the others constant. This process is called **partial differentiation,** and the result is referred to as the **partial derivative** of f with respect to the chosen independent variable.

For functions of two variables, we give the following definition.

Definition of First Partial Derivatives

If $z = f(x, y)$, then the **first partial derivatives of f with respect to x and to y** are the functions $\partial f/\partial x$ and $\partial f/\partial y$ defined as

$$\frac{\partial f}{\partial x} = \lim_{\Delta x \to 0} \frac{f(x + \Delta x, y) - f(x, y)}{\Delta x} \qquad (y \text{ is held constant})$$

$$\frac{\partial f}{\partial y} = \lim_{\Delta y \to 0} \frac{f(x, y + \Delta y) - f(x, y)}{\Delta y} \qquad (x \text{ is held constant})$$

Remark

Note that this definition indicates that partial derivatives are determined by temporarily treating a function of two variables as a function of one variable by considering the other variable to be fixed. For instance, if $z = f(x, y)$, then to find $\partial f/\partial x$ we consider y to be a constant and differentiate with respect to x. To find $\partial f/\partial y$ we consider x fixed and differentiate with respect to y.

Example 1

Finding Partial Derivatives

Find $\partial f/\partial x$ and $\partial f/\partial y$ where $f(x, y) = 3x - x^2y^2 + 2x^3y$.

Solution Considering y to be constant, and differentiating with respect to x, we have

$$\frac{\partial f}{\partial x} = 3 - 2xy^2 + 6x^2y$$

Now, considering x to be constant and differentiating with respect to y gives us

$$\frac{\partial f}{\partial y} = -2x^2y + 2x^3$$

□

There are several notations for first partial derivatives. We list the common ones here, along with the notation for a partial derivative evaluated at some point (a, b).

Notation for First Partial Derivatives

If $z = f(x, y)$, then

$$\frac{\partial f}{\partial x} = f_x = z_x = \frac{\partial z}{\partial x} = \frac{\partial}{\partial x} f(x, y) = f_x(x, y)$$

and

$$\frac{\partial f}{\partial y} = f_y = z_y = \frac{\partial z}{\partial y} = \frac{\partial}{\partial y} f(x, y) = f_y(x, y)$$

The first partials evaluated at the point (a, b) are denoted by

$$\left.\frac{\partial f}{\partial x}\right|_{(a, b)} \quad \text{or} \quad f_x(a, b)$$

and

$$\left.\frac{\partial f}{\partial y}\right|_{(a, b)} \quad \text{or} \quad f_y(a, b)$$

The partial derivatives of a function of two variables, $z = f(x, y)$, have simple geometric interpretations. If y is fixed, say, $y = y_0$, then $z = f(x, y_0)$ represents a curve that is the intersection of the plane $y = y_0$ and the surface $z = f(x, y)$. (See Figure 8.20.) Thus,

$$f_x(x_0, y_0)$$

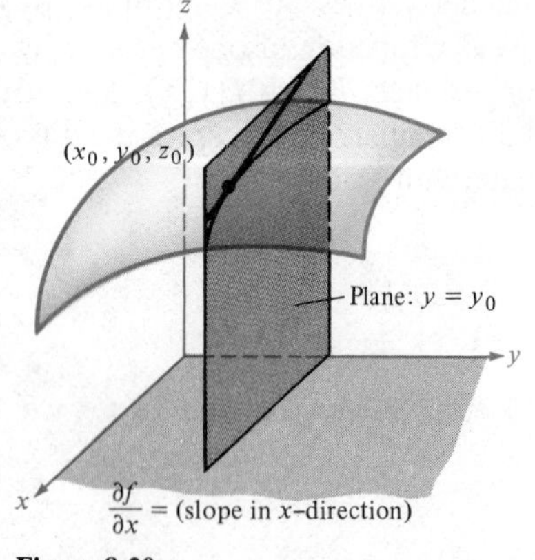

Figure 8.20

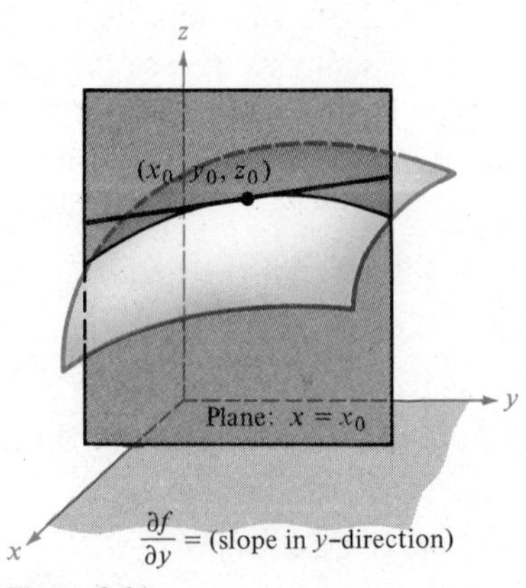

Figure 8.21

represents the slope of this curve at the point (x_0, y_0, z_0). (Note that both the curve and the tangent line lie in the plane $y = y_0$.) Similarly,

$$f_y(x_0, y_0)$$

represents the slope of the intersection of $z = f(x, y)$ and the plane $x = x_0$ at (x_0, y_0, z_0). (See Figure 8.21.)

Example 2

Finding the Slope of a Surface in the x and y Directions

Find the slope of the surface given by

$$f(x, y) = -\frac{x^2}{2} - y^2 + \frac{25}{8}$$

at the point $(\frac{1}{2}, 1, 2)$:
(a) in the x direction (b) in the y direction

Solution

(a) To find the slope in the x direction, we hold y constant and differentiate with respect to x to obtain

$$f_x(x, y) = -x$$

Thus, at the point $(\frac{1}{2}, 1, 2)$, the slope in the x direction is

$$f_x(\tfrac{1}{2}, 1) = -\tfrac{1}{2}$$

as shown in Figure 8.22.

(b) To find the slope in the y direction, we hold x constant and differentiate with respect to y to obtain

$$f_y(x, y) = -2y$$

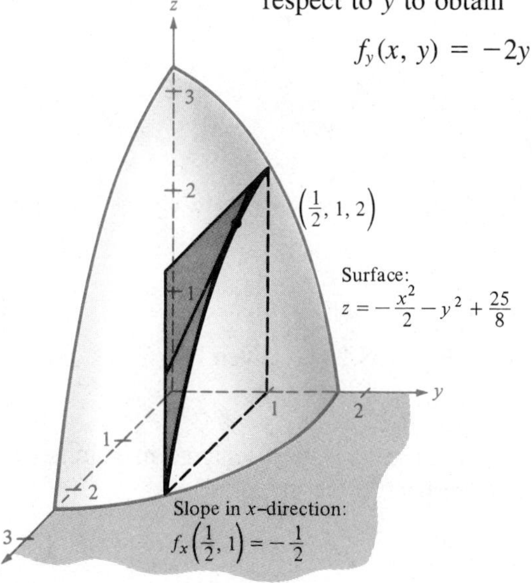

Figure 8.22

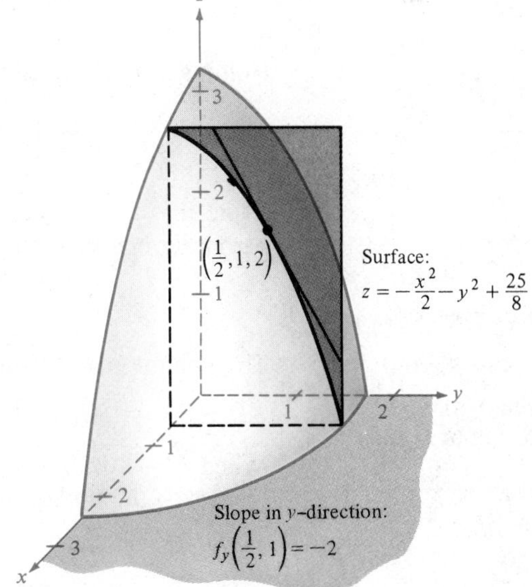

Figure 8.23

Thus, at the point $(\frac{1}{2}, 1, 2)$, the slope in the y direction is

$$f_y(\tfrac{1}{2}, 1) = -2$$

as shown in Figure 8.23. □

All the definitions and notations of this section can easily be extended to functions of three or more variables

$$u = f(x, y, z) \qquad \text{or} \qquad w = f(x_1, x_2, \ldots, x_n)$$

even though there is no simple geometric interpretation for partial derivatives of functions of three or more variables. However, we can interpret the partial derivatives of functions of several variables as *rates of change,* no matter how many variables are involved.

Example 3

Finding Partial Derivatives for a Function of Three Variables

Find $\partial w/\partial x$, $\partial w/\partial y$, and $\partial w/\partial z$ where $w = xe^{xy+2z}$.

Solution Holding y and z constant, we have

$$\begin{aligned}\frac{\partial w}{\partial x} &= x\frac{\partial}{\partial x}[e^{xy+2z}] + e^{xy+2z}\frac{\partial}{\partial x}[x] \\ &= x(ye^{xy+2z}) + e^{xy+2z}(1) \\ &= (xy + 1)e^{xy+2z}\end{aligned}$$

Holding x and z constant, we have

$$\frac{\partial w}{\partial y} = x(x)e^{xy+2z} = x^2e^{xy+2z}$$

Holding x and y constant, we have

$$\frac{\partial w}{\partial z} = x(2)e^{xy+2z} = 2xe^{xy+2z}$$

(Note that it was necessary to apply the Product Rule only when finding the partial derivative with respect to x.) □

Example 4

Using Partial Derivatives to Find Rates of Change

The volume of a frustum of a cone, Figure 8.24, is given by

$$V = f(R, r, h) = \tfrac{1}{3}\pi h(R^2 + Rr + r^2)$$

When $R = 10$, $r = 4$, and $h = 6$, what is the rate of change in volume with respect to the upper radius r? With respect to the height h?

Solution Since

$$\frac{\partial V}{\partial r} = f_r = \frac{1}{3}\pi h(0 + R + 2r)$$

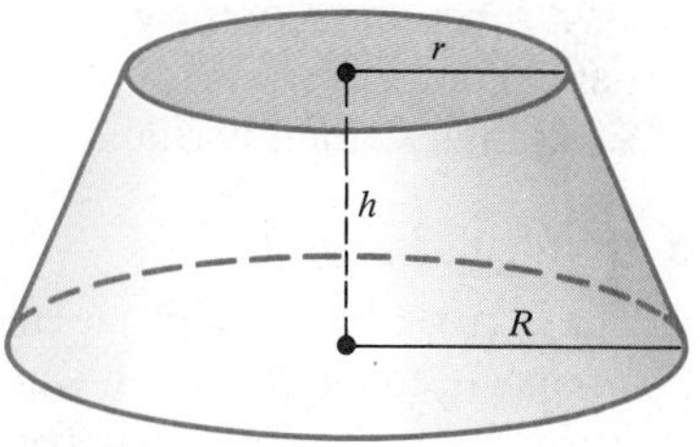

Frustum of a Cone

Figure 8.24

we have for $R = 10$, $r = 4$, and $h = 6$

$$f_r(10, 4, 6) = \tfrac{1}{3}\pi(6)(18) = 36\pi$$

Furthermore, since

$$\frac{\partial V}{\partial h} = f_h = \frac{1}{3}\pi(R^2 + Rr + r^2)$$

we have

$$f_h(10, 4, 6) = \tfrac{1}{3}\pi(100 + 40 + 16) = 52\pi$$ □

As is true for ordinary derivatives, it is possible to take a second, third, and so forth, partial derivative of a function of several variables. For instance, if $z = f(x, y)$, we can take the following second partial derivatives:

1. both with respect to x,

$$\frac{\partial}{\partial x}\left(\frac{\partial f}{\partial x}\right) = \frac{\partial^2 f}{\partial x^2} = \frac{\partial^2 z}{\partial x^2} = f_{xx} = z_{xx}$$

2. both with respect to y,

$$\frac{\partial}{\partial y}\left(\frac{\partial f}{\partial y}\right) = \frac{\partial^2 f}{\partial y^2} = \frac{\partial^2 z}{\partial y^2} = f_{yy} = z_{yy}$$

3. first with respect to x, then y,

$$\frac{\partial}{\partial y}\left(\frac{\partial f}{\partial x}\right) = \frac{\partial^2 f}{\partial y\, \partial x} = \frac{\partial^2 z}{\partial y\, \partial x} = f_{xy} = z_{xy}$$

4. first with respect to y, then x,

$$\frac{\partial}{\partial x}\left(\frac{\partial f}{\partial y}\right) = \frac{\partial^2 f}{\partial x\, \partial y} = \frac{\partial^2 z}{\partial x\, \partial y} = f_{yx} = z_{yx}$$

Partials 3 and 4 are referred to as "mixed" partial derivatives. Note that the symbol $\partial^2 f/\partial y\, \partial x$ means the partial derivative with respect to x first, then with respect to y.

Observe that for a function of two variables, there are two first partials and four second partials. For a function of three variables, there are three first partials, f_x, f_y, f_z, and nine second partials,

$$f_{xx}, f_{xy}, f_{xz}, f_{yx}, f_{yy}, f_{yz}, f_{zx}, f_{zy}, f_{zz}$$

six of which are mixed partials. We will see in the next two examples that some of these mixed partials are equal. In fact, it can be shown that if f has continuous second partials, then the order in which the partial derivatives are taken is immaterial. That is

$$f_{xy} = f_{yx}, \qquad f_{xz} = f_{zx}$$

and so on.

To take partial derivatives of order three and higher, we follow this same pattern. For instance, if $z = f(x, y)$, we have

$$z_{xxx} = \frac{\partial}{\partial x}\left(\frac{\partial^2 f}{\partial x^2}\right) = \frac{\partial^3 f}{\partial x^3} \quad \text{and} \quad z_{xxy} = \frac{\partial}{\partial y}\left(\frac{\partial^2 f}{\partial x^2}\right) = \frac{\partial^3 f}{\partial y\, \partial x^2}$$

Example 5

Finding Second Partial Derivatives

Find the second partial derivatives of $z = 3xy^2 - 2y + 5x^2y^2$. Determine the value of $z_{xy}(-1, 2)$.

Solution Since

$$\frac{\partial z}{\partial x} = 3y^2 + 10xy^2 \quad \text{and} \quad \frac{\partial z}{\partial y} = 6xy - 2 + 10x^2y$$

we have

$$\frac{\partial^2 z}{\partial y\, \partial x} = 6y + 20xy \quad \text{and} \quad \frac{\partial^2 z}{\partial x\, \partial y} = 6y + 20xy$$

Furthermore,

$$\frac{\partial^2 z}{\partial x^2} = 10y^2 \quad \text{and} \quad \frac{\partial^2 z}{\partial y^2} = 6x + 10x^2$$

Finally,

$$z_{xy}(-1, 2) = \left.\frac{\partial^2 z}{\partial y\, \partial x}\right|_{(-1,2)} = 12 - 40 = -28$$

□

Example 6

Finding Second Partial Derivatives

Find the second partial derivatives of $f(x, y, z) = ye^x + x \ln z$.

Solution Since

$$f_x = ye^x + \ln z$$
$$f_y = e^x$$
$$f_z = x\left(\frac{1}{z}\right) = \frac{x}{z}$$

it follows that

$$f_{xx} = ye^x, \quad f_{xy} = e^x, \quad f_{xz} = \frac{1}{z}$$
$$f_{yx} = e^x, \quad f_{yy} = 0, \quad f_{yz} = 0$$
$$f_{zx} = \frac{1}{z}, \quad f_{zy} = 0, \quad f_{zz} = -\frac{x}{z^2}$$

□

Section Exercises (8.3)

In Exercises 1–25, find all the first partial derivatives and evaluate each one at the indicated point (when given).

1. $f(x, y) = 2x - 3y + 5$

2. $f(x, y) = x^2 - 3y^2 + 7$; $(1, 3)$

3. $g(x, y) = xy$; $(3, 4)$

4. $f(x, y) = \dfrac{x}{y}$

5. $z = x^2 - 3xy + y^2$; $(-1, 2)$

6. $z = x\sqrt{y}$; $(2, 9)$

7. $z = x^2e^{2y}$

8. $z = xe^{x/y}$; $(0, 1)$

9. $z = \ln(x^2 + y^2)$

10. $z = \ln\sqrt{xy}$

11. $z = \ln\left(\dfrac{x+y}{x-y}\right)$

12. $z = \sqrt{20}$

13. $h(x, y) = e^{-(x^2 + y^2)}$

14. $h(x, y) = \dfrac{x^2}{2y} + \dfrac{4y^2}{x}$

15. $z = \sqrt{x^2 + y^2}$; $(-3, 4)$

16. $z = \ln\sqrt{x^2 + y^2}$; $(-2, 0)$

17. $z = \dfrac{xy}{x - y}$; $(2, -2)$

18. $f(x, y) = \dfrac{4xy}{x^2 + y^2}$

19. $f(x, y) = xe^y + ye^x$

20. $f(x, y) = \dfrac{xy}{\sqrt{x^2 + y^2}}$

21. $w = \sqrt{x^2 + y^2 + z^2}$; $(2, -1, 2)$

22. $w = \dfrac{xy}{x + y + z}$; $(1, 2, 0)$

23. $F(x, y, z) = \ln\sqrt{x^2 + y^2 + z^2}$

24. $G(x, y, z) = \dfrac{1}{\sqrt{1 - x^2 - y^2 - z^2}}$

25. $H(x, y, z) = 3x^2y - 5xyz + 10yz^2$

In Exercises 26–34, find the second partial derivatives

$$\frac{\partial^2 z}{\partial x^2}, \quad \frac{\partial^2 z}{\partial y^2}, \quad \frac{\partial^2 z}{\partial x\,\partial y}, \quad \text{and} \quad \frac{\partial^2 z}{\partial y\,\partial x}$$

26. $z = 3x^2 - xy + 2y^3$

27. $z = x^3 + 3x^2y - 5y^2$

28. $z = \dfrac{x}{x + y}$

29. $z = \dfrac{xy}{x - y}$

30. $z = 2e^{xy}$

31. $z = x^4 - 3x^2y^2 + y^4$

32. $z = \ln(x - y)$

33. $z = 9 + 4x - 6y - x^2 - y^2$

34. $z = \sqrt{9 - x^2 - y^2}$

35. Show that each of the following functions satisfies *Laplace's equation*

$$\frac{\partial^2 z}{\partial x^2} + \frac{\partial^2 z}{\partial y^2} = 0$$

(a) $z = 5xy$

(b) $z = x^2 - y^2$

(c) $z = x^3 - 3xy^2$

(d) $z = \dfrac{y}{x^2 + y^2}$

36. Find the slope of the paraboloid $z = x^2 + 4y^2$ at the point (2, 1, 8):
(a) in the x direction (b) in the y direction

37. Find the slope of the paraboloid $z = 9x^2 - y^2$ at the point (1, 3, 0):
(a) in the x direction (b) in the y direction

38. Find the slope of the hemisphere $z = \sqrt{25 - x^2 - y^2}$ at the point (3, 0, 4):
(a) in the x direction (b) in the y direction

39. Find the slope of the plane $z = 2x - 3y + 5$ at any point on its surface:
(a) in the x direction (b) in the y direction

40. Find the slope of the hyperbolic paraboloid $z = x^2 - 4y^2$ at the point (3, 1, 5):
(a) in the x direction (b) in the y direction

Ⓒ**41.** A company manufactures two types of wood-burning stoves: a free-standing model and a fireplace-insert model. The cost function for producing x free-standing stoves and y fireplace-insert stoves is

$$C = 32\sqrt{xy} + 175x + 205y + 1050$$

Find the marginal costs ($\partial C/\partial x$ and $\partial C/\partial y$) when $x = 80$ and $y = 20$.

42. Let N be the number of applicants to a university, p the charge for food and housing at the university, and t the tuition. If $N = f(p, t)$, explain why $\partial N/\partial p < 0$ and $\partial N/\partial t < 0$.

43. Let $x = 1000$ and $y = 500$ in the *Cobb-Douglas* production function:

$$f(x, y) = 100x^{0.6}y^{0.4}$$

(See Example 4, Section 8.2.)
(a) Find the marginal productivity of labor, $\partial f/\partial x$.
(b) Find the marginal productivity of capital, $\partial f/\partial y$.

44. Repeat Exercise 43 for the production function

$$f(x, y) = 100x^{0.75}y^{0.25}$$

Ⓒ**45.** The temperature at any point (x, y) in a steel plate is given by

$$T = 500 - 0.6x^2 - 1.5y^2$$

where x and y are measured in feet. At the point (2, 3), find the rate of change of the temperature with respect to the distance moved along the plate in the directions of the x- and y-axes, respectively.

Ⓒ**46.** A corporation has two plants that produce the same product. If q_1 and q_2 are the number of units produced in plant 1 and plant 2, respectively, then the total revenue for the product is given by

$$R(q_1, q_2) = 200q_1 + 200q_2 - 4q_1^2 - 8q_1q_2 - 4q_2^2$$

If $q_1 = 4$ and $q_2 = 12$, find:
(a) the marginal revenue for plant 1, $\partial R/\partial q_1$
(b) the marginal revenue for plant 2, $\partial R/\partial q_2$

Ⓒ**47.** Using the notation of the Introductory Example for this section, we let

x_1 = demand for product 1

x_2 = demand for product 2

p_1 = price of product 1

p_2 = price of product 2

Determine if the following demand functions describe complementary or competitive product relationships. (For the definitions of complementary and competitive demand functions, refer to the Introductory Example for this section.)

(a) $x_1 = 150 - 2p_1 - \frac{5}{2}p_2$
$x_2 = 350 - \frac{3}{2}p_1 - 3p_2$

(b) $x_1 = 150 - 2p_1 + 1.8p_2$
$x_2 = 350 + 0.75p_1 - 1.9p_2$

(c) $x_1 = \dfrac{1000}{\sqrt{p_1p_2}}$
$x_2 = \dfrac{750}{p_2\sqrt{p_1}}$

Section 8.4

Extrema of Functions of Two Variables

Introductory Example

The Maximum Profit for Two Competitive Products

A certain company markets two competitive products whose demand equations are

$$\text{demand for product 1} = x_1 = 200(p_2 - p_1)$$

$$\text{demand for product 2} = x_2 = 500 + 100p_1 - 180p_2$$

where p_1 and p_2 are the product prices in dollars. (Note that x_1 increases as p_2 increases and x_2 increases as p_1 increases; this is consistent with our claim that the two products have a competitive demand relationship.)

The costs of producing the two products are \$0.50 and \$0.75 per unit, respectively, and hence the total cost; the total revenue, and the profit are

$$C = 0.5x_1 + 0.75x_2$$

$$R = p_1x_1 + p_2x_2$$

$$\begin{aligned} P = R - C &= p_1x_1 + p_2x_2 - 0.5x_1 - 0.75x_2 \\ &= (p_1 - 0.5)x_1 + (p_2 - 0.75)x_2 \\ &= (p_1 - 0.5)200(p_2 - p_1) \\ &\quad + (p_2 - 0.75)(500 + 100p_1 - 180p_2) \\ &= -200p_1^2 + 300p_1p_2 - 180p_2^2 + 25p_1 \\ &\quad + 535p_2 - 375 \end{aligned}$$

The **maximum profit** occurs when the two first partial derivatives of P are zero:

$$\frac{\partial P}{\partial p_1} = -400p_1 + 300p_2 + 25 = 0$$

$$\frac{\partial P}{\partial p_2} = 300p_1 - 360p_2 + 535 = 0$$

The solutions to this system of equations are $p_1 = \$3.14$ and $p_2 = \$4.10$. Table 8.3 lists the profit for varying pairs of prices, and Figure 8.25 reinforces our conclusion that there actually exists a pair of values for p_1 and p_2 that produces a maximum profit.

Table 8.3

p_1 (price for product 1)	p_2 (price for product 2) \$3.00	\$3.50	\$4.00	\$4.50	\$5.00
\$2.00	\$660.00	\$642.50	\$535.00	\$337.50	\$50.00
\$2.50	\$672.50	\$730.00	\$697.50	\$575.00	\$362.50
\$3.00	\$585.00	\$717.50	\$760.00	\$712.50	\$575.00
\$3.50	\$397.50	\$605.00	\$722.50	\$750.00	\$687.50
\$4.00	\$110.00	\$392.50	\$585.00	\$687.50	\$700.00

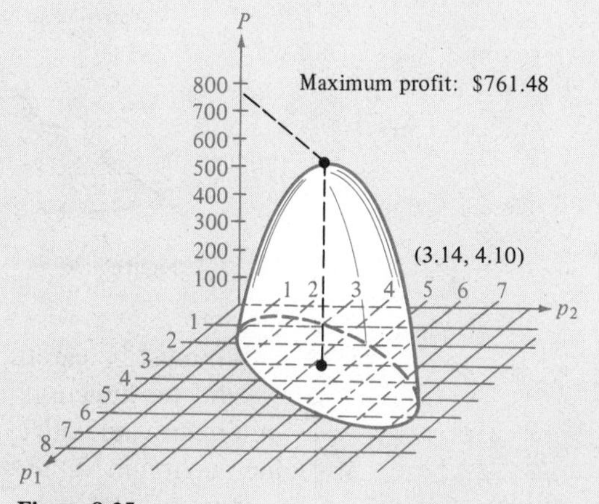

Figure 8.25

Section Objectives: ***To define the relative extrema of a function of two variables ▪ To define the critical points of a function of two variables ▪ To discuss the Second-Partials Test for relative extrema.***

A considerable portion of our study of the derivative of a function of one variable dealt with the finding and testing of the extreme values of a function. In this section, we take up this subject for functions of several variables, and we will find the ideas to be much the same. Although most theorems and definitions will be given in terms of functions of two variables, analogous ones may be stated for functions of three or more variables.

We define first the relative extrema of a function of two variables.

Definition of Relative Extrema

If f is a function defined at (x_0, y_0), then $f(x_0, y_0)$ is called:

1. a **relative maximum** of f if there is a circular region R centered at (x_0, y_0) such that $f(x, y) \leq f(x_0, y_0)$ for all (x, y) in R
2. a **relative minimum** of f if there is a circular region R centered at (x_0, y_0) such that $f(x, y) \geq f(x_0, y_0)$ for all (x, y) in R

Informally, we say that a point is a relative maximum of a surface if it is at least as high as all "nearby" points on the surface. Similarly, we say that a point is a relative minimum if it is at least as low as all "nearby" points on the surface. Several relative maximum and relative minimum points are pictured in Figure 8.26.

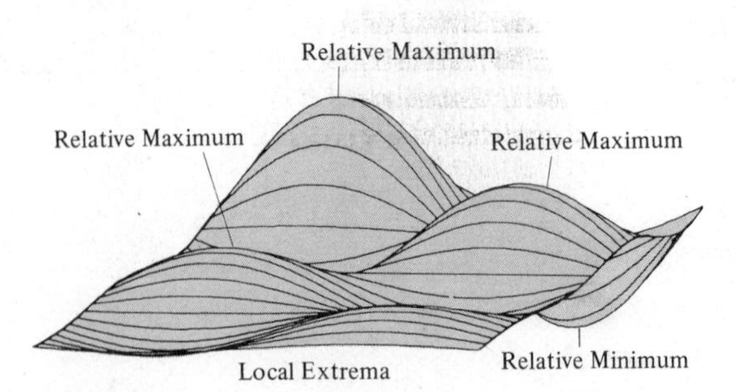

Figure 8.26

We must be careful to distinguish between relative and absolute extrema as we did for functions of a single variable. For instance, in Figure 8.27, the minimum value of $f(x, y)$ over the region R is not a relative minimum, whereas the maximum value of $f(x, y)$ over the region R happens also to be a relative maximum.

To locate the relative extrema of a function of two variables, we can use a procedure that is similar to the First-Derivative Test used for functions of a single variable. For functions of two variables, we call this test the **First-Partials Test,** and it is described as follows. (See Figure 8.28.)

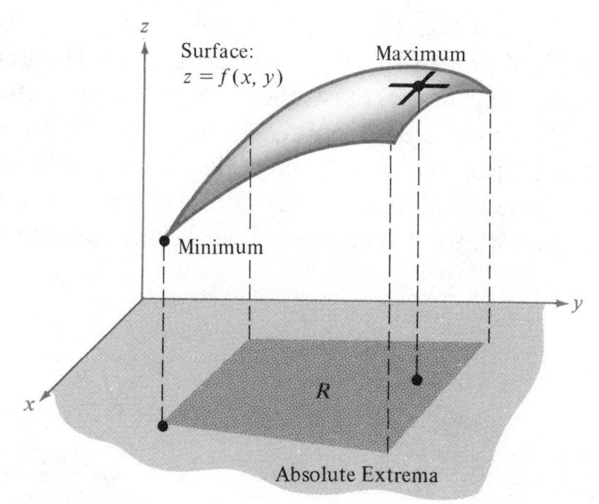

Figure 8.27

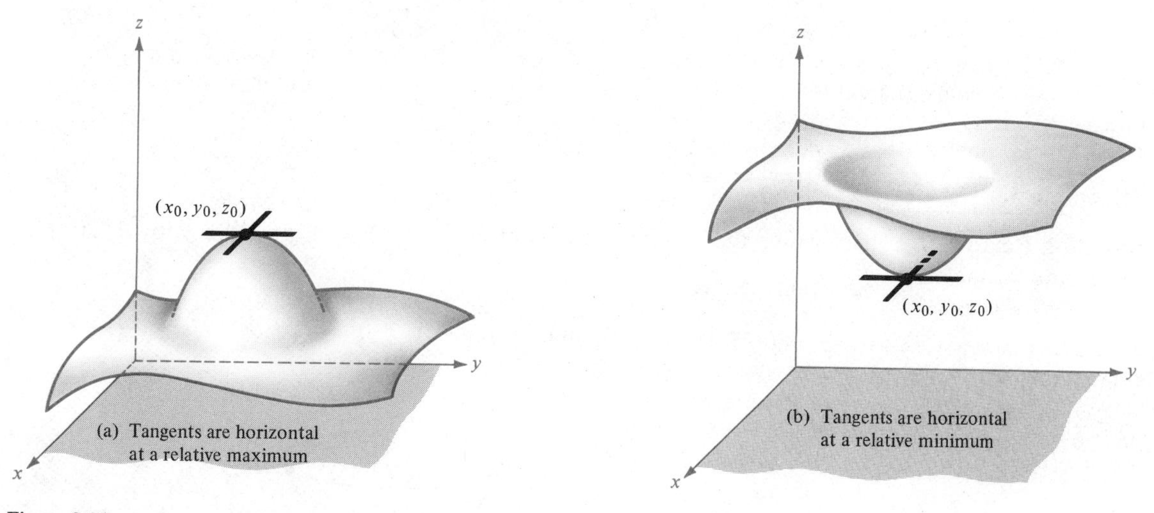

Figure 8.28

Theorem 8.1 (First-Partials Test for Relative Extrema)

If $f(x_0, y_0)$ is a relative extremum of f on an open region R in the xy-plane, and the first partial derivatives of f exist in R, then

$$f_x(x_0, y_0) = 0 = f_y(x_0, y_0)$$

Remark

By a *closed region R* in the xy-plane, we mean all points within R as well as all points on the boundary of R. This is comparable to a closed interval $[a, b]$ of real numbers, which includes the endpoints a and b. Similarly, an *open region* is comparable to an open interval and does *not* include the points on the boundary.

A point (x_0, y_0) for which both $f_x(x_0, y_0) = 0$ and $f_y(x_0, y_0) = 0$ is called a **critical point** of f. The contrapositive of Theorem 8.1 suggests that, if the first partial derivatives exist, then to find local extrema we need only examine values of $f(x, y)$ at critical points. However, as is true for a function of one variable, the critical points of a function of two variables do not always yield relative maxima or minima. For example, some critical points yield **saddle points** (see Figure 8.29), which are neither relative maxima nor relative minima.

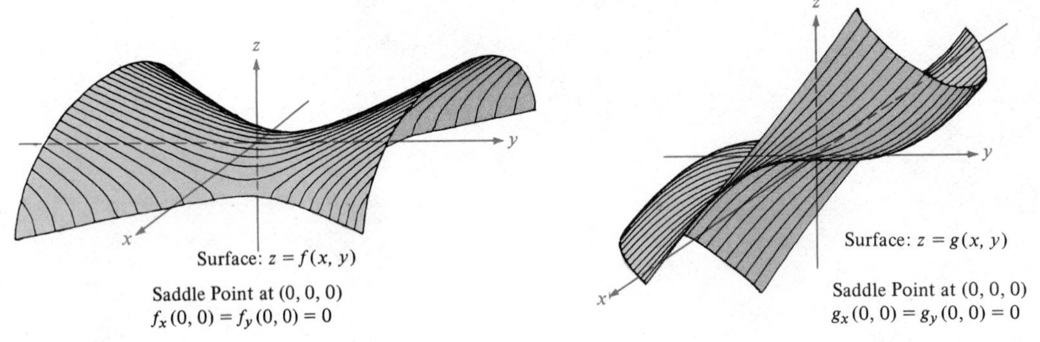

Figure 8.29

Example 1

Finding Relative Extrema

Determine the relative extrema of $f(x, y) = 2x^2 + y^2 + 8x - 6y + 20$.

Solution Since

$$f_x = 4x + 8 \qquad \text{and} \qquad f_y = 2y - 6$$

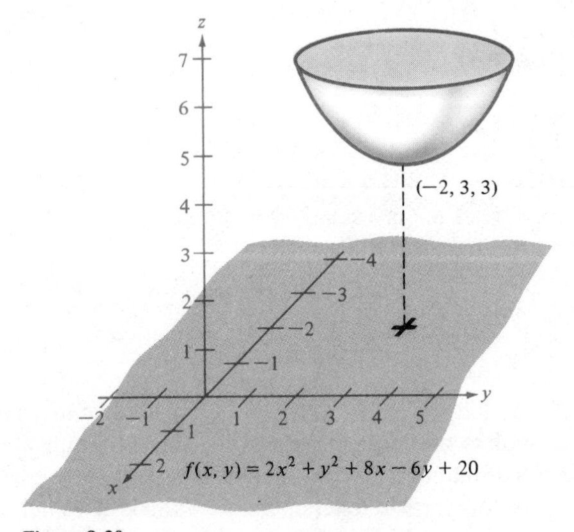

Figure 8.30

we solve the system of equations

$$4x + 8 = 0 \qquad \text{and} \qquad 2y - 6 = 0$$

to obtain the solution $(-2, 3)$. Now, for all $(x, y) \neq (-2, 3)$, completing the square shows us that

$$f(x, y) = 2(x + 2)^2 + (y - 3)^2 + 3 > 3$$

Hence a relative *minimum* of f occurs at $(-2, 3)$ and its value is $f(-2, 3) = 3$. (See Figure 8.30.) □

For a function like the one in Example 1, it is relatively easy to determine the *type* of extrema at the critical points. This can be done by algebraic arguments or by sketching a graph of the function. For more complicated functions such procedures are generally not so fruitful, and hence we seek a test that gives conditions under which a critical point will yield a relative maximum, a relative minimum, or neither.

The basic test for determining the relative maxima and minima of a function of two variables is the Second-Partials Test, which is the counterpart of the Second-Derivative Test for functions of one variable.

Theorem 8.2 (Second-Partials Test)

If f has continuous first and second partial derivatives on an open region and there exists a point (a, b) in the region such that

$$f_x(a, b) = 0 \qquad \text{and} \qquad f_y(a, b) = 0$$

then the quantity

$$d = f_{xx}(a, b)f_{yy}(a, b) - [f_{xy}(a, b)]^2$$

can be used as follows:

i. $f(a, b)$ is a **relative minimum** if

$$d > 0 \qquad \textit{and} \qquad f_{xx}(a, b) > 0$$

ii. $f(a, b)$ is a **relative maximum** if

$$d > 0 \qquad \textit{and} \qquad f_{xx}(a, b) < 0$$

iii. $(a, b, f(a, b))$ is a **saddle point** if

$$d < 0$$

Remark

If $d = 0$ in the Second-Partials Test, the test gives no information, and we must rely on a sketch or a method similar to the one used in Example 1 to determine the nature of the critical point.

Example 2

Applying the Second-Partials Test

Find the relative extrema of $f(x, y) = x^3 - 4xy + 2y^2$, if any exist.

Solution Since

$$f_x(x, y) = 3x^2 - 4y \qquad \text{and} \qquad f_y(x, y) = -4x + 4y$$

we obtain the system

$$3x^2 - 4y = 0$$

$$-4x + 4y = 0 \qquad \text{or} \qquad x = y$$

Thus, by substitution,

$$3(y)^2 - 4y = 0$$

$$y(3y - 4) = 0$$

and $\quad y = 0 \quad$ or $\quad y = \frac{4}{3}$

Therefore, the critical points are $(0, 0)$ and $(\frac{4}{3}, \frac{4}{3})$. Since

$$f_{xx}(x, y) = 6x, \qquad f_{yy}(x, y) = 4, \qquad f_{xy}(x, y) = -4$$

it follows that

$$f_{xx}(0, 0)f_{yy}(0, 0) - [f_{xy}(0, 0)]^2 = 0 - 16 < 0$$

and by part iii of Theorem 8.2, we conclude that $(0, 0)$ yields a saddle point of f. Furthermore, since $f_{xx}(\frac{4}{3}, \frac{4}{3}) = 8 > 0$ and

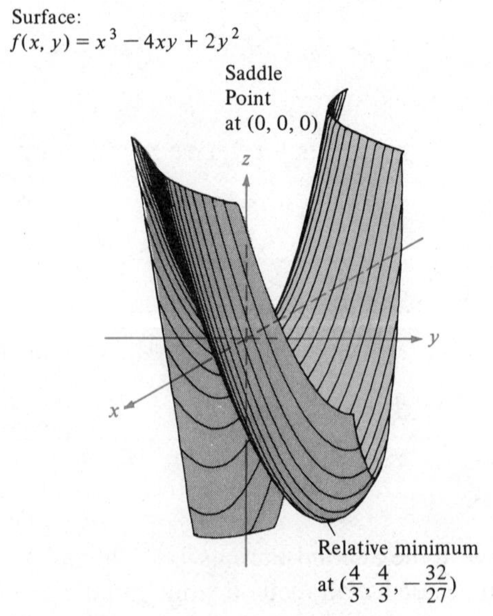

Figure 8.31

$$f_{xx}(\tfrac{4}{3}, \tfrac{4}{3})f_{yy}(\tfrac{4}{3}, \tfrac{4}{3}) - [f_{xy}(\tfrac{4}{3}, \tfrac{4}{3})]^2 = 8(4) - 16 = 16 > 0$$

we conclude by part i of Theorem 8.2 that $f(\frac{4}{3}, \frac{4}{3}) = -\frac{32}{27}$ is a relative minimum of f. (See Figure 8.31.) □

Example 3

Finding the Maximum Volume of a Box

A rectangular box is resting on the xy-plane with one vertex at the origin. Find the maximum volume of the box if its vertex opposite the origin lies in the plane $6x + 4y + 3z = 24$ (see Figure 8.32).

Solution Since one vertex of the box lies in the plane

$$6x + 4y + 3z = 24 \qquad \text{or} \qquad z = \tfrac{1}{3}(24 - 6x - 4y)$$

we can denote the volume of the box by

$$V = xyz = \tfrac{1}{3}xy(24 - 6x - 4y) = \tfrac{1}{3}(24xy - 6x^2y - 4xy^2)$$

Now, from the system

$$V_x = \frac{1}{3}(24y - 12xy - 4y^2) = \frac{y}{3}(24 - 12x - 4y) = 0$$

$$V_y = \frac{1}{3}(24x - 6x^2 - 8xy) = \frac{x}{3}(24 - 6x - 8y) = 0$$

we obtain the solutions $x = 0$, $y = 0$ and $x = \frac{4}{3}$, $y = 2$. Since the volume is zero for $x = 0$ and $y = 0$, we test the values $x = \frac{4}{3}$ and $y = 2$ to see if they yield a maximum volume. From

$$V_{xx}(x, y) = -4y, \quad V_{yy}(x, y) = \frac{-8x}{3}, \quad V_{xy}(x, y) = \frac{1}{3}(24 - 12x - 8y)$$

and

$$V_{xx}(\tfrac{4}{3}, 2)V_{yy}(\tfrac{4}{3}, 2) - [V_{xy}(\tfrac{4}{3}, 2)]^2 = (-8)(\tfrac{-32}{9}) - [\tfrac{-8}{3}]^2 = \tfrac{64}{3} > 0$$

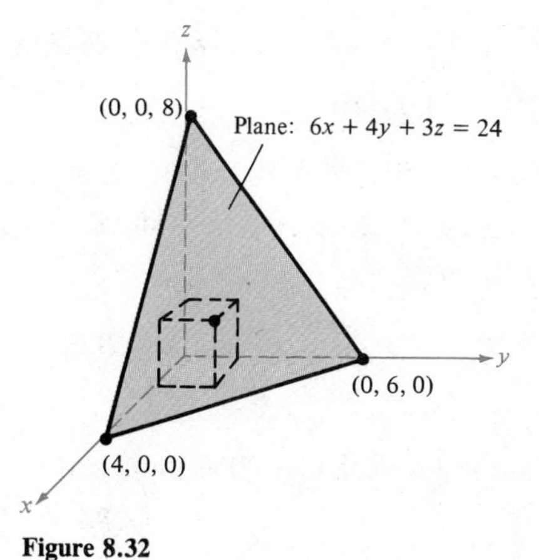

Figure 8.32

we conclude by Theorem 8.2 that V is maximum for $x = \frac{4}{3}$ and $y = 2$, and this maximum volume is

$$V = xyz = \tfrac{4}{3}(2)(\tfrac{1}{3})(24 - 8 - 8) = \tfrac{64}{9} \text{ cubic units}$$ □

In Example 3, we were able to convert a function of three variables to a function of two variables to obtain a solution. In general, this is not necessary since the definitions of local extrema and critical points can be extended to functions of three or more variables. Specifically, if all first partials of $w = f(x, y, z, \ldots)$ exist, then it can be shown that a local maximum or minimum can occur at $(x, y, z, \ldots)$ only if

$$f_x = 0, \qquad f_y = 0, \qquad f_z = 0, \qquad \ldots$$

(compare to Theorem 8.1), which means the critical points are obtained by solving this system of equations. The extension of Theorem 8.2 to three or more variables is also possible, though we will not consider such an extension in this text.

Example 4

Finding the Maximum Profit

The profit obtained by producing x units of product 1 and y units of product 2 is given by

$$P(x, y) = 8x + 10y - (0.001)(x^2 + xy + y^2) - 10{,}000$$

Find the production level that produces a maximum profit.

Solution Since

$$P_x = 8 - (0.001)(2x + y)$$

and

$$P_y = 10 - (0.001)(x + 2y)$$

we conclude that $P_x = 0$ and $P_y = 0$ if

$$8 - (0.001)(2x + y) = 0 \Rightarrow 2x + y = 8{,}000$$

$$10 - (0.001)(x + 2y) = 0 \Rightarrow x + 2y = 10{,}000$$

Solving this system, we have

$$x = 2000 \qquad \text{and} \qquad y = 4000$$

Since

$$P_{xx}P_{yy} - (P_{xy})^2 = (-0.002)(-0.002) - (-0.001)^2 > 0$$

and

$$P_{xx} < 0$$

the production level of $x = 2000$ units and $y = 4000$ units yields a maximum profit. □

Section Exercises (8.4)

In Exercises 1–12, examine each function for relative extrema and saddle points.

1. $f(x, y) = 2x^2 + 3y^2 - 4x - 12y + 13$

2. $f(x, y) = 5 + 3x - 4y - 3x^2 - 2y^2$

3. $f(x, y) = x^2 - y^2 - 2x - 4y - 4$

4. $f(x, y) = x^2 - 3xy - y^2$

5. $f(x, y) = xy$

6. $f(x, y) = 120x + 120y - xy - x^2 - y^2$

7. $f(x, y) = x^3 - 3xy + y^3$

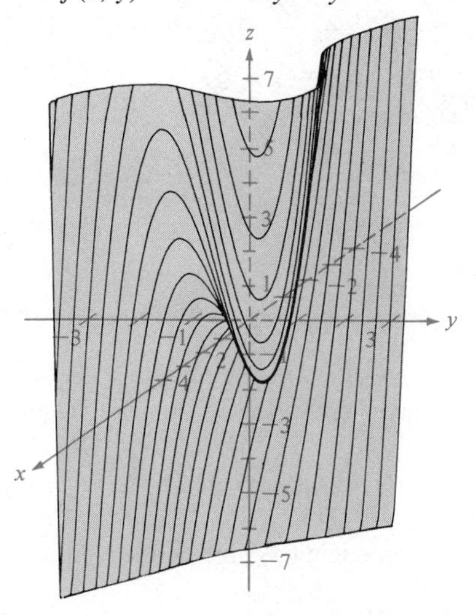

8. $f(x, y) = xy^2 - x^2y + x - y$

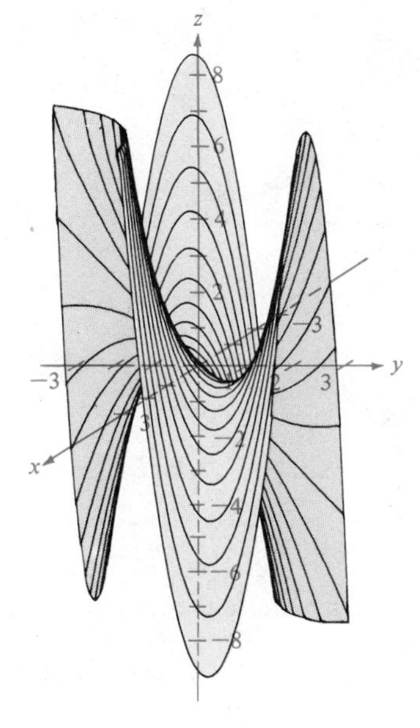

9. $f(x, y) = xy + \dfrac{1}{x} + \dfrac{1}{y}$

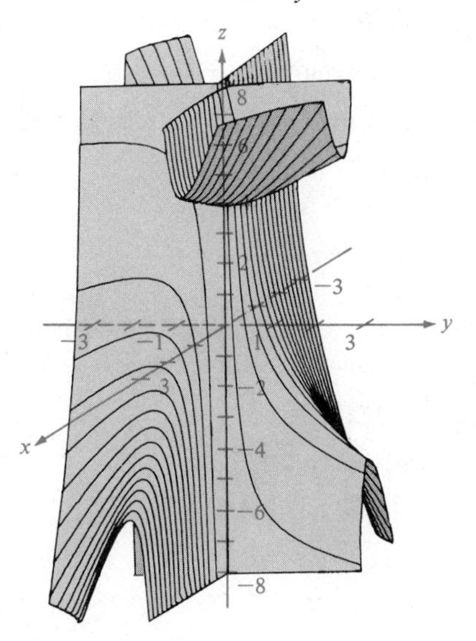

10. $f(x, y) = y^3 - 3yx^2 - 3y^2 - 3x^2 + 1$

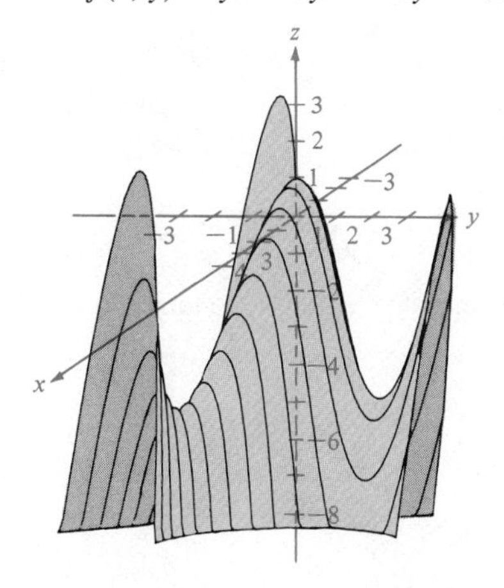

11. $f(x, y) = e^{xy}$

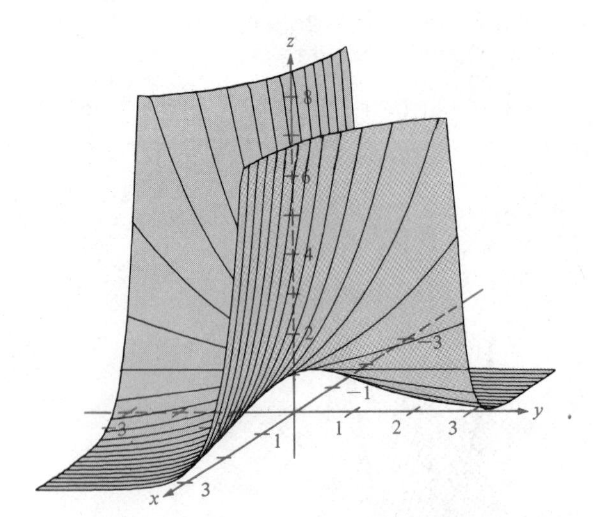

12. $f(x, y) = \dfrac{-4x}{x^2 + y^2 + 1}$

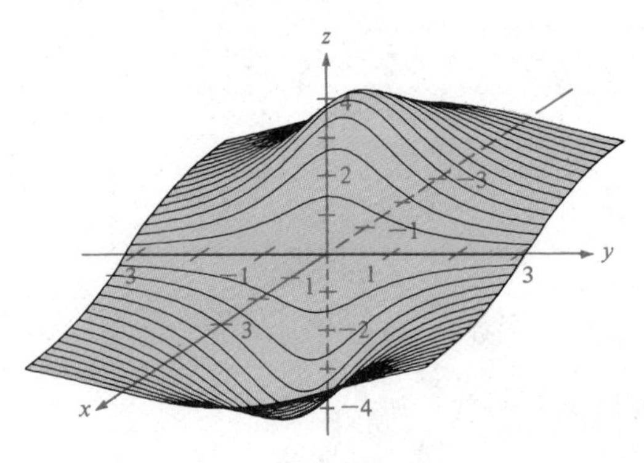

Ⓒ**13.** A company manufactures two products. The total revenue from selling x_1 units of product 1 and x_2 units of product 2 is

$$R(x_1, x_2) = 5x_1^2 - 8x_2^2 - 2x_1x_2 + 42x_1 + 102x_2$$

Find x_1 and x_2 so as to maximize the revenue R.

Ⓒ**14.** A retail outlet sells two competitive products the prices of which are p_1 and p_2. Find p_1 and p_2 so as to maximize the total revenue function

$$R(p_1, p_2) = 500p_1 + 800p_2 + 1.5p_1p_2 - 1.5p_1^2 - p_2^2$$

Ⓒ**15.** A retail outlet sells two competitive products having the demand functions

$$x_1 = 1000 - 2p_1 + p_2$$
$$x_2 = 1500 + 2p_1 - 1.5p_2$$

Find p_1 and p_2 so as to maximize the total revenue function

$$R(p_1, p_2) = x_1p_1 + x_2p_2$$

Ⓒ**16.** Repeat Exercise 15 for the demand functions

$$x_1 = 1000 - 4p_1 + 2p_2$$
$$x_2 = 900 + 4p_1 - 3p_2$$

Ⓒ**17.** A corporation manufactures a product at two locations. The cost of producing x_1 units at location 1 is

$$C_1 = 0.02x_1^2 + 4x_1 + 500$$

and the cost of producing x_2 units at location 2 is

$$C_2 = 0.05x_2^2 + 4x_2 + 275$$

If the product sells for \$15 per unit, find the quantity produced (at each location) that will maximize the profit

$$P(x_1, x_2) = 15(x_1 + x_2) - C_1 - C_2$$

Ⓒ**18.** Repeat Exercise 17 for a product that sells for \$50 per unit with cost functions given by

$$C_1 = 0.1x_1^3 + 20x_1 + 150$$
$$C_2 = 0.05x_2^3 + 20.6x_2 + 125$$

Ⓒ**19.** A corporation manufactures a product at two locations. The cost functions for producing x_1 units at location 1 and x_2 units at location 2 are given by

$$C_1 = 0.05x_1^2 + 15x_1 + 5400$$
$$C_2 = 0.03x_2^2 + 15x_2 + 6100$$

The demand function is given by

$$p = 225 - 0.4(x_1 + x_2)$$

Ⓒ Calculator may be helpful.

and, therefore, the total revenue function is

$$R(x_1, x_2) = [225 - 0.4(x_1 + x_2)](x_1 + x_2)$$

Find the production levels (at the two locations) that will maximize the profit

$$P(x_1, x_2) = R - C_1 - C_2$$

20. The material for constructing the base of an open box costs 1.5 times as much as the material for the sides. For a fixed amount of money C_0, find the dimensions of the box of largest volume that can be made in this manner.

21. Find the dimensions of a rectangular package of largest volume that may be sent by parcel post if the sum of the length and the girth (perimeter of a cross section) cannot exceed 108 in.

22. Find three positive numbers, x, y, and z, whose sum is 32 and for which $P = xy^2z$ is a maximum.

23. Find three positive numbers whose sum is 30 and whose product is maximum.

24. Find three positive numbers whose sum is 30 and whose sum of squares is minimum.

Section 8.5

Lagrange Multipliers and Constrained Optimization

Introductory Example

The Maximum Profit in a Limited Market

In the Introductory Example of Section 8.4, we looked at the maximum profit for two competitive products with the demand equations

$$x_1 = 200(p_2 - p_1)$$
$$\text{and } x_2 = 500 + 100p_1 - 180p_2$$

Thus, the total demand for the two products is

$$\begin{aligned} x_1 + x_2 &= 200(p_2 - p_1) + 500 + 100p_1 - 180p_2 \\ &= -100p_1 + 20p_2 + 500 \end{aligned}$$

Note that the total demand is determined solely by the prices p_1 and p_2. In many situations this assumption is overly simplistic. For example, regardless of the prices of the competing brands, the annual total demand for toothpaste is relatively constant. In such situations we say that the total market is *limited*, and variations in price do not affect the total market as much as they affect the market share of the competing brands.

Recall that in the Introductory Example of Section 8.4 the maximum profit for two products in an unlimited market was realized when $p_1 = \$3.14$ and $p_2 = \$4.10$. This corresponds to a total demand of

$$x_1 + x_2 = -100p_1 + 20p_2 + 500 = 268 \text{ units}$$

Now, let us suppose that the market for these two products is limited to 200 units (per year). Thus, we have

$$\begin{aligned} x_1 + x_2 = -100p_1 + 20p_2 + 500 &= 200 \\ -5p_1 + p_2 + 15 &= 0 \end{aligned}$$

We call this additional condition on p_1 and p_2 a **constraint.** In this particular case the graph of the constraint equation is a vertical plane, as shown in Figure 8.33. Now, to find the maximum profit

$$\begin{aligned} P = -200p_1^2 + 300p_1p_2 - 180p_2^2 + 25p_1 \\ + 535p_2 - 375 \end{aligned}$$

subject to our limited market constraint, we consider only those points that lie on *both* the surface representing the profit and the plane representing the constraint. In this section, we will see that by using a **Lagrange multiplier** we can determine that the maximum profit occurs when $p_1 = \$3.94$ and $p_2 = \$4.70$. (This corresponds to a profit of \$712.48.) Recall that the maximum profit in an unlimited market was $P = \$761.48$. Hence, just as the name implies, the addition of a constraint equation *reduced* the maximum profit available for the two products.

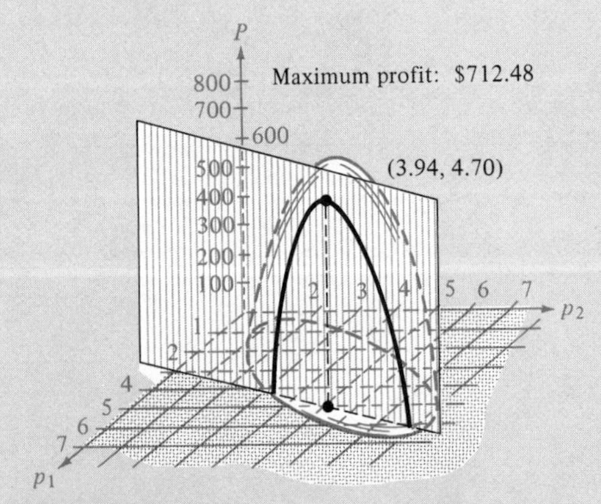

Figure 8.33

Section Objective: ***To introduce the method of Lagrange multipliers to solve optimization problems.***

The maximum and minimum problems in the preceding section are examples of a general type of problem called **optimization problems.** In Example 3 of that section, we looked at an important type of optimization problem called **constrained optimization.** Recall that in Example 3 we wanted to find the dimensions of the rectangular box of maximum volume that would fit in the first octant beneath the plane $6x + 4y + 3z = 24$. (See Figure 8.34.) That is, we wanted to find the maximum of

$$V = xyz \qquad \text{(Optimization Equation)}$$

subject to the constraint

$$6x + 4y + 3z = 24 \qquad \text{(Constraint Equation)}$$

In Section 8.4, we solved this problem by solving for z in the constraint equation and then rewriting V as a function of two variables.

In this section, we look at a different (and often better) way to solve constrained optimization problems. This new method involves the use of variables called **Lagrange multipliers.** We first outline this procedure and then demonstrate its use in the examples in this section.

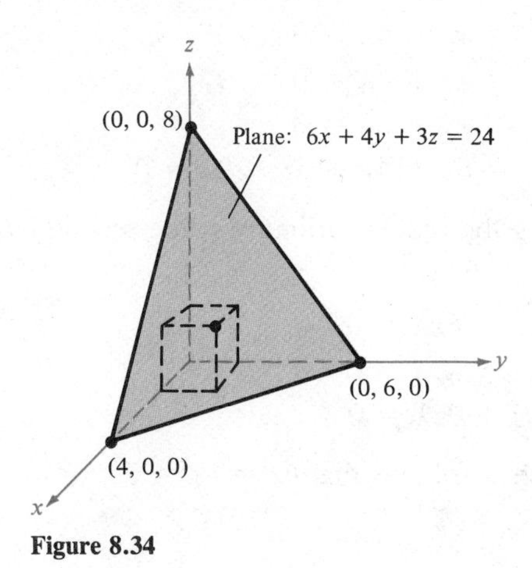

Figure 8.34

Method of Lagrange Multipliers

To find the local extrema of $f(x, y, z)$ subject to the constraint $g(x, y, z) = 0$, find the critical numbers for the new function F defined by

$$F(x, y, z, \lambda) = f(x, y, z) + \lambda g(x, y, z)$$

The variable λ (the lower-case Greek letter lambda) is called a **Lagrange multiplier.**

Remark As we present it, the method of Lagrange multipliers gives us a way of finding critical points, but does not tell us whether these points yield minimums or maximums or neither. To make this distinction, we rely on the context of the problem.

Example 1

Using Lagrange Multipliers for Optimization with One Constraint

Find the maximum value of

$$V = xyz, \qquad x > 0, y > 0, z > 0$$

subject to the constraint

$$6x + 4y + 3z = 24$$

Solution First, we let

$$f(x, y, z) = xyz \qquad \text{and} \qquad g(x, y, z) = 6x + 4y + 3z - 24$$

Then we define a new function F by

$$\begin{aligned} F(x, y, z, \lambda) &= f(x, y, z) + \lambda g(x, y, z) \\ &= xyz + \lambda(6x + 4y + 3z - 24) \end{aligned}$$

From this equation, we obtain the system

$$\begin{aligned} F_x &= yz + 6\lambda = 0 \\ F_y &= xz + 4\lambda = 0 \\ F_z &= xy + 3\lambda = 0 \\ F_\lambda &= 6x + 4y + 3z - 24 = 0 \end{aligned}$$

Multiplying the first equation by x, the second by y, and the third by z gives us the system

$$\begin{aligned} xyz + 6x\lambda &= 0 \\ xyz + 4y\lambda &= 0 \\ xyz + 3z\lambda &= 0 \end{aligned}$$

from which it follows that $6x = 4y = 3z$, or

$$y = \tfrac{3}{2}x \qquad \text{and} \qquad z = 2x$$

Substituting these values for y and z into $F_\lambda = 0$, we obtain

$$\begin{aligned} F_\lambda = 6x + 4(\tfrac{3}{2}x) + 3(2x) - 24 &= 0 \\ 18x &= 24 \\ x &= \tfrac{4}{3} \end{aligned}$$

Finally, the critical values are

$$x = \tfrac{4}{3}, \qquad y = 2, \qquad z = \tfrac{8}{3}$$

Thus, the maximum is

$$V = xyz = (\tfrac{4}{3})(2)(\tfrac{8}{3}) = \tfrac{64}{9}$$

□

Remark

Note that in Example 1 we gave no particular interpretation to the value of λ that produced a maximum and in fact we were even able to solve for x, y, and z without actually solving for λ. In Example 3, we look at a problem in which it is useful to solve for λ.

To optimize functions of two variables with one constraint, we follow a procedure similar to that for three variables. This is demonstrated in the next example.

Example 2

A Business Application

The Cobb-Douglas production function (see Example 4, Section 8.2) for a particular manufacturer is given by

$$f(x, y) = 100x^{3/4}y^{1/4}$$

where x represents the units of labor and y represents the units of capital. If units of labor and capital cost \$150 and \$250 each, respectively, and the total expense on labor and capital is limited to \$50,000, then we have the constraint

$$150x + 250y = 50{,}000$$

Find the maximum production level for this manufacturer.

Solution Let

$$F(x, y, \lambda) = 100x^{3/4}y^{1/4} + \lambda(50{,}000 - 150x - 250y)$$

Then

$$\frac{\partial F}{\partial x} = 75x^{-1/4}y^{1/4} - 150\lambda = 0$$

$$\frac{\partial F}{\partial y} = 25x^{3/4}y^{-3/4} - 250\lambda = 0$$

$$\frac{\partial F}{\partial \lambda} = 50{,}000 - 150x - 250y = 0$$

Solving for λ in the first equation,

$$\lambda = \frac{75x^{-1/4}y^{1/4}}{150} = \frac{x^{-1/4}y^{1/4}}{2}$$

and substituting into the second equation, we have

$$25x^{3/4}y^{-3/4} - 250\left(\frac{x^{-1/4}y^{1/4}}{2}\right) = 0$$

Multiplying by $x^{1/4}y^{3/4}$, we have

$$25x - 125y = 0$$

$$x = \frac{125y}{25} = 5y$$

Finally, substituting this value of x into the third equation, we have

$$
\begin{aligned}
50{,}000 - 150(5y) - 250y &= 0 \\
1{,}000y &= 50{,}000 \\
y &= 50 \text{ units of capital} \\
x &= 250 \text{ units of labor}
\end{aligned}
$$

Thus, the maximum production is

$$f(250, 50) = 100(250^{3/4})(50)^{1/4} \approx 16{,}719 \text{ product units}$$

□

Remark

In Example 2, we could just as easily have used the equation

$$F(x, y, \lambda) = 100x^{3/4}y^{1/4} + \lambda(150x + 250y - 50{,}000)$$

However, in many business applications it is convenient to obtain positive values for λ, and thus economists normally use the form

$$\lambda(50{,}000 - 150x - 250y)$$

as shown in Example 2.

In Example 2, the Lagrange multiplier λ is called the **marginal productivity of money.** Thus, for $x = 250$ and $y = 50$, we have

$$\lambda = \frac{x^{-1/4}y^{1/4}}{2} = \frac{(250^{-1/4})(50^{1/4})}{2} \approx 0.334$$

This means that, if one additional dollar can be spent on production, then 0.334 additional units of the product can be produced. We use this concept in the next example.

Example 3

Using Lagrange Multipliers to Find the Marginal Productivity of Money

Suppose that \$70,000 is available for labor and capital for the Cobb-Douglas production function in Example 2. What is the maximum number of units that can be produced?

Solution We could rework the entire problem following the procedure used in Example 2. However, since the only change in the problem is the availability of additional money to spend on labor and capital, we use the fact that the marginal productivity of money is

$$\lambda \approx 0.334$$

Since an additional \$20,000 is available and since the maximum production in Example 2 was 16,719 units, we conclude that the maximum production is now

$$16{,}719 + (0.334)(20{,}000) = 23{,}406 \text{ product units}$$

We suggest that you test this procedure by working out this example by the procedure in Example 2 to see if you arrive at the same production level. □

For optimization problems with *two* constraints, we introduce a second Lagrange multiplier, as is illustrated in the next example.

Example 4

Using Lagrange Multipliers for Optimization with Two Constraints

If $T(x, y, z) = 20 + 2x + 2y + z^2$ represents the temperature at each point on the hemisphere $x^2 + y^2 + z^2 = 11$ $(z > 0)$, find the maximum temperature on the curve of intersection of the plane $x + y + z - 3 = 0$ and the hemisphere.

Solution In this case we have two constraints,

$$g(x, y, z) = x^2 + y^2 + z^2 - 11 = 0 \qquad \text{(hemisphere)}$$

and

$$h(x, y, z) = x + y + z - 3 = 0 \qquad \text{(plane)}$$

Let

$$F(x, y, z, \lambda, \mu) = 20 + 2x + 2y + z^2 + \lambda(x^2 + y^2 + z^2 - 11) + \mu(x + y + z - 3)$$

(μ is the lowercase Greek letter mu). Then we have the system

$$\begin{aligned} F_x &= 2 + 2x\lambda + \mu = 0 \\ F_y &= 2 + 2y\lambda + \mu = 0 \\ F_z &= 2z + 2z\lambda + \mu = 0 \\ F_\lambda &= x^2 + y^2 + z^2 - 11 = 0 \\ F_\mu &= x + y + z - 3 = 0 \end{aligned}$$

Subtracting F_y from F_x yields

$$2\lambda(x - y) = 0$$

which means that $\lambda = 0$ or $x = y$.

Case 1: If $\lambda = 0$, then the first equation yields $\mu = -2$, and, subsequently, the third equation yields $z = 1$. The fourth and fifth equations now become

$$x^2 + y^2 - 10 = 0 \quad \text{and} \quad x + y - 2 = 0$$

and by substitution we obtain

$$\begin{aligned} x^2 + (2 - x)^2 - 10 &= 0 \\ 2x^2 - 4x - 6 &= 0 \\ x^2 - 2x - 3 &= 0 \end{aligned}$$

with solutions $x = 3$ and $x = -1$. The corresponding y-values are $y = -1$ and $y = 3$, respectively. Thus, for $\lambda = 0$ the critical points are $(3, -1, 1)$ and $(-1, 3, 1)$.

Case 2: If $x = y$, then the fourth and fifth equations become

$$2x^2 + z^2 = 11 \quad \text{and} \quad 2x + z = 3$$

and by substitution we obtain

$$\begin{aligned} 2x^2 + (3 - 2x)^2 &= 11 \\ 6x^2 - 12x - 2 &= 0 \\ 3x^2 - 6x - 1 &= 0 \end{aligned}$$

The solutions are $x = (3 \pm 2\sqrt{3})/3$, and we discard $x = (3 + 2\sqrt{3})/3$ since $z = 3 - 2x$ would then be negative. Thus, for $x = y$ the critical point is $((3 - 2\sqrt{3})/3, (3 - 2\sqrt{3})/3, (3 + 4\sqrt{3})/3)$.

Finally, since

$$T(3, -1, 1) = T(-1, 3, 1) = 25$$

and
$$T\left(\frac{3 - 2\sqrt{3}}{3}, \frac{3 - 2\sqrt{3}}{3}, \frac{3 + 4\sqrt{3}}{3}\right) = \frac{91}{3} \approx 30.33$$

we conclude that $T = 30.33$ is the maximum temperature on the curve of intersection. □

Remark We can see from Examples 1, 2, and 4 that the system of equations that arises in the method of Lagrange multipliers is not, in general, a linear system. Because of this nonlinearity, the solution of the system often requires some ingenuity, and we encourage you to tailor your method of solution to each individual system. Of course, if the system happens to be linear, we can resort to familiar methods, as is demonstrated in the next example.

Example 5

Using Lagrange Multipliers to Optimize Functions of Four Variables

Find the minimum of the function

$$f(x, y, z, w) = x^2 + y^2 + z^2 + w^2$$

subject to the constraint $3x + 2y - 4z + w = 3$.

Solution Let

$$F(x, y, z, w, \lambda) = x^2 + y^2 + z^2 + w^2 + \lambda(3x + 2y - 4z + w - 3)$$

Then we obtain the linear system

$$\begin{aligned} F_x &= 2x + 3\lambda = 0 & x &= -\tfrac{3}{2}\lambda \\ F_y &= 2y + 2\lambda = 0 & y &= -\lambda \\ F_z &= 2z - 4\lambda = 0 & z &= 2\lambda \\ F_w &= 2w + \lambda = 0 & w &= -\tfrac{1}{2}\lambda \\ F_\lambda &= 3x + 2y - 4z + w - 3 = 0 & 3x + 2y - 4z + w &= 3 \end{aligned}$$

Substituting for x, y, z, and w into the last equation yields

$$\begin{aligned} 3(-\tfrac{3}{2}\lambda) + 2(-\lambda) - 4(2\lambda) + (-\tfrac{1}{2}\lambda) &= 3 \\ -15\lambda &= 3 \\ \lambda &= -\tfrac{1}{5} \end{aligned}$$

Therefore, the critical numbers are

$$x = \tfrac{3}{10}, \qquad y = \tfrac{1}{5}, \qquad z = -\tfrac{2}{5}, \qquad w = \tfrac{1}{10}$$

and the minimum value of f subject to the given constraint is

$$f\left(\tfrac{3}{10}, \tfrac{1}{5}, -\tfrac{2}{5}, \tfrac{1}{10}\right) = \tfrac{9}{100} + \tfrac{1}{25} + \tfrac{4}{25} + \tfrac{1}{100} = \tfrac{3}{10}$$

□

For our last example, we return to the concepts discussed in the Introductory Example of this section. Note that the system of equations obtained in this problem is also linear.

Example 6

Finding the Maximum Profit in a Limited Market

Referring to the Introductory Example of this section, find the prices yielding a maximum profit

$$P = -200p_1^2 + 300p_1p_2 - 180p_2^2 + 25p_1 + 535p_2 - 375$$

subject to the constraint

$$-5p_1 + p_2 + 15 = 0$$

Solution Let

$$F(p_1, p_2, \lambda) = -200p_1^2 + 300p_1p_2 - 180p_2^2 + 25p_1 + 535p_2 - 375 + \lambda(-5p_1 + p_2 + 15)$$

Then we obtain the linear system

$$\frac{\partial F}{\partial p_1} = -400p_1 + 300p_2 + 25 - 5\lambda = 0$$

$$\frac{\partial F}{\partial p_2} = 300p_1 - 360p_2 + 535 + \lambda = 0$$

$$\frac{\partial F}{\partial \lambda} = -5p_1 + p_2 + 15 = 0$$

Now, to solve this system, we divide the first equation by 5 and add the result to the second equation to obtain

$$\begin{array}{rcl} -80p_1 + 60p_2 + 5 - \lambda &=& 0 \\ 300p_1 - 360p_2 + 535 + \lambda &=& 0 \\ \hline 220p_1 - 300p_2 + 540 &=& 0 \end{array}$$

Finally, dividing this result by 20 and multiplying the third equation by 15, we have

$$\begin{array}{rcl} 11p_1 - 15p_2 + 27 &=& 0 \\ -75p_1 + 15p_2 + 225 &=& 0 \\ \hline -64p_1 + 252 &=& 0 \end{array}$$

$$p_1 = \tfrac{252}{64} = \$3.94$$

$$p_2 = 5p_1 - 15 = \$4.70$$

□

Section Exercises (8.5)

In Exercises 1–20, solve the given problem by using Lagrange multipliers. (In each exercise, assume that x, y, and z are positive.)

1. Maximize $f(x, y) = xy$ subject to $x + y = 10$.
2. Maximize $f(x, y) = xy$ subject to $2x + y = 4$.
3. Minimize $f(x, y) = x^2 + y^2$ subject to $x + y - 4 = 0$.
4. Minimize $f(x, y) = x^2 + y^2$ subject to $2x - 4y + 5 = 0$.
5. Maximize $f(x, y) = x^2 - y^2$ subject to $y - x^2 = 0$.
6. Maximize $f(x, y) = x^2 - y^2$ subject to $x - 2y + 6 = 0$.
7. Maximize $f(x, y) = 2x + 2xy + y$ subject to $2x + y = 100$.
8. Minimize $f(x, y) = 3x + y + 10$ subject to $2 \ln x + \ln y - 6 = 0$.
9. Maximize $f(x, y) = \sqrt{6 - x^2 - y^2}$ subject to $x + y - 2 = 0$.
10. Minimize $f(x, y) = \sqrt{x^2 + y^2}$ subject to $2x + 4y - 15 = 0$.
11. Maximize $f(x, y) = e^{xy}$ subject to $x^2 + y^2 - 8 = 0$.
12. Minimize $f(x, y) = 2x + y$ subject to $xy = 32$.
13. Minimize $f(x, y, z) = x^2 + y^2 + z^2$ subject to $x + y + z - 6 = 0$.
14. Maximize $f(x, y, z) = xyz$ subject to $x + y + z - 6 = 0$.
15. Maximize $f(x, y, z) = xyz$ subject to $x + y + z = 32$ and $x - y + z = 0$.
16. Minimize $f(x, y, z) = x^2 + y^2 + z^2$ subject to $x + 2z = 4$ and $x + y = 8$.
17. Minimize $f(x, y, z) = x^2 + y^2 + z^2$ subject to $x + y + z = 1$.
18. Maximize $f(x, y, z) = xy + yz$ subject to $x + 2y = 6$ and $x - 3z = 0$.
19. Maximize $f(x, y, z) = xyz$ subject to $x^2 + z^2 = 5$ and $x - 2y = 0$.
20. Minimize $x^2 - 8x + y^2 - 12y + 48$ subject to $x + y = 8$.
21. Find the dimensions of the rectangular package of largest volume subject to the constraint that the sum of the length and the girth (the perimeter of a cross section) cannot exceed 108 in. (In other words, maximize $V = xyz$ subject to the constraint $x + 2y + 2z = 108$.) (See Figure 8.35.)

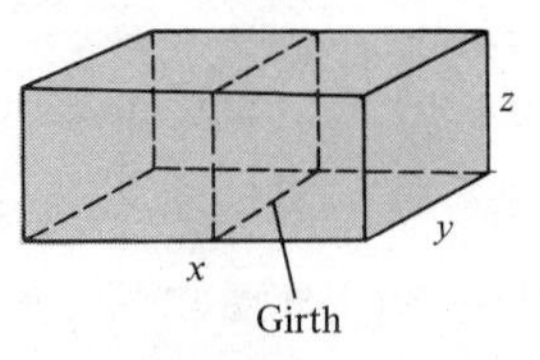

Figure 8.35

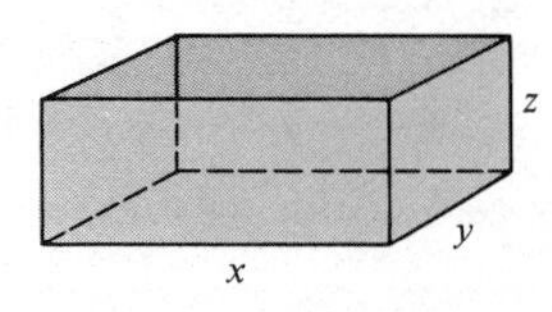

Figure 8.36

22. The material for constructing the base of an open box costs 1.5 times as much as the material for the sides. For a fixed amount of money C, find the dimensions of the box of largest volume that can be made in this manner. (In other words, maximize $V = xyz$ subject to the constraint $1.5xy + 2xz + 2yz = C$.) (See Figure 8.36.)
23. [C] Find the minimum distance from the origin to the line $2x + 3y = -1$. (That is, minimize $d = \sqrt{x^2 + y^2}$ subject to the constraint $2x + 3y = -1$.)
24. [C] Find the minimum distance from the point (2, 1, 1) to the plane $x + y + z = 1$. (That is, minimize $d = \sqrt{(x - 2)^2 + (y - 1)^2 + (z - 1)^2}$ subject to the constraint $x + y + z = 1$.)
25. [C] The production function for a firm is

$$f(x, y) = 100x^{0.25}y^{0.75}$$

where x is the number of units of labor and y is the number of units of capital. Suppose that labor costs \$48 per unit and capital costs \$36 per unit and that management sets a production goal of 20,000 units.

(a) Find the number of units of labor and capital needed to meet the production goal while minimizing the cost.

(b) Show that the conditions of part (a) are met when

$$\frac{\text{marginal productivity of labor}}{\text{marginal productivity of capital}} = \frac{\text{unit price of labor}}{\text{unit price of capital}}$$

This proportion is called the **Least-Cost Rule.**

26. [C] Repeat Exercise 25 for the production function

$$f(x, y) = 100x^{0.6}y^{0.4}$$

[C] Calculator may be helpful.

Ⓒ **27.** The production function for a corporation is

$$f(x, y) = 100x^{0.25}y^{0.75}$$

where x is the number of units of labor and y is the number of units of capital. Each unit of labor costs \$48, and each unit of capital costs \$36. The total cost of labor and capital is limited to \$100,000.

(a) Find the maximum production level for this manufacturer.

(b) Find the marginal productivity of money.

(c) Use the marginal productivity of money to find the maximum number of units that can be produced if \$125,000 is available for labor and capital.

Ⓒ **28.** Repeat Exercise 27 for the production function

$$f(x, y) = 100x^{0.6}y^{0.4}$$

Ⓒ **29.** Repeat Exercise 27 for the production function

$$f(x, y) = 4x + xy + 2y$$

Assume that the total amount available for labor and capital is \$2000 and that units of labor and capital cost \$20 and \$4, respectively.

Ⓒ **30.** A manufacturer has an order for 1000 units that it can produce at two locations. Let x_1 and x_2 be the number of units produced at the two plants. Find the number of units that should be produced at each plant to minimize the cost if the cost function is given by

$$C = 0.25x_1^2 + 10x_1 + 0.15x_2^2 + 12x_2$$

Section 8.6

The Method of Least Squares

Introductory Example

Egg Consumption in the United States

Although eating patterns in the United States follow fairly stable patterns, the past 20 years have witnessed some gradual shifts away from certain types of foods to other types. For example, two protein foods that have shown significant gains in popularity are chicken and cheese, while eggs (another source of protein) have decreased in popularity. Figure 8.37 shows the average annual consumption (per person) in the United States for these three protein products.

Part of the reason for the loss of popularity of eggs was the "cholesterol scare" of the 1960s and 1970s. However, as is reflected in Figure 8.37, cheese consumption continued to climb, even though cheese is also high in cholesterol. Whatever the reason for the decline, the nation's egg producers finally decided to invest in a national advertising campaign to push "the incredible edible egg" as an inexpensive and nutritious source of protein. By 1978, this advertising effort seemed to be paying off.

Using the **method of least squares,** we can determine that until 1977 the average egg consumption roughly followed the linear model.

$$y = -3.68t + 299.67$$

where $t = 0$ represents 1970. Had this pattern continued through 1979, the average consumption would have dropped to 267. (See Figure 8.37.) Since consumption in 1979 was actaully 283, we can conclude that (probably due to effective advertising) the "average" American consumed 16 eggs that year that would not have been consumed had the pre-1977 pattern continued. At a dollar a dozen, this increase amounted to additional retail sales of about $300,000,000 in 1979.

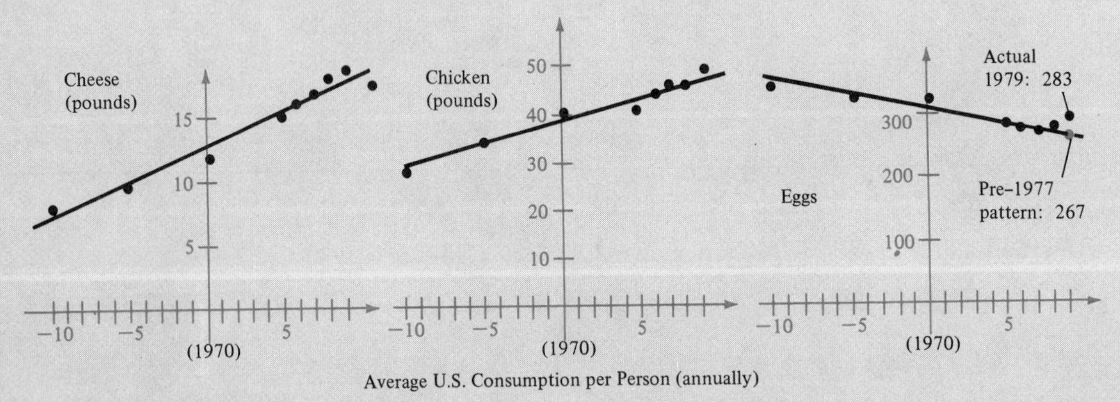

Figure 8.37

Section Objective: ***To introduce the method of least squares to fit a model to a set of points in the plane.***

We have already spent some time discussing the concept of a **mathematical model.** We pointed out that it is not always feasible to find an equation that *exactly* describes a particular phenomenon. In such cases we try to find a model that is both as *simple* and as *accurate* as possible. For instance, in Figure 8.38, a simple (linear) model for the given points is

$$y = 1.8566x - 5.0246$$

However, in Figure 8.39, it appears that by choosing the slightly more complicated (quadratic) model

$$y = 0.1996x^2 - 0.7281x + 1.3749$$

we can achieve significantly greater accuracy.

As a measure of how well the model $y = f(x)$ fits the collection of points

$$\{(x_1, y_1), (x_2, y_2), \ldots, (x_n, y_n)\}$$

we sum the squares of the differences between the actual y-values and the y-values given by the model. We call this the **sum of the squares error** and denote it by S.

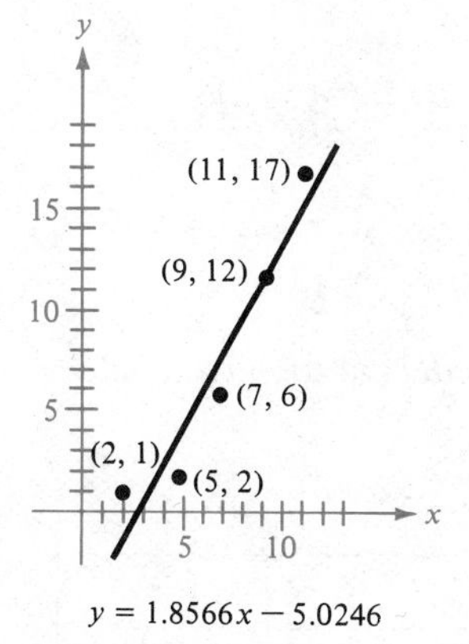

$y = 1.8566x - 5.0246$

Figure 8.38

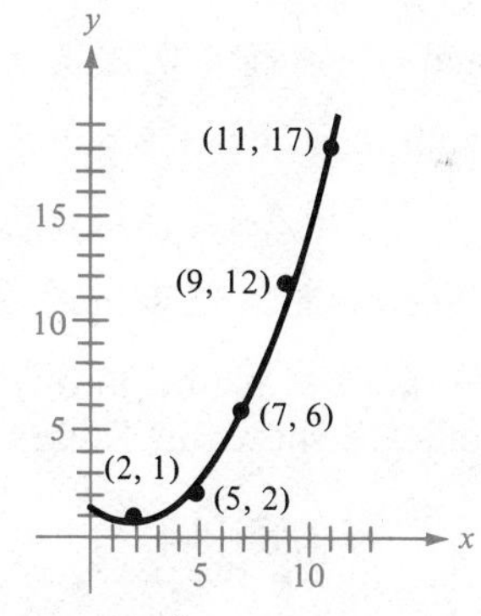

$y = 0.1996x^2 - 0.7281x + 1.3749$

Figure 8.39

Definition of the Sum of the Squares Error

The **sum of the squares error** for the model $y = f(x)$ with respect to the points

$$\{(x_1, y_1), (x_2, y_2), \ldots, (x_n, y_n)\}$$

is given by

$$S = [f(x_1) - y_1]^2 + [f(x_2) - y_2]^2 + \cdots + [f(x_n) - y_n]^2$$

Graphically, S is the sum of the squares of the vertical distances between the graph of f and the given points in the plane (see Figure 8.40).

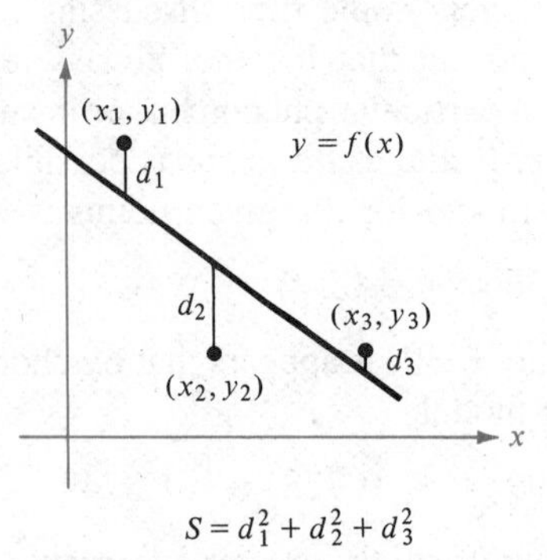

Figure 8.40

Example 1

Finding the Sum of the Squares Error

Find the sum of the squares error for the models

$$f(x) = 1.8566x - 5.0246$$

and $$g(x) = 0.1996x^2 - 0.7281x + 1.3749$$

with respect to the points

$$\{(2, 1), (5, 2), (7, 6), (9, 12), (11, 17)\}$$

(See Figures 8.38 and 8.39.)

Solution We begin by evaluating each of the two models at the given values of x, as shown in Table 8.4.

Table 8.4

x	2	5	7	9	11
y	1	2	6	12	17
$f(x)$	−1.3114	4.2584	7.9716	11.6848	15.3980
$g(x)$	0.7171	2.7244	6.0586	10.9896	17.5174

Now, for the linear model f, we find the sum of the squares error:

$$S = (-1.3114 - 1)^2 + (4.2584 - 2)^2 + (7.9716 - 6)^2 + (11.6848 - 12)^2 + (15.3980 - 17)^2$$

$$\approx 16.9959$$

Similarly, we compute the sum of the squares error for the quadratic model:

$$S = (0.7171 - 1)^2 + (2.7244 - 2)^2 + (6.0586 - 6)^2 + (10.9896 - 12)^2 + (17.5174 - 17)^2$$
$$\approx 1.8969$$

Note that the sum of the squares error for the quadratic model is less than the sum of the squares error for the linear model, which confirms our initial claim that the quadratic model provides a better fit. □

At this point, you might well be asking "What does all of this have to do with functions of several variables?" The answer is that we can use the optimization techniques of this chapter to find models with minimum least squares errors. Before looking at a general outline of this procedure, we look next at a simple example. Study this example carefully.

Example 2

Using Partial Derivatives to Find the Best Linear Model

Find the values of a and b such that the linear model

$$f(x) = ax + b$$

has a minimum sum of the squares error for the points

$$\{(-3, 0), (-1, 1), (0, 2), (2, 3)\}$$

Solution Table 8.5 lists the x-values, the actual y-values, and the estimated y-values for these four points.

Table 8.5

x	-3	-1	0	2
y	0	1	2	3
$f(x)$	$-3a + b$	$-a + b$	b	$2a + b$

The sum of the squares error is

$$S = (-3a + b)^2 + (-a + b - 1)^2 + (b - 2)^2 + (2a + b - 3)^2$$

To minimize this error, we set the partial derivatives of S (with respect to a and b) equal to zero as follows:

$$\frac{\partial S}{\partial a} = 2(-3)(-3a + b) + 2(-1)(-a + b - 1) + 2(2)(2a + b - 3) = 0$$
$$18a - 6b + 2a - 2b + 2 + 8a + 4b - 12 = 0$$
$$28a - 4b - 10 = 0$$
$$14a - 2b - 5 = 0$$

$$\frac{\partial S}{\partial b} = 2(-3a + b) + 2(-a + b - 1) + 2(b - 2) + 2(2a + b - 3) = 0$$

$$-6a + 2b - 2a + 2b - 2 + 2b - 4 + 4a + 2b - 6 = 0$$

$$-4a + 8b - 12 = 0$$

$$-a + 2b - 3 = 0$$

Now, we can solve for a and b as follows:

$$\begin{array}{r} 14a - 2b - 5 = 0 \\ -a + 2b - 3 = 0 \\ \hline 13a \qquad - 8 = 0 \\ a = \frac{8}{13} \end{array}$$

Since $-a + 2b - 3 = 0$, we have

$$b = \tfrac{1}{2}(a + 3) = \tfrac{1}{2}\left(\tfrac{8}{13} + 3\right) = \tfrac{47}{26}$$

Finally, we conclude that the line that best fits the four given points is

$$f(x) = \tfrac{8}{13}x + \tfrac{47}{26}$$

(See Figure 8.41.)

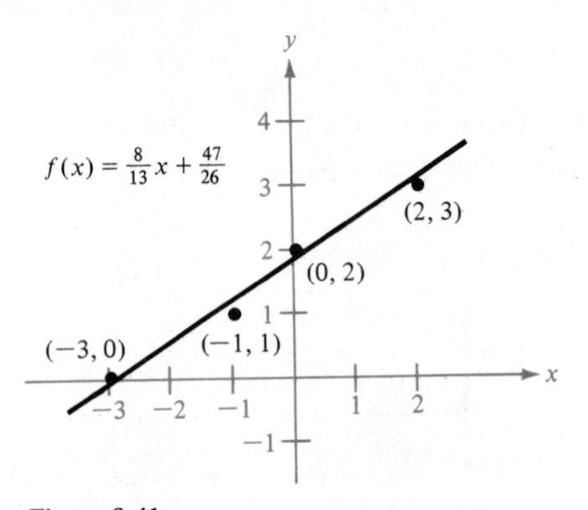

Figure 8.41

□

We call the line in Example 2 the **least squares regression line** for the given points. Before discussing a general procedure for finding such lines, we introduce the following notation for sums:

$$\sum_{i=1}^{n} x_i = x_1 + x_2 + \cdots + x_n$$

(Σ is the upper case Greek letter sigma.) Now, using this shorthand notation for the linear model

$$f(x) = ax + b$$

and the points

$$\{(x_1, y_1), (x_2, y_2), \ldots, (x_n, y_n)\}$$

we have

$$S = \sum_{i=1}^{n} [f(x_i) - y_i]^2 = \sum_{i=1}^{n} (ax_i + b - y_i)^2$$

To minimize S, we set its partial derivatives (with respect to a and b) equal to zero and solve for a and b. We omit the details of this procedure (they are similar to the steps in Example 2) and simply summarize the result in the following theorem.

Theorem 8.3 (The Least Squares Regression Line)

The **least squares regression line** for the points

$$\{(x_1, y_1), (x_2, y_2), \ldots, (x_n, y_n)\}$$

is $\quad y = ax + b$

where

$$a = \frac{n \sum_{i=1}^{n} x_i y_i - \sum_{i=1}^{n} x_i \sum_{i=1}^{n} y_i}{n \sum_{i=1}^{n} x_i^2 - \left(\sum_{i=1}^{n} x_i\right)^2} \quad \text{and} \quad b = \frac{1}{n}\left(\sum_{i=1}^{n} y_i - a \sum_{i=1}^{n} x_i\right)$$

Remark

In Theorem 8.3, note that if the x-values are symmetrically spaced about zero, then

$$\sum_{i=1}^{n} x_i = 0$$

and the formulas for a and b simplify to

$$a = \frac{\sum_{i=1}^{n} x_i y_i}{\sum_{i=1}^{n} x_i^2} \qquad b = \frac{1}{n} \sum_{i=1}^{n} y_i$$

Remark

Note also that only the *development* of Theorem 8.3 involves partial derivatives. The *application* of Theorem 8.3 is simply a matter of computing the values of a and b. This is demonstrated in the next example.

Example 3

Finding the Least Squares Regression Line by Formula

The total number of degrees (bachelor's, master's, and doctoral) conferred each year in the United States from 1971 to 1975 is given in the following table.

Year	1971	1972	1973	1974	1975
Number of degrees (in millions)	1.148	1.224	1.280	1.321	1.316

Find the least squares regression line for this data and use the result to predict the number of degrees conferred in 1976.

Solution To simplify the calculations in this problem, we let $x = -2$ represent 1971, $x = -1$ represent 1972, and so on. Table 8.6 summarizes the calculations involved in applying Theorem 8.3.

Table 8.6

x	y	xy	x^2
-2	1.148	-2.296	4
-1	1.224	-1.224	1
0	1.280	0	0
1	1.321	1.321	1
2	1.316	2.632	4
$\sum_{i=1}^{n} x_i = 0$	$\sum_{i=1}^{n} y_i = 6.289$	$\sum_{i=1}^{n} x_i y_i = 0.433$	$\sum_{i=1}^{n} x_i^2 = 10$

Now, since the sum of the x-values is zero, we apply the simplified version of Theorem 8.3 to obtain

$$a = \frac{\sum_{i=1}^{n} x_i y_i}{\sum_{i=1}^{n} x_i^2} = \frac{0.433}{10} = 0.0433$$

$$b = \frac{1}{n} \sum_{i=1}^{n} y_i = \frac{1}{5}(6.289) = 1.2578$$

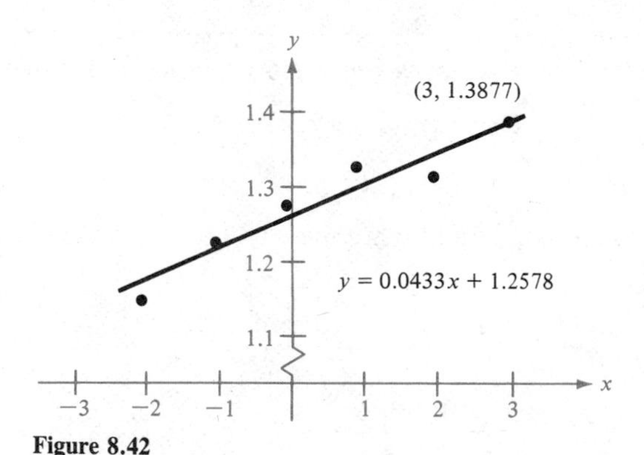

Figure 8.42

Thus, the least squares regression line is

$$y = 0.0433x + 1.2578$$

Finally, for 1976 we have $x = 3$, and we estimate the number of degrees conferred in that year to be

$$\begin{aligned} y &= 0.0433(3) + 1.2578 \\ &= 1.3877 \text{ million} \end{aligned}$$

(See Figure 8.42.) □

Before concluding this section, we must emphasize that the least squares regression line provides the best *linear* model for a given set of points. It does not, however, necessarily provide the best possible model. For example, it seems from Figure 8.42 that in 1975 ($x = 2$) a downward trend in the number of degrees began to occur. Thus, for a prediction for 1976, we might do better to use a quadratic model of the form

$$y = ax^2 + bx + c$$

Theorem 8.4 summarizes the formulas for finding the least squares regression quadratic. As with Theorem 8.3, the proof of this theorem involves the partial derivatives of the sum of the squares error

$$S = \sum_{i=1}^{n} (ax_i^2 + bx_i + c - y_i)^2$$

We omit the details of the rather lengthy proof and simply state the theorem as follows.

Theorem 8.4 (Least Squares Quadratic Fit)

The **least squares regression quadratic** for the points

$$\{(x_1, y_1), (x_2, y_2), \ldots, (x_n, y_n)\}$$

is

$$y = ax^2 + bx + c$$

where a, b, and c are the solutions to the system

$$a\sum_{i=1}^{n} x_i^4 + b\sum_{i=1}^{n} x_i^3 + c\sum_{i=1}^{n} x_i^2 = \sum_{i=1}^{n} x_i^2 y_i$$

$$a\sum_{i=1}^{n} x_i^3 + b\sum_{i=1}^{n} x_i^2 + c\sum_{i=1}^{n} x_i = \sum_{i=1}^{n} x_i y_i$$

$$a\sum_{i=1}^{n} x_i^2 + b\sum_{i=1}^{n} x_i + cn = \sum_{i=1}^{n} y_i$$

Example 4

Finding the Best Quadratic Fit for a Set of Points

Find the least squares regression quadratic for the data in Example 3 and use the result to estimate the number of degrees conferred in the United States in 1976.

Solution As in Example 3, we let $x = -2$ represent 1971. Then applying Theorem 8.4, we have

$$\sum_{i=1}^{n} x_i = 0 \qquad \text{(from Example 3)}$$

$$\sum_{i=1}^{n} x_i^2 = 10 \qquad \text{(from Example 3)}$$

$$\sum_{i=1}^{n} x_i^3 = (-2)^3 + (-1)^3 + (0)^3 + (1)^3 + (2)^3 = 0$$

$$\sum_{i=1}^{n} x_i^4 = (-2)^4 + (-1)^4 + (0)^4 + (1)^4 + (2)^4 = 34$$

$$\sum_{i=1}^{n} y_i = 6.289 \qquad \text{(from Example 3)}$$

$$\sum_{i=1}^{n} x_i y_i = 0.433 \qquad \text{(from Example 3)}$$

$$\sum_{i=1}^{n} x_i^2 y_i = (-2)^2(1.148) + (-1)^2(1.224) + (0)^2(1.280) + (1)^2(1.321) + (2)^2(1.316) = 12.401$$

Thus, we have

$$\begin{aligned} 34a + 10c &= 12.401 \\ 10b &= 0.433 \\ 10a + 5c &= 6.289 \end{aligned}$$

Solving this system, we have

$$\begin{aligned} a &= -0.01264 \\ b &= 0.0433 \\ c &= 1.2831 \end{aligned}$$

and the best quadratic fit is given by

$$y = -0.01264x^2 + 0.0433x + 1.2831$$

For 1976, we have $x = 3$ and

$$y = -0.01264(3^2) + 0.0433(3) + 1.2831 = 1.2992 \text{ million}$$

(See Figure 8.43.)

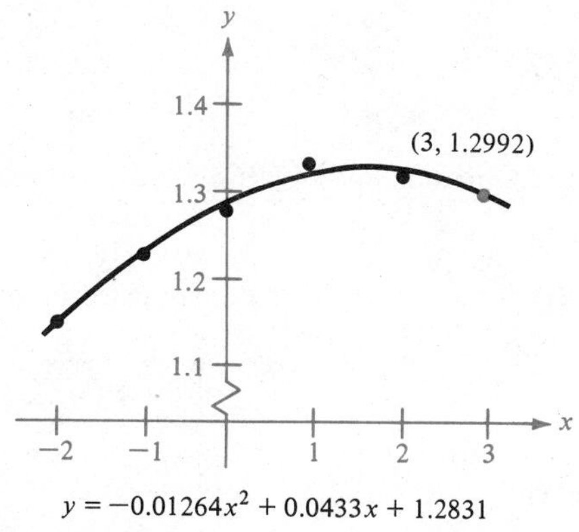

Figure 8.43

□

Remark The quadratic model in Example 4 has a sum of the squares error of $S \approx 0.00009$, while the linear model in Example 3 has a sum of the squares error of $S \approx 0.00232$.

Section Exercises (8.6)

In Exercises 1–4, (a) mentally draw a best-fitting line to the given points, (b) use the method of least squares to find the least squares regression line, and (c) calculate the sum of the squares error.

1.

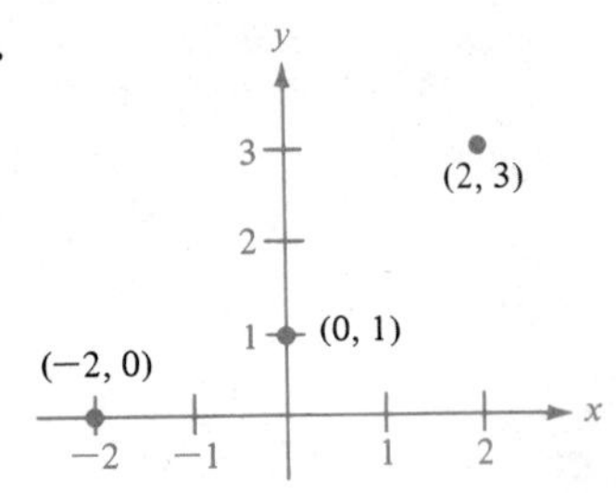

2.

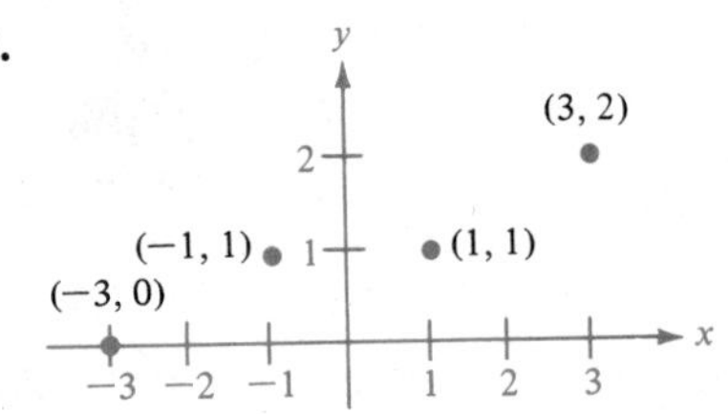

3.

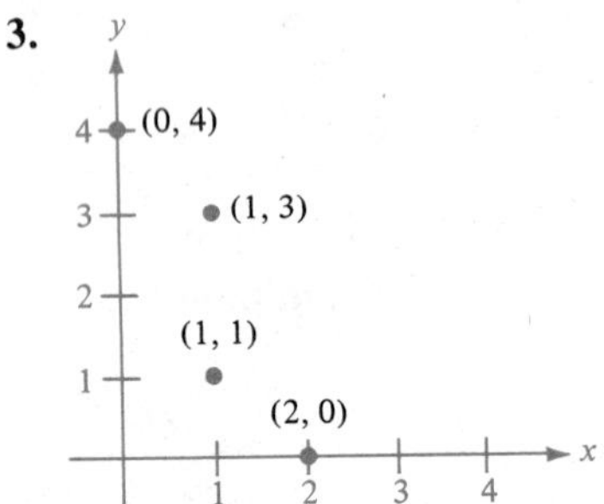

4.

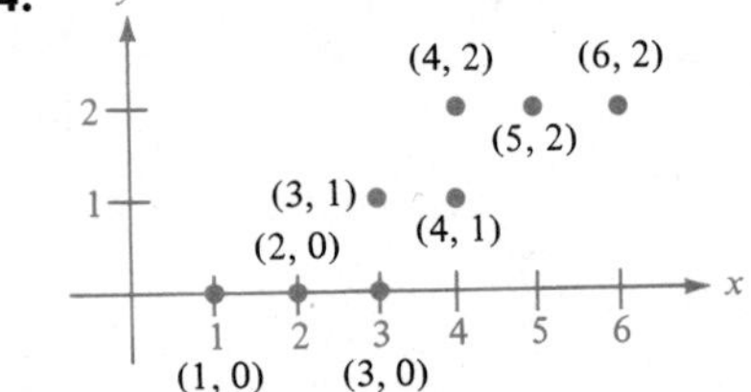

In Exercises 5–14, find the least squares regression line for the given points.

ⓒ **5.** $(-2, 0)$, $(-1, 1)$, $(0, 1)$, $(1, 2)$, $(2, 3)$

ⓒ **6.** $(-4, -1)$, $(-2, 0)$, $(2, 4)$, $(4, 5)$

ⓒ **7.** $(-3, 0)$, $(1, 4)$, $(2, 6)$

ⓒ **8.** $(-5, 1)$, $(1, 3)$, $(2, 3)$, $(2, 5)$

ⓒ **9.** $(-3, 4)$, $(-1, 2)$, $(1, 1)$, $(3, 0)$

ⓒ **10.** $(-10, 10)$, $(-5, 8)$, $(3, 6)$, $(7, 4)$, $(5, 0)$

ⓒ **11.** $(0, 0)$, $(1, 1)$, $(3, 4)$, $(4, 2)$, $(5, 5)$

ⓒ **12.** $(1, 0)$, $(3, 3)$, $(5, 6)$

ⓒ **13.** $(0, 6)$, $(4, 3)$, $(5, 0)$, $(8, -4)$, $(10, -5)$

ⓒ **14.** $(5, 2)$, $(0, 0)$, $(2, 1)$, $(7, 4)$, $(10, 6)$, $(12, 6)$

ⓒ **15.** A store manager wants to know the demand for a certain product as a function of the price. The daily sales for three different prices of the product are given in the following table.

Price (x)	\$1.00	\$1.25	\$1.50
Demand (y)	450	375	330

(a) Find the least squares regression line for this data.
(b) Estimate the demand when the price is \$1.40.

ⓒ **16.** A hardware retailer wants to know the demand for a certain tool as a function of the price. The monthly sales for four different prices of the tool are listed in the following table.

Price (x)	\$25	\$30	\$35	\$40
Demand (y)	82	75	67	55

(a) Find the least squares regression line for this data.
(b) Estimate the demand when the price is \$32.95.

ⓒ **17.** A farmer used four test plots to determine the relationship between the wheat yield (in bushels per acre) and the amount of fertilizer (in hundreds of pounds per acre). The results are in the following table.

Fertilizer (x)	1.0	1.5	2.0	2.5
Yield (y)	32	41	48	53

(a) Find the least squares regression line for this data.

ⓒ Calculator may be helpful.

(b) Estimate the yield for a fertilizer application of 160 lb per acre.

ⓒ**18.** A sociologist who wants to analyze the increase in the number of aggravated assaults per 1,000,000 persons in the United States obtains the following data from governmental records.

Year (x)	1967	1970	1973	1976	1979
Number of assaults (y)	13.0	16.5	20.1	22.9	27.9

Let $x = 0$ represent the year 1973.

(a) Find the least squares regression line for this data.

(b) Estimate the number of aggravated assaults in 1985.

ⓒ**19.** The number of cars imported from Japan has increased dramatically in the past several years, as indicated in the following table (in thousands).

Year (x)	1970	1975	1977	1979
Number of imports (y)	381	696	1342	1617

Let $x = 0$ represent the year 1970.

(a) Find the least squares regression line for this data.

(b) Estimate the number of imports for the year 1978.

ⓒ**20.** The Consumer Price Index (CPI) for all items for four different years is given in the following table.

Year (x)	1970	1972	1974	1976
CPI (y)	116.3	125.3	147.7	170.5

Let $x = 0$ represent the year 1970.

(a) Find the least squares regression line for this data.

(b) Estimate the CPI for 1982.

In Exercises 21–24, find the values of a and b such that the linear model

$$f(x) = ax + b$$

has a minimum sum of the squares error for the given points. (Use partial derivatives, following the pattern of Example 2 in this section.)

21. $(-2, 0), (0, 1), (2, 3)$ (See Exercise 1.)

22. $(-3, 0), (-1, 1), (1, 1), (3, 2)$ (See Exercise 2.)

23. $(0, 4), (1, 1), (1, 3), (2, 0)$ (See Exercise 3.)

24. $(1, 0), (2, 0), (3, 0), (3, 1), (4, 1), (4, 2), (5, 2)$ $(6, 2)$ (See Exercise 4.)

In Exercises 25–28, find the least squares regression quadratic for the given points.

ⓒ**25.** $(-2, 0), (-1, 0), (0, 1), (1, 2), (2, 5)$

ⓒ**26.** $(-4, 5), (-2, 6), (2, 6), (4, 2)$

ⓒ**27.** $(0, 0), (2, 2), (3, 6), (4, 12)$

ⓒ**28.** $(0, 10), (1, 9), (2, 6), (3, 0)$

ⓒ**29.** The following table gives the world population (in billions) in four different years.

Year (x)	1960	1970	1974	1976
Population (y)	3.0	3.6	4.0	4.1

Let $x = 0$ represent the year 1970.

(a) Find the least squares regression quadratic for this data.

(b) Use this quadratic to estimate the world population in 1985.

ⓒ**30.** The following table gives the U.S. energy consumption in quads (quadrillion British thermal units).

Year (x)	1960	1965	1970	1972	1973
Quads (y)	44.1	53.0	66.8	71.6	74.6

Let $x = 0$ represent the year 1968.

(a) Find the least squares regression line for this data.

(b) Use this equation to estimate the energy consumption in 1975. (The actual usage was 70.7 quads. The dramatic decrease in consumption was due to OPEC's oil embargo and indicates how difficult it is to make accurate forecasts on politically sensitive issues.)

ⓒ Calculator may be helpful.

Section 8.7

Double Integrals and Area in the Plane

Introductory Example

Population Density of a City

Demographers have discovered that most cities have population densities that are greatest at the city's center. As one moves farther and farther from the city center, the population becomes less and less dense. The most common model to represent this changing density (in number of people per square mile) is density $= Ce^{-ar}$, where a and C are constants and r is the distance (in miles) from the city center. To determine the total population of a city with a given density function, we assume that the center of the city lies at the origin, as shown in Figure 8.44. Since the distance from the origin to the point (x, y) is $r = \sqrt{x^2 + y^2}$, we can rewrite the density function as a function of x and y as follows:

$$\text{density} = f(x, y) = Ce^{-a\sqrt{x^2 + y^2}}$$

Now, the total population of the city can be determined by integrating the density function over the plane region representing the city's limits. We do this by setting up a **double integral** to represent the area of the plane region. For example, a circular city with a radius of 5 mi has an area represented by the double integral

$$\text{area} = \int_{-5}^{5} \int_{-\sqrt{25-x^2}}^{\sqrt{25 - x^2}} dy\, dx$$

Thus, if the density function for this circular city is

$$f(x, y) = (100{,}000)e^{-\sqrt{x^2+y^2}}$$

then the total population for the city would be given by the double integral

$$\text{population} = \int_{-5}^{5} \int_{-\sqrt{25-x^2}}^{\sqrt{25-x^2}} (100{,}000e)^{-\sqrt{x^2+y^2}}\, dy\, dx$$

Although evaluation of this *particular* double integral is beyond the scope of this text (the population for this model turns out to be 602,917), you will be able to evaluate a wide variety of double integrals after studying the techniques presented in this section.

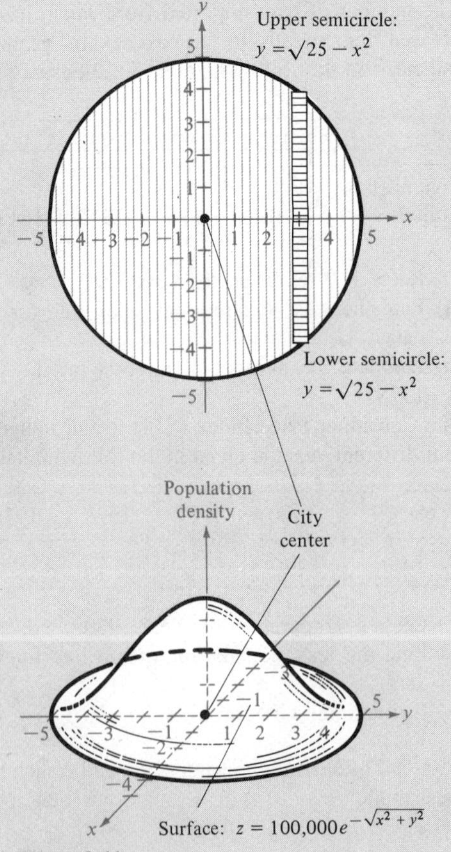

Figure 8.44

Section Objectives: ***To introduce double integrals and demonstrate their evaluation ▪ To represent areas of a plane region by double integrals.***

In Section 8.3, we showed that it is meaningful to differentiate functions of more than one variable by differentiating with respect to one variable at a time while holding the other variables constant. It should not be surprising to learn that we can *integrate* functions of two or more variables by a similar procedure. For example, if we are given the partial derivative $f_x(x, y) = 2xy$, then by holding y constant, we can integrate with respect to x to obtain

$$f(x, y) = x^2y + C(y)$$

This procedure is sometimes referred to as **partial integration with respect to** x. Note that the "constant of integration," $C(y)$, is assumed to be a function of y since y is fixed while integrating with respect to x.

To evaluate the definite integral of a function of two or more variables, we can use the Fundamental Theorem of Calculus (Section 4.3) with one variable while holding the others constant. For instance,

$$\int_1^{2y} 2xy\,dx = x^2y\Big]_1^{2y} = (2y)^2y - (1)^2y = 4y^3 - y$$

x is the variable of integration and y is fixed

replace x by the limits of integration

the result is a function of y

Note that we omit the constant of integration just as we do for the definite integral of a function of one variable.

A partial integral with respect to y can be evaluated in a similar manner, as is demonstrated in Example 1.

Example 1

Partial Integration

Evaluate the integrals

$$\int_1^x (2x^2y^{-2} + 2y)\,dy \qquad \text{and} \qquad \int_y^{5y} \sqrt{x - y}\,dx$$

Solution Considering x to be constant while integrating with respect to y, we obtain

$$\begin{aligned}\int_1^x (2x^2y^{-2} + 2y)\,dy &= \left[\frac{-2x^2}{y} + y^2\right]_1^x \\ &= \left(\frac{-2x^2}{x} + x^2\right) - \left(\frac{-2x^2}{1} + 1\right) \\ &= 3x^2 - 2x - 1\end{aligned}$$

Similarly, considering y to be constant, we have

$$\begin{aligned}\int_y^{5y} \sqrt{x - y}\,dx &= \frac{2}{3}(x - y)^{3/2}\Big]_y^{5y} = \frac{2}{3}[(5y - y)^{3/2} - (y - y)^{3/2}] \\ &= \tfrac{2}{3}(4y)^{3/2} = \tfrac{16}{3}y^{3/2}\end{aligned}$$

□

Note that in Example 1 the integral

$$\int_1^x (2x^2y^{-2} + 2y)\, dy = 3x^2 - 2x - 1$$

defines a function of one variable x and, as such, can *itself* be integrated, as shown in the next example.

Example 2

Evaluating a Double Integral

Evaluate the integral

$$\int_0^1 \left[\int_1^x (2x^2y^{-2} + 2y)\, dy\right] dx$$

Solution Using the results from Example 1, we have

$$\int_0^1 \left[\int_1^x (2x^2y^{-2} + 2y)\, dy\right] dx = \int_0^1 (3x^2 - 2x - 1)\, dx$$

$$= \left[x^3 - x^2 - x\right]_0^1 = -1$$

□

We call integrals of the form

$$\int_a^b \left[\int_{g_1(x)}^{g_2(x)} f(x, y)\, dy\right] dx \quad \text{or} \quad \int_c^d \left[\int_{h_1(y)}^{h_2(y)} f(x, y)\, dx\right] dy$$

double integrals, and we normally shorten the notation by omitting the brackets. Thus, by definition,

$$\int_a^b \int_{g_1(x)}^{g_2(x)} f(x, y)\, dy\, dx = \int_a^b \left[\int_{g_1(x)}^{g_2(x)} f(x, y)\, dy\right] dx$$

One of the simplest applications of a double integral is in finding the area of a plane region. For instance, if R is the region bounded by $a \leq x \leq b$ and $g_1(x) \leq y \leq g_2(x)$, then by the methods of Section 4.4, we know that the area of R is given by

$$\int_a^b [g_2(x) - g_1(x)]\, dx$$

(See Figure 8.45.) This same area is also given by the double integral

$$\int_a^b \int_{g_1(x)}^{g_2(x)} dy\, dx$$

because

$$\int_a^b \int_{g_1(x)}^{g_2(x)} dy\, dx = \int_a^b y\Big]_{g_1(x)}^{g_2(x)} dx = \int_a^b [g_2(x) - g_1(x)]\, dx$$

Figure 8.45 shows the two basic types of plane region whose area can be determined by a double integral.

In Figure 8.45, note that the position (vertical or horizontal) of the narrow rectangle indicates the order of integration. The "outer" variable of integration always corresponds to the width (thickness) of the rectangle.

Area in the Plane by Double Integrals

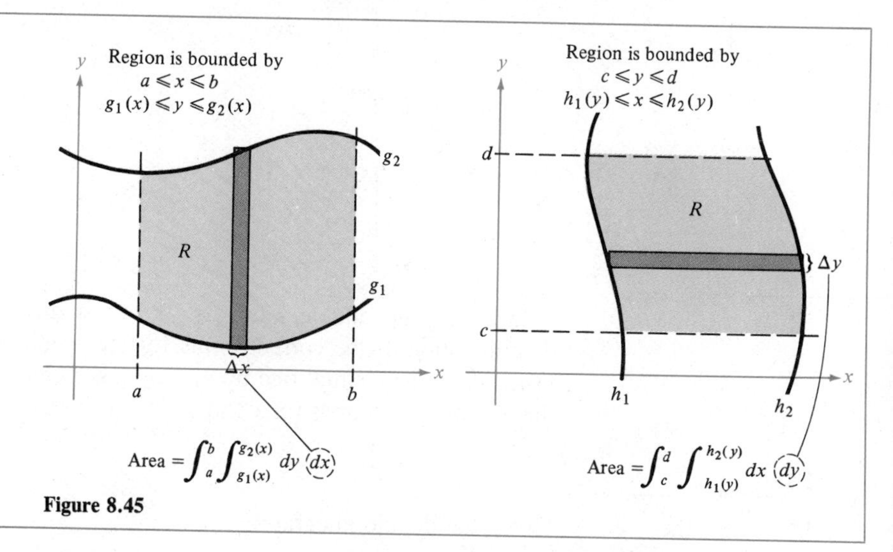

Figure 8.45

(Compare this with the procedures of Section 4.4 for the area between two curves.) Note also that the outer limits of integration for a double integral are constant, whereas the inner limits may be functions of the outer variable.

Example 3

Finding Area by a Double Integral

Set up a double integral for the area of the region bounded by $0 \leq x \leq 1$ and $x^3 \leq y \leq x^2$. (See Figure 8.46.)

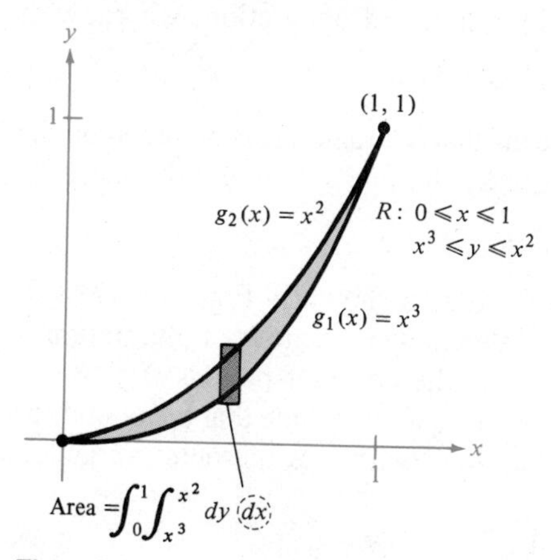

Figure 8.46

Solution Since the limits for x are constant, we let x be the outer variable and write

$$\begin{aligned}\text{area} &= \int_0^1 \int_{x^3}^{x^2} dy\, dx \\ &= \int_0^1 y\Big]_{x^3}^{x^2} dx \\ &= \int_0^1 (x^2 - x^3)\, dx \\ &= \left[\frac{x^3}{3} - \frac{x^4}{4}\right]_0^1 \\ &= \tfrac{1}{3} - \tfrac{1}{4} = \tfrac{1}{12}\end{aligned}$$

□

In setting up double integrals, the most difficult task is likely to be the determination of the correct limits for the specified order of integration. This task can often be simplified by making a sketch of the region R and identifying the appropriate bounds for x and y. The next example illustrates this procedure.

Example 4

Changing the Order of Integration

Given the double integral

$$\int_0^2 \int_{y^2}^4 dx\, dy$$

(a) Sketch the region R whose area is given by this integral.
(b) Rewrite the integral so that x is the outer variable.
(c) Show that both orders of integration yield the same value.

Solution

(a) From the limits of integration, we know that

$$y^2 \leq x \leq 4$$

which means that the region R is bounded on the left by the parabola $y^2 = x$ and on the right by the line $x = 4$. Furthermore, since

$$0 \leq y \leq 2$$

we have the region shown in Figure 8.47.
(b) If we interchange the order of integration so that x is the outer variable, we see that x has the constant bounds $0 \leq x \leq 4$, and, by solving for y in the equation $y^2 = x$, we conclude that the bounds for y are $0 \leq y \leq \sqrt{x}$ (see Figure 8.48). Therefore, with x as the outer variable, we obtain the integral

$$\int_0^4 \int_0^{\sqrt{x}} dy\, dx$$

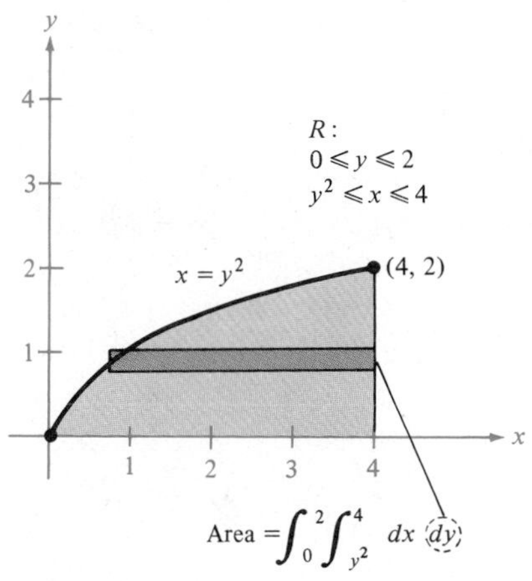

Figure 8.47

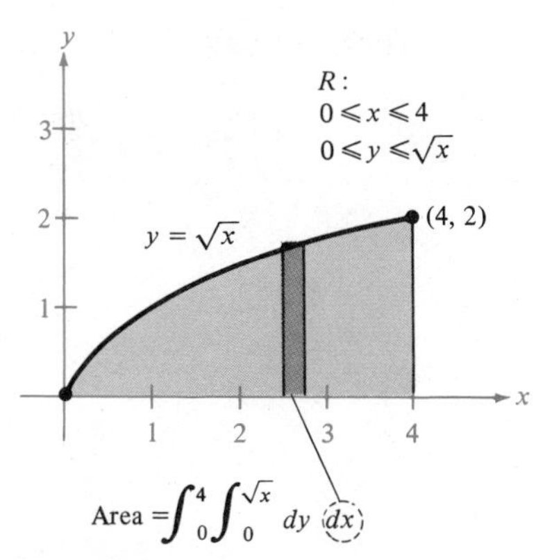

Figure 8.48

(c) Evaluating both integrals, we obtain

$$\int_0^2 \int_{y^2}^4 dx\, dy = \int_0^2 x \Big]_{y^2}^4 dy = \int_0^2 (4 - y^2)\, dy = \left[4y - \frac{y^3}{3}\right]_0^2 = \frac{16}{3}$$

and

$$\int_0^4 \int_0^{\sqrt{x}} dy\, dx = \int_0^4 y \Big]_0^{\sqrt{x}} dx = \int_0^4 \sqrt{x}\, dx = \frac{2}{3}x^{3/2}\Big]_0^4 = \frac{16}{3}$$

□

To designate a double integral or an area of a region without specifying a particular order of integration, we use the symbol

$$\iint_R dA$$

where $dA = dx\, dy$ or $dA = dy\, dx$.

Example 5

Finding Area by a Double Integral

Use a double integral to calculate the area denoted by

$$\iint_R dA$$

where R is the region bounded by $y = x$ and $y = x^2 - x$.

Solution From a sketch (Figure 8.49) of region R, we can see that vertical rectangles of width dx are more convenient than horizontal ones. Therefore, x is the outer variable of integration, and its constant bounds are $0 \le x \le 2$.

Thus, we have

$$\iint_R dA = \int_0^2 \int_{x^2-x}^{x} dy\,dx = \int_0^2 y\Big]_{x^2-x}^{x} dx$$

$$= \int_0^2 [x - (x^2 - x)]\,dx = \int_0^2 (2x - x^2)\,dx$$

$$= \left[x^2 - \frac{x^3}{3}\right]_0^2 = 4 - \frac{8}{3} = \frac{4}{3}$$

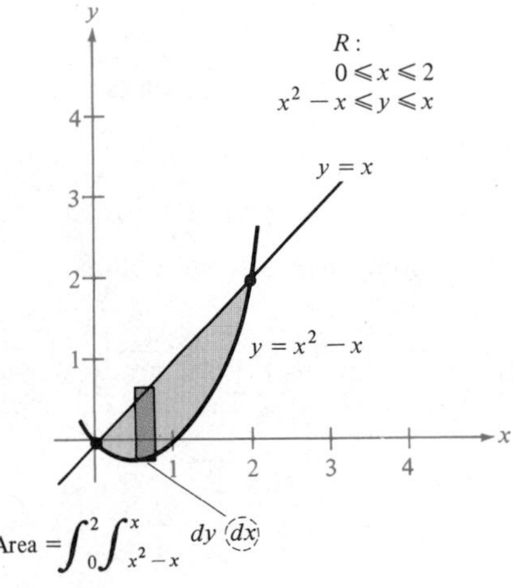

Figure 8.49

□

At this point you may be wondering why we need double integrals since we can already find the area between curves in the plane by single integrals. The real need for double integrals will be demonstrated in Section 8.8 where they are used to find volumes and average functional values. When working the exercises in this section, remember that we have chosen to introduce double integrals by way of areas in the plane so that we can give primary attention to the procedures for finding the limits of integration. As the examples of this section show, this task is greatly simplified by making a sketch of the region under consideration.

Section Exercises (8.7)

In Exercises 1–10, evaluate the specified integral.

1. $\int_0^x (2x - y)\,dy$

2. $\int_x^{x^2} \frac{y}{x}\,dy$

3. $\int_1^{2y} \frac{y}{x}\,dx$

4. $\int_0^{e^y} y\,dx$

5. $\int_0^{\sqrt{4-x^2}} x^2y\,dy$

6. $\int_{x^2}^{\sqrt{x}} (x^2 + y^2)\,dy$

7. $\int_{e^y}^{y} \frac{y \ln x}{x}\,dx$

8. $\int_{-\sqrt{1-y^2}}^{\sqrt{1-y^2}} (x^2 + y^2)\,dx$

9. $\displaystyle\int_0^{x^3} ye^{-y/x}\, dy$

10. $\displaystyle\int_y^3 \frac{xy}{\sqrt{x^2+1}}\, dx$

In Exercises 11–20, evaluate the specified double integral.

11. $\displaystyle\int_0^1 \int_0^2 (x+y)\, dy\, dx$

12. $\displaystyle\int_0^1 \int_0^x \sqrt{1-x^2}\, dy\, dx$

13. $\displaystyle\int_1^2 \int_0^4 (x^2 - 2y^2 + 1)\, dx\, dy$

14. $\displaystyle\int_0^1 \int_y^{2y} (1 + 2x^2 + 2y^2)\, dx\, dy$

15. $\displaystyle\int_0^1 \int_0^{\sqrt{1-y^2}} (x+y)\, dx\, dy$

16. $\displaystyle\int_0^2 \int_{3y^2-6y}^{2y-y^2} 3y\, dx\, dy$

17. $\displaystyle\int_0^2 \int_0^{\sqrt{4-y^2}} \frac{2}{\sqrt{4-y^2}}\, dx\, dy$

18. $\displaystyle\int_0^4 \int_0^x \frac{2}{(x+1)(y+1)}\, dy\, dx$

19. $\displaystyle\int_0^2 \int_0^{4-x^2} x^3\, dy\, dx$

20. $\displaystyle\int_0^\infty \int_0^\infty xye^{-(x^2+y^2)}\, dx\, dy$

In Exercises 21–27, sketch the region R whose area is given by the double integral, switch the order of integration, and show that both orders yield the same area.

21. $\displaystyle\int_0^1 \int_0^2 dy\, dx$

22. $\displaystyle\int_1^2 \int_2^4 dx\, dy$

23. $\displaystyle\int_0^4 \int_0^{\sqrt{x}} dy\, dx$

24. $\displaystyle\int_0^1 \int_{2y}^2 dx\, dy$

25. $\displaystyle\int_0^2 \int_{x/2}^1 dy\, dx$

26. $\displaystyle\int_0^4 \int_{\sqrt{x}}^2 dy\, dx$

27. $\displaystyle\int_0^1 \int_{y^2}^{\sqrt[3]{y}} dx\, dy$

In Exercises 28–36, use a double integral to find the area of the specified region.

28.

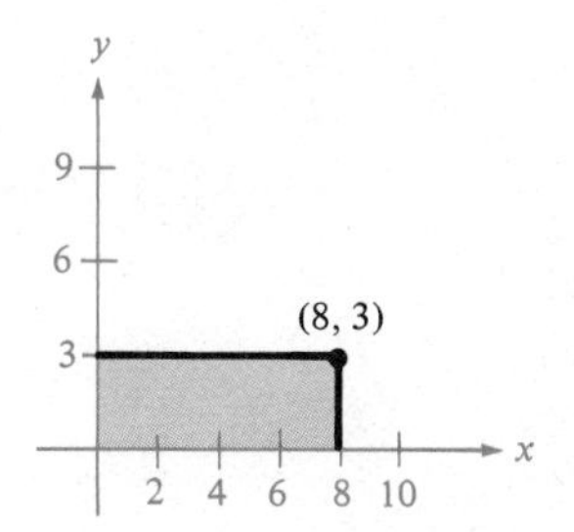

29.

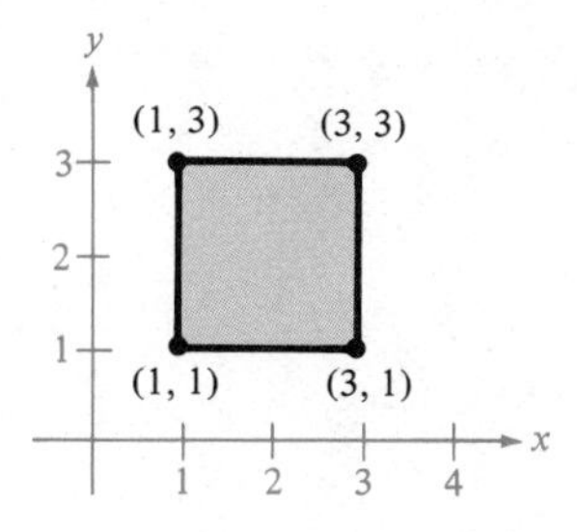

30.

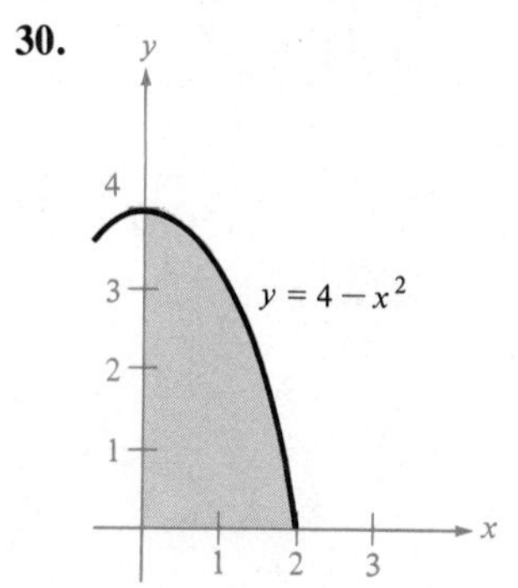

31.

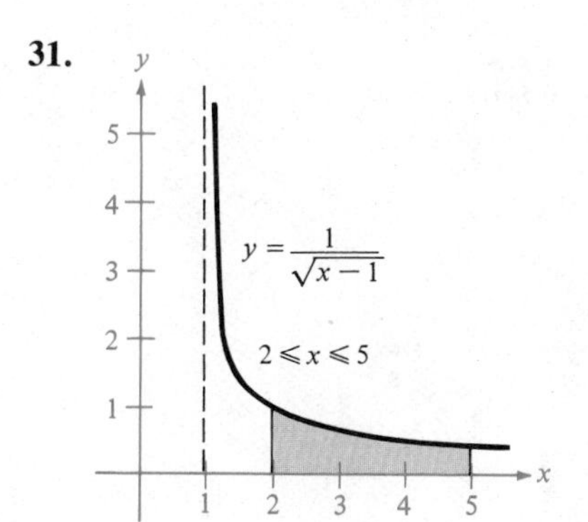

32.

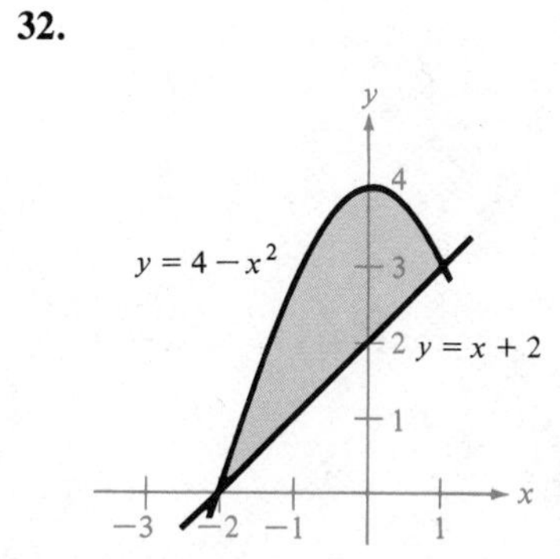

33.

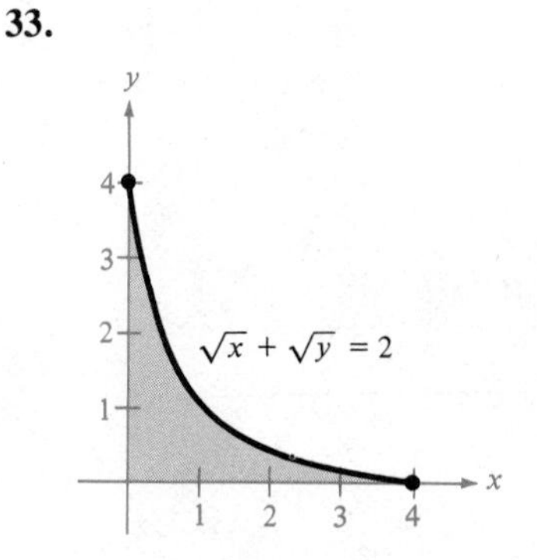

34. R is bounded by $y = x^{3/2}$ and $y = x$.

35. R is the triangle with vertices (0, 0), (5, 0), and (3,2).

36. R is bounded by $xy = 9$, $x = 1$, $y = 0$, and $x = 9$.

Section 8.8

Applications of Double Integrals

Introductory Example

Average Elevation

The highest point in the United States is Mt. McKinley in Alaska, with an elevation of 20,320 ft (above sea level). The lowest point is in Death Valley, California, where the elevation drops to -282 ft (below sea level). Does this mean that the average elevation in the United States is $(20{,}320 - 282)/2 = 10{,}019$ feet? Quite clearly, the answer to this question is no! Actually, the average elevation of the United States is estimated to be about 2,500 ft. The reason the average elevation is so much less than the mean of the two extremes is that considerably more land lies at lower elevations than at higher ones.

To see how the average elevation can be found, let us consider a simple example. Suppose that we want to find the average elevation of a triangular coastal county. (See Figure 8.50.) The elevation (in miles above sea level) at the point (x, y) is given by

$$\text{elevation} = f(x, y) = 0.25 - 0.025x - 0.01y$$

where x and y are measured in miles. Now, to find the average elevation of this county, we consider the (first octant) solid bounded above by the plane $z = f(x, y)$ and below by the xy-plane. The average elevation is given by the **double integral**

$$\text{average elevation} = \frac{\text{volume of solid}}{\text{area of base}} = \frac{1}{A}\iint_R f(x, y)\, dA$$

where A is the area of the base. Since the base is triangular, its area is

$$\text{area of base} = \left(\tfrac{1}{2}\right)(10)(25) = 125 \text{ mi}^2$$

Thus, the average elevation is

average elevation

$$= \frac{1}{125}\int_0^{10}\int_0^{25-2.5x} (0.25 - 0.025x - 0.01y)\, dy\, dx$$

$$= \frac{1}{125}\int_0^{10}\left[0.25y - 0.025xy - 0.005y^2\right]_0^{25-2.5x} dx$$

$$= \frac{1}{125}\int_0^{10}(0.03125x^2 - 0.625x + 3.125)\, dx$$

$$= \frac{1}{125}\left[\frac{0.03125x^3}{3} - \frac{0.625x^2}{2} + 3.125x\right]_0^{10}$$

$$= 0.0833 \text{ mi} = 440 \text{ ft above sea level}$$

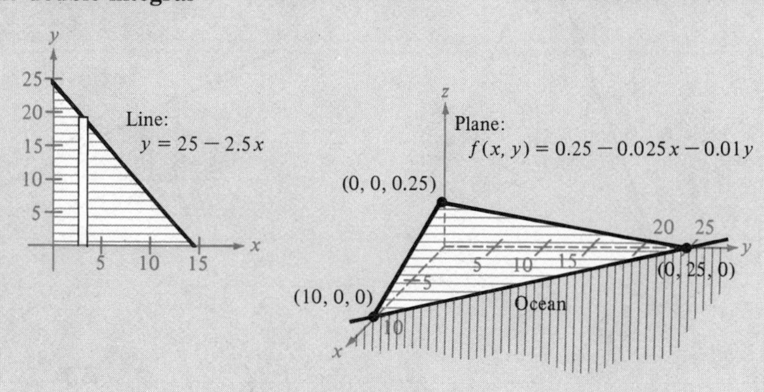

Figure 8.50

Section Objectives: *To calculate the volume of solid regions by double integrals ▪ To find the average value of a function of two variables over a region in the plane.*

In the preceding section, we demonstrated the use of a double integral as an alternative way to find the area of a plane region. In this section, we discuss more imperative uses of double integrals, namely, to find the volume of a solid region and to find the average value of a function.

Specifically, if f is continuous and nonnegative over a region R, then the volume between $z = f(x, y)$ and the region R in the xy-plane is given by the *double integral*

$$\iint_R f(x, y)\, dA$$

where $dA = dy\, dx$ or $dA = dx\, dy$.

Volume by Double Integrals

If R is a bounded region in the xy-plane and f is continuous and nonnegative over R, then the **volume of the solid** between the surface $z = f(x, y)$ and R is given by

$$\iint_R f(x, y)\, dA$$

Example 1

Finding Volume by a Double Integral

Find the volume of the solid bounded in the first octant by the plane $z = 2 - x - 2y$.

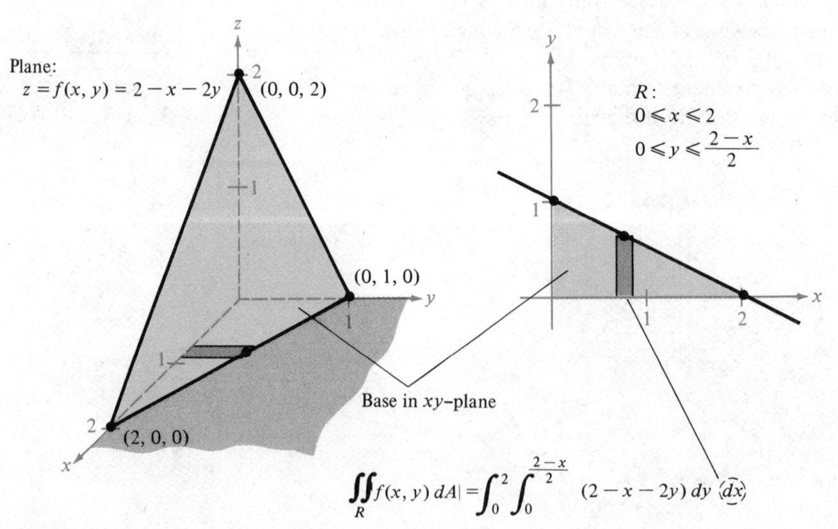

Figure 8.51

Solution To set up the double integral for the volume, it is helpful to sketch both the solid and the region R in the xy-plane. From Figure 8.51, we can see that if $z = 0$, then R is bounded in the xy-plane by the lines $x = 0$, $y = 0$, and

$$y = \frac{2 - x}{2}$$

For a rectangle of width dx, we obtain

constant bounds for x: $0 \leq x \leq 2$

variable bounds for y: $0 \leq y \leq \dfrac{2 - x}{2}$

Thus, the volume of the solid region is

$$\begin{aligned}
\int_0^2 \int_0^{(2-x)/2} (2 - x - 2y)\, dy\, dx &= \int_0^2 \left[(2 - x)y - y^2 \right]_0^{(2-x)/2} dx \\
&= \int_0^2 \left[(2 - x)\left(\frac{2 - x}{2}\right) - \left(\frac{2 - x}{2}\right)^2 \right] dx \\
&= \frac{1}{4} \int_0^2 (2 - x)^2\, dx \\
&= \frac{-1}{12} (2 - x)^3 \Big]_0^2 \\
&= \frac{8}{12} \\
&= \frac{2}{3}
\end{aligned}$$

□

Following the pattern of Example 1, we outline a procedure for determining the volume of a solid region:

1. Write the equation of the surface in the form $z = f(x, y)$ and sketch the solid region.
2. Sketch region R in the xy-plane and determine the order and limits of integration.
3. Evaluate the double integral,

$$\iint_R f(x, y)\, dA$$

using the order and limits determined in the second step.

In Example 1, the order of integration was arbitrary. We could have used y as the outer variable, as shown in Figure 8.52. There are, however, some occasions in which one order of integration is much more convenient than the other. Example 2 shows such a case.

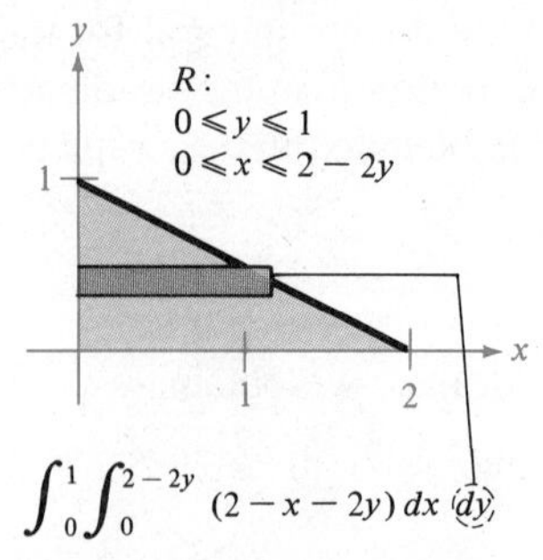

Reversing the Order of Integration
(Compare with Figure 8.51.)

Figure 8.52

Example 2

Comparing Different Orders of Integration

Find the volume under the cylindrical surface $f(x, y) = e^{-x^2}$ bounded by the xz-plane and the planes $y = x$ and $x = 1$.

Solution In the xy-plane the bounds of region R are the lines $y = 0$, $x = 1$, and $y = x$ (see Figure 8.53). The two possible orders of integration are given in Figure 8.54. In attempting to evaluate these two double integrals, we discover that the one on the right requires the antiderivative $\int e^{-x^2}\, dx$, which we know is not an elementary function. On the other hand, we can evaluate the integral on the left in the following manner:

$$\int_0^1 \int_0^x e^{-x^2}\, dy\, dx = \int_0^1 e^{-x^2} y \Big]_0^x dx = \int_0^1 x e^{-x^2}\, dx$$

$$= -\frac{1}{2} e^{-x^2} \Big]_0^1 = -\frac{1}{2}\left(\frac{1}{e} - 1\right) = \frac{e - 1}{2e}$$

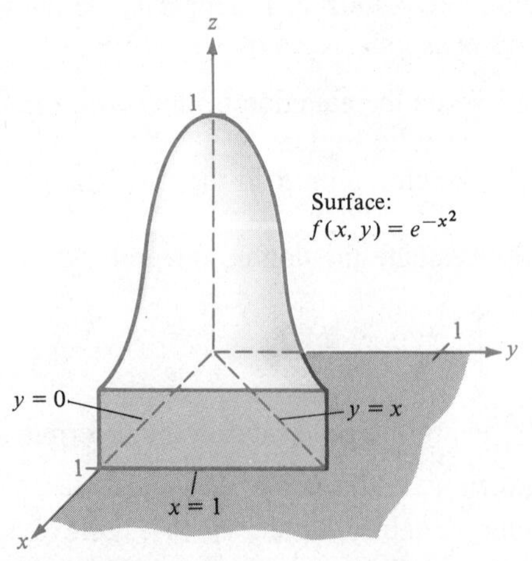

Figure 8.53

In Examples 1 and 2, we provided the sketches of *both* the solid whose volume we were finding *and* the plane region that formed the base of the solid. However, keep in mind that you do not have to be an artist to be able to find the volume of three-dimensional solids. If you are not able to sketch a particular solid, you can still find its volume by referring to a sketch of its base in the xy-plane. We demonstrate this in Example 3.

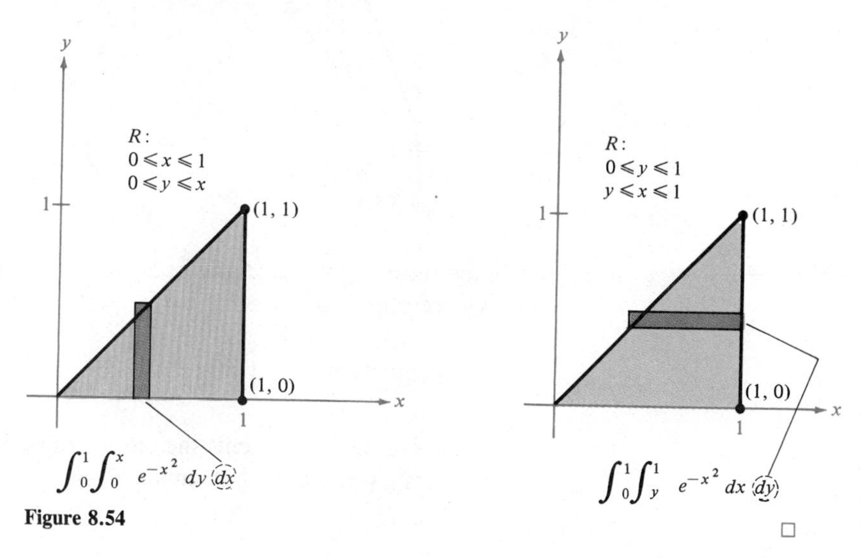

Figure 8.54

□

Example 3

Finding Volume by a Double Integral

Find the volume of the solid bounded above by the surface

$$f(x, y) = 6x^2 - 2xy$$

and below by the plane region R shown in Figure 8.55.

Solution Since the region R is bounded above by the parabola $y = 3x - x^2$ and below by the line $y = x$, y has the limits $x \le y \le 3x - x^2$ (see Figure 8.55). The value of the integral is

$$\int_0^2 \int_x^{3x-x^2} (6x^2 - 2xy)\,dy\,dx$$

$$= \int_0^2 \left[6x^2y - xy^2 \right]_x^{3x-x^2} dx$$

$$= \int_0^2 [(18x^3 - 6x^4 - 9x^3 + 6x^4 - x^5) - (6x^3 - x^3)]\,dx$$

$$= \int_0^2 (4x^3 - x^5)\,dx = \left[x^4 - \frac{x^6}{6} \right]_0^2 = \frac{16}{3}$$

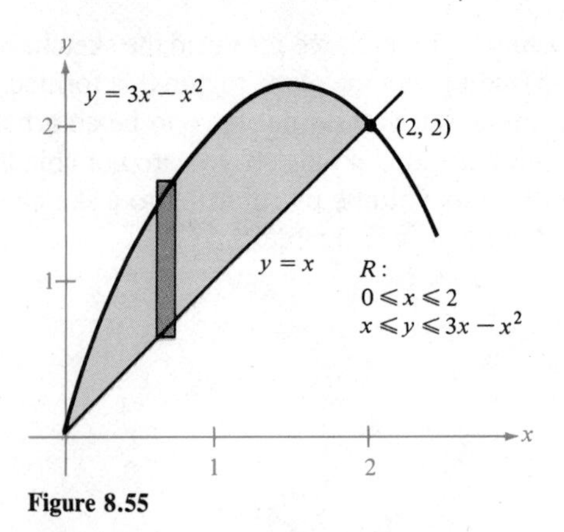

Figure 8.55

□

In the Introductory Example of Section 8.7, we saw that the population of a city whose population density is $f(x, y)$ is given by

$$\text{population} = \iint_R f(x, y)\, dA$$

where the region R represents the city's limits. We demonstrate an application of this formula in the next example.

Example 4

Finding the Population of a City

Find the population of a city whose density function is

$$f(x, y) = \frac{50{,}000}{x + |y| + 1}$$

and whose boundary is shown in Figure 8.56.

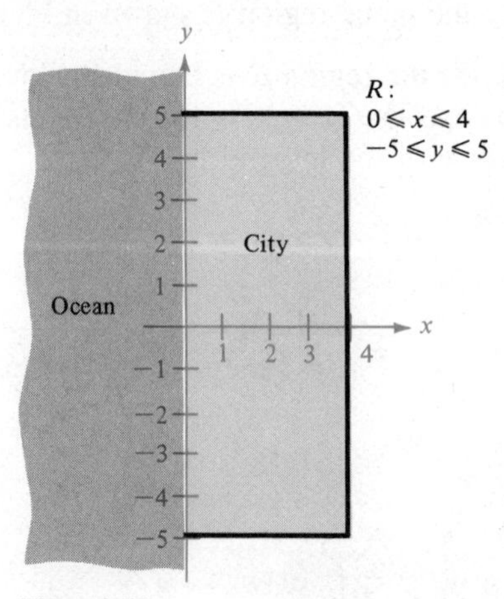

Figure 8.56

Solution To begin with, we note (from the absolute value of y in the density function) that the population density at the point (x, y) is the same as that at the point $(x, -y)$. Thus, we can conclude that the population in the first quadrant equals the population in the fourth quadrant. In other words, we can find the total population by doubling the population in the first quadrant as follows:

$$\begin{aligned}
&\text{population}\\
&= 2\int_0^4\int_0^5 \frac{50{,}000}{x+y+1}\,dy\,dx\\
&= 100{,}000\int_0^4 \Big[\ln(x+y+1)\Big]_0^5\,dx\\
&= 100{,}000\int_0^4 \Big[\ln(x+6)-\ln(x+1)\Big]\,dx\\
&= 100{,}000\Big[(x+6)\ln(x+6)-(x+6)-(x+1)\ln(x+1)+(x+1)\Big]_0^4\\
&= 100{,}000\Big[(x+6)\ln(x+6)-(x+1)\ln(x+1)-5\Big]_0^4\\
&= 100{,}000\,[10\ln(10)-5\ln(5)-5-6\ln(6)+5]\\
&\approx 422{,}810 \text{ people}
\end{aligned}$$

□

Recall from Chapter 4 that one important application of single integrals is finding the **average value** of the function $f(x)$ over the interval $[a, b]$. We can generalize this concept to functions of two variables as follows.

Average Value of $f(x, y)$ Over the Region R

If f is integrable over the plane region R, then its **average value** over R is given by

$$\text{average value of } f(x, y) \text{ over } R = \frac{1}{A}\iint_R f(x, y)\,dA$$

where A is the area of the region R.

Example 5

Finding the Average Profit

A manufacturer determines that the profit for selling x units of one product and y units of a second product is

$$P = -(x-200)^2 - (y-100)^2 + 5{,}000$$

The weekly sales for product 1 vary between 150 and 200 units, while the weekly sales for product 2 vary between 80 and 100 units. Estimate the weekly profit for these two products.

Solution Since we are given $150 \le x \le 200$ and $80 \le y \le 100$, we estimate the weekly profit to be the average of the profit function over the rectangular region shown in Figure 8.57. Since the area of this region is

$$\text{area} = (50)(20) = 1{,}000$$

we have

$$\begin{aligned}
\text{average profit} &= \frac{1}{1{,}000}\int_{150}^{200}\int_{80}^{100} [-(x-200)^2 - (y-100)^2 + 5{,}000]\, dy\, dx \\
&= \frac{1}{1{,}000}\int_{150}^{200} \left[-(x-200)^2 y - \frac{(y-100)^3}{3} + 5{,}000y\right]_{80}^{100} dx \\
&= \frac{1}{1{,}000}\int_{150}^{200} [-20(x-200)^2 + \frac{308{,}000}{3}]\, dx \\
&= \frac{1}{3{,}000}\left[-20(x-200)^3 + 308{,}000x\right]_{150}^{200} \\
&= \frac{12{,}900{,}000}{3{,}000} = \$4{,}300
\end{aligned}$$

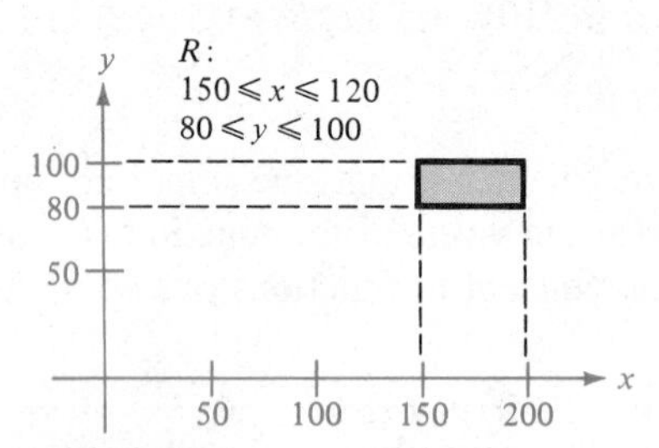

Figure 8.57

□

Section Exercises (8.8)

In Exercises 1–4, sketch the region R and evaluate the integral.

1. $\int_0^2 \int_0^1 (1 + 2x + 2y)\, dy\, dx$

2. $\int_0^6 \int_{y/2}^3 (x + y)\, dx\, dy$

3. $\int_0^1 \int_y^{\sqrt{y}} x^2y^2\, dx\, dy$

4. $\int_{-a}^{a} \int_{-\sqrt{a^2-x^2}}^{\sqrt{a^2-x^2}} (x + y)\, dy\, dx$

In Exercises 5–8, set up the integrals for both *dx dy* and *dy dx* and use the most convenient order to evaluate the given double integral over the region R.

5. $\iint_R xy\, dA$; R: rectangle with vertices (0, 0), (0, 5), (3, 5), (3, 0)

6. $\iint_R \frac{y}{x^2 + y^2}\, dA$; R: triangle bounded by $y = x$, $y = 2x$, $x = 2$.

7. $\displaystyle\iint_R \frac{y}{1+x^2}\,dA$; R: region bounded by $y = 0$, $y = \sqrt{x}$, $x = 4$.

8. $\displaystyle\iint_R x\,dA$; R: semicircle bounded by $y = \sqrt{25 - x^2}$ and $y = 0$

In Exercises 9–20, use a double integral to find the volume of the specified solid.

9.

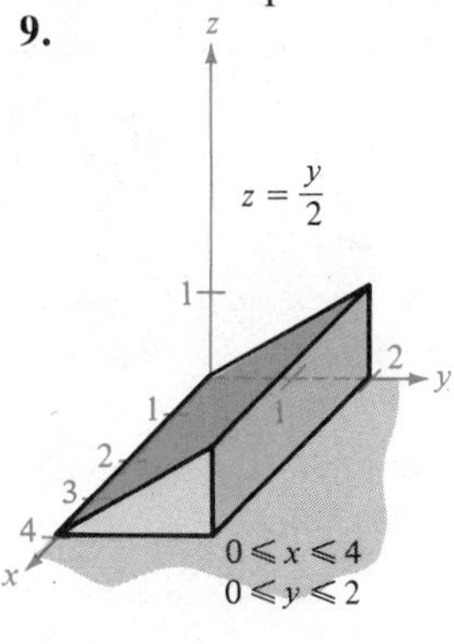

10.

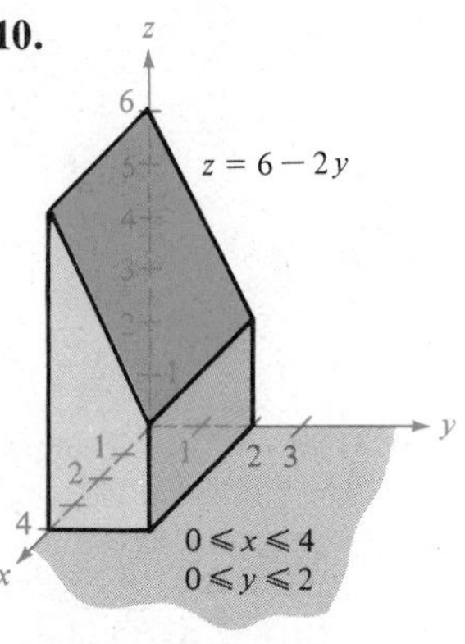

11.

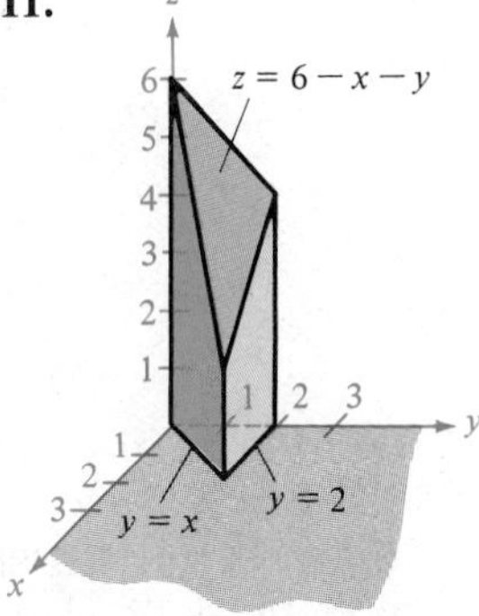

12.

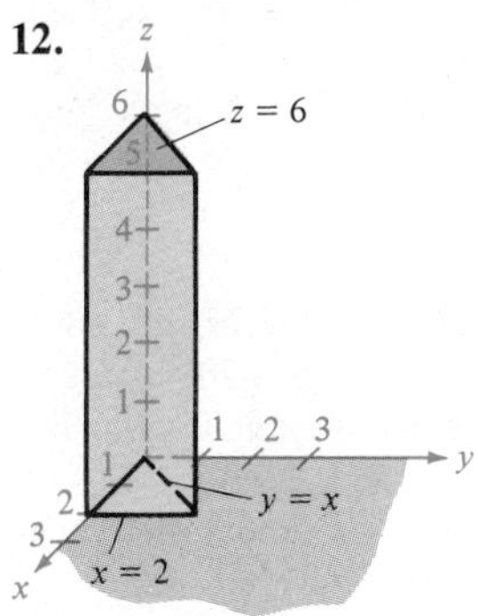

13.

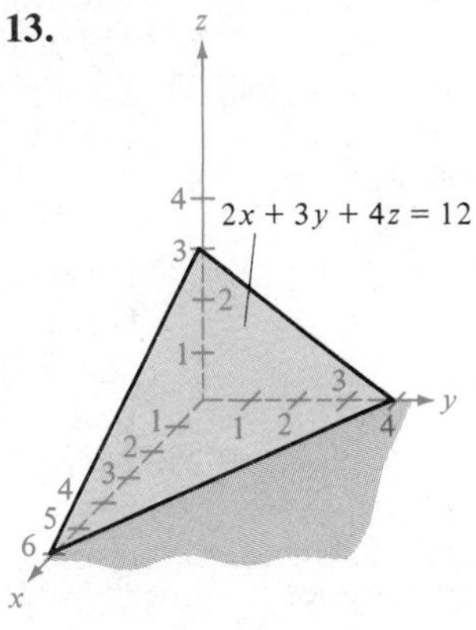

14.

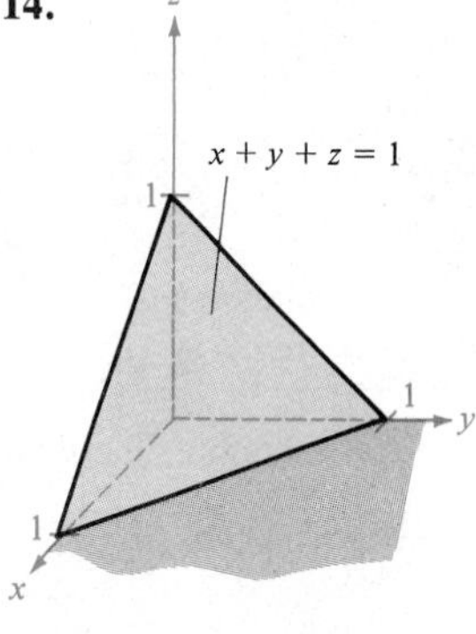

15.

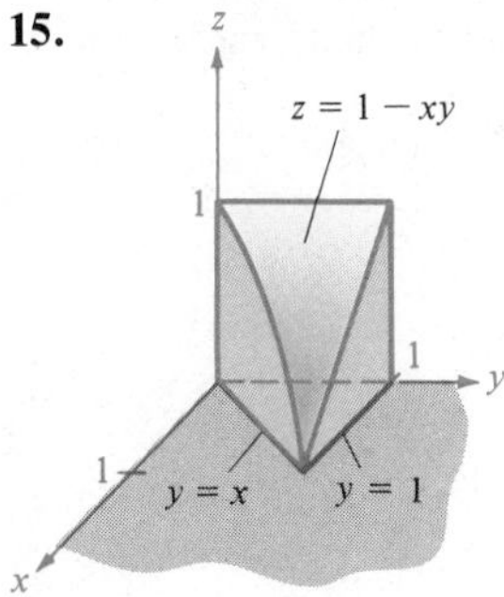

16.

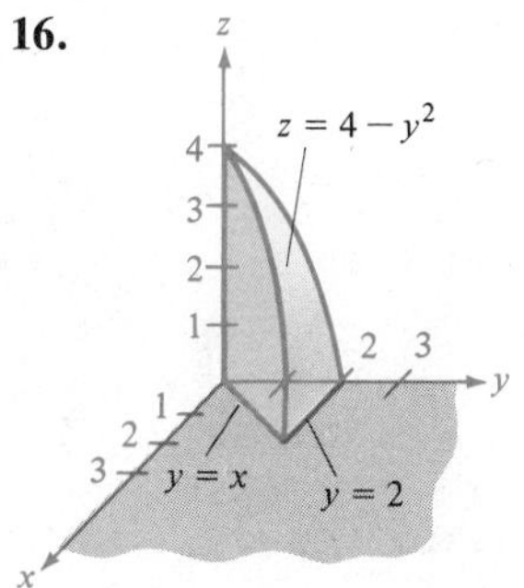

17.

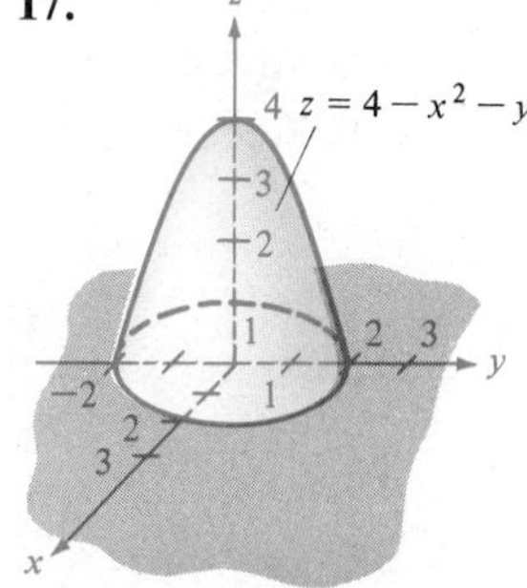

18.

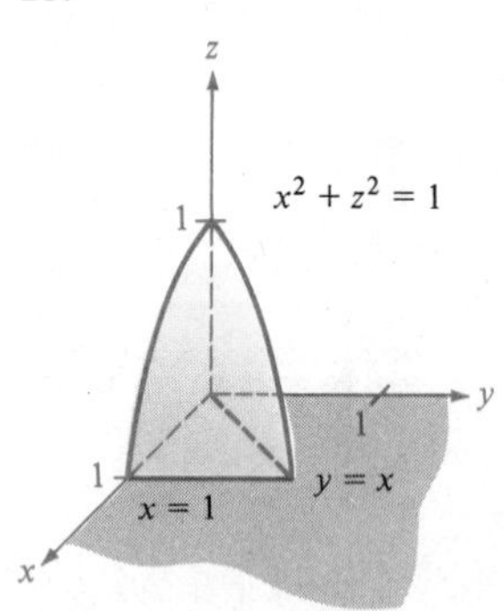

19. Improper integral

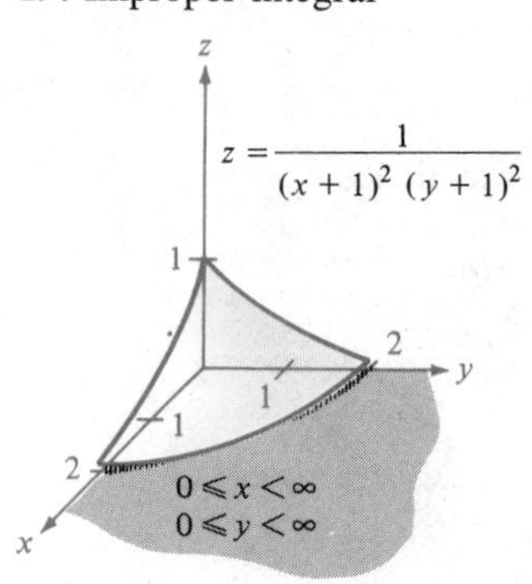

20. Improper integral

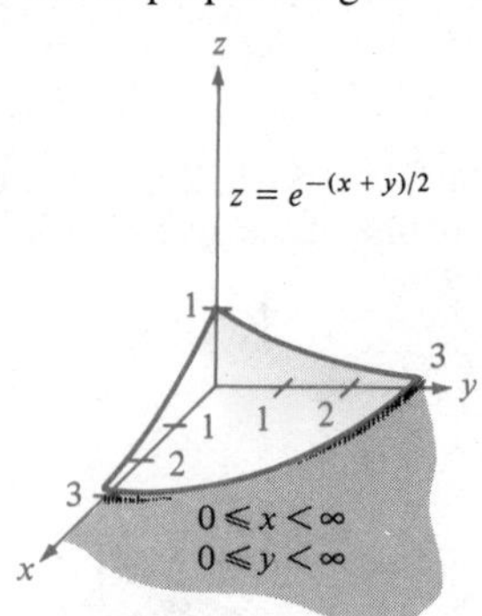

In Exercises 21–24, find the average value of the given function over the region R.

21. $f(x, y) = x$; R: rectangle with vertices $(0, 0)$, $(4, 0)$, $(4, 2)$, $(0, 2)$

22. $f(x, y) = xy$; R: rectangle with vertices $(0, 0)$, $(4, 0)$, $(4, 2)$, $(0, 2)$

23. $f(x, y) = x^2 + y^2$; R: square with vertices $(0, 0)$, $(2, 0)$, $(2, 2)$, $(0, 2)$

24. $f(x, y) = e^{x+y}$; R: triangle with vertices $(0, 0)$, $(0, 1)$, $(1, 1)$

Ⓒ**25.** A company sells two products whose demand functions are

$$x_1 = 500 - 3p_1$$
$$x_2 = 750 - 2.4p_2$$

and, therefore, the total revenue is given by

$$R = x_1p_1 + x_2p_2$$

Estimate the average revenue if the price p_1 varies between \$50 and \$75 and the price p_2 varies between \$100 and \$150.

Ⓒ**26.** A corporation produces a product at two locations. Cost functions at plants 1 and 2 are

$$C_1 = 0.03x_1^2 + 8x_1 + 5000$$
$$C_2 = 0.06x_2^2 + 6x_2 + 8000$$

Therefore, the total cost for producing the product is

$$C = C_1 + C_2$$

Estimate the average cost if the production level at plant 1 varies between 200 and 225 units and the production level at plant 2 varies between 100 and 150 units.

Ⓒ**27.** For a particular company, the Cobb-Douglas production function is

$$f(x, y) = 100x^{0.6}y^{0.4}$$

Estimate the average production level if the number of units of labor varies between 200 and 250 units and the number of units of capital varies between 300 and 325 units.

Ⓒ**28.** A firm's profit in marketing two products is given by

$$P = 192x_1 + 576x_2 - x_1^2 - 5x_2^2 - 2x_1x_2 - 5000$$

where x_1 and x_2 represent the number of units of each product. Estimate the average weekly profit if x_1 varies between 40 and 50 units and x_2 varies between 45 and 50 units.

Ⓒ Calculator may be helpful.

Chapter 9

Series and Taylor Polynomials

Section 9.1

Sequences

Introductory Example

Newton's Method

We call a function that is defined only on the positive integers a **sequence**, and we usually denote its variable by n rather than x.

A common example of a sequence is the approximation of an x-intercept of a function by Newton's Method.* For this method, the sequence

$$x_1, x_2, x_3, x_4, \ldots, x_n, \ldots$$

consists of successively better approximations to a particular x-intercept of $y = f(x)$. To use this approximation method, we choose an initial estimate x_1 of the intercept and then calculate each successive term in the sequence by the rule

$$x_{n+1} = x_n - \frac{f(x_n)}{f'(x_n)}$$

For example, if we want to find the x-intercept of $f(x) = x^3 + x + 1$ (see Figure 9.1), we have

$$x_{n+1} = x_n - \frac{x_n^{\ 3} + x_n + 1}{3x_n^{\ 2} + 1}$$

Now, suppose that we guess that f crosses the x-axis at $x_1 = 0$. This is *not* the actual x-intercept, and we apply Newton's Method to obtain a better estimate:

$$x_2 = 0 - \frac{0^3 + 0 + 1}{3(0^2) + 1} = -1$$

[Geometrically, we interpret x_2 to be the x-intercept of the tangent line to the graph of f at the point $(x_1, f(x_1))$.] Next, using $x_2 = -1$, we can obtain an even better estimate:

$$x_3 = -1 - \frac{(-1)^3 - 1 + 1}{3(-1)^2 + 1} = -1 + \frac{1}{4}$$

$$= -0.75$$

By continuing this process, we obtain the sequence shown in Table 9.1. If a is the actual x-intercept of $f(x) = x^3 + x + 1$, then we say this sequence **converges** to a, and we write

$$\lim_{n \to \infty} x_n = a$$

Table 9.1

x_1	x_2	x_3	x_4	x_5	x_6
0	−1	−0.75	−0.68605	−0.68234	−0.68233

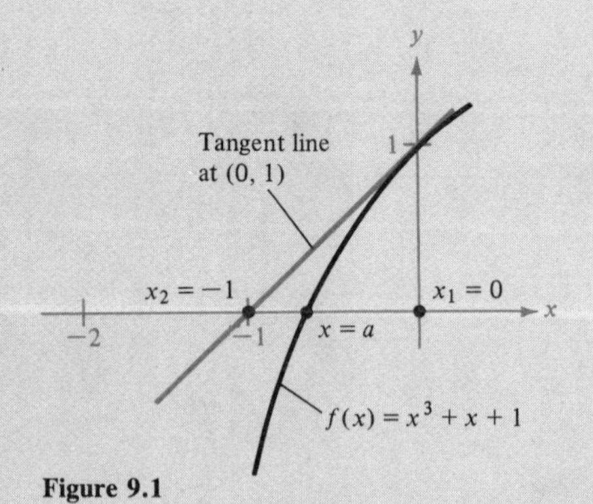

Figure 9.1

*We will study Newton's Method in detail in Section 9.5.

Section Objectives: *To define a sequence and the limit of a sequence ▪ To discuss some techniques for determining the general term of a sequence.*

In mathematics the word "sequence" is used in much the same way as it is in ordinary English. When we say that a collection of objects or events is "in sequence," we usually mean that the collection is ordered so that it has an identified first member, second member, third member, and so on. We define a sequence mathematically as a function whose domain is the set of positive integers. For instance, the equation

$$a(n) = \frac{1}{2^n}$$

defines the sequence

$$\frac{1}{2}, \frac{1}{4}, \frac{1}{8}, \frac{1}{16}, \ldots, \frac{1}{2^n}, \ldots$$

the terms of which correspond respectively to

$$a(1), a(2), a(3), a(4), \ldots, a(n), \ldots$$

We will generally write a sequence such as this in the more convenient *subscript form,*

$$a_1, a_2, a_3, a_4, \ldots, a_n, \ldots$$

or we will denote it by the symbol $\{a_n\}$, where a_n is the nth term of the sequence.

Definition of a Sequence

A **sequence** $\{a_n\}$ is a function whose domain is the set of positive integers. The functional values $a_1, a_2, a_3, \ldots, a_n, \ldots$ are called the **terms** of the sequence.

Remark

It is occasionally convenient to begin subscripting a sequence with zero. In such cases we have

$$a_0, a_1, a_2, \ldots, a_n, \ldots$$

Example 1

Finding Terms in a Sequence

List the first four terms of the sequences given by the following equations. (In each case assume n begins with 1.)

(a) $a_n = 3 + (-1)^n$

(b) $b_n = \dfrac{2n}{1 + n}$

(c) $c_n = \dfrac{2^n}{2^n - 1}$

Solution

(a) $\{a_n\} = \{3 + (-1)^n\}$

$$= \{3 + (-1)^1, 3 + (-1)^2, 3 + (-1)^3, 3 + (-1)^4, \ldots\}$$

$$= \{2, 4, 2, 4, \ldots\}$$

(b) $\{b_n\} = \left\{\dfrac{2n}{1 + n}\right\} = \left\{\dfrac{2(1)}{1 + 1}, \dfrac{2(2)}{1 + 2}, \dfrac{2(3)}{1 + 3}, \dfrac{2(4)}{1 + 4}, \ldots\right\}$

$$= \left\{\frac{2}{2}, \frac{4}{3}, \frac{6}{4}, \frac{8}{5}, \ldots\right\}$$

(c) $\{c_n\} = \left\{\dfrac{2^n}{2^n - 1}\right\} = \left\{\dfrac{2^1}{2^1 - 1}, \dfrac{2^2}{2^2 - 1}, \dfrac{2^3}{2^3 - 1}, \dfrac{2^4}{2^4 - 1}, \ldots\right\}$

$$= \left\{\frac{2}{1}, \frac{4}{3}, \frac{8}{7}, \frac{16}{15}, \ldots\right\}$$

□

In this chapter, our primary interest is in sequences whose terms approach a (unique) limiting value. Such sequences are said to **converge.** (Sequences having no limit are said to **diverge.**) For instance, the terms of the sequence

$$\left\{\frac{1}{2}, \frac{1}{4}, \frac{1}{8}, \frac{1}{16}, \ldots, \frac{1}{2^n}, \ldots\right\}$$

approach 0 as n increases, and we write

$$\lim_{n\to\infty} a_n = \lim_{n\to\infty} \frac{1}{2^n} = 0$$

Similarly, for the sequence $\{b_n\}$ in Example 1, we have

$$\lim_{n\to\infty} \frac{2n}{1 + n} = 2 \lim_{n\to\infty} \frac{n}{1 + n} = 2(1) = 2$$

Although there are technical differences, for our purposes we can operate with limits of sequences in the same way as we operated with limits of continuous functions in Section 3.6. In other words, when we are asked to evaluate the limit of the sequence

$$\lim_{n\to\infty} \frac{2n}{1 + n}$$

we can imagine that n is replaced by x to obtain

$$\lim_{x\to\infty} \frac{2x}{1 + x} = 2$$

Again we conclude that the limit of the sequence is 2.

Example 2

Finding the Limit of a Sequence

Determine the convergence or divergence of the sequences given by the following equations.

$$\text{(a) } a_n = 3 + (-1)^n \qquad \text{(b) } b_n = \frac{n}{1-2n} \qquad \text{(c) } c_n = \frac{2^n}{2^n - 1}$$

Solution

(a) Since the sequence

$$\{a_n\} = \{3 + (-1)^n\} = \{2, 4, 2, 4, \ldots\}$$

oscillates between 2 and 4, we know that $\lim_{n\to\infty} a_n$ does not exist, and we conclude that $\{a_n\}$ diverges.

(b) For

$$\{b_n\} = \left\{\frac{n}{1-2n}\right\}$$

we divide the numerator and denominator by n to obtain

$$\lim_{n\to\infty} \frac{n}{1-2n} = \lim_{n\to\infty} \left[\frac{1}{(1/n) - 2}\right] = \frac{1}{-2}$$

and $\{b_n\}$ converges to $-\frac{1}{2}$.

(c) For

$$\{c_n\} = \left\{\frac{2^n}{2^n - 1}\right\}$$

we divide the numerator and denominator by 2^n to obtain

$$\lim_{n\to\infty} \frac{2^n}{2^n - 1} = \lim_{n\to\infty} \frac{1}{1 - (1/2^n)} = 1$$

which implies that $\{c_n\}$ converges to 1. □

Some of the most important sequences we will be dealing with involve factorials. If n is a positive integer, then n **factorial** is defined as

$$n! = 1 \cdot 2 \cdot 3 \cdot 4 \cdots (n-1) \cdot n$$

Moreover, for $n = 0$ we let $0! = 1$. Thus, we have

$$0! = 1$$
$$1! = 1$$
$$2! = 1 \cdot 2 = 2$$
$$3! = 1 \cdot 2 \cdot 3 = 6$$
$$4! = 1 \cdot 2 \cdot 3 \cdot 4 = 24$$

Factorials follow the same conventions for order of operation as exponents do. That is, just as $2x^3$ and $(2x)^3$ imply different orders of operations, $2n!$ and $(2n)!$ imply the following orders.

$$2n! = 2(n!) = 2(1 \cdot 2 \cdot 3 \cdot 4 \cdots n)$$
$$(2n)! = 1 \cdot 2 \cdot 3 \cdot 4 \cdots n \cdot (n + 1) \cdots (2n)$$

Example 3

Finding the Limit of a Sequence

Find the limit of the sequence

$$\left\{\frac{(-1)^n}{n!}\right\}$$

Solution The first several terms of this sequence are listed in the following table.

n	1	2	3	4	5	6
$\left\{\frac{(-1)^n}{n!}\right\}$	$\frac{-1}{1}$	$\frac{1}{2}$	$-\frac{1}{6}$	$\frac{1}{24}$	$-\frac{1}{120}$	$\frac{1}{720}$

It is clear that the denominator is growing without bound while the numerator is bounded, and we have

$$\lim_{n\to\infty} \frac{(-1)^n}{n!} = 0$$

□

It is sometimes the case that the terms of a sequence are generated by some rule that does not explicitly identify the nth term of the sequence. Under such circumstances we are required to discover a *pattern* in the sequence and come up with a description of the nth term. Once the nth term is specified, then we can discuss the convergence or divergence of the sequence. This is demonstrated in the next example.

Example 4

Finding a Pattern for a Sequence

Given $f(x) = e^{x/3}$, determine the convergence or divergence of the sequence given by

$$a_n = f^{(n-1)}(0)$$

where $f^{(0)} = f$ and $f^{(n)}$ is the nth derivative of f.

Solution We determine the first several terms of the sequence as shown in Table 9.2.

Table 9.2

n	1	2	3	4	5	...	n
$f^{(n-1)}(x)$	$e^{x/3}$	$\frac{e^{x/3}}{3}$	$\frac{e^{x/3}}{3^2}$	$\frac{e^{x/3}}{3^3}$	$\frac{e^{x/3}}{3^4}$	...	$\frac{e^{x/3}}{3^{n-1}}$
$f^{(n-1)}(0)$	1	$\frac{1}{3}$	$\frac{1}{3^2}$	$\frac{1}{3^3}$	$\frac{1}{3^4}$	...	$\frac{1}{3^{n-1}}$

The pattern in the sequence $\{f^{(n-1)}(0)\}$ suggests that the nth term is

$$a_n = \frac{1}{3^{n-1}}$$

and, therefore, the sequence converges to 0 because

$$\lim_{n\to\infty} a_n = \lim_{n\to\infty}\left(\frac{1}{3^{n-1}}\right) = 0 \qquad \square$$

Without a specific rule for generating the terms of a sequence or some knowledge of the context in which the terms of the sequence are obtained, it is not possible to determine the convergence or divergence of the sequence merely from its first several terms. For instance, the first three terms of the four sequences given next are identical, yet from the description of their individual nth terms, we see that the first two sequences converge to 0, the third sequence converges to $\frac{1}{9}$, and the fourth one diverges.

$$\{a_n\} = \left\{\frac{1}{2}, \frac{1}{4}, \frac{1}{8}, \frac{1}{16}, \ldots, \frac{1}{2^n}, \ldots\right\}$$

$$\{b_n\} = \left\{\frac{1}{2}, \frac{1}{4}, \frac{1}{8}, \frac{1}{15}, \ldots, \frac{6}{(n+1)(n^2-n+6)}, \ldots\right\}$$

$$\{c_n\} = \left\{\frac{1}{2}, \frac{1}{4}, \frac{1}{8}, \frac{7}{62}, \ldots, \frac{n^2-3n+3}{9n^2-25n+18}, \ldots\right\}$$

$$\{d_n\} = \left\{\frac{1}{2}, \frac{1}{4}, \frac{1}{8}, 0, \ldots, \frac{-n(n+1)(n-4)}{6(n^2+3n-2)}, \ldots\right\}$$

Thus, if only the first several terms of a sequence are given, there are many forms for an nth term of the sequence. In such a situation we can only determine the convergence or divergence of the sequence on the basis of our choice for the nth term.

Example 5

Finding the Pattern for a Sequence

Determine an nth term for the sequence

$$\left\{-\frac{1}{1}, \frac{3}{2}, -\frac{7}{6}, \frac{15}{24}, -\frac{31}{120}, \ldots\right\}$$

Solution Observe that the numerators are one less than 2^n for $n = 1, 2, 3, 4, 5, \ldots$. Hence, we can generate the numerators by the rule $2^n - 1$. Now, if we factor the denominators successively, we have

$$1,\ 1 \cdot 2,\ 1 \cdot 2 \cdot 3,\ 1 \cdot 2 \cdot 3 \cdot 4,\ 1 \cdot 2 \cdot 3 \cdot 4 \cdot 5,\ \ldots$$

This suggests that the denominators are represented by $n!$. Furthermore, since the signs alternate, we can write

$$a_n = (-1)^n\left(\frac{2^n - 1}{n!}\right)$$

as an nth term for the given sequence. □

As an aid in finding patterns for sequences, we provide the following table summarizing the most common sequence patterns.

Patterns for sequences	n	1	2	3	4	5	6
Changes in sign	$(-1)^n$	-1	1	-1	1	-1	1
	$(-1)^{n+1}$	1	-1	1	-1	1	-1
Arithmetic sequences	$2n$	2	4	6	8	10	12
	$2n - 1$	1	3	5	7	9	11
	$an + b$	$a + b$	$2a + b$	$3a + b$	$4a + b$	$5a + b$	$6a + b$
Binary sequence	2^{n-1}	1	2	4	8	16	32
Geometric sequence	ar^{n-1}	a	ar	ar^2	ar^3	ar^4	ar^5
Power sequences	n^2	1	4	9	16	25	36
	n^p	1	2^p	3^p	4^p	5^p	6^p
Factorial sequence	$n!$	1	2	6	24	120	720
Sequences of products	$2^n n!$	2	$2 \cdot 4$	$2 \cdot 4 \cdot 6$	$2 \cdot 4 \cdot 6 \cdot 8$	$2 \cdot 4 \cdot 6 \cdot 8 \cdot 10$	$2 \cdot 4 \cdot 6 \cdot 8 \cdot 10 \cdot 12$
	$\frac{(2n)!}{2^n n!}$	1	$3 \cdot 5$	$3 \cdot 5 \cdot 7$	$3 \cdot 5 \cdot 7 \cdot 9$	$3 \cdot 5 \cdot 7 \cdot 9 \cdot 11$	$3 \cdot 5 \cdot 7 \cdot 9 \cdot 11 \cdot 13$

Example 6

A Business Application

A deposit of \$1000 is made in an account that earns 12% compounded monthly. Find a sequence to represent the balance in this account after n months ($n = 1, 2, 3, \ldots$).

Solution Since an annual interest rate of 12% corresponds to a monthly rate of 1%, the balance after 1 month is

$$A_1 = 1000 + 1000(0.01) = 1000(1.01)$$

After 2 months the balance is

$$A_2 = 1000(1.01) + 1000(1.01)(0.01) = 1000(1.01)^2$$

By continuing this pattern, we find the sequence representing the monthly balances to be

$$\{A_n\} = \{1000(1.01)^n\} = \{1000(1.01), 1000(1.01)^2, 1000(1.01)^3, \ldots\}$$

Note that this sequence is of the form $\{ar^n\}$ and is therefore a geometric sequence. □

Section Exercises (9.1)

In Exercises 1–8, write out the first five terms of the specified sequence.

1. $a_n = 2^n$

2. $a_n = \dfrac{n}{n+1}$

3. $a_n = \left(-\dfrac{1}{2}\right)^n$

4. $a_n = \dfrac{n^2-1}{n^2+2}$

5. $a_n = \dfrac{3^n}{n!}$

6. $a_n = 5 - \dfrac{1}{n} + \dfrac{1}{n^2}$

7. $a_n = \dfrac{(-1)^n}{n^2}$

8. $a_n = \dfrac{3n!}{(n-1)!}$

In Exercises 9–24, write an expression for the nth term of the sequence.

9. $\{1, 4, 7, 10, \ldots\}$

10. $\{3, 7, 11, 15, \ldots\}$

11. $\{-1, 2, 7, 14, 23, \ldots\}$

12. $\{1, \frac{1}{4}, \frac{1}{9}, \frac{1}{16}, \ldots\}$

13. $\{\frac{2}{3}, \frac{3}{4}, \frac{4}{5}, \frac{5}{6}, \ldots\}$

14. $\{2, \frac{3}{3}, \frac{4}{5}, \frac{5}{7}, \frac{6}{9}, \ldots\}$

15. $\{2, -1, \frac{1}{2}, -\frac{1}{4}, \frac{1}{8}, \ldots\}$

16. $\{\frac{1}{2}, \frac{1}{3}, \frac{2}{9}, \frac{4}{27}, \frac{8}{81}, \ldots\}$

17. $\{2, 1 + \frac{1}{2}, 1 + \frac{1}{3}, 1 + \frac{1}{4}, 1 + \frac{1}{5}, \ldots\}$

18. $\{1 + \frac{1}{2}, 1 + \frac{3}{4}, 1 + \frac{7}{8}, 1 + \frac{15}{16}, 1 + \frac{31}{32}, \ldots\}$

19. $\left\{\dfrac{1}{2\cdot 3}, \dfrac{2}{3\cdot 4}, \dfrac{3}{4\cdot 5}, \dfrac{4}{5\cdot 6}, \ldots\right\}$

20. $\left\{1, \dfrac{1}{2}, \dfrac{1}{6}, \dfrac{1}{24}, \dfrac{1}{120}, \ldots\right\}$

21. $\left\{1, \dfrac{1}{1\cdot 3}, \dfrac{1}{1\cdot 3\cdot 5}, \dfrac{1}{1\cdot 3\cdot 5\cdot 7}, \ldots\right\}$

22. $\{2, -4, 6, -8, 10, \ldots\}$

23. $\{1, -1, -1, 1, 1, -1, -1, \ldots\}$

24. $\left\{1, x, \dfrac{x^2}{2}, \dfrac{x^3}{6}, \dfrac{x^4}{24}, \dfrac{x^5}{120}, \ldots\right\}$

In Exercises 25–38, determine the convergence or divergence of each sequence. If the sequence converges, find its limit.

25. $a_n = \dfrac{n+1}{n}$

26. $a_n = \dfrac{1}{n^{3/2}}$

27. $a_n = (-1)^n\left(\dfrac{n}{n+1}\right)$

28. $a_n = \dfrac{n-1}{n} - \dfrac{n}{n-1}, \quad n \geq 2$

29. $a_n = \dfrac{3n^2 - n + 4}{2n^2 + 1}$

30. $a_n = \dfrac{\sqrt{n}}{\sqrt{n} + 1}$

31. $a_n = \dfrac{n^2 - 1}{n + 1}$

32. $a_n = 1 + (-1)^n$

33. $a_n = \dfrac{1 + (-1)^n}{n}$

34. $a_n = \dfrac{n!}{n}$

35. $a_n = 3 - \dfrac{1}{2^n}$

36. $a_n = \dfrac{n}{\sqrt{n^2 + 1}}$

37. $a_n = \dfrac{3^n}{4^n}$

38. $a_n = \dfrac{(n - 2)!}{n!}$

Ⓒ**39.** A deposit of \$5000 is made in an account that earns 8% interest compounded quarterly. The balance in the account after n quarters is given by

$$A_n = 5000\left(1 + \frac{0.08}{4}\right)^n, \qquad n = 1, 2, 3, \ldots$$

(a) Compute the first eight terms of this sequence.
(b) Find the balance in this account after 10 years by computing the 40th term of the sequence.

Ⓒ**40.** A deposit of \$100 is made *each* month in an account that earns 12% interest compounded monthly. The balance in the account after n months is given by

$$A_n = 100(101)[(1.01)^n - 1], \qquad n = 1, 2, 3, \ldots$$

(a) Compute the first six terms of this sequence.
(b) Find the balance after 5 years by computing the 60th term of the sequence.
(c) Find the balance after 20 years by computing the 240th term of the sequence.

41. The sum of the first n positive integers is given by

$$S_n = \frac{n(n + 1)}{2}, \qquad n = 1, 2, 3, \ldots$$

(a) Compute the first five terms of this sequence and check to see that each term is the correct sum.
(b) Find the sum of the first 50 positive integers by computing S_{50}.

42. A ball is dropped from a height of 12 ft, and on each rebound it rises to $\frac{2}{3}$ of its previous height.
(a) Write an expression for the height of the nth rebound.
(b) Determine the convergence or divergence of this sequence. If it converges, find the limit.

Ⓒ**43.** A government program that currently costs taxpayers \$2.5 billion per year is to be cut back by 20% per year.
(a) Write an expression for the amount budgeted for this program after n years.
(b) Compute the budgets for the first 4 years.
(c) Determine the convergence or divergence of the sequence of reduced budgets. If the sequence converges, find its limit.

Ⓒ**44.** If the average price of a new car increases $5\frac{1}{2}$% per year and the average price is currently \$9000, then the average price after n years is

$$P_n = 9000(1.055)^n, \qquad n = 1, 2, 3, \ldots$$

Compute the average price for the first 5 years of increases.

Ⓒ**45.** Consider an idealized population with the characteristic that each population member produces 1 offspring at the end of every time period. If each population member has a life span of 3 time periods and the population begins with 10 newborn members, then the following table gives the population during the first 5 time periods.

Age bracket	Time period 1	2	3	4	5
0–1	10	10	20	40	70
1–2		10	10	20	40
2–3			10	10	20
Total	10	20	40	70	130

The sequence for the total population has the property that

$$S_n = S_{n-1} + S_{n-2} + S_{n-3}, \qquad n > 3$$

Find the total population during the next 5 time periods.

Ⓒ Calculator may be helpful.

Section 9.2

Series and Convergence

Introductory Example

Market Stabilization

One of the many uses of sequences is in stabilization problems. To see how this works, let us consider a manufacturer who sells 10,000 units of a certain product each year. Suppose that in any given year each unit of this product (regardless of its age) has a 10% chance of breaking. In other words, after a year we would expect only 9,000 of the previous year's 10,000 units to be still in use. During the next year this number would drop by an additional 10% to 8,100, and so on. Assuming that 10,000 new units are sold every year, how many units would be in use after n years?

To solve this problem, we set up a geometric sequence that represents the declining number of items (out of one year's sales of 10,000) in use after n years:

Year	1	2	. . .	n
Number of items	10,000	10,000(0.9)	. . .	$10{,}000(0.9)^{n-1}$

Now, since each year witnesses the introduction of 10,000 new units, we have the following totals in use:

1st Year: 10,000

2nd Year: $10{,}000 + 10{,}000(0.9)$

3rd Year: $10{,}000 + 10{,}000(0.9) + 10{,}000(0.9)^2$

⋮

nth Year: $10{,}000 + 10{,}000(0.9) + 10{,}000(0.9)^2 + 10{,}000(0.9)^{n-1}$

These totals are graphically represented in Figure 9.2.

In this section, we will see that the sum of the **geometric series** representing the nth year's total is

$$\sum_{i=0}^{n-1} 10{,}000(0.9)^i = \frac{10{,}000[1 - (0.9)^n]}{1 - 0.9}$$

$$= 100{,}000[1 - (0.9)^n]$$

As n increases, we have

$$\lim_{n\to\infty} (0.9)^n = 0$$

and we see that the stabilization point is 100,000 units.

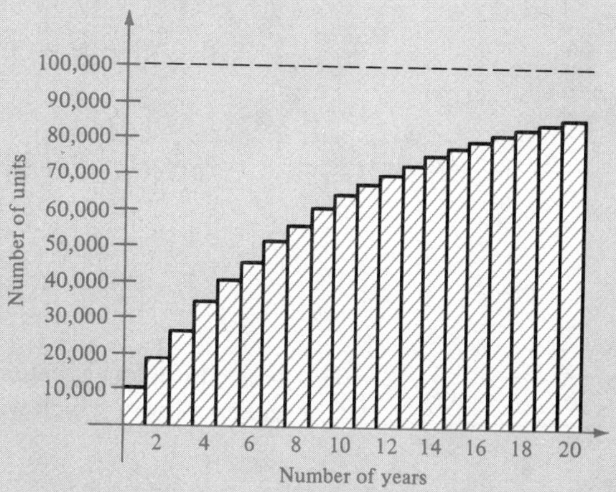

Figure 9.2

Section Objectives: *To define a series and the convergence of a series* ▪ *To introduce the* nth-*Term Test for divergence* ▪ *To find the sum of a geometric series.*

In this section, we investigate an important application of infinite sequences, namely, their use in representing infinite summations. As a simple illustration, suppose we write the decimal representation of $\frac{1}{3}$ as

$$\frac{1}{3} = 0.33333\ldots = \frac{3}{10} + \frac{3}{10^2} + \frac{3}{10^3} + \frac{3}{10^4} + \frac{3}{10^5} + \cdots$$

We consider this representation to be an *infinite summation* whose value is $\frac{1}{3}$.

To facilitate writing sums involving many terms, we use the standard **sigma notation** for sums, as introduced in Section 8.6. Some illustrations of the use of sigma notation are given in the following example.

Example 1

Examples of Sigma Notation

Sum	*Σ Notation*
(a) $1 + 2 + 3 + 4 + 5 + 6$	$\sum_{i=1}^{6} i$

(Note that for this particular sum the **index of summation** begins at $i = 1$. In general, we can use any variable as the index, and we can begin at any integer.)

(b) $3^2 + 4^2 + 5^2 + 6^2 + 7^2$	$\sum_{n=3}^{7} n^2$
(c) $3(1) + 3\left(\frac{1}{2}\right) + 3\left(\frac{1}{2}\right)^2 + \cdots + 3\left(\frac{1}{2}\right)^n$	$\sum_{k=0}^{n} 3\left(\frac{1}{2}\right)^k$
(d) $a_1 + a_2 + a_3 + \cdots + a_n$	$\sum_{i=1}^{n} a_i$

□

Now, to obtain a better picture of what an infinite summation is, let $\{a_n\}$ be a sequence from which we form another sequence $\{S_n\}$ in the following manner:

$$S_0 = a_0$$
$$S_1 = a_0 + a_1$$
$$S_2 = a_0 + a_1 + a_2$$
$$\vdots$$
$$S_n = a_0 + a_1 + a_2 + \cdots + a_n = \sum_{i=0}^{n} a_i$$

As $n \to \infty$, the infinite summation

$$a_0 + a_1 + a_2 + a_3 + a_4 + \cdots$$

is denoted by the expression

$$\sum_{n=0}^{\infty} a_n$$

and is called an **infinite series.** The sequence $\{S_n\}$ is called a **sequence of partial sums,** and we have the following:

1. If

   $$\lim_{n \to \infty} S_n = S$$

   the series $\Sigma_{n=0}^{\infty} a_n$ **converges,** and S is called the **sum of the series.**
2. If $\lim_{n\to\infty} S_n$ does not exist, the series $\Sigma_{n=0}^{\infty} a_n$ **diverges.**

The following properties are useful in determining the sum of an infinite series.

Properties of an Infinite Series

If c, A, and B are real numbers such that

$$\sum_{n=0}^{\infty} a_n = A, \qquad \sum_{n=0}^{\infty} b_n = B$$

then

1. $$\sum_{n=0}^{\infty} ca_n = c\sum_{n=0}^{\infty} a_n = cA$$

2. $$\sum_{n=0}^{\infty} (a_n \pm b_n) = \sum_{n=0}^{\infty} a_n \pm \sum_{n=0}^{\infty} b_n = A \pm B$$

The two primary questions regarding an infinite series are as follows:

1. Does the series converge, or does it diverge?
2. If the series converges, to what value does it converge?

We begin our pursuit for answers to these questions with a simple test for **divergence.**

Theorem 9.1 (*n*th-Term Test for Divergence)

If

$$\lim_{n \to \infty} a_n \neq 0$$

then the series $\Sigma_{n=0}^{\infty} a_n$ **diverges.**

Remark

Note that this theorem does *not* state that a series converges if $\lim_{n\to\infty} a_n = 0$, but rather that a series diverges if $\lim_{n\to\infty} a_n \neq 0$. In other words, the condition that the nth term of a series approaches zero as $n \to \infty$ is *necessary* for convergence to occur, but it is *not sufficient* to guarantee convergence.

Example 2

Testing for Divergence

Determine which, if any, of the following series can be said to diverge by Theorem 9.1.

$$\text{(a)} \sum_{n=0}^{\infty} 2^n \qquad \text{(b)} \sum_{n=0}^{\infty} \frac{1}{2^n} \qquad \text{(c)} \sum_{n=1}^{\infty} \frac{n!}{2n! + 1}$$

Solution

(a) The series

$$\sum_{n=0}^{\infty} 2^n = 1 + 2 + 4 + 8 + 16 + \cdots$$

diverges since

$$\lim_{n\to\infty} 2^n = \infty$$

(b) The nth-Term Test tells us *nothing* about the series

$$\sum_{n=0}^{\infty} \frac{1}{2^n} = 1 + \frac{1}{2} + \frac{1}{4} + \frac{1}{8} + \frac{1}{16} + \cdots$$

since

$$\lim_{n\to\infty} \frac{1}{2^n} = 0$$

(Later in this section, we will see that this particular series happens to converge. The point here is that we cannot deduce this from the nth-Term Test.)

(c) The series

$$\sum_{n=1}^{\infty} \frac{n!}{2n! + 1} = \frac{1}{3} + \frac{2}{5} + \frac{6}{13} + \frac{24}{49} + \frac{120}{241} + \cdots$$

diverges since

$$\lim_{n\to\infty} \frac{n!}{2n! + 1} = \frac{1}{2}$$

□

We have already pointed out that the nth-Term Test is a test for divergence, not convergence. In the remaining portion of this section, we discuss a special type of series, called a geometric series, for which we will develop a test for convergence (as well as for divergence).

Definition of Geometric Series

The series given by

$$\sum_{n=0}^{\infty} ar^n = a + ar + ar^2 + \cdots + ar^n + \cdots \qquad a \neq 0$$

is called a **geometric series** with ratio r.

Remark

Note that the first term in this geometric series is $ar^0 = a$. If the index had begun with $n = 1$, the first term would be $ar^1 = ar$. For a geometric series whose index begins at $n = 0$, we have the following formula for its nth partial sum.

Theorem 9.2 (nth Partial Sum of a Geometric Series)

The nth partial sum of the geometric series

$$\sum_{n=0}^{\infty} ar^n$$

is given by

$$S_n = \frac{a(1 - r^{n+1})}{1 - r}$$

Proof Let

$$S_n = a + ar + ar^2 + \cdots + ar^n$$

Then multiplication by r yields

$$rS_n = ar + ar^2 + ar^3 + \cdots + ar^{n+1}$$

By subtracting these two equations, we obtain

$$\begin{aligned} S_n &= a + ar + ar^2 + \cdots + ar^n \\ -rS_n &= -ar - ar^2 - \cdots - ar^n - ar^{n+1} \\ S_n - rS_n &= a - ar^{n+1} \end{aligned}$$

Therefore,

$$S_n(1 - r) = a(1 - r^{n+1})$$

or

$$S_n = \frac{a(1 - r^{n+1})}{1 - r}$$

□

When applying Theorem 9.2, be sure you check to see if the index begins at $n = 0$. If it does not, you will have to adjust S_n accordingly. For instance, by comparing the two geometric series

$$\sum_{n=1}^{10} ar^n = ar + ar^2 + ar^3 + \cdots + ar^{10}$$

and

$$\sum_{n=0}^{10} ar^n = a + ar + ar^2 + ar^3 + \cdots + ar^{10}$$

we see that by subtracting a from the second series we obtain the first. In other words,

$$\sum_{n=1}^{10} ar^n = \left(\sum_{n=0}^{10} ar^n\right) - a = \frac{a(1 - r^{11})}{1 - r} - a$$

We illustrate this procedure in the next example.

Example 3

Finding the Balance in an Increasing Annuity

A deposit of \$50 is made every month for two years in a savings account that pays 12% compounded monthly. What is the balance in the account at the end of the two years?

Solution The money that was deposited the first month will have a balance of

$$A_{24} = 50\left(1 + \frac{0.12}{12}\right)^{24} = 50(1.01)^{24}$$

after 24 months. Similarly, the money deposited in the second month will have a balance of

$$A_{23} = 50(1.01)^{23}$$

after 23 months. By continuing this process, the total balance resulting from the 24 deposits will be

$$A = A_1 + A_2 + \cdots + A_{24} = \sum_{n=1}^{24} A_n = \sum_{n=1}^{24} 50(1.01)^n$$

Now, noting that the index begins at $n = 1$, we apply Theorem 9.2 to determine

$$S_{24} = \sum_{n=0}^{24} 50(1.01)^n = \frac{50(1 - 1.01^{25})}{1 - 1.01}$$

$$= \frac{50(1 - 1.01^{25})}{-0.01}$$

Finally, accounting for the difference in the two indexes, we have a balance of

$$A = \frac{50(1 - 1.01^{25})}{-0.01} - 50 = \$1362.16$$

□

Now, since we have a formula for the nth partial sum of a geometric series, we have the necessary tool to determine which *infinite* geometric series converge and which ones diverge. The following theorem tells us that the only geometric series that converge are those in which the ratio r lies between 1 and -1.

Theorem 9.3 (Convergence of a Geometric Series)

The geometric series

$$\sum_{n=0}^{\infty} ar^n = a + ar + ar^2 + \cdots$$

has these properties

i. If $|r| \geq 1$, it *diverges*.
ii. If $|r| < 1$, it *converges* and

$$\sum_{n=0}^{\infty} ar^n = \frac{a}{1-r}$$

Proof This proof hinges on the fact that the only numbers whose magnitudes decrease with repeated multiplication are numbers lying in the interval, $-1 < r < 1$. For instance, if $r = -\frac{1}{2}$ or $r = \frac{1}{4}$, then the sequences

$$(-\tfrac{1}{2})^0, (-\tfrac{1}{2})^1, (-\tfrac{1}{2})^2, (-\tfrac{1}{2})^3, (-\tfrac{1}{2})^4, \ldots$$

and $$(\tfrac{1}{4})^0, (\tfrac{1}{4})^1, (\tfrac{1}{4})^2, (\tfrac{1}{4})^3, (\tfrac{1}{4})^4, \ldots$$

both converge to 0. But if $r = 1$ or $r = -2$, then neither of the sequences

$$(1)^0, (1)^1, (1)^2, (1)^3, (1)^4, \ldots$$

or $$(-2)^0, (-2)^1, (-2)^2, (-2)^3, (-2)^4, \ldots$$

converge to 0.

In other words, if $|r| \geq 1$, then

$$\lim_{n\to\infty} ar^n \neq 0$$

and by the nth-Term Test the sequence diverges.

On the other hand, if $|r| < 1$, then using the formula for the nth partial sum we have

$$\lim_{n\to\infty} S_n = \lim_{n\to\infty} \frac{a(1-r^{n+1})}{1-r} = \frac{a}{1-r}\lim_{n\to\infty}(1-r^{n+1}) = \frac{a}{1-r}$$

and the series converges to

$$\frac{a}{1-r}$$

□

Example 4

Finding the Sum of an Infinite Geometric Series

Determine the convergence or divergence of the following geometric series.

$$\text{(a)} \sum_{n=0}^{\infty} \left(-\frac{1}{2}\right)^n \qquad \text{(b)} \sum_{n=0}^{\infty} \left(\frac{3}{2}\right)^n \qquad \text{(c)} \sum_{n=1}^{\infty} \frac{4}{3^n}$$

Solution

(a) Since $a = 1$, $r = -\frac{1}{2}$, and $\left|-\frac{1}{2}\right| < 1$, the series converges to

$$\frac{a}{1-r} = \frac{1}{1-(-\frac{1}{2})} = \frac{1}{\frac{3}{2}} = \frac{2}{3}$$

(b) Since $r = \frac{3}{2} > 1$, the series diverges.

(c) By rewriting the series as

$$\sum_{n=1}^{\infty} 4\left(\frac{1}{3}\right)^n$$

we see that $a = 4$ and $r = \frac{1}{3}$. Therefore, for the series beginning at $n = 0$, we have

$$\sum_{n=0}^{\infty} 4\left(\frac{1}{3}\right)^n = \frac{4}{1-\frac{1}{3}} = \frac{4}{\frac{2}{3}} = 6$$

Finally, for the given series (beginning at $n = 1$), we have

$$\sum_{n=1}^{\infty} 4\left(\frac{1}{3}\right)^n = \sum_{n=0}^{\infty} 4\left(\frac{1}{3}\right)^n - 4 = 6 - 4 = 2$$

□

Example 5

Finding the Total Distance Traveled by a Bouncing Ball

A ball is dropped from a height of 6 ft and begins bouncing. Suppose that the height of each bounce is $\frac{3}{4}$ of that of the previous bounce. Find the total distance traveled by the ball. (See Figure 9.3.)

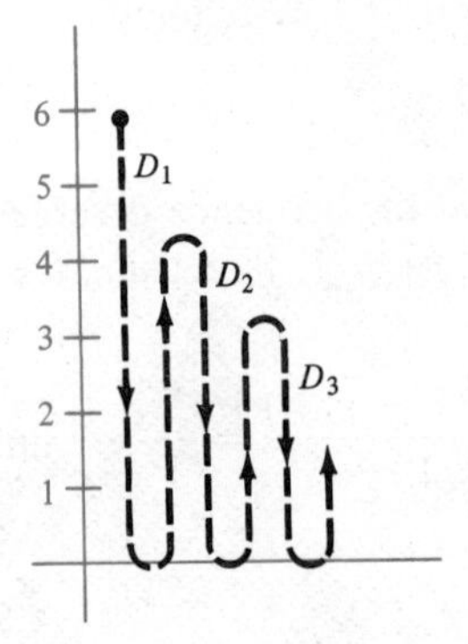

Figure 9.3

Solution When the ball hits the ground for the first time, it has traveled a distance of

$$D_1 = 6$$

Between the first and second times it hits the ground, it has traveled an additional distance of

$$D_2 = \underbrace{6\left(\tfrac{3}{4}\right)}_{up} + \underbrace{6\left(\tfrac{3}{4}\right)}_{down} = 12\left(\tfrac{3}{4}\right)$$

Between the second and third times it hits the ground, it has traveled an additional distance of

$$D_3 = \underbrace{6\left(\tfrac{3}{4}\right)\left(\tfrac{3}{4}\right)}_{up} + \underbrace{6\left(\tfrac{3}{4}\right)\left(\tfrac{3}{4}\right)}_{down} = 12\left(\tfrac{3}{4}\right)^2$$

Continuing this process, we have a total distance traveled of

$$\begin{aligned} D = 6 + 12\left(\tfrac{3}{4}\right) + 12\left(\tfrac{3}{4}\right)^2 + \cdots &= -6 + \sum_{n=0}^{\infty} 12\left(\frac{3}{4}\right)^n \\ &= -6 + \frac{12}{1 - \left(\frac{3}{4}\right)} \\ &= -6 + 48 = 42 \text{ ft} \end{aligned}$$

□

Section Exercises (9.2)

In Exercises 1–10, verify that the infinite series diverges.

1. $\displaystyle\sum_{n=1}^{\infty} \frac{n}{n+1} = \frac{1}{2} + \frac{2}{3} + \frac{3}{4} + \frac{4}{5} + \cdots$

2. $\displaystyle\sum_{n=1}^{\infty} \frac{n}{2n+3} = \frac{1}{5} + \frac{2}{7} + \frac{3}{9} + \frac{4}{11} + \cdots$

3. $\displaystyle\sum_{n=1}^{\infty} \frac{n^2}{n^2+1} = \frac{1}{2} + \frac{4}{5} + \frac{9}{10} + \frac{16}{17} + \cdots$

4. $\displaystyle\sum_{n=1}^{\infty} \frac{n}{\sqrt{n^2+1}} = \frac{1}{\sqrt{2}} + \frac{2}{\sqrt{5}} + \frac{3}{\sqrt{10}} + \frac{4}{\sqrt{17}} + \cdots$

5. $\displaystyle\sum_{n=0}^{\infty} 3\left(\frac{3}{2}\right)^n = 3 + \frac{9}{2} + \frac{27}{4} + \frac{81}{8} + \cdots$

6. $\displaystyle\sum_{n=0}^{\infty} \left(\frac{4}{3}\right)^n = 1 + \frac{4}{3} + \frac{16}{7} + \frac{64}{27} + \cdots$

7. $\displaystyle\sum_{n=0}^{\infty} 1000(1.055)^n = 1000 + 1055 + 1113.025 + \cdots$

8. $\displaystyle\sum_{n=0}^{\infty} 2(-1.03)^n = 2 - 2.06 + 2.1218 - \cdots$

9. $\displaystyle\sum_{n=0}^{\infty} 5\left(-\frac{3}{2}\right)^n = 5 - \frac{15}{2} + \frac{45}{4} - \frac{135}{8} + \cdots$

10. $\sum_{n=1}^{\infty} \sqrt{\frac{n}{4n+1}} = \sqrt{\frac{1}{5}} + \sqrt{\frac{2}{9}} + \sqrt{\frac{3}{13}} + \sqrt{\frac{4}{17}} + \cdots$

In Exercises 11–20 find the sum of the (convergent) geometric series.

11. $\sum_{n=0}^{\infty} \left(\frac{1}{2}\right)^n = 1 + \frac{1}{2} + \frac{1}{4} + \frac{1}{8} + \cdots$

12. $\sum_{n=0}^{\infty} 2\left(\frac{2}{3}\right)^n = 2 + \frac{4}{3} + \frac{8}{9} + \frac{16}{27} + \cdots$

13. $\sum_{n=0}^{\infty} \left(-\frac{1}{2}\right)^n = 1 - \frac{1}{2} + \frac{1}{4} - \frac{1}{8} + \cdots$

14. $\sum_{n=0}^{\infty} 2\left(-\frac{2}{3}\right)^n = 2 - \frac{4}{3} + \frac{8}{9} - \frac{16}{27} + \cdots$

15. $\sum_{n=0}^{\infty} 2\left(\frac{1}{\sqrt{2}}\right)^n = 2 + \sqrt{2} + 1 + \frac{1}{\sqrt{2}} + \cdots$

16. $4 + 1 + \frac{1}{4} + \frac{1}{16} + \cdots$

17. $1 + 0.1 + 0.01 + 0.001 + \cdots$

18. $8 + 6 + \frac{9}{2} + \frac{27}{8} + \cdots$

19. $3 - 1 + \frac{1}{3} - \frac{1}{9} + \cdots$

20. $4 - 2 + 1 - \frac{1}{2} + \cdots$

21. Find the sum of the infinite series

$$\sum_{n=0}^{\infty} \left(\frac{1}{2^n} - \frac{1}{3^n}\right)$$

22. Find the sum of the infinite series

$$\sum_{n=0}^{\infty} [(0.7)^n + (0.9)^n]$$

Ⓒ**23.** A company produces a new product and estimates the annual sales will be 8000 units. Suppose that in any given year 10% of the units (regardless of age) will become inoperative.
(a) How many units will be in use after n years?
(b) Find the market stabilization level for this product.

Ⓒ**24.** Repeat Exercise 23 with the assumption that 25% of the units will become inoperative each year.

Ⓒ**25.** A ball is dropped from a height of 16 ft. Each time it drops h feet, it rebounds $0.81h$ feet. Find the total distance traveled by the ball.

Ⓒ**26.** The ball in Exercise 25 takes the following times for each fall:

$$\begin{aligned} s_1 &= -16t^2 + 16, & s_1 &= 0 \text{ if } t = 1 \\ s_2 &= -16t^2 + 16(0.81), & s_2 &= 0 \text{ if } t = 0.9 \\ s_3 &= -16t^2 + 16(0.81)^2, & s_3 &= 0 \text{ if } t = (0.9)^2 \\ s_4 &= -16t^2 + 16(0.81)^3, & s_4 &= 0 \text{ if } t = (0.9)^3 \\ &\vdots \\ s_n &= -16t^2 + 16(0.81)^{n-1}, & s_n &= 0 \text{ if } t = (0.9)^{n-1} \end{aligned}$$

Beginning with s_2, the ball takes the same amount of time to bounce up as it takes to fall, and thus the total time elapsed before the ball comes to rest is

$$t = 1 + 2\sum_{n=1}^{\infty} (0.9)^n$$

Find this total.

27. Write each of the infinite repeating decimals as a rational number by considering each to be the sum of an infinite geometric series.

(a) $0.363636\ldots = 0.36 + 0.0036 + 0.000036 + \ldots = 0.36 + 0.36(0.01) + 0.36(.01)^2 + \ldots$

(b) $0.212121\ldots = 0.21 + 0.0021 + 0.000021 + \ldots = 0.21 + 0.21(0.01) + 0.21(0.01)^2 + \ldots$

In Exercises 28–32, use the formula for the nth partial sum of a geometric series:

$$a + ar + ar^2 + \cdots + ar^n = \frac{a(1 - r^{n+1})}{1 - r}$$

Ⓒ**28.** A deposit of \$100 is made each month for five years in an account that pays 10% compounded monthly. What is the balance A in the account at the end of the five years?

$$A = 100\left(1 + \frac{0.10}{12}\right) + 100\left(1 + \frac{0.10}{12}\right)^2 + \cdots + 100\left(1 + \frac{0.10}{12}\right)^{60}$$

Ⓒ Calculator may be helpful.

Ⓒ**29.** A deposit of \$50 is made each month in an account that pays 12% interest compounded monthly. What is the balance A in the account at the end of ten years?

$$A = 50\left(1 + \frac{0.12}{12}\right) + 50\left(1 + \frac{0.12}{12}\right)^2 + \cdots + 50\left(1 + \frac{0.12}{12}\right)^{120}$$

30. A deposit of P dollars is made every month for T years in an account that pays R percent interest compounded monthly. Let $N = 12T$ be the total number of deposits. The balance A after T years is

$$A = P\left(1 + \frac{R}{12}\right) + P\left(1 + \frac{R}{12}\right)^2 + \cdots + P\left(1 + \frac{R}{12}\right)^N$$

Show that this sum is given by

$$A = P\left[\left(1 + \frac{R}{12}\right)^N - 1\right]\left(1 + \frac{12}{R}\right)$$

Ⓒ**31.** Use the formula in Exercise 29 to find the amount in an account earning 9% compounded monthly after monthly deposits of \$50 have been made for 40 years.

32. The number of direct ancestors a person has had is given by

$$2 + 2^2 + 2^3 + 2^4 + \cdots + 2^n + \cdots$$

(*parents*, *grandparents*, *great-grandparents*, *great-great-grandparents*)

This formula is valid *provided* the person has had no common ancestors. [A common ancestor is one to whom you are related in more than one way. For example, one of your great-grandmothers on your father's side might also be one of your great-grandmothers on your mother's side, as shown in Figure 9.4.] How many direct ancestors have you had who lived since the year 0 A.D.? Assume that the average time between generations was 30 years (resulting in 66 generations) so that the total is given by

$$2 + 2^2 + 2^3 + 2^4 + \cdots + 2^{66}$$

Considering your total, is it reasonable to assume that you have had no common ancestors in the past 2000 years?

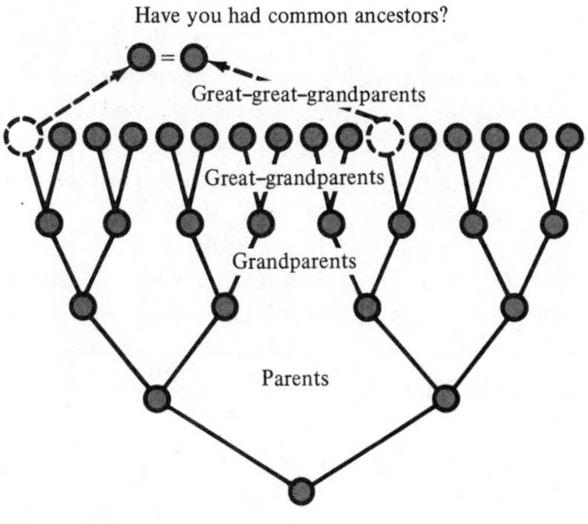

Have you had common ancestors?

Great-great-grandparents

Great-grandparents

Grandparents

Parents

Figure 9.4

Ⓒ Calculator may be helpful.

Section 9.3

Power Series and Taylor's Theorem

Introductory Example

Evaluating the Exponential Function on a Calculator

How does a calculator (or a computer) compute values for the exponential function? Does it simply retrieve the values from a long list stored in memory or does it actually compute values such as the following?

$$e^{0.25} \approx 1.2840254$$
$$e^{-3.5} \approx 0.03019738$$

The answer is that it actually computes these values each time the exponential key is activated. (We don't have to think very long about the possibility of storing such a list to realize that this is not a feasible solution to the problem.)

When you enter e^x in a calculator, you will notice that there is a pause before the result is displayed. During this pause, the calculator is summing up a series representation for e^x. In this section, we will see that the **power series** for the exponential function is

$$e^x = 1 + x + \frac{x^2}{2} + \frac{x^3}{3!} + \frac{x^4}{4!} + \frac{x^5}{5!} + \cdots$$

Remember that, although this series is infinite, a calculator only displays eight (or ten) digits, and, thus, it only sums enough terms to produce this degree of accuracy.

For example, suppose we want to use this series to evaluate the first eight digits of $e^{0.25}$. We would have

$$e^{1/4} \approx 1 + \frac{1}{4} + \frac{1}{2(4^2)} + \frac{1}{3!(4^3)} + \frac{1}{4!(4^4)} + \frac{1}{5!(4^5)} + \cdots$$

The following equations indicate that once we get past the eighth term in this series we are no longer adding enough to change the first eight digits. Thus, the first eight terms are sufficient to give us an eight-digit representation of $e^{0.25}$.

$$1 = 1.00000000 \qquad \frac{1}{4!\,(4^4)} = 0.00016276$$
$$\frac{1}{4} = 0.25000000 \qquad \frac{1}{5!\,(4^5)} = 0.00000814$$
$$\frac{1}{2(4^2)} = 0.03125000 \qquad \frac{1}{6!\,(4^6)} = 0.00000034$$
$$\frac{1}{3!\,(4^3)} = 0.00260417 \qquad \frac{1}{7!\,(4^7)} = 0.00000001$$

$$\left.\begin{array}{r} 1.00000000 \\ 0.25000000 \\ 0.03125000 \\ 0.00260417 \\ 0.00016276 \\ 0.00000814 \\ 0.00000034 \\ +\ 0.00000001 \end{array}\right\} e^{0.25} \approx 1.28402542$$
$$\overline{1.28402542}$$

Section Objectives: *To define a power series and determine its interval of convergence ▪ To derive and use Taylor's Theorem for generating the power series for a given function.*

Up to this point we have been dealing with series whose terms are constants. Now we want to consider series whose terms are variable. In particular, if x is a variable, then a series of the form

$$\sum_{n=0}^{\infty} a_n x^n = a_0 + a_1 x + a_2 x^2 + a_3 x^3 + \cdots + a_n x^n + \cdots$$

is called a **power series.** More generally, we call a series of the form

$$\sum_{n=0}^{\infty} a_n (x - c)^n = a_0 + a_1(x - c) + a_2(x - c)^2 + \cdots + a_n(x - c)^n + \cdots$$

a **power series centered at** c. Note that we begin a power series at $n = 0$, and to simplify the nth term, we agree that $(x - c)^0 = 1$ even if $x = c$.

Since a power series has variable terms, it can be viewed as a function of x,

$$f(x) = \sum_{n=0}^{\infty} a_n (x - c)^n$$

where the *domain of f* is the set of all x for which the power series converges. The determination of this *domain of convergence* is one of the primary problems associated with power series.

Quite obviously, every power series converges at its center c, since for $x = c$

$$f(c) = \sum_{n=0}^{\infty} a_n (c - c)^n = a_0(1) + 0 + 0 + \cdots + 0 + \cdots = a_0$$

Thus, $x = c$ always lies in the domain of f. It may happen that the domain of f consists *only* of the single point $x = c$. In fact, this is one of the three basic types of domains for power series. All three types are described in the following theorem.

Theorem 9.4 (Convergence of a Power Series)

For a power series centered at c, precisely one of the following is true:

i. The series converges only for $x = c$.
ii. The series converges for all x.
iii. There exists an $R > 0$ such that the series converges for $|x - c| < R$, and diverges for $|x - c| > R$.

In Theorem 9.4, part iii, we call R the **radius of convergence** of the power series, and in parts i and ii, we consider the radius of convergence to be 0 and

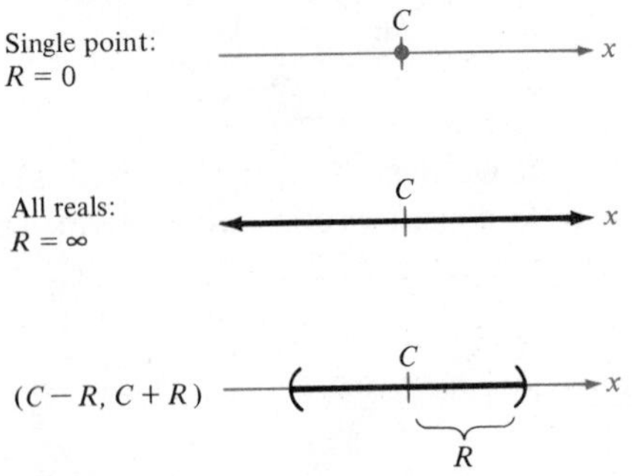

Single point: $R = 0$

All reals: $R = \infty$

$(C - R, C + R)$

Three Types of Domains

Figure 9.5

∞, respectively. Furthermore, since Theorem 9.4 states that the domain of convergence of a power series is always an interval, we call this domain the **interval of convergence.** (See Figure 9.5.)

Note that Theorem 9.4 says nothing about the convergence of the series at the endpoints of its interval of convergence. Determining the convergence or divergence at the endpoints can be a very difficult problem. In our presentation of power series, we will focus primarily on determination of the radius of convergence and (except for simple cases) leave the endpoint question open. To find the radius of convergence of a power series, we use the following theorem, which is called the Power Series Ratio Test.

Theorem 9.5 (Power Series Ratio Test)

The **radius of convergence** of the power series

$$f(x) = \sum_{n=0}^{\infty} a_n (x - c)^n$$

is given by

$$\lim_{n\to\infty} \left| \frac{a_n}{a_{n+1}} \right| = R, \qquad 0 \le R \le \infty$$

Example 1

Finding the Radius of Convergence

Find the radius of convergence for the series

$$\sum_{n=0}^{\infty} \frac{x^n}{n!}$$

Solution Since $a_n = 1/n!$, we have

$$\lim_{n\to\infty} \left| \frac{a_n}{a_{n+1}} \right| = \lim_{n\to\infty} \left| \frac{1/n!}{1/(n+1)!} \right| = \lim_{n\to\infty} \left| \frac{(n+1)!}{n!} \right|$$

$$= \lim_{n\to\infty} (n + 1) = \infty$$

and the Ratio Test shows that the series converges for all x ($R = \infty$). □

Example 2

Finding the Radius of Convergence

Find the radius of convergence for the series

$$\sum_{n=0}^{\infty} \frac{(-1)^n(x+1)^n}{2^n}$$

Solution Since

$$\lim_{n\to\infty}\left|\frac{a_n}{a_{n+1}}\right| = \lim_{n\to\infty}\left|\frac{(-1)^n/2^n}{(-1)^{n+1}/2^{n+1}}\right| = \lim_{n\to\infty}\left|\frac{2^{n+1}}{2^n}\right| = 2$$

the radius of convergence is $R = 2$. Since the series is centered at $x = -1$, it will converge in the interval $(-3, 1)$. □

Why are we interested in power series? To begin with, power series share many of the desirable properties of polynomials. For example, like polynomials, power series can be easily integrated. From our previous work, we are well aware that there are many functions (such as $y = e^{x^2}$) that are not easily integrated. Thus, if we could find a power series to represent one of these difficult functions, we could integrate the series to obtain a power series representation of the antiderivative. The point is that we are interested in power series because the power series representation of a particular function may enable us to perform some otherwise difficult operation on the function.

This brings us to our next question. How can we find a power series for a given function? The answer to this question is given by **Taylor's Theorem,** which gives the power series for a function f in terms of the first (and higher-order) derivatives of f.

Theorem 9.6 (Taylor's Theorem)

If f is differentiable (at all orders), then the power series for f centered at $x = c$ is given by

$$f(x) = f(c) + f'(c)(x-c) + \frac{f''(c)}{2!}(x-c)^2 + \frac{f'''(c)}{3!}(x-c)^3 + \cdots$$

$$= \sum_{n=0}^{\infty} \frac{f^{(n)}(c)}{n!}(x-c)^n$$

Proof Suppose that the function f is represented by a power series, centered at $x = c$,

$$f(x) = \sum_{n=0}^{\infty} a_n(x-c)^n$$

Thus, by successive differentiation we have

$$
\begin{aligned}
f^{(0)}(x) &= a_0 + a_1(x-c) + a_2(x-c)^2 + a_3(x-c)^3 + \cdots \\
f^{(1)}(x) &= a_1 + 2a_2(x-c) + 3a_3(x-c)^2 + 4a_4(x-c)^3 + \cdots \\
f^{(2)}(x) &= 2a_2 + 3!a_3(x-c) + 4\cdot 3a_4(x-c)^2 \\
&\quad + 5\cdot 4a_5(x-c)^3 + \cdots \\
f^{(3)}(x) &= 3!a_3 + 4!a_4(x-c) + 5\cdot 4\cdot 3a_5(x-c)^2 \\
&\quad + 6\cdot 5\cdot 4(x-c)^3 + \cdots \\
&\vdots \\
f^{(n)}(x) &= n!a_n + (n+1)!a_{n+1}(x-c) + \cdots
\end{aligned}
$$

Now, evaluating each of these derivatives at $x = c$ yields

$$
\begin{aligned}
f^{(0)}(c) &= a_0 = 0!a_0 \\
f^{(1)}(c) &= a_1 = 1!a_1 \\
f^{(2)}(c) &= 2a_2 = 2!a_2 \\
f^{(3)}(c) &= 3!a_3 \\
&\vdots \\
f^{(n)}(c) &= n!a_n
\end{aligned}
$$

By solving for a_n in the last equation, we find that the nth coefficient of the power series representation of $f(x)$ is

$$a_n = \frac{f^{(n)}(c)}{n!}$$

□

The series obtained in Theorem 9.6 is often referred to as a **Taylor series** for $f(x)$ at $x = c$. If the series is centered at $x = 0$, it is called a **Maclaurin series.**

Example 3

Finding a Taylor Series

Find the power series centered at $x = 1$ for

$$f(x) = \frac{1}{x}$$

Solution Successive differentiation of $f(x)$ yields

$$
\begin{aligned}
f(x) &= (x)^{-1} & f(1) &= 1 = 0! \\
f'(x) &= -(x)^{-2} & f'(1) &= -1 = --(1!) \\
f''(x) &= 2(x)^{-3} & f''(1) &= 2 = 2! \\
f'''(x) &= -6(x)^{-4} & f'''(1) &= -6 = -(3!) \\
f^{(4)}(x) &= 24(x)^{-5} & f^{(4)}(1) &= 24 = 4! \\
f^{(5)}(x) &= -120(x)^{-6} & f^{(5)}(1) &= -120 = -(5!)
\end{aligned}
$$

Following this pattern, we see that

$$f^{(n)}(1) = (-1)^n(n!)$$

Therefore, by Taylor's Theorem we have

$$\begin{aligned}\frac{1}{x} &= f(1) + f'(1)(x-1) + \frac{f''(1)(x-1)^2}{2!} + \frac{f'''(1)(x-1)^3}{3!} + \cdots \\ &= 1 - (x-1) + \frac{2!\,(x-1)^2}{2!} - \frac{3!\,(x-1)^3}{3!} + \frac{4!\,(x-1)^4}{4!} - \cdots \\ &= 1 - (x-1) + (x-1)^2 - (x-1)^3 + (x-1)^4 - \cdots \\ &= \sum_{n=0}^{\infty} (-1)^n(x-1)^n\end{aligned}$$

□

Note that the radius of convergence of the series in Example 3 is $R = 1$. Thus, the series converges in the interval $(0, 2)$. Furthermore, for any fixed value of x this series is geometric, and by Theorem 9.3 we know it diverges at the endpoints $x = 0$ and $x = 2$. Figure 9.6 compares the graphs of $f(x) = 1/x$ and this particular series representations. Remember that this series is valid *only* within the interval $(0, 2)$.

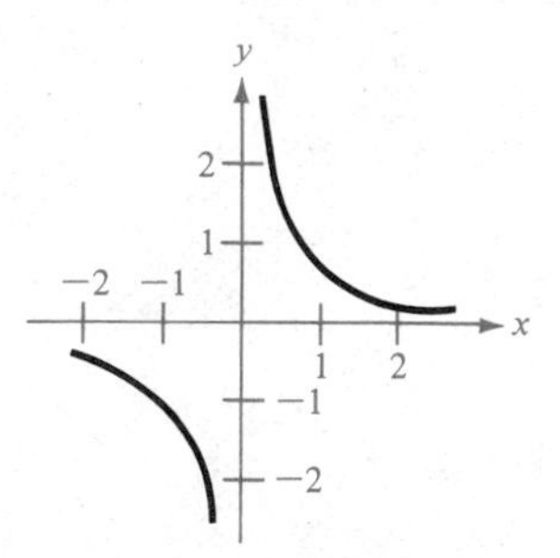

$f(x) = \frac{1}{x}$, Domain: all $x \neq 0$

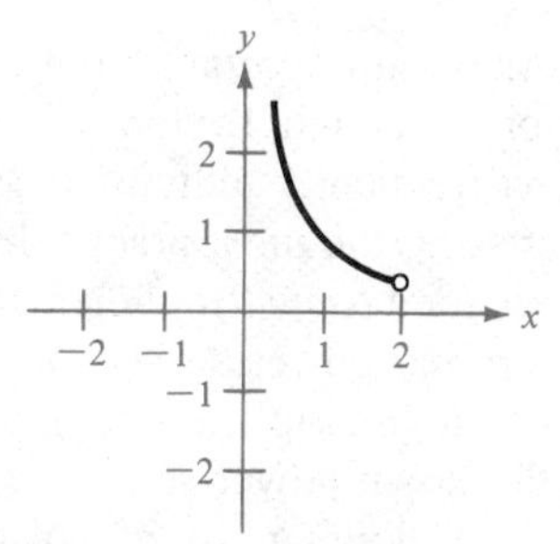

$f(x) = \sum_{n=0}^{\infty} (-1)^n (x-1)^n$, Domain: $0 < x < 2$

Figure 9.6

Example 4

Finding a Maclaurin Series

Find a power series, centered at $x = 0$, for $f(x) = e^{x^2}$.

Solution To use Taylor's Theorem, we must calculate successive derivatives of $f(x) = e^{x^2}$. By calculating just the first two,

$$f'(x) = 2xe^{x^2} \qquad \text{and} \qquad f''(x) = (4x^2 + 2)e^{x^2}$$

we recognize this to be a rather cumbersome task. Fortunately, there is an alternative. Suppose we first consider the power series for $g(x) = e^x$. Then we have

$$\begin{aligned} g(x) &= e^x & g(0) &= 1 \\ g'(x) &= e^x & g'(0) &= 1 \\ g''(x) &= e^x & g''(0) &= 1 \\ &\vdots \\ g^{(n)}(x) &= e^x & g^{(n)}(0) &= 1 \end{aligned}$$

Therefore,

$$g(x) = e^x = \sum_{n=0}^{\infty} \frac{g^{(n)}(0)}{n!} x^n = \sum_{n=0}^{\infty} \frac{1}{n!} x^n$$

Now, since $e^{x^2} = g(x^2)$, we have

$$\begin{aligned} e^{x^2} = g(x^2) &= \sum_{n=0}^{\infty} \frac{1}{n!}(x^2)^n \\ &= 1 + x^2 + \frac{x^4}{2!} + \frac{x^6}{3!} + \frac{x^8}{4!} + \cdots \end{aligned}$$

as the power series for e^{x^2}. The radius of convergence for this series is infinite, and hence this series converges for all x. □

Remark

Example 4 illustrates an important point in determining power series representations of functions. Though Taylor's Theorem is applicable to a wide variety of functions, it is frequently tedious to use because of the complexity of the derivatives. Therefore, the most practical use of Taylor's Theorem is in developing power series for a *basic list* of elementary functions. Then, from this basic list, we can determine power series for other functions by the operations of addition, subtraction, multiplication, division, differentiation, integration, or composition with known power series.

We provide the following list of power series for some basic elementary functions.

Power Series for Elementary Functions

$$\frac{1}{x} = 1 - (x-1) + (x-1)^2 - (x-1)^3 + (x-1)^4 - \cdots + (-1)^n(x-1)^n + \cdots, \quad 0 < x < 2$$

$$\frac{1}{1+x} = 1 - x + x^2 - x^3 + x^4 - x^5 + \cdots + (-1)^n x^n + \cdots, \quad -1 < x < 1$$

$$\ln x = (x-1) - \frac{(x-1)^2}{2} + \frac{(x-1)^3}{3} - \frac{(x-1)^4}{4} + \frac{(x-1)^5}{5} - \cdots + \frac{(-1)^{n-1}(x-1)^n}{n} + \cdots, \quad 0 < x \le 2$$

$$e^x = 1 + x + \frac{x^2}{2!} + \frac{x^3}{3!} + \frac{x^4}{4!} + \frac{x^5}{5!} + \cdots + \frac{x^n}{n!} + \cdots, \quad -\infty < x < \infty$$

$$(1+x)^k = 1 + kx + \frac{k(k-1)x^2}{2!} + \frac{k(k-1)(k-2)x^3}{3!} + \frac{k(k-1)(k-2)(k-3)x^4}{4!} + \cdots, \quad -1 < x < 1^*$$

$$(1+x)^{-k} = 1 - kx + \frac{k(k+1)x^2}{2!} - \frac{k(k+1)(k+2)x^3}{3!} + \frac{k(k+1)(k+2)(k+3)x^4}{4!} - \cdots, \quad -1 < x < 1^*$$

The last two series in this basic list are called **binomial series.** In the final two examples in this section, we demonstrate how to use the basic list of power series to obtain series for functions that are related to those in the list.

Example 5

Using the Basic List of Power Series

Determine the power series centered at $x = 0$ for $g(x) = e^{2x+1}$.

Solution First, we consider $f(x)$ to be $e^{2x+1} = e^{2x}e$. Then, using the power series

$$f(x) = e^x = 1 + x + \frac{x^2}{2!} + \frac{x^3}{3!} + \frac{x^4}{4!} + \cdots$$

*The convergence at $x = \pm 1$ depends on the value k.

we write

$$ee^{2x} = e[f(2x)] = e\left[1 + 2x + \frac{(2x)^2}{2!} + \frac{(2x)^3}{3!} + \frac{(2x)^4}{4!} + \cdots\right]$$

$$= e\sum_{n=0}^{\infty}\frac{(2x)^n}{n!} = e\sum_{n=0}^{\infty}\frac{2^n}{n!}x^n, \qquad -\infty < x < \infty$$

□

Example 6

Using the Basic List of Power Series

Find the power series centered at $x = 0$ for $g(x) = \sqrt[3]{1 + x}$.

Solution Using the binomial series

$$(1 + x)^k = 1 + kx + \frac{k(k-1)x^2}{2!} + \frac{k(k-1)(k-2)x^3}{3!} + \cdots$$

we let $k = \frac{1}{3}$ and write

$$(1 + x)^{1/3} = 1 + \frac{x}{3} - \frac{2x^2}{3^2 2!} + \frac{2 \cdot 5x^3}{3^3 3!} - \frac{2 \cdot 5 \cdot 8x^4}{3^4 4!} + \cdots$$

which converges for $-1 < x < 1$.

□

Section Exercises (9.3)

In Exercises 1–20, find the radius of convergence for the given power series.

1. $\sum_{n=0}^{\infty}\left(\frac{x}{2}\right)^n$

2. $\sum_{n=0}^{\infty}\left(\frac{x}{k}\right)^n$

3. $\sum_{n=1}^{\infty}\frac{(-1)^n x^n}{n}$

4. $\sum_{n=0}^{\infty}(-1)^{n+1}nx^n$

5. $\sum_{n=0}^{\infty}\frac{x^n}{n!}$

6. $\sum_{n=0}^{\infty}\frac{(3x)^n}{(2n)!}$

7. $\sum_{n=0}^{\infty}(2n)!\left(\frac{x}{2}\right)^n$

8. $\sum_{n=0}^{\infty}\frac{(-1)^n x^n}{(n+1)(n+2)}$

9. $\sum_{n=1}^{\infty}\frac{(-1)^{n+1}x^n}{4^n}$

10. $\sum_{n=0}^{\infty}\frac{(-1)^n n!\,(x-4)^n}{3^n}$

11. $\sum_{n=1}^{\infty}\frac{(-1)^{n+1}(x-5)^n}{n5^n}$

12. $\sum_{n=0}^{\infty}\frac{(x-2)^{n+1}}{(n+1)3^{n+1}}$

13. $\sum_{n=0}^{\infty}\frac{(-1)^{n+1}(x-1)^{n+1}}{n+1}$

14. $\sum_{n=1}^{\infty}\frac{(-1)^{n+1}(x-c)^n}{nc^n}$

15. $\sum_{n=1}^{\infty}\frac{(x-c)^{n-1}}{c^{n-1}}$

16. $\sum_{n=1}^{\infty}\frac{(-1)^{n+1}x^{2n-1}}{2n-1}$

17. $\sum_{n=1}^{\infty}\frac{n}{n+1}(-2x)^{n-1}$

18. $\sum_{n=0}^{\infty}\frac{(-1)^n x^{2n}}{n!}$

19. $\sum_{n=0}^{\infty}\frac{x^{2n+1}}{(2n+1)!}$

20. $\sum_{n=1}^{\infty}\frac{n!\,x^n}{(2n)!}$

In Exercises 21–26, apply Taylor's Theorem directly to find the power series (centered at c) for the given function, and find the radius of convergence.

21. $f(x) = \ln x,\ c = 1$
22. $f(x) = \sqrt{x},\ c = 1$
23. $f(x) = e^{2x},\ c = 0$
24. $f(x) = \sqrt{x},\ c = 4$
25. $f(x) = \sqrt[3]{x},\ c = 1$
26. $f(x) = \frac{1}{x+1},\ c = 0$

In Exercises 27–30, apply Taylor's Theorem directly to find the binomial series (centered at $c = 0$) for the given function, and find the radius of convergence.

27. $f(x) = \dfrac{1}{(1 + x)^2}$ **28.** $f(x) = \sqrt{1 + x}$

29. $f(x) = \dfrac{1}{\sqrt{1 + x}}$ **30.** $f(x) = \sqrt[3]{1 + x}$

In Exercises 31–38, use the basic list of power series for elementary functions provided in this section.

31. Use the power series for e^x to find the power series for

$$f(x) = e^{-x^2}$$

32. Use the power series for e^x to find the power series for

$$f(x) = \frac{e^x + e^{-x}}{2}$$

33. Use the power series for $1/(1 + x)$ to find the power series for

$$f(x) = \frac{1}{1 + x^2}$$

34. Use the power series for $1/(1 + x)$ to find the power series for

$$f(x) = \frac{2x}{1 + x^2}$$

35. Use the power series for $(1 + x)^{-k}$ to find the power series for

$$f(x) = \frac{1}{\sqrt{1 + x^2}}$$

36. Differentiate the power series for e^x term by term and use the resulting series to show that

$$\frac{d}{dx}[e^x] = e^x$$

37. Integrate the power series for $1/x$ to find the power series for

$$f(x) = \ln x$$

Compare this result with the series obtained in Exercise 21.

38. Integrate the power series for $1/(1 + x)$ to find the power series for

$$f(x) = \ln(1 + x)$$

39. Integrate the power series for $2x/(1 + x^2)$ in Exercise 34 to find the power series for

$$f(x) = \ln(1 + x^2)$$

40. Integrate the power series for $(e^x + e^{-x})/2$ in Exercise 32 to find the power series for

$$f(x) = \frac{e^x - e^{-x}}{2}$$

In Chapter 10, we will encounter the trigonometric functions sine and cosine. The power series for these functions are

$$\sin x = x - \frac{x^3}{3!} + \frac{x^5}{5!} - \frac{x^7}{7!} + \cdots + (-1)^n \frac{x^{2n+1}}{(2n + 1)!} + \cdots$$

$$\cos x = 1 - \frac{x^2}{2!} + \frac{x^4}{4!} - \frac{x^6}{6!} + \cdots + (-1)^n \frac{x^{2n}}{(2n)!} + \cdots$$

Use these two series in Exercises 41–46.

41. Find the radius of convergence for the power series representation of $\sin x$.

42. Find the radius of convergence for the power series representation of $\cos x$.

43. Use the power series for $\sin x$ to find the power series for

$$f(x) = \sin x^2$$

44. Use the power series for $\cos x$ to find the power series for

$$f(x) = \cos 2x$$

45. Differentiate the power series for $\sin x$ to show that

$$\frac{d}{dx}[\sin x] = \cos x$$

46. Differentiate the power series for $\cos x$ to show that

$$\frac{d}{dx}[\cos x] = -\sin x$$

Section 9.4

Taylor Polynomials

Introductory Example

A Decimal Approximation of π

From elementary geometry, we know that the area of a circle of radius r is $A = \pi r^2$. Thus, the area of a circle whose radius is 1 is π. We can use this information to find a decimal approximation for π as follows. (Recall that $\pi \approx 3.1416$.)

We know that the area of the quarter-circle pictured in Figure 9.7 is $\pi/4$, and thus we have

$$4 \int_0^1 \sqrt{1 - x^2}\, dx = \pi$$

Using the Taylor series of the preceding section, we find that

$$(1 - x^2)^{1/2} = 1 - \frac{x^2}{2} - \frac{x^4}{2^2 2!} - \frac{3x^6}{2^3 3!} - \frac{3 \cdot 5x^8}{2^4 4!} - \frac{3 \cdot 5 \cdot 7x^{10}}{2^5 5!} - \cdots$$

We call the nth partial sum of this series the nth-degree **Taylor polynomial** for $f(x) = \sqrt{1 - x^2}$, and we can obtain a decimal approximation for π by substituting one of these polynomials into the above integral. Moreover, the higher the degree of the polynomial we use, the better the approximation will be. To see this, let us begin with the second-degree Taylor polynomial for $\sqrt{1 - x^2}$. Thus,

$$\pi \approx 4 \int_0^1 \left(1 - \frac{x^2}{2}\right) dx = 4\left[x - \frac{x^3}{6}\right]_0^1 = 4\left(1 - \frac{1}{6}\right) \approx 3.33$$

A better approximation is obtained by using the fourth degree Taylor polynomial.

$$\pi \approx 4 \int_0^1 \left(1 - \frac{x^2}{2} - \frac{x^4}{8}\right) dx = 4\left[x - \frac{x^3}{6} - \frac{x^5}{40}\right]_0^1 = 4\left(1 - \frac{1}{6} - \frac{1}{40}\right) \approx 3.23$$

This process could be continued to obtain increasingly better approximations for π. [Actually, this particular sequence converges to π quite slowly. That is, Taylor polynomials of large degree are needed to obtain close approximations. However, there are other Taylor series (involving functions we will not cover in this text) that converge to π quite rapidly, and the principles demonstrated in this example can be applied to those series to obtain more accurate approximations of π.]

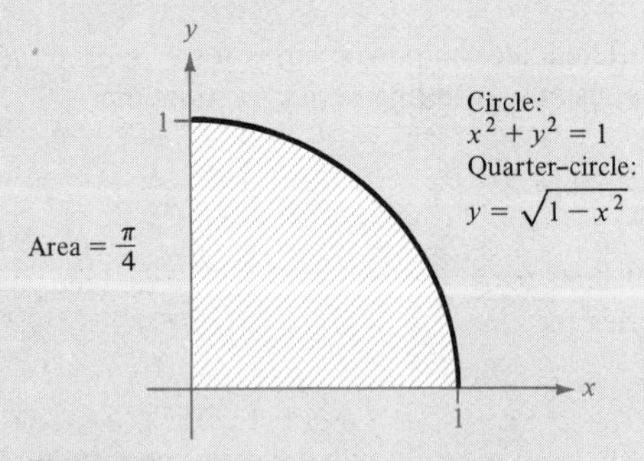

Figure 9.7

Section Objectives: ***To introduce the use of Taylor polynomials in approximating functions ▪ To use Taylor's Remainder Theorem to estimate the error involved in using Taylor polynomials as approximations.***

In Section 9.3, we saw that it is sometimes possible to obtain an *exact* power series representation for a function. For example, the function $f(x) = e^{-x}$ can be represented exactly by the power series

$$e^{-x} = \sum_{n=0}^{\infty} \frac{(-1)^n}{n!} x^n$$

The problem with using this power series is that the exactness of its representation depends on the summation of an infinite number of terms. Since this is usually not feasible, we must (in practice) content ourselves with a finite summation that approximates the given function rather than representing it exactly.

To obtain a better understanding of just how this particular series can be used to approximate e^{-x}, consider the following sequence of partial sums:

$$S_0(x) = 1$$

$$S_1(x) = 1 - x$$

$$S_2(x) = 1 - x + \frac{x^2}{2}$$

$$S_3(x) = 1 - x + \frac{x^2}{2} - \frac{x^3}{3!}$$

$$\vdots$$

$$S_n(x) = 1 - x + \frac{x^2}{2} - \frac{x^3}{3!} + \cdots + \frac{(-1)^n x^n}{n!}$$

The members of this sequence are called the **Taylor polynomials** for e^{-x}. As n approaches infinity, the graphs of these Taylor polynomials become closer and closer approximations to the graph of e^{-x}. For example, the graphs of four of these polynomials are shown in Figure 9.8. Note that the graphs of S_1, S_2, S_3, and S_4 are successively better approximations to the graph of e^{-x}. Furthermore, from Figure 9.8 it appears that the closer x is to the center of convergence ($x = 0$ in this case), the better the polynomial $S_n(x)$ approximates e^{-x}.

To reinforce this conclusion, consider the values shown in Table 9.3. Using the Ratio Test we see that the power series for e^{-x} converges for *all* x. However, from Table 9.3 and Figure 9.8, we see that the further x is from 0, the more terms we need to obtain a good approximation.

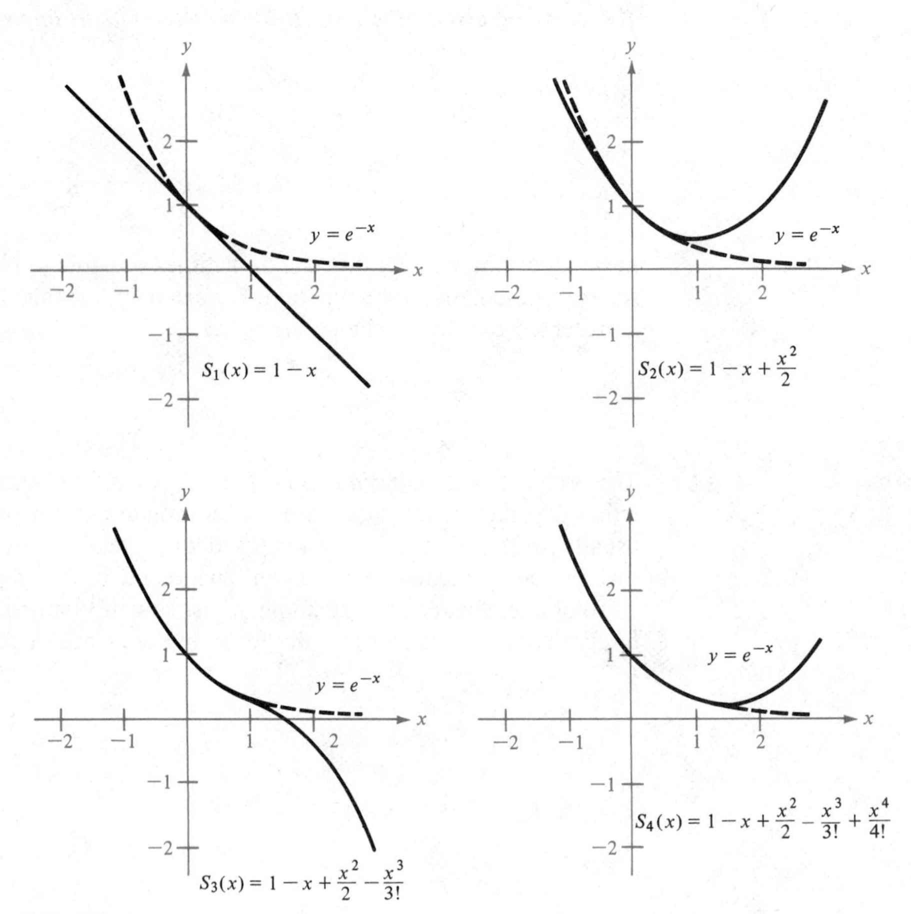

Figure 9.8

Table 9.3

x	0	0.5	1.0	1.5	2.0
$S_1 = 1 - x$	1	0.5000	0	−0.5000	−1.0
$S_2 = 1 - x + \frac{x^2}{2}$	1	0.6250	0.5000	0.6250	1.0
$S_3 = 1 - x + \frac{x^2}{2} - \frac{x^3}{3!}$	1	0.6042	0.3333	0.0625	−0.3333
$S_4 = 1 - x + \frac{x^2}{2} - \frac{x^3}{3!} + \frac{x^4}{4!}$	1	0.6068	0.3750	0.2734	0.3333
$S_5 = 1 - x + \frac{x^2}{2} - \frac{x^3}{3!} + \frac{x^4}{4!} - \frac{x^5}{5!}$	1	0.6065	0.3667	0.2102	0.0667
e^{-x}	1	0.6065	0.3679	0.2231	0.1353

A formal description of how closely a Taylor polynomial approximates a function f is given in the following theorem.

Theorem 9.7 (Taylor's Remainder Theorem)

Let f be a function such that $f^{(n+1)}(x)$ exists for every x in an interval I containing c. Then for all x in I,

$$f(x) = f(c) + f'(c)(x - c) + \frac{f''(c)}{2!}(x - c)^2 + \cdots + \frac{f^{(n)}(c)}{n!}(x - c)^n + R_n$$

where

$$R_n = \frac{f^{(n+1)}(z)}{(n + 1)!}(x - c)^{n+1}$$

for some number z between c and x.

Although this theorem appears to give a formula for the exact remainder, note that the theorem does not specify which value of z should be used to find R_n. In other words, the practical application of this theorem lies not in calculating R_n but in finding bounds for R_n. Once we have found the bounds for R_n, we can tell how closely the Taylor polynomial of degree n approximates the function $f(x)$. The following three examples demonstrate this concept.

Example 1

Finding Bounds for the Error in a Taylor Polynomial Approximation

Approximate $e^{-0.75}$ by a fourth-degree Taylor polynomial and determine bounds for the accuracy of this approximation.

Solution The fourth-degree Taylor polynomial for e^{-x} is

$$e^{-x} \approx 1 - x + \frac{x^2}{2} - \frac{x^3}{3!} + \frac{x^4}{4!}$$

Now, from Theorem 9.7 we know that the error in using this polynomial to approximate $e^{-0.75}$ is

$$R_4 = \frac{e^{-z}}{5!}(0.75)^5, \qquad 0 \leq z \leq 0.75$$

Furthermore, since we know that e^{-z} has a maximum value of 1 in the interval $[0, 0.75]$, we can determine that the maximum possible error for this approximation is

$$|R_4| \leq \frac{1}{5!}(0.75)^5 = \frac{1}{5!}\left(\frac{3}{4}\right)^5 = \frac{243}{(120)(1024)} \approx 0.002$$

Finally, we have

$$e^{-0.75} \approx 1 - (0.75) + \frac{(0.75)^2}{2} - \frac{(0.75)^3}{3!} + \frac{(0.75)^4}{4!}$$

$$\approx 0.474$$

This approximation is off by at most 0.002; therefore, we know that the exact value of $e^{-0.75}$ lies in the interval

$$0.472 \leq e^{-0.75} \leq 0.476$$ □

Example 2

Using a Taylor Polynomial to Approximate a Function

What degree Taylor polynomial must be used to approximate $f(x) = e^x$ in the interval $[-2, 2]$ to an accuracy of ± 0.001?

Solution For $f(x) = e^x$, we have

$$f^{(n+1)}(x) = e^x$$

and thus the maximum value of $f^{(n+1)}(x)$ in the interval $[-2, 2]$ is

$$e^2 \approx (2.71828)^2 \approx 7.389 \leq 7.4$$

Therefore, in the interval $[-2, 2]$ we know that the error in using the nth degree Taylor polynomial to approximate e^x is bounded by

$$|R_n| \leq \left| \frac{7.4}{(n+1)!} x^{n+1} \right|, \qquad -2 \leq x \leq 2$$

$$|R_n| \leq \frac{7.4}{(n+1)!} 2^{n+1}$$

Since we want this error to be less than 0.001, we use trial and error to determine that

$$\frac{7.4}{10!}(2^{10}) = \frac{7.4(1{,}024)}{3{,}628{,}800} \approx 0.002$$

$$\frac{7.4}{11!}(2^{11}) = \frac{7.4(2{,}048)}{39{,}916{,}800} \approx 0.0004$$

Thus, since $n + 1 = 11$, we have $n = 10$ and the tenth-degree Taylor polynomial for e^x is sufficient to approximate e^x (to the nearest 0.001) for *any* x in the interval $[-2, 2]$. □

Example 3

Using a Taylor Polynomial to Approximate a Function

Use a Taylor polynomial to approximate $\ln \frac{3}{2}$. Choose the degree of the polynomial so that your approximation is accurate to within ± 0.001.

Solution The nth derivative of $f(x) = \ln x$ is found as follows

$$f(x) = \ln x$$

$$f'(x) = \frac{1}{x}$$

$$f''(x) = -\frac{1}{x^2}$$

$$f'''(x) = \frac{2}{x^3}$$

$$f^{(4)}(x) = -\frac{3!}{x^4}$$

$$\vdots$$

$$f^{(n)}(x) = \frac{(-1)^{n-1}(n-1)!}{x^n}$$

Now, the Taylor series (centered at $x = 1$) for $\ln x$ is

$$\ln x = (x-1) - \frac{(x-1)^2}{2} + \frac{(x-1)^3}{3} - \frac{(x-1)^4}{4} + \cdots$$

Furthermore, the maximum value of $|f^{(n+1)}(x)|$ in the interval $[1, \frac{3}{2}]$ is

$$|f^{(n+1)}(1)| = \frac{n!}{(1)^{n+1}} = n!$$

Thus, we know that the remainder in approximating $\ln \frac{3}{2}$ by an nth-degree Taylor polynomial is bounded by

$$|R_n| \leq \frac{n!}{(n+1)!}\left(\frac{3}{2} - 1\right)^{n+1} = \frac{1}{(n+1)(2^{n+1})}$$

Since we want this error to be less than 0.001, we use trial and error to determine that

$$\frac{1}{7(2^7)} = \frac{1}{7(128)} \approx 0.0011$$

$$\frac{1}{8(2^8)} = \frac{1}{8(256)} \approx 0.0005$$

Thus, since $n + 1 = 8$, we have $n = 7$ and the seventh-degree Taylor polynomial for $\ln x$ is sufficient to approximate $\ln \frac{3}{2}$. Finally, we have

$$\ln \tfrac{3}{2} \approx \frac{1}{2} - \frac{(\frac{1}{2})^2}{2} + \frac{(\frac{1}{2})^3}{3} - \frac{(\frac{1}{2})^4}{4} + \frac{(\frac{1}{2})^5}{5} - \frac{(\frac{1}{2})^6}{6} + \frac{(\frac{1}{2})^7}{7}$$

$$\approx 0.406$$

(Using a calculator, we find that $\ln \frac{3}{2} \approx 0.405465108$, and thus our approximation is within 0.001 as specified.) □

In our last example, we demonstrate the use of a Taylor polynomial to evaluate an integral whose antiderivative we cannot find.

Example 4

Using a Taylor Polynomial to Approximate a Definite Integral

Use the eighth-degree Taylor polynomial for e^{-x^2} to approximate

$$\int_0^1 e^{-x^2}\,dx$$

Solution By replacing x with $-x^2$ in the series for e^x, we have

$$e^{-x^2} = 1 - x^2 + \frac{x^4}{2!} - \frac{x^6}{3!} + \frac{x^8}{4!} - \cdots$$

Now, integrating the eighth-degree Taylor polynomial for e^{-x^2}, we have

$$\int_0^1 e^{-x^2}\,dx \approx \left[x - \frac{x^3}{3} + \frac{x^5}{5 \cdot 2!} - \frac{x^7}{7 \cdot 3!} + \frac{x^9}{9 \cdot 4!}\right]_0^1$$

$$\approx 1 - \tfrac{1}{3} + \tfrac{1}{10} - \tfrac{1}{42} + \tfrac{1}{216}$$

Summing these terms, we have

$$\int_0^1 e^{-x^2}\,dx \approx 0.747$$

□

Section Exercises (9.4)

1. For the function $f(x) = e^x$, give the Taylor polynomial (centered at $c = 0$) of degree:
   (a) 1 (b) 2
   (c) 3 (d) 4
   (e) 5
2. For the function $f(x) = \ln(x + 1)$, give the Taylor polynomial (centered at $c = 0$) of degree:
   (a) 1 (b) 2
   (c) 3 (d) 4
   (e) 5
3. For the function $f(x) = \sqrt{x}$, give the Taylor polynomial (centered at $c = 1$) of degree:
   (a) 1 (b) 2
   (c) 3 (d) 4
4. For the function $f(x) = e^{-x^2}$, give the Taylor polynomial (centered at $c = 0$) of degree:
   (a) 2 (b) 4
   (c) 6 (d) 8
5. For the function

   $$f(x) = \frac{1}{(x + 1)^2}$$

   give the Taylor polynomial (centered at $c = 0$) of degree:
   (a) 1 (b) 2
   (c) 3 (d) 4
6. For the function

   $$f(x) = \frac{1}{1 + x^2}$$

   give the Taylor polynomial (centered at $c = 0$) of degree:
   (a) 2 (b) 4
   (c) 6 (d) 8

Ⓒ **7.** Complete the following table using Taylor polynomials as approximations to the function $f(x) = e^{x/2}$.

x	$e^{x/2}$	$1+\frac{x}{2}$	$1+\frac{x}{2}+\frac{x^2}{8}$
0	1,0000		
0.25	1.1331		
0.50	1.2840		
0.75	1.4550		
1.0	1.6487		

x	$e^{x/2}$	$1+\frac{x}{2}+\frac{x^2}{8}+\frac{x^3}{48}$	$1+\frac{x}{2}+\frac{x^2}{8}+\frac{x^3}{48}+\frac{x^4}{384}$
0	1.0000		
0.25	1.1331		
0.50	1.2840		
0.75	1.4550		
1.0	1.6487		

Ⓒ **8.** Complete the following table using Taylor polynomials as approximations to the function $f(x) = \ln(x^2+1)$.

x	$\ln(x^2+1)$	x^2	$x^2-\frac{x^4}{2}$
0	0.00000		
0.25	0.06062		
0.50	0.22314		
0.75	0.44629		

x	$\ln(x^2+1)$	$x^2-\frac{x^4}{2}+\frac{x^6}{3}$	$x^2-\frac{x^4}{2}+\frac{x^6}{3}-\frac{x^8}{4}$
0	0.00000		
0.25	0.06062		
0.50	0.22314		
0.75	0.44629		

Ⓒ **9.** What degree Taylor polynomial must be used to approximate $f(x) = e^x$ in the interval $[-1, 1]$ to an accuracy of ± 0.001?

Ⓒ **10.** What degree Taylor polynomial centered at $c = 1$ must be used to approximate $f(x) = 1/x$ in the interval $[1, \frac{3}{2}]$ to an accuracy of ± 0.001?

Ⓒ **11.** The fifth-degree Taylor polynomial

$$S_5(x) = 1 - x + \frac{x^2}{2!} - \frac{x^3}{3!} + \frac{x^4}{4!} - \frac{x^5}{5!}$$

is used to approximate the function $f(x) = e^{-x}$ in the interval $[0, 1]$. What is the maximum error (guaranteed by Taylor's remainder) for this approximation?

Ⓒ **12.** The fifth-degree Taylor polynomial

$$S_5(x) = 1 - (x-1) + (x-1)^2 - (x-1)^3 + (x-1)^4 - (x-1)^5$$

is used to approximate the function $f(x) = 1/x$ in the interval $[1, \frac{3}{2}]$. What is the maximum error (guaranteed by Taylor's remainder) for this approximation?

Ⓒ **13.** Use the sixth-degree Taylor polynomial for e^{-x^2} to approximate

$$\int_0^{1/2} e^{-x^2}\,dx$$

Ⓒ **14.** The normal probability density function is

$$f(x) = \frac{1}{\sqrt{2\pi}} e^{-x^2/2}$$

and the probability that x falls between a and b is

$$P(a \le x \le b) = \frac{1}{\sqrt{2\pi}} \int_a^b e^{-x^2/2}\,dx$$

Approximate $P(0 \le x \le 1)$ using the sixth-degree Taylor polynomial for $f(x)$.

Ⓒ Calculator may be helpful.

15. Use the sixth-degree Taylor polynomial for

$$\frac{1}{\sqrt{1 + x^2}}$$

to approximate

$$\int_0^1 \frac{1}{\sqrt{1 + x^2}}\, dx$$

16. It can be shown that

$$4 \int_0^1 \frac{1}{1 + x^2}\, dx = \pi$$

(a) Approximate π by using this integral and the Taylor polynomial for

$$\frac{4}{1 + x^2}$$

of degree:
(i) 6 (ii) 8 (iii) 10 (iv) 12

(b) Note that the approximations obtained in part (a) are alternately too large and then too small. Average the last two approximations to obtain a better estimate for π.

Section 9.5

Newton's Method

Introductory Example

Creating Examples for a Text

One of the greatest difficulties in writing a mathematics text is choosing appropriate examples to illustrate a particular mathematical concept or technique. The problem lies basically in choosing the appropriate level of difficulty for the example. If the example is too difficult, the point one is attempting to illustrate becomes obscured. On the other hand, if the example is too simple, the new technique may seem unnecessary since the reader may be able to anticipate the solution without having to resort to the technique being illustrated.

As a case in point, consider the optimization problem discussed in Example 3 of Section 3.4. In that example we were asked to find the minimum distance between the point $(0, 2)$ and the parabola $y = 4 - x^2$. To solve the problem we set up the distance equation

$$d = \sqrt{(x-0)^2 + (y-2)^2}$$
$$= \sqrt{x^2 + (4 - x^2 - 2)^2} = \sqrt{x^4 - 3x^2 + 4}$$

By differentiating the term inside the radical and setting the result equal to zero, we obtained the equation

$$4x^3 - 6x = 2x(2x^2 - 3) = 0$$

and, by factoring, we concluded that the minimum distance occurred when $x = \sqrt{\frac{3}{2}}$.

Suppose we must find the minimum distance between the point $(0, 5)$ and the parabola $y = 4 - x^2$. We could go through the same steps, but it is clear from Figure 9.9 that the closest point on the parabola is its vertex and the minimum distance is 1. Thus, this example makes the optimization technique seem unnecessary.

Finally, suppose we are asked to find the minimum distance between the point $(1, 0)$ and the same parabola. In this case, we have

$$d = \sqrt{(x-1)^2 + (y-0)^2}$$
$$= \sqrt{(x-1)^2 + (4-x^2)^2}$$
$$= \sqrt{x^4 - 7x^2 - 2x + 17}$$

By differentiating and setting equal to zero, we obtain

$$4x^3 - 14x - 2 = 2(2x^3 - 7x - 1) = 0$$

This particular cubic cannot be solved by elementary factoring techniques, and thus this example is too difficult to be a good illustration of the optimization technique. In Section 9.5, we will introduce a technique (**Newton's Method**) that can be used to solve this cubic.

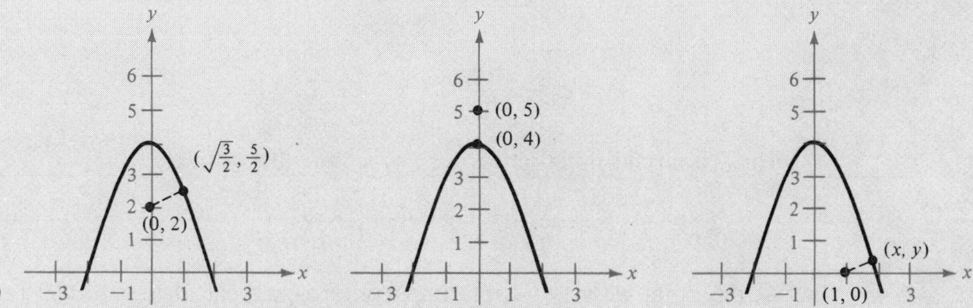

Figure 9.9

Section Objective: ***To develop Newton's Method for approximating zeros of functions.***

In the preceding chapters of this text, we frequently needed to find the zeros of a function. [The zeros of f are those values of x for which $f(x) = 0$.] Until now our functions have been carefully chosen so that elementary algebraic techniques suffice for finding their zeros. For example, the zeros of

$$f(x) = x^2 - 6x + 8$$
$$g(x) = 2x^2 - 3x - 7$$
$$h(x) = x^3 - 2x^2 - x + 2$$

can all be found by factoring or by the quadratic formula. However, in practice we frequently encounter functions whose zeros are more difficult to find. For example, the zeros of a function as simple as

$$f(x) = x^3 - x + 1$$

cannot be found by elementary algebraic methods. In such cases, the first-degree Taylor polynomial for the function may serve as a convenient tool for approximating the zeros.

For example, the first-degree Taylor polynomial (centered at $x = x_1$) for $f(x)$ is

$$p(x) = f(x_1) + f'(x_1)(x - x_1)$$

Note that this linear polynomial is simply the equation of the tangent line to f at the point $(x_1, f(x_1))$. (See Figure 9.10.) Newton's Method for approximating zeros is based on the assumption that f and the tangent to f at $(x_1, f(x_1))$ both cross the x-axis at about the same point. To illustrate this, suppose that the function f crosses the x-axis at some point in the interval (a, b). If we cannot find this point, we guess its value to be x_1. Now, since we can easily calculate the x-intercept for the tangent line, we use it as our second (and hopefully better) estimate for the zero of f. In other words, we use our initial guess, x_1, to find a better guess, x_2. To find x_2, we let $p(x) = 0$ and solve for x as follows:

$$0 = f(x_1) + f'(x_1)(x - x_1)$$

$$x - x_1 = -\frac{f(x_1)}{f'(x_1)}$$

$$x = x_1 - \frac{f(x_1)}{f'(x_1)}$$

Thus, from our initial guess we arrive at a new estimate,

$$x_2 = x_1 - \frac{f(x_1)}{f'(x_1)}$$

At this point we may wish to improve on x_2 and calculate yet a third estimate,

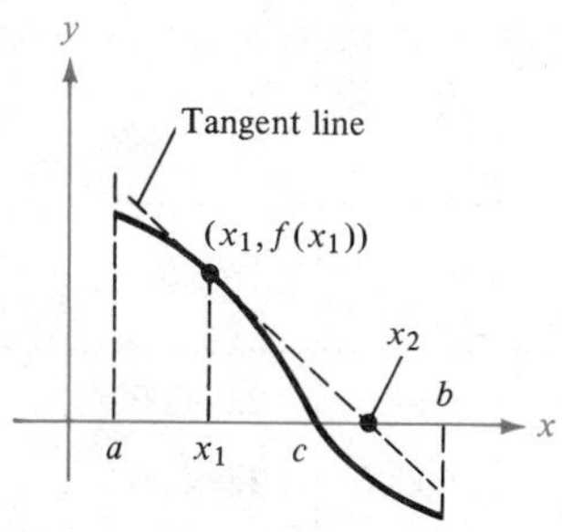

The x-intercept of the tangent line approximates the zero of f.

Figure 9.10

$$x_3 = x_2 - \frac{f(x_2)}{f'(x_2)}$$

Repeated application of this process is called **Newton's Method.**

Newton's Method for Approximating the Zeros of a Function

Let f be differentiable on (a, b) and $f(c) = 0$, where c is in (a, b). Then to approximate c we do the following:

1. Make an initial estimate x_1 "close" to c.
2. Determine a new approximation by the formula

$$x_{n+1} = x_n - \frac{f(x_n)}{f'(x_n)}$$

3. If $|x_n - x_{n+1}|$ is less than the desired accuracy, let x_{n+1} serve as our final approximation. Otherwise, return to step 2 and calculate a new approximation.

The process of calculating each successive approximation is called an **iteration.**

Example 1

Using Newton's Method

Calculate three iterations of Newton's Method to approximate a zero of $f(x) = x^2 - 2$. Use $x_1 = 1$ as the initial guess.

Solution Since

$$f(x) = x^2 - 2$$

we have

$$f'(x) = 2x$$

and the iterative process is given by the formula

$$x_{n+1} = x_n - \frac{f(x_n)}{f'(x_n)} = x_n - \frac{x_n^2 - 2}{2x_n}$$

The calculations for three iterations are shown in Table 9.4.

Table 9.4

n	x_n	$f(x_n)$	$f'(x_n)$	$\frac{f(x_n)}{f'(x_n)}$	$x_n - \frac{f(x_n)}{f'(x_n)}$
1	1.000000	−1.000000	2.000000	−0.500000	1.500000
2	1.500000	0.250000	3.000000	0.083333	1.416667
3	1.416667	0.006945	2.833334	0.002451	1.414216
4	1.414216				

Of course, in this example we know that the two zeros of the function are $\pm\sqrt{2}$. To six decimal places, $\sqrt{2} = 1.414214$. Thus, after only three iterations of Newton's Method, we have obtained an approximation (1.414216) that is within 0.000002 of the actual root. The first iteration of this process is pictured in Figure 9.11.

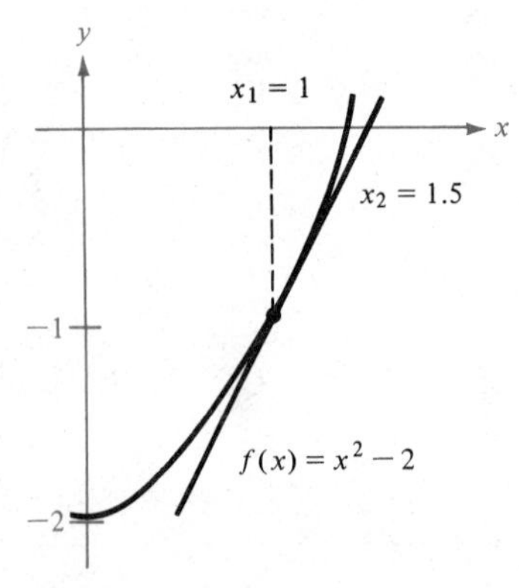

Figure 9.11 □

As you might expect, Newton's Method has become more popular since the emergence of electronic calculators.

Example 2

Using Newton's Method

Use Newton's Method to approximate the zeros of $f(x) = 2x^3 + x^2 - x + 1$. Continue the iterations until two successive approximations differ by less than 0.0001.

Solution We begin by sketching a graph of f and observing that it has only one zero, which occurs near $x = -1.2$ (see Figure 9.12). The calculations are shown in Table 9.5.

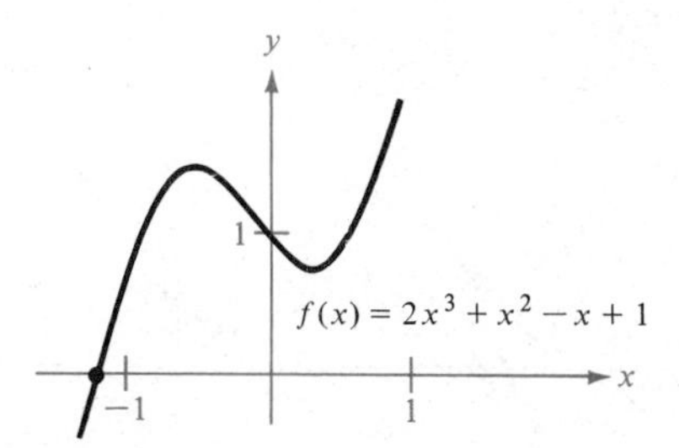

Figure 9.12

Table 9.5

n	x_n	$f(x_n)$	$f'(x_n)$	$\dfrac{f(x_n)}{f'(x_n)}$	$x_n - \dfrac{f(x_n)}{f'(x_n)}$
1	−1.20000	0.18400	5.24000	0.03511	−1.23511
2	−1.23511	−0.00771	5.68276	−0.00136	−1.23375
3	−1.23375	0.00001	5.66533	0.00000	−1.23375
4	−1.23375				

Thus, we estimate the zero of f to be -1.23375, since two successive approximations differ by less than the required 0.0001. □

When, as in Examples 1 and 2, the approximations approach a zero of the function, we say that the method **converges.** It is important to realize that Newton's Method does not always converge. Two ways in which this may happen are as follows:

1. If $f'(x_n) = 0$ for some n (see Figure 9.13)
2. If $\lim_{n\to\infty} x_n$ does not exist (see Figure 9.14)

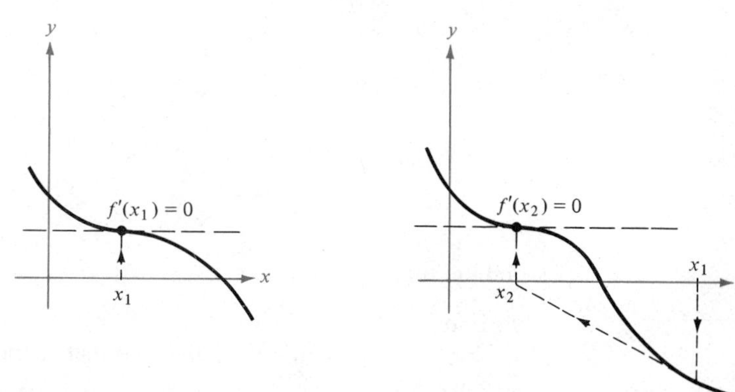

Figure 9.13

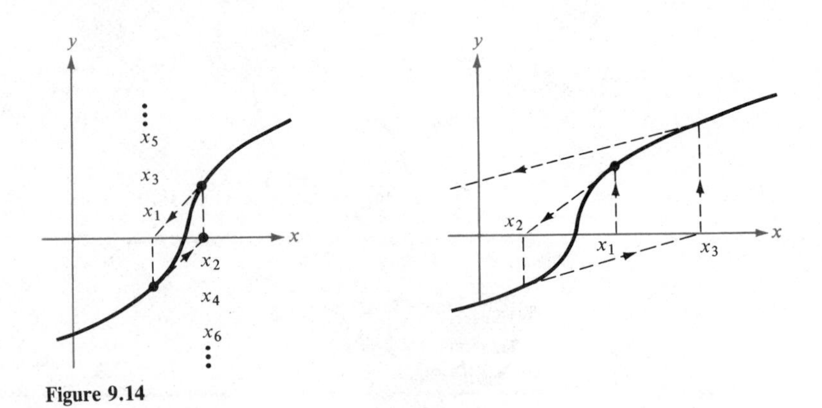

Figure 9.14

The type of problem encountered in Figure 9.13 can usually be overcome with a better choice for x_1. However, the problem illustrated in Figure 9.14 is usually more serious and may be independent of the choice of x_1. For example, Newton's Method does not converge for any choice of x_1 (other than the actual zero) for the function $f(x) = x^{1/3}$.

Example 3

An Example in Which Newton's Method Fails

Refer to Figure 9.15. Using $x_1 = 0.1$, show that $\lim_{n\to\infty} x_n$ does not exist for the function $f(x) = x^{1/3}$.

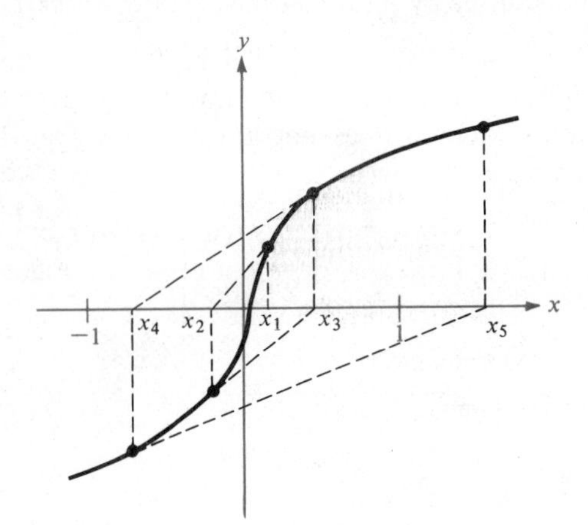

Figure 9.15

Solution

Table 9.6 and Figure 9.15 indicate that x_n increases in magnitude as $n \to \infty$, and thus $\lim_{n\to\infty} x_n$ does not exist.

Table 9.6

n	x_n	$f(x_n)$	$f'(x_n)$	$\frac{f(x_n)}{f'(x_n)}$	$x_n - \frac{f(x_n)}{f'(x_n)}$
1	0.10000	0.46416	1.54720	0.30000	−0.20000
2	−0.20000	−0.58480	0.97467	−0.60000	0.40000
3	0.40000	0.73681	0.61401	1.20000	−0.80000
4	−0.80000	−0.92832	0.38680	−2.40000	1.60000

□

Example 4

Using Newton's Method to Find a Point of Intersection

Estimate the point of intersection of the two curves given by

$$y = e^{-x^2} \quad \text{and} \quad y = x$$

as shown in Figure 9.16. (Use Newton's Method and continue the iterations until two successive approximations differ by less than 0.0001.)

Solution We seek the value of x such that

$$e^{-x^2} = x$$

and
$$0 = x - e^{-x^2}$$

To use Newton's Method, we let

$$f(x) = x - e^{-x^2}$$

and the iterative formula is

$$x_{n+1} = x_n - \frac{f(x_n)}{f'(x_n)} = x_n - \frac{x_n - e^{-x_n^2}}{1 + 2x_n e^{-x_n^2}}$$

The calculations are shown in Table 9.7, beginning with an initial guess of $x_1 = 0.5$.

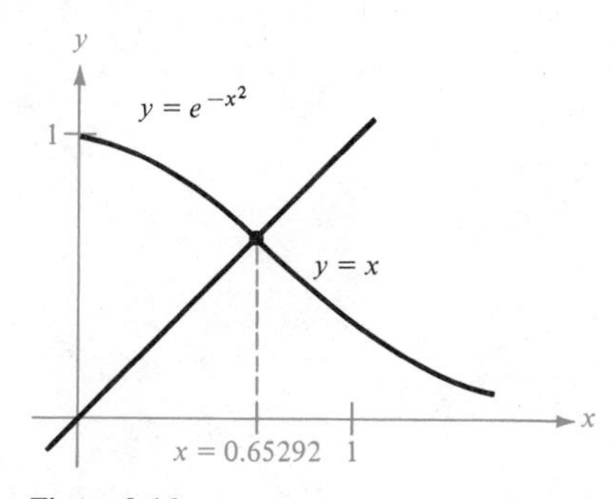

Figure 9.16

Table 9.7

n	x_n	$f(x_n)$	$f'(x_n)$	$\frac{f(x_n)}{f'(x_n)}$	$x_n - \frac{f(x_n)}{f'(x_n)}$
1	0.50000	−0.27880	1.77880	−0.15674	0.65674
2	0.65674	0.00707	1.85331	0.00382	0.65292
3	0.65292	0.00000	1.85261	0.00000	0.65292
4	0.65292				

Thus, the two curves intersect when $x \approx 0.65292$. □

Section Exercises (9.5)

1. Complete one iteration of Newton's Method on the function

$f(x) = x^2 - 3$

using $x_1 = 1.7$ as the initial guess.

2. Complete one iteration of Newton's Method on the function

$f(x) = 3x^2 - 2$

using $x_1 = 1$ as the initial guess.

In Exercises 3–11, approximate the indicated zero(s) of the function. Use Newton's method and continue the process until you are correct to three decimal places.

Ⓒ **3.** $f(x) = x^3 + x - 1$

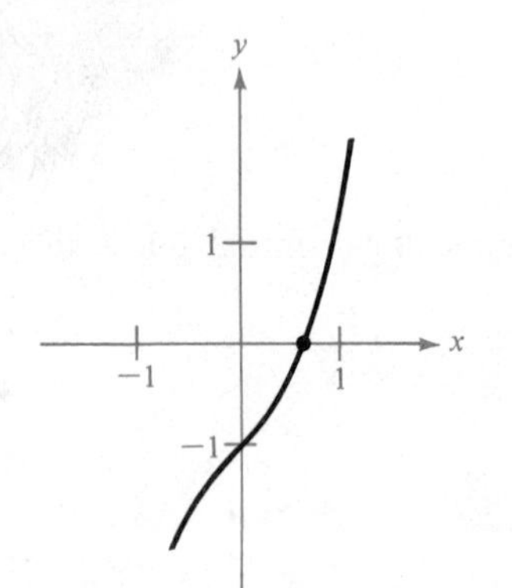

Ⓒ **4.** $f(x) = x^5 + x - 1$

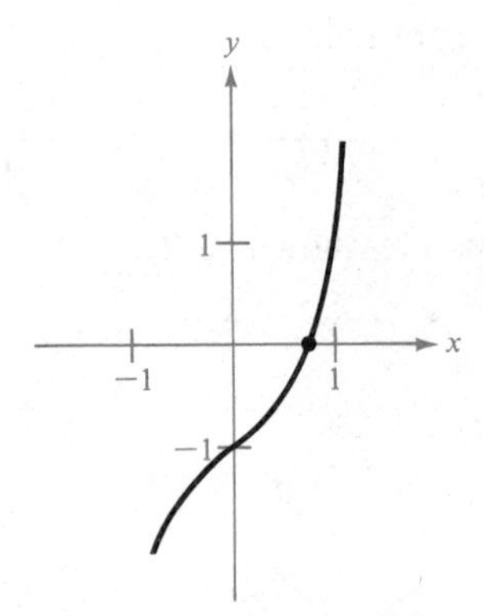

Ⓒ **5.** $f(x) = 3\sqrt{x - 1} - x$

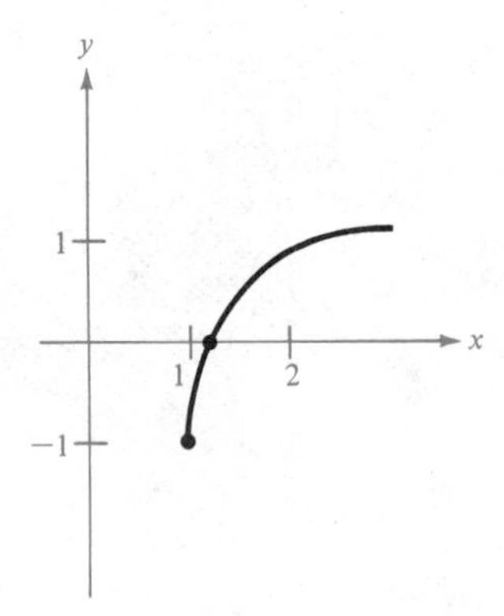

Ⓒ Calculator may be helpful.

ⓒ 6. $f(x) = x^3 - 3.9x^2 + 4.79x - 1.881$

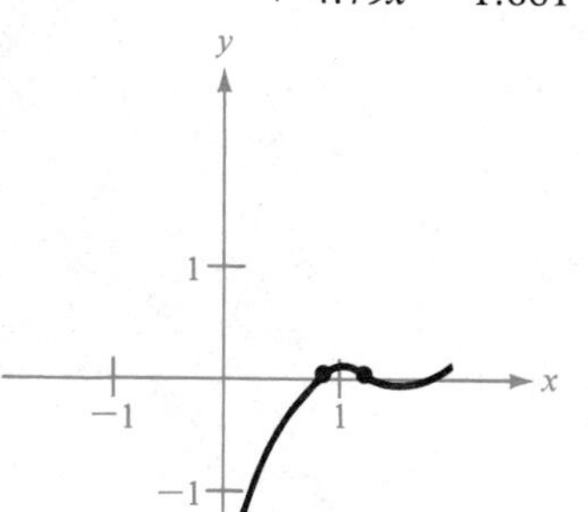

ⓒ 7. $f(x) = x^4 - 10x^2 - 11$

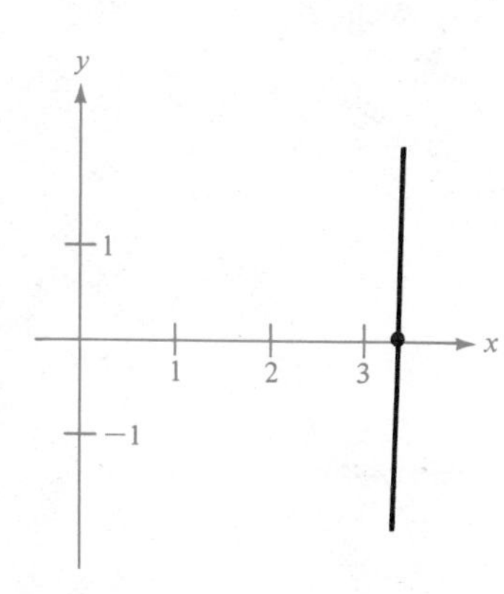

ⓒ 8. $f(x) = x^3 + 3$

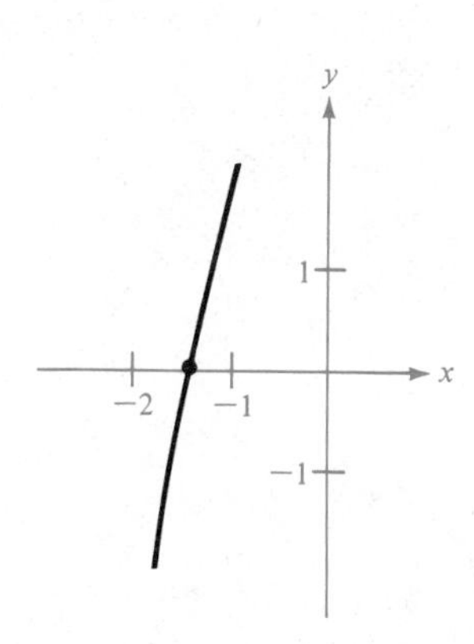

ⓒ 9. $f(x) = \ln x + x$

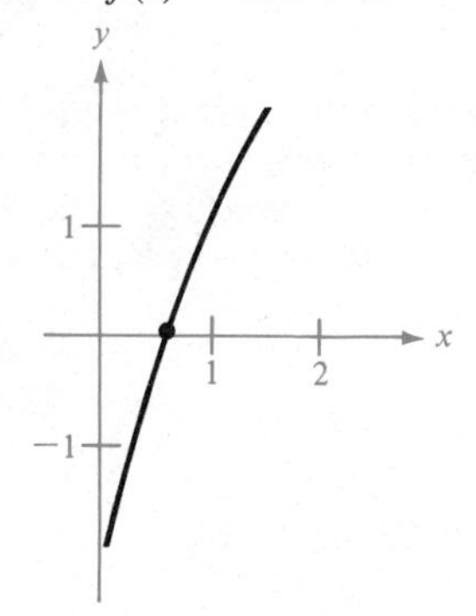

ⓒ10. $f(x) = -2 + e^{3x}(4 - 2x)$

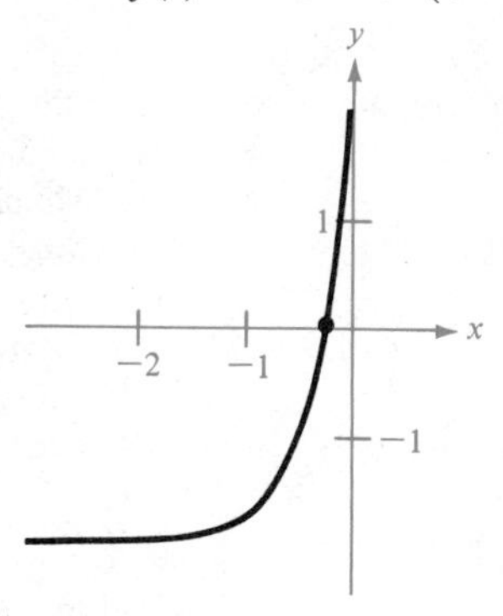

ⓒ11. $f(x) = e^{-x^2} - x^2$

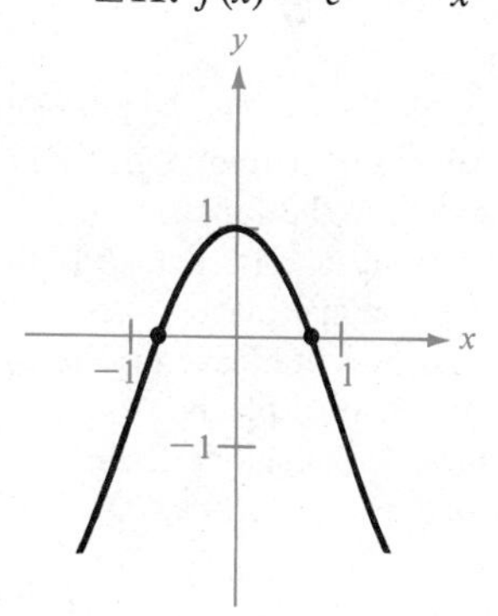

In Exercises 12–14, apply Newton's Method to approximate (to two decimal places) the x-value of the point of intersection of the two graphs.

ⓒ12. $f(x) = 2x + 1$; $g(x) = \sqrt{x + 4}$

ⓒ13. $f(x) = 3 - x$; $g(x) = \ln x$

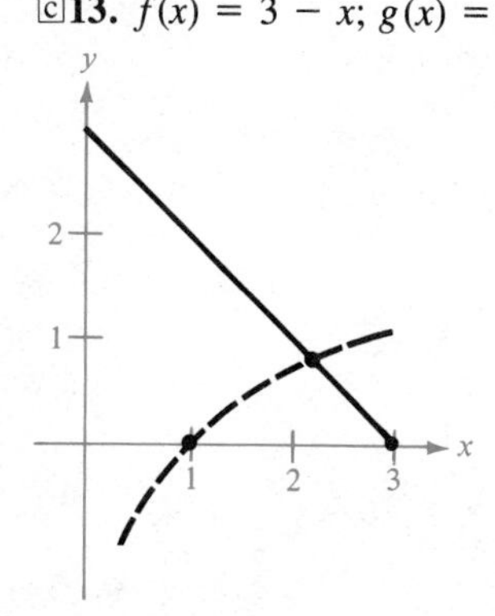

ⓒ14. $f(x) = x$; $g(x) = e^{-x}$

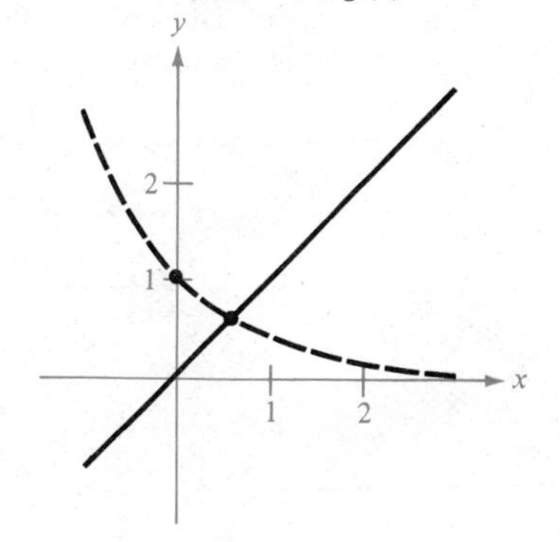

ⓒ Calculator may be helpful.

In Exercises 15–17, apply Newton's Method (using the indicated initial guess) and explain why the method fails to converge.

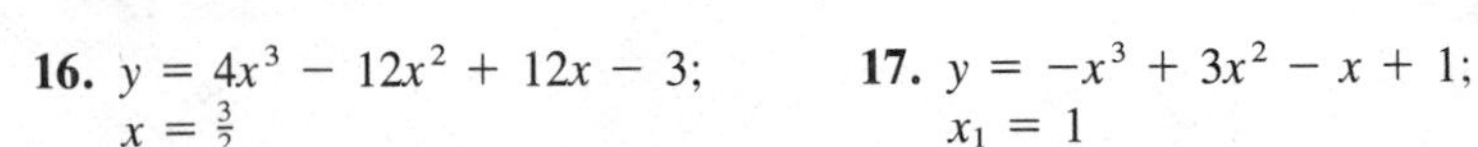

15. $y = 2x^3 - 6x^2 + 6x - 1$; $x_1 = 1$

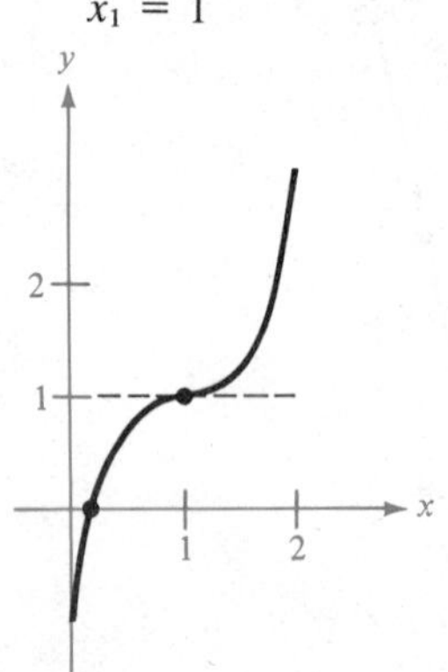

16. $y = 4x^3 - 12x^2 + 12x - 3$; $x = \frac{3}{2}$

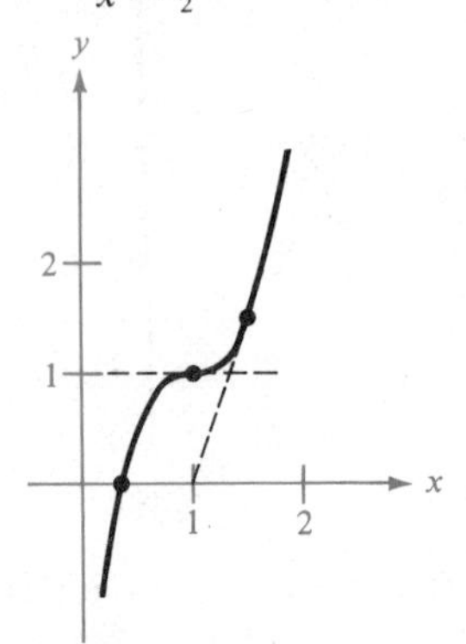

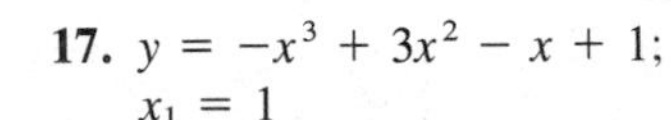

17. $y = -x^3 + 3x^2 - x + 1$; $x_1 = 1$

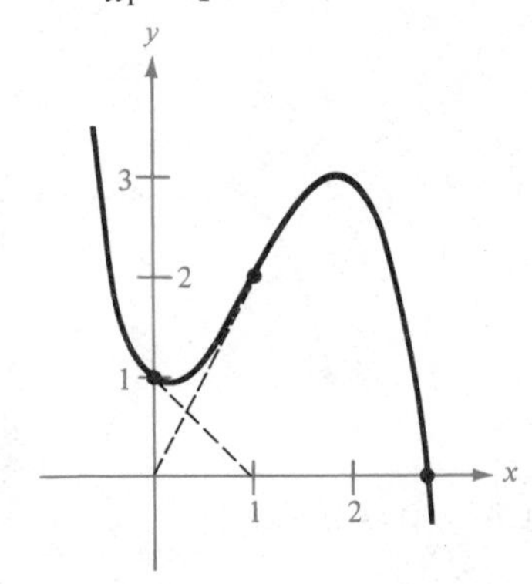

18. Use Newton's Method to obtain a general formula for approximating $\sqrt[n]{a}$. [Hint: Apply Newton's Method to the function $f(x) = x^n - a$.]

ⓒ**19.** Use the result of Exercise 18 to approximate $\sqrt[3]{7}$ to three decimal places.

ⓒ**20.** Use the result of Exercise 18 to approximate $\sqrt[4]{6}$ to three decimal places.

ⓒ**21.** In the Introductory Example of this section, the equation

$$2x^3 - 7x - 1 = 0$$

must be solved to find the coordinates of the point on the curve $y = 4 - x^2$ that is closest to the point (1, 0). Use Newton's Method to solve for x (to three decimal places). What is the y-coordinate of the point?

ⓒ**22.** Find the coordinates of the point on the curve $y = x^2$ that is closest to the point $(4, -3)$.

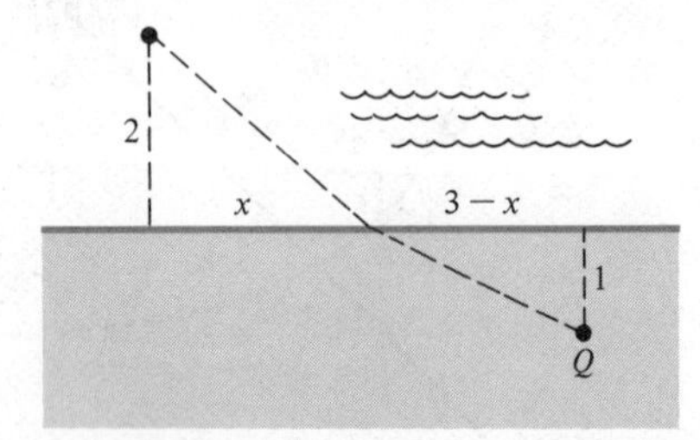

Figure 9.17

ⓒ**23.** A man is in a boat 2 mi from the nearest point on the coast. He is to go to a point Q, which is 3 mi down the coast and 1 mi inland. If he can row at 3 mi/h and walk at 4 mi/h, toward what point on the coast should he row in order to reach point Q in the least time? (See Figure 9.17.) (Note: compare this problem to Exercise 29 in Section 3.4.)

ⓒ**24.** A company estimates that the cost (in dollars) of producing x units of a certain product is given by the model

$$C = 800 + 0.4x + 0.02x^2 + 0.0001x^3$$

Find the production level that minimizes the average cost per unit.

ⓒ**25.** The concentration C of a certain chemical in the bloodstream t hours after injection into muscle tissue is given by

$$C = \frac{3t^2 + t}{50 + t^3}$$

When is the concentration greatest? (Note: Compare this problem to Exercise 42 in Section 3.2.)

ⓒ**26.** The ordering and transportation costs C of the components used in manufacturing a certain product is given by

$$C = 100\left(\frac{200}{x^2} + \frac{x}{x + 30}\right), \quad 1 \le x$$

where C is measured in thousands of dollars and x is the order size in hundreds. Find the order size that minimizes the costs.

ⓒ Calculator may be helpful.

Chapter 10

The Trigonometric Functions

Section 10.1

Radian Measure of Angles

Introductory Example

Winding a Pulley

One way to lift objects is by means of a simple pulley. As the cylindrical drum of the pulley is revolved, the pulley cable winds around the circumference of the drum. The length of cable wound around the drum corresponds precisely to the height the object is raised off the ground. (See Figure 10.1.)

To illustrate this concept, suppose that we are asked to write a computer program to operate a pulley in an automated factory. If the radius of the pulley's drum is 1 ft, how many degrees should the drum be revolved in order to lift the object s feet off the ground? The answer can be seen by observing that the entire circumference of this particular drum is

$$\text{circumference} = (2\pi)(\text{radius}) = (2\pi)(1) = 2\pi$$

Thus, when the drum is turned one full revolution (an angle of $\theta = 360°$), the object is lifted 2π ft. (Since $\pi \approx 3.1416$, one revolution lifts the object about 6.2832 ft.) Now, since 360° corresponds to 2π ft, we have

$$360° \Leftrightarrow 2\pi \text{ ft}$$

$$\frac{180°}{\pi} = \frac{360°}{2\pi} \Leftrightarrow 1 \text{ ft}$$

and, by multiplying by s, we have

$$\left(\frac{180°}{\pi}\right)s \Leftrightarrow s \text{ feet}$$

In other words, if we want to lift the object 2 ft off the ground, we should revolve the drum

$$\left(\frac{180°}{\pi}\right)(2) \approx 114.6°$$

Similarly, if we want to lift the object 7 ft off the ground, we should revolve the drum

$$\left(\frac{180°}{\pi}\right)(7) \approx 401.1°$$

Note that there is a one-to-one correspondence between the number of degrees revolved and the number of feet of cable wound around the drum. In this section, we make use of this correspondence to develop an alternative to the standard degree measure of an angle. This alternative measure is called **radian measure,** and in this particular example (since the radius is 1) a revolution of θ degrees corresponds to a radian measure of s radians. (See Figure 10.1.)

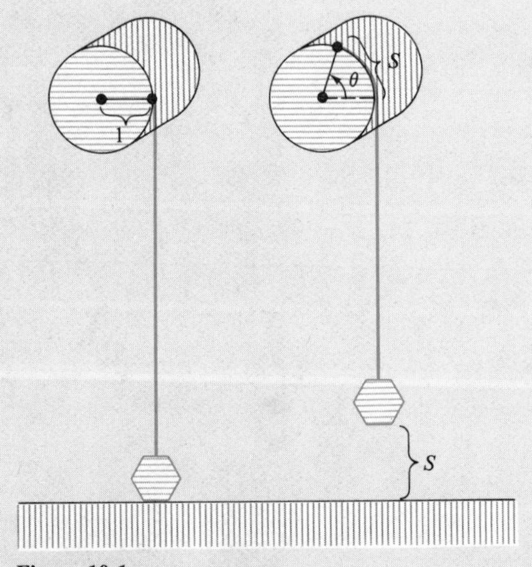

Figure 10.1

Section Objectives: ***To review the properties of angles and their degree measure ▪ To introduce the radian measure of angles ▪ To convert between radian and degree measures.***

The concept of an angle is central to the study of trigonometry, and we devote this section to a discussion of angles and their measure. As shown in Figure 10.2, an **angle** has three parts: an **initial ray,** a **terminal ray,** and a **vertex** (the point of intersection of these two rays). We say that an angle is in **standard position** if its initial ray coincides with the positive x-axis and its vertex is at the origin.

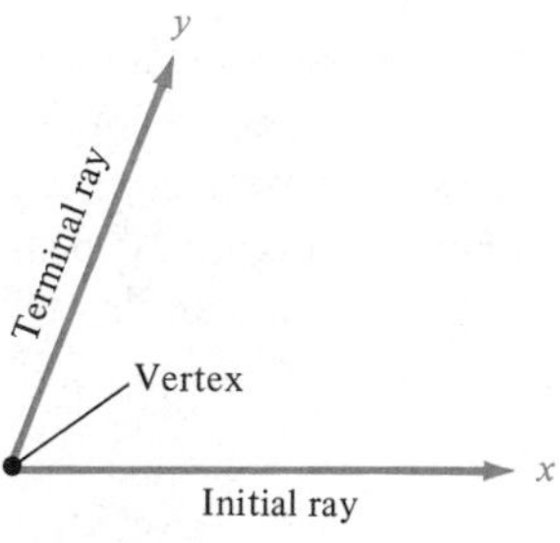

Figure 10.2

We assume that you are somewhat familiar with the measurement of angles in terms of degrees. For example, Figure 10.3 shows several common angles with their corresponding degree measure. Note that we use θ (the lowercase Greek letter theta) to represent the measure of an angle. (Actually, it is common practice to use θ to represent both the angle and its measure.)

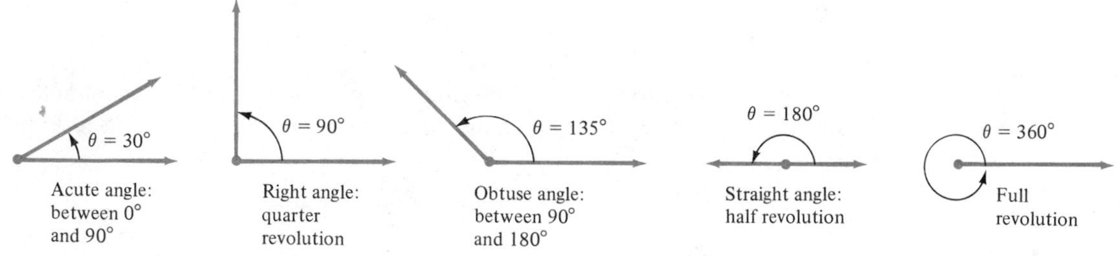

Figure 10.3

Also note in Figure 10.3 that we classify angles between 0° and 90° as **acute** and angles between 90° and 180° as **obtuse.** Positive angles are measured *counterclockwise* beginning with the initial ray. Negative angles are measured *clockwise*. For instance, Figure 10.4 shows an angle whose measure is −45°.

Be sure you understand from Figure 10.4 that we cannot assign a measure to an angle merely by knowing where its initial and terminal rays are located; to

Figure 10.4

Figure 10.5

measure an angle we must also know how the terminal ray was revolved. For example, Figure 10.5 shows that the angle measuring $-45°$ has the same terminal ray as the angle measuring 315°. We call such angles **coterminal.**

Although it may seem strange to consider angles that are larger than 360°, we will see that such angles have very useful applications in trigonometry. An angle that is larger than 360° is one whose terminal ray has revolved more than one full revolution counterclockwise. Figure 10.6 shows two angles measuring more than 360°. Similarly, we can generate angles whose measure is less than $-360°$ by revolving a terminal ray more than one full revolution clockwise.

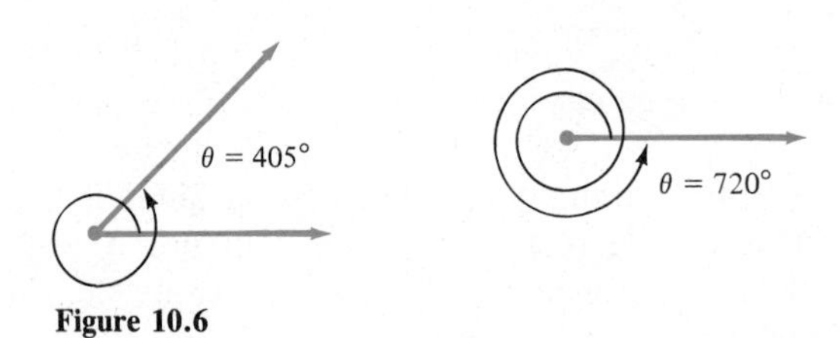

Figure 10.6

Example 1

Finding Coterminal Angles

For each of the following angles, find a coterminal angle θ such that $0° \leq \theta < 360°$.

(a) 450° (b) 750° (c) $-160°$ (d) $-390°$

Solution To find the required coterminal angle, we must add (or subtract) the proper multiple of 360° to bring the result between 0° and 360°. (See Figure 10.7.)

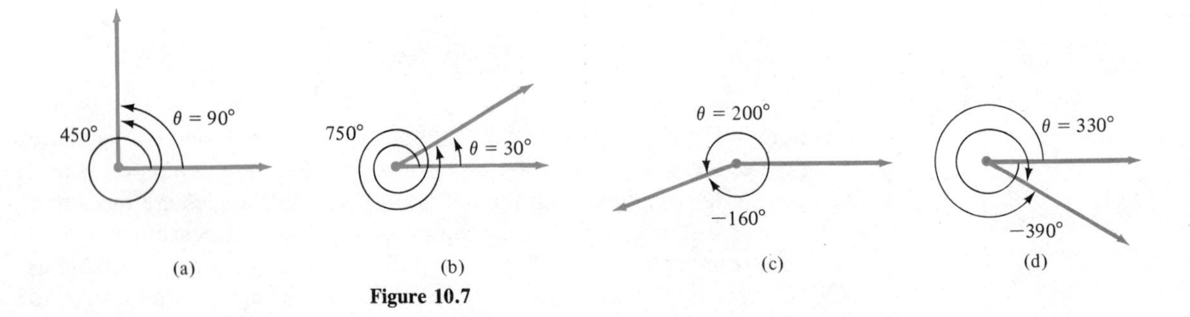

Figure 10.7

(a) $450° - 360° = 90°$

(b) $750° - 2(360°) = 750° - 720° = 30°$

(c) $-160° + 360° = 200°$

(d) $-390° + 2(360°) = -390° + 720° = 330°$ □

One of the most common applications of angle measure occurs in problems dealing with triangles. Many rules from geometry concern triangles. You need to be familiar with several of these, and for convenience we summarize them next.

Triangles: A Summary of Some Rules

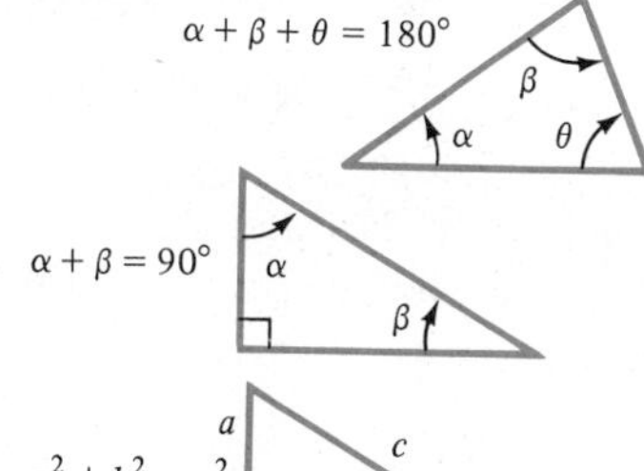

1. The *sum of the angles* of a triangle is 180°.

2. The *sum of the two acute angles* of a right triangle is 90°.

3. *Pythagorean Theorem:* The sum of the squares of the (perpendicular) sides of a right triangle is equal to the square of the hypotenuse.

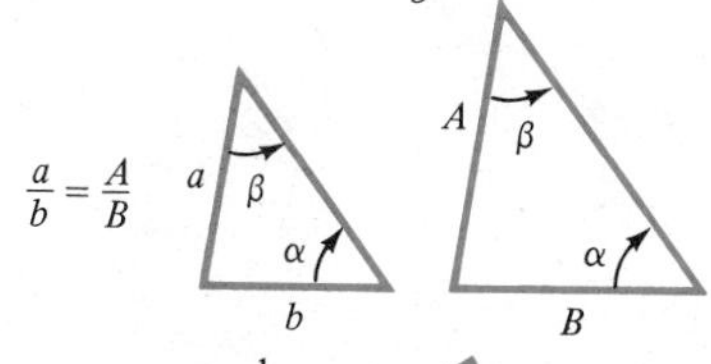

4. *Similar Triangles:* If two triangles are similar (have the same angle measures) then the ratios of corresponding sides are equal.

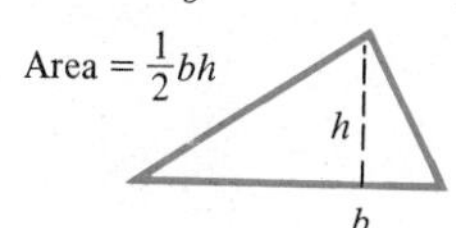

5. The *area* of a triangle is equal to one-half the base times the height.

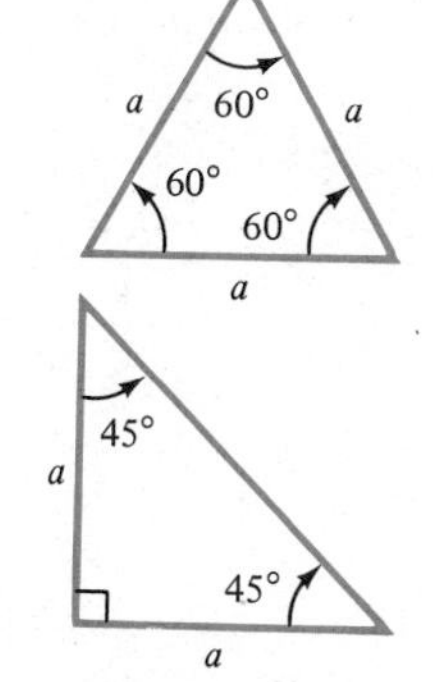

6. Each of the angles in an *equilateral triangle* measures 60°.

7. Each of the acute angles in an *isosceles right triangle* measures 45°.

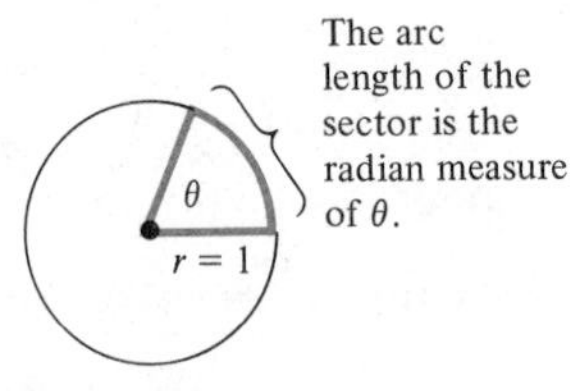

Figure 10.8

A second way to measure angles is in terms of radians (rad). This second type of angle measure turns out to be very useful in calculus. (We will see why radian measure is preferable to degree measure in calculus in Section 10.3.) To assign a radian measure to an angle θ, we consider θ to be the central angle of a circular sector of radius 1, as shown in Figure 10.8. The radian measure of θ is then defined to be the length of the arc of this sector. Recall that the total circumference of a circle of radius 1 is

$$\text{circumference} = (2\pi)(\text{radius})$$

Thus, the circumference of a circle of radius 1 is simply 2π, and we may conclude that the radian measure of an angle measuring 360° is 2π. In other words,

$$360° = 2\pi \text{ rad}$$

Figure 10.9 gives the radian measure for several common angles.

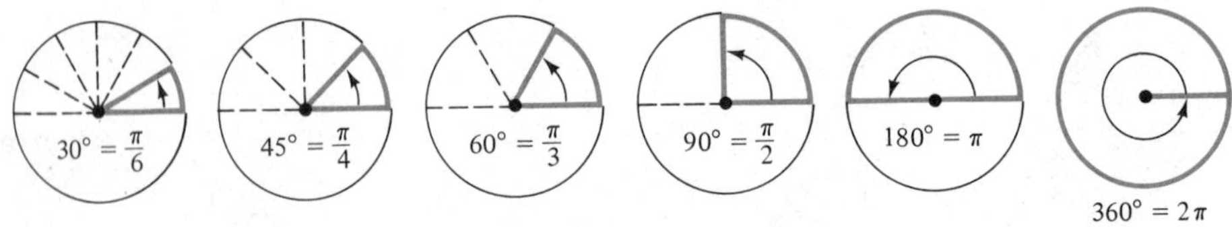

Figure 10.9

It is important for you to become adept at converting back and forth between the degree and radian measures of angles. To do this, we suggest that you memorize the common conversions (such as those pictured in Figure 10.9). For other conversions, you can use one of the following conversion rules.

Conversion Rules: Degrees ⟷ Radians

$$180° = \pi \text{ rad}$$

$$1° = \frac{\pi}{180} \text{ rad}$$

$$1 \text{ rad} = \frac{180°}{\pi}$$

Example 2

Converting from Degrees to Radians

Convert the following degree measures to radian measures.

(a) $135°$ (b) $40°$ (c) $540°$ (d) $-270°$

Solution

(a) $135° = (135)\left(\dfrac{\pi}{180}\right) = \dfrac{3\pi}{4}$ rad

(b) $40° = (40)\left(\dfrac{\pi}{180}\right) = \dfrac{2\pi}{9}$ rad

(c) $540° = (540)\left(\dfrac{\pi}{180}\right) = 3\pi$ rad

(d) $-270° = (-270)\left(\dfrac{\pi}{180}\right) = -\dfrac{3\pi}{2}$ rad □

Example 3

Converting from Radians to Degrees

Convert the following radian measures to degree measures.

(a) $-\dfrac{\pi}{2}$ (b) $\dfrac{7\pi}{4}$ (c) $\dfrac{11\pi}{6}$ (d) $\dfrac{9\pi}{2}$

Solution

(a) $-\dfrac{\pi}{2}$ rad $= \left(-\dfrac{\pi}{2}\right)\left(\dfrac{180}{\pi}\right) = -90°$

(b) $\dfrac{7\pi}{4}$ rad $= \left(\dfrac{7\pi}{4}\right)\left(\dfrac{180}{\pi}\right) = 315°$

(c) $\dfrac{11\pi}{6}$ rad $= \left(\dfrac{11\pi}{6}\right)\left(\dfrac{180}{\pi}\right) = 330°$

(d) $\dfrac{9\pi}{2}$ rad $= \left(\dfrac{9\pi}{2}\right)\left(\dfrac{180}{\pi}\right) = 810°$ □

Section Exercises (10.1)

In Exercises 1–8, determine two coterminal angles (one positive and one negative) for the given angle. Give your answers in degrees.)

1. **2.**

3.

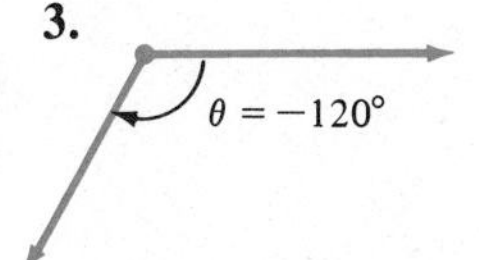

4.

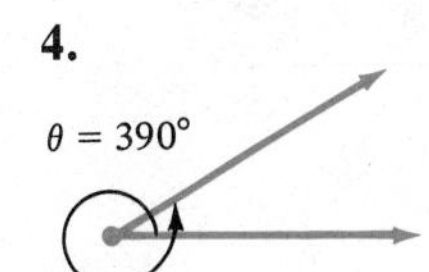

5.

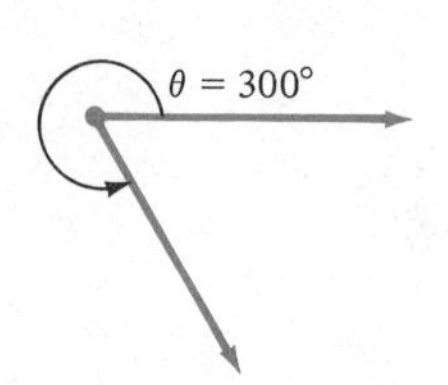

6.

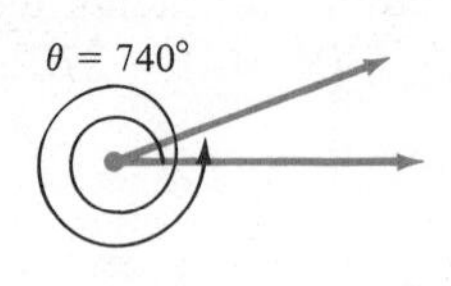

7.

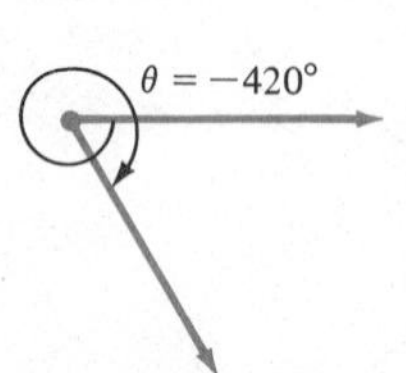

8.

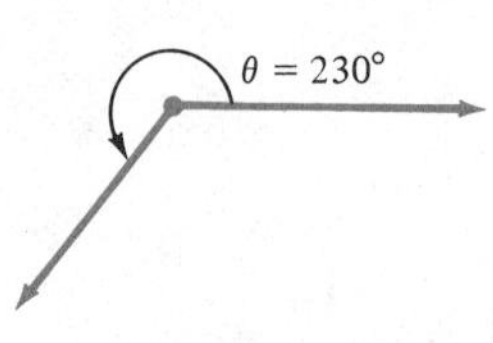

In Exercises 9–16, determine two coterminal angles (one positive and one negative) for the given angle. (Give your answers in radians.)

9.

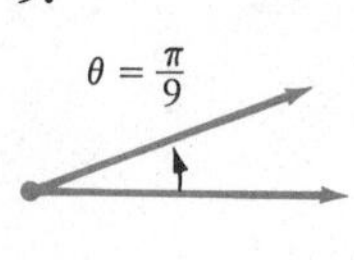

10.

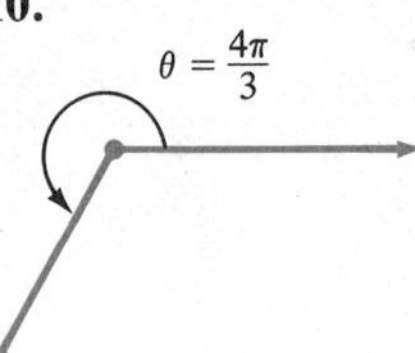

11.

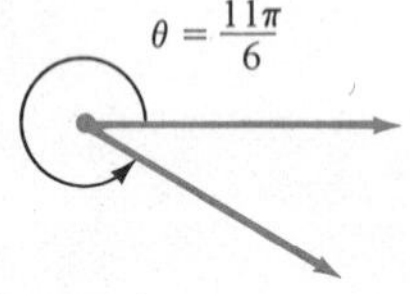
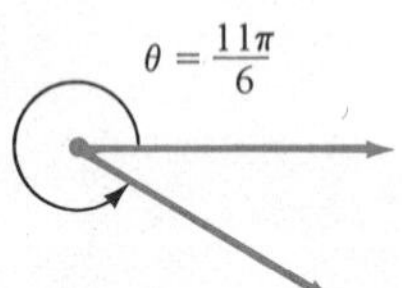

12.

$\theta = -\frac{7\pi}{6}$

13.

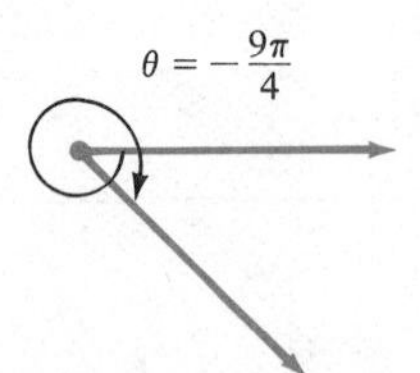

14.

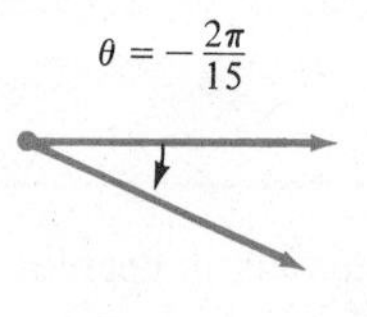

15.

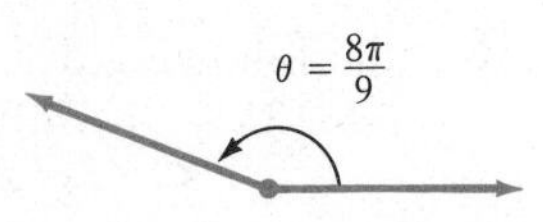

16.

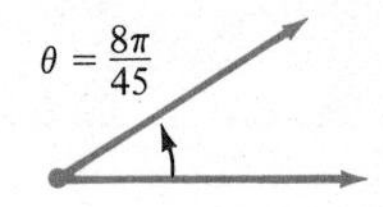

In Exercises 17–24, express the given angle in radian measure as a multiple of π.

17. $30°$ **18.** $150°$
19. $315°$ **20.** $120°$
21. $-20°$ **22.** $-240°$
23. $-270°$ **24.** $144°$

In Exercises 25–32, express the given angle in degree measure.

25. $\frac{3\pi}{2}$ **26.** $\frac{7\pi}{6}$
27. $-\frac{7\pi}{12}$ **28.** $\frac{\pi}{9}$
29. $\frac{7\pi}{3}$ **30.** $-\frac{11\pi}{30}$
31. $\frac{11\pi}{6}$ **32.** $\frac{34\pi}{15}$

In Exercises 33–38, solve the triangle for the indicated side and/or angle.

33.

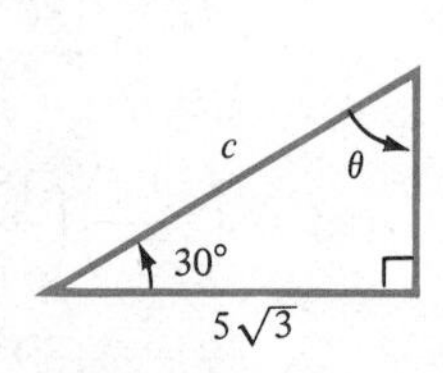

34.

288
θ
a
45°
a

35.

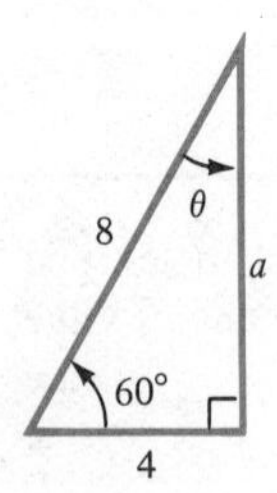

36.

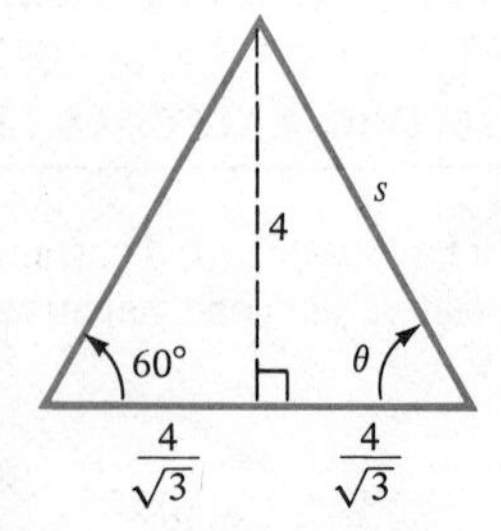

37.

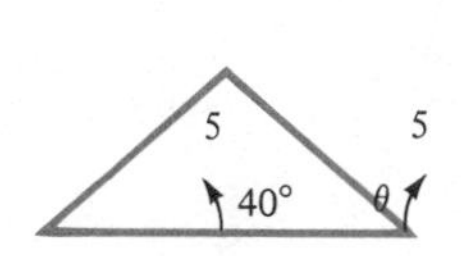

38.

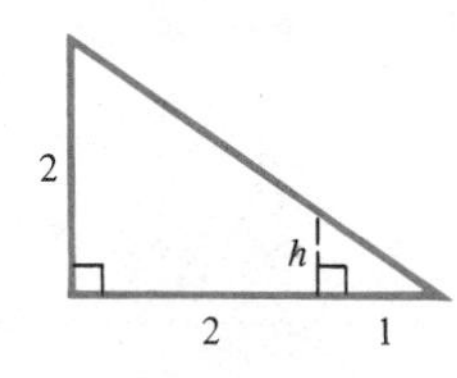

[C] **39.** A 6-ft person standing 12 ft from a streetlight casts an 8-ft shadow. (See Figure 10.10.) What is the height of the streetlight?

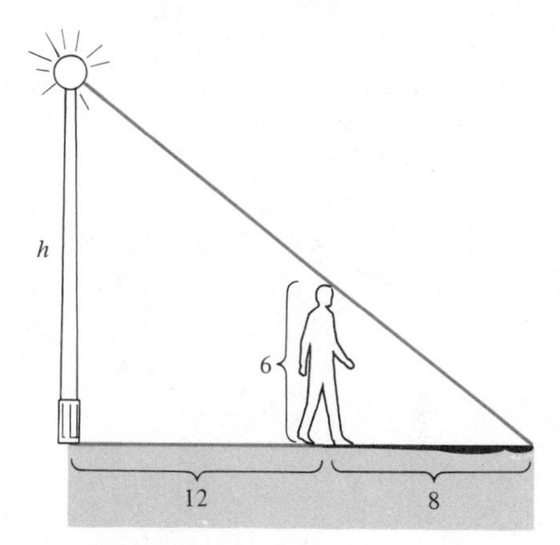

Figure 10.10

[C] **40.** A guy wire is stretched from a broadcasting tower at a point 200 ft above the ground to an anchor 125 ft from the base. (See Figure 10.11.) How long is the wire?

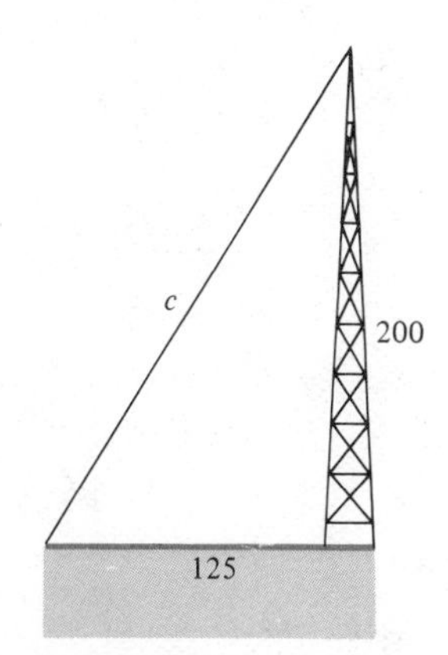

Figure 10.11

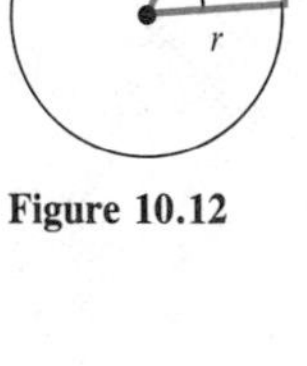

Figure 10.12

[C] **41.** Let r represent the radius of a circle, θ the central angle (measured in radians), and s the length of the arc subtended by the angle, as shown in Figure 10.12. Use the relationship

$$\theta = \frac{s}{r}$$

to complete the following table.

r	8 ft	15 in	85 cm		
s	12 ft			96 in	8642 mi
θ		1.6	$\frac{3\pi}{4}$	4	$\frac{2\pi}{3}$

[C] **42.** The minute hand on a clock is $3\frac{1}{2}$ in long. (See Figure 10.13.) Through what distance does the tip of the minute hand move in 25 min?

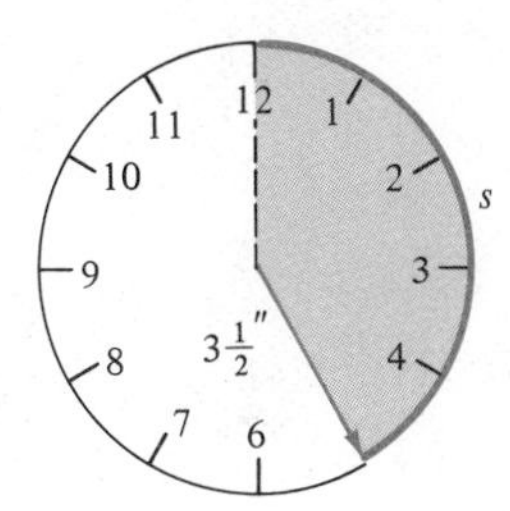

Figure 10.13

[C] **43.** A man bends his elbow through 75°. The distance from his elbow to the top of his index finger is 18.75 in. (See Figure 10.14.)

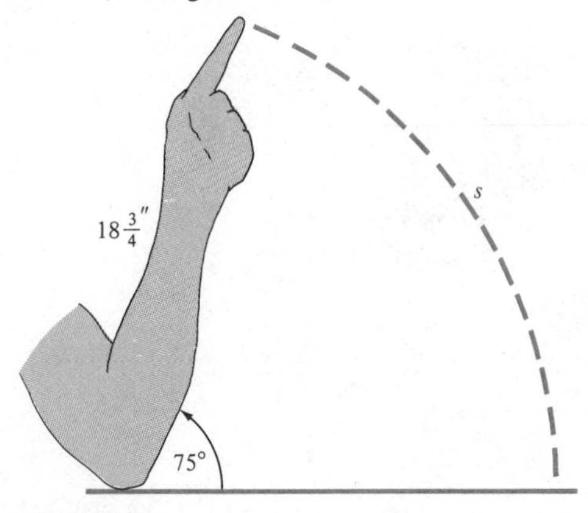

Figure 10.14

[C] Calculator may be helpful.

(a) Find the radian measure of this angle.
(b) Find the distance the tip of the index finger moves.

Ⓒ **44.** A tractor tire, 5 ft in diameter, is partially filled with a liquid ballast for additional traction. To check the air pressure, the tractor operator rotates the tire until the valve stem is at the top so that the liquid will not enter the gauge. On a given occasion, the operator notes that the tire must be rotated 80° to have the stem in the proper position. (See Figure 10.15.)
(a) Find the radian measure of this rotation.
(b) How far must the tractor be moved to get the valve stem in the proper position?

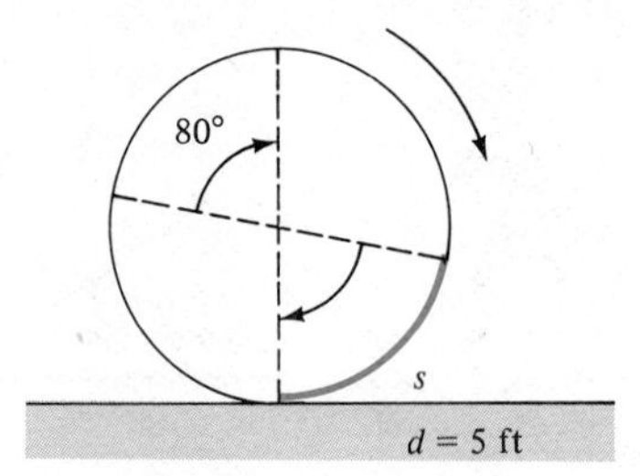

Figure 10.15

Ⓒ Calculator may be helpful.

Section 10.2

The Trigonometric Functions

Introductory Example

Peripheral Vision

When you are looking straight ahead, how great an arc can you see? How can we measure this angle of peripheral vision?

One way would be to stand at the center of a large eye-level circle that is marked off in degrees, as shown in Figure 10.16. While looking straight ahead, you could turn until your right eye could just barely see the 0° marker. At that point, the largest degree marker your left eye could read would represent the angle of your peripheral vision. This solution requires little mathematics, but, unfortunately, it does require the construction of a fairly elaborate structure.

A simpler solution (at least simpler physically) would be to stand facing the corner of a room and measure how far back along the wall you could see. Suppose that a person standing 1 ft from the corner can see an object located 2 ft down the wall, as shown in Figure 10.17. Using the Pythagorean Theorem, we can determine the values of x and y, as shown in Figure 10.18.

In this section, we will see that we can determine either of the acute angles of a right triangle by knowing the ratio of any two sides of the triangle. The study of these ratios is called **trigonometry.** In this particular example, we call the ratio of y to x the **tangent** of the angle θ and write

$$\tan \theta = \frac{y}{x}$$

Since we know the values of x and y, we can solve the problem by solving the equation

$$\tan \theta = \frac{\sqrt{2}}{\sqrt{2} - 1} \approx 3.414$$

The solution to this equation is $\theta \approx 73.7°$, which means that $\alpha/2 \approx 106.3°$ and $\alpha \approx 212.6°$. Thus, the person's angle of peripheral vision is about 212.6°.

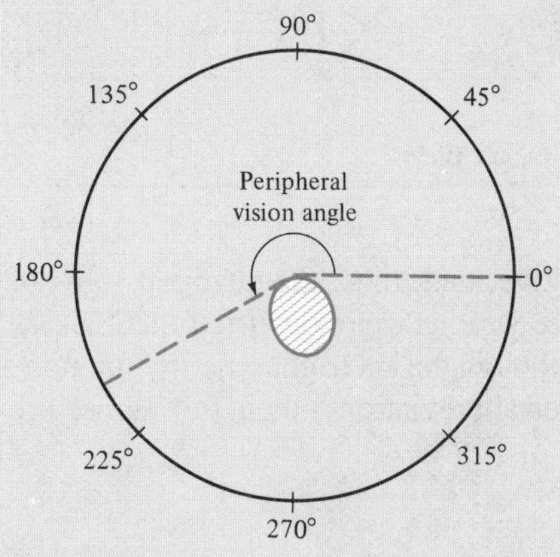

Figure 10.16

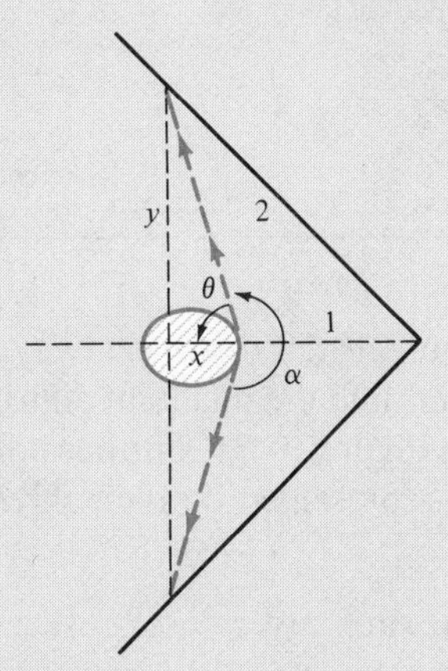

Figure 10.17

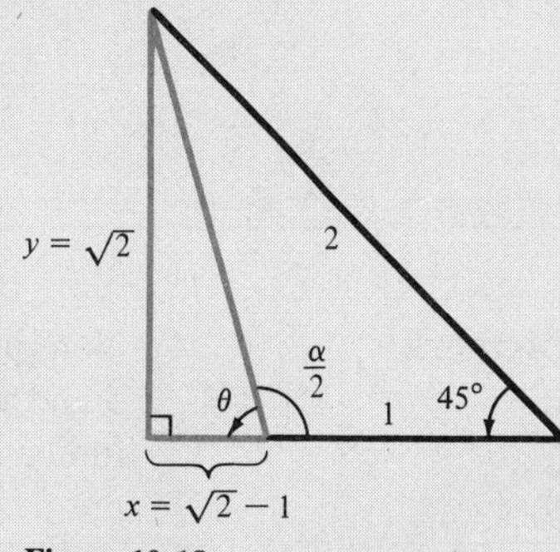

Figure 10.18

Section Objectives: *To evaluate the six trigonometric functions ▪ To introduce some common trigonometric identities ▪ To solve equations involving trigonometric functions.*

There are two common approaches to the study of trigonometry. In one case the trigonometric functions are defined as ratios of two sides of a right triangle. In the other case these functions are defined in terms of a point on the terminal side of an arbitrary angle. The first approach is the one generally used in surveying, navigation, and astronomy, where a typical problem involves a fixed triangle having three of its six parts (sides and angles) known and three to be determined. The second approach is the one normally used in physics, electronics, and biology, where the periodic nature of the trigonometric functions is emphasized. In the following definition we define the six trigonometric functions from both viewpoints.

Definition of the Six Trigonometric Functions

Refer to Figure 10.19.

$$\sin\theta = \frac{\text{opp.}}{\text{hyp.}} \qquad \csc\theta = \frac{\text{hyp.}}{\text{opp.}}$$

$$\cos\theta = \frac{\text{adj.}}{\text{hyp.}} \qquad \sec\theta = \frac{\text{hyp.}}{\text{adj.}}$$

$$\tan\theta = \frac{\text{opp.}}{\text{adj.}} \qquad \cot\theta = \frac{\text{adj.}}{\text{opp.}}$$

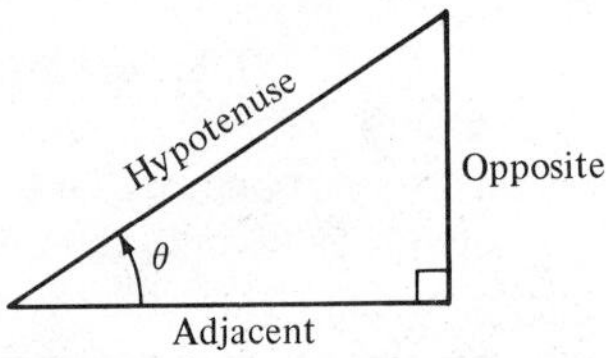

Hypotenuse
Opposite
Adjacent
θ

Figure 10.19

Refer to Figure 10.20.

$$\sin\theta = \frac{y}{r} \qquad \csc\theta = \frac{r}{y}$$

$$\cos\theta = \frac{x}{r} \qquad \sec\theta = \frac{r}{x}$$

$$\tan\theta = \frac{y}{x} \qquad \cot\theta = \frac{x}{y}$$

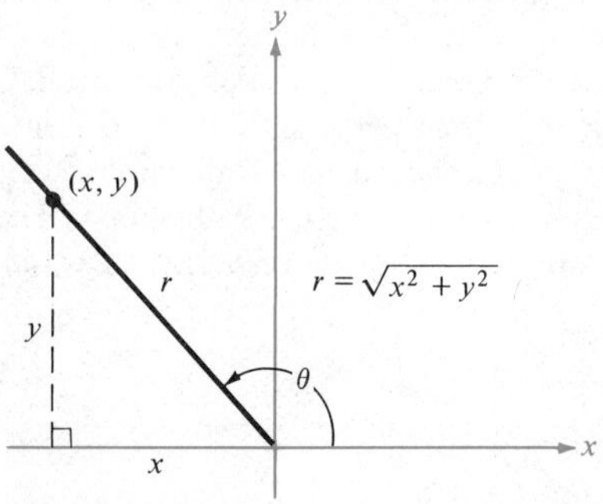

(x, y)
r
$r = \sqrt{x^2 + y^2}$
y
θ
x

Figure 10.20

Remark

Note in Figure 10.20 that r is always positive, and, thus, the quadrant signs of x and y determine the quadrant signs of the various trigonometric functions, as indicated in Figure 10.21. Furthermore, although the six trigonometric functions are usually represented by their three-letter abbreviations, their full names are actually

sin: sine	csc: cosecant
cos: cosine	sec: secant
tan: tangent	cot: cotangent

The following formulas are direct consequences of the definitions.

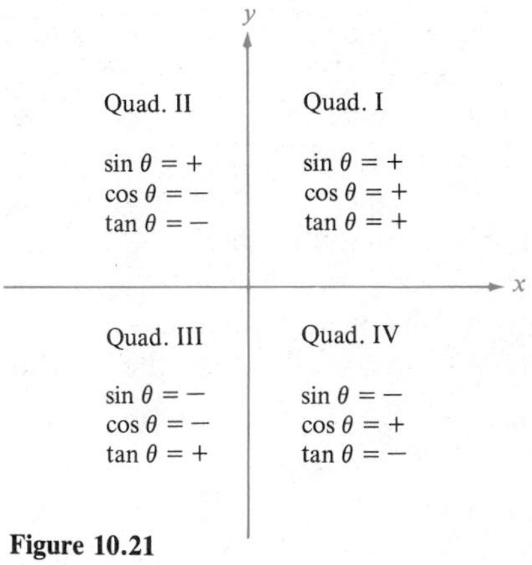

Quad. II	Quad. I
$\sin\theta = +$ $\cos\theta = -$ $\tan\theta = -$	$\sin\theta = +$ $\cos\theta = +$ $\tan\theta = +$
Quad. III	**Quad. IV**
$\sin\theta = -$ $\cos\theta = -$ $\tan\theta = +$	$\sin\theta = -$ $\cos\theta = +$ $\tan\theta = -$

Figure 10.21

$$\csc\theta = \frac{1}{\sin\theta} \qquad \cot\theta = \frac{1}{\tan\theta} \qquad \cot\theta = \frac{\cos\theta}{\sin\theta}$$

$$\sec\theta = \frac{1}{\cos\theta} \qquad \tan\theta = \frac{\sin\theta}{\cos\theta}$$

Furthermore, since

$$\sin^2\theta + \cos^2\theta = \left(\frac{y}{r}\right)^2 + \left(\frac{x}{r}\right)^2 = \frac{x^2 + y^2}{r^2} = \frac{r^2}{r^2} = 1$$

we can readily obtain the Pythagorean Identity

$$\sin^2\theta + \cos^2\theta = 1$$

[Note that we use $\sin^2\theta$ to mean $(\sin\theta)^2$.] Additional trigonometric identities are listed next, without proof.

Trigonometric Identities

Pythagorean Identities:

$\sin^2\theta + \cos^2\theta = 1$

$\tan^2\theta + 1 = \sec^2\theta$

$\cot^2\theta + 1 = \csc^2\theta$

Reduction Formula:

$\sin(-\theta) = -\sin\theta$

$\cos(-\theta) = \cos\theta$

$\tan(-\theta) = -\tan\theta$

$\sin\theta = -\sin(\theta - \pi)$

$\cos\theta = -\cos(\theta - \pi)$

$\tan\theta = \tan(\theta - \pi)$

Sum or Difference of Two Angles:*

$\sin(\theta \pm \phi) = \sin\theta\cos\phi \pm \cos\theta\sin\phi$

$\cos(\theta \pm \phi) = \cos\theta\cos\phi \mp \sin\theta\sin\phi$

$$\tan(\theta \pm \phi) = \frac{\tan\theta \pm \tan\phi}{1 \mp \tan\theta\tan\phi}$$

Double Angle:

$\sin 2\theta = 2\sin\theta\cos\theta$

$\cos 2\theta = 2\cos^2\theta - 1 = 1 - 2\sin^2\theta$

Half Angle:

$\sin^2\theta = \frac{1}{2}(1 - \cos 2\theta)$

$\cos^2\theta = \frac{1}{2}(1 + \cos 2\theta)$

*Note that ϕ is the lowercase Greek letter phi.

Remark

Although an angle can be measured in either degrees or radians, we have already mentioned that in calculus radian measure is preferred. Thus, all angles in the remainder of this chapter are measured in radians *unless stated otherwise*. In other words, when we write sin 3, we mean the sine of three radians, and when we write sin 3°, we mean the sine of three degrees.

There are two common methods of evaluating trigonometric functions: decimal approximations with a calculator (or a table of trigonometric values) and exact evaluations using trigonometric identities and formulas from geometry. The next three examples illustrate the second method. Study these examples carefully.

Example 1

Evaluating Trigonometric Functions

Evaluate the sine, cosine, and tangent of $\pi/3$.

Solution We begin by drawing the angle $\theta = \pi/3$ in the standard position, as shown in Figure 10.22. Then, since $\pi/3$ rad $= 60°$, we imagine an equilateral triangle whose sides have a length of 1 and with θ as one of its angles. Since the altitude of this triangle bisects its base, we know that

$$x = \frac{1}{2}$$

Now, using the Pythagorean Theorem, we have

$$y = \sqrt{r^2 - x^2} = \sqrt{1 - \left(\frac{1}{2}\right)^2} = \sqrt{\frac{3}{4}} = \frac{\sqrt{3}}{2}$$

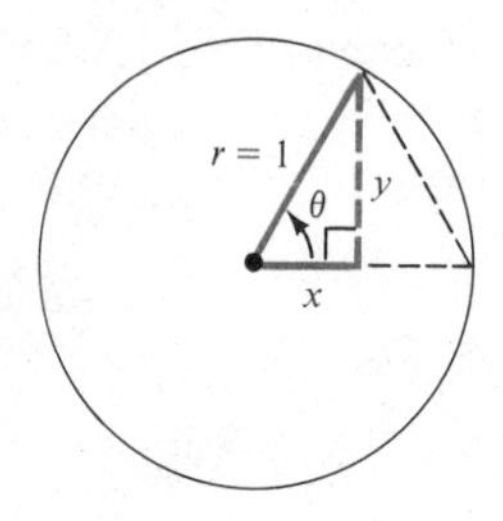

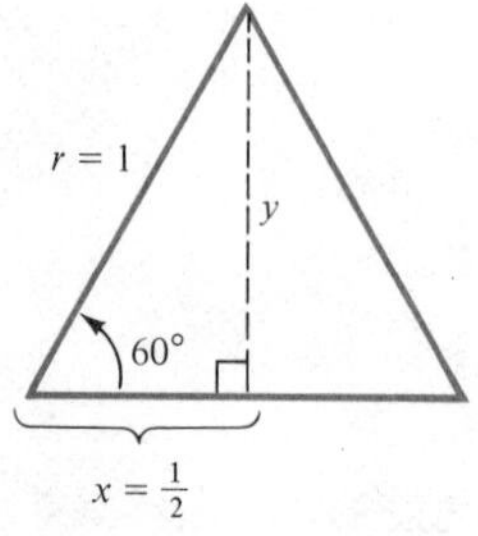

Figure 10.22

Finally, we have

$$\sin \frac{\pi}{3} = \frac{y}{r} = \frac{\sqrt{3}/2}{1} = \frac{\sqrt{3}}{2}$$

$$\cos \frac{\pi}{3} = \frac{x}{r} = \frac{\frac{1}{2}}{1} = \frac{1}{2}$$

$$\tan \frac{\pi}{3} = \frac{y}{x} = \frac{\sqrt{3}/2}{\frac{1}{2}} = \sqrt{3}$$

□

The degree and radian measures of several common angles are given in Table 10.1 along with the corresponding values of sine, cosine, and tangent.

Table 10.1

Degrees	0	30°	45°	60°	90°
Radians	0	$\frac{\pi}{6}$	$\frac{\pi}{4}$	$\frac{\pi}{3}$	$\frac{\pi}{2}$
$\sin\theta$	0	$\frac{1}{2}$	$\frac{\sqrt{2}}{2}$	$\frac{\sqrt{3}}{2}$	1
$\cos\theta$	1	$\frac{\sqrt{3}}{2}$	$\frac{\sqrt{2}}{2}$	$\frac{1}{2}$	0
$\tan\theta$	0	$\frac{1}{\sqrt{3}}$	1	$\sqrt{3}$	undef.

As an aid to memorizing the values in Table 10.1, note the pattern for $\sin\theta$ and $\cos\theta$ shown in the following table.

θ	0°	30°	45°	60°	90°
$\sin\theta$	$\frac{\sqrt{0}}{2}$	$\frac{\sqrt{1}}{2}$	$\frac{\sqrt{2}}{2}$	$\frac{\sqrt{3}}{2}$	$\frac{\sqrt{4}}{2}$
$\cos\theta$	$\frac{\sqrt{4}}{2}$	$\frac{\sqrt{3}}{2}$	$\frac{\sqrt{2}}{2}$	$\frac{\sqrt{1}}{2}$	$\frac{\sqrt{0}}{2}$

Using the values in the table, you can determine the corresponding values for the tangent by using the identity

$$\tan\theta = \frac{\sin\theta}{\cos\theta}$$

To extend the use of the table to angles in quadrants other than the first quadrant, you can use the concept of a reference angle (as shown in Figure

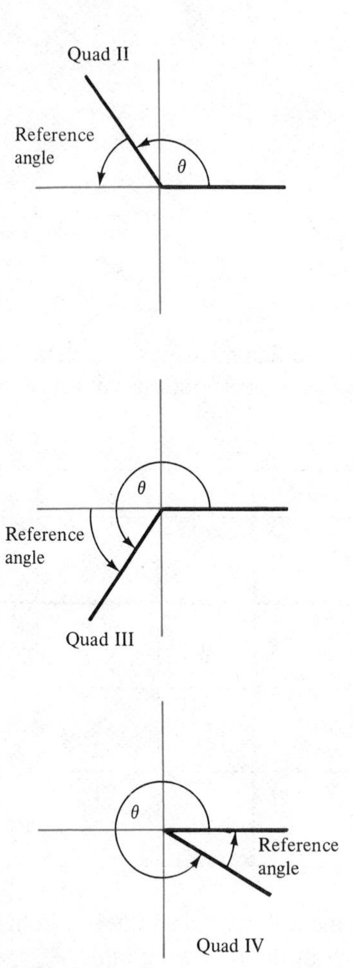

Figure 10.23

10.23), together with the appropriate quadrant sign. For instance, the reference angle for 135° is 45° since 135° + 45° = 180°, and the reference angle for 210° is 30° since 210° − 30° = 180°.

Example 2

Evaluating Trigonometric Functions in the Second, Third, and Fourth Quadrants

Evaluate the following trigonometric functions.

$$\text{(a) } \sin \frac{3\pi}{4} \qquad \text{(b) } \tan 330° \qquad \text{(c) } \cos \frac{7\pi}{6}$$

Solution

(a) Since the reference angle for $3\pi/4$ is $\pi/4$ and the sine is positive in the second quadrant (see Figure 10.24), we have

$$\sin\frac{3\pi}{4} = +\sin\frac{\pi}{4} = \frac{\sqrt{2}}{2}$$

(b) Since the reference angle for 330° is 30° and the tangent is negative in the fourth quadrant, we have

$$\tan 330° = -\tan 30° = -\frac{1}{\sqrt{3}} = -\frac{\sqrt{3}}{3}$$

(c) Since the reference angle for $7\pi/6$ is $\pi/6$ and the cosine is negative in the third quadrant, we have

$$\cos\frac{7\pi}{6} = -\cos\frac{\pi}{6} = -\frac{\sqrt{3}}{2}$$

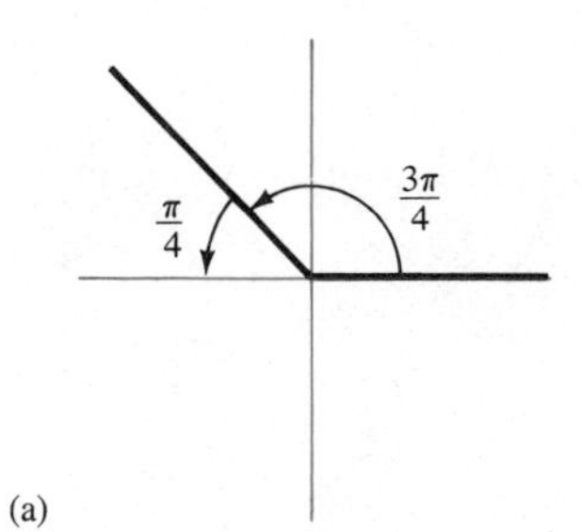

(a)

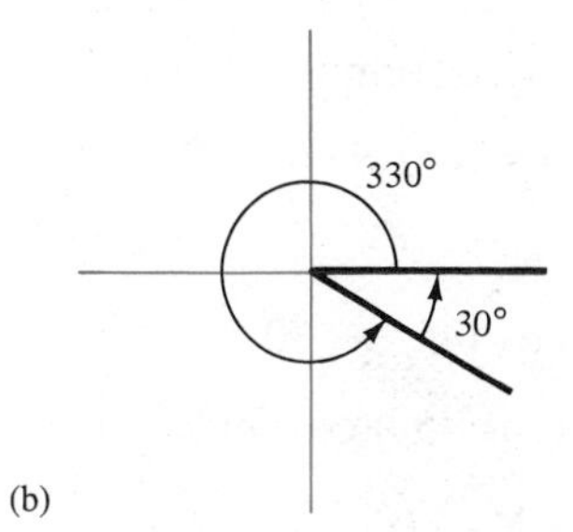

(b)

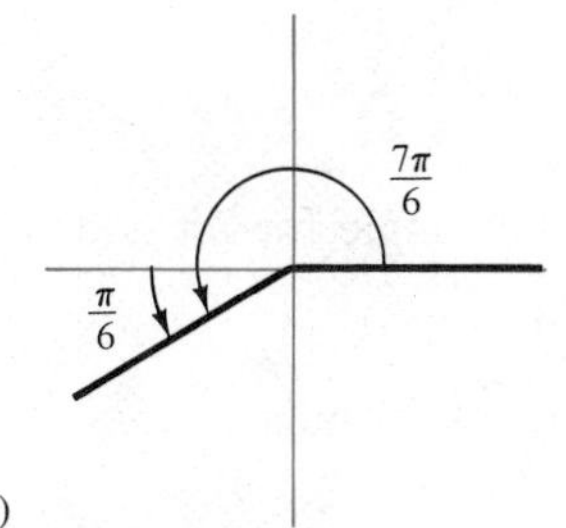

(c)

Figure 10.24

□

Example 3

Using Trigonometric Identities to Evaluate Trigonometric Functions

Evaluate the following trigonometric functions.

(a) $\sin\left(-\frac{\pi}{3}\right)$ (b) $\sec 60°$ (c) $\cos 15°$

(d) $\sin 2\pi$ (e) $\cot 0°$ (f) $\tan \frac{9\pi}{4}$

Solution

(a) Using the reduction formula

$$\sin(-\theta) = -\sin\theta$$

we have

$$\sin\left(-\frac{\pi}{3}\right) = -\sin\frac{\pi}{3} = -\frac{\sqrt{3}}{2}$$

(b) Using the reciprocal formula

$$\sec\theta = \frac{1}{\cos\theta}$$

we have

$$\sec 60° = \frac{1}{\cos 60°} = \frac{1}{\frac{1}{2}} = 2$$

(c) Using the difference formula

$$\cos(\theta - \phi) = \cos\theta\cos\phi + \sin\theta\sin\phi$$

we have

$$\begin{aligned}\cos 15° &= \cos(45° - 30°)\\ &= (\cos 45°)(\cos 30°) + (\sin 45°)(\sin 30°)\\ &= \left(\frac{\sqrt{2}}{2}\right)\left(\frac{\sqrt{3}}{2}\right) + \left(\frac{\sqrt{2}}{2}\right)\left(\frac{1}{2}\right)\\ &= \frac{\sqrt{6} + \sqrt{2}}{4}\end{aligned}$$

(d) Since the reference angle for 2π is 0, we have

$$\sin 2\pi = \sin 0 = 0$$

(e) Using the reciprocal formula

$$\cot\theta = \frac{1}{\tan\theta}$$

we have

$$\cot 0° = \frac{1}{\tan 0°} = \frac{1}{0} \qquad \text{(undefined)}$$

(f) Since the reference angle for $9\pi/4$ is $\pi/4$ and the tangent is positive in the first quadrant, we have

$$\tan \frac{9\pi}{4} = \tan \frac{\pi}{4} = 1$$

□

Up to this point, we have restricted our trigonometric evaluations to "standard" angles. How can we evaluate trigonometric functions of nonstandard angles? For example, what is the sine of 17° or the cosine of $\pi/11$? In such cases, it is best to use decimal approximations obtained with a calculator. When doing this, be sure to remember to distinguish between degree and radian measure. Most calculators have a degree/radian key to distinguish between these two types of angle measure. Furthermore, most calculators have only three trigonometric function keys: sin, cos, and tan. To evaluate the other three functions, we combine one of these keys with the reciprocal key. For example, we can find the secant of 10° by hitting the following sequence of keys:

Problem	*Calculator Steps*	*Display*
To find sec 10°	[10] [cos] [$\frac{1}{x}$]	1.015426611

We demonstrate an additional use of a calculator in the following example.

Example 4

Using Trigonometric Functions to Solve Triangles

A surveyor is standing 50 ft away from the base of a large tree. The surveyor measures the angle of elevation to the top of the tree as 71.5°. How tall is the tree?

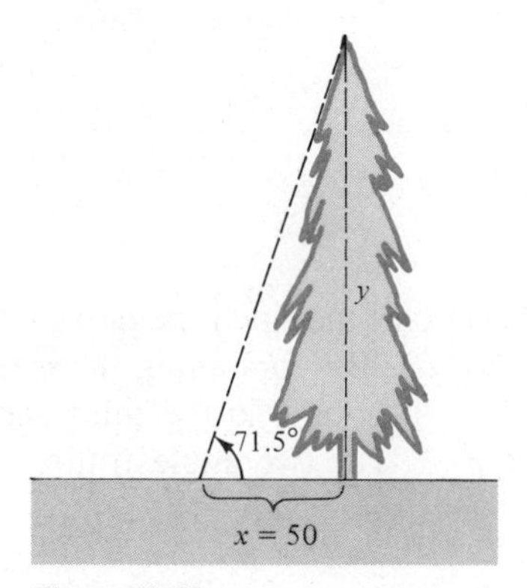

Figure 10.25

Solution By referring to Figure 10.25, we see that

$$\tan 71.5° = \frac{y}{x}$$

where $x = 50$ and y is the height of the tree. Thus, we can determine the height as follows:

$$\begin{aligned} y &= (x)(\tan 71.5°) \\ &\approx (50)(2.98868) \\ &\approx 149.4 \text{ ft} \end{aligned}$$

□

To conclude this section, we look at the solution of trigonometric equations. For example, consider the equation

$$\sin \theta = 0$$

We know $\theta = 0$ is one solution. Also, in Example 3, part (d), we saw that $\theta = 2\pi$ is another solution. But these are not the only solutions. Actually, there are infinitely many solutions to this equation! Any one of the following values will work:

$$\ldots, -3\pi, -2\pi, -\pi, 0, \pi, 2\pi, 3\pi, \ldots$$

To simplify this situation, we usually restrict our search for solutions to the interval $0 \le \theta \le 2\pi$, as shown in the following example.

Example 5

Solving Trigonometric Equations

Solve for θ in the following equations.

(a) $\sin \theta = -\dfrac{\sqrt{3}}{2}, \quad 0 \le \theta \le 2\pi$

(b) $\cos \theta = 1, \quad 0 \le \theta \le 2\pi$

(c) $\tan \theta = 1, \quad 0 \le \theta \le 2\pi$

Solution

(a) To solve the equation

$$\sin \theta = -\frac{\sqrt{3}}{2}$$

we make two observations: the sine is negative in the third and fourth quadrants, *and* $\sin(\pi/3) = \sqrt{3}/2$. By combining these two observations, we conclude that we are seeking values of θ in the third and fourth quadrants that have a reference angle of $\pi/3$. The two angles fitting these criteria are

$$\theta = \pi + \frac{\pi}{3} = \frac{4\pi}{3}$$

and $\theta = 2\pi - \frac{\pi}{3} = \frac{5\pi}{3}$

(b) To solve $\cos \theta = 1$, we observe that $\cos 0 = 1$ and note that in the interval $[0, 2\pi]$ the only angles whose reference angle is 0 are 0, π, and 2π. Since $\cos \pi = -1$ and $\cos 2\pi = 1$, we conclude that there are two solutions:

$$\theta = 0 \quad \text{or} \quad \theta = 2\pi$$

(c) Since $\tan(\pi/4) = 1$ and the tangent is positive in the first and third quadrants, we have

$$\theta = \frac{\pi}{4} \quad \text{or} \quad \theta = \pi + \frac{\pi}{4} = \frac{5\pi}{4}$$

□

Example 6

Solving Trigonometric Equations

Solve the following equation for θ.

$$\cos 2\theta = 2 - 3 \sin \theta, \qquad 0 \le \theta \le 2\pi$$

Solution Using the double angle identity

$$\cos 2\theta = 1 - 2 \sin^2 \theta$$

we have the following polynomial (in $\sin \theta$):

$$1 - 2\sin^2 \theta = 2 - 3 \sin \theta$$

$$0 = 2 \sin^2 \theta - 3 \sin \theta + 1$$

$$0 = (2 \sin \theta - 1)(\sin \theta - 1)$$

If $2 \sin \theta - 1 = 0$, we have $\sin \theta = \frac{1}{2}$ and

$$\theta = \frac{\pi}{6} \quad \text{or} \quad \frac{5\pi}{6}$$

If $\sin \theta - 1 = 0$, we have $\sin \theta = 1$ and

$$\theta = \frac{\pi}{2}$$

Thus, for $0 \le \theta \le 2\pi$, there are three solutions to the given equation:

$$\theta = \frac{\pi}{6}, \frac{\pi}{2}, \text{ or } \frac{5\pi}{6}$$

□

Remark

In Example 6, we factored the quadratic equation involving $\sin \theta$ just as if the sine of θ were a single unknown variable. For instance, if we let $\sin \theta = x$, then the factorization would have taken the standard algebraic form

$$2x^2 - 3x + 1 = (2x - 1)(x - 1)$$

Example 7

Solving Trigonometric Equations

Solve the following equation for θ.

$$-1 = \sin \theta + \cos \theta, \qquad 0 \le \theta \le 2\pi$$

Solution To convert this equation to a polynomial in $\sin \theta$, we write

$$-\sin \theta - 1 = \cos \theta$$

Now, in order to apply the Pythagorean identity for $\cos^2 \theta$, we square both sides of this equation to obtain

$$(-\sin \theta - 1)^2 = \cos^2 \theta$$

$$\sin^2 \theta + 2 \sin \theta + 1 = 1 - \sin^2 \theta$$

$$2 \sin^2 \theta + 2 \sin \theta = 0$$

$$2 \sin \theta (\sin \theta + 1) = 0$$

If $\sin \theta = 0$, we have

$$\theta = 0, \ \pi, \text{ or } 2\pi$$

If $\sin \theta + 1 = 0$, we have $\sin \theta = -1$, and

$$\theta = \frac{3\pi}{2}$$

However, by checking these four values of θ, we see that the only ones satisfying the original equation are $\theta = \pi$ and $\theta = 3\pi/2$. The other two values are called *extraneous roots,* and in this particular solution they were introduced by squaring both sides of the equation. □

Section Exercises (10.2)

In Exercises 1–6, determine all six trigonometric functions for the given angle θ.

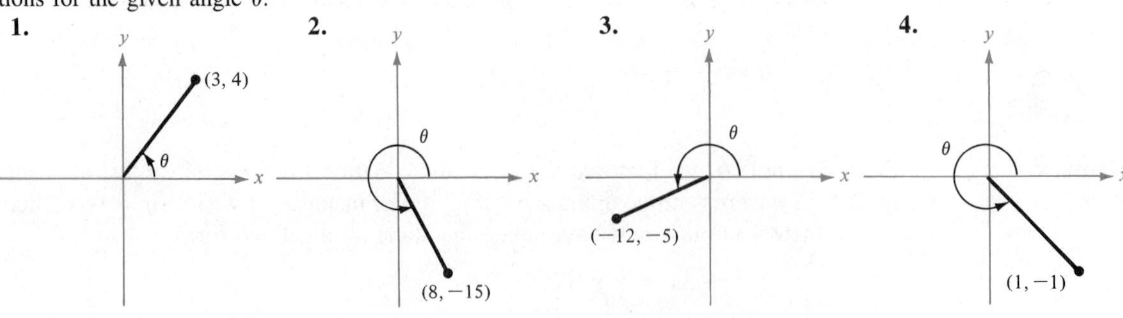

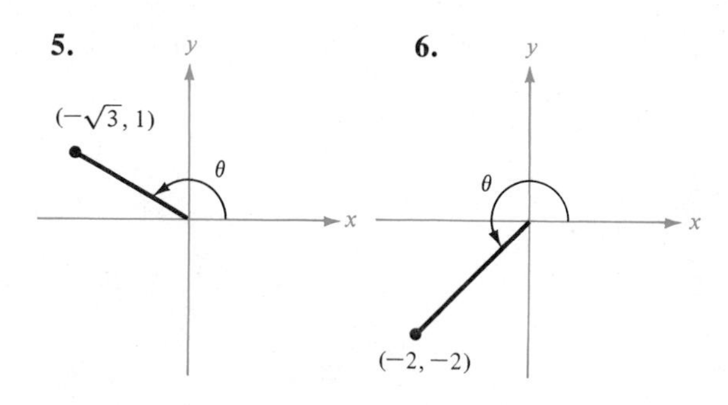

In Exercises 7–12, determine the quadrant in which θ lies.

7. $\sin\theta < 0$ and $\cos\theta < 0$
8. $\sin\theta > 0$ and $\cos\theta < 0$
9. $\sin\theta > 0$ and $\cos\theta > 0$
10. $\sin\theta < 0$ and $\cos\theta > 0$
11. $\sin\theta > 0$ and $\tan\theta < 0$
12. $\cos\theta > 0$ and $\tan\theta < 0$

In Exercises 13–18, find the indicated trigonometric function from the given one.

13. Given: $\sin\theta = \frac{1}{2}$
Find: $\csc\theta$

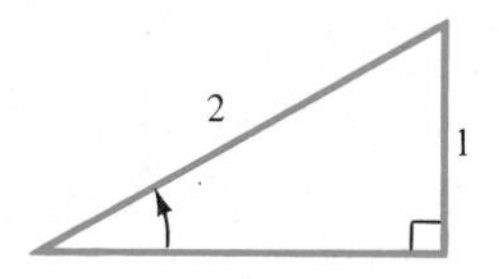

14. Given: $\sin\theta = \frac{1}{3}$
Find: $\tan\theta$

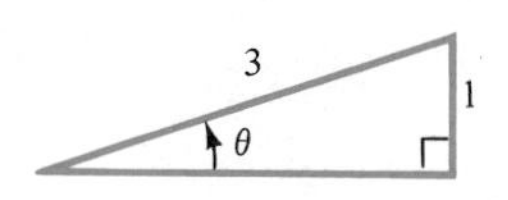

15. Given: $\cos\theta = \frac{4}{5}$
Find: $\cot\theta$

16. Given: $\sec\theta = \frac{13}{5}$
Find: $\cot\theta$

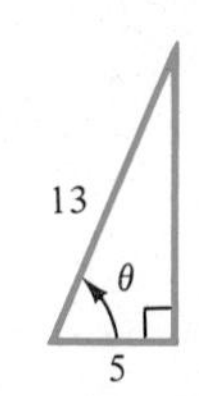

17. Given: $\cot\theta = \frac{15}{8}$
Find: $\sec\theta$

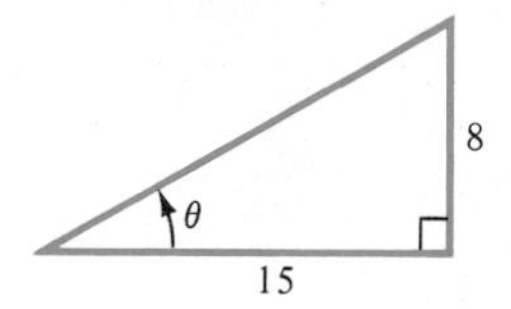

18. Given: $\tan\theta = \frac{1}{2}$
Find: $\sin\theta$

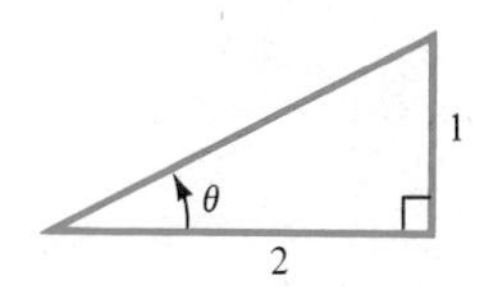

In Exercises 19–26, evaluate the sine, cosine, and tangent of the given angles *without* using a calculator or tables.

19. (a) $60°$ (b) $\dfrac{2\pi}{3}$
20. (a) $\dfrac{\pi}{4}$ (b) $\dfrac{5\pi}{4}$
21. (a) $-\dfrac{\pi}{6}$ (b) $150°$
22. (a) $-\dfrac{\pi}{2}$ (b) $\dfrac{\pi}{2}$
23. (a) $225°$ (b) $-225°$
24. (a) $300°$ (b) $330°$
25. (a) $750°$ (b) $510°$
26. (a) $\dfrac{10\pi}{3}$ (b) $\dfrac{17\pi}{3}$

In Exercises 27–34, use a calculator to evaluate the given trigonometric functions to four decimal places.

[c] **27.** (a) $\sin 10°$ (b) $\csc 10°$
[c] **28.** (a) $\sec 225°$ (b) $\sec 135°$
[c] **29.** (a) $\tan\dfrac{\pi}{9}$ (b) $\tan\dfrac{10\pi}{9}$
[c] **30.** (a) $\cot(1.35)$ (b) $\tan(1.35)$
[c] **31.** (a) $\cos(-110°)$ (b) $\cos 250°$
[c] **32.** (a) $\sin(-0.65)$ (b) $\sin(5.63)$
[c] **33.** (a) $\tan 240°$ (b) $\cot 210°$
[c] **34.** (a) $\csc(2.62)$ (b) $\csc 150°$

In Exercises 35–40, find two values of θ corresponding to the given functions. List the measure of θ in degrees $(0 \le \theta < 360°)$ *and* radians $(0 \le \theta < 2\pi)$. Do not use a calculator or tables.

35. (a) $\sin\theta = \frac{1}{2}$ (b) $\sin\theta = -\frac{1}{2}$
36. (a) $\cos\theta = \dfrac{\sqrt{2}}{2}$ (b) $\cos\theta = -\dfrac{\sqrt{2}}{2}$
37. (a) $\csc\theta = \dfrac{2\sqrt{3}}{3}$ (b) $\cot\theta = -1$
38. (a) $\sec\theta = 2$ (b) $\sec\theta = -2$
39. (a) $\tan\theta = 1$ (b) $\cot\theta = -\sqrt{3}$
40. (a) $\sin\theta = \dfrac{\sqrt{3}}{2}$ (b) $\sin\theta = -\dfrac{\sqrt{3}}{2}$

In Exercises 41–46, use the tables found in the Appendix of this text to estimate two values of θ corresponding to the given functions. List the measure of θ in

[c] Calculator may be helpful.

degrees ($0 \le \theta < 360°$) *and* radians ($0 \le \theta < 2\pi$).

41. (a) $\sin\theta = 0.8191$ (b) $\sin\theta = -0.2589$
42. (a) $\sin\theta = 0.0175$ (b) $\sin\theta = -0.6691$
43. (a) $\cos\theta = 0.9848$ (b) $\cos\theta = -0.5890$
44. (a) $\cos\theta = 0.8746$ (b) $\cos\theta = -0.2419$
45. (a) $\tan\theta = 1.192$ (b) $\tan\theta = -8.144$
46. (a) $\cot\theta = 5.671$ (b) $\cot\theta = -1.280$

In Exercises 47–56, solve the given equation for θ. ($0 \le \theta < 2\pi$) (Note: For some of the equations, you should use the trigonometric identities listed in this section.)

47. $2\sin^2\theta = 1$
48. $\tan^2\theta = 3$
49. $\tan^2\theta - \tan\theta = 0$
50. $2\cos^2\theta - \cos\theta = 1$
51. $\sin 2\theta + 2\sin\theta = 0$
52. $\sec\theta\csc\theta = 2\csc\theta$
53. $\sin\theta = \cos\theta$
54. $\cos 2\theta + 3\cos\theta + 2 = 0$
55. $\cos^2\theta + \sin\theta = 1$
56. $\cos\dfrac{\theta}{2} - \cos\theta = 1$

In Exercises 57–64, solve for x, y, or r, as indicated.

57. Solve for y.

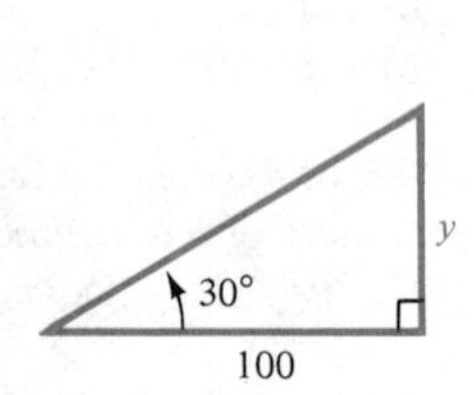

58. Solve for x.

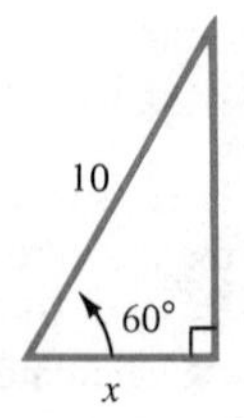

59. Solve for x.

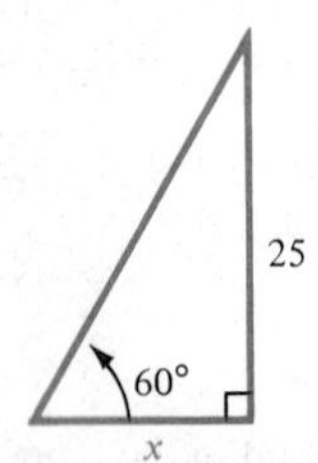

60. Solve for r.

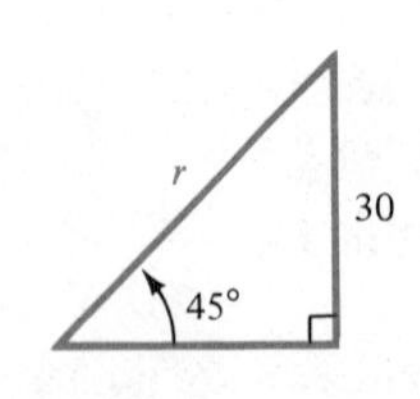

Ⓒ**61.** Solve for r.

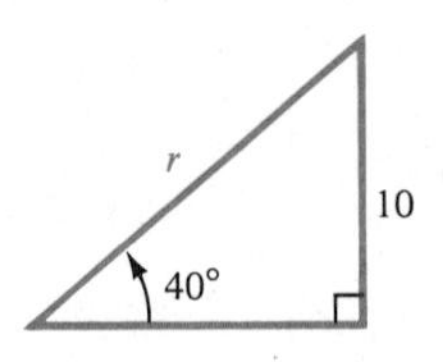

Ⓒ**62.** Solve for x.

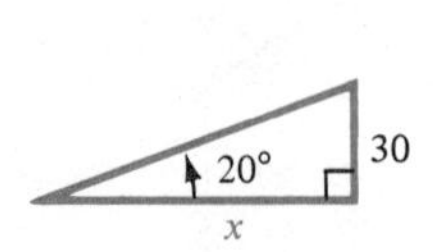

Ⓒ**63.** Solve for y.

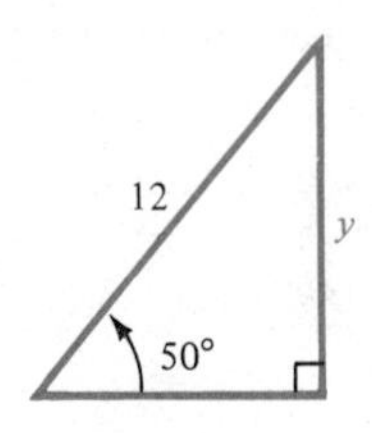

Ⓒ**64.** Solve for r.

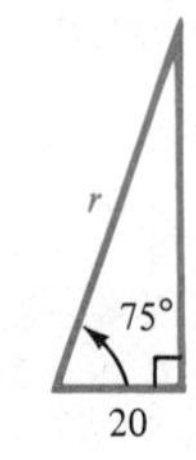

Ⓒ**65.** A 20-ft ladder leaning against the side of a house makes a 70° angle with the ground. (See Figure 10.26.) How far up the side of the house does the ladder reach?

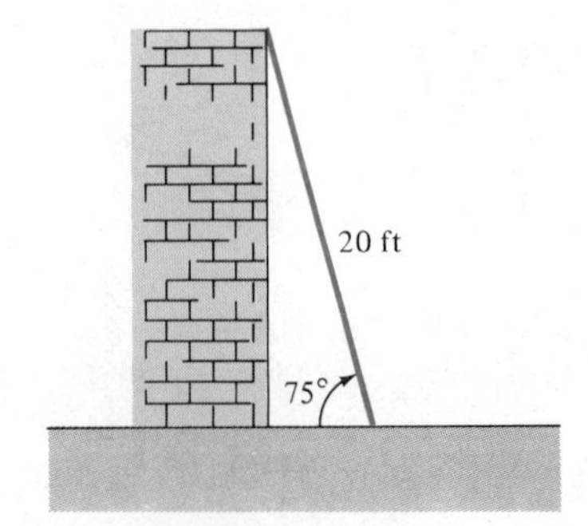

Figure 10.26

Ⓒ**66.** A biologist wants to know the width w of a river in order to properly set instruments to study the pollutants in the water. From point A, he walks downstream 100 ft and sights to point C. From this sighting, he determines that $\theta = 50°$. (See Figure 10.27.) How wide is the river?

Ⓒ Calculator may be helpful.

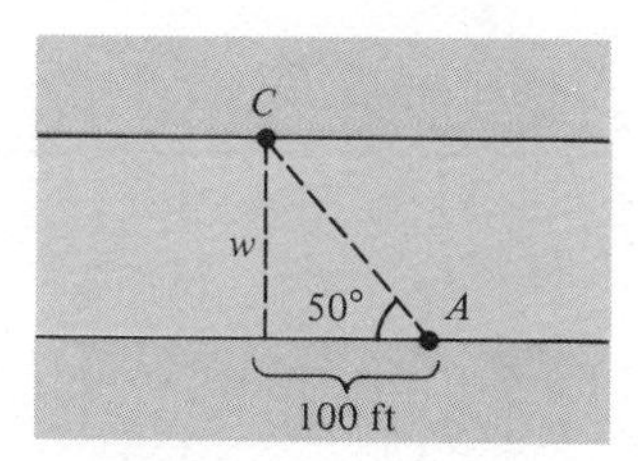

Figure 10.27

Ⓒ**67.** From a 150-ft observation tower on the coast, a Coast Guard officer sights a boat in difficulty. The angle of depression of the boat is 4°. (See Figure 10.28.) How far is the boat from the shoreline?

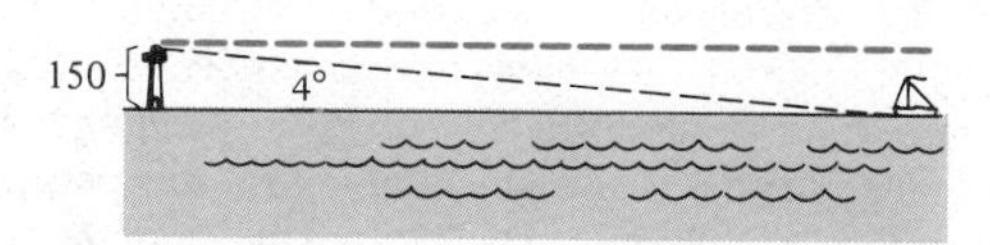

Figure 10.28

Ⓒ**68.** A ramp $17\frac{1}{2}$ ft in length rises to a loading platform that is $3\frac{1}{3}$ ft off the ground. (See Figure 10.29.)

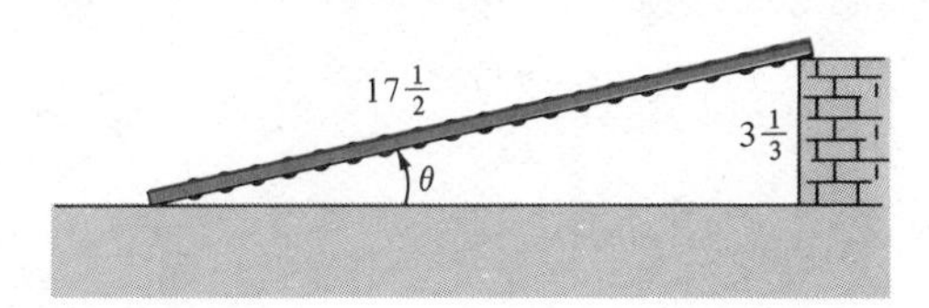

Figure 10.29

Find the angle that the ramp makes with the ground. (Hint: Find the sine of θ and then use the table of trigonometric values in the Appendix to estimate θ.)

Ⓒ**69.** The average daily temperature (in degrees Fahrenheit) for a certain city is given by

$$T = 45 - 23 \cos\left[\frac{2\pi}{365}(t - 32)\right]$$

where t is the time in days with $t = 1$ corresponding to January 1. Find the average temperature on the following days.
(a) January 1
(b) July 4 ($t = 185$)
(c) October 18 ($t = 291$)

Ⓒ**70.** A company that produces a seasonal product forecasts monthly sales over the next two years to be

$$S = 23.1 + 0.442t + 4.3 \sin\left(\frac{\pi t}{6}\right)$$

where S is measured in thousands of units and t is the time in months, with $t = 1$ representing January 1984. Predict the sales for the following months.
(a) February 1984 (b) February 1985
(c) September 1984 (d) September 1985

Ⓒ Calculator may be helpful.

Section 10.3

Graphs of Trigonometric Functions

Introductory Example

Biorythms

A popular theory that attempts to explain the ups and downs of everyday life states that each of us has three cycles, which begin at birth. These three *biorythm cycles* have different periods.

At birth these cycles are at their average energy levels and begin increasing until they reach their peak energy levels during the 6th, 7th, and 8th days of life. Thereafter, the cycles begin dropping until they reach their lowest energy levels from the 16th through the 25th days of life. As the cycles move more out of phase, a person will have days in which one cycle is up while the other two are down, days in which two cycles are up and one is down, and so on.

The energy levels for each cycle can be represented as a sine function:

$$\text{physical (23 days):} \quad P(t) = \sin \frac{2\pi t}{23}$$

$$\text{emotional (28 days):} \quad E(t) = \sin \frac{2\pi t}{28}$$

$$\text{intellectual (33 days):} \quad I(t) = \sin \frac{2\pi t}{33}$$

where t represents the number of days since birth. For example, Figure 10.30 shows the September 1984 biorythm **graphs** for a person born on 20 July 1964. Note that September has a dismal beginning for this person, but by the 18th through the 20th of the month, the person's cycles are all at high energy levels. To compute the biorythm energy levels for any particular day, we need to know that day's t-value. For instance, 1 September 1984 is the 7349th day of this person's life, and we can calculate the energy levels as follows:

$$P(7349) = \sin \frac{2\pi(7349)}{23} \approx -0.14$$

$$E(7349) = \sin \frac{2\pi(7349)}{28} \approx 0.22$$

$$I(7349) = \sin \frac{2\pi(7349)}{33} \approx -0.94$$

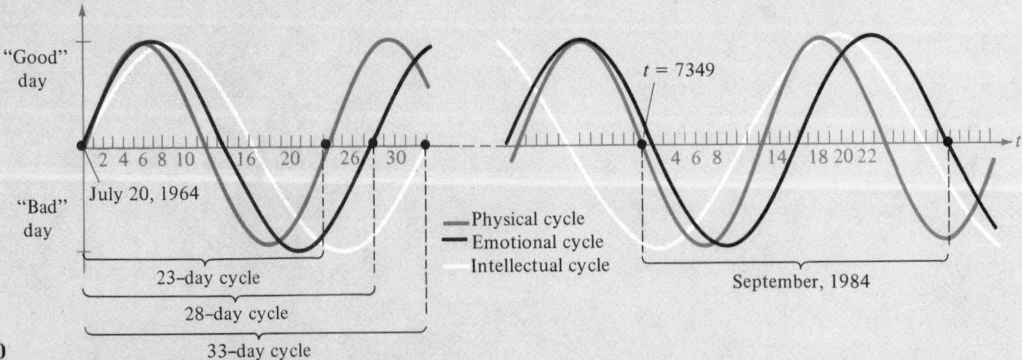

Figure 10.30

Section Objective: ***To sketch graphs of trigonometric functions.***

In the beginning of Section 10.2, we mentioned that there are two common approaches to the study of trigonometry: the right triangle approach and the periodic function approach. In Section 10.2, we stressed the right triangle approach, and we followed the standard convention of using θ to represent one of the acute angles of a right triangle. In this section, we look at the second approach, which stresses the periodic nature of the six trigonometric functions. Since we are emphasizing the functional nature of trigonometry, we use x (rather than θ) to represent the angle and write

$$f(x) = \sin x$$

where x is measured in radians.

We begin by examining the graph of the sine function. Table 10.2 gives the values for the sine for several values of x between 0 and 2π.

Table 10.2

x	0	$\frac{\pi}{6}$	$\frac{\pi}{4}$	$\frac{\pi}{3}$	$\frac{\pi}{2}$	$\frac{4\pi}{3}$	$\frac{3\pi}{4}$	$\frac{5\pi}{6}$	π
$\sin x$	0	$\frac{1}{2}$	$\frac{\sqrt{2}}{2}$	$\frac{\sqrt{3}}{2}$	1	$\frac{\sqrt{3}}{2}$	$\frac{\sqrt{2}}{2}$	$\frac{1}{2}$	0

x	$\frac{7\pi}{6}$	$\frac{5\pi}{4}$	$\frac{4\pi}{3}$	$\frac{3\pi}{2}$	$\frac{5\pi}{3}$	$\frac{7\pi}{4}$	$\frac{11\pi}{6}$	2π
$\sin x$	$-\frac{1}{2}$	$-\frac{\sqrt{2}}{2}$	$-\frac{\sqrt{3}}{2}$	-1	$-\frac{\sqrt{3}}{2}$	$-\frac{\sqrt{2}}{2}$	$-\frac{1}{2}$	0

Now, by plotting these points and connecting them with a smooth curve, we get the graph shown in Figure 10.31.

Note in Figure 10.31 that the maximum value of $\sin x$ is 1 and the minimum value is -1. The **amplitude** of the sine function (or of the cosine function) is defined to be half of the difference between its maximum and minimum values. Thus, the amplitude of $f(x) = \sin x$ is 1. To realize the periodic nature of the sine function, we observe that as x increases beyond 2π, the graph repeats itself

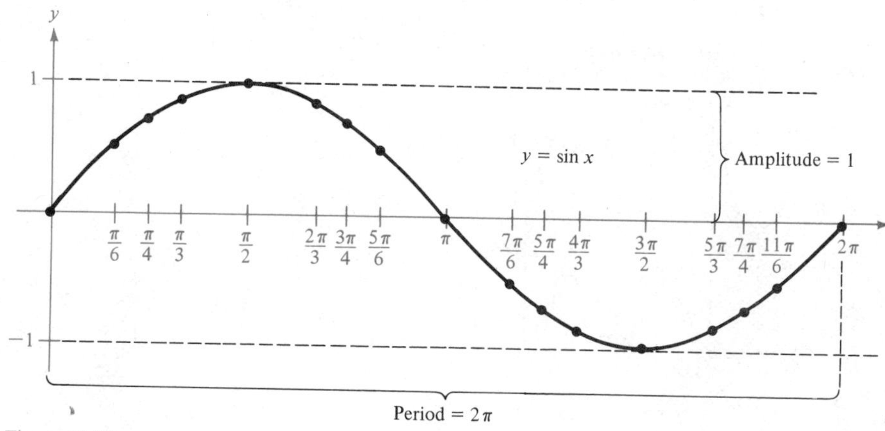

Figure 10.31

over and over, continuously oscillating about the x-axis. The **period** of the function is the distance (measured on the x-axis) between successive cycles. Thus, the period of $f(x) = \sin x$ is 2π.

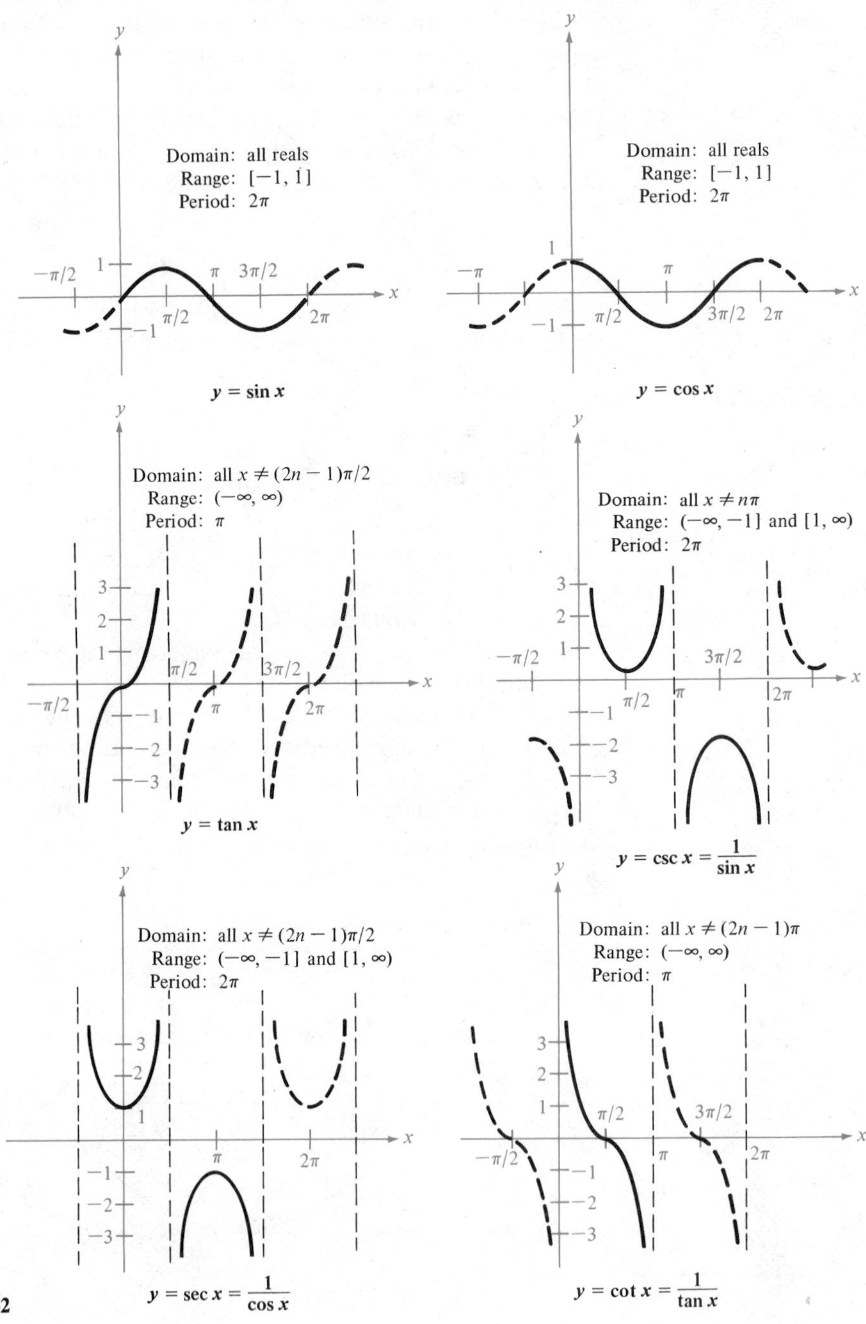

Figure 10.32

Figure 10.32 gives the graphs of at least one cycle of all six trigonometric functions. Study these graphs carefully and try to memorize their basic features.

Familiarity with the graphs of the six basic trigonometric functions enables us to sketch graphs of more general functions, such as

$$y = a \sin bx \qquad \text{or} \qquad y = a \cos bx$$

Note that the function $y = a \sin bx$ oscillates between $-a$ and a and hence has an amplitude of $|a|$. Furthermore, since $bx = 0$ when $x = 0$ and $bx = 2\pi$ when $x = 2\pi/b$, we may conclude that the function $y = a \sin bx$ has a period of $2\pi/|b|$. Table 10.3 summarizes the amplitudes and periods for some of the more general types of trigonometric functions.

Table 10.3

Function	Period	Amplitude
$y = a \sin bx$ or $y = a \cos bx$	$\frac{2\pi}{\|b\|}$	$\|a\|$
$y = a \tan bx$ or $y = a \cot bx$	$\frac{\pi}{\|b\|}$	—
$y = a \sec bx$ or $y = a \csc bx$	$\frac{2\pi}{\|b\|}$	—

Example 1

Sketching the Graph of a Trigonometric Function

Sketch the graph of $f(x) = 3 \cos 2x$.

Solution The graph of $f(x) = 3 \cos 2x$ has the following characteristics:

amplitude: 3

period: $\frac{2\pi}{2} = \pi$

Several cycles of the graph are shown in Figure 10.33, starting with the maximum point (0, 3).

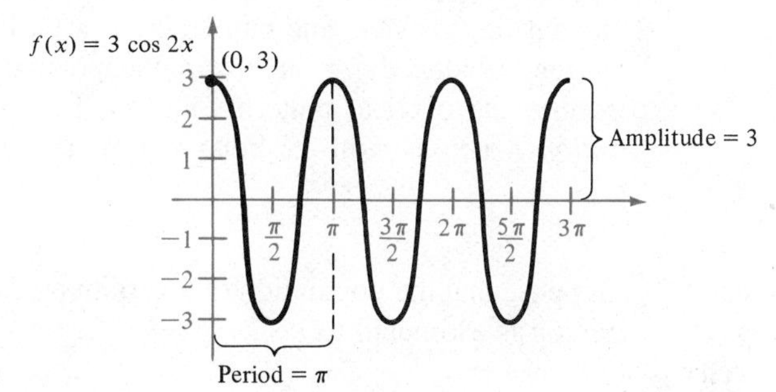

Figure 10.33

□

Example 2

Sketching the Graph of a Trigonometric Function

Sketch the graph of $f(x) = -2 \tan 3x$.

Solution The graph of this function has a period of $\pi/3$. The vertical asymptotes of this particular tangent function occur at

$$x = \cdots \underbrace{-\frac{\pi}{6}, \frac{\pi}{6}}_{\text{period} = \frac{\pi}{3}}, \frac{\pi}{2}, \frac{5\pi}{6}, \cdots$$

Several cycles of the graph are shown in Figure 10.34, starting with the vertical asymptote $x = -\pi/6$.

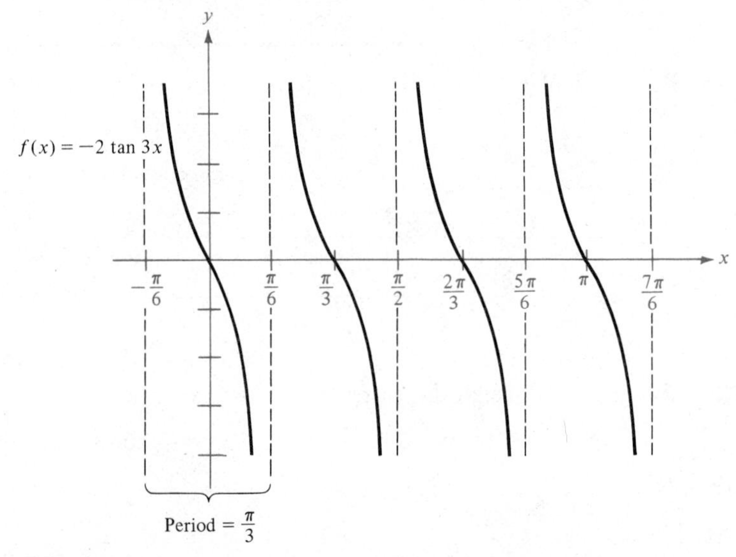

Figure 10.34

□

There are many examples of periodic phenomena in both business and biology. Many businesses have cyclical sales patterns; plant growth is affected by the day/night cycle; and human brain waves can be measured by the oscillating current produced by an electroencephalograph. The following example describes the cyclical pattern followed by many types of *predator-prey populations,* such as wolf-caribou, owl-mouse, and bird-insect.

Example 3

An Example of Predator-Prey Population Cycles

Suppose that the population of a certain predator at time t (in months) in a given region is estimated to be

$$P(t) = 10{,}000 + 3{,}000 \sin\left(\frac{2\pi t}{24}\right)$$

and the population of its primary food source (its prey) is estimated to be

$$p(t) = 15{,}000 + 5{,}000 \cos\left(\frac{2\pi t}{24}\right)$$

Sketch both of these functions on the same graph and explain the oscillations in the size of each population.

Solution Both of these functions have a period of

$$\frac{2\pi}{2\pi/24} = 24 \text{ months}$$

The predator's population has an amplitude of 3,000 and oscillates about the line $y = 10{,}000$, whereas the prey's population has an amplitude of 5,000 and oscillates about the line $y = 15{,}000$. The graphs of these two functions are given in Figure 10.35.

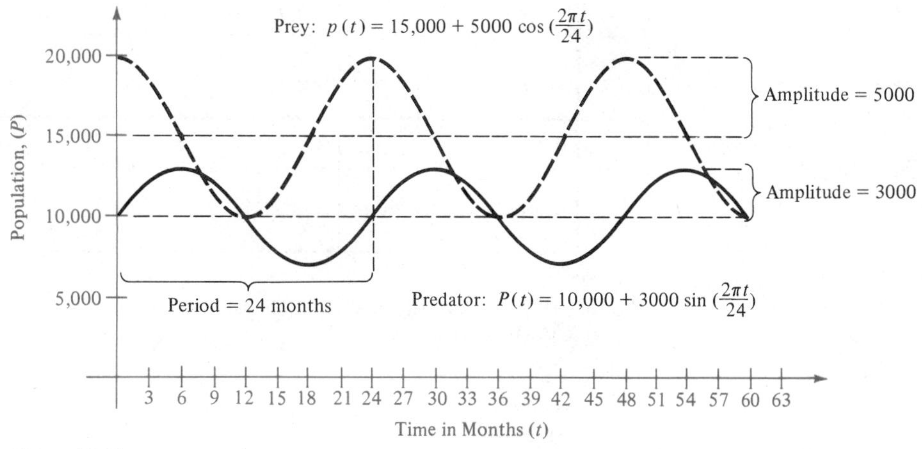

Figure 10.35

We can explain the cycles of this predator-prey population by noting the following cause and effect pattern:

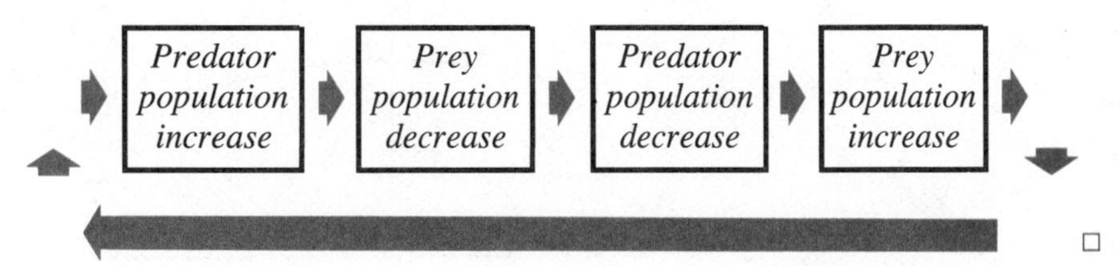

□

Our last example in this section examines the graph of a function that we will encounter again in Section 10.4.

Example 4

Using a Calculator to Graph a Trigonometric Function

Use a calculator to evaluate the function

$$f(x) = \frac{\sin x}{x}$$

at several points in the interval $[-1, 1]$ and then use these points to sketch the function's graph. From your graph, estimate the limit

$$\lim_{x\to 0} \frac{\sin x}{x}$$

Solution Table 10.4 shows the computations for intervals of 0.1 from -1 to 1. Note that the function is undefined when $x = 0$.

Table 10.4

x	-1.0	-0.9	-0.8	-0.7	-0.6	-0.5	-0.4	-0.3	-0.2	-0.1
$\frac{\sin x}{x}$	0.841	0.870	0.897	0.920	0.941	0.959	0.974	0.985	0.993	0.998
x	0.1	0.2	0.3	0.4	0.5	0.6	0.7	0.8	0.9	1.0
$\frac{\sin x}{x}$	0.998	0.993	0.985	0.974	0.959	0.941	0.920	0.897	0.870	0.841

Figure 10.36 shows the result of plotting these points. It appears from the graph that the limit of this function as x approaches 0 (from either side) is 1, and we have

$$\lim_{x\to 0} \frac{\sin x}{x} = 1$$

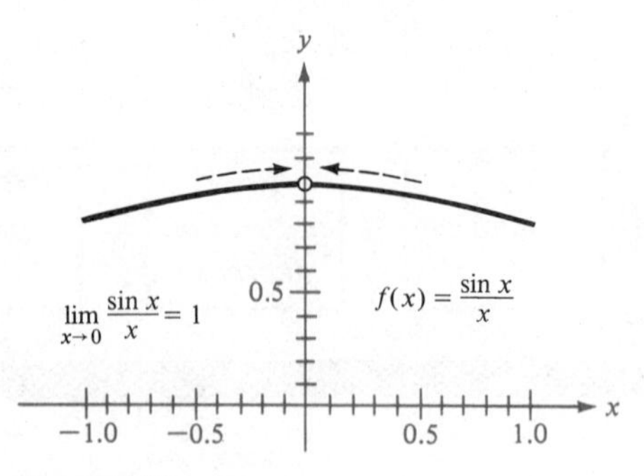

Figure 10.36

□

Section Exercises (10.3)

In Exercises 1–14, determine the period and amplitude of the given function.

1. $y = 2 \sin 2x$

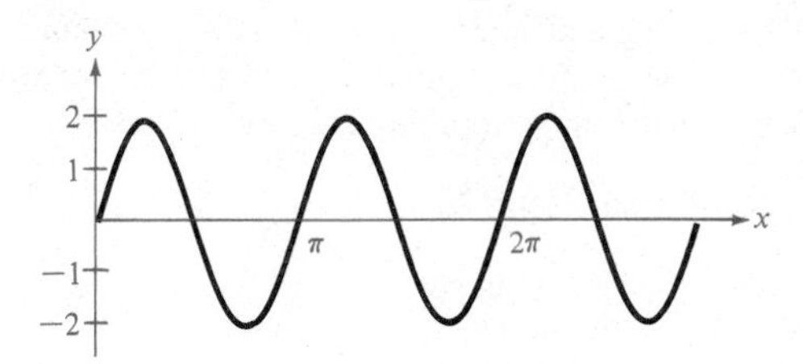

2. $y = 3 \cos 3x$

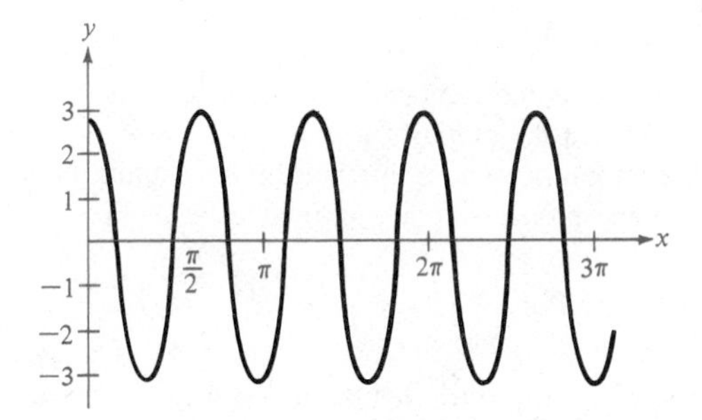

3. $y = \dfrac{3}{2} \cos \dfrac{x}{2}$

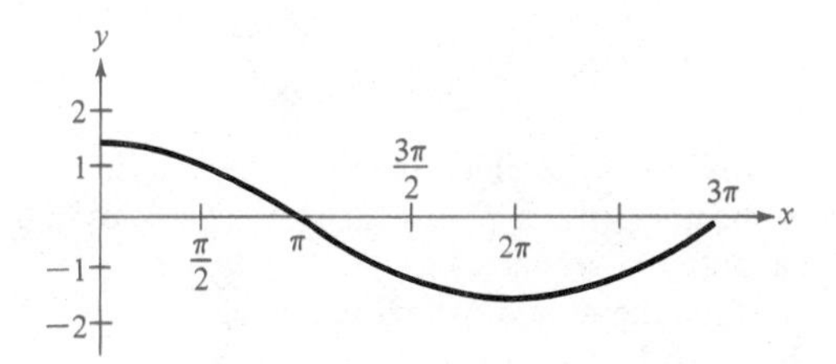

4. $y = -2 \sin \dfrac{x}{3}$

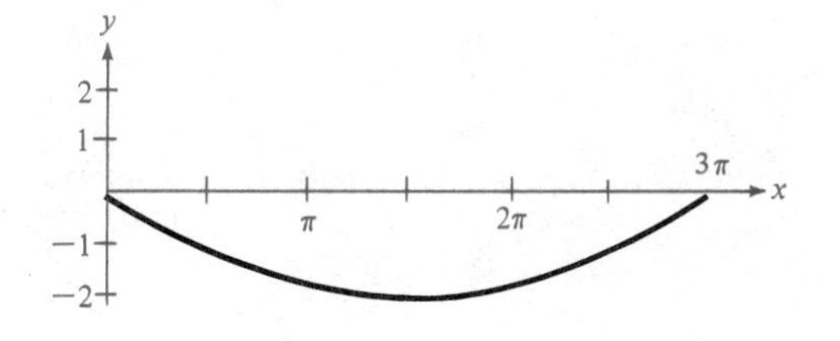

5. $y = \frac{1}{2} \sin \pi x$

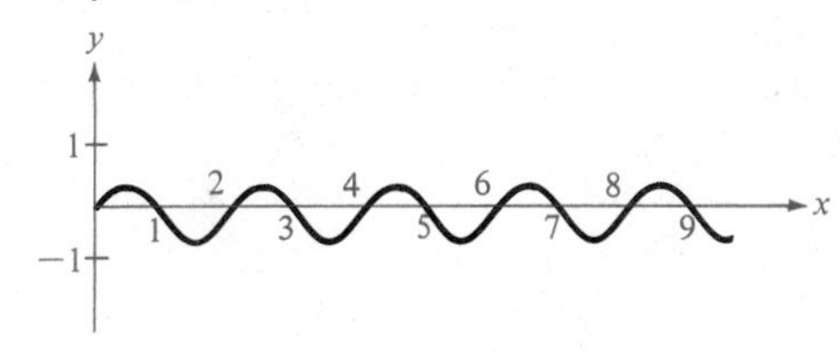

6. $y = \dfrac{5}{2} \cos \dfrac{\pi x}{2}$

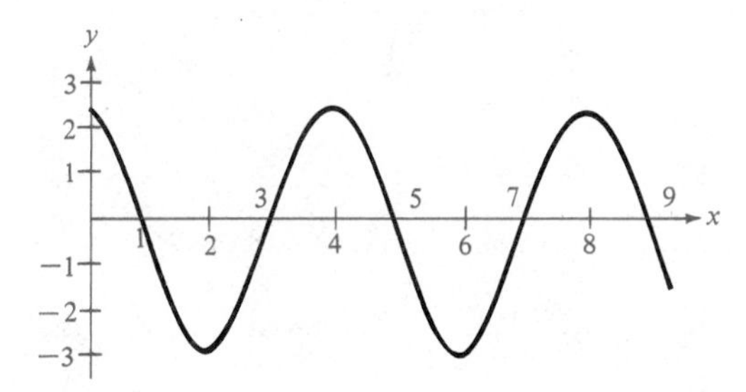

7. $y = 2 \sin x$

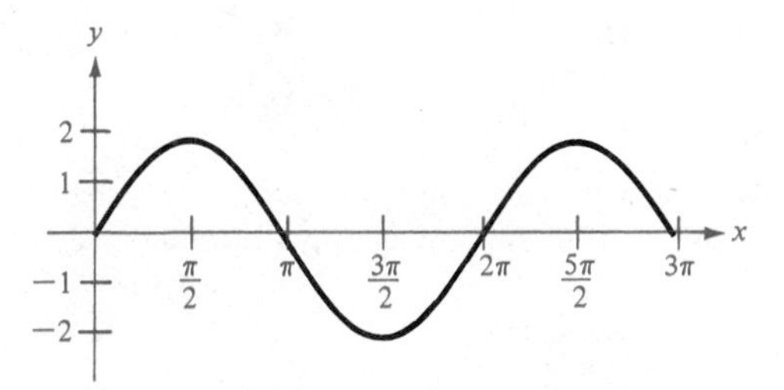

8. $y = -\cos \dfrac{2x}{3}$

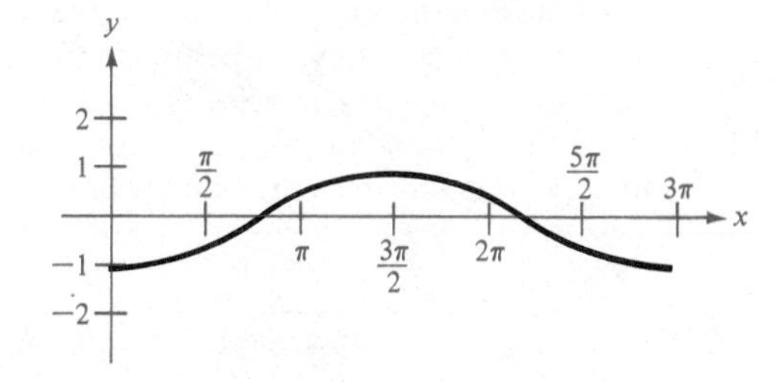

9. $y = -2 \sin 10x$

10. $y = \frac{1}{3} \sin 8x$

11. $y = \frac{1}{2} \cos \frac{2x}{3}$

12. $y = \frac{5}{2} \cos \frac{x}{4}$

13. $y = 3 \sin 4\pi x$

14. $y = \frac{2}{3} \cos \frac{\pi x}{10}$

In Exercises 15–20, find the period of the given function.

15. $y = 5 \tan 2x$

16. $y = \frac{7}{2} \tan 2\pi x$

17. $y = \frac{1}{2} \sec 5x$

18. $y = 4.2 \csc 4x$

19. $y = \cot \frac{\pi x}{3}$

20. $y = 5 \tan \frac{2\pi x}{3}$

In Exercises 21–40, sketch the graph of the given function. (Include two full periods.)

21. $y = \sin \frac{x}{2}$

22. $y = 4 \sin \frac{x}{3}$

23. $y = 2 \cos 2x$

24. $y = \frac{3}{2} \cos \frac{2x}{3}$

25. $y = -2 \sin 6x$

26. $y = -3 \cos 4x$

27. $y = \cos 2\pi x$

28. $y = \frac{3}{2} \sin \frac{\pi x}{4}$

29. $y = -\sin \frac{2\pi x}{3}$

30. $y = 10 \cos \frac{\pi x}{6}$

31. $y = 2 \tan x$

32. $y = 2 \cot x$

33. $y = \csc \frac{x}{2}$

34. $y = 3 \tan \pi x$

35. $y = \tan 2x$

36. $y = \csc \frac{x}{3}$

37. $y = 2 \sec 2x$

38. $y = \sec \pi x$

39. $y = \csc 2\pi x$

40. $y = -\tan x$

ⓒ **41.** For a person at rest, the velocity v (in liters per second) of air flow during a respiratory cycle is

$$v = 0.85 \sin \frac{\pi t}{3}$$

where t is the time in seconds. (Inhalation occurs when $v > 0$ and exhalation occurs when $v < 0$.)
(a) Find the time for one full respiratory cycle.
(b) Find the number of cycles per minute.
(c) Sketch the graph of the velocity function.

ⓒ **42.** After exercising for a few minutes, a person has a respiratory cycle for which the velocity of air flow is approximated by

$$v = 1.75 \sin \frac{\pi t}{2}$$

Use this model to repeat Exercise 41.

ⓒ **43.** When tuning a piano, a technician strikes a tuning fork for the A above middle C and sets up wave motion that can be approximated by

$$y = 0.001 \sin 880\pi t$$

where t is the time in seconds.
(a) What is the period p of this function?
(b) The frequency f is given by

$$f = \frac{1}{p}$$

What is the frequency of this note?
(c) Sketch the graph of this function.

ⓒ **44.** The monthly sales S in thousands of units of a seasonal product are approximated by

$$S = 74.50 + 43.75 \sin \frac{\pi t}{6}$$

where t is the time in months with $t = 1$ corresponding to January. Sketch the graph of this sales function over one year.

ⓒ **45.** The function

$$P = 100 - 20 \cos \frac{5\pi t}{3}$$

approximates the blood pressure P (in millimeters of mercury) for a person at rest. (The time t is measured in seconds.)
(a) Find the period of this function.
(b) Find the number of heartbeats per minute for this individual.
(c) Sketch the graph of this pressure function.

ⓒ **46.** Complete the following table for the function

$$f(x) = \frac{\sin 2x}{\sin 3x}$$

Use the table to graph this function in the interval

ⓒ Calculator may be helpful.

[−1, 1] and estimate the limit

$$\lim_{x \to 0} \frac{\sin 2x}{\sin 3x}$$

x	−1.0	−0.8	−0.6	−0.4	−0.2	−0.1	−0.01	−0.001
$\frac{\sin 2x}{\sin 3x}$								
x	1.0	0.8	0.6	0.4	0.2	0.1	0.01	0.001
$\frac{\sin 2x}{\sin 3x}$								

Ⓒ**47.** Repeat Exercise 46 using the function

$$f(x) = \frac{1 - \cos x}{x}$$

Ⓒ Calculator may be helpful.

Section 10.4

Derivatives of Trigonometric Functions

Introductory Example

The Construction of a Honeycomb

A honeycomb is constructed of two offset rows of hexagonal cells that fit together perfectly, with no spaces between the cells. Each cell has a hexagonal base and three rhombic upper faces which meet the altitude of the cell at an angle of θ. (See Figures 10.37 and 10.38.)

Important geometrical properties of a honeycomb are:

1. As long as each cell in the honeycomb has the *same angle* θ, the cells will fit together so that each interior wall serves two cells.
2. The volume of each cell is

$$\text{volume} = V = \frac{3\sqrt{3}}{2}s^2h$$

3. The surface area of each cell is

$$\text{surface area} = S = 6hs + \frac{3s^2}{2}\left(\frac{\sqrt{3} - \cos\theta}{\sin\theta}\right)$$

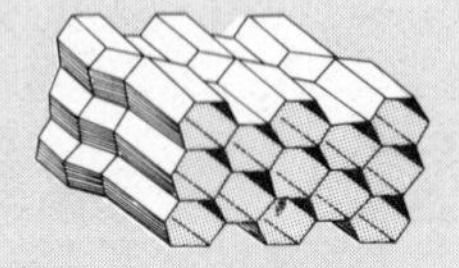

Figure 10.37

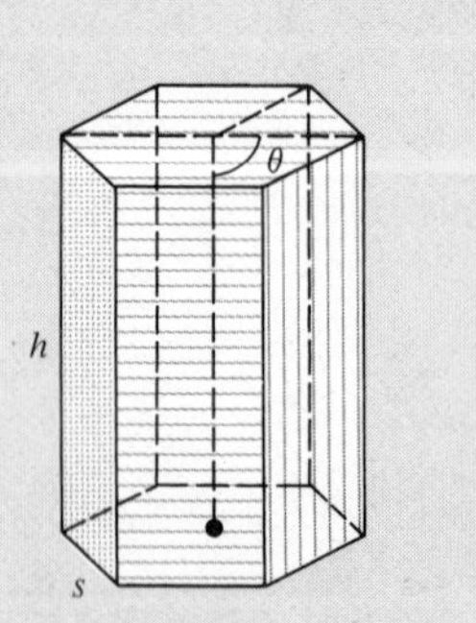

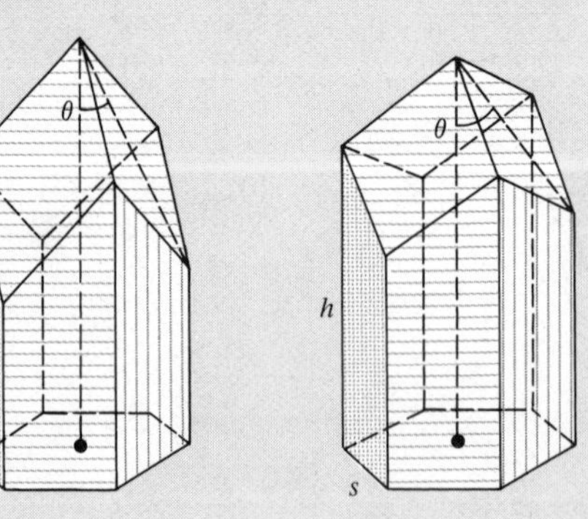

Figure 10.38

From these properties, we note the value of θ does not affect the fit or the volume *but* does affect the surface area. Table 10.5 lists the surface area for several values of θ. It appears that we can minimize the surface area (and consequently minimize the amount of wax needed to build the honeycomb) by choosing values of θ between 50° and 60°. Using the differentiation formulas in this section, we will see that we can minimize the surface area by taking its **derivative** with respect to θ:

$$\frac{dS}{d\theta} = \frac{3s^2}{2}\left[\frac{\sin^2\theta - (\sqrt{3} - \cos\theta)\cos\theta}{\sin^2\theta}\right]$$

$$= \frac{3s^2}{2}\left(\frac{1 - \sqrt{3}\cos\theta}{\sin^2\theta}\right)$$

Finally, by setting $dS/d\theta = 0$, we have

$$\cos\theta = \frac{1}{\sqrt{3}} \approx 0.57735$$

The value of θ that satisfies this equation is $\theta \approx 54.74°$. It is not surprising that this value of θ is the one actually used by honeybees.

Table 10.5

θ	Surface Area
90°	$6hs + 2.598s^2$
80°	$6hs + 2.374s^2$
70°	$6hs + 2.219s^2$
60°	$6hs + 2.134s^2$
50°	$6hs + 2.133s^2$
40°	$6hs + 2.254s^2$
30°	$6hs + 2.598s^2$
20°	$6hs + 3.475s^2$
10°	$6hs + 6.455s^2$

Section Objective: ***To introduce differentiation formulas for the six trigonometric functions and demonstrate their use.***

In the preceding section (Example 4 and Exercise 47), we looked at the following two limits:

$$\lim_{\Delta x \to 0} \frac{\sin \Delta x}{\Delta x} = 1 \quad \text{and} \quad \lim_{\Delta x \to 0} \frac{1 - \cos \Delta x}{\Delta x} = 0$$

and

$$\lim_{\Delta x \to 0} \frac{1 - \cos \Delta x}{\Delta x} = 0$$

These two limits turn out to be crucial in the *theoretical development* of the derivative of the sine function.

$$\begin{aligned}
\frac{d}{dx}[\sin x] &= \lim_{\Delta x \to 0} \frac{\sin (x + \Delta x) - \sin x}{\Delta x} \\
&= \lim_{\Delta x \to 0} \frac{\sin x \cos \Delta x + \cos x \sin \Delta x - \sin x}{\Delta x} \\
&= \lim_{\Delta x \to 0} \frac{\cos x \sin \Delta x - \sin x (1 - \cos \Delta x)}{\Delta x} \\
&= \lim_{\Delta x \to 0} \left[\cos x \frac{\sin \Delta x}{\Delta x} - \sin x \frac{1 - \cos \Delta x}{\Delta x}\right] \\
&= \cos x \left[\lim_{\Delta x \to 0} \frac{\sin \Delta x}{\Delta x}\right] - \sin x \left[\lim_{\Delta x \to 0} \frac{1 - \cos \Delta x}{\Delta x}\right] \\
&= (\cos x)(1) - (\sin x)(0) \\
&= \cos x
\end{aligned}$$

This differentiation formula is shown graphically in Figure 10.39. Note that the *slope* of the sine curve determines the *value* of the cosine curve.

Recall from our previous work with derivatives that we usually listed two versions of each differentiation rule: a simple version and a general (Chain Rule) version. The versions of the sine rule are

Simple Version	*Chain Rule Version*
$\frac{d}{dx}[\sin x] = \cos x$	$\frac{d}{dx}[\sin u] = \cos u \frac{du}{dx}$

We omit the development of the other five differentiation rules for trigonometric functions and simply summarize the results as follows.

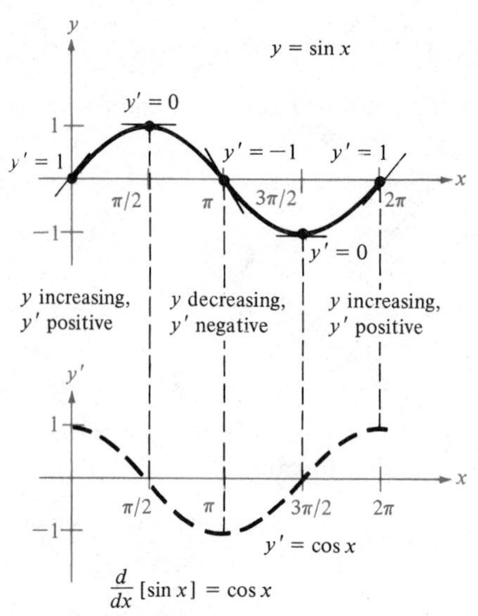

Figure 10.39

Derivatives of Trigonometric Functions

$$\frac{d}{dx}[\sin u] = \cos u \frac{du}{dx} \qquad \frac{d}{dx}[\cos u] = -\sin u \frac{du}{dx}$$

$$\frac{d}{dx}[\tan u] = \sec^2 u \frac{du}{dx} \qquad \frac{d}{dx}[\cot u] = -\csc^2 u \frac{du}{dx}$$

$$\frac{d}{dx}[\sec u] = \sec u \tan u \frac{du}{dx} \qquad \frac{d}{dx}[\csc u] = -\csc u \cot u \frac{du}{dx}$$

Remark As an aid to memorization, note that the co-functions (cosine, cotangent, and cosecant) require a negative sign as part of their derivatives.

Example 1

Differentiating Trigonometric Functions

Differentiate the following trigonometric functions.

(a) $y = \sin 2x$

(b) $y = \cos (x - 1)$

(c) $y = \tan 3x$

Solution

(a) Considering $u = 2x$, we have

$$y' = \cos u \frac{du}{dx} = \cos 2x \frac{d}{dx}[2x]$$

$$= (\cos 2x)(2)$$

$$= 2 \cos 2x$$

(b) Considering $u = x - 1$, we see that $du/dx = 1$, and thus the derivative is simply

$$y' = -\sin (x - 1)$$

(c) Considering $u = 3x$, we have $du/dx = 3$, and it follows that

$$y' = 3 \sec^2 3x$$

□

Example 2

Differentiating Trigonometric Functions

Differentiate $y = \cos 3x^2$.

Solution We consider $u = 3x^2$, then

$$y' = -\sin u \frac{du}{dx} = -\sin 3x^2 \frac{d}{dx}[3x^2]$$

$$= -(\sin 3x^2)(6x)$$

$$= -6x \sin 3x^2$$

□

Example 3

Differentiating Trigonometric Functions

Differentiate $y = \tan^4 3x$.

Solution By the Power Rule, we have

$$\frac{d}{dx}\left[(\tan 3x)^4\right] = 4(\tan 3x)^3 \frac{d}{dx}[\tan 3x]$$

$$= 4(\tan^3 3x)(3)(\sec^2 3x)$$

$$= 12 \tan^3 3x \sec^2 3x$$

□

Example 4

Differentiating Trigonometric Functions

Differentiate $y = \csc \frac{x}{2}$

Solution

$$y' = -\csc \frac{x}{2} \cot \frac{x}{2} \frac{d}{dx}\left[\frac{x}{2}\right]$$

$$= -\frac{1}{2} \csc \frac{x}{2} \cot \frac{x}{2}$$

□

Example 5

Differentiating Trigonometric Functions

Differentiate $f(t) = \sqrt{\sin 4t}$.

Solution By the Power Rule, we have

$$f(t) = (\sin 4t)^{1/2}$$

$$f'(t) = \left(\frac{1}{2}\right)(\sin 4t)^{-1/2}\frac{d}{dx}[\sin 4t]$$

$$= \left(\frac{1}{2}\right)(\sin 4t)^{-1/2}(4 \cos 4t)$$

$$= \frac{2 \cos 4t}{\sqrt{\sin 4t}}$$

□

Example 6

Differentiating Trigonometric Functions

Differentiate $y = x \sin x$.

Solution By the Product Rule, we have

$$\frac{dy}{dx} = x\frac{d}{dx}[\sin x] + \sin x \frac{d}{dx}[x]$$

$$= x \cos x + \sin x$$

□

Example 7

Finding Relative Extrema for Trigonometric Functions

Determine the relative extrema of the graph of

$$y = \frac{1 + \cos x}{1 + \sin x}$$

on the interval $(-\pi/2, 3\pi/2)$.

Solution By the Quotient Rule, we have

$$y' = \frac{(1 + \sin x)(-\sin x) - (1 + \cos x)(\cos x)}{(1 + \sin x)^2}$$

$$= \frac{-\sin x - \cos x - (\sin^2 x + \cos^2 x)}{(1 + \sin x)^2}$$

$$= \frac{-1 - \sin x - \cos x}{(1 + \sin x)^2}$$

Now, setting this derivative equal to zero, we have

$$0 = -1 - \sin x - \cos x$$

$$-1 = \sin x + \cos x$$

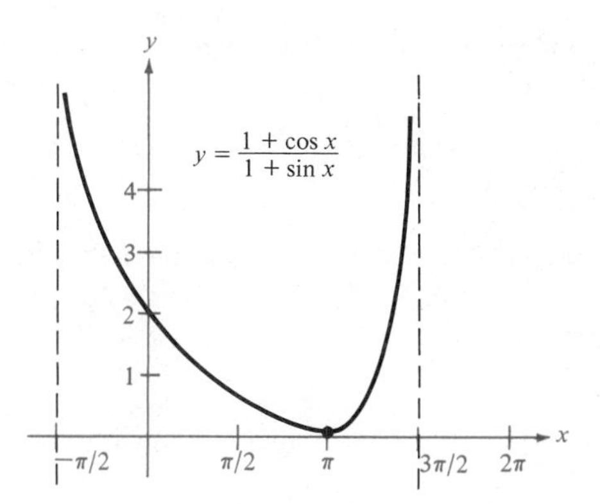

Figure 10.40

Finally, on the interval $(-\pi/2, 3\pi/2)$ this equation has only one solution, $x = \pi$. (See Example 7, Section 10.2.) Thus, by the First-Derivative Test, the point $(\pi, 0)$ is a minimum, as shown in Figure 10.40. □

Example 8

Finding Relative Extrema for Trigonometric Functions

Find the relative extrema and sketch the graph of $f(x) = 2 \sin x - \cos 2x$ on the interval $[0, 2\pi]$.

Solution By setting $f'(x)$ equal to zero, we have

$$f'(x) = 2 \cos x + 2 \sin 2x = 0$$

Since $\sin 2x = 2 \cos x \sin x$, we have

$$2 \cos x + 4 \cos x \sin x = 0$$

$$2(\cos x)(1 + 2 \sin x) = 0$$

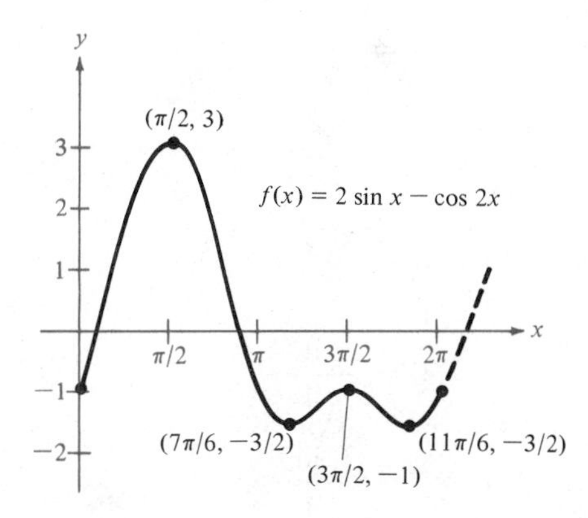

Figure 10.41

This equation has solutions when $\cos x = 0$ or when $\sin x = -\frac{1}{2}$. Thus, $f'(x) = 0$ if

$$x = \frac{\pi}{2}, \frac{3\pi}{2} \qquad \text{or} \qquad x = \frac{7\pi}{6}, \frac{11\pi}{6}$$

By the Second-Derivative Test, we can determine that $(\pi/2, 3)$ and $(3\pi/2, -1)$ are relative maxima, and $(7\pi/6, -3/2)$ and $(11\pi/6, -3/2)$ are relative minima (Figure 10.41). □

Example 9

A Business Application

A fertilizer manufacturer finds that the national sales of fertilizer roughly follow the cyclical pattern

$$F = 100{,}000\left[1 + \sin\frac{2\pi(t - 60)}{365}\right]$$

where F is measured in pounds and t is measured in days, with $t = 1$ representing January 1.* (See Figure 10.42.) On which day of the year is the maximum amount of fertilizer sold?

Solution By taking the derivative, we have

$$\frac{dF}{dt} = 100{,}000\left[\frac{2\pi}{365}\cos\frac{2\pi(t - 60)}{365}\right]$$

Now, by setting this derivative equal to zero, we have

$$\cos\frac{2\pi(t - 60)}{365} = 0$$

Since the cosine is zero at $\pi/2$ and $3\pi/2$, we have

$$\frac{2\pi(t - 60)}{365} = \frac{\pi}{2}$$

$$t - 60 = \frac{365}{4}$$

$$t = \frac{365}{4} + 60 \approx 151 \text{ (May 31)}$$

and

$$\frac{2\pi(t - 60)}{365} = \frac{3\pi}{2}$$

$$t - 60 = \frac{3(365)}{4}$$

$$t = \frac{3(365)}{4} + 60 \approx 334 \text{ (November 30)}$$

*For simplicity's sake, we will follow the convention of saying January 1 is representerd by $t = 1$ even though what we actually mean is that January 1 is represented by the *interval* $[0, 1]$.

From Figure 10.42, we see that the maximum sales occur on May 31 (near the Memorial Day weekend).

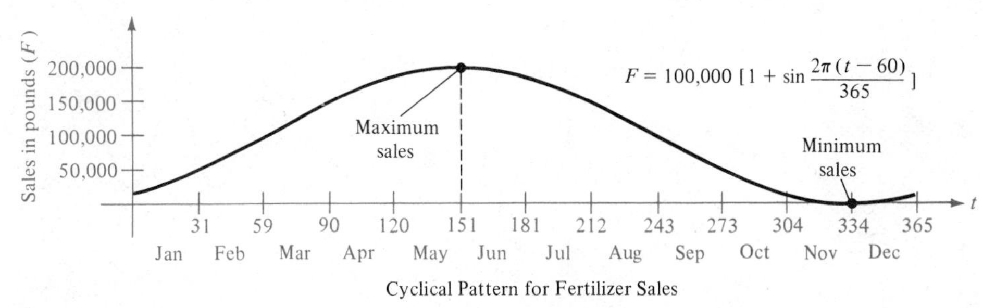

Figure 10.42

□

Example 10

An Application to Temperature Change

The temperature during a given 24-hour period is approximated by the model

$$T = 70 + 15 \sin \frac{\pi(t - 8)}{12}$$

where T is measured in degrees Fahrenheit and t is measured in hours, with $t = 0$ representing midnight. (See Figure 10.43.) Find the rate at which the temperature is changing at 6 A.M.

Solution The rate of change of the temperature is given by the derivative

$$\frac{dT}{dt} = \frac{15\pi}{12} \cos \frac{\pi(t - 8)}{12}$$

Since 6 A.M. corresponds to $t = 6$, the rate of change at 6 A.M. is

$$\frac{dT}{dt} = \frac{15\pi}{12} \cos \left(\frac{-2\pi}{12}\right) = \frac{5\pi}{4} \cos \left(-\frac{\pi}{6}\right)$$

$$= \frac{5\pi}{4}\left(\frac{\sqrt{3}}{2}\right) \approx 3.4°/h$$

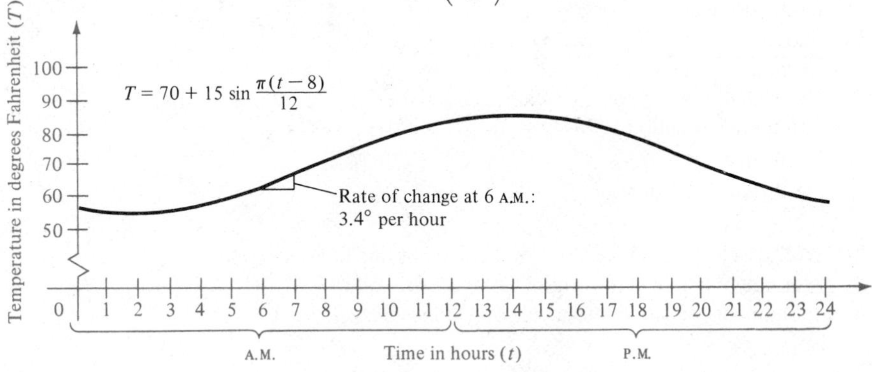

Figure 10.43

□

Section Exercises (10.4)

In Exercises 1–20, find dy/dx.

1. $y = \sin 4x$
2. $y = \cos 3x$
3. $y = 2 \cos \dfrac{x}{2}$
4. $y = 3 \tan 4x$
5. $y = \sec (x^2)$
6. $y = \dfrac{1}{\sec x}$
7. $y = x \cos x$
8. $y = \sin \pi x$
9. $y = \cos \pi x$
10. $y = \sec^3 2x$
11. $y = \dfrac{\cos x + 1}{x}$
12. $y = \dfrac{1 + \sin x}{1 - \sin x}$
13. $y = \frac{1}{2} \csc 2x$
14. $y = \csc^2 x$
15. $y = \tan \dfrac{\pi(x - 1)}{2}$
16. $y = \cot \dfrac{x + \pi}{4}$
17. $y = x \sin \dfrac{1}{x}$
18. $y = x^2 \sin \dfrac{1}{x}$
19. $y = \tan^2 e^x$
20. $y = \sin e^x$

In Exercises 21–36, find dy/dx and then simplify your answer by using the trigonometric identities listed in Section 10.2.

21. $y = \cos^2 x$
22. $y = \frac{1}{4} \sin^2 2x$
23. $y = \cos^2 x - \sin^2 x$
24. $y = \sin x \cos x$
25. $y = \dfrac{\cos x}{\sin x}$
26. $y = \dfrac{x}{2} + \dfrac{\sin 2x}{4}$
27. $y = \ln (\cot x)$
28. $y = \ln (\tan x)$
29. $y = \tan x - x$
30. $y = \dfrac{\sec^7 x}{7} - \dfrac{\sec^5 x}{5}$
31. $y = \ln (\csc x - \cot x)$
32. $y = \ln (\sec x + \tan x)$
33. $y = \frac{1}{2}(x \tan x - \sec x)$
34. $y = \frac{2}{3} \sin^{3/2} x - \frac{2}{7} \sin^{7/2} x$
35. $y = \sin 3x \cos^2 3x$
36. $y = \ln (\sin^2 x)$
37. Show that $y = 2 \sin x + 3 \cos x$ satisfies the differential equation $y'' + y = 0$.
38. Show that $y = (10 - \cos x)/x$ satisfies the differential equation $xy' + y = \sin x$.
39. Show that $y = e^x(\cos 2x + \sin 2x)$ satisfies the differential equation $y'' + 4y = 0$.
40. Show that $y = e^x(\cos \sqrt{2}x + \sin \sqrt{2}x)$ satisfies the differential equation $y'' - 2y' + 3y = 0$.

In Exercises 41–46, find the slope of the tangent line to the given sine function at the origin. Compare this value to the number of complete cycles in the interval $[0, 2\pi]$.

41. $y = \sin 3x$

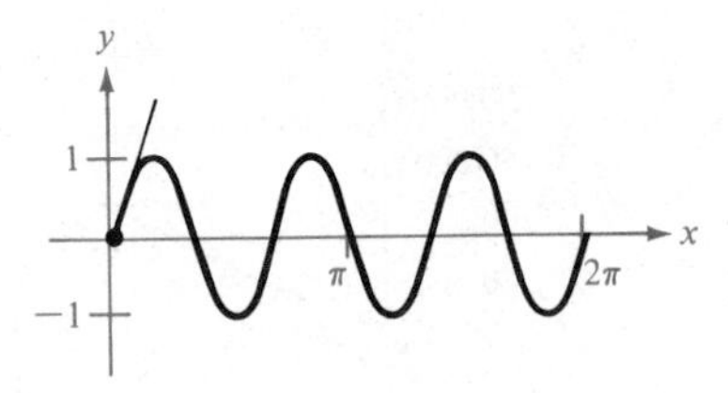

42. $y = \sin \dfrac{5x}{2}$

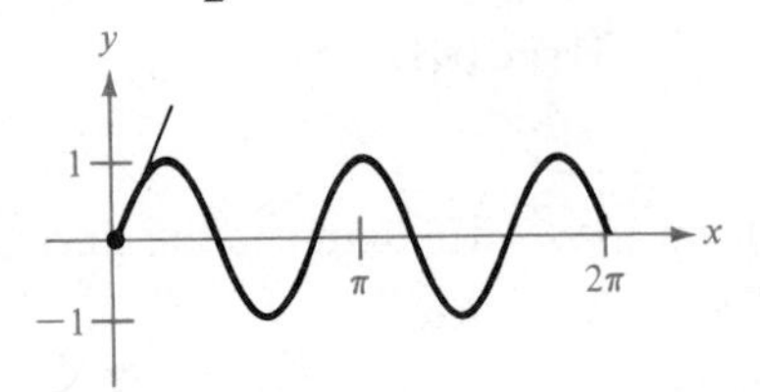

43. $y = \sin 2x$

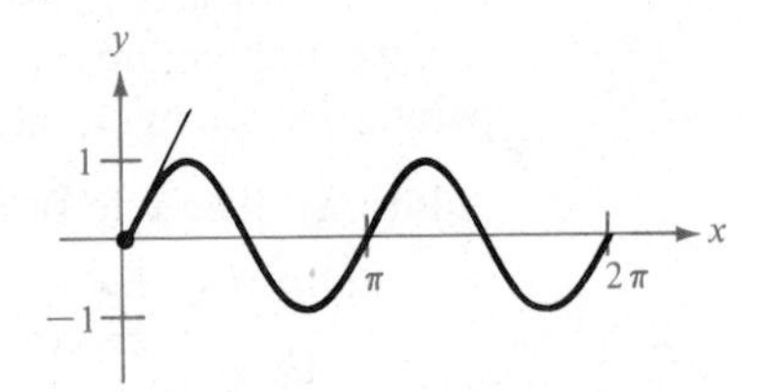

44. $y = \sin \dfrac{3x}{2}$

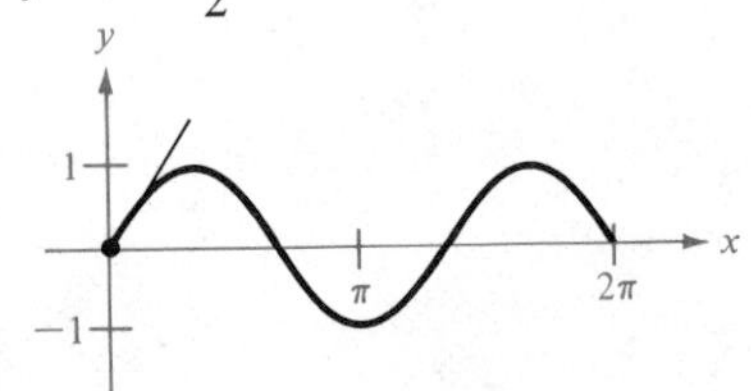

45. $y = \sin x$

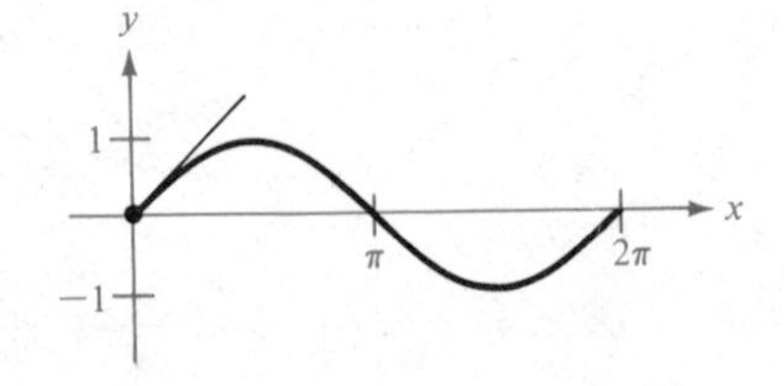

46. $y = \sin \frac{x}{2}$

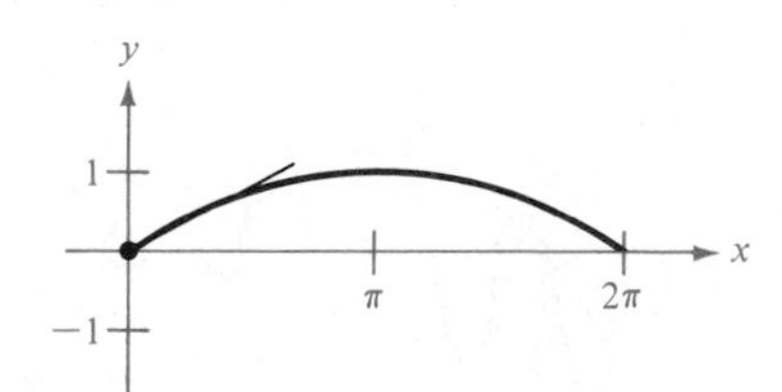

In Exercises 47–50, find the relative extrema and sketch the graph of the given function on the indicated interval.

47. $f(x) = 2 \sin x + \sin 2x$; $[0, 2\pi]$

48. $f(x) = 2 \sin x + \cos 2x$; $[0, 2\pi]$

49. $f(x) = x - 2 \sin x$; $[0, 4\pi]$

50. $f(x) = e^{-x} \sin x$; $[0, 2\pi]$

51. Find the Taylor series (centered at $c = 0$) for the function

$$f(x) = \sin x$$

What is the radius of convergence for this series?

52. Find the Taylor series (centered at $c = 0$) for the function

$$f(x) = \cos x$$

What is the radius of convergence for this series?

53. Use the series in Exercise 51 to find the Taylor series (centered at $c = 0$) for the function

$$f(x) = \begin{cases} \dfrac{\sin x}{x}, & x \neq 0 \\ 1, & x = 0 \end{cases}$$

What is the radius of convergence for this series?

54. Use the series in Exercise 52 to find the Taylor series (centered at $c = 0$) for the function

$$f(x) = \cos (x^2)$$

What is the radius of convergence for this series?

Ⓒ **55.** The normal average daily temperature (in degrees Fahrenheit) for a certain city is given by

$$T = 45 - 23 \cos \frac{2\pi(t - 32)}{365}$$

where t is the time in days with $t = 1$ corresponding to January 1. Find the expected date of:

(a) the warmest day

(b) the coldest day

Ⓒ **56.** Domestic energy consumption in the United States is seasonal and in 1979 can be approximated by the model

$$Q = 6.9 + \cos \frac{\pi(2t - 1)}{12}$$

where Q is the total consumption in quads (quadrillion BTUs) and t is the time in months, with $t = 1$ corresponding to January. (Remember that since t is measured in months, January is actually represented by the *interval* $[0, 1]$, with 0 representing the first of the month, 0.5 representing the middle of the month, and 1 representing the end of the month.) (See Figure 10.44.)

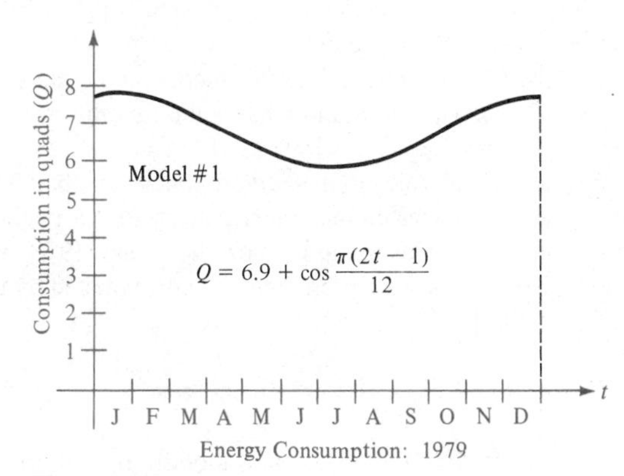

Figure 10.44

(a) On which day does this model predict the greatest consumption, and what is the monthly rate for that day?

(b) On which day does this model predict the least consumption, and what is the monthly rate for that day?

Ⓒ **57.** The model for energy consumption in Exercise 56 does not account for the increase in energy consumption due to air conditioning during the summer months. A model that does account for this additional use of energy is given by

$$Q = 6.9 + \frac{2}{3} \sin \frac{\pi(2t + 5)}{12} - \frac{1}{3} \cos \frac{\pi(2t + 5)}{6}$$

(See Figure 10.45.)

Ⓒ Calculator may be helpful.

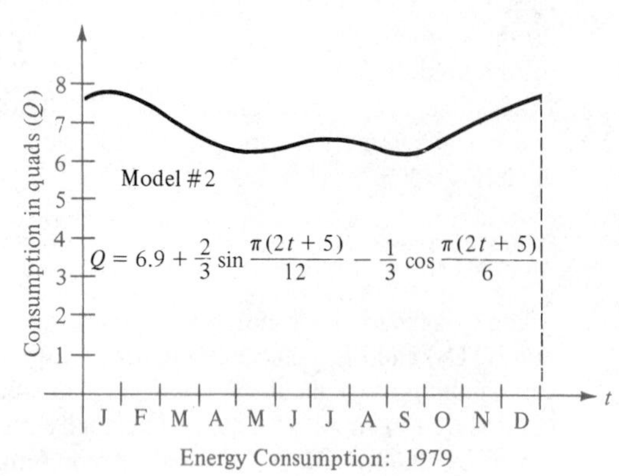

Figure 10.45

(a) On which day does this model predict the greatest consumption, and what is the monthly rate for that day?

(b) On which days does this model predict the least consumption, and what is the monthly rate for these days?

Ⓒ**58.** Electricity sales in the United States have had both an increasing annual sales pattern *and* a cyclical monthly sales pattern. For the years 1978 and 1979, the sales pattern can be approximated by the model

$$S = 164.68 + 0.56t + 13.60 \cos \frac{\pi t}{3}$$

where S is the sales (per month) in billions of kilowatt hours and t is the time in months, with $t = 1$ corresponding to January, 1978. (See Figure 10.46.)

(a) Find the relative extrema of this function for the years 1978 and 1979.

(b) Use this model to predict the sales in August 1980. (Use $t = 31.5$.)

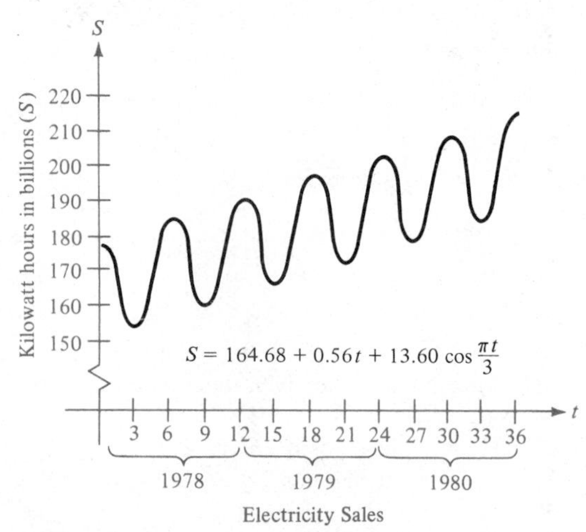

Figure 10.46

Ⓒ**59.** Plants do not grow at constant rates during a normal 24-h period since their growth is affected by sunlight. Suppose that the growth of a certain plant species in a controlled environment is given by the model

$$h = 0.2t + 0.03 \sin 2\pi t$$

where h is the height of the plant in inches and t is the time in hours, with $t = 0$ corresponding to midnight. (See Figure 10.47.)

(a) During what time of day is the *rate of growth* of this plant the greatest?

(b) During what time of day is the *rate of growth* of this plant the least?

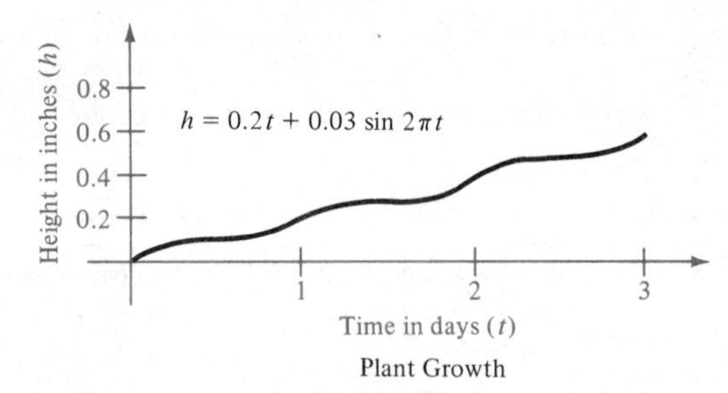

Figure 10.47

Ⓒ Calculator may be helpful.

Section 10.5

Integrals of Trigonometric Functions

Introductory Example

Probability of a Needle Touching a Line

If a 2-in needle is tossed randomly onto a floor ruled with parallel lines that are 2 in apart, what is the probability that the needle will touch one of the lines? At first glance, this problem may seem to be very difficult. However, by utilizing trigonometry and integration, the solution becomes quite manageable.

We begin by letting θ represent the angle between the needle and the lines on the floor. Without loss of generality, we can assume $0 \leq \theta \leq \pi/2$. Furthermore, we assume that each possible value of θ is equally likely. Now, for a given value of θ, the needle will touch one of the lines if its center falls within $\sin\theta$ inches of one of the lines. (See Figure 10.48). In other words, there is a "touching region" of width $2\sin\theta$ inches between each pair of lines on the floor. Since the distance between each pair of lines is 2 in, the probability (for a given θ) of touching one of the lines is

$$\text{Probability of touching} = \frac{\text{Width of "touching region"}}{\text{distance between lines}} = \frac{2\sin\theta}{2} = \sin\theta$$

(Note that for $\theta = 0$ the only way the needle could touch a line would be to fall exactly on one of the lines, and we consider this probability to be 0. Similarly, for $\theta = \pi/2$, we consider the probability of touching to be 1 since the only way such a needle could not touch would be for its center to fall exactly halfway between two of the lines.)

Finally, we define the total probability of the needle touching the line to be the average of the probabilities as θ ranges between 0 and $\pi/2$. Using the **integration techniques** discussed in this section, we can determine the probability of touching to be

$$\text{Probability of touching} = \frac{1}{(\pi/2) - 0}\int_0^{\pi/2} \sin\theta\, d\theta = \frac{2}{\pi}\Big[-\cos\theta\Big]_0^{\pi/2} = \frac{2}{\pi} \approx 63.7\%$$

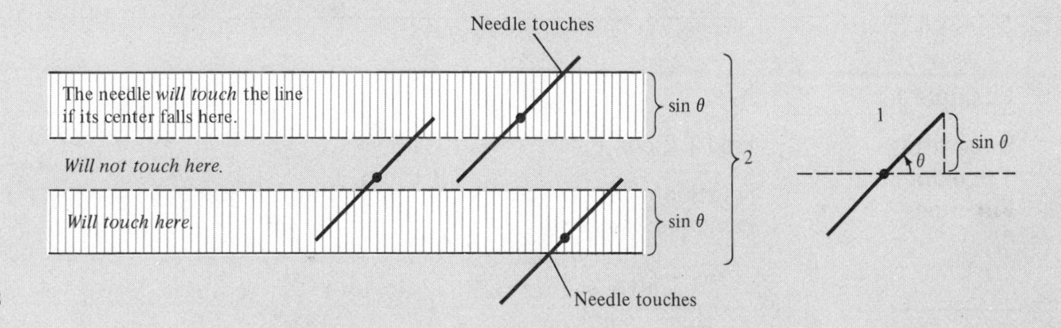

Figure 10.48

Section Objective: ***To derive formulas for integrating trigonometric functions.***

Corresponding to each formula for differentiating a trigonometric function is an integration formula. For instance, the formula

$$\frac{d}{dx}[\cos u] = -(\sin u)u'$$

corresponds to the formula

$$\int (\sin u)u'\,dx = -\cos u + C$$

We list next the integration formulas corresponding to the derivatives of the six basic trigonometric functions.

Trigonometric Functions

Integrals	*Derivatives*
$\int (\cos u)u'\,dx = \sin u + C$	$\frac{d}{dx}[\sin u] = (\cos u)u'$
$\int (\sin u)u'\,dx = -\cos u + C$	$\frac{d}{dx}[\cos u] = (-\sin u)u'$
$\int (\sec^2 u)u'\,dx = \tan u + C$	$\frac{d}{dx}[\tan u] = (\sec^2 u)u'$
$\int (\sec u \tan u)u'\,dx = \sec u + C$	$\frac{d}{dx}[\sec u] = (\sec u \tan u)u'$
$\int (\csc^2 u)u'\,dx = -\cot u + C$	$\frac{d}{dx}[\cot u] = (-\csc^2 u)u'$
$\int (\csc u \cot u)u'\,dx = -\csc u + C$	$\frac{d}{dx}[\csc u] = (-\csc u \cot u)u'$

Example 1

Integrating Trigonometric Functions

Find $\int 2\cos x\,dx$.

Solution Let $u = x$; then $u' = 1$. Since 2 is not a necessary part of the integrand, we have

$$\begin{aligned}\int 2\cos x\,dx = 2\int \cos x\,dx &= 2\int (\cos u)u'\,dx \\ &= 2\sin u + C \\ &= 2\sin x + C\end{aligned}$$

□

Example 2

Integrating Trigonometric Functions

Find $\int 3x^2 \sin x^3 \, dx$.

Solution We let $u = x^3$, and thus $u' = 3x^2$. Therefore, we write

$$\begin{aligned}\int 3x^2 \sin x^3 \, dx &= \int (\sin x^3)(3x^2) \, dx \\ &= \int (\sin u) \, u' \, dx \\ &= -\cos u + C = -\cos x^3 + C\end{aligned}$$

□

Example 3

Integrating Trigonometric Functions

Evaluate $\int \sec (3x + 1) \tan (3x + 1) \, dx$.

Solution Let $u = 3x + 1$; then $u' = 3$, and we write

$$\begin{aligned}\int \sec (3x + 1) \tan (3x + 1) \, dx &= \tfrac{1}{3} \int \sec (3x + 1) \tan (3x + 1)(3) \, dx \\ &= \tfrac{1}{3} \sec (3x + 1) + C\end{aligned}$$

□

Example 4

Integrating Trigonometric Functions

Evaluate

$$\int \frac{\sec^2 \sqrt{x}}{\sqrt{x}} \, dx$$

Solution Consider $u = \sqrt{x}$: then $u' = 1/(2\sqrt{x})$. Thus, we write

$$\begin{aligned}\int \frac{\sec^2 \sqrt{x}}{\sqrt{x}} \, dx &= \int (\sec^2 \sqrt{x}) \left(\frac{1}{\sqrt{x}} \right) dx \\ &= 2 \int (\sec^2 \sqrt{x}) \left(\frac{1}{2\sqrt{x}} \right) dx \\ &= 2(\tan \sqrt{x}) + C = 2 \tan \sqrt{x} + C\end{aligned}$$

□

Example 5

Finding Area by Integration

Find the area of the region bounded by the x-axis and one arc of the sine curve $y = \sin x$.

Solution As indicated in Figure 10.49, this area is given by

$$\text{area} = \int_0^{\pi} \sin x \, dx = -\cos x \Big]_0^{\pi} = -(-1) - (-1) = 2$$

and we conclude that the area is 2.

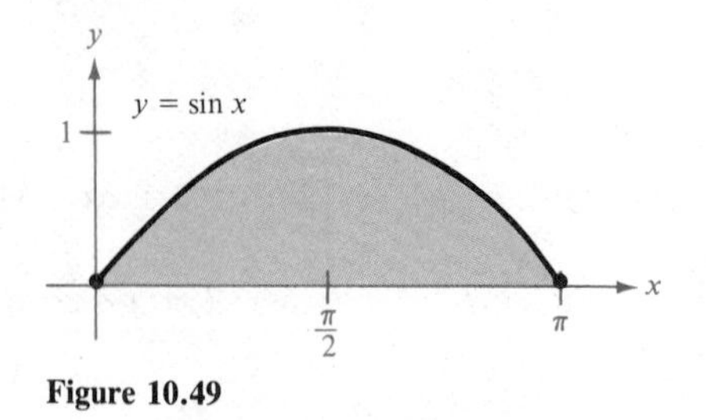

Figure 10.49

□

Example 6

Finding the Average Value of a Function Over an Interval

The temperature during a 24-h period is given by

$$T = 72 + 18 \sin \frac{\pi(t - 8)}{12}$$

where T is measured in degrees Fahrenheit and t is measured in hours, with $t = 0$ representing midnight. Find the average temperature during the 4-h period from noon to 4 P.M.

Solution The average temperature during this 4-h period is given by the following integral.

$$\begin{aligned}
\text{average temperature} &= \frac{1}{4}\int_{12}^{16}\left[72 + 18 \sin \frac{\pi(t-8)}{12}\right] dt \\
&= \frac{1}{4}\left[72t + 18\left(\frac{12}{\pi}\right)\left(-\cos \frac{\pi(t-8)}{12}\right)\right]_{12}^{16} \\
&= \frac{1}{4}\left[72(16) + 18\left(\frac{12}{\pi}\right)\left(\frac{1}{2}\right) - 72(12)\right. \\
&\qquad \left. + 18\left(\frac{12}{\pi}\right)\left(\frac{1}{2}\right)\right] \\
&= \frac{1}{4}\left[288 + \frac{216}{\pi}\right] \\
&= 72 + \frac{54}{\pi} \approx 89.2^\circ
\end{aligned}$$

(See Figure 10.50.)

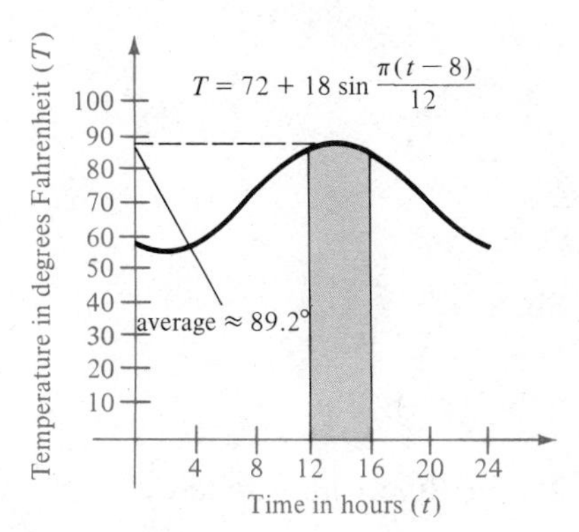

Figure 10.50

□

We frequently use the Power and Log Rules to evaluate integrals containing trigonometric functions. The key to their use lies in identifying, as part of the integrand, the derivative of one of the functions involved. The next three examples show how these rules are used. Recall that the Power and Log Rules for integration are

Power Rule

$$\int u^n u'\,dx = \frac{u^{n+1}}{n+1} + C, \qquad n \neq -1$$

Log Rule

$$\int \frac{u'}{u}\,dx = \ln|u| + C$$

Example 7

Using the Power Rule on Trigonometric Integrals

Evaluate $\int \sin^2 4x \cos 4x\,dx$.

Solution The integrand does not fit any of the six basic trigonometric integration formulas. However, by letting $u = \sin 4x$, we have

$$u' = \frac{d}{dx}[\sin 4x] = 4 \cos 4x$$

Thus, using the Power rule, we have

$$\begin{aligned}\int \sin^2 4x \cos 4x\,dx &= \frac{1}{4}\int \overbrace{(\sin 4x)^2}^{u^n}\overbrace{(4\cos 4x)}^{u'}\,dx \\ &= \frac{1}{4}\,\frac{(\sin 4x)^3}{3} + C \\ &= \tfrac{1}{12}\sin^3 4x + C\end{aligned}$$

□

Example 8

Using the Power Rule on Trigonometric Integrals

Evaluate

$$\int \frac{\sec^2 x \, dx}{\sqrt{\tan x}}$$

Solution Since

$$\frac{d}{dx}[\tan x] = \sec^2 x$$

we use the Power Rule with $u = \tan x$.

$$\int \frac{\sec^2 x}{\sqrt{\tan x}} \, dx = \int \overbrace{(\tan x)^{-1/2}}^{u^n} \overbrace{(\sec^2 x)}^{u'} \, dx$$

$$= \frac{(\tan x)^{1/2}}{\frac{1}{2}} + C$$

$$= 2\sqrt{\tan x} + C$$

□

Example 9

Using the Log Rule on Trigonometric Integrals

Find

$$\int \frac{\sin x}{\cos x} \, dx$$

Solution Knowing that

$$\frac{d}{dx}[\cos x] = -\sin x$$

we apply the Log Rule

$$\int \frac{u'}{u} \, dx = \ln |u| + C$$

and write

$$\int \frac{\sin x}{\cos x} \, dx = -\int \frac{(-\sin x)}{\cos x} \, dx = -\ln |\cos x| + C$$

Since

$$\tan x = \frac{\sin x}{\cos x}$$

we have the following formula:

$$\int \tan x \, dx = -\ln |\cos x| + C$$

□

We listed integration formulas for sin x and cos x in the beginning of this section. Now, using the result of Example 9, we also have an integration formula for tan x. We omit the development of the integration formulas for the other three trigonometric functions and simply summarize all six formulas as follows.

Integrals of the Six Basic Trigonometric Functions

$$\int (\sin u)u'\,dx = -\cos u + C \qquad \int (\cos u)u'\,dx = \sin u + C$$

$$\int (\tan u)u'\,dx = -\ln|\cos u| + C \qquad \int (\cot u)u'\,dx = \ln|\sin u| + C$$

$$\int (\sec u)u'\,dx = \ln|\sec u + \tan u| + C \qquad \int (\csc u)u'\,dx = \ln|\csc u - \cot u| + C$$

Section Exercises (10.5)

In Exercises 1–30, evaluate the given integral

1. $\int \sin 2x\,dx$

2. $\int x \sin x^2\,dx$

3. $\int x \cos x^2\,dx$

4. $\int \cos 6x\,dx$

5. $\int_{\pi/2}^{2\pi/3} \sec^2 \frac{x}{2}\,dx$

6. $\int_{\pi/3}^{\pi/2} \csc^2 \frac{x}{2}\,dx$

7. $\int_0^{\pi/9} \tan 3x\,dx$

8. $\int \tan \frac{\pi - 8x}{4}\,dx$

9. $\int \cot 2x\,dx$

10. $\int_0^{\pi/8} \sin 2x \cos 2x\,dx$

11. $\int_0^1 \sec(1 - x)\tan(1 - x)\,dx$

12. $\int_{\pi/12}^{\pi/4} \csc 2x \cot 2x\,dx$

13. $\int \frac{\sec^2 x}{\tan x}\,dx$

14. $\int \frac{\sin x}{\cos^2 x}\,dx$

15. $\int \cot^2 x \csc^2 x\,dx$

16. $\int \sec \frac{3x}{2}\,dx$

17. $\int \tan^4 x \sec^2 x\,dx$

18. $\int \sqrt{\cot x}\,\csc^2 x\,dx$

19. $\int_0^{\pi/4} \sec^4 x \tan x\,dx$

20. $\int_0^{\pi/2} (x + \cos x)\,dx$

21. $\int \frac{\sec x \tan x}{\sec x - 1}\,dx$

22. $\int \frac{\sin \sqrt{x}}{\sqrt{x}}\,dx$

23. $\int \frac{\sin x}{1 + \cos x}\,dx$

24. $\int \frac{\csc^2 x}{\sqrt{\cot x - 1}}\,dx$

25. $\int \sin\theta \sqrt{1 + \cos\theta}\,d\theta$

26. $\int \frac{1 - \cos\theta}{\theta - \sin\theta}\,d\theta$

27. $\int e^x \cos e^x\,dx$

28. $\int e^x \sin e^x\,dx$

29. $\int \frac{\cos x}{\sqrt{\sin x}}\,dx$

30. $\int \frac{\sin x}{\cos^3 x}\,dx$

In Exercises 31–34, evaluate the given integral, using the integration by parts formula

$$\int uv'\,dx = uv - \int u'v\,dx$$

31. $\int x \cos x\,dx$

32. $\int x \sin x\,dx$

33. $\int x \sec^2 x\,dx$

34. $\int \theta \sec\theta \tan\theta\,d\theta$

In Exercises 35–40, find the area of the given region.

35. $y = \cos \dfrac{x}{2}$

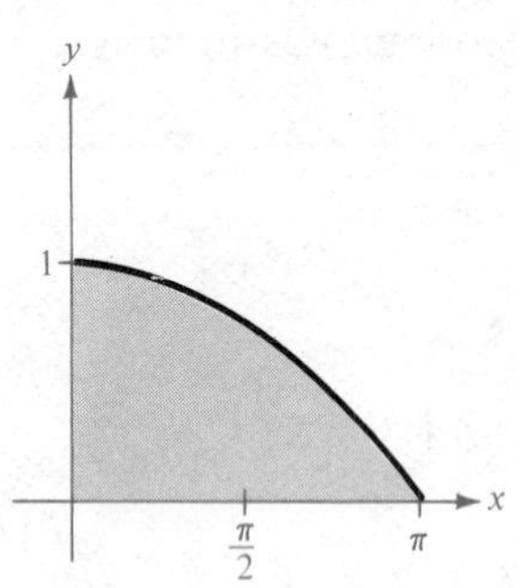

36. $y = x + \sin x$

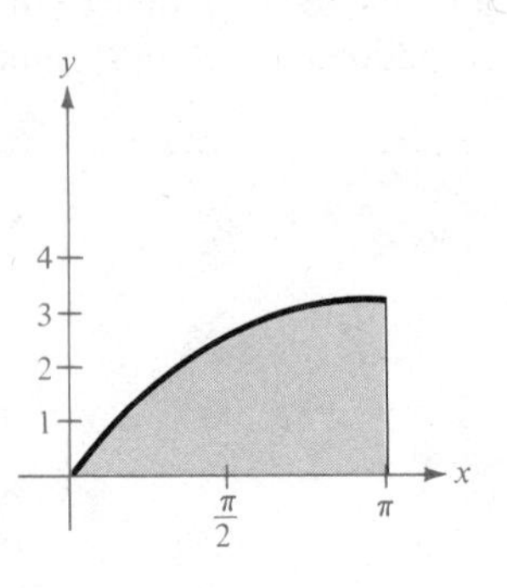

37. $y = \tan x$

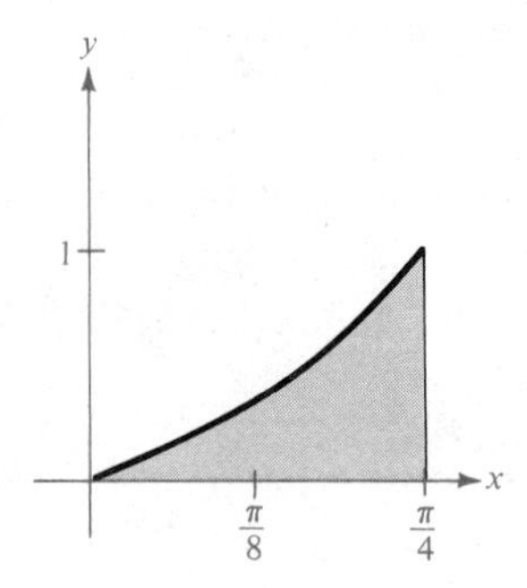

38. $y = 2 \sin x + \sin 2x$

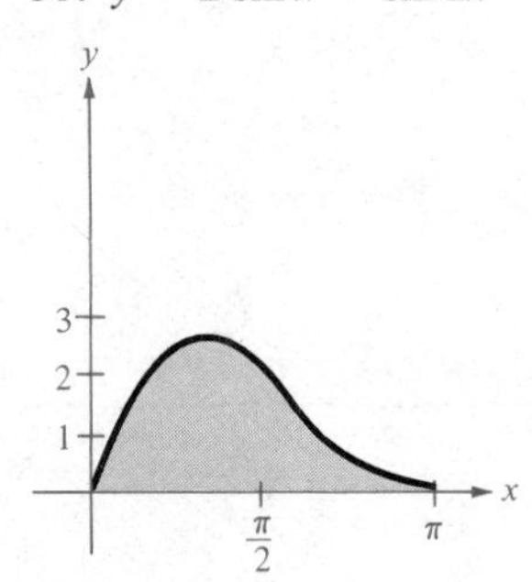

39. $y = \sin x + \cos 2x$

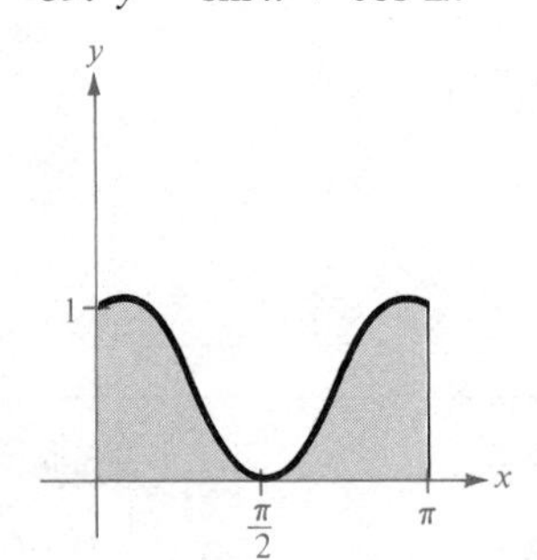

40. $y = x \sin x$

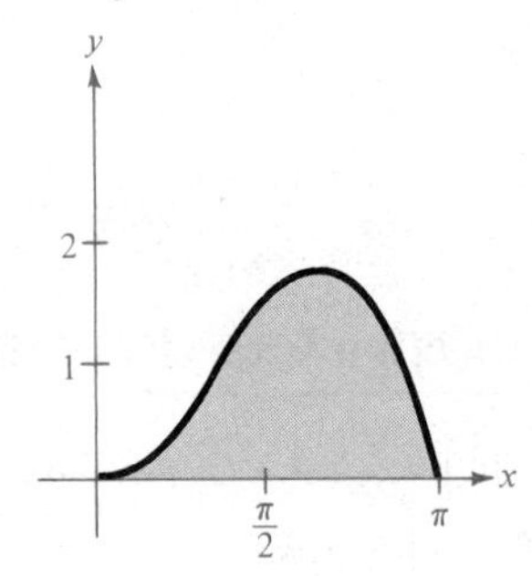

ⓒ**41.** Use the Taylor polynomial

$$\frac{\sin x}{x} \approx 1 - \frac{x^2}{6} + \frac{x^4}{120} - \frac{x^6}{5040}$$

to approximate the integral

$$\int_0^{\pi/2} \frac{\sin x}{x}\, dx$$

ⓒ**42.** Use the Taylor polynomial

$$\cos x^2 \approx 1 - \frac{x^4}{2} + \frac{x^8}{24} - \frac{x^{12}}{720}$$

to approximate the integral

$$\int_0^1 \cos x^2\, dx$$

ⓒ**43.** The minimum acceptable level of stockpiled gasoline in the United States can be approximated by the model

$$Q = 217 + 13 \cos \frac{\pi(t - 3)}{6}$$

where Q is measured in millions of barrels of gasoline and t is the time in months, with $t = 1$ corresponding to January. Find the average minimum level given by this model during:
(a) the first quarter of the year ($0 \le t \le 3$)
(b) the second quarter of the year ($3 \le t \le 6$)
(c) the entire year ($0 \le t \le 12$)

ⓒ**44.** The sales of a seasonal product are given by the model

$$S = 74.50 + 43.75 \sin \frac{\pi t}{6}$$

where S is the sales in thousands of units and t is the time in months, with $t = 1$ corresponding to January. Find the average sales during:
(a) the first quarter of the year ($0 \le t \le 3$)
(b) the second quarter of the year ($3 \le t \le 6$)
(c) the entire year ($0 \le t \le 12$)

ⓒ Calculator may be helpful.

Ⓒ**45.** For a person at rest, the velocity v (in liters per second) of air flow during a respiratory cycle is

$$v = 0.85 \sin \frac{\pi t}{3}$$

where t is the time in seconds. Find the volume (in liters) of air inhaled during one cycle by integrating this function over the interval $[0, 3]$.

Ⓒ**46.** After exercising a few minutes, a person has a respiratory cycle for which the velocity of air flow is approximated by

$$v = 1.75 \sin \frac{\pi t}{2}$$

How much does the lung capacity of a person increase as a result of exercising? In other words, how much more air is inhaled during a cycle after exercising than was inhaled during a cycle at rest in Exercise 45? (Remember that the cycle is shorter and you should integrate from 0 to 2 to find the volume of air inhaled.)

Ⓒ**47.** Suppose that the temperature (in degrees Fahrenheit) is given by

$$T = 72 + 12 \sin \frac{\pi(t-8)}{12}$$

where t is the time in hours, with $t = 0$ representing midnight. Furthermore, suppose that it costs \$0.10 to cool a particular house 1° for 1 h.

(a) Find the cost C of cooling this house if its thermostat is set at 72° by evaluating the integral

$$C = 0.1 \int_{8}^{20} \left[72 + 12 \sin \frac{\pi(t-8)}{12} - 72\right] dt$$

(See Figure 10.51.)

(b) Find the savings in resetting the thermostat to 78° by evaluating the integral

$$C = 0.1 \int_{10}^{18} \left[72 + 12 \sin \frac{\pi(t-8)}{12} - 78\right] dt$$

(See Figure 10.52.)

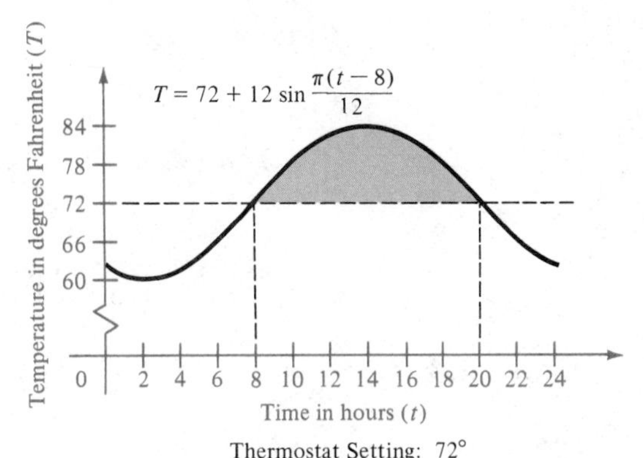

Figure 10.51

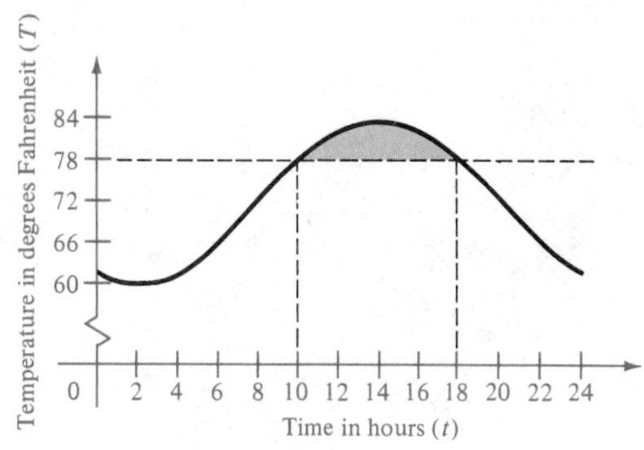

Figure 10.52

Ⓒ**48.** In Example 9 of Section 10.4, we looked at the seasonal sales model

$$F = 100{,}000\left[1 + \sin \frac{2\pi(t-60)}{365}\right]$$

where F is measured in pounds and t is the time in days, with $t = 1$ representing January 1. The manufacturer of this product wants to set up a manufacturing schedule to produce a *uniform amount* each day. What should this amount be? (Assume that there are 200 production days during the year.)

Ⓒ Calculator may be helpful.

Appendix A

TABLE OF SQUARE ROOTS AND CUBE ROOTS

n	$\sqrt{n}$	$\sqrt[3]{n}$
1	1.00000	1.00000
2	1.41421	1.25992
3	1.73205	1.44225
4	2.00000	1.58740
5	2.23607	1.70998
6	2.44949	1.81712
7	2.64575	1.91293
8	2.82843	2.00000
9	3.00000	2.08008
10	3.16228	2.15443
11	3.31662	2.22398
12	3.46410	2.28943
13	3.60555	2.35133
14	3.74166	2.41014
15	3.87298	2.46621
16	4.00000	2.51984
17	4.12311	2.57128
18	4.24264	2.62074
19	4.35890	2.66840
20	4.47214	2.71442
21	4.58258	2.75892
22	4.69042	2.80204
23	4.79583	2.84387
24	4.89898	2.88450
25	5.00000	2.92402
26	5.09902	2.96250
27	5.19615	3.00000
28	5.29150	3.03659
29	5.38516	3.07232
30	5.47723	3.10723

n	$\sqrt{n}$	$\sqrt[3]{n}$
31	5.56776	3.14138
32	5.65685	3.17480
33	5.74456	3.20753
34	5.83095	3.23961
35	5.91608	3.27107
36	6.00000	3.30193
37	6.08276	3.33222
38	6.16441	3.36198
39	6.24500	3.39121
40	6.32456	3.41995
41	6.40312	3.44822
42	6.48074	3.47603
43	6.55744	3.50340
44	6.63325	3.53035
45	6.70820	3.55689
46	6.78233	3.58305
47	6.85565	3.60883
48	6.92820	3.63424
49	7.00000	3.65931
50	7.07107	3.68403
51	7.14143	3.70843
52	7.21110	3.73251
53	7.28011	3.75629
54	7.34847	3.77976
55	7.41620	3.80295
56	7.48331	3.82586
57	7.54983	3.84850
58	7.61577	3.87088
59	7.68115	3.89300
60	7.74597	3.91487

n	$\sqrt{n}$	$\sqrt[3]{n}$
61	7.81025	3.93650
62	7.87401	3.95789
63	7.93725	3.97906
64	8.00000	4.00000
65	8.06226	4.02073
66	8.12404	4.04124
67	8.18535	4.06155
68	8.24621	4.08166
69	8.30662	4.10157
70	8.36660	4.12129
71	8.42615	4.14082
72	8.48528	4.16017
73	8.54400	4.17934
74	8.60233	4.19834
75	8.66025	4.21716
76	8.71780	4.23582
77	8.77496	4.25432
78	8.83176	4.27266
79	8.88819	4.29084
80	8.94427	4.30887
81	9.00000	4.32675
82	9.05539	4.34448
83	9.11043	4.36207
84	9.16515	4.37952
85	9.21954	4.39683
86	9.27362	4.41400
87	9.32738	4.43105
88	9.38083	4.44796
89	9.43398	4.46475
90	9.48683	4.48140

TABLE OF SQUARE ROOTS AND CUBE ROOTS (Continued)

n	$\sqrt{n}$	$\sqrt[3]{n}$
91	9.53939	4.49794
92	9.59166	4.51436
93	9.64365	4.53065
94	9.69536	4.54684
95	9.74679	4.56290
96	9.79796	4.57886
97	9.84886	4.59470
98	9.89949	4.61044
99	9.94987	4.62606
100	10.0000	4.64159
101	10.0499	4.65701
102	10.0995	4.67233
103	10.1489	4.68755
104	10.1980	4.70267
105	10.2470	4.71769
106	10.2956	4.73262
107	10.3441	4.74746
108	10.3923	4.76220
109	10.4403	4.77686
110	10.4881	4.79142
111	10.5357	4.80590
112	10.5830	4.82028
113	10.6301	4.83459
114	10.6771	4.84881
115	10.7238	4.86294
116	10.7703	4.87700
117	10.8167	4.89097
118	10.8628	4.90487
119	10.9087	4.91868
120	10.9545	4.93242
121	11.0000	4.94609
122	11.0454	4.95968
123	11.0905	4.97319
124	11.1355	4.98663
125	11.1803	5.00000
126	11.2250	5.01330
127	11.2694	5.02653

n	$\sqrt{n}$	$\sqrt[3]{n}$
128	11.3137	5.03968
129	11.3578	5.05277
130	11.4018	5.06580
131	11.4455	5.07875
132	11.4891	5.09164
133	11.5326	5.10447
134	11.5758	5.11723
135	11.6190	5.12993
136	11.6619	5.14256
137	11.7047	5.15514
138	11.7473	5.16765
139	11.7898	5.18010
140	11.8322	5.19249
141	11.8743	5.20483
142	11.9164	5.21710
143	11.9583	5.22932
144	12.0000	5.24148
145	12.0416	5.25359
146	12.0830	5.26564
147	12.1244	5.27763
148	12.1655	5.28957
149	12.2066	5.30146
150	12.2474	5.31329
151	12.2882	5.32507
152	12.3288	5.33680
153	12.3693	5.34848
154	12.4097	5.36011
155	12.4499	5.37169
156	12.4900	5.38321
157	12.5300	5.39469
158	12.5698	5.40612
159	12.6095	5.41750
160	12.6491	5.42884
161	12.6886	5.44012
162	12.7279	5.45136
163	12.7671	5.46256
164	12.8062	5.47370

n	$\sqrt{n}$	$\sqrt[3]{n}$
165	12.8452	5.48481
166	12.8841	5.49586
167	12.9228	5.50688
168	12.9615	5.51785
169	13.0000	5.52877
170	13.0384	5.53966
171	13.0767	5.55050
172	13.1149	5.56130
173	13.1529	5.57205
174	13.1909	5.58277
175	13.2288	5.59344
176	13.2665	5.60408
177	13.3041	5.61467
178	13.3417	5.62523
179	13.3791	5.63574
180	13.4164	5.64622
181	13.4536	5.65665
182	13.4907	5.66705
183	13.5277	5.67741
184	13.5647	5.68773
185	13.6015	5.69802
186	13.6382	5.70827
187	13.6748	5.71848
188	13.7113	5.72865
189	13.7477	5.73879
190	13.7840	5.74890
191	13.8203	5.75897
192	13.8564	5.76900
193	13.8924	5.77900
194	13.9284	5.78896
195	13.9642	5.79889
196	14.0000	5.80879
197	14.0357	5.81865
198	14.0712	5.82848
199	14.1067	5.83827
200	14.1421	5.84804

Appendix B

BLE OF EXPONENTIAL AND LOGARITHMIC FUNCTIONS

x	e^x	e^{-x}	$\ln x$
0.0	1.0000	1.0000	———
0.1	1.1052	0.9048	−2.3026
0.2	1.2214	0.8187	−1.6094
0.3	1.3499	0.7408	−1.2040
0.4	1.4918	0.6703	−0.9163
0.5	1.6487	0.6065	−0.6932
0.6	1.8221	0.5488	−0.5108
0.7	2.0138	0.4966	−0.3567
0.8	2.2255	0.4493	−0.2231
0.9	2.4596	0.4066	−0.1054
1.0	2.7183	0.3679	0.0000
1.1	3.0042	0.3329	0.0953
1.2	3.3201	0.3012	0.1823
1.3	3.6693	0.2725	0.2624
1.4	4.0552	0.2466	0.3365
1.5	4.4817	0.2231	0.4055
1.6	4.9530	0.2019	0.4700
1.7	5.4739	0.1827	0.5306
1.8	6.0496	0.1653	0.5878
1.9	6.6859	0.1496	0.6419
2.0	7.3891	0.1353	0.6931
2.1	8.1662	0.1225	0.7419
2.2	9.0250	0.1108	0.7885
2.3	9.9742	0.1003	0.8329
2.4	11.023	0.0907	0.8755

x	e^x	e^{-x}	$\ln x$
2.5	12.182	0.0821	0.9163
2.6	13.464	0.0743	0.9555
2.7	14.880	0.0672	0.9933
2.8	16.445	0.0608	1.0296
2.9	18.174	0.0550	1.0647
3.0	20.086	0.0498	1.0986
3.1	22.198	0.0450	1.1314
3.2	24.533	0.0408	1.1632
3.3	27.113	0.0369	1.1939
3.4	29.964	0.0334	1.2238
3.5	33.115	0.0302	1.2528
3.6	36.598	0.0273	1.2809
3.7	40.447	0.0247	1.3083
3.8	44.701	0.0224	1.3350
3.9	49.402	0.0202	1.3610
4.0	54.598	0.0183	1.3863
4.1	60.340	0.0166	1.4110
4.2	66.686	0.0150	1.4351
4.3	73.700	0.0136	1.4586
4.4	81.451	0.0123	1.4816
4.5	90.017	0.0111	1.5041
4.6	99.484	0.0101	1.5261
4.7	109.95	0.0091	1.5476
4.8	121.51	0.0082	1.5686
4.9	134.29	0.0074	1.5892

TABLE OF EXPONENTIAL AND LOGARITHMIC FUNCTIONS (Continued)

x	e^x	e^{-x}	$\ln x$
5.0	148.41	0.0067	1.6094
5.1	164.02	0.0061	1.6292
5.2	181.27	0.0055	1.6487
5.3	200.34	0.0050	1.6677
5.4	221.41	0.0045	1.6864
5.5	244.69	0.0041	1.7048
5.6	270.43	0.0037	1.7228
5.7	298.87	0.0033	1.7405
5.8	330.30	0.0030	1.7579
5.9	365.04	0.0027	1.7750
6.0	403.43	0.0025	1.7918
6.1	445.86	0.0022	1.8083
6.2	492.75	0.0020	1.8246
6.3	544.57	0.0018	1.8406
6.4	601.85	0.0017	1.8563
6.5	665.14	0.0015	1.8718
6.6	735.10	0.0014	1.8871
6.7	812.41	0.0012	1.9021
6.8	897.85	0.0011	1.9169
6.9	992.27	0.0010	1.9315
7.0	1096.63	0.0009	1.9459
7.1	1211.97	0.0008	1.9601
7.2	1339.43	0.0007	1.9741
7.3	1480.30	0.0007	1.9879
7.4	1635.98	0.0006	2.0015

x	e^x	e^{-x}	$\ln x$
7.5	1808.04	0.0006	2.0149
7.6	1998.20	0.0005	2.0282
7.7	2208.35	0.0005	2.0412
7.8	2440.60	0.0004	2.0541
7.9	2697.28	0.0004	2.0669
8.0	2980.96	0.0003	2.0794
8.1	3294.47	0.0003	2.0919
8.2	3640.95	0.0003	2.1041
8.3	4023.87	0.0002	2.1163
8.4	4447.07	0.0002	2.1282
8.5	4914.77	0.0002	2.1401
8.6	5431.66	0.0002	2.1518
8.7	6002.91	0.0002	2.1633
8.8	6634.24	0.0002	2.1748
8.9	7331.97	0.0001	2.1861
9.0	8103.08	0.0001	2.1972
9.1	8955.29	0.0001	2.2083
9.2	9897.13	0.0001	2.2192
9.3	10938.02	0.0001	2.2300
9.4	12088.38	0.0001	2.2407
9.5	13359.73	0.0001	2.2513
9.6	14764.78	0.0001	2.2618
9.7	16317.61	0.0001	2.2721
9.8	18033.74	0.0001	2.2824
9.9	19930.37	0.0001	2.2925
10.0	22026.47	0.0000	2.3026

Appendix C

TRIGONOMETRIC TABLES

1 degree ≈ 0.01745 radians
1 radian ≈ 57.29578 degrees

For $0 \leqslant \theta \leqslant 45$, read from upper left.
For $45 \leqslant \theta \leqslant 90$, read from lower right.
For $90 \leqslant \theta \leqslant 360$, use the identities:

θ	Quadrant II	Quadrant III	Quadrant IV
$\sin\theta$	$\sin(180-\theta)$	$-\sin(\theta-180)$	$-\sin(360-\theta)$
$\cos\theta$	$-\cos(180-\theta)$	$-\cos(\theta-180)$	$\cos(360-\theta)$
$\tan\theta$	$-\tan(180-\theta)$	$\tan(\theta-180)$	$-\tan(360-\theta)$
$\cot\theta$	$-\cot(180-\theta)$	$\cot(\theta-180)$	$-\cot(360-\theta)$

Degrees	Radians	sin	cos	tan	cot		
0°00′	.0000	.0000	1.0000	.0000	–	1.5708	90°00′
10	.0029	.0029	1.0000	.0029	343.774	1.5679	50
20	.0058	.0058	1.0000	.0058	171.885	1.5650	40
30	.0087	.0087	1.0000	.0087	114.589	1.5621	30
40	.0116	.0116	.9999	.0116	85.940	1.5592	20
50	.0145	.0145	.9999	.0145	68.750	1.5563	10
1°00′	.0175	.0175	.9998	.0175	57.290	1.5533	89°00′
10	.0204	.0204	.9998	.0204	49.104	1.5504	50
20	.0233	.0233	.9997	.0233	42.964	1.5475	40
30	.0262	.0262	.9997	.0262	38.188	1.5446	30
40	.0291	.0291	.9996	.0291	34.368	1.5417	20
50	.0320	.0320	.9995	.0320	31.242	1.5388	10
2°00′	.0349	.0349	.9994	.0349	28.636	1.5359	88°00′
10	.0378	.0378	.9993	.0378	26.432	1.5330	50
20	.0407	.0407	.9992	.0407	24.542	1.5301	40
30	.0436	.0436	.9990	.0437	22.904	1.5272	30
40	.0465	.0465	.9989	.0466	21.470	1.5243	20
50	.0495	.0494	.9988	.0495	20.206	1.5213	10
3°00′	.0524	.0523	.9986	.0524	19.081	1.5184	87°00′
10	.0553	.0552	.9985	.0553	18.075	1.5155	50
20	.0582	.0581	.9983	.0582	17.169	1.5126	40
30	.0611	.0610	.9981	.0612	16.350	1.5097	30
40	.0640	.0640	.9980	.0641	15.605	1.5068	20
50	.0669	.0669	.9978	.0670	14.924	1.5039	10
		cos	sin	cot	tan	Radians	Degrees

Degrees	Radians	sin	cos	tan	cot		
4°00′	.0698	.0698	.9976	.0699	14.301	1.5010	86°00′
10	.0727	.0727	.9974	.0729	13.727	1.4981	50
20	.0756	.0756	.9971	.0758	13.197	1.4952	40
30	.0785	.0785	.9969	.0787	12.706	1.4923	30
40	.0814	.0814	.9967	.0816	12.251	1.4893	20
50	.0844	.0843	.9964	.0846	11.826	1.4864	10
5°00′	.0873	.0872	.9962	.0875	11.430	1.4835	85°00′
10	.0902	.0901	.9959	.0904	11.059	1.4806	50
20	.0931	.0929	.9957	.0934	10.712	1.4777	40
30	.0960	.0958	.9954	.0963	10.385	1.4748	30
40	.0989	.0987	.9951	.0992	10.078	1.4719	20
50	.1018	.1016	.9948	.1022	9.788	1.4690	10
6°00′	.1047	.1045	.9945	.1051	9.514	1.4661	84°00′
10	.1076	.1074	.9942	.1080	9.255	1.4632	50
20	.1105	.1103	.9939	.1110	9.010	1.4603	40
30	.1134	.1132	.9936	.1139	8.777	1.4573	30
40	.1164	.1161	.9932	.1169	8.556	1.4544	20
50	.1193	.1190	.9929	.1198	8.345	1.4515	10
7°00′	.1222	.1219	.9925	.1228	8.144	1.4486	83°00′
10	.1251	.1248	.9922	.1257	7.953	1.4457	50
20	.1280	.1276	.9918	.1287	7.770	1.4428	40
30	.1309	.1305	.9914	.1317	7.596	1.4399	30
40	.1338	.1334	.9911	.1346	7.429	1.4370	20
50	.1367	.1363	.9907	.1376	7.269	1.4341	10
		cos	sin	cot	tan	Radians	Degrees

TRIGONOMETRIC TABLES (Continued)

Degrees	Radians	sin	cos	tan	cot		
8°00′	.1396	.1392	.9903	.1405	7.115	1.4312	82°00′
10	.1425	.1421	.9899	.1435	6.968	1.4283	50
20	.1454	.1449	.9894	.1465	6.827	1.4254	40
30	.1484	.1478	.9890	.1495	6.691	1.4224	30
40	.1513	.1507	.9886	.1524	6.561	1.4195	20
50	.1542	.1536	.9881	.1554	6.435	1.4166	10
9°00′	.1571	.1564	.9877	.1584	6.314	1.4137	81°00′
10	.1600	.1593	.9872	.1614	6.197	1.4108	50
20	.1629	.1622	.9868	.1644	6.084	1.4079	40
30	.1658	.1650	.9863	.1673	5.976	1.4050	30
40	.1687	.1679	.9858	.1703	5.871	1.4021	20
50	.1716	.1708	.9853	.1733	5.769	1.3992	10
10°00′	.1745	.1736	.9848	.1763	5.671	1.3963	80°00′
10	.1774	.1765	.9843	.1793	5.576	1.3934	50
20	.1804	.1794	.9838	.1823	5.485	1.3904	40
30	.1833	.1822	.9833	.1853	5.396	1.3875	30
40	.1862	.1851	.9827	.1883	5.309	1.3846	20
50	.1891	.1880	.9822	.1914	5.226	1.3817	10
11°00′	.1920	.1908	.9816	.1944	5.145	1.3788	79°00′
10	.1949	.1937	.9811	.1974	5.066	1.3759	50
20	.1978	.1965	.9805	.2004	4.989	1.3730	40
30	.2007	.1994	.9799	.2035	4.915	1.3701	30
40	.2036	.2022	.9793	.2065	4.843	1.3672	20
50	.2065	.2051	.9787	.2095	4.773	1.3643	10
12°00′	.2094	.2079	.9781	.2126	4.705	1.3614	78°00′
10	.2123	.2108	.9775	.2156	4.638	1.3584	50
20	.2153	.2136	.9769	.2186	4.574	1.3555	40
30	.2182	.2164	.9763	.2217	4.511	1.3526	30
40	.2211	.2193	.9757	.2247	4.449	1.3497	20
50	.2240	.2221	.9750	.2278	4.390	1.3468	10
13°00′	.2269	.2250	.9744	.2309	4.331	1.3439	77°00′
10	.2298	.2278	.9737	.2339	4.275	1.3410	50
20	.2327	.2306	.9730	.2370	4.219	1.3381	40
30	.2356	.2334	.9724	.2401	4.165	1.3352	30
40	.2385	.2363	.9717	.2432	4.113	1.3323	20
50	.2414	.2391	.9710	.2462	4.061	1.3294	10
14°00′	.2443	.2419	.9703	.2493	4.011	1.3265	76°00′
10	.2473	.2447	.9696	.2524	3.962	1.3235	50
20	.2502	.2476	.9689	.2555	3.914	1.3206	40
30	.2531	.2504	.9681	.2586	3.867	1.3177	30
40	.2560	.2532	.9674	.2617	3.821	1.3148	20
50	.2589	.2560	.9667	.2648	3.776	1.3119	10
15°00′	.2618	.2588	.9659	.2679	3.732	1.3090	75°00′
10	.2647	.2616	9652	.2711	3.689	1.3061	50
20	.2676	.2644	.9644	.2742	3.647	1.3032	40
30	.2705	.2672	.9636	.2773	3.606	1.3003	30
40	.2734	.2700	.9628	.2805	3.566	1.2974	20
50	.2763	.2728	.9621	.2836	3.526	1.2945	10
16°00′	.2793	.2756	.9613	.2867	3.487	1.2915	74°00′
10	.2822	.2784	.9605	.2899	3.450	1.2886	50
20	.2851	.2812	.9596	.2931	3.412	1.2857	40
30	.2880	.2840	.9588	.2962	3.376	1.2828	30
40	.2909	.2868	.9580	.2994	3.340	1.2799	20
50	.2938	.2896	.9572	.3026	3.305	1.2770	10
17°00′	.2967	.2924	.9563	.3057	3.271	1.2741	73°00′
10	.2996	.2952	.9555	.3089	3.237	1.2712	50
20	.3025	.2979	.9546	.3121	3.204	1.2683	40
30	.3054	.3007	.9537	.3153	3.172	1.2654	30
40	.3083	.3035	.9528	.3185	3.140	1.2625	20
50	.3113	.3062	.9520	.3217	3.108	1.2595	10
		cos	sin	cot	tan	Radians	Degrees

Degrees	Radians	sin	cos	tan	cot		
18°00′	.3142	.3090	.9511	.3249	3.078	1.2566	72°0
10	.3171	.3118	.9502	.3281	3.047	1.2537	5
20	.3200	.3145	.9492	.3314	3.018	1.2508	4
30	.3229	.3173	.9483	.3346	2.989	1.2479	3
40	.3258	.3201	.9474	.3378	2.960	1.2450	2
50	.3287	.3228	.9465	.3411	2.932	1.2421	1
19°00′	.3316	.3256	.9455	.3443	2.904	1.2392	71°0
10	.3345	.3283	.9446	.3476	2.877	1.2363	5
20	.3374	.3311	.9436	.3508	2.850	1.2334	4
30	.3403	.3338	.9426	.3541	2.824	1.2305	3
40	.3432	.3365	.9417	.3574	2.798	1.2275	2
50	.3462	.3393	.9407	.3607	2.773	1.2246	1
20°00′	.3491	.3420	.9397	.3640	2.747	1.2217	70°0
10	.3520	.3448	.9387	.3673	2.723	1.2188	5
20	.3549	.3475	.9377	.3706	2.699	1.2159	4
30	.3578	.3502	.9367	.3739	2.675	1.2130	3
40	.3607	.3529	.9356	.3772	2.651	1.2101	2
50	.3636	.3557	.9346	.3805	2.628	1.2072	1
21°00′	.3665	.3584	.9336	.3839	2.605	1.2043	69°0
10	.3694	.3611	.9325	.3872	2.583	1.2014	5
20	.3723	.3638	.9315	.3906	2.560	1.1985	4
30	.3752	.3665	.9304	.3939	2.539	1.1956	3
40	.3782	.3692	.9293	.3973	2.517	1.1926	2
50	.3811	.3719	.9283	.4006	2.496	1.1897	1
22°00′	.3840	.3746	.9272	.4040	2.475	1.1868	68°0
10	.3869	.3773	.9261	.4074	2.455	1.1839	5
20	.3898	.3800	.9250	.4108	2.434	1.1810	4
30	.3927	.3827	.9239	.4142	2.414	1.1781	3
40	.3956	.3854	.9228	.4176	2.394	1.1752	2
50	.3985	.3881	.9216	.4210	2.375	1.1723	1
23°00′	.4014	.3907	.9205	.4245	2.356	1.1694	67°0
10	.4043	.3934	.9194	.4279	2.337	1.1665	5
20	.4072	.3961	.9182	.4314	2.318	1.1636	4
30	.4102	.3987	.9171	.4348	2.300	1.1606	3
40	.4131	.4014	.9159	.4383	2.282	1.1577	2
50	.4160	.4041	.9147	.4417	2.264	1.1548	1
24°00′	.4189	.4067	.9135	.4452	2.246	1.1519	66°0
10	.4218	.4094	.9124	.4487	2.229	1.1490	5
20	.4247	.4120	.9112	.4522	2.211	1.1461	4
30	.4276	.4147	.9100	.4557	2.194	1.1432	3
40	.4305	.4173	.9088	.4592	2.177	1.1403	2
50	.4334	.4200	.9075	.4628	2.161	1.1374	1
25°00′	.4363	.4226	.9063	.4663	2.145	1.1345	65°0
10	.4392	.4253	.9051	.4699	2.128	1.1316	5
20	.4422	.4279	.9038	.4734	2.112	1.1286	4
30	.4451	.4305	.9026	.4770	2.097	1.1257	3
40	.4480	.4331	.9013	.4806	2.081	1.1228	2
50	.4509	.4358	.9001	.4841	2.066	1.1199	1
26°00′	.4538	.4384	.8988	.4877	2.050	1.1170	64°0
10	.4567	.4410	.8975	.4913	2.035	1.1141	5
20	.4596	.4436	.8962	.4950	2.020	1.1112	4
30	.4625	.4462	.8949	.4986	2.006	1.1083	3
40	.4654	.4488	.8936	.5022	1.991	1.1054	2
50	.4683	.4514	.8923	.5059	1.977	1.1025	1
27°00′	.4712	.4540	.8910	.5095	1.963	1.0996	63°0
10	.4741	.4566	.8897	.5132	1.949	1.0966	5
20	.4771	.4592	.8884	.5169	1.935	1.0937	4
30	.4800	.4617	.8870	.5206	1.921	1.0908	3
40	.4829	.4643	.8857	.5243	1.907	1.0879	2
50	.4858	.4669	.8843	.5280	1.894	1.0850	1
		cos	sin	cot	tan	Radians	Degr

IGONOMETRIC TABLES (Continued)

egrees	Radians	sin	cos	tan	cot		
8°00′	.4887	.4695	.8829	.5317	1.881	1.0821	62°00′
10	.4916	.4720	.8816	.5354	1.868	1.0792	50
20	.4945	.4746	.8802	.5392	1.855	1.0763	40
30	.4974	.4772	.8788	.5430	1.842	1.0734	30
40	.5003	.4797	.8774	.5467	1.829	1.0705	20
50	.5032	.4823	.8760	.5505	1.816	1.0676	10
9°00′	.5061	.4848	.8746	.5543	1.804	1.0647	61°00′
10	.5091	.4874	.8732	.5581	1.792	1.0617	50
20	.5120	.4899	.8718	.5619	1.780	1.0588	40
30	.5149	.4924	.8704	.5658	1.767	1.0559	30
40	.5178	.4950	.8689	.5696	1.756	1.0530	20
50	.5207	.4975	.8675	.5735	1.744	1.0501	10
0°00′	.5236	.5000	.8660	.5774	1.732	1.0472	60°00′
10	.5265	.5025	.8646	.5812	1.720	1.0443	50
20	.5294	.5050	.8631	.5851	1.709	1.0414	40
30	.5323	.5075	.8616	.5890	1.698	1.0385	30
40	.5325	.5100	.8601	.5930	1.686	1.0356	20
50	.5381	.5125	.8587	.5969	1.675	1.0327	10
1°00′	.5411	.5150	.8572	.6009	1.664	1.0297	59°00′
10	.5440	.5175	.8557	.6048	1.653	1.0268	50
20	.5469	.5200	.8542	.6088	1.643	1.0239	40
30	.5498	.5225	.8526	.6128	1.632	1.0210	30
40	.5527	.5250	.8511	.6168	1.621	1.0181	20
50	.5556	.5275	.8496	.6208	1.611	1.0152	10
2°00′	.5585	.5299	.8480	.6249	1.600	1.0123	58°00′
10	.5614	.5324	.8465	.6289	1.590	1.0094	50
20	.5643	.5348	.8450	.6330	1.580	1.0065	40
30	.5672	.5373	.8434	.6371	1.570	1.0036	30
40	.5701	.5398	.8418	.6412	1.560	1.0007	20
50	.5730	.5422	.8403	.6453	1.550	.0977	10
3°00′	.5760	.5446	.8387	.6494	1.540	.9948	57°00′
10	.5789	.5471	.8371	.6536	1.530	.9919	50
20	.5818	.5495	.8355	.6577	1.520	.9890	40
30	.5847	.5519	.8339	.6619	1.511	.9861	30
40	.5876	.5544	.8323	.6661	1.501	.9832	20
50	.5905	.5568	.8307	.6703	1.492	.9803	10
4°00′	.5934	.5592	.8290	.6745	1.483	.9774	56°00′
10	.5963	.5616	.8274	.6787	1.473	.9745	50
20	.5992	.5640	.8258	.6830	1.464	.9716	40
30	.6021	.5664	.8241	.6873	1.455	.9687	30
40	.6050	.5688	.8225	.6916	1.446	.9657	20
50	.6080	.5712	.8208	.6959	1.437	.9628	10
5°00′	.6109	.5736	.8192	.7002	1.428	.9599	55°00′
10	.6138	.5760	.8175	.7046	1.419	.9570	50
20	.6167	.5783	.8158	.7089	1.411	.9541	40
30	.6196	.5807	.8141	.7133	1.402	.9512	30
40	.6225	.5831	.8124	.7177	1.393	.9483	20
50	.6254	.5854	.8107	.7221	1.385	.9454	10
°00′	.6283	.5878	.8090	.7265	1.376	.9425	54°00′
10	.6312	.5901	.8073	.7310	1.368	.9396	50
20	.6341	.5925	.8056	.7355	1.360	.9367	40
30	.6370	.5948	.8039	.7400	1.351	.9338	30
40	.6400	.5972	.8021	.7445	1.343	.9308	20
50	.6429	.5995	.8004	.7490	1.335	.9279	10
		cos	sin	cot	tan	Radians	Degrees

Degrees	Radians	sin	cos	tan	cot		
37°00′	.6458	.6018	.7986	.7536	1.327	.9250	53°00′
10	.6487	.6041	.7969	.7581	1.319	.9221	50
20	.6516	.6065	.7951	.7627	1.311	.9192	40
30	.6545	.6088	.7934	.7673	1.303	.9163	30
40	.6574	.6111	.7916	.7720	1.295	.9134	20
50	.6603	.6134	.7898	.7766	1.288	.9105	10
38°00′	.6632	.6157	.7880	.7813	1.280	.9076	52°00′
10	.6661	.6180	.7862	.7860	1.272	.9047	50
20	.6690	.6202	.7844	.7907	1.265	.9018	40
30	.6720	.6225	.7826	.7954	1.257	.8988	30
40	.6749	.6248	.7808	.8002	1.250	.8959	20
50	.6778	.6271	.7790	.8050	1.242	.8930	10
39°00′	.6807	.6293	.7771	.8098	1.235	.8901	51°00′
10	.6836	.6316	.7753	.8146	1.228	.8872	50
20	.6865	.6338	.7735	.8195	1.220	.8843	40
30	.6894	.6361	.7716	.8243	1.213	.8814	30
40	.6923	.6383	.7698	.8292	1.206	.8785	20
50	.6952	.6406	.7679	.8342	1.199	.8756	10
40°00′	.6981	.6428	.7660	.8391	1.192	.8727	50°00′
10	.7010	.6450	.7642	.8441	1.185	.8698	50
20	.7039	.6472	.7623	.8491	1.178	.8668	40
30	.7069	.6494	.7604	.8541	1.171	.8639	30
40	.7098	.6517	.7585	.8591	1.164	.8610	20
50	.7127	.6539	.7566	.8642	1.157	.8581	10
41°00′	.7156	.6561	.7547	.8693	1.150	.8552	49°00′
10	.7185	.6583	.7528	.8744	1.144	.8523	50
20	.7214	.6604	.7509	.8796	1.137	.8494	40
30	.7243	.6626	.7490	.8847	1.130	.8465	30
40	.7272	.6648	.7470	.8899	1.124	.8436	20
50	.7301	.6670	.7451	.8952	1.117	.8407	10
42°00′	.7330	.6691	.7431	.9004	1.111	.8378	48°00′
10	.7359	.6713	.7412	.9057	1.104	.8348	50
20	.7389	.6734	.7392	.9110	1.098	.8319	40
30	.7418	.6756	.7373	.9163	1.091	.8290	30
40	.7447	.6777	.7353	.9217	1.085	.8261	20
50	.7476	.6799	.7333	.9271	1.079	.8232	10
43°00′	.7505	.6820	.7314	.9325	1.072	.8203	47°00′
10	.7534	.6841	.7294	.9380	1.066	.8174	50
20	.7563	.6862	.7274	.9435	1.060	.8145	40
30	.7592	.6884	.7254	.9490	1.054	.8116	30
40	.7621	.6905	.7234	.9545	1.048	.8087	20
50	.7650	.6926	.7214	.9601	1.042	.8058	10
44°00′	.7679	.6947	.7193	.9657	1.036	.8029	46°00′
10	.7709	.6967	.7173	.9713	1.030	.7999	50
10	.7738	.6988	.7153	.9770	1.024	.7970	40
30	.7767	.7009	.7133	.9827	1.018	.7941	30
40	.7796	.7030	.7112	.9884	1.012	.7912	20
50	.7825	.7050	.7092	.9942	1.006	.7883	10
45°00′	.7854	.7071	.7071	1.0000	1.000	.7854	45°00′
		cos	sin	cot	tan	Radians	Degrees

Answers to Odd-Numbered Exercises

Chapter 0

Section 0.1

1. $[-2, 0)$ $\;-2 \leqslant x < 0$

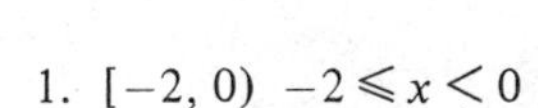

3. $[3, 11/2]$ $\;3 \leqslant x \leqslant 11/2$

5. $[100, \infty)$ $\;100 \leqslant x$

7. $(\sqrt{2}, 8]$ $\;\sqrt{2} < x \leqslant 8$

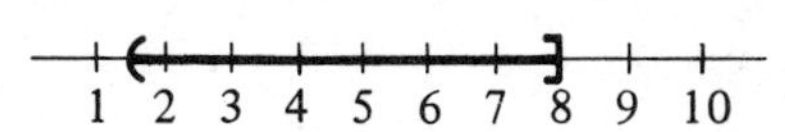

9. $x \geqslant 12$

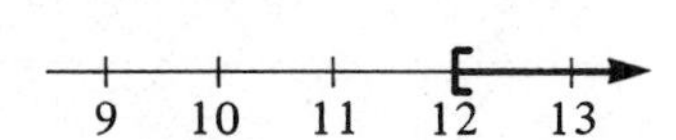

11. $x < -1/2$

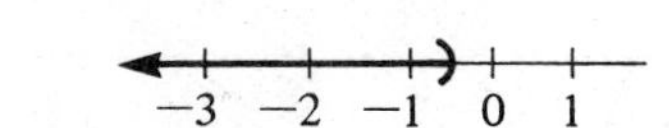

13. $1/2 \leqslant x$

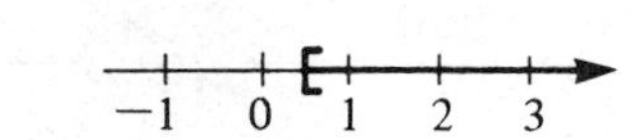

15. $1/2 < x$

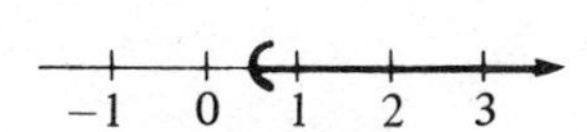

17. $-1/2 < x < 7/2$

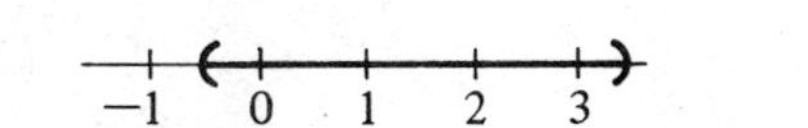

19. $-3/4 < x < -1/4$

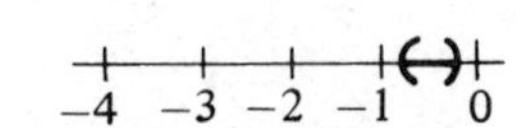

21. (a) Yes (b) No (c) Yes (d) No

23. (a) Yes (b) No (c) No (d) Yes

25. $r > 12.5\%$

27. $x > 35.80$, or at least 36 units must be sold.

29. (a) 355/113 (b) 22/7

Section 0.2

1. (a) 4 (b) −4 (c) 4

3. (a) 23/4 (b) −23/4 (c) 23/4

5. (a) −51 (b) 51 (c) 51

7. (a) −14.99 (b) 14.99 (c) 14.99

9. 1

11. −3.25

13. 14

15. 1.25

17. $-5 < x < 5$

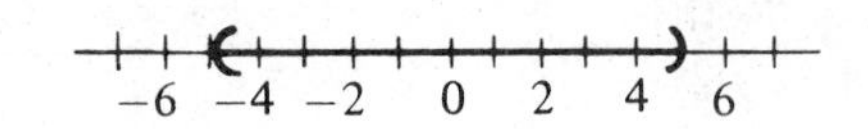

19. $x < -6$ or $x > 6$

21. $-7 < x < 3$

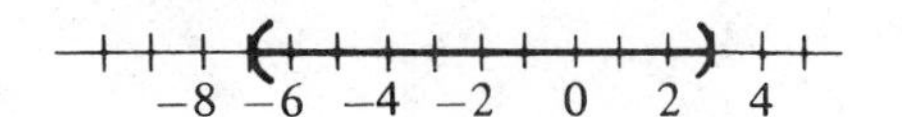

23. $x \leqslant -7$ or $x \geqslant 13$

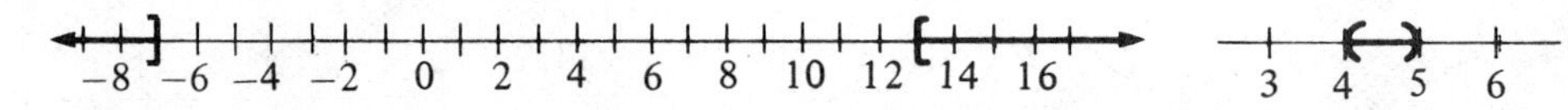

25. $4 < x < 5$

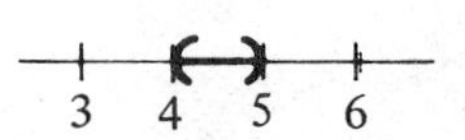

27. $|x| \leqslant 2$

29. $|x| > 2$

31. $|x - 4| \leqslant 2$

33. $|x - 2| > 2$

35. $|x - 12| < 10$

37. $65.8 \leqslant h \leqslant 71.2$

39. $2{,}125{,}000 < p < 2{,}375{,}000$

Section 0.3

1. −24

3. 1/2

5. 3

7. 44

9. 5

11. 9

13. 1/2

15. 1/4

17. 908.3484

19. −5.3601

21. $5x^6$

23. $24y^{10}$

25. $10x^4$

27. $7x^5$

29. $(4/3)(x + y)^2$

31. $3x$

33. $4x^4$

35. $y(y + \sqrt{3})(y - \sqrt{3}$

37. $2x\sqrt{x}\,(2\sqrt{x} - 3)$

39. $x^{1/2}(3 + 4x)$

41. $\dfrac{3 + 4x^2}{\sqrt{x}}$

43. $\dfrac{3x+2}{2\sqrt{x+1}}$

45. $\dfrac{5x^2+3}{3(x^2+1)^{2/3}}$

47. $\dfrac{2-x}{2x^2\sqrt{x-1}}$

49. $\dfrac{(x^2+1)(4x^2-4x-1)}{(x+1)^2}$

51. $x \geqslant 1$

53. $(-\infty, \infty)$

55. $(-\infty, 1)$ and $(1, \infty)$

57. $x > 3$

59. $1 \leqslant x \leqslant 5$

Section 0.4

1. $1/2, -1/3$

3. $3/2$

5. $-2 \pm \sqrt{3}$

7. $\dfrac{1 \pm \sqrt{7}}{3}$

9. $\dfrac{7 \pm \sqrt{17}}{4}$

11. No real roots

13. $\dfrac{-1 \pm \sqrt{13}}{2}$

15. $1, -2$

17. $2, 3$

19. $-1/2, 1$

21. $0, 5$

23. ± 3

25. $\pm\sqrt{3}$

27. $3 \pm 2\sqrt{2}$

29. $-6, 8$

31. $x \leqslant 3$ or $x \geqslant 4$

33. $-2 \leqslant x \leqslant 2$

35. $-4 \leqslant x \leqslant 3$

37. $(-\infty, \infty)$

39. $(x+2)(x^2-2x+4)$

41. $(x-1)(2x^2+x-1)$

43. $(x-3)(x^3+5x^2+9x+9)$

45. $1, 1, -1$

47. $1, 2, 3$

49. $-1/2, 1/2, 1$

51. 4

53. $-2, -3 \pm \sqrt{10}$

55. $\pm 2, \pm 3$

57. $1, \pm 2$

59. $-2, -1/2, 1$

Section 0.5

1. $\dfrac{x+5}{x-1}$

3. $\dfrac{5x-1}{x^2+2}$

5. $\dfrac{4x-3}{x^2}$

7. $\dfrac{x-6}{x^2-4}$

9. $\dfrac{2}{x-3}$

11. $-\dfrac{x^2+3}{(x+1)(x-2)(x-3)}$

13. $\dfrac{(A + B)x + 3(A - 2B)}{(x + 3)(x - 6)}$

15. $\dfrac{(A + B)x^2 + (10A + C)x + 5(5A - 5B - C)}{(x + 5)^2 (x - 5)}$

17. $-\dfrac{(x - 1)^2}{x(x^2 + 1)}$

19. $\dfrac{x + 2}{(x + 1)^{3/2}}$

21. $-\dfrac{3t}{2\sqrt{1 + t}}$

23. $\dfrac{2x^2 + 5}{\sqrt{x^2 + 5}}$

25. $\dfrac{x(x^2 + 2)}{(x^2 + 1)^{3/2}}$

27. $-\dfrac{2x + 3}{3x^2(x + 1)^{2/3}}$

29. $1/\sqrt{x^2 + 1}$

31. $\sqrt{21}/7$

33. $\dfrac{2}{3\sqrt{2}}$

35. $\dfrac{x\sqrt{x - 4}}{x - 4}$

37. $\dfrac{y}{6\sqrt{y}}$

39. $\dfrac{49\sqrt{x^2 - 9}}{x + 3}$

41. $\sqrt{6} - \sqrt{5}$

43. $\dfrac{\sqrt{x^2 - 2} + \sqrt{x}}{x - 2}$

45. $\dfrac{1}{x(\sqrt{3} + \sqrt{2})}$

47. $\dfrac{8x(\sqrt{17x} + 1)}{17x - 1}$

49. $\dfrac{2x - 1}{2x + \sqrt{4x - 1}}$

Section 1.1

1. 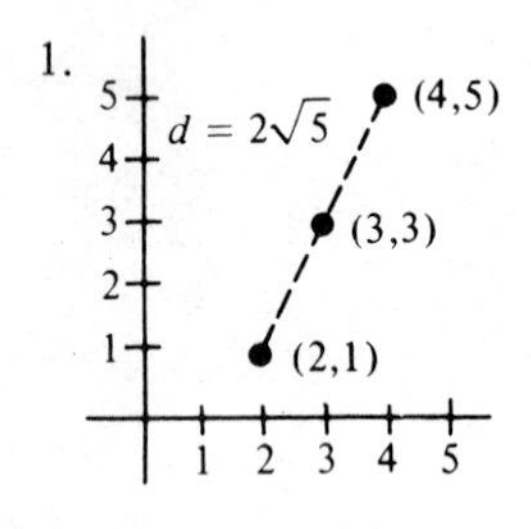

3. 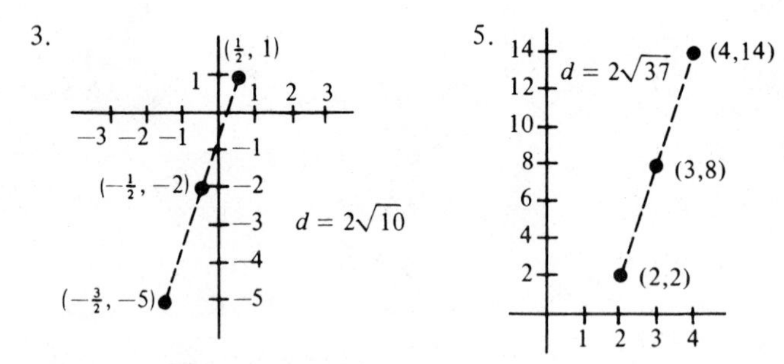

5. 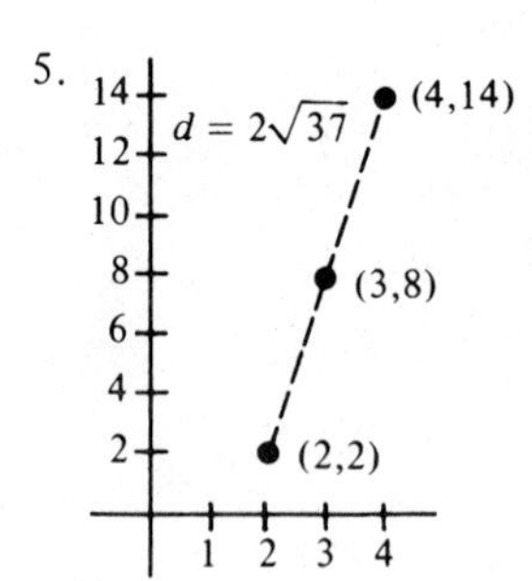

7.
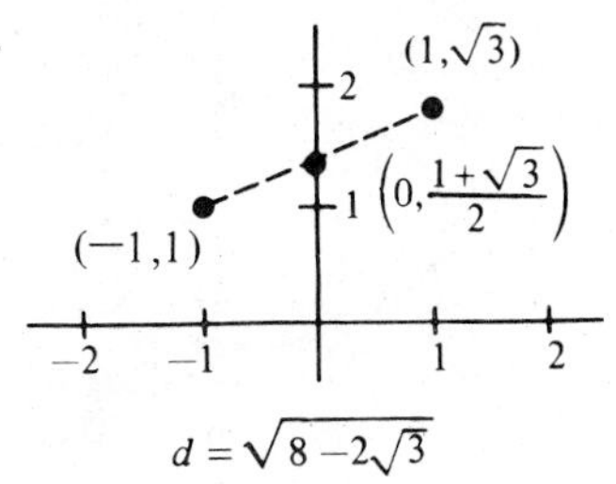

$d = \sqrt{8 - 2\sqrt{3}}$

9.
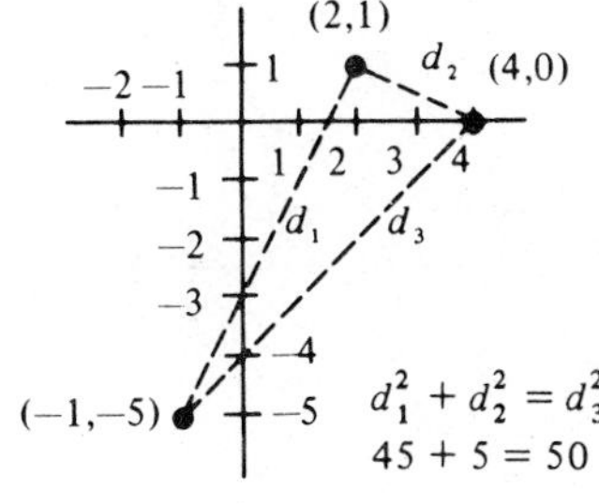

$d_1^2 + d_2^2 = d_3^2$
$45 + 5 = 50$

11.
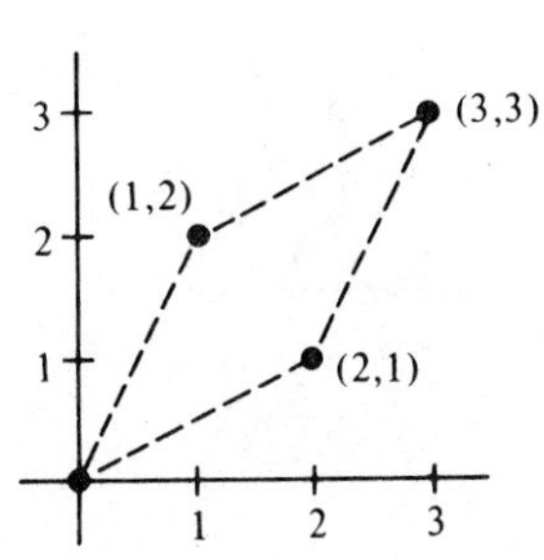

The length of each side is $\sqrt{5}$

13. (3,2), (2,0), (0,−4); d_1, d_2, d_3

Collinear since $d_1 + d_2 = d_3$

$2\sqrt{5} + \sqrt{5} = 3\sqrt{5}$

15. (−2,1), (−1,0), (2,−2)

$d_1 = \sqrt{2}$
$d_2 = \sqrt{13}$
$d_3 = 5$

Not on a line since $d_1 + d_2 > d_3$

17. $x = \pm 3$

19. $2y = 3x - 1$

21. $\left(\dfrac{3x_1 + x_2}{4}, \dfrac{3y_1 + y_2}{4}\right), \left(\dfrac{x_1 + x_2}{2}, \dfrac{y_1 + y_2}{2}\right), \left(\dfrac{x_1 + 3x_2}{4}, \dfrac{y_1 + 3y_2}{4}\right)$

25. (a) 15.25% (b) 14.8% (c) 12.5% (d) 15.5%

27. (a) 175 (b) 800 (c) 280 (d) 750

29. 1972, 1973, 1976

Section 1.2

1. (c)

3. (b)

5. (a)

7. (0, −3), (3/2, 0)

9. (0, −2), (−2, 0) (1, 0)

11. (0, 0), (−3, 0), (3, 0)

13. (0, 1/2), (1,0)

15. (0, 0)

17. $y = x$, (0,0)

19.
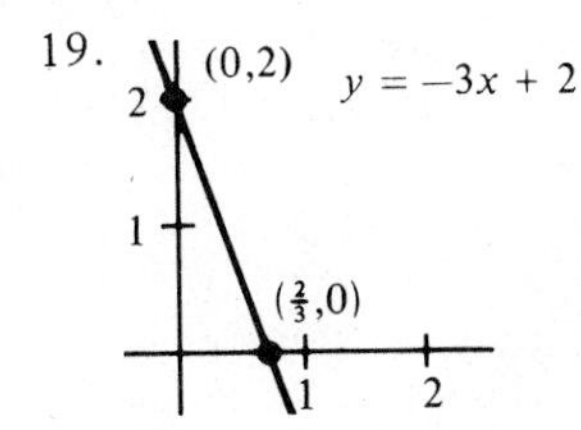

21. $y = 1 - x^2$
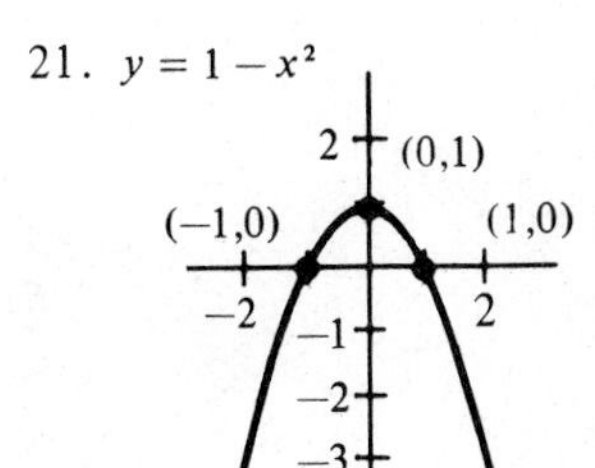

23. $y = -2x^2 + x + 1$

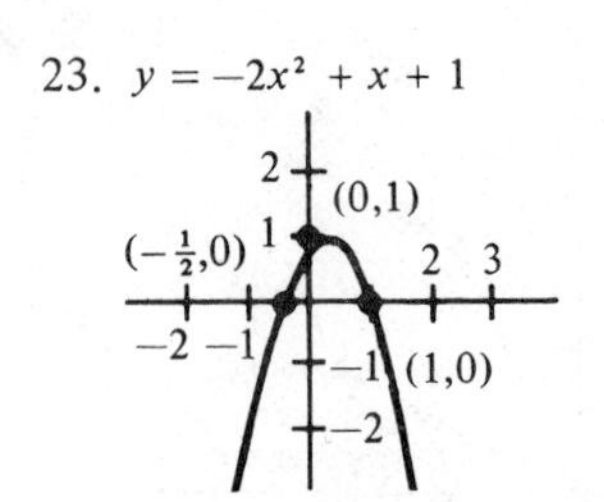

25. $y = x^3 + 2$

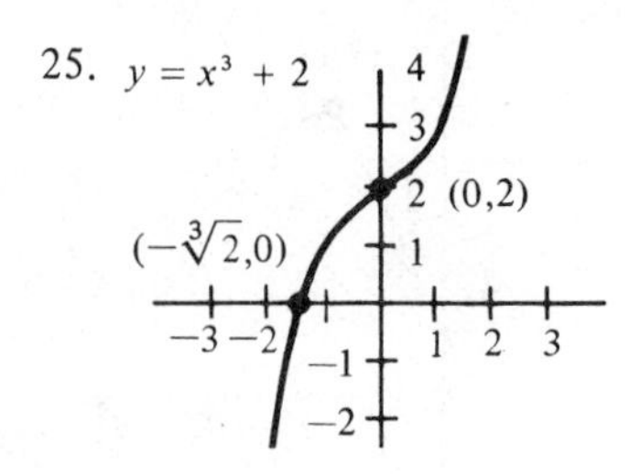

27. $x^2 + 4y^2 = 4$

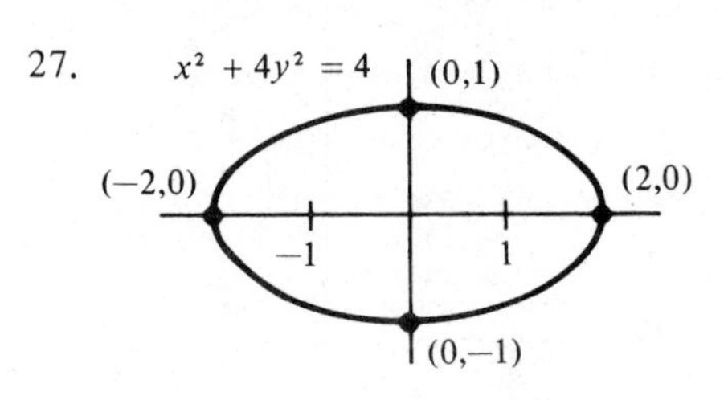

29. $y = (x + 2)^2$

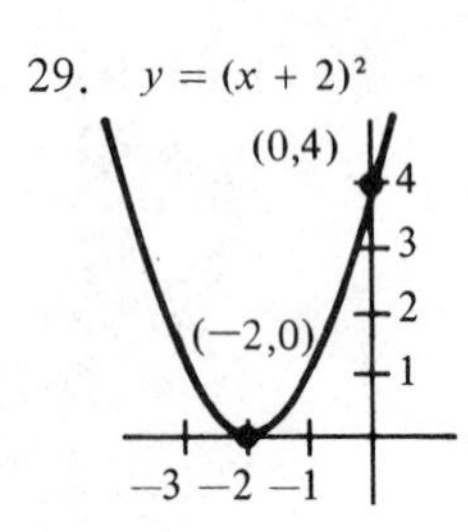

31.

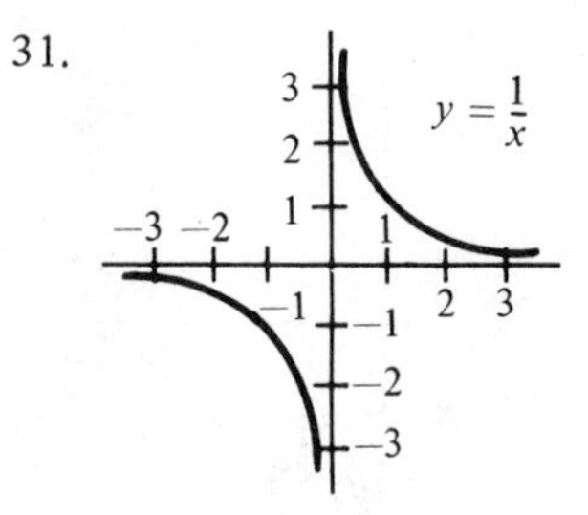

33. $y = |x - 2|$

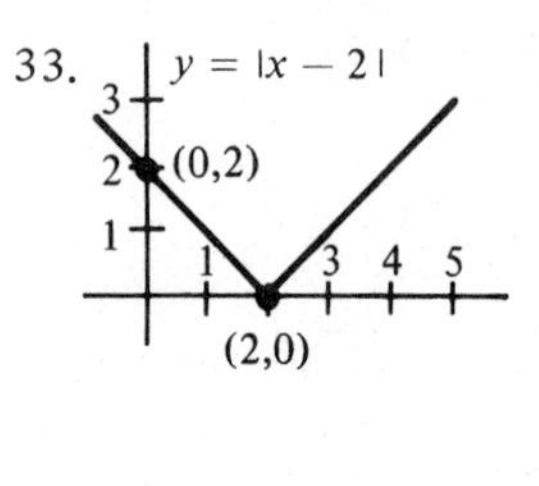

35. $y = \sqrt{x - 3}$

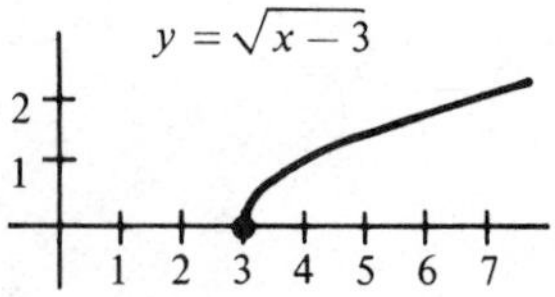

37. $x^2 + y^2 - 9 = 0$

39. $x^2 + y^2 - 4x + 2y - 11 = 0$

41. $x^2 + y^2 + 2x - 4y = 0$

43. $(x - 1)^2 + (y + 3)^2 = 4$

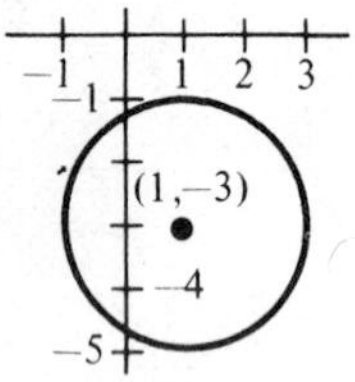

(1, −3)

45. $(x - 1)^2 + (y + 3)^2 = 0$

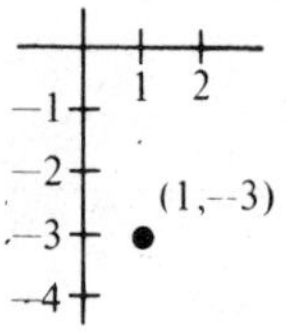

(1, −3)

47. $\left(x - \frac{1}{2}\right)^2 + \left(y - \frac{1}{2}\right)^2 = 2$

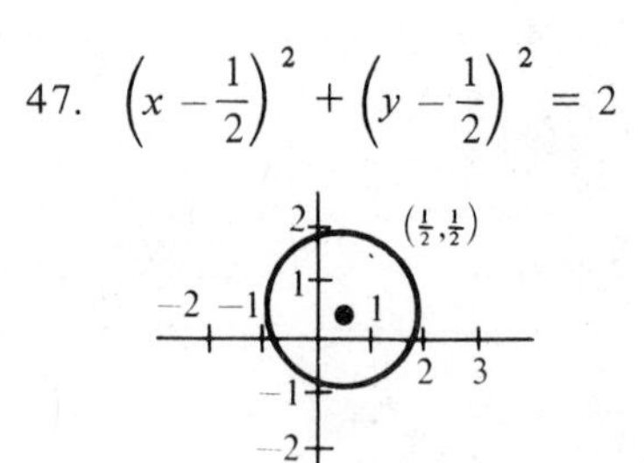

49. $\left(x + \frac{1}{2}\right)^2 + \left(y + \frac{5}{4}\right)^2 = \frac{9}{4}$

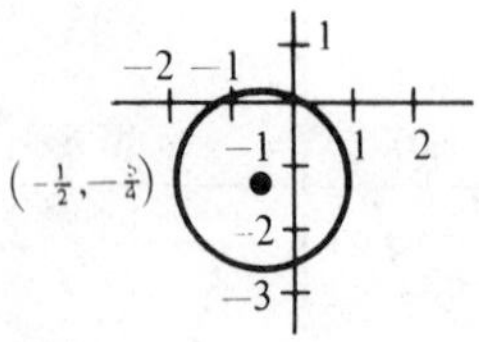

$(-\frac{1}{2},-\frac{5}{4})$

51. (1, 1)

53. (5, 2)

55. (2, 1), (−1, −2)

57. (−1, −1), (0, 0), (1, 1)

59. (−1, −5), (0, −1), (2, 1)

61. (1, 2) is not on the graph
(1, −1) is on the graph
(4, 5) is on the graph

63. (1, 1/5) is on the graph
(2, 1/2) is on the graph
(−1, −2) is not on the graph

65. (a) $C = 21.60x + 5000$ (b) $R = 34.10x$ (c) 400 units

67. $x \approx 192.3$. Therefore, 193 units must be sold.

69. (a)

Year	1970	1971	1972	1973	1974	1975	1976	1977	1978	1979
t	0	1	2	3	4	5	6	7	8	9
CPI	114.4	120.8	128.3	136.9	146.6	157.4	169.3	182.3	196.4	211.6

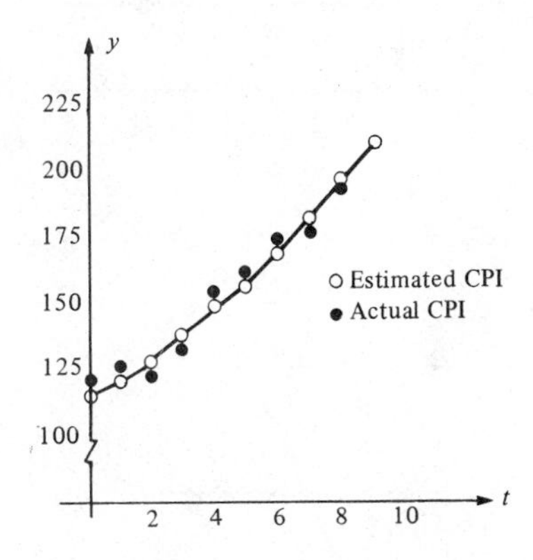

(b) 325.91

71. (a)

Year	1950	1955	1960	1965	1970	1975	1979
t	0	5	10	15	20	25	29
Percentage	20.4	11.3	7.8	6.0	4.8	4.1	3.6

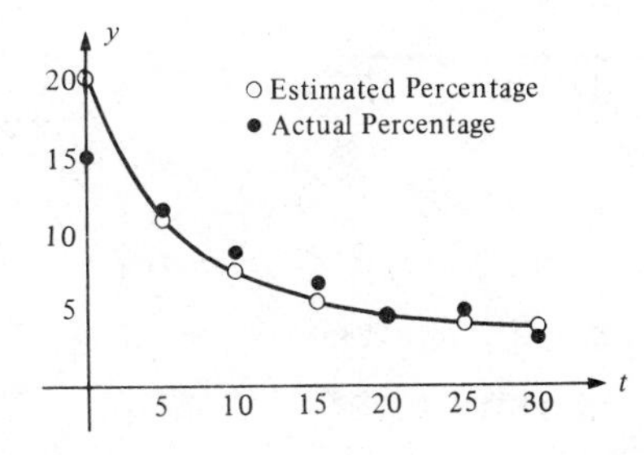

(b) 2.74

Section 1.3

1. 1
3. 0
5. −3
7. $m = 3$

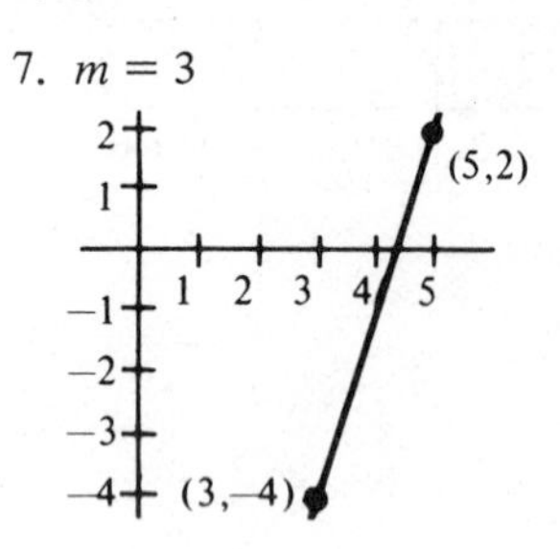

9. $m = 0$

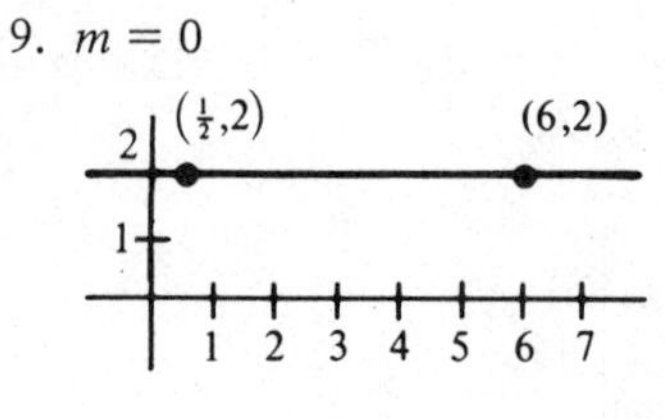

11. m is undefined

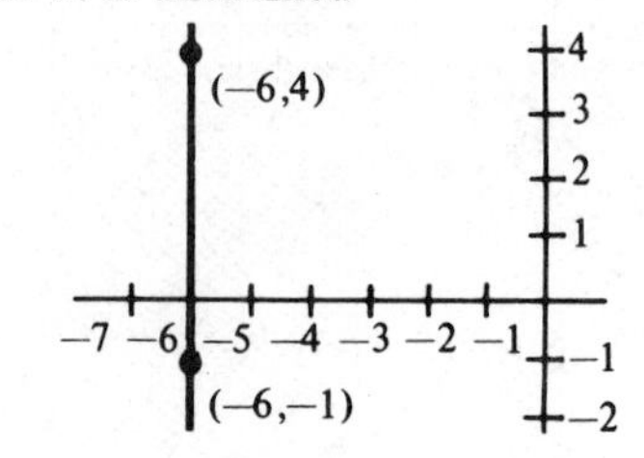

13. $m = 4/3$

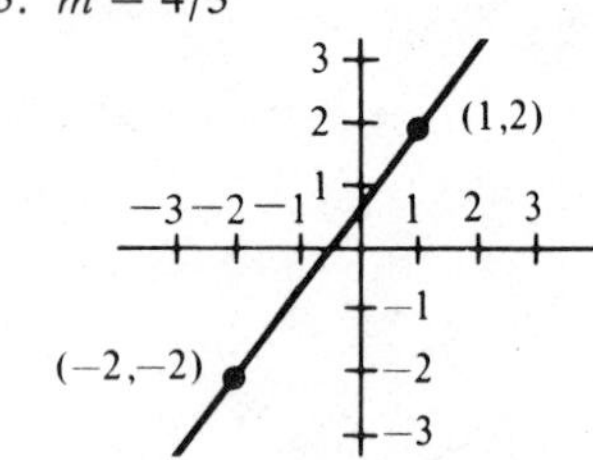

15. $2x - y - 3 = 0$

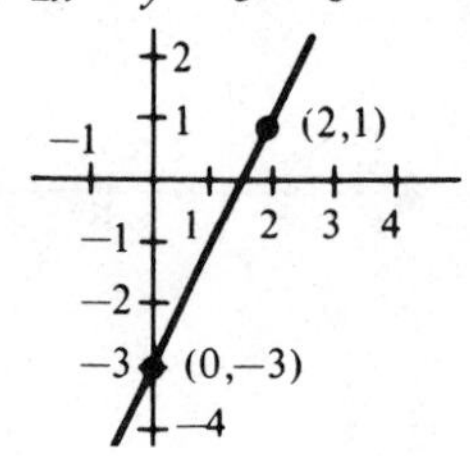

17. $3x + y = 0$

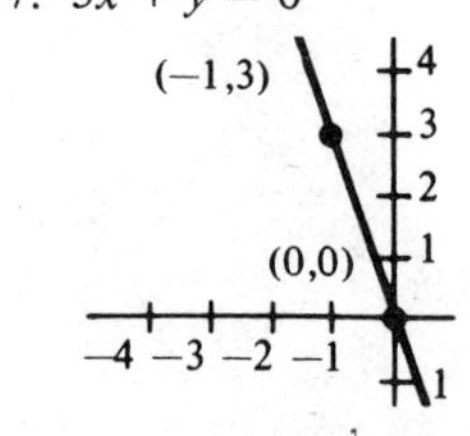

19. $x - 2 = 0$

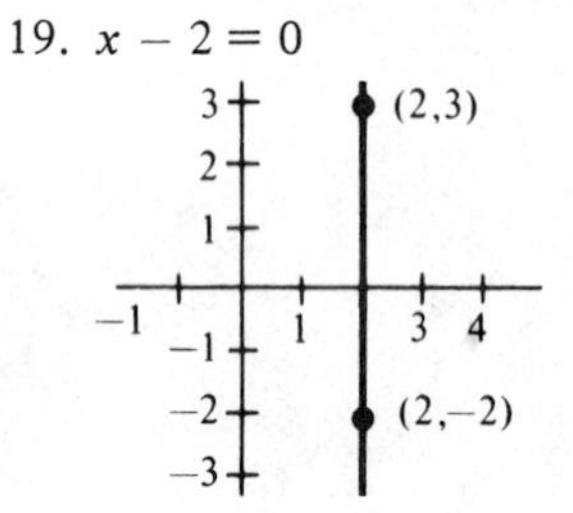

21. $y + 2 = 0$

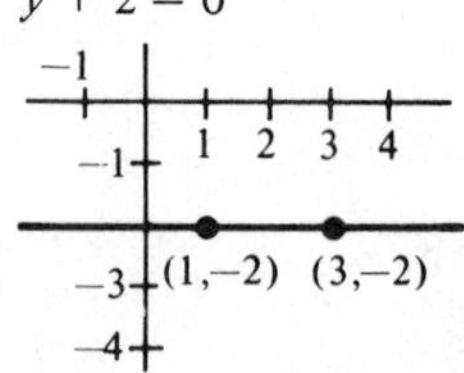

23. $3x - 4y + 12 = 0$

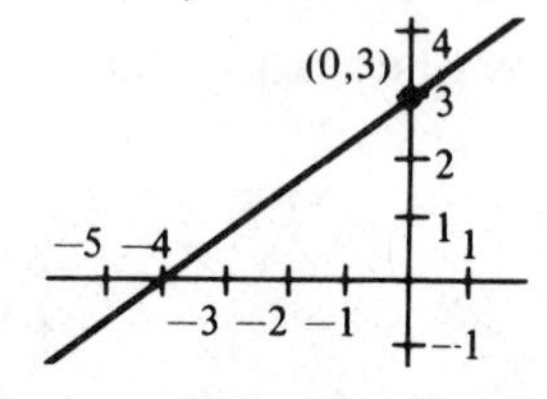

25. $2x - 3y = 0$

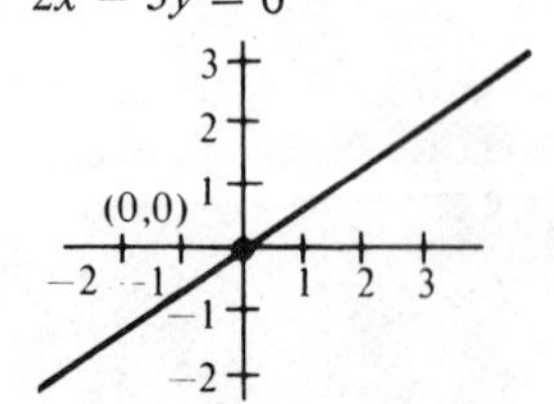

27. $2x + y - 5 = 0$

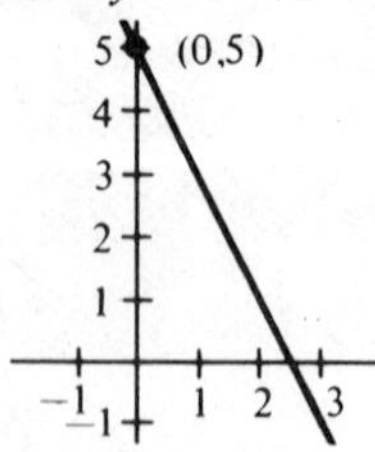

29. $4x - y + 2 = 0$

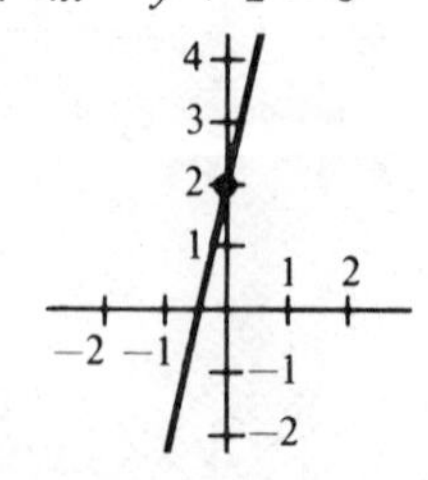

31. $9x - 12y + 8 = 0$

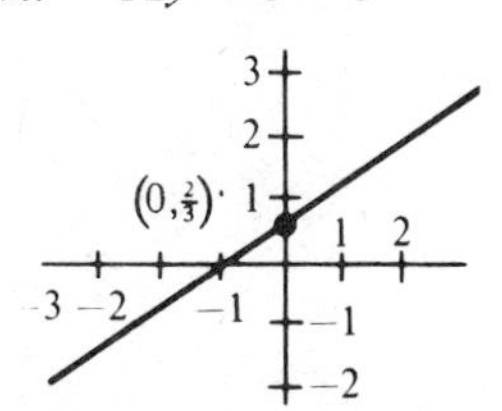

33. $x - 3 = 0$

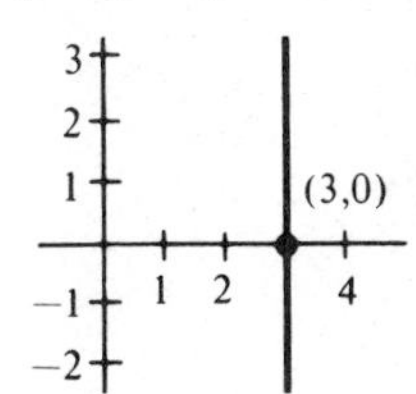

35. (a) $x + y + 1 = 0$ (b) $x - y + 5 = 0$

37. (a) $3x + 4y + 2 = 0$ (b) $4x - 3y + 36 = 0$

39. (a) $y = 0$ (b) $x + 1 = 0$

41.

°C	$-17.\overline{77}$	-10	10	20	$32.\overline{22}$	177
°F	0	14	50	68	90	350.6

43. $V = 875 - 175t$

45. (a) $x = \dfrac{1}{15}(1030 - p)$ (b) 45 (c) 49

47. (a) $C = 105x + 739.40$ (b) \$1369.40

49. $C = 0.22x + 75$

Section 1.4

1. (a) -1 (b) -3 (c) -9 (d) $2b - 3$ (e) $2x - 5$ (f) $-5/2$

3. (a) 0 (b) 1 (c) $\sqrt{3}$ (d) 3 (e) $\sqrt{x + \Delta x + 3}$ (f) $\sqrt{c + 3}$

5. (a) 1 (b) -1 (c) -1 (d) 1 (e) 1 (f) $|x - 1|/(x - 1)$

7. $3 + \Delta x$

9. $3x^2 + 3x\,\Delta x + (\Delta x)^2$

11. $\dfrac{-1}{\sqrt{x - 1}\,(1 + \sqrt{x - 1})}$

13. (a) -1 (b) 2 (c) 56 (d) 17 (e) $-7/4$ (f) $9x^2 - 9x + 2$

15. (a) $1/3$ (b) $-3/4$ (c) -2 (d) 1 (e) $(1 - x^2)/x^2$ (f) $1/(x^2 - 1)$

17. ± 3

19. $10/7$

21. Domain: $[1, \infty)$ Range: $[0, \infty)$

23. Domain: $(-\infty, \infty)$ Range: $[0, \infty)$

25. Domain: $[-3, 3]$ Range: $[0, 3]$

27. Domain: $(-\infty, 0)$ and $(0, \infty)$ Range: $(0, \infty)$

29. Domain: $(-\infty, 0)$ and $(0, \infty)$ Range: -1 and 1

31. y is not a function of x

33. y is a function of x

35. y is a function of x

37. y is not a function of x

39. y is a function of x

41. $f(x) = 2x - 3, f^{-1}(x) = (x + 3)/2$

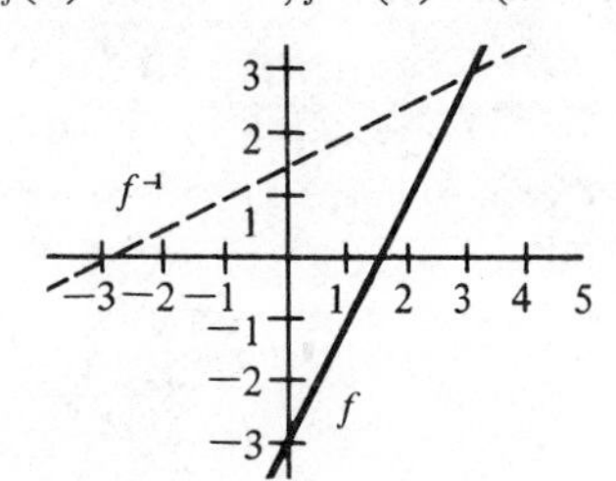

43. $f(x) = x^3, f^{-1}(x) = \sqrt[3]{x}$

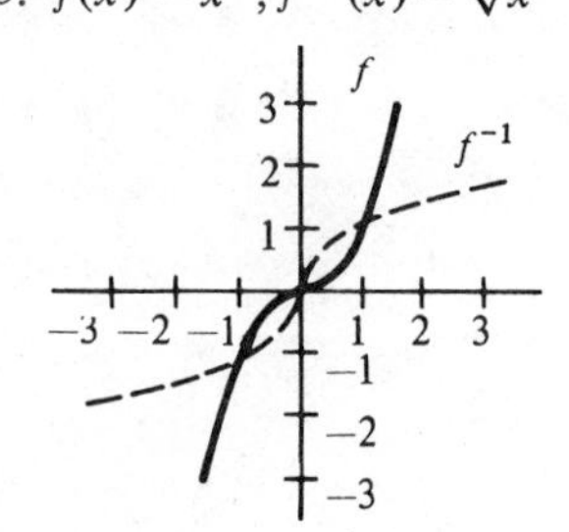

45. $f(x) = \sqrt{x}, f^{-1}(x) = x^2, 0 \leqslant x$

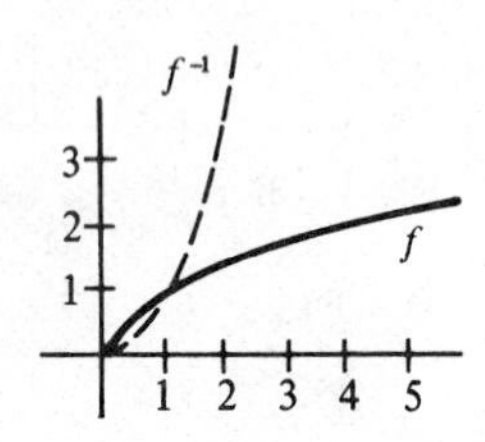

47. $f(x) = 1/x, f^{-1}(x) = 1/x$

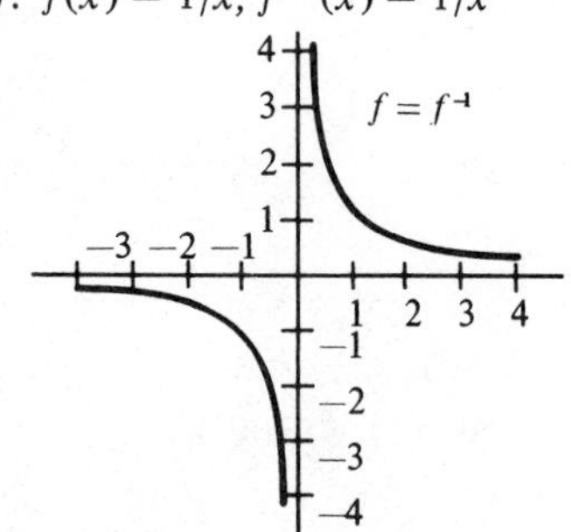

49. Yes, as long as $a \neq 0$

51. No

53. No

55. $V = 1{,}750a + 500{,}000; a > 0$

57. (a) $C = 12.30x + 98{,}000$ (b) $R = 17.98x$ (c) $P = 5.68x - 98{,}000$

59. (a) $x = \dfrac{100(14.75 - p)}{p}$ (b) $x = 47.5$

61. $C = 15\sqrt{x^2 + \frac{1}{4}} + 10(3 - x)$

63. $P = kr^2(0.18 - r)$

Section 1.5

1. 4
3. 4
5. -2
7. 1/6
9. Limit doesn't exist
11. 2
13. 12
15. Limit doesn't exist
17. 2
19. $2x - 2$
21. 1/10
23. 3/2
25. Limit doesn't exist
27. $\sqrt{3}/6$
29. $-1/4$
31. (a) -1 (b) 6 (c) 2/3

33.

x	1.5	1.9	1.99	2	2.01	2.1	2.5
$f(x)$	11.5	13.5	13.95	14	14.05	14.5	16.5

$\lim_{x \to 2} (5x + 4) = 14$

35.

x	1.9	1.99	1.999	2	2.001	2.01	2.1
$f(x)$	0.256	0.251	0.250	undefined	0.250	0.249	0.244

$$\lim_{x \to 2} \frac{x-2}{x^2-4} = \frac{1}{4}$$

37.

x	−0.1	−0.01	−0.001	0	0.001	0.01	0.1
$f(x)$	0.358	0.354	0.354	undefined	0.354	0.353	0.349

$$\lim_{x \to 0} \frac{\sqrt{x+2}-\sqrt{2}}{x} = \frac{1}{2\sqrt{2}} \approx 0.354$$

39.

x	−0.1	−0.01	−0.001	0	0.001	0.01	0.1
$f(x)$	−0.263	−0.251	−0.250	undefined	−0.250	−0.249	−0.238

$$\lim_{x \to 0} \frac{\frac{1}{2+x}-\frac{1}{2}}{x} = -\frac{1}{4}$$

41. (a) 1 (b) 1 (c) 1

43. (a) 0 (b) 0 (c) 0

45. (a) 3 (b) −3 (c) Limit doesn't exist

Section 1.6

1. Continuous
3. Removable discontinuity at $x = -1$
5. Removable discontinuity at $x = 1$
7. Continuous
9. Nonremovable discontinuity at $x = 1$
11. Continuous
13. Removable discontinuity at $x = -2$
    Nonremovable discontinuity at $x = 5$
15. Continuous
17. Nonremovable discontinuity at $x = 2$
19. Nonremovable discontinuity at $x = -2$
21. Continuous
23. Nonremovable discontinuity at $x = n$, where n is any integer
25. Nonremovable discontinuity at $x = 1$
27.

Year	1982	1983	1984	1985	1986	1987
t	0	1	2	3	4	5
$S(t)$	28,500	31,065	33,861	36,908	40,230	43,851

Discontinuous at every positive integer

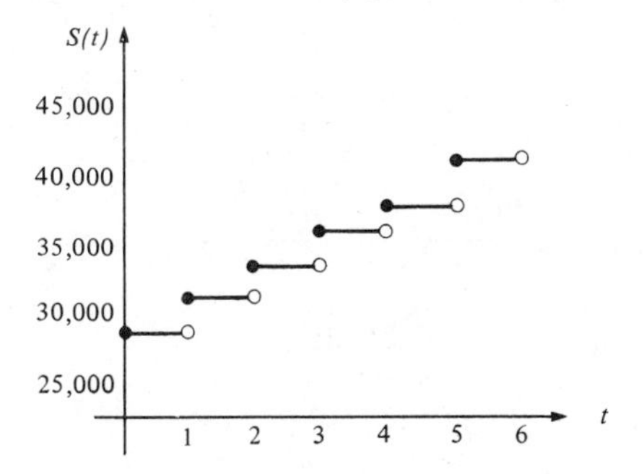

29.

t	0	1	1.8	2	3	3.8
$N(t)$	50	25	5	50	25	5

Discontinuous at every positive even integer. The company replenishes its inventory every two months.

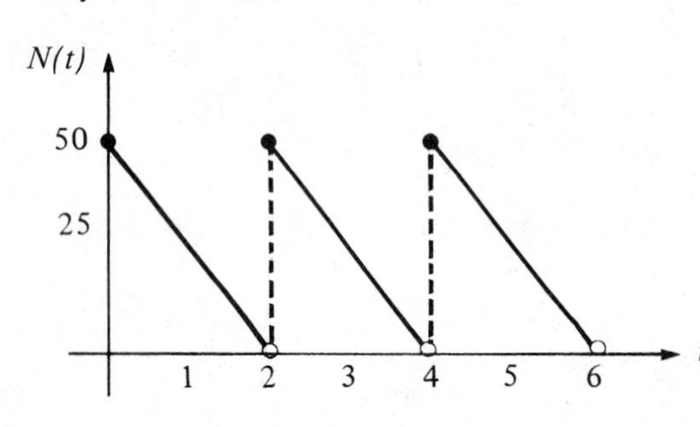

31.

x	0	20	40	60	80	90
$C(x)$	0	0.5	$1.33\overline{3}$	3	8	18

Domain: $[0, 100)$. $C(x)$ is continuous in its domain.

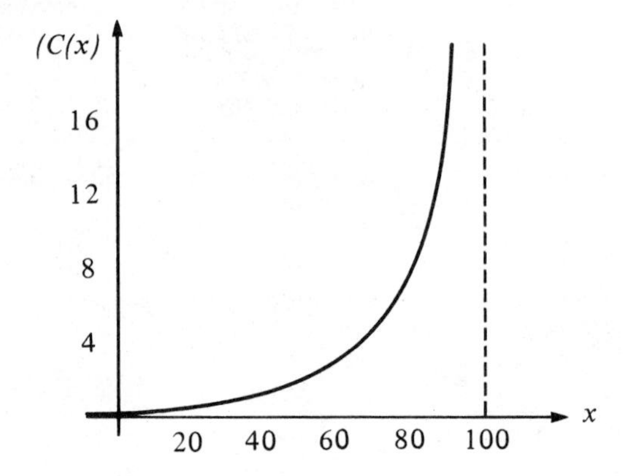

Chapter 2

Section 2.1

1.

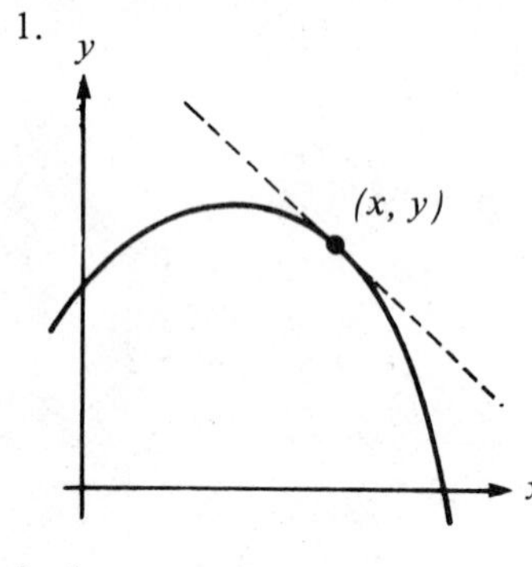

3.

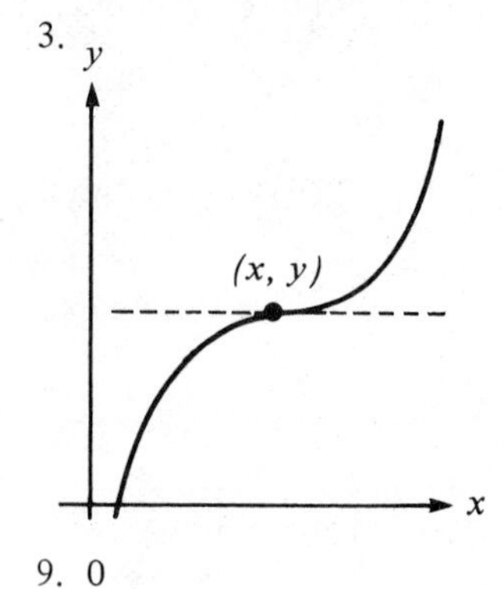

5.

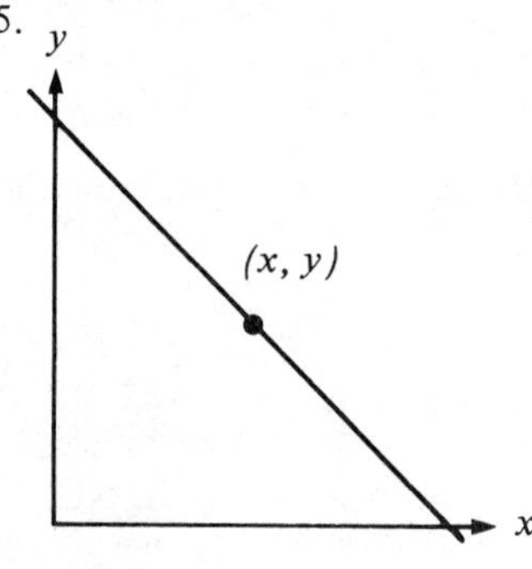

7. 1

9. 0

11. $-1/3$

13.
$$f(x) = 3$$
(i) $$f(x + \Delta x) = 3$$
(ii) $$f(x + \Delta x) - f(x) = 0$$
(iii) $$\frac{f(x + \Delta x) - f(x)}{\Delta x} = 0$$
(iv) $$\lim_{\Delta x \to 0} \frac{f(x + \Delta x) - f(x)}{\Delta x} = 0$$

15.
$$f(x) = -5x$$
(i) $$f(x + \Delta x) = -5x - 5\,\Delta x$$
(ii) $$f(x + \Delta x) - f(x) = -5\,\Delta x$$
(iii) $$\frac{f(x + \Delta x) - f(x)}{\Delta x} = -5$$
(iv) $$\lim_{\Delta x \to 0} \frac{f(x + \Delta x) - f(x)}{\Delta x} = -5$$

17.
$$f(x) = 2x^2 + x - 1$$

(i) $$f(x + \Delta x) = 2x^2 + 4x\,\Delta x + 2(\Delta x)^2 + x + \Delta x - 1$$

(ii) $$f(x + \Delta x) - f(x) = 4x\,\Delta x + 2(\Delta x)^2 + \Delta x$$

(iii) $$\frac{f(x + \Delta x) - f(x)}{\Delta x} = 4x + 2\,\Delta x + 1$$

(iv) $$\lim_{\Delta x \to 0} \frac{f(x + \Delta x) - f(x)}{\Delta x} = 4x + 1$$

19.
$$f(x) = \frac{1}{x - 1}$$

(i) $$f(x + \Delta x) = \frac{1}{x + \Delta x - 1}$$

(ii) $$f(x + \Delta x) - f(x) = \frac{-\Delta x}{(x + \Delta x - 1)(x - 1)}$$

(iii) $$\frac{f(x + \Delta x) - f(x)}{\Delta x} = \frac{-1}{(x + \Delta x - 1)(x - 1)}$$

(iv) $$\lim_{\Delta x \to 0} \frac{f(x + \Delta x) - f(x)}{\Delta x} = \frac{-1}{(x - 1)^2}$$

21.
$$f(t) = t^3 - 12t$$

(i) $$f(t + \Delta t) = t^3 + 3t^2\,\Delta t + 3t(\Delta t)^2 + (\Delta t)^3 - 12t - 12\,\Delta t$$

(ii) $$f(t + \Delta t) - f(t) = 3t^2\,\Delta t + 3t(\Delta r)^2 + (\Delta t)^3 - 12\,\Delta t$$

(iii) $$\frac{f(t + \Delta t) - f(t)}{\Delta t} = 3t^2 + 3t\,\Delta t + (\Delta t)^2 - 12$$

(iv) $$\lim_{\Delta t \to 0} \frac{f(t + \Delta t) - f(t)}{\Delta t} = 3t^2 - 12$$

23. $f'(x) = 2x$

Tangent line: $y = 4x - 3$

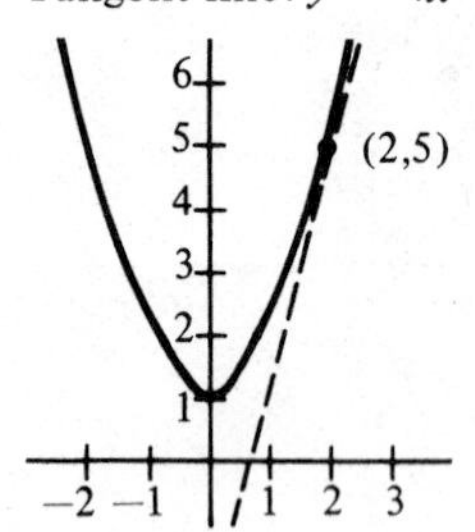

25. $f'(x) = 3x^2$

Tangent line: $y = 12x - 16$

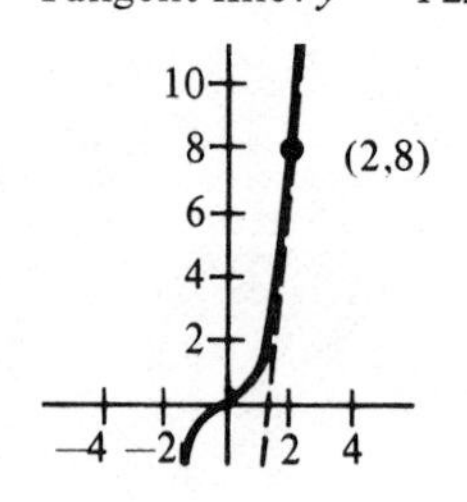

27. $f'(x) = \dfrac{1}{2\sqrt{x+1}}$

Tangent line: $4y = x + 5$

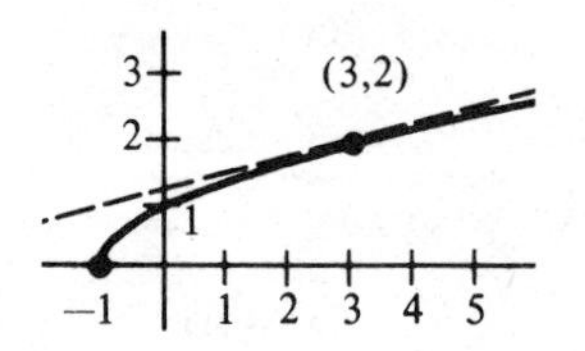

29. $f'(x) = -1/x^2$
Tangent line: $y = -x + 2$

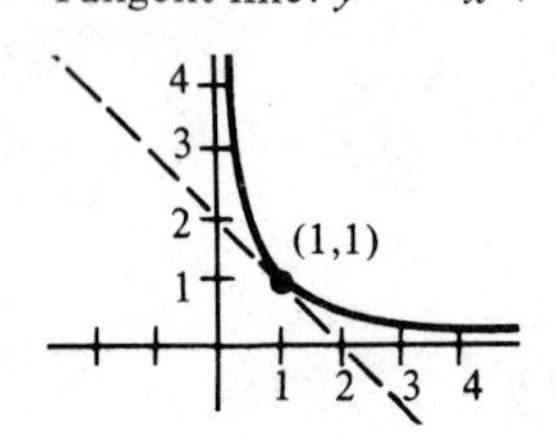

31. $y = 3x - 2$ or $y = 3x + 2$

33. $y = 2x + 1$ and $y = -2x + 9$

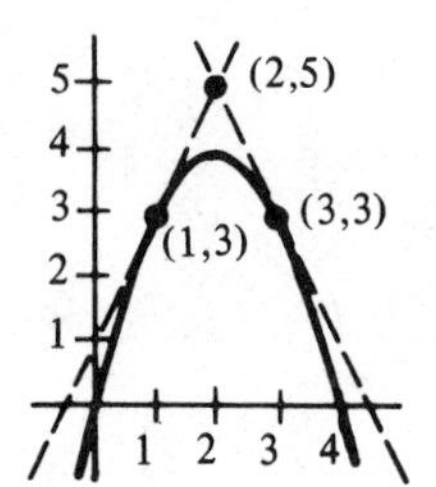

Section 2.2

1. $y' = 0$

3. $f'(x) = 1$

5. $g'(x) = 2x$

7. $f'(t) = -4t + 3$

9. $s'(t) = 3t^2 - 2$

11. $f'(x) = -1/x^2; -1$

13. $f'(t) = 3/5t^2; 5/3$

15. $y' = -1/x^4; -1$

17. $y' = 8x + 4; 4$

19. $f'(x) = 32x - 128x^3$

21. $f'(x) = 2x + \dfrac{4}{x^2}$

23. $f'(x) = 3x^2 - 3 + \dfrac{8}{x^5}$

25. $f'(x) = 1 - \dfrac{8}{x^3}$

27. $f'(x) = -2\pi/9x^3$

29. $f'(x) = 3x^2 - 1$

31. $f'(x) = \dfrac{4}{5x^{1/5}}$

33. $f'(x) = \dfrac{1}{3x^{2/3}} + \dfrac{1}{5x^{4/5}}$

35. $f'(x) = \dfrac{-1}{2x^{3/2}} - \dfrac{2}{x^3} - \dfrac{4}{x^5}$

37. 1/2

39. 3/2

41. 3

43. −1

45. −2

47. $y = 4x + 2$

49. No horizontal tangents

Section 2.3

1. (a) \$0.22 (b) \$0.264 (c) \$0.316 (d) \$0.325

3. (a) 17 (b) 9

5. (a) 50 vibrations/s (b) $33.33\overline{3}$ vibrations/s

7. (a) 4/9 (b) 1/3 (c) 0 (d) −5/9

9. (a) 0 ft/s (b) 15 ft/s (c) 30 ft/s (d) 45 ft/s

11. (a) 200 ft/s (b) 104 ft/s

13. (a) \$2.00 (b) \$2.00

15. (a)

x	10	15	20	23	25
P	902.00	1308.00	1820.00	1970.91	1940.00

(b)

x	10	15	20	23	25
$\frac{dP}{dx}$	37.40	108.40	79.80	14.83	−48.40

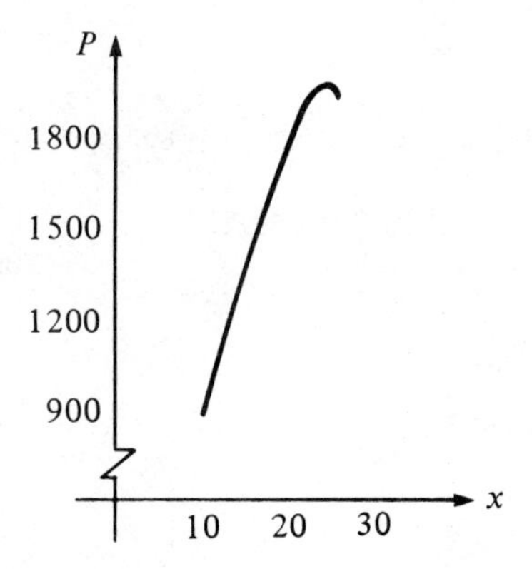

17. (b)

x	10	15	20	25	30	35	40
C	2625.00	1750.00	1312.50	1050.00	875.00	750.00	656.25

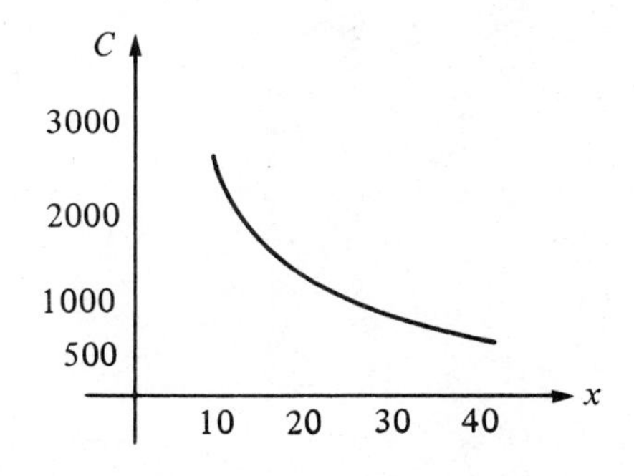

(c)

x	10	15	20	25	30	35	40
$\frac{dC}{dx}$	−262.50	−116.67	−65.63	−42.00	−29.17	−21.43	−16.41

(d) −238.64

(e) −28.23

19. (a) $R = xp = \dfrac{x(1100 - x)}{400}$ (b) $\dfrac{dR}{dx} = \dfrac{550 - x}{200}$

x	300	400	500	550	600	700
$\frac{dR}{dx}$	1.25	0.75	0.25	0	−0.25	−0.75
p	2.00	1.75	1.50	1.38	1.25	1.00

21. (a) $P = \dfrac{x(1100 - x)}{400} - (65 + 1.25x)$ (b) $\dfrac{dP}{dx} = \dfrac{550 - x}{200} - 1.25$

$P(250) = \$153.75$ $P(300) = \$160.00$
$P'(250) = \$0.25$ $P'(300) = 0$

23. (a) $P = \dfrac{x(1100 - x)}{400} - (65 + 1.50x)$ (b) $\dfrac{dP}{dx} = \dfrac{550 - x}{200} - 1.50$

$P(250) = \$91.25$ $P(300) = \$85.00$
$P'(250) = 0$ $P'(300) = -\$0.25$

Section 2.4

1. $2x^2; 0$

3. $-7/x^4; -7$

5. $(x - 1)^2 (5x^2 + 2x + 2); 0$

7. $(3x - 5)(x - 1); 5$

9. $3\left(2x^5 + x^2 - 2x + \dfrac{1}{x^2}\right); 6$

11. $-5/(2x - 3)^2$

13. $2/(x + 1)^2$

15. $6x/(4 - 3x^2)^2$

17. $\dfrac{5}{6x^{1/6}} + \dfrac{1}{x^{2/3}}$

19. $\dfrac{-t^2 - 2t}{(t^2 + 2t + 2)^2}$

21. $6s^2(s^3 - 2)$

23. $\dfrac{2x^2 + 8x - 1}{(x + 2)^2}$

25. $15x^4 - 48x^3 - 33x^2 - 32x - 20$

27. $-4c^2x/(c^2 + x^2)^2$

29. $y = -x + 4$

31. $y = -x - 2$

33. $(0, 0)$ and $(2, 4)$

35. $P'(t) = \dfrac{2000(50 - t^2)}{(50 + t^2)^2}$; $P'(2) = 31.55$ bacteria/h

37. $T'(t) = \dfrac{-1400(t + 2)}{(t^2 + 4t + 10)^2}$; (a) $T'(1) = -18.67°/\text{h}$
(b) $T'(3) = -7.28°/\text{h}$
(c) $T'(5) = -3.24°/\text{h}$
(d) $T'(10) = -0.75°/\text{h}$

Section 2.5

1. $6(2x - 7)^2$
3. $12x(x^2 - 1)^2$
5. $-1/(x - 2)^2$
7. $-2/(t - 3)^3$
9. $-9x^2/(x^3 - 4)^2$
11. $2x(x - 2)^3(3x - 2)$
13. $(x + 3)^2(4x - 3)$
15. $\dfrac{1}{2\sqrt{t + 1}}$
17. $\dfrac{t + 1}{\sqrt{t^2 + 2t - 1}}$
19. $6x/(9x^2 + 4)^{2/3}$
21. $\dfrac{2x}{\sqrt{x^2 + 4}}$
23. $\dfrac{4x}{3(x^2 - 9)^{1/3}}$
25. $\dfrac{-1}{2(x + 2)^{3/2}}$
27. $-3x^2/(x^3 - 1)^{4/3}$
29. $\dfrac{1}{2\sqrt{x}(\sqrt{x} + 1)^2}$
31. $12x(x^2 - 1)^2$
33. $\dfrac{6x}{(9x^2 + 4)^{2/3}}$
35. $\dfrac{1}{(x^2 + 1)^{3/2}}$
37. $\dfrac{x + 2}{(x + 1)^{3/2}}$
39. $\dfrac{1 - 3x^2 - 4x^{3/2}}{2\sqrt{x}(x^2 + 1)^2}$
41. $-5/(t - 1)^2$
43. $\dfrac{3t(t^2 + 3t - 2)}{(t^2 + 2t - 1)^{3/2}}$
45. $\dfrac{5}{6(t + 1)^{1/6}}$
47. $\dfrac{-1}{2x^{3/2}\sqrt{x + 1}}$
49. $\dfrac{t}{\sqrt{1 + t}}$
51. $\dfrac{t^2 + 4}{4t^{3/2}\sqrt{t^2 - 4}}$
53. $9x - 5y - 2 = 0$
55. $y = 0$
57. $\dfrac{dC}{dt} = 9.12\sqrt{1.52t + 10}$

t	1	2	3	4	5
$\dfrac{dC}{dt}$	30.95	32.93	34.80	36.57	38.26

59. (a) 6 board ft/in (b) 18 board ft/in (c) 30 board ft/in (d) 48 board ft/in

61. $\dfrac{dN}{dt} = \dfrac{4800t}{(t^2 + 2)^3}$

(a)

t	0	1	2	3	4
$\dfrac{dN}{dt}$	0	177.78	44.44	10.82	3.29

(b) 400

Section 2.6

1. $f''(x) = 2$
3. $f''(x) = 0$
5. $f'''(x) = 24x - 12$
7. $f^{(4)}(x) = -24/x^5$
9. $f''(t) = 4/9t^{7/3}$
11. $f'''(x) = -9/2x^5$
13. $f'''(x) = 12$
15. $f''(x) = 24(x^2 - 1)(5x^2 - 1)$
17. $f''(x) = 4/(x - 1)^3$
19. $h''(s) = 2(s - 1)(10s^2 - 2s + 1)$
21. $g'''(x) = \dfrac{-3}{8(4 - x)^{5/2}}$
23. $f''(x) = 6(x - 3) = 0$ when $x = 3$
25. $f''(x) = 6(x - 2) = 0$ when $x = 2$
27. $f''(x) = 12(x - 3)(x - 1) = 0$ when $x = 1$ or $x = 3$
29. $f''(x) = \dfrac{2x(x^2 - 9)}{(x^2 + 3)^3} = 0$ when $x = 0$, $x = -3$, or $x = 3$
31. (a) $v(t) = -32t + 48$ (b) 1.5 s (c) 36 ft
    $a(t) = -32$
33.

t	0	10	20	30	40	50	60
$\dfrac{ds}{dt}$	0	45	60	67.5	72	75	77.14
$\dfrac{d^2s}{dt^2}$	9.00	2.25	1.00	0.56	0.36	0.25	0.18

Section 2.7

1. $-x/y$
3. $-y/x$
5. $-\sqrt{y/x}$
7. $\dfrac{y - 3x^2}{2y - x}$
9. $\dfrac{36x}{2y(x^2 + 9)^2}$; undefined at (3,0)
11. $\dfrac{1 - 3x^2y^3}{3x^3y^2 - 1}$; -1
13. $-\sqrt[3]{\dfrac{y}{x}}$; $-1/2$
15. $\dfrac{4xy - 3x^2 - 3y^2}{2x(3y - x)}$; $-15/28$

17. (a) $y = \pm\sqrt{16 - x^2}$

(b) $y' = \dfrac{\mp x}{\sqrt{16 - x^2}}$

(c) $y' = -x/y$

(d) $y = \sqrt{16 - x^2}$ (upper semicircle)

$y = -\sqrt{16 - x^2}$ (lower semicircle)

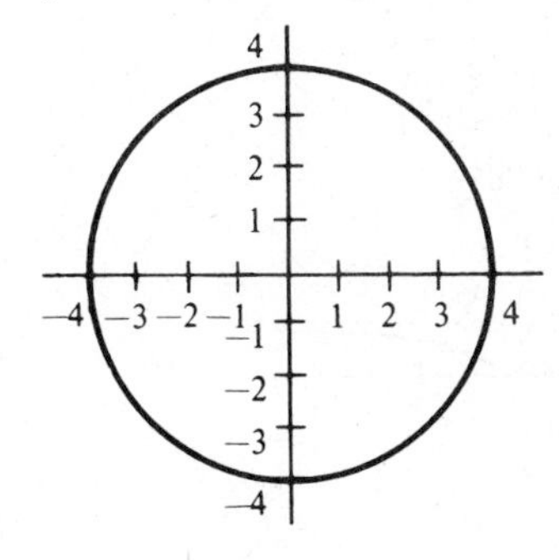

19. (a) $y = \pm\dfrac{1}{4}\sqrt{144 - 9x^2}$

(b) $y' = \mp\dfrac{9}{4}\dfrac{x}{\sqrt{144 - 9x^2}}$

(c) $y' = -9x/16y$

(d) $y = \dfrac{1}{4}\sqrt{144 - 9x^2}$ (upper semi-ellipse)

$y = -\dfrac{1}{4}\sqrt{144 - 9x^2}$ (lower semi-ellipse)

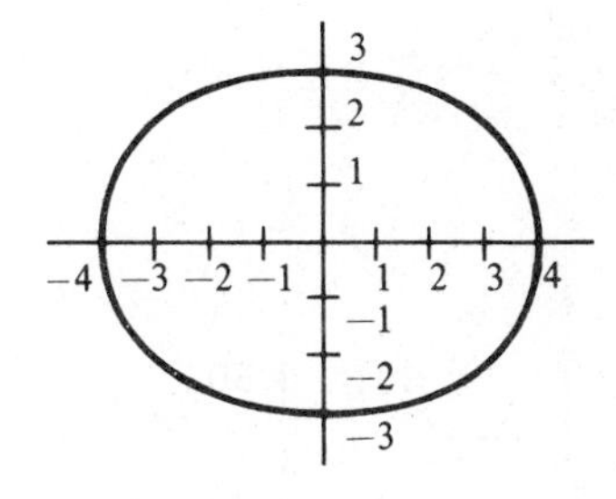

21. $\dfrac{2(x + y)}{x^2} = \dfrac{10}{x^3}$

23. $-16/y^3$

25. $3x/4y$

Chapter 3

Section 3.1

1. Decreasing on $(-\infty, 3)$
   Increasing on $(3, \infty)$

3. Increasing on $(-\infty, -2)$ and $(2, \infty)$
   Decreasing on $(-2, 2)$

5. Increasing on $(-\infty, 0)$
   Decreasing on $(0, \infty)$

7. No critical numbers
Increasing on $(-\infty, \infty)$

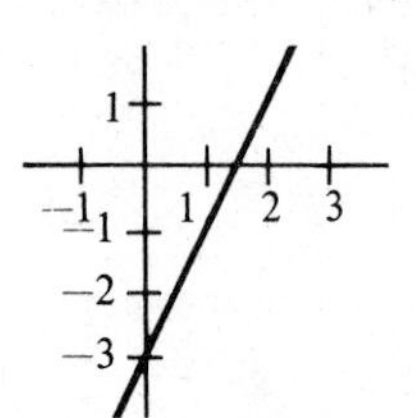

9. Critical number: $x = 1$
Decreasing on $(-\infty, 1)$
Increasing on $(1, \infty)$

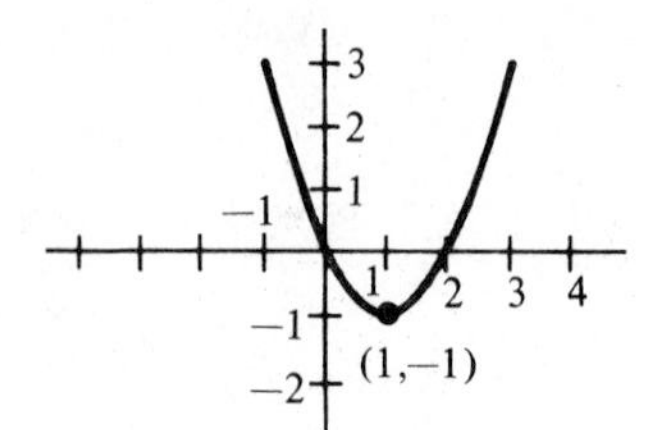

11. Critical numbers: $x = 0, 2$
Increasing on $(-\infty, 0)$
Decreasing on $(0, 2)$
Increasing on $(2, \infty)$

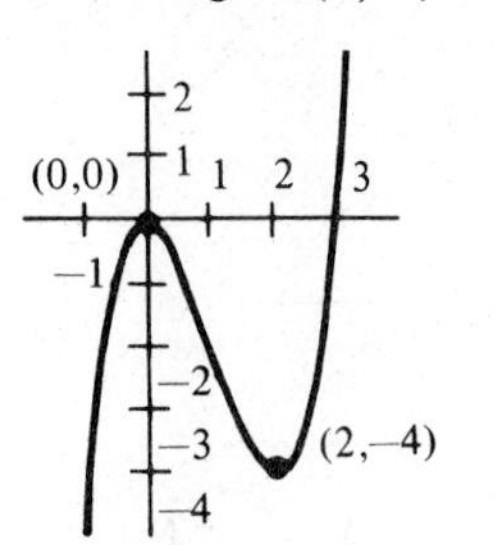

13. Critical numbers: $x = \pm 2$
Decreasing on $(-\infty, -2), (2, \infty)$
Increasing on $(-2, 2)$

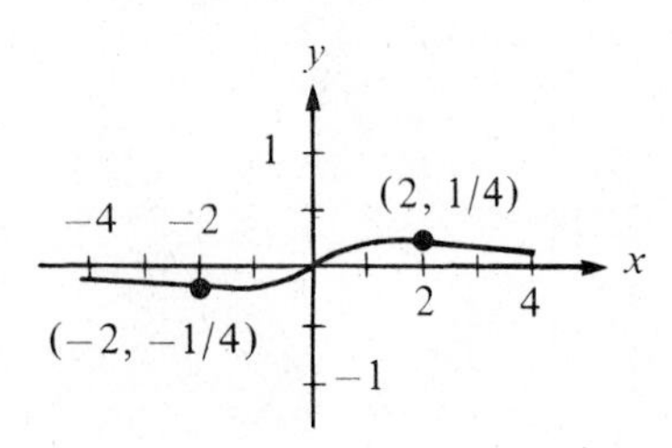

15. Critical number: $x = 0$
Decreasing on $(-\infty, 0)$
Increasing on $(0, \infty)$

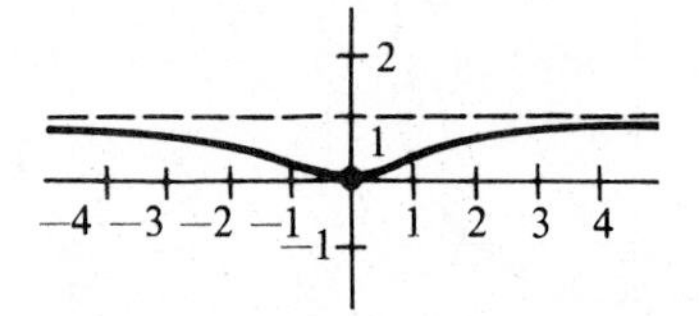

17. Domain: $(-\infty, -1), (1, \infty)$
Critical numbers: $x = \pm 1$
Decreasing on $(-\infty, -1)$
Increasing on $(1, \infty)$

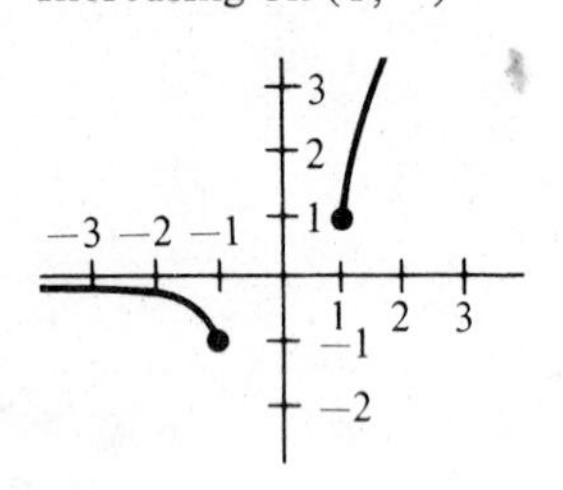

19. Critical number: $x = 0$
Decreasing on $(-\infty, 0)$
Increasing on $(0, \infty)$

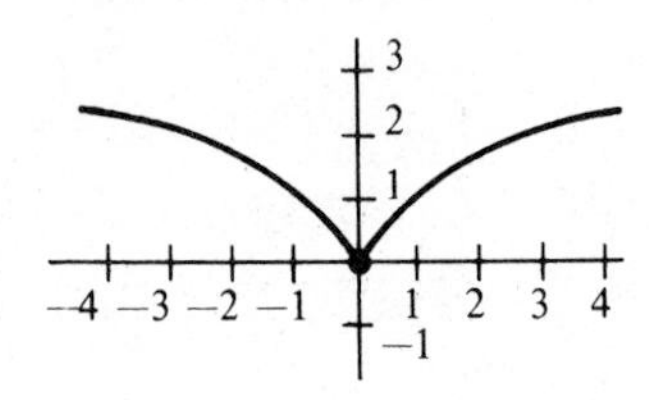

21. Domain: $(-\infty, 6]$
Critical numbers: $x = 4, 6$
Increasing on $(-\infty, 4)$
Decreasing on $(4, 6]$

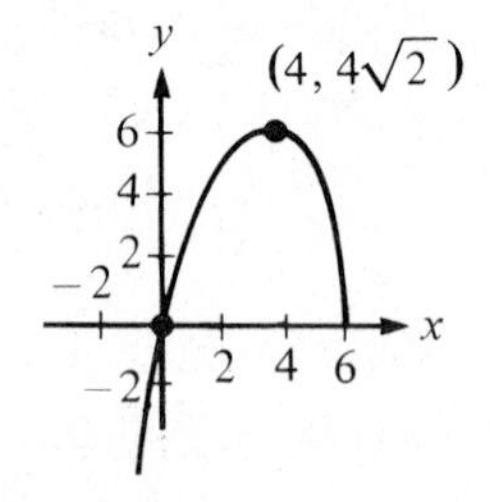

23. Critical number: $x = 0$
Increasing on $(-\infty, 0)$
Decreasing on $(0, \infty)$

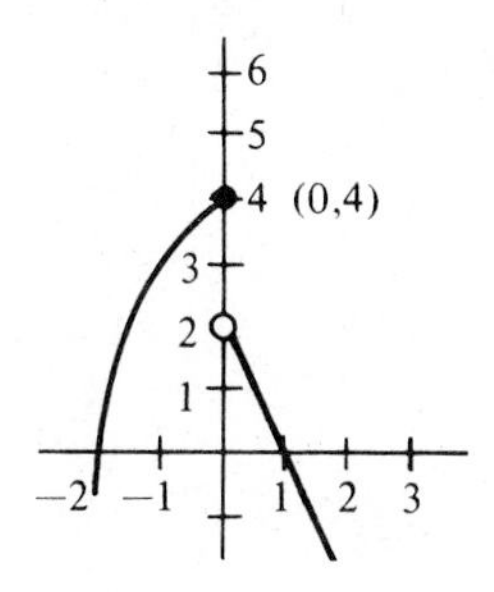

25. Increasing on $[0, 84.34)$
Decreasing on $(84.34, 120]$

27. Decreasing on $[0, 6.0)$
Increasing on $(6.0, 14]$

Section 3.2

1. $(1, 5)$, relative maximum

3. $(3, -9)$, relative minimum

5. $(-2, 20)$, relative maximum
$(1, -7)$, relative minimum

7. $(0, 15)$ relative maximum
$(4, -17)$, relative minimum

9. $(3/2, -27/16)$, relative minimum

11. No relative extrema

13. $(1, 2)$, relative minimum
$(-1, -2)$, relative maximum

15. $(0, 0)$, relative maximum

17. $(2, -44)$, relative minimum

19. $(-3, -8)$, relative maximum
$(1, 0)$, relative minimum

21. $(2, 2)$, minimum
$(-1, 8)$, maximum

23. $(0, 0)$, minimum
$(2, 4)$, maximum

25. $(-1, -4)$, and $(2, -4)$, minima
$(0, 0)$ and $(3, 0)$, maxima

27. $(0, 0)$, minimum
$(-1, 5)$, maximum

29. $(1, -1)$, minimum
$(0, -1/2)$, maximum

31. Yes

33. Yes

35. No

37. No

39. (a) 96 ft/s (b) 144 ft (c) Downward (d) 80 ft

41. $r = 2R/3$

45. $T = 10^\circ\text{C}$

45. $|f''(0)| = 2$ maximum

47. $\left| f''\left(\sqrt[3]{\dfrac{-20+\sqrt{432}}{2}}\right)\right| \cong 1.96$ maximum

49. $|f^{(4)}(1/2)| = 360$ maximum

51. $|f^{(4)}(0)| = 56/81$ maximum

Section 3.3

1. Concave upward on $(-\infty, \infty)$

3. Concave upward on $(-\infty, -2)$ and $(2, \infty)$
   Concave downward on $(-2, 2)$

5. Concave upward on $(-\infty, -1)$ and $(1, \infty)$
   Concave downward on $(-1, 1)$

7. $(3, 9)$, relative maximum

9. $(5, 0)$, relative minimum

11. $(0, 3)$, relative maximum
    $(2, -1)$, relative minimum

13. $(3, -25)$, relative minimum

15. $(0, -3)$, relative minimum

17. $(-2, -4)$, relative maximum
    $(2, 4)$ relative minimum

19. $(2, 0)$, point of inflection

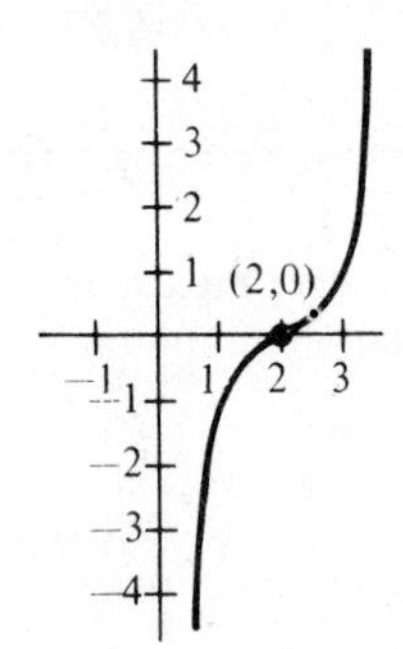

21. $(2, -16)$, relative minimum
    $(-2, 16)$, relative maximum
    $(0, 0)$, point of inflection

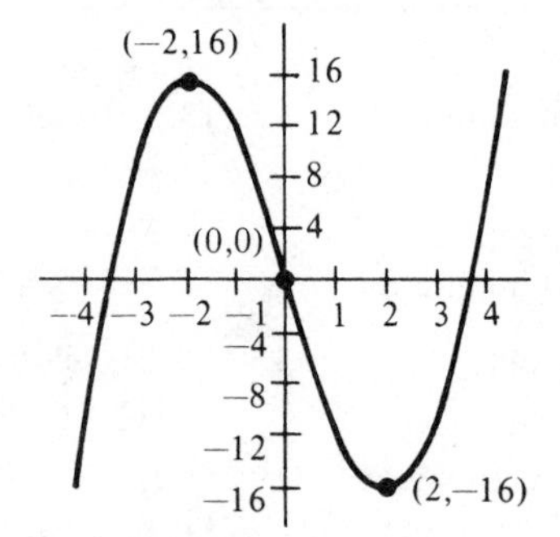

23. $(-2, -4)$, relative minimum
    $(0, 0)$, relative maximum
    $(2, -4)$, relative minimum
    $(2/\sqrt{3}, -20/9)$ point of inflection
    $(-2/\sqrt{3}, -20/9)$ point of inflection

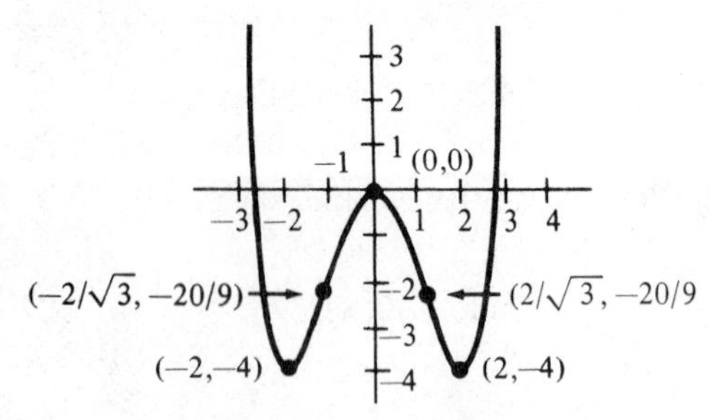

25. $(0, 4)$, relative maximum
    $(-\sqrt{3}/3, 3)$, point of inflection
    $(\sqrt{3}/3, 3)$, point of inflection

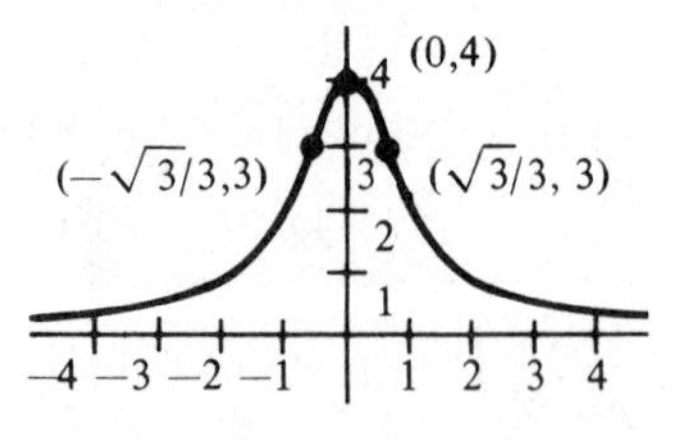

27. $(-2, -2)$, relative minimum

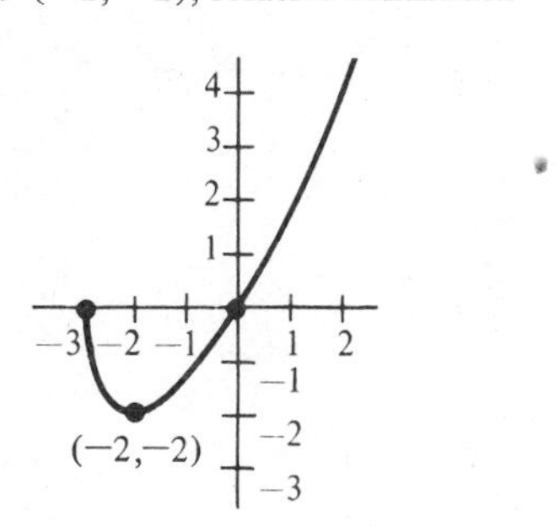

29. $\left(\frac{7}{4}, -\frac{15^4}{3(4^4)}\right)$, relative minimum

$(-2, 0)$, point of inflection

$\left(\frac{1}{2}, -\left(\frac{5}{2}\right)^4\right)$, point of inflection

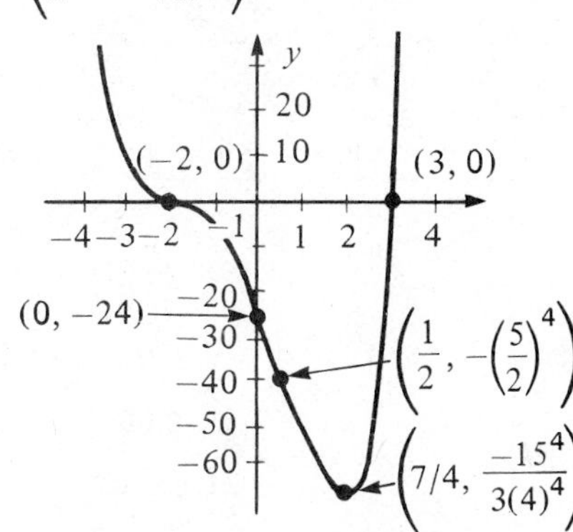

31. $(-3, 0)$, relative maximum
$(-1, -\sqrt[3]{4})$, relative minimum
$(0, 0)$, point of inflection

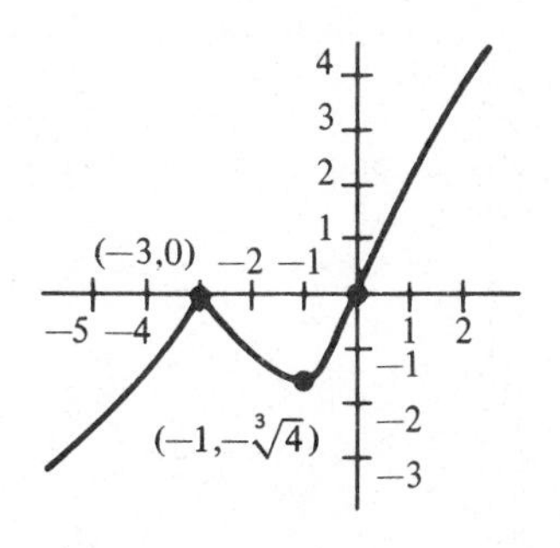

33. 100 units

35. $x = \dfrac{15 - \sqrt{33}}{16} L \approx 0.578L$ ft

Section 3.4

1. $l = w = 25$ ft

3. 12 and 6

5. 55 and 55

7. $\sqrt{192}$ and $\sqrt{192}$

9. $x = 25, y = 100/3$

11. $(7/2, \sqrt{7/2})$

13. $V = 128$ when $x = 2$

15. 0.392 ft × 1.215 ft × 2.215 ft

17. 50 m × $100/\pi$ m

19. $r \approx 1.51$ in, $h = 2r \approx 3.02$ in

21. $x = 3, y = 3/2$

23. $x = y = 5\sqrt{2}/2 \approx 3.54$

25. 50/3 in X 50/3 in X 100/3 in

27. Radius: $8/(4+\pi)$; side of square: $16/(4+\pi)$

29. 1 mi down coast from nearest point

31. (4, 0) and (0, 6)

33. $w = 8\sqrt{3}, h = 8\sqrt{6}$

35. $32\pi r^3/81$

Section 3.5

1. 4500
3. 300
5. 200
7. 200
9. 30
11. 1500
13. 20
15. 200
17. Line should run from the power station to a point across the river $3/(2\sqrt{7})$ mi downstream.
19. 8%
21. $x = 3$
23. (a) 6.5% (b) −1.3 (c) −4/3 (d) $x = 40/3, p = \sqrt{10/3}$
25. $p = 4\sqrt{3}/3, x = 32/3$
27. (a) $80.00 (b) $99.29
29. $92.50

Section 3.6

1. $x = 0$, even
3. $x = -1$, odd; $x = 2$, odd
5. $x = -1$, odd; $x = 1$, odd
7. 2/3
9. 0
11. $-\infty$
13. ∞
15. 5
17.

x	1	10	10^2	10^4	10^6
$f(x)$	2.000	0.348	0.101	0.010	0.001

$$\lim_{x\to\infty} \frac{x+1}{x\sqrt{x}} = 0$$

19.

x	1	10	10^2	10^4	10^6
$f(x)$	1.0	5.1	50.1	5,000.1	500,000.2

$$\lim_{x\to\infty} x^2 - x\sqrt{x(x-1)} = \infty$$

21. $y = \dfrac{2 + x}{1 - x}$

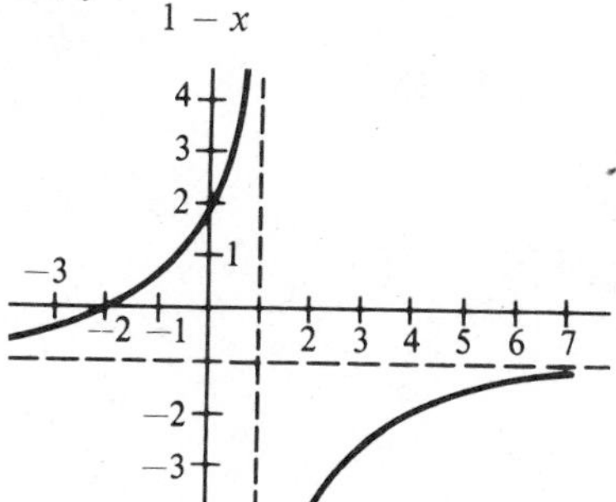

23. $y = \dfrac{x^2}{x^2 + 9}$

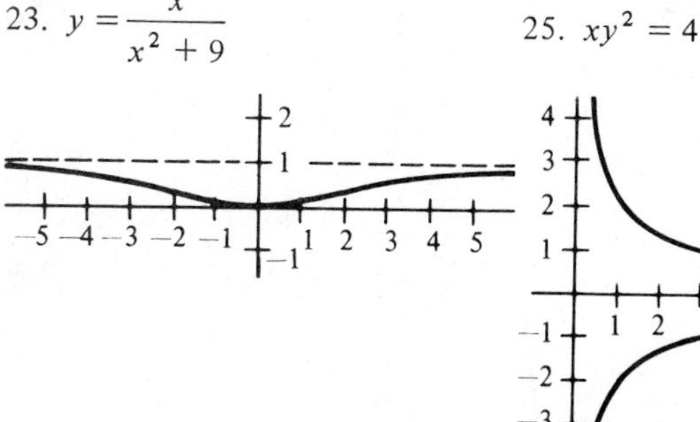

25. $xy^2 = 4$

27. $y = \dfrac{2x}{1 - x}$

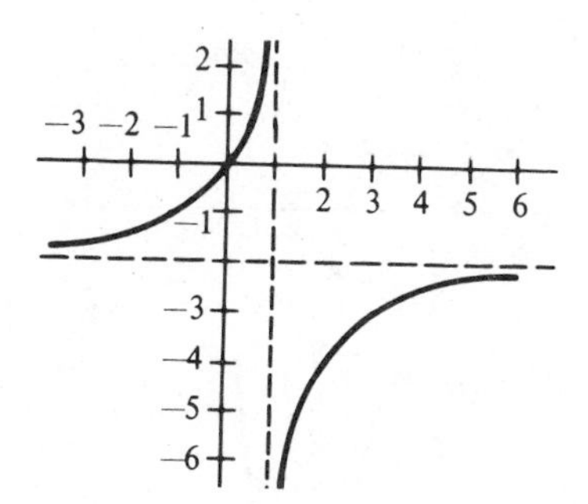

29. $y = 2 - \dfrac{3}{x^2}$

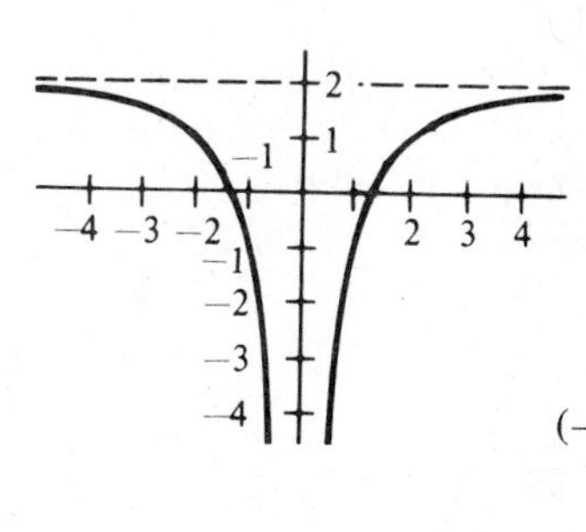

31. $y = \dfrac{x^3}{\sqrt{x^2 - 4}}$

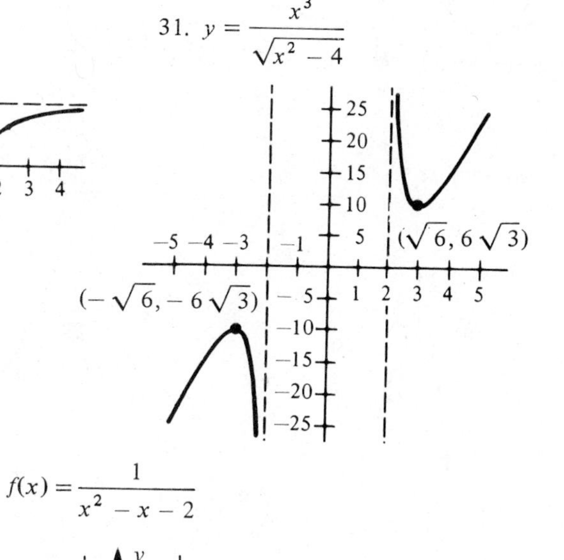

33. $f(x) = \dfrac{x^2}{x^2 - 1}$

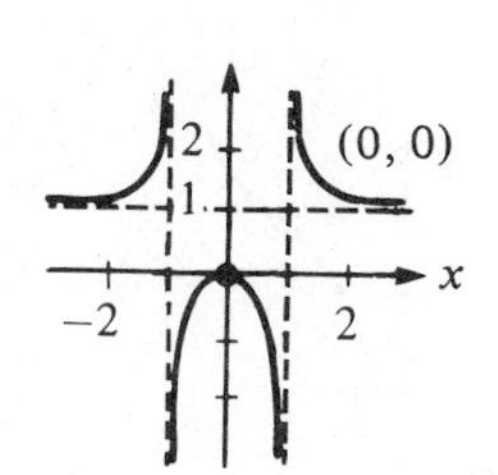

35. $f(x) = \dfrac{1}{x^2 - x - 2}$

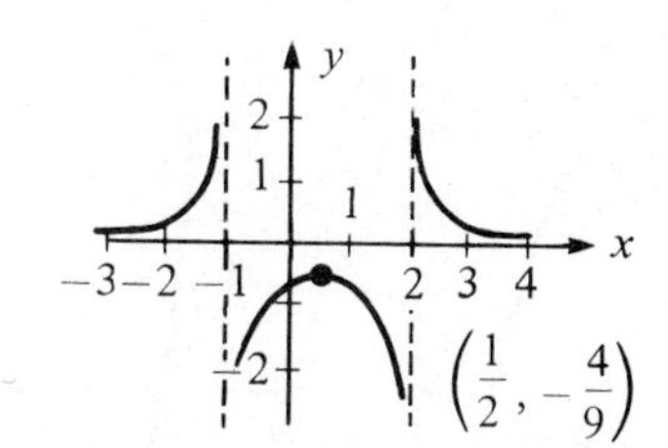

37. (a) (i) \$176 million
(ii) \$528 million
(iii) \$1,584 million
(b) ∞, no
(c)

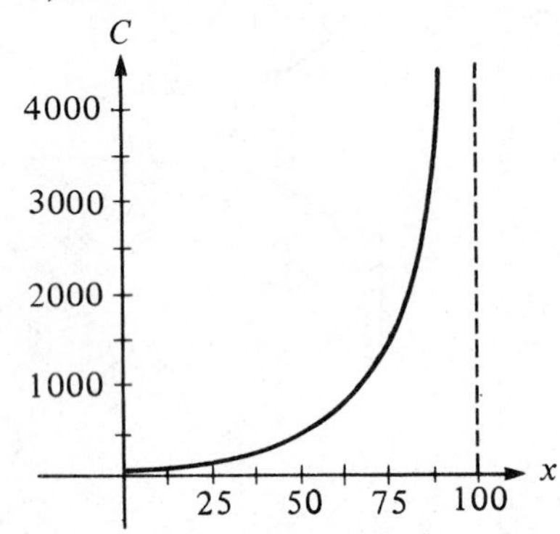

39. a

41. (a) (i) \$47.05
(ii) \$5.92
(b) \$1.35

Section 3.7

1.

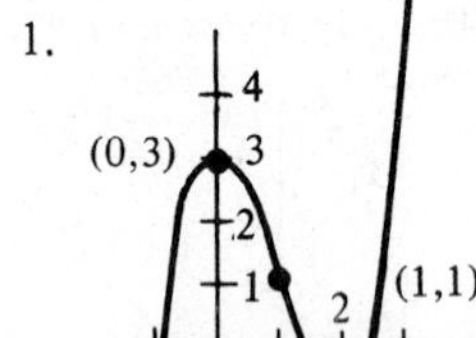
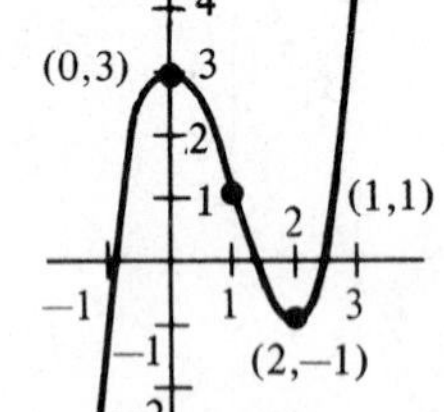

3.

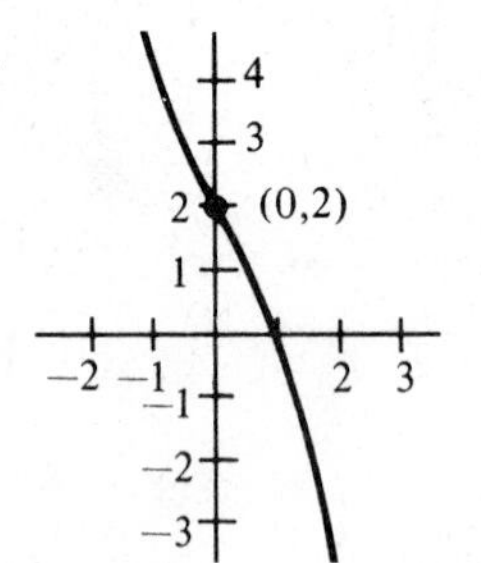

5.

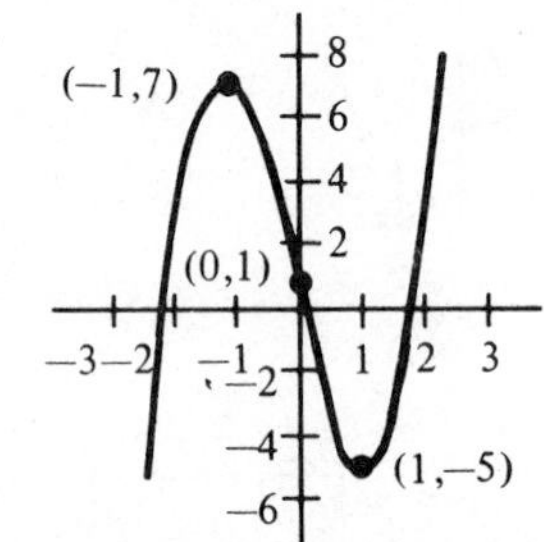

7.

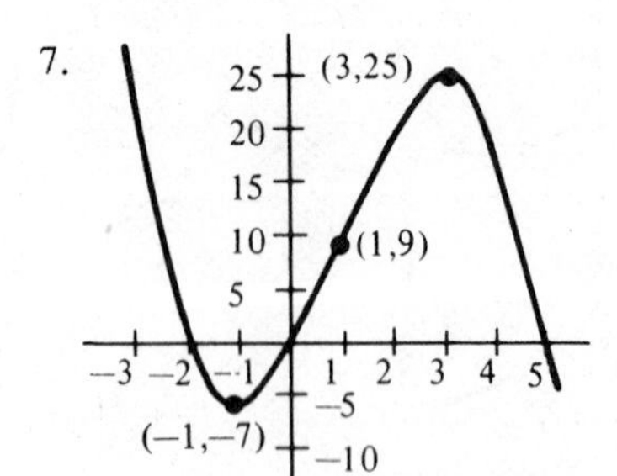

9.

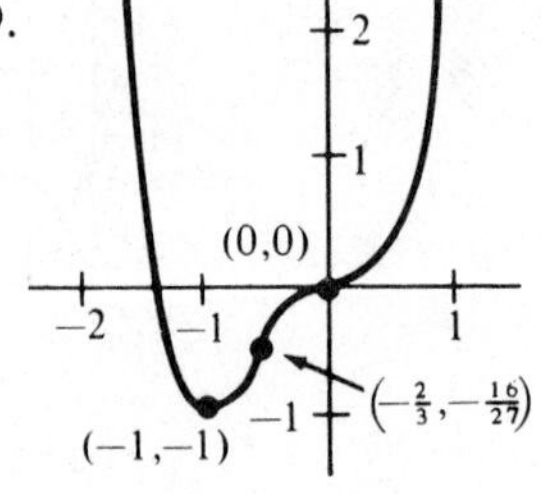

11.

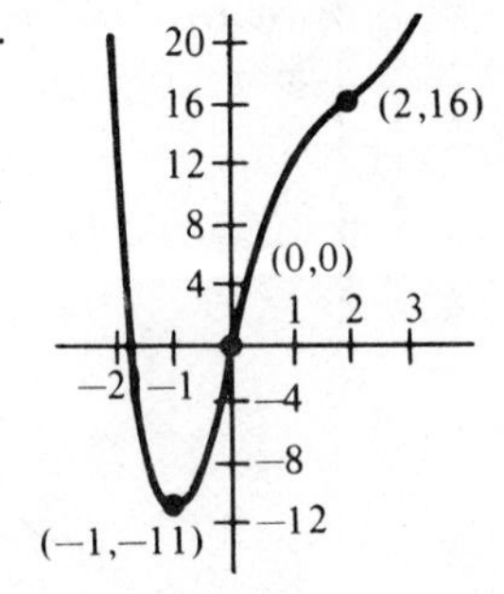

13.

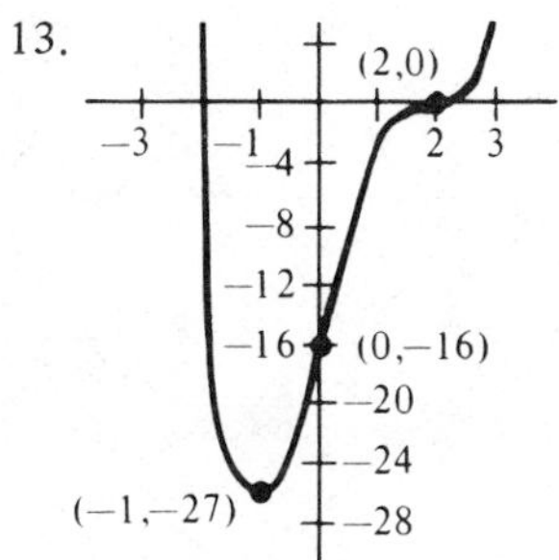

15.

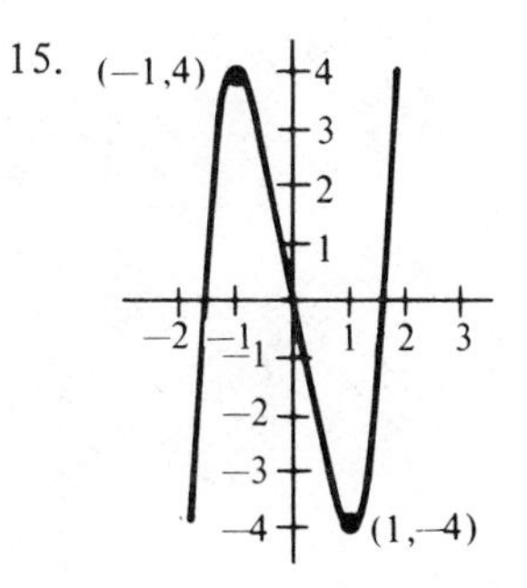

17.

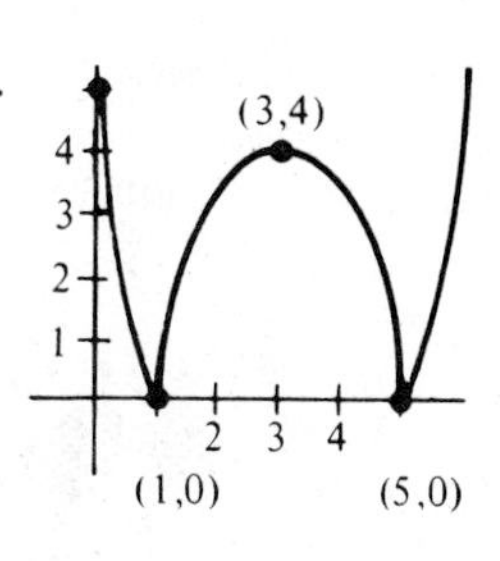

19.

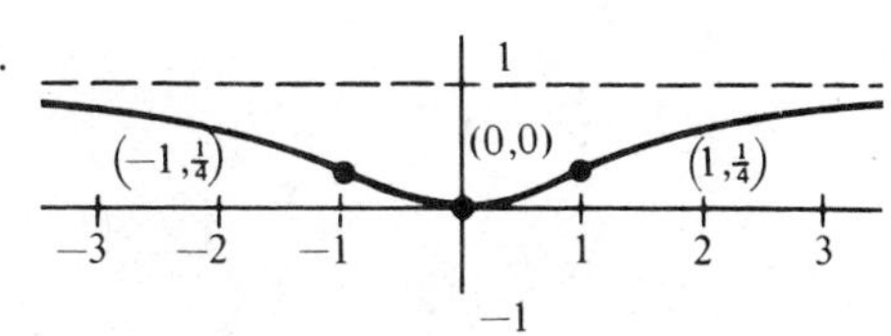

21.

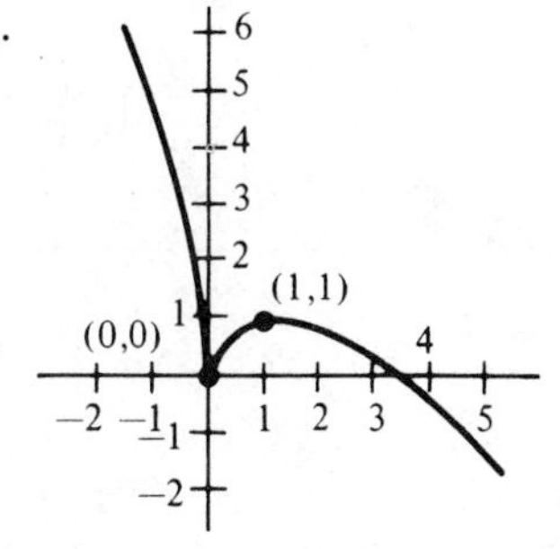

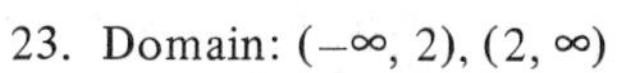
23. Domain: $(-\infty, 2), (2, \infty)$

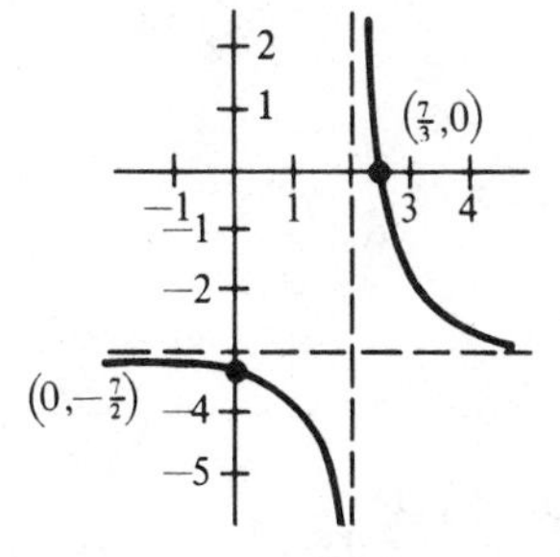

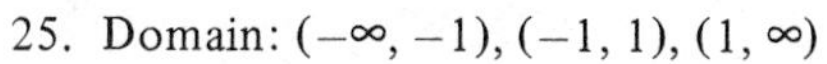
25. Domain: $(-\infty, -1), (-1, 1), (1, \infty)$

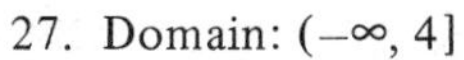
27. Domain: $(-\infty, 4]$

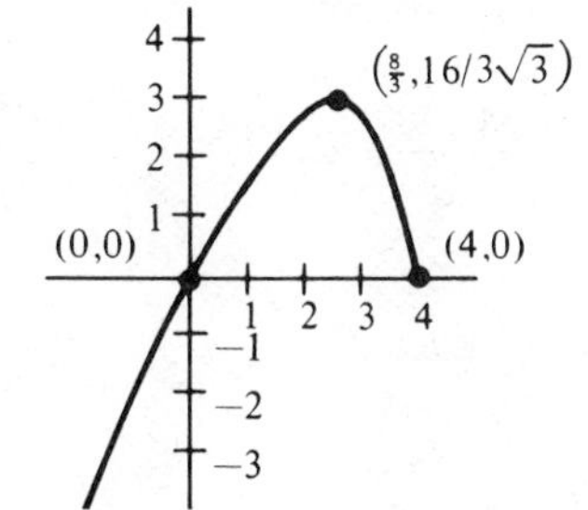

29. Domain: $(-\infty, 0), (0, \infty)$

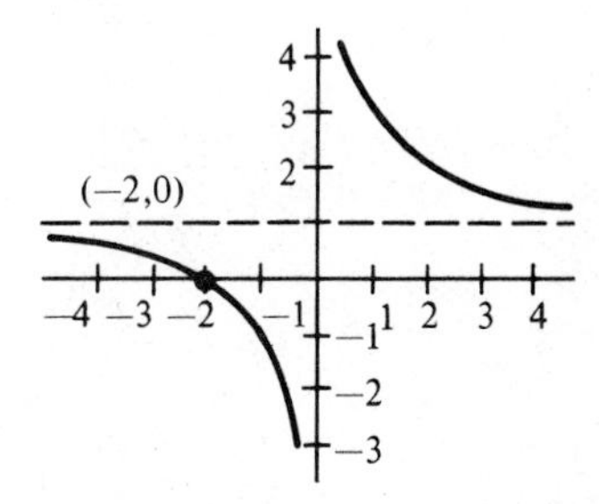

Chapter 4

Section 4.1

1. $\frac{x^4}{4} + 2x + C$

3. $\frac{2x^{5/2}}{5} + x^2 + x + C$

5. $\frac{3x^{5/3}}{5} + C$

7. $\frac{-1}{2x^2} + C$

9. $\frac{-1}{4x} + C$

11. $\frac{2\sqrt{x}}{15}(3x^2 + 5x + 15) + C$

13. $x^3 + \frac{x^2}{2} - 2x + C$

15. $t - \frac{2}{t} + C$

17. $\frac{2y^{7/2}}{7} + C$

19. $x + C$

21. $\frac{t^5}{5} - \frac{4t^3}{3} + 4t + C$

23. $\frac{2y^{3/2}}{5}(15 - y) + C$

25. $f(x) = x^2 + x + 4$

27. $f(x) = -4\sqrt{x} + 3x$

29. (a) $C = x^2 - 12x + 125$
(b) $\overline{C} = x - 12 + 125/x$
(c) \$2025

31. (a) $R = 100x - \frac{5}{2}x^2$
(b) \$990
(c) $p = 100 - \frac{5}{2}x$

33. (a) $h(t) = \frac{t^2}{4} + 2t + 5$
(b) 26 in

35. 56.25 ft

Section 4.2

1. $\frac{(1 + 2x)^5}{5} + C$

3. $\frac{(x^3 - 1)^5}{15} + C$

5. $\frac{(x^2 - 1)^8}{16} + C$

7. $4\sqrt{1 + x^2} + C$

9. $\frac{15(1 + x^2)^{4/3}}{8} + C$

11. $-3\sqrt{2x + 3} + C$

13. $\frac{-1}{2(x^2 + 2x - 3)} + C$

15. $\frac{-2}{1 + \sqrt{x}} + C$

17. $\frac{-1}{3(1 + x^3)} + C$

19. $\frac{1}{2}\sqrt{1 + x^4} + C$

21. $\sqrt{x} + C$

23. $\sqrt{2x} + C$

25. (a) $W = \frac{3}{2}(\sqrt{16t + 9} - 3)$ (b) 55.67 lb

27. (a) $\frac{3}{2}(12x + 1)^{2/3} + 56.48$ (b)

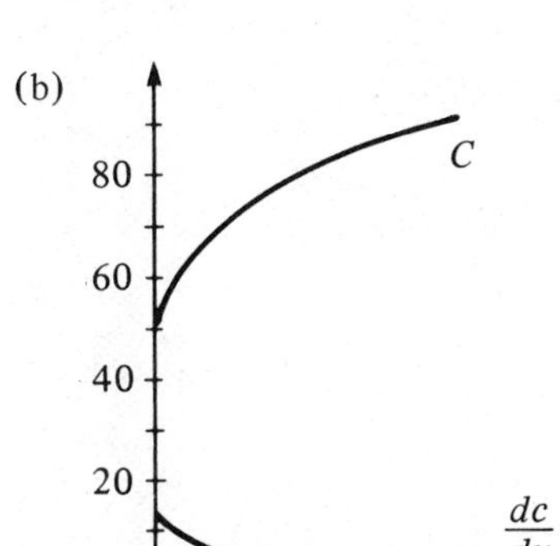

Section 4.3

1. 1
3. $-5/2$
5. $-10/3$
7. $1/3$
9. $1/2$
11. 36
13. -4
15. $2/3$
17. $-1/18$
19. 1
21. 2
23. 0
25. $1/6$
27. $8/5$
29. 6
31. $A = 6$ 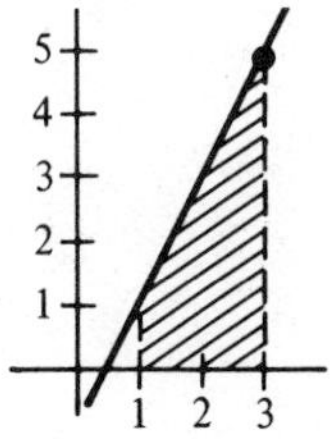
33. $A = 10/3$
35. $A = 1/4$

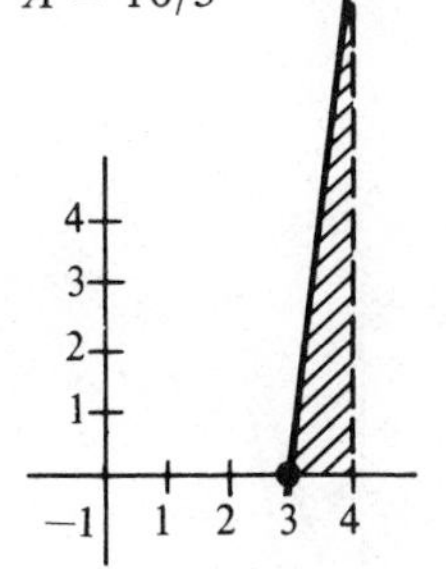

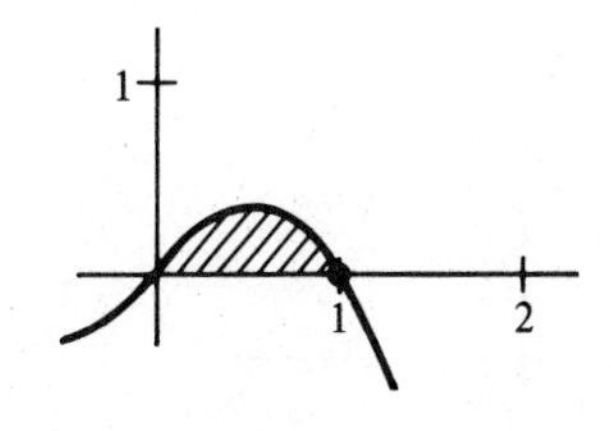

37. Average = 8/3

$x = \pm 2\sqrt{3}/3 \approx \pm 1.155$

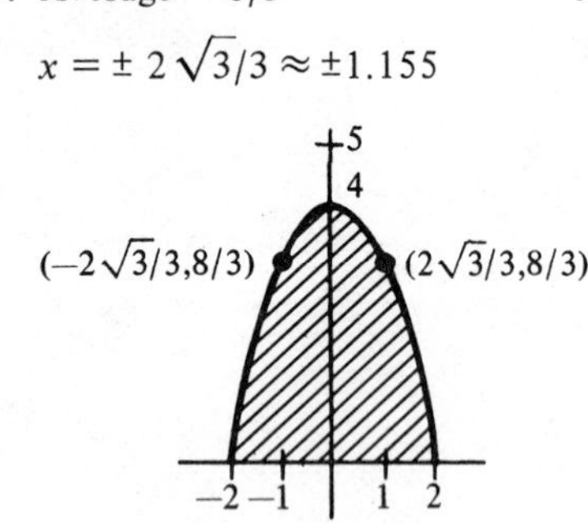

39. Average = 4/3

$$x = \sqrt{2 + \frac{2\sqrt{5}}{3}} \approx 1.868$$

$$x = \sqrt{2 - \frac{2\sqrt{5}}{3}} \approx 0.714$$

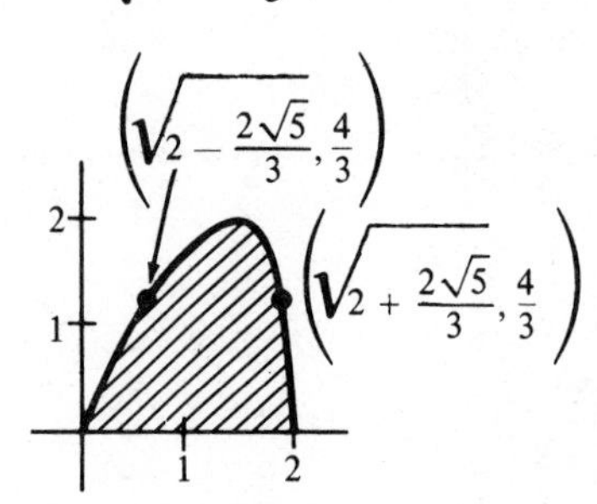

41. Average = −2/3

$$x = \frac{4 + 2\sqrt{3}}{3} \approx 2.488$$

$$x = \frac{4 - 2\sqrt{3}}{3} \approx 0.179$$

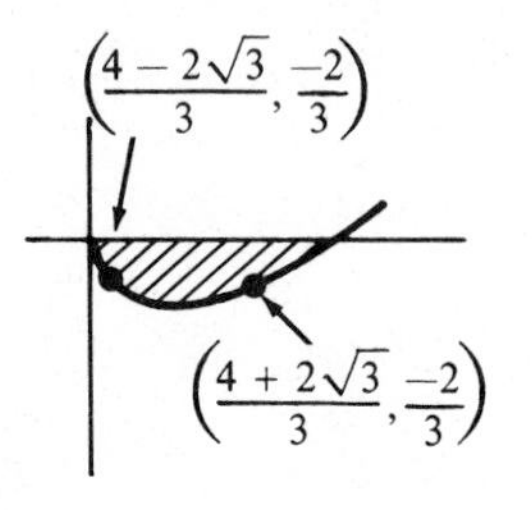

43. 0.53183 liter

45. (a) 64.4°F (b) 68.6°F

Section 4.4

1. 36

3. 9

5. 1/12

7. $A = 32/3$

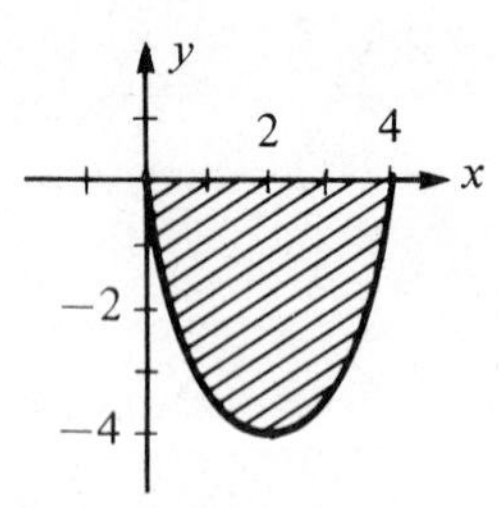

9. $A = 9/2$

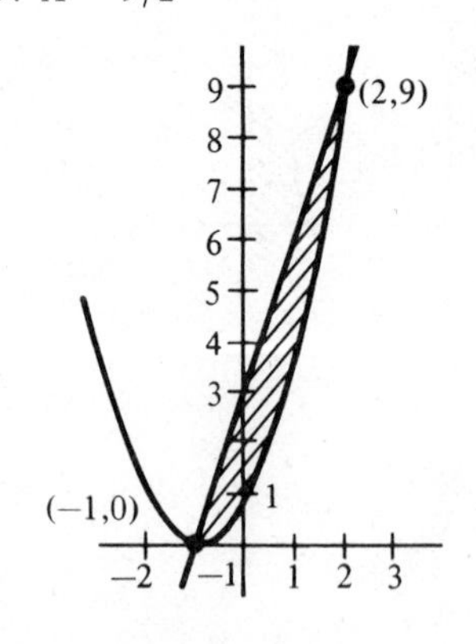

11. $A = 1$

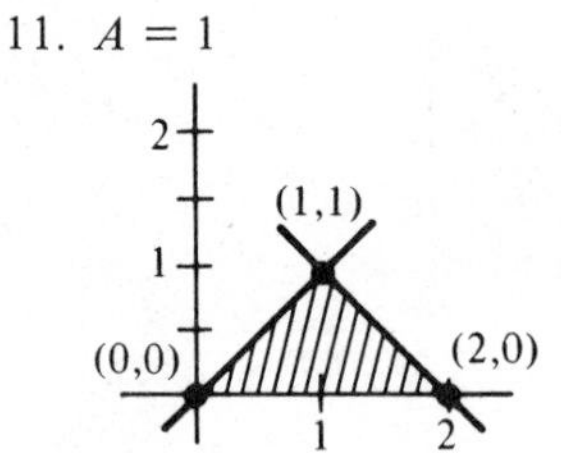

13. $A = 500/27 \cong 18.52$

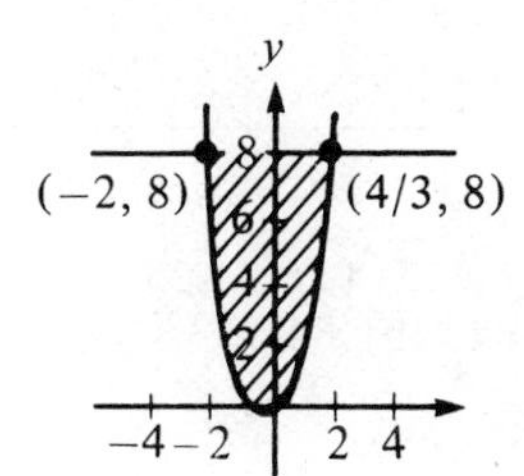

15. $A = 2$

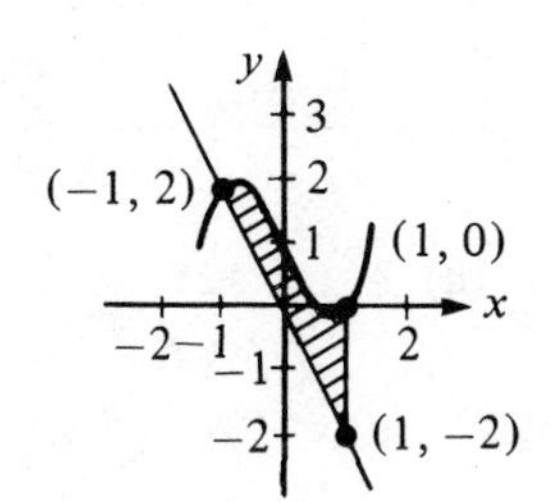

17. $A = 3/2$

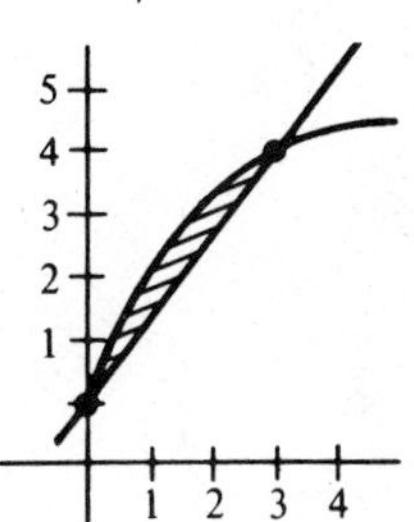

19. $A = 64/3$

21. 41.2633 billion lb

23. \$2.4167 billion

25. Point of equilibrium: (80, 10)
Consumer Surplus: 1600
Producer Surplus: 400

27. Point of equilibrium: (300, 500)
Consume Surplus: 50,000
Producer Surplus: $\dfrac{10{,}000}{3}(4\sqrt{10} - 5) \approx 25,497$

Section 4.5

1. $\pi/3$
3. $16\pi/3$
5. $15\pi/2$
7. $2\pi/35$
9. $128\pi/5$
11. $243\pi/5$
13. 8π
15. $\pi/4$
19. 60π
21. 1,134,115 cu. ft.
2,268 fish
23. Approx. 2.26 quarts

Chapter 5

Section 5.1

1. (a) 625 (b) 9 (c) 8 (d) 9 (e) 125

3. (a) 3125 (b) 1/5 (c) 625 (d) 1/125 (e) 1/125 (f) 25 (g) 4 (h) 64

5. (a) e^6 (b) e^{12} (c) $1/e^6$ (d) e^2 (e) e^2 (f) $1/e^3$

7. $g(x) = 5^x$

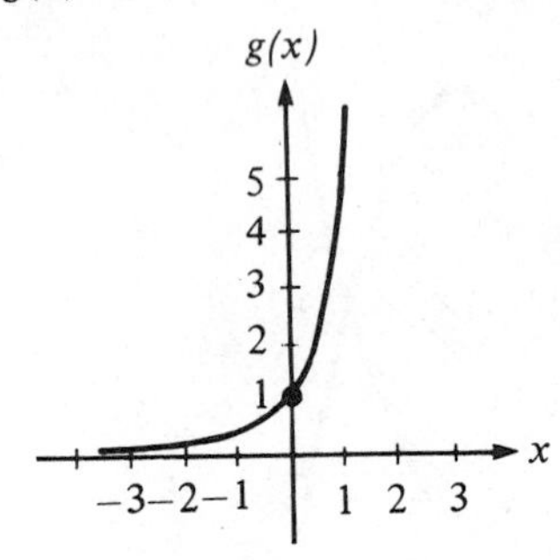

9. $f(x) = 5^{-x}$

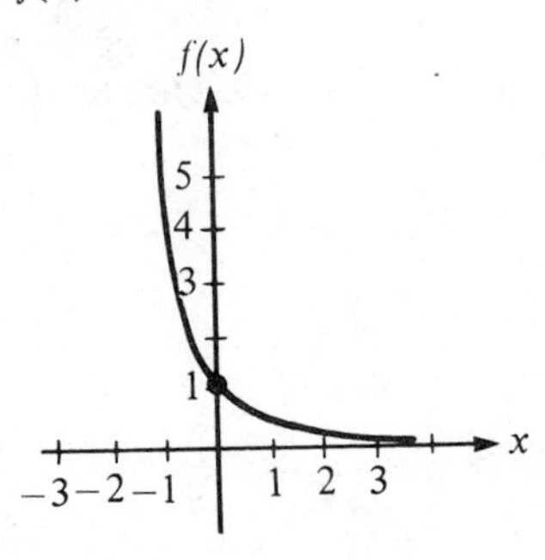

11. Same graph as Exercise 7

13. $y = 3^{-x^2}$

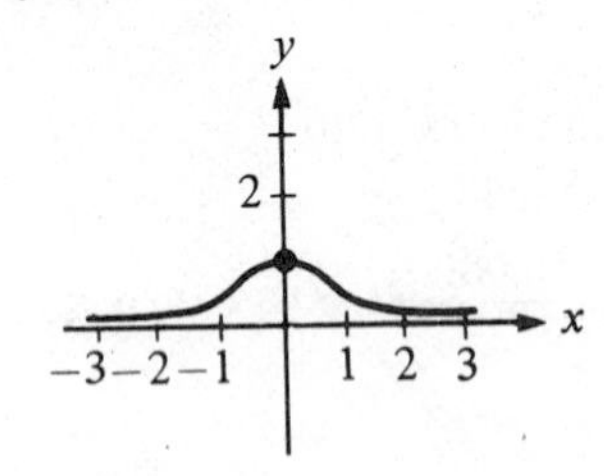

15. $y = 3^{-|x|}$

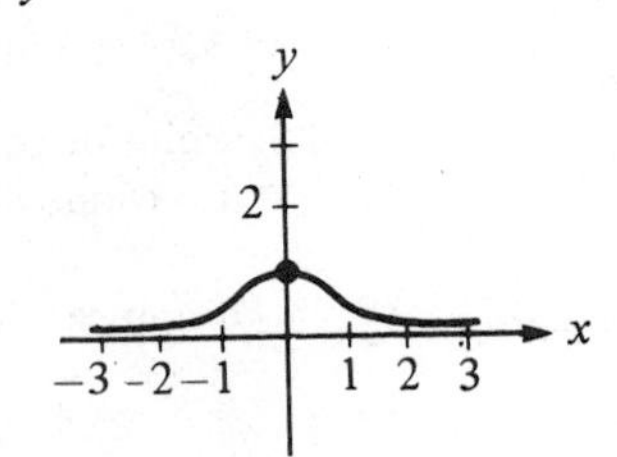

17. $s(t) = 3^{-t}/4$

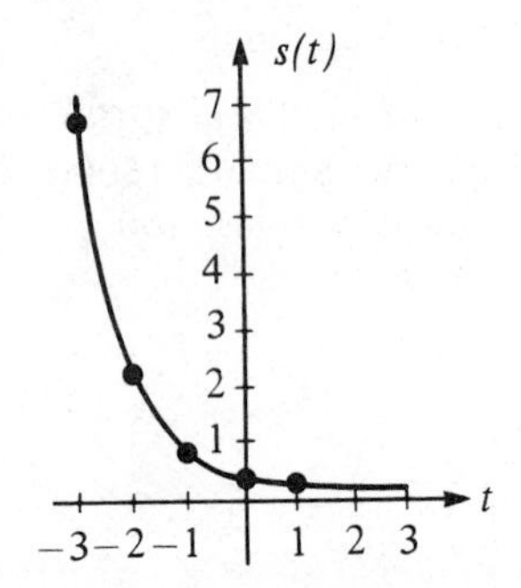

19. $h(x) = e^{x-2}$

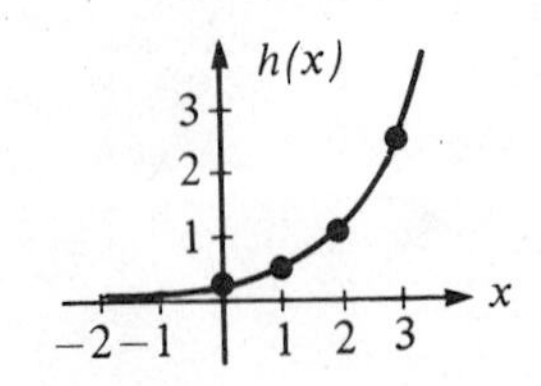

21. $N(t) = 1000\,e^{-0.2t}$

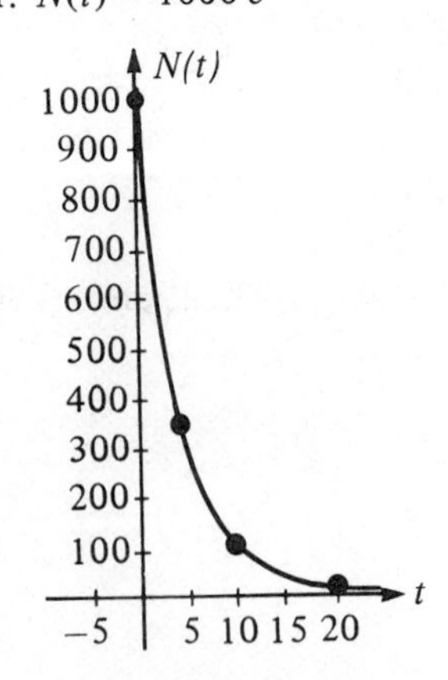

23. (a) \$2593.74 (b) \$2653.30 (c) \$2685.06 (d) \$2707.04 (e) \$2717.91 (f) \$2718.28

25. (a) \$849.53 (b) \$421.12

27. (a) 850
(b)

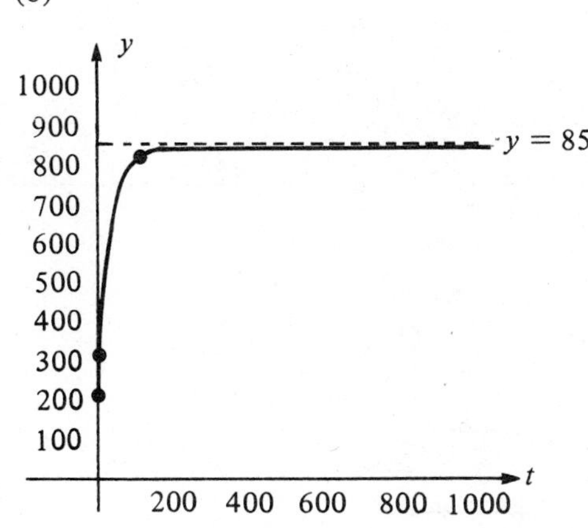

29. (a) 0.731 (b) 0.830

31. (a) \$88,692.04 (b) \$30,119.42 (c) \$9,071.80
(d) \$247.88

Section 5.2

1. 3

3. 1

5. -2

7. $2e^{2x}$

9. $2(x-1)e^{1-2x+x^2}$

11. $\dfrac{e^{\sqrt{x}}}{2\sqrt{x}}$

13. $3(e^x+e^{-x})^2\,(e^x-e^{-x})$

15. $\dfrac{-2(e^x-e^{-x})}{(e^x+e^{-x})^2}$

17. xe^x

19. $\frac{1}{2}(1-e^{-2}) \approx 0.432$

21. $\frac{e}{3}(e^2-1) \approx 5.789$

23. $\frac{1}{2a}\,e^{ax^2}+C$

25. $\frac{e}{3}(e^2-1) \approx 5.789$

27. $e^x+2x-e^{-x}+C$

29. $-\frac{2}{3}(1-e^x)^{3/2}+C$

31.

2 $y = 2$ (0, 1) x
−4 −2 2 4

33.

5 4 3 2 1 (2,4/e^2) (0,0)
−2 −1 1 2 3 4 5
$x = 2 \pm \sqrt{2}$

35. 113.51; 403.20; $t = 5 \ln 19 \approx 14.7$ months

37. (a) 0.058 million cubic feet per year
(b) 0.073 million cubic feet per year
(c) 0.040 million cubic feet per year

39. (a) \$433.31 per year
(b) \$890.22 per year
(c) \$21,839.26 per year

41. $\frac{1}{2}(1 - e^{-16}) \approx 0.5$

43. (a) 70.57%
(b) 38.60%

Section 5.3

1. $e^{0.6931\ldots} = 2$

3. $e^{-0.6931\ldots} = 0.5$

5. $\ln 1 = 0$

7. $\ln 0.1353\ldots = -2$

9. $y = \ln(x - 1)$

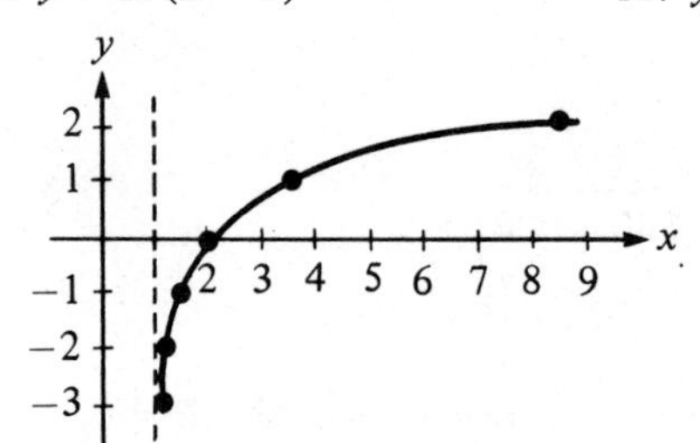

11. $y = \ln 2x$

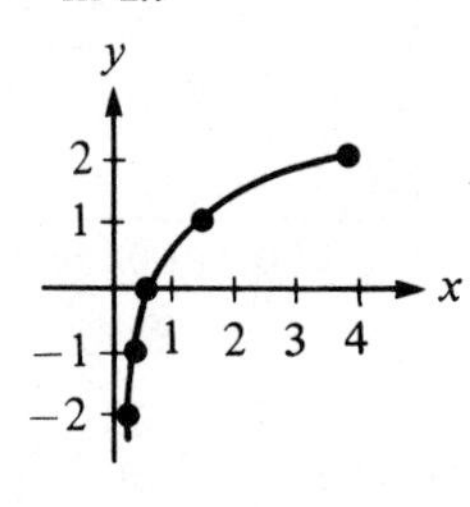

13. $y = 3 \ln x$

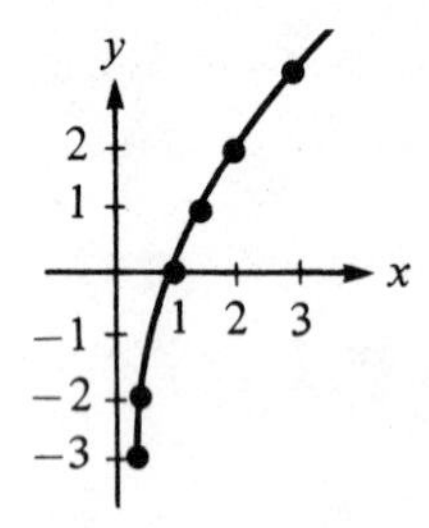

15. $f(x) = e^{2x}$; $g(x) = \ln \sqrt{x}$

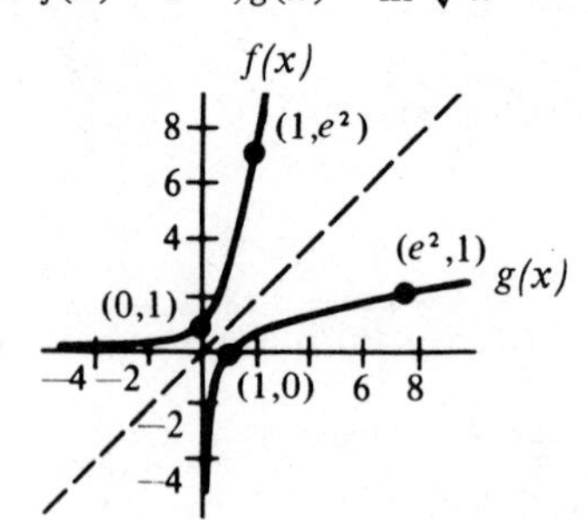

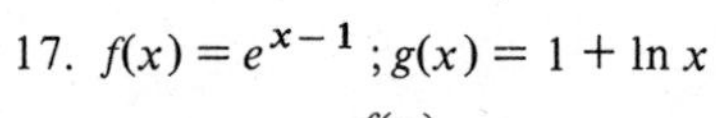
17. $f(x) = e^{x-1}$; $g(x) = 1 + \ln x$

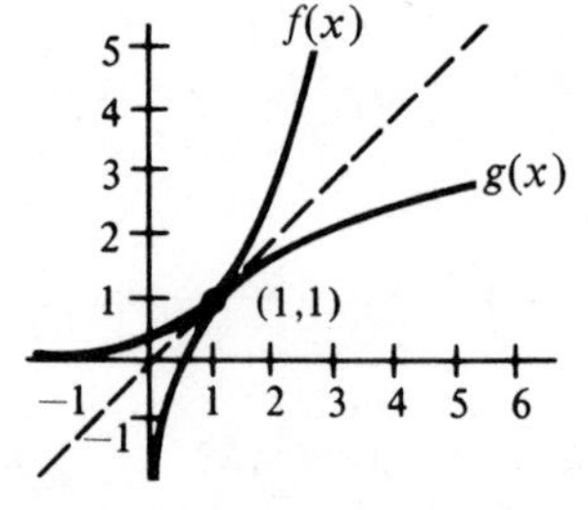

19. x^2

21. $5x + 2$

23. $\sqrt{x}$

25. $\frac{1}{2}\ln(a - 1)$

27. $3[\ln(x + 1) + \ln(x - 1) - 3\ln x]$

29. -1

31. $\ln\left(\frac{x^3y^2}{z^4}\right)$

33. $\ln\left(\frac{x}{x^2 - 1}\right)^2$

35. $x = 4$

37. $x = 1$

39. $x = (\ln 4) - 1 \approx 0.3863$

41. $x = \frac{4}{3}\ln 2.197225 \approx 1.0496$

43. $x = \frac{\ln 6 - \ln 5}{0.11} \approx 1.6575$

45. $t = \frac{\ln 0.3}{0.0315} \approx 38.2214$

47. $x = \frac{\ln 15}{2\ln 5} \approx 0.8413$

49. $t = \frac{\ln 2}{\ln 1.07} \approx 10.2448$

51. (a) 6.64 years (b) 6.33 years (c) 6.30 years (d) 6.30 years

53.

r	2%	4%	6%	8%	10%	12%
t	54.93	27.47	18.31	13.73	10.99	9.16

55. (a) $x = \dfrac{\ln 300}{0.004} \approx 1426$ (b) $x = \dfrac{\ln 400}{0.004} \approx 1498$

Section 5.4

1. 3

3. 2

5. 1

7. $\dfrac{2}{x}$

9. $\dfrac{2(x^3 - 1)}{x(x^3 - 4)}$

11. $\dfrac{4(\ln x)^3}{x}$

13. $\dfrac{2x^2 - 1}{x(x^2 - 1)}$

15. $\dfrac{1 - x^2}{x(x^2 + 1)}$

17. $\dfrac{1 - 2\ln x}{x^3}$

19. $\dfrac{-4}{x(x^2 + 4)}$

21. $1/(1 - x^2)$

23. $e^{-x}\left(\dfrac{1}{x} - \ln x\right)$

25. $2x$

27. $\ln|x + 1| + C$

29. $-\dfrac{1}{2}\ln|3 - 2x| + C$

31. $\ln\sqrt{x^2 + 1} + C$

33. $4\ln 2 - \dfrac{3}{2} \approx 1.273$

35. $-\ln 3$

37. $7/3$

39. $-[x + \ln(x - 1)]$

41. $2\sqrt{x + 1} + C$

43. $\ln|e^x - e^{-x}| + C$

45. $x - \ln(1 + e^x) + C$

47. $(1, 1)$, relative minimum

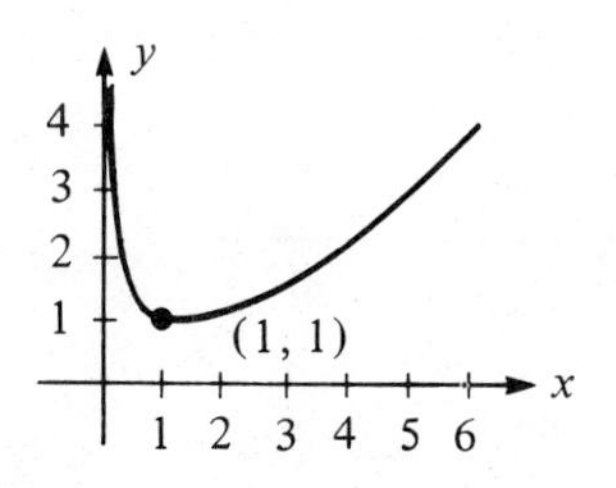

49. $\left(e, \dfrac{1}{e}\right)$, relative maximum

$\left(e^{3/2}, \dfrac{3}{2e^{3/2}}\right)$, point of inflection

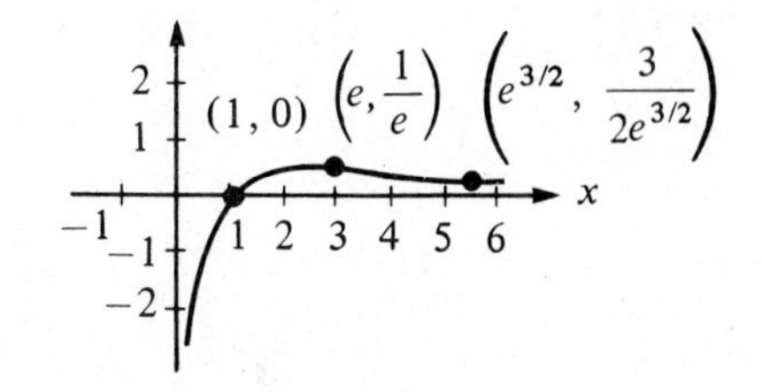

51. $\left(\frac{1}{\sqrt{e}}, \frac{-1}{2e}\right)$, relative minimum

$\left(\frac{1}{e^{3/2}}, \frac{-3}{2e^3}\right)$, point of inflection

53. $4 + 5 \ln 5 \approx 0.227$

55. $3000 \ln \frac{55}{52} = \$168.27$

Section 5.5

1. $y = 2e^{0.1014t}$

3. $y = 4e^{-0.4159t}$

5. $y = 0.6687e^{0.4024t}$

7. (a) $r = \frac{\ln 2}{7.75} \approx 9\%$

(b) $A = 750e^{10 \ln 2/7.75} = \$1834.37$

9. $y = 742{,}000\, e^{-0.0802(3)} = \$583{,}327.77$

11. (a) $N = 30(1 - e^{-0.0502t})$

(b) $t = \frac{\ln 6}{0.0502} \approx 36$ days

13. (a) $S = 30e^{-1.7918/t}$

(b) $S = 30e^{-1.7918/5} \approx 20.9646$, or 20,965 units

(c)

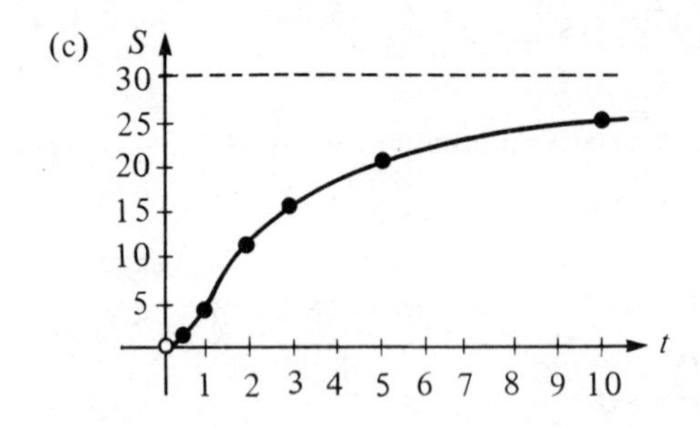

15. 900

17. 6015

19. 95.76%

Chapter 6

Section 6.1

1. $-\dfrac{2x+1}{2(x+1)^2}+C$

3. $\dfrac{1}{9}\left(\ln|3x-1|-\dfrac{1}{3x-1}\right)+C$

5. $-\dfrac{1}{30}(1-x)^5(1+5x)+C$

7. $\dfrac{1}{2}\ln|x^2-2x|+C$

9. $\dfrac{2}{5}(x-3)^{3/2}(x+2)+C$

11. $-\dfrac{2}{105}(1-x)^{3/2}(15x^2+12x+8)+C$

13. $\dfrac{\sqrt{2x-1}}{15}(3x^2+2x-13)+C$

15. $2(2-\sqrt{3})\sqrt{t}+C$

17. $2[\sqrt{x}-\ln(1+\sqrt{x})]+C$

19. $\dfrac{1}{9}[3\sqrt{2x}-\ln(3\sqrt{2x}+1]+C$

21. $\ln 2-1/2$

23. $\dfrac{13}{320}$

25. $\dfrac{144}{5}$

27. $\dfrac{1209}{28}$

29. $2-\ln 3$

31. $4/15$

33. 21.56%

35. (a) 0.353 (b) 0.586

37. $24,658.75

39. 0.22

Section 6.2

1. $\dfrac{1}{4}e^{2x}(2x-1)+C$

3. $\dfrac{1}{2}e^{x^2}+C$

5. $-\dfrac{1}{4}e^{-2x}(2x+1)+C$

7. $e^x(x^3-3x^2+6x-6)+C$

9. $\dfrac{x^4}{16}(4\ln x-1)+C$

11. $\dfrac{1}{4}[2(t^2-1)\ln(t+1)+t(2-t)]+C$

13. $x[(\ln x)^2-2\ln x+2]+C$

15. $\dfrac{(\ln x)^3}{3}+C$

17. $\dfrac{e^{2x}}{4(2x+1)}+C$

19. $\dfrac{2}{15}(x-1)^{3/2}(3x+2)+C$

21. $(x-1)^2e^x+C$

23. $\dfrac{2}{5}(2x-3)^{3/2}(x+1)+C$

25. $\dfrac{1}{3}\sqrt{x^2+4}(x^2-8)+C$

27. $\frac{2}{75}\sqrt{5x+4}\,(5x-8)+C$ 29. $\frac{e^{5x}}{125}(25x^2-10x+2)+C$ 31. $e-2$

33. $4\ln 4-3$ 35. (a) $3.2\ln 2-0.2=2.0181$
(b) $12.8\ln 4-7.2\ln 3-1.8=8.0346$

Section 6.3

1. $\frac{1}{2}\ln\left|\frac{x-1}{x+1}\right|+C$ 3. $\ln\left|\frac{x}{x+1}\right|+C$ 5. $\ln\left|\frac{x}{2x+1}\right|+C$

7. $\ln\left|\frac{x-1}{x+2}\right|+C$ 9. $\frac{3}{2}\ln|2x-1|-2\ln|x+1|+C$

11. $5\ln|x-2|-\ln|x+2|-3\ln|x|+C$ 13. $x^2+\frac{3}{2}\ln|x-4|-\frac{1}{2}\ln|x+2|+C$

15. $\frac{1}{2}\ln\left|\frac{(x-4)^3}{x}\right|+C$ 17. $2\ln|x-1|+\frac{1}{x-1}+C$

19. $\frac{1}{2}\left(5\ln|x+1|-\ln|x|+\frac{3}{x+1}\right)+C$ 21. $\frac{1}{4}\ln\frac{5}{3}$

23. $1-\ln 3$ 25. $\frac{1}{5}\ln\left|\frac{e^x-1}{e^x+4}\right|+C$ 27. $\frac{1}{4}\ln\left|\frac{\sqrt{4+x^2}-2}{\sqrt{4+x^2}+2}\right|+C$

29. $-\frac{2}{\sqrt{x}+1}$ 35. $x=\frac{n[e^{(n+1)kt}-1]}{e^{(n+1)kt}+n}$

Section 6.4

1. $\ln\left|\frac{x}{x+1}\right|+C$ 3. $-\ln\left|\frac{1+\sqrt{x^2+1}}{x}\right|+C$ 5. $\frac{x^2}{4}(2\ln x-1)+C$

7. $-\frac{1}{2}\ln\left|\frac{2+\sqrt{4-x^2}}{x}\right|+C$ 9. $-\left[e^{-x}+\ln\left(\frac{e^x}{e^x+1}\right)\right]+C$

11. $\frac{1}{4}[x^2\sqrt{x^4-9} - 9\ln(x^2+\sqrt{x^4-9})] + C$

13. $\frac{1}{27}\left(\frac{4}{2+3t} - \frac{4}{2(2+3t)^2} + \ln|2+3t|\right) + C$

15. $\frac{1}{\sqrt{3}}\ln\left|\frac{\sqrt{3+s}-\sqrt{3}}{\sqrt{3+s}+\sqrt{3}}\right| + C$

17. $\frac{x}{2}(x-2) + \ln|x+1| + C$

19. $-\frac{\sqrt{1-x^2}}{x} + C$

21. $\frac{x^3}{9}(3\ln x - 1) + C$

23. $\frac{1}{27}\left(3x - \frac{25}{3x-5} + 10\ln|3x-5|\right) + C$

25. $x - \frac{1}{2}\ln(1+e^{2x}) + C$

27. $\frac{1}{4}(2\ln x - 3\ln|3+2\ln x|) + C$

29 $\frac{1}{8}\left((2x-3)[(2x-3)^2+2]\sqrt{(2x-3)^2+4} - 8\ln|(2x-3)+\sqrt{(2x-3)^2+4}|\right) + C$

31. (a) $(x+3)^2 - 9$
(b) $(x-4)^2 - 7$
(c) $(x^2+1)^2 - 6$

33. $\frac{1}{4}\ln\left|\frac{x-3}{x+1}\right| + C$

35. $-\ln\left|\frac{1+\sqrt{x^2-2x+2}}{x-1}\right| + C$

37. $\frac{1}{8}\ln\left|\frac{x-3}{x+1}\right| + C$

39. $\frac{1}{2}\ln|x^2+1+\sqrt{x^4+2x^2+2}| + C$

41. 40/3 square units

43. $42.57 \approx 43$

45. Consumer surplus = 17.92
Producer surplus = 24.00

Section 6.5

	Exact	*Trapezoidal*	*Simpson's*
1.	2.6667	2.7500	2.6667
3.	4.0000	4.2500	4.0000
5.	4.0000	4.0625	4.0000
7.	0.5000	0.5090	0.5004
9.	0.6931	0.6970	0.6933
11.		3.41	3.22
13.		0.342	0.372
15.		0.749	0.771
17.		0.772	0.780
19.		0.286	0.274

21. (a) 1/2 (b) 0

23. (a) $5e/64 \approx 0.212$ (b) $13e/1024 \approx 0.035$

25. $n = 10$

27. (a) 0.3414 (b) 0.4772

29. 22.12 acres

Section 6.6

1. 4
3. 6
5. 1
7. 6
9. Diverges
11. $\ln(2+\sqrt{3})$
13. 0
15. 1
17. Diverges
19. Diverges
21. (a)

x	1	10	25	50
$x e^{-x}$	0.3679	0.0005	0	0

(b)

x	1	10	25	50
$x^2 e^{-x}$	0.3579	0.0045	0	0

(c)

x	1	10	25	50
$x^5 e^{-x}$	0.3679	4.5400	0.0001	0

23. 2
25. (b) $e^{-5/7} \approx 0.49$
27. (a) \$743,997.58
    (b) \$795,584.54
    (c) \$858,333.33

Chapter 7

Section 7.1

15. $y = 3e^{-2x}$
17. $y = 5 + \ln\sqrt{x}$
19. $y = -\frac{4}{11}e^{6x} + \frac{4}{11}e^{-5x}$
21. $x^2 + y^2 = C$

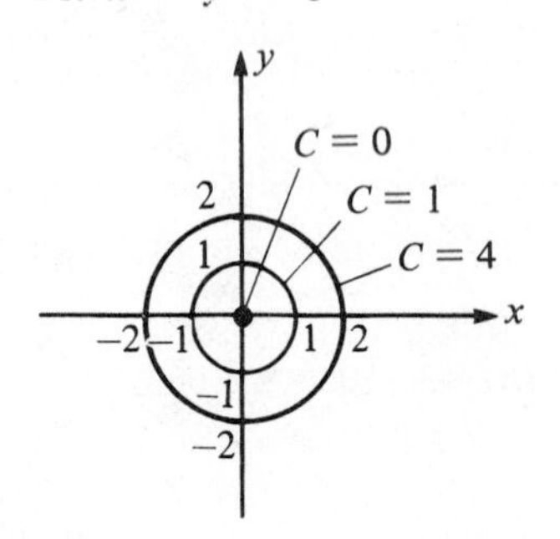

23. $y = C(x + 2)^2$

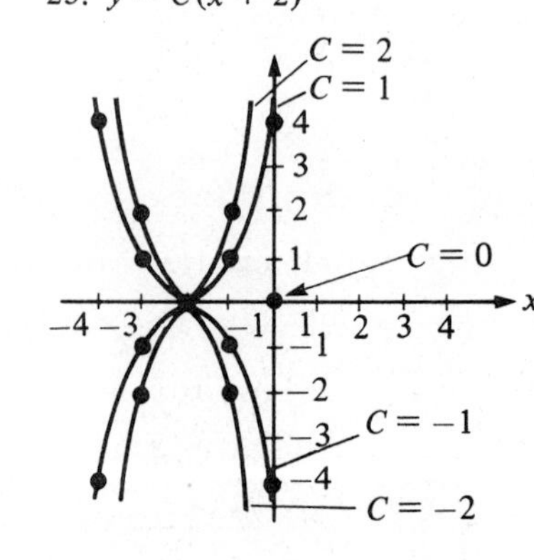

25. (a) $N = 750 - 650e^{-0.0484t}$
(b)

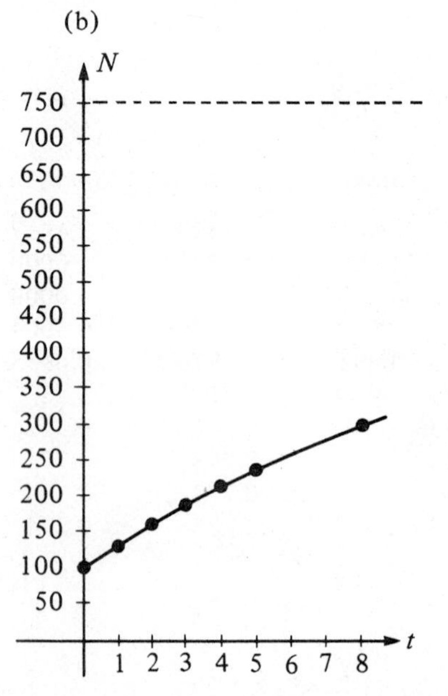

Section 7.2

1. $y = x^2 + C$

3. $y = \sqrt[3]{x + C}$

5. $y = Ce^{x^2/2}$

7. $y = C(1 + x)^2$

9. $y = \ln |t^3 + t + C|$

11. $x^2 - y^2 = C$

13. $y = C(2 + x)^2$

15. $y = Ce^{(\ln x)^2/2}$

17. $y^2 = 2e^x + 14$

19. $y = e^{-x(x+2)/2}$

21. $P = P_0 e^{kt}$

23. $9x^2 + 16y^2 = 25$

25. $v = 34.56(1 - e^{-0.1t})$; 34.56 ft/s

27. $t = \dfrac{\ln 0.25}{\ln 0.8} \approx 6.2\ h$

29. $T = 70\, e^{-0.3773t}$; $t = 5.16\ h$

Section 7.3

1. $A = \dfrac{2000}{17t + 20}$

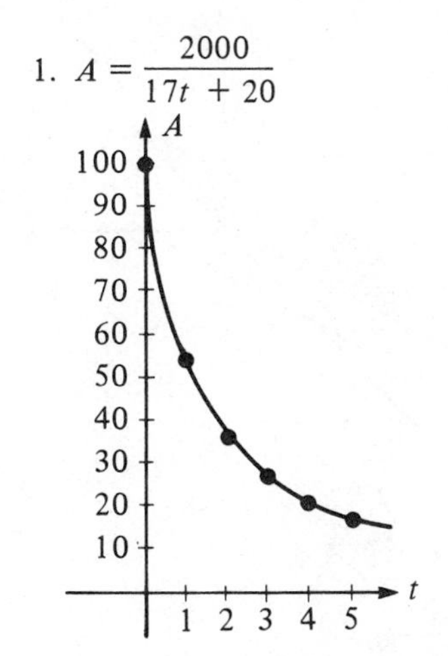

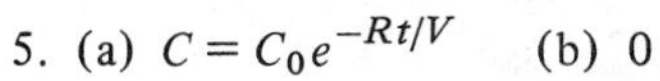

3. $y = 500e^{-1.6094e^{-0.1451t}}$

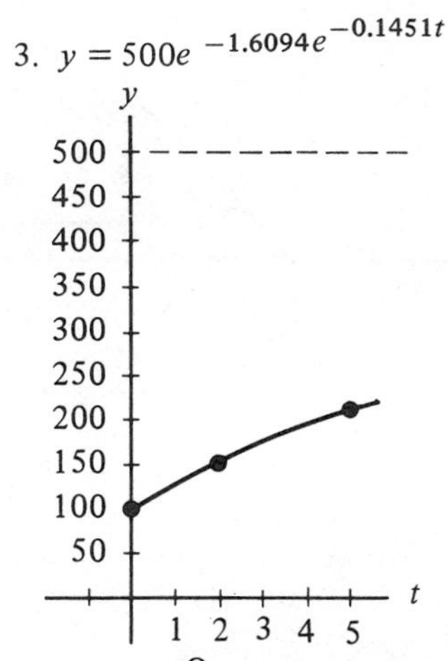

5. (a) $C = C_0 e^{-Rt/V}$ (b) 0

7. (a) $C = \dfrac{Q}{R}(1 - e^{-Rt/V})$ (b) Q/R

9. $s = 25 - \dfrac{13 \ln (h/2)}{\ln 3}$; $2 < h < 15$

11. $\dfrac{dP}{dn} = kP(L - P)$; $P = \dfrac{CL}{e^{-kLn} + C}$

13. $y = Cx$

15. $A = \dfrac{P}{r}(e^{rt} - 1)$

17. $7,305,295.15

Chapter 8

Section 8.1

1.

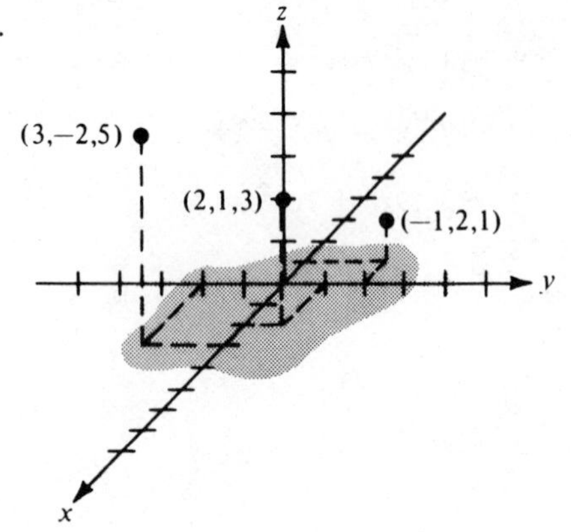

3. $(3/2, -3, 5)$

5. $(0.65, 1.1, -0.8)$

7. $3\sqrt{2}$

9. $\sqrt{206}$

11. $x^2 + y^2 + z^2 + 4x - 2y - 2z + 2 = 0$

13. Center: $(1, -3, -4)$; radius: 5

15. Center: $(1/3, -1, 0)$; radius: 2

17. $3x + 3y + 5z = 15$

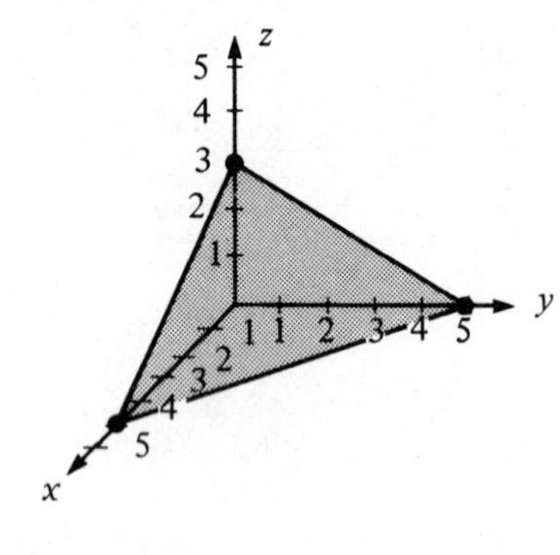

19. $z = 3$

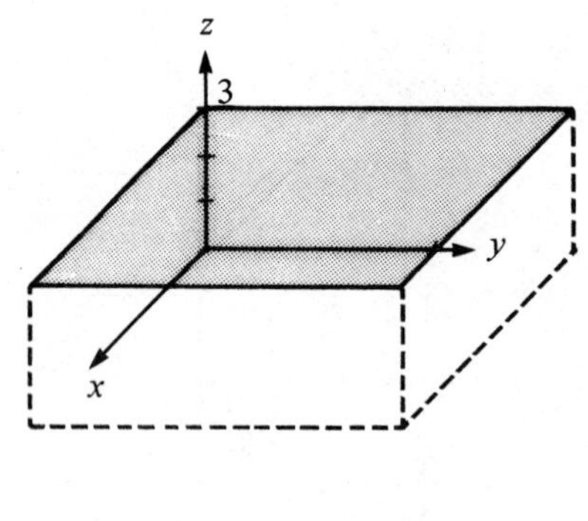

21. $x + y - z = 0$

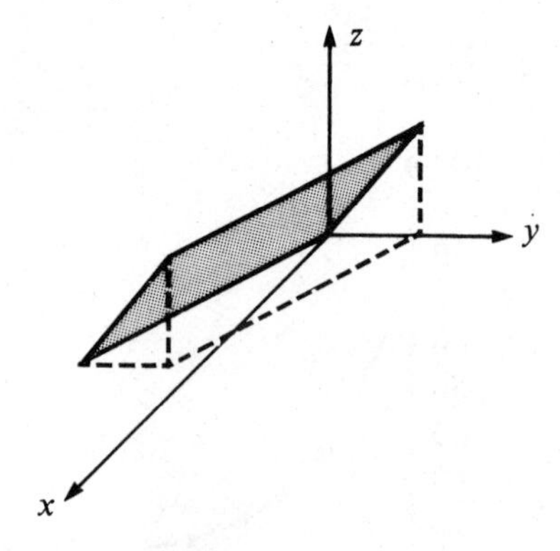

23. (c)

25. (f)

27. (d)

29. (a)

Section 8.2

1. (a) $3/2$ (b) $-1/4$ (c) 6
   (d) $5/y$ (e) $x/2$ (f) $5/t$

3. (a) 5 (b) $3e^2$ (c) $2/e$
   (d) $5e^y$ (e) xe^2 (f) te^t

5. (a) $2/3$ (b) Undefined

7. (a) 90π (b) 50π

9. (a) 4 (b) 6

11. Domain: All points inside and on the boundary of the circle $x^2 + y^2 = 4$
    Range: $[0, 2]$

13. Domain: All points on the xy-plane
    Range: $(-\infty, 4]$

15. Domain: The half-plane below the line $y = -x + 4$
Range: $(-\infty, \infty)$

17. Domain: All points in the xy-plane except those on the x and y axes
Range: All real numbers except 0

19. Domain: All points in the xy-plane where y is nonnegative
Range: $(-\infty, \infty)$

21. The level curves are circles of radius 5 or less centered at the origin.

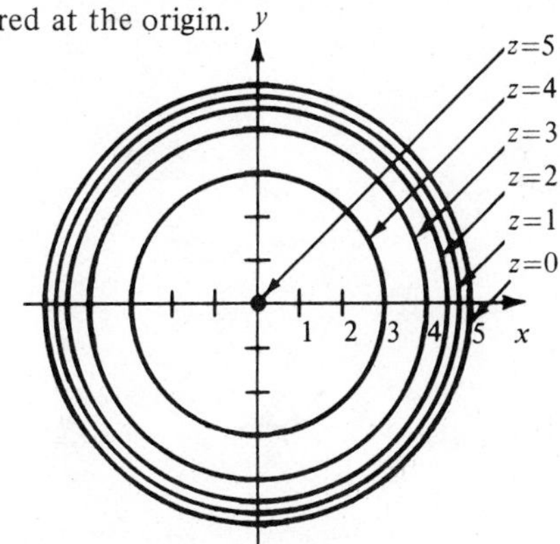

23. The level curves are hyperbolas centered at the origin with the x and y axes as asymptotes.

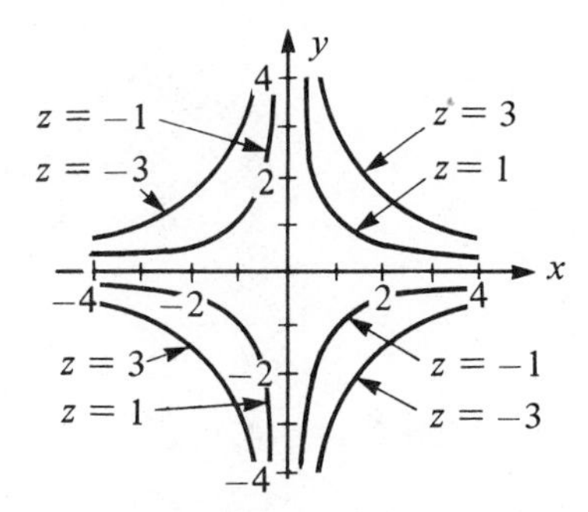

25. The level curves are lines of slope 1 passing through the fourth quadrant.

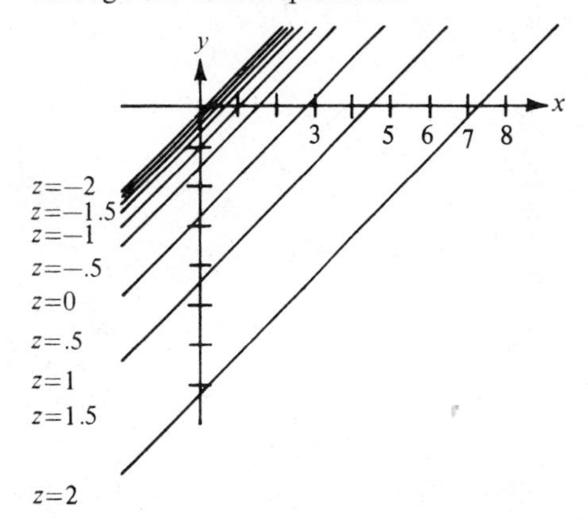

27. (a) 243 board-feet (b) 507 board-feet

29. (a) $\frac{2}{15}$ h = 8 min (b) $\frac{1}{4}$ h = 15 min
(c) $\frac{1}{12}$ h = 5 min

Section 8.3

1. $\frac{\partial f}{\partial x} = 2$

   $\frac{\partial f}{\partial y} = -3$

3. $\frac{\partial g}{\partial x} = y; 4$

   $\frac{\partial g}{\partial y} = x; 3$

5. $\frac{\partial z}{\partial x} = 2x - 3y; -8$

   $\frac{\partial z}{\partial y} = -3x + 2y; 7$

7. $\frac{\partial z}{\partial x} = 2xe^{2y}$

   $\frac{\partial z}{\partial y} = 2x^2e^{2y}$

9. $\frac{\partial z}{\partial x} = \frac{2x}{x^2 + y^2}$

   $\frac{\partial z}{\partial y} = \frac{2y}{x^2 + y^2}$

11. $\frac{\partial z}{\partial x} = \frac{-2y}{x^2 - y^2}$

    $\frac{\partial z}{\partial y} = \frac{2x}{x^2 - y^2}$

13. $\frac{\partial h}{\partial x} = -2xe^{-(x^2 + y^2)}$

    $\frac{\partial h}{\partial y} = -2ye^{-(x^2 + y^2)}$

15. $\frac{\partial z}{\partial x} = x/\sqrt{x^2 + y^2}; -3/5$

    $\frac{\partial z}{\partial y} = y/\sqrt{x^2 + y^2}; 4/5$

17. $\frac{\partial z}{\partial x} = -y^2/(x - y)^2; -1/4$

    $\frac{\partial z}{\partial y} = x^2/(x - y)^2; 1/4$

19. $\frac{\partial f}{\partial x} = e^y + ye^x$

    $\frac{\partial f}{\partial y} = xe^y + e^x$

Section 8.4

1. Relative minimum: (1, 2, −1)
3. Saddle point: (1, −2, −1)
5. Saddle point: (0, 0, 0)
7. Relative minimum: (1, 1, −1); Saddle Point: (0, 0, 0)
9. Relative minimum: (1, 1, 3)
11. Saddle Point (0, 0, 1)
13. $x_1 = 3, x_2 = 6$
15. $p_1 = 2500, p_2 = 3000$
17. $x_1 = 275, x_2 = 110$
19. $x_1 \approx 94, x_2 \approx 157$
21. 18 × 18 × 36 in
23. 10, 10, 10

Section 8.5

1. $f(5, 5) = 25$
3. $f(2, 2) = 8$
5. $f\left(\frac{\sqrt{2}}{2}, \frac{1}{2}\right) = \frac{1}{4}$
7. $f(25, 50) = 2600$
9. $f(1, 1) = 2$
11. $f(2, 2) = e^4$
13. $f(2, 2, 2) = 12$
15. $f(8, 16, 8) = 1024$
17. $f\left(\frac{1}{3}, \frac{1}{3}, \frac{1}{3}\right) = \frac{1}{3}$

19. $f\left(\sqrt{\frac{10}{3}}, \frac{1}{2}\sqrt{\frac{10}{3}}, \sqrt{\frac{5}{3}}\right) = \frac{5\sqrt{15}}{9}$

21. $36 \times 18 \times 18$ in

23. $\sqrt{13}/13$

25. (a) $x = 50\sqrt{2} \approx 71$, $y = 200\sqrt{2} \approx 283$

27. (a) $f\left(\frac{3{,}125}{6}, \frac{6{,}250}{3}\right) = 147{,}314$

(b) 1.473

(c) 184,142

29. (a) $f(49.4, 253) = 13{,}202$

(b) 12.85

(c) 14,487

Section 8.6

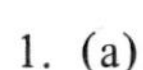

1. (a)

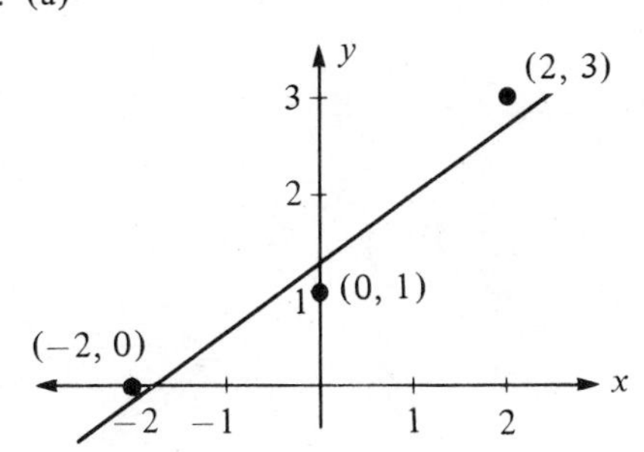

(b) $y = \frac{3}{4}x + \frac{4}{3}$

(c) 1/6

3. (a)

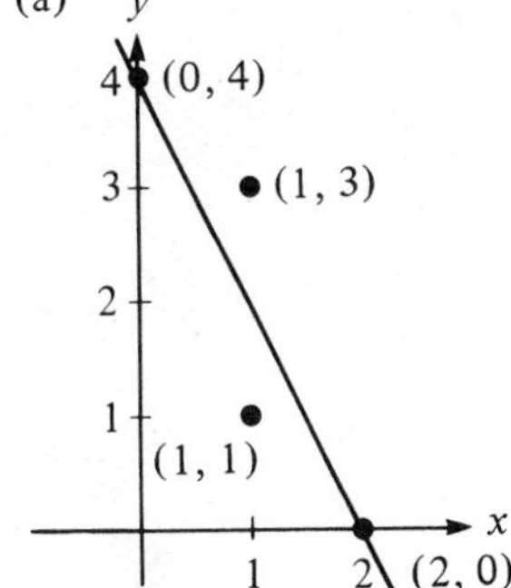

(b) $y = -2x + 4$

(c) 2

5. $y = \frac{7}{10}x + \frac{7}{5}$

7. $y = \frac{8}{7}x + \frac{10}{3}$

9. $y = -\frac{13}{20}x + \frac{7}{4}$

11. $y = \frac{37}{43}x + \frac{7}{43}$

13. $y = -\frac{175}{148}x + \frac{945}{148}$

15. (a) $y = -240x + 685$ (b) 349

17. (a) $y = 14x + 19$
(b) 41.4 bushels per acre

19. (a) $y = 139.40x + 277.17$
(b) 1392 thousand imports

21. $y = \frac{3}{4}x + \frac{4}{3}$

23. $y = -2x + 4$

25. $y = \frac{3}{7}x^2 + \frac{6}{5}x + \frac{26}{35}$

27. $y = x^2 - x$

29. (a) $y = 0.00136x^2 + 0.07617x + 3.62343$
(b) 5.1 billion

Section 8.7

1. $3x^2/2$

3. $y \ln(2y)$

5. $\dfrac{4x^2 - x^4}{2}$

7. $\dfrac{y}{2}(\ln^2 y - y^2)$

9. $x^2\left(1 - e^{-x^2} - x^2 e^{-x^2}\right)$

11. 3

13. 20/3

15. 2/3

17. 4

19. 16/3

21. $\displaystyle\int_0^1 \int_0^2 dy\, dx = \int_0^2 \int_0^1 dx\, dy = 2$

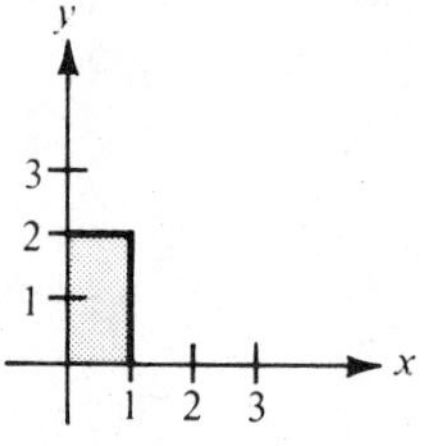

23. $\displaystyle\int_0^4 \int_0^{\sqrt{x}} dy\, dx = \int_0^2 \int_{y^2}^4 dx\, dy = 16/3$

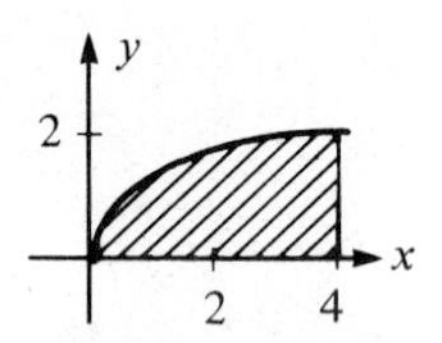

25. $\displaystyle\int_0^2 \int_{x/2}^1 dy\, dx = \int_0^1 \int_0^{2y} dx\, dy = 1$

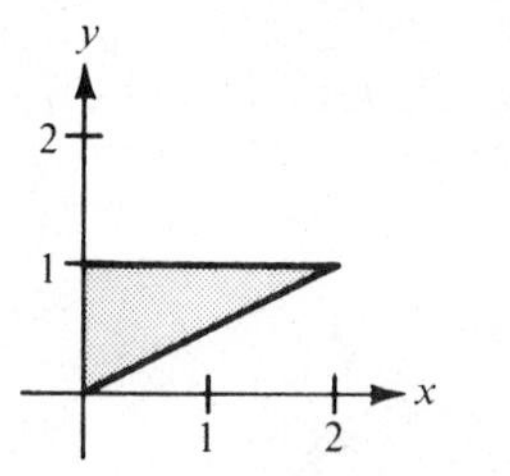

27. $\displaystyle\int_0^1 \int_{y^2}^{\sqrt[3]{y}} dx\, dy = \int_0^1 \int_{x^3}^{\sqrt{x}} dy\, dx = 5/12$

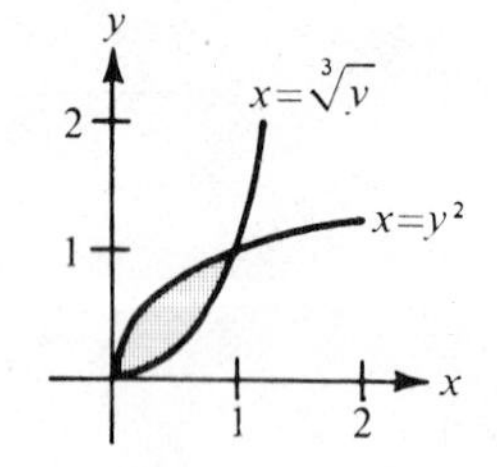

29. 4

31. 2

33. 8/3

35. 5

Section 8.8

1. 8

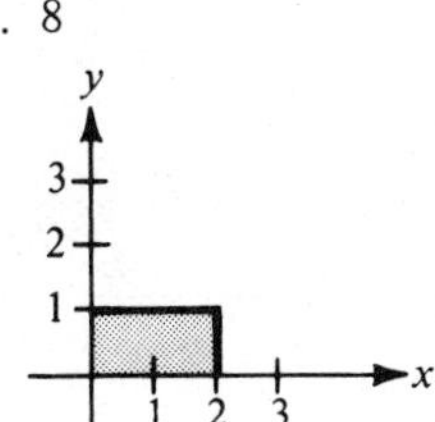

3. 1/54

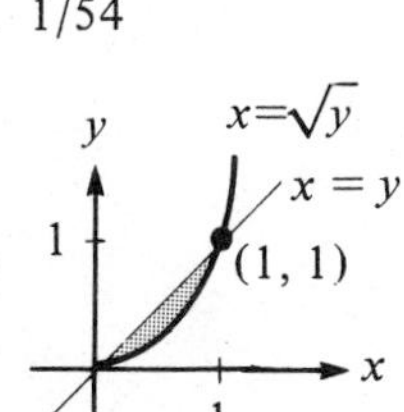

5. $\int_0^3 \int_0^5 xy\, dy\, dx = \int_0^5 \int_0^3 xy\, dx\, dy = 225/4$

7. $\int_0^4 \int_0^{\sqrt{x}} \frac{y}{1+x^2}\, dy\, dx = \int_0^2 \int_{y^2}^4 \frac{y}{1+x^2}\, dx\, dy = \frac{1}{4} \ln 17$

9. 4

11. 8

13. 12

15. 3/8

17. 8π

19. 1

21. 2

23. 8/3

25. $75,125

27. 25,645.24

29. (a) $(e^{-1/3} - 1)(e^{-1/2} - 1) \times 0.1115$
(b) $e^{-1/3} \times 0.7165$
(c) $1 - e^{-1} \times 0.6321$

Chapter 9

Section 9.1

1. 2, 4, 8, 16, 32

3. $-\frac{1}{2}, \frac{1}{4}, -\frac{1}{8}, \frac{1}{16}, -\frac{1}{32}$

5. $3, \frac{9}{2}, \frac{27}{6}, \frac{81}{24}, \frac{243}{120}$

7. $-1, -\frac{1}{4}, \frac{1}{9}, \frac{1}{16}, -\frac{1}{25}$

9. $3n - 2$

11. $n^2 - 2$

13. $\frac{n+1}{n+2}$

15. $\frac{(-1)^{n-1}}{2^{n-2}}$

17. $\frac{n+1}{n}$

19. $\frac{n}{(n+1)(n+2)}$

21. $\frac{2^n n!}{(2n)!}$

23. $(-1)^{n(n-1)/2}$

25. Converges to 1

27. Diverges

29. Converges to 3/2

31. Diverges

33. Converges to 0

35. Converges to 3

37. Converges to 0

39. (a) $A_1 = \$5100.00$ $A_5 = \$5520.40$
$A_2 = \$5202.00$ $A_6 = \$5630.81$
$A_3 = \$5306.04$ $A_7 = \$5743.43$
$A_4 = \$5412.16$ $A_8 = \$5858.30$
(b) $A_{40} = \$11{,}040.20$

41. (a) $S_1 = 1$ $S_4 = 10$
$S_2 = 3$ $S_5 = 15$
$S_3 = 6$
(b) $S_{50} = 1275$

43. (a) $2(0.8)^{n-1}$
(b) \$2 billion, \$1.6 billion, \$1.28 billion \$1.024 billion
(c) Converges to 0

45. $S_6 = 240$, $S_7 = 440$, $S_8 = 810$, $S_9 = 1490$, $S_{10} = 2740$

Section 9.2

1. Use Theorem 9.1:
$$\lim_{n\to\infty} a_n = \lim_{n\to\infty} \frac{n}{n+1} = 1 \neq 0$$

3. Use Theorem 9.1:
$$\lim_{n\to\infty} a_n = \lim_{n\to\infty} \frac{n^2}{n^2+1} = 1 \neq 0$$

5. Use Theorem 9.3:
$$a_n = 3\left(\frac{3}{2}\right)^n; a = 3, r = \frac{3}{2} > 1$$

7. Use Theorem 9.3:
$$a_n = 1000(1.055)^n; a = 1000, r = 1.055 > 1$$

9. Use Theorem 9.3:
$$a_n = 5\left(-\frac{3}{2}\right)^n; a = 5, |r| = \left|-\frac{3}{2}\right| > 1$$

11. 2

13. 2/3

15. $4 + 2\sqrt{2}$

17. 10/9

19. 9/4

21. 1/2

23. (a) $80{,}000(1 - 0.9^n)$
(b) 80,000

25. 152.42 ft

27. (a) 4/11
(b) 7/33

29. \$11,616.95

31. \$235,821.51

Section 9.3

1. $R = 2$

3. $R = 1$

5. $R = \infty$

7. $R = 0$

9. $R = 4$

11. $R = 5$

13. $R = 1$

15. $R = c$

17. $R = 1/2$

19. $R = \infty$

21. $\sum_{n=1}^{\infty} \frac{(-1)^{n-1}}{n}(x-1)^n; R = 1$

23. $\sum_{n=0}^{\infty} \frac{(2x)^n}{n!}; R = \infty$

25. $1 + \frac{1}{3}(x-1) + \sum_{n=2}^{\infty} \frac{(-1)^{n-1} 2 \cdot 5 \cdot 8 \cdots (3n-4)}{3^n \cdot n!}(x-1)^n; R = 1$

27. $\sum_{n=0}^{\infty} (-1)^n (n+1)x^n; R = 1$

29. $1 + \sum_{n=1}^{\infty} \frac{(-1)^n 1 \cdot 3 \cdot 5 \cdot 7 \cdots (2n-1)}{2^n \cdot n!} x^n; R = 1$

31. $\sum_{n=0}^{\infty} \frac{(-1)^n}{n!} x^{2n}$

33. $\sum_{n=0}^{\infty} (-1)^n x^{2n}$

35. $1 + \sum_{n=1}^{\infty} \frac{(-1)^n 1 \cdot 3 \cdot 5 \cdot 7 \cdots (2n-1)}{2^n \cdot n!} x^{2n}$

37. $x + \sum_{n=2}^{\infty} \frac{(-1)^{n+1}}{n} (x-1)^n$

39. $2 \sum_{n=0}^{\infty} (-1)^n \frac{x^{2n+3}}{2n+3}$

41. $R = \infty$

43. $\sum_{n=0}^{\infty} (-1)^n \frac{x^{4n+2}}{(2n+1)!}$

Section 9.4

1. (a) $S_1(x) = 1 + x$

(b) $S_2(x) = 1 + x + \frac{x^2}{2}$

(c) $S_3(x) = 1 + x + \frac{x^2}{2} + \frac{x^3}{6}$

(d) $S_4(x) = 1 + x + \frac{x^2}{2} + \frac{x^3}{6} + \frac{x^4}{24}$

(e) $S_5(x) = 1 + x + \frac{x^2}{2} + \frac{x^3}{6} + \frac{x^4}{24} + \frac{x^5}{120}$

3. (a) $S_1(x) = 1 + \frac{1}{2}(x-1)$

(b) $S_2(x) = 1 + \frac{1}{2}(x-1) - \frac{1}{2^2 \cdot 2!}(x-1)^2$

(c) $S_3(x) = 1 + \frac{1}{2}(x-1) - \frac{1}{2^2 \cdot 2!}(x-1)^2 + \frac{3}{2^3 \cdot 3!}(x-1)^3$

(d) $S_4(x) = 1 + \frac{1}{2}(x-1) - \frac{1}{2^2 \cdot 2!}(x-1)^2 + \frac{3}{2^3 \cdot 3!}(x-1)^3 - \frac{3 \cdot 5}{2^4 \cdot 4!}(x-1)^4$

5. (a) $S_1(x) = 1 - 2x$

(b) $S_2(x) = 1 - 2x + 3x^2$

(c) $S_3(x) = 1 - 2x + 3x^2 - 4x^3$

(d) $S_4(x) = 1 - 2x + 3x^2 - 4x^3 + 5x^4$

7.

x	$e^{x/2}$	$1+\frac{x}{2}$	$1+\frac{x}{2}+\frac{x^2}{8}$	$1+\frac{x}{2}+\frac{x^2}{8}+\frac{x^3}{48}$	$1+\frac{x}{2}+\frac{x^2}{8}+\frac{x^3}{48}+\frac{x^4}{384}$
0	1.0000	1.000	1.0000	1.0000	1.0000
0.25	1.1331	1.125	1.1328	1.1331	1.1331
0.50	1.2840	1.250	1.2813	1.2839	1.2841
0.75	1.4556	1.375	1.4453	1.4541	1.4549
1.0	1.6487	1.500	1.6250	1.6458	1.6484

9. $n = 8$

11. $\frac{1}{6!}$

13. 0.4613

15. 0.8637

Section 9.5

1. 1.732

3. 0.682

5. 1.146

7. 3.317

9. 0.567

11. ± 0.753

13. 2.21

15. $f'(x_1) = 0$

17. $\lim_{n\to\infty} x_n$ does not exist.

19. 1.913

21. Point nearest the origin: (1.939, 0.240)

23. 1.56 mi

25. 4.486 h after injection

Chapter 10

Section 10.1

1. $396°, -324°$

3. $240°, -480°$

5. $660°, -60°$

7. $300°, -60°$

9. $19\pi/9, -17\pi/9$

11. $23\pi/6, -\pi/6$

13. $7\pi/4, -\pi/4$

15. $26\pi/9, -10\pi/9$

17. $\pi/6$

19. $7\pi/4$

21. $-\pi/9$

23. $-3\pi/2$

25. $270°$

27. $-105°$

29. $420°$

31. $330°$

33. $c = 10, \theta = 60°$

35. $a = 4\sqrt{3}, \theta = 30°$

37. $\theta = 40°$

39. $h = 15$ ft

41.

r	8 ft	15 in	85 cm	24 in	$\frac{12963}{\pi}$ mi
s	12 ft	24 in	63.75 cm	96 in	8642 mi
θ	1.5	1.6	$3\pi/4$	4	$2\pi/3$

43. (a) $5\pi/12$ (b) 7.8125π in

Section 10.2

1. $\sin\theta = 4/5$, $\csc\theta = 5/4$; $\cos\theta = 3/5$, $\sec\theta = 5/3$; $\tan\theta = 4/3$, $\cot\theta = 3/4$

3. $\sin\theta = -5/13$, $\csc\theta = -13/5$; $\cos\theta = -12/13$, $\sec\theta = -13/12$; $\tan\theta = 5/12$, $\cot\theta = 12/5$

5. $\sin\theta = 1/2$, $\csc\theta = 2$; $\cos\theta = -\sqrt{3}/2$, $\sec\theta = -2\sqrt{3}/3$; $\tan\theta = -\sqrt{3}/3$, $\cot\theta = -\sqrt{3}$

7. Quadrant III

9. Quadrant I

11. Quadrant II

13. $\csc\theta = 2$

15. $\cot\theta = 4/3$

17. $\sec\theta = 17/15$

19. (a) $\sin 60^\circ = \sqrt{3}/2$, $\cos 60^\circ = 1/2$, $\tan 60^\circ = \sqrt{3}$
(b) $\sin\frac{2\pi}{3} = \sqrt{3}/2$, $\cos\frac{2\pi}{3} = -1/2$, $\tan\frac{2\pi}{3} = -\sqrt{3}$

21. (a) $\sin\left(-\frac{\pi}{6}\right) = -1/2$, $\cos\left(-\frac{\pi}{6}\right) = \sqrt{3}/2$, $\tan\left(-\frac{\pi}{6}\right) = -\sqrt{3}/3$
(b) $\sin 150^\circ = 1/2$, $\cos 150^\circ = -\sqrt{3}/2$, $\tan 150^\circ = -\sqrt{3}/3$

23. (a) $\sin 225^\circ = -\sqrt{2}/2$, $\cos 225^\circ = -\sqrt{2}/2$, $\tan 225^\circ = 1$
(b) $\sin(-225^\circ) = \sqrt{2}/2$, $\cos(-225^\circ) = -\sqrt{2}/2$, $\tan(-225^\circ) = -1$

25. (a) $\sin 750^\circ = 1/2$, $\cos 750^\circ = \sqrt{3}/2$, $\tan 750^\circ = \sqrt{3}/3$
(b) $\sin 510^\circ = 1/2$, $\cos 510^\circ = -\sqrt{3}/2$, $\tan 510^\circ = -\sqrt{3}/3$

27. (a) $\sin 10^\circ = 0.1736$ (b) $\csc 10^\circ = 5.7588$

29. (a) $\tan\frac{\pi}{9} = 0.3640$ (b) $\tan\frac{10\pi}{9} = 0.3640$

31. (a) $\cos(-110^\circ) = -0.3420$ (b) $\cos 250^\circ = -0.3420$

33. (a) $\tan 240^\circ = 1.7321$ (b) $\cot 210^\circ = 1.7321$

		Degrees	*Radians*
35.	(a)	30°, 150°	$\pi/6$, $5\pi/6$
	(b)	210°, 330°	$7\pi/6$, $11\pi/6$
37.	(a)	60°, 120°	$\pi/3$, $2\pi/3$
	(b)	135°, 315°	$3\pi/4$, $7\pi/4$
39.	(a)	45°, 225°	$\pi/4$, $5\pi/4$
	(b)	150°, 330°	$5\pi/6$, $11\pi/6$
41.	(a)	55°, 125°	$11\pi/36$, $25\pi/36$
	(b)	195°, 345°	$13\pi/12$, $23\pi/12$
43.	(a)	10°, 350°	$\pi/18$, $35\pi/18$
	(b)	126°, 234°	$7\pi/10$, $13\pi/10$
45.	(a)	50°, 230°	$5\pi/18$, $23\pi/18$
	(b)	97°, 227°	$97\pi/180$, $277\pi/180$

47. $\pi/4$, $3\pi/4$, $5\pi/4$, $7\pi/4$

49. 0, $\pi/4$, $5\pi/4$

51. 0, π

53. $\pi/4$, $5\pi/4$

55. 0, $\pi/2$, π

57. $100\sqrt{3}/3$

59. $25\sqrt{3}/3$

61. 15.5572

63. 9.1925

65. 18.7939 ft

67. 2145.1 ft

69. (a) 25.2°F (b) 65.1°F (c) 50.8°F

Section 10.3

1. Period: π
   Amplitude: 2

3. Period: 4π
   Amplitude: $3/2$

5. Period: 2
   Amplitude: $1/2$

7. Period: 2π
   Amplitude: 2

9. Period: $\pi/5$
   Amplitude: 2

11. Period: 3π
    Amplitude: $1/2$

13. Period: $1/2$
    Amplitude: 3

15. $\pi/2$

17. $2\pi/5$

19. 3

21.

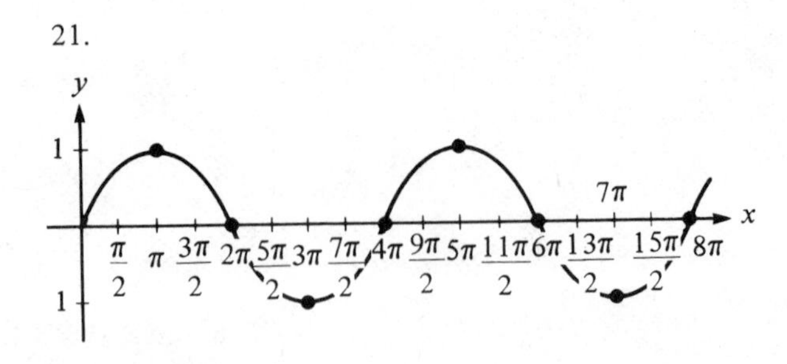

23.

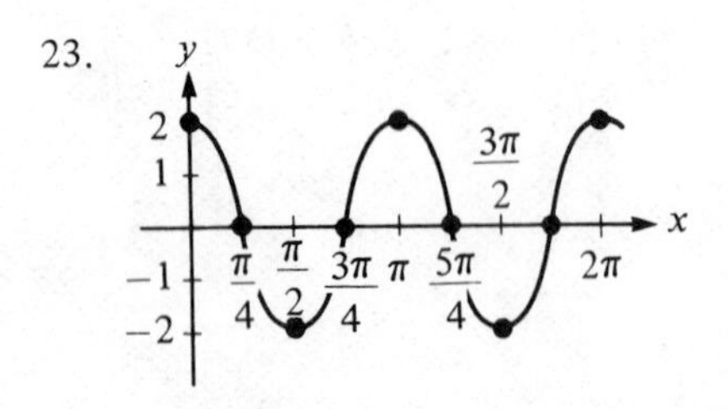

25.

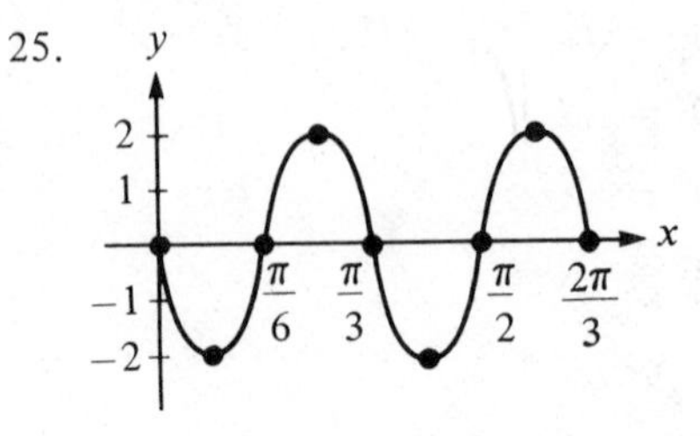

27. 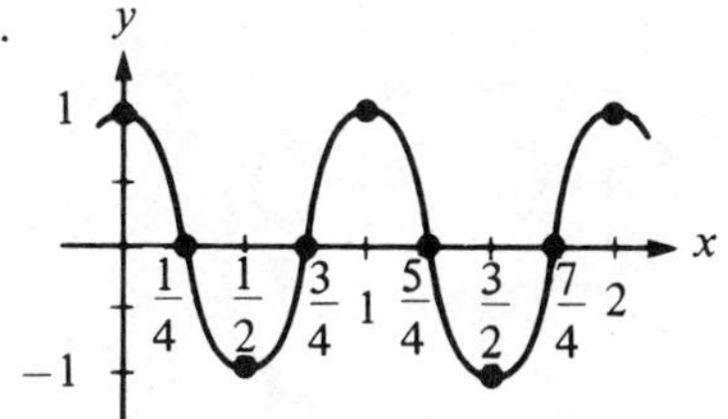

29. 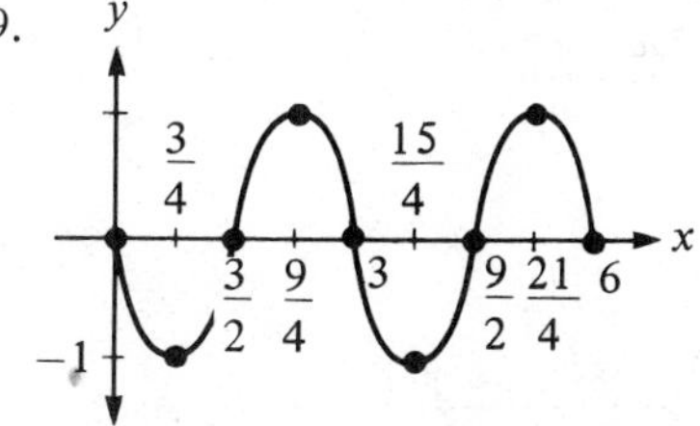

31. 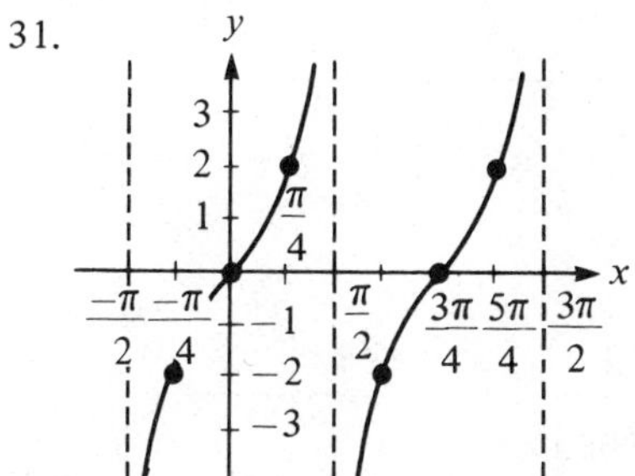

33. 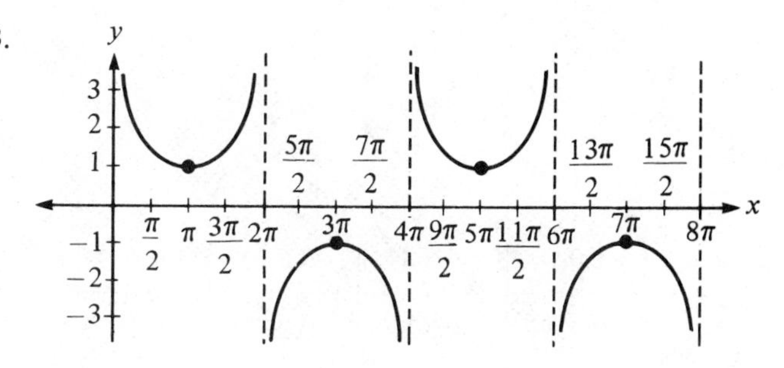

35. 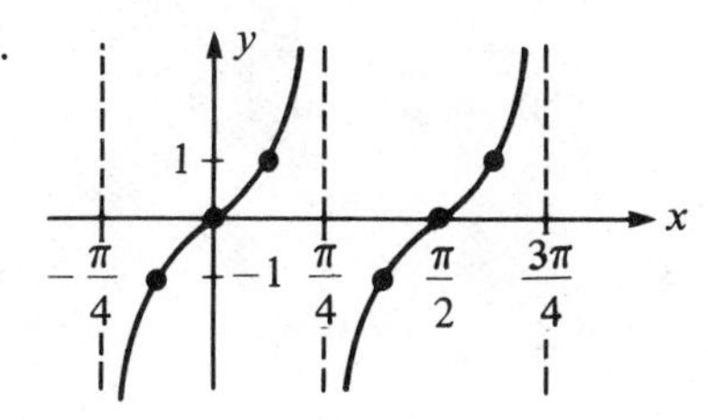

37. 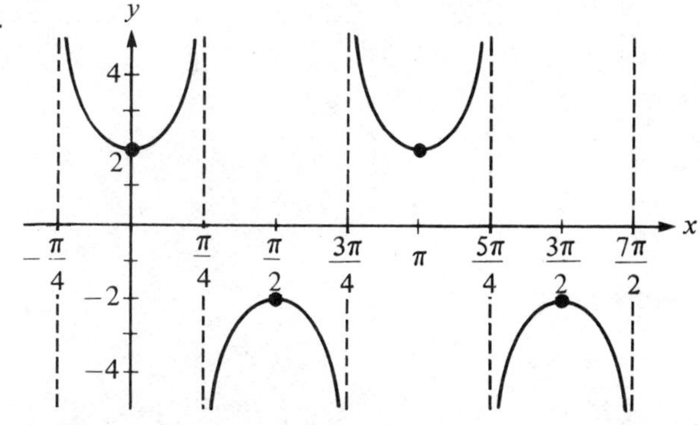

39. 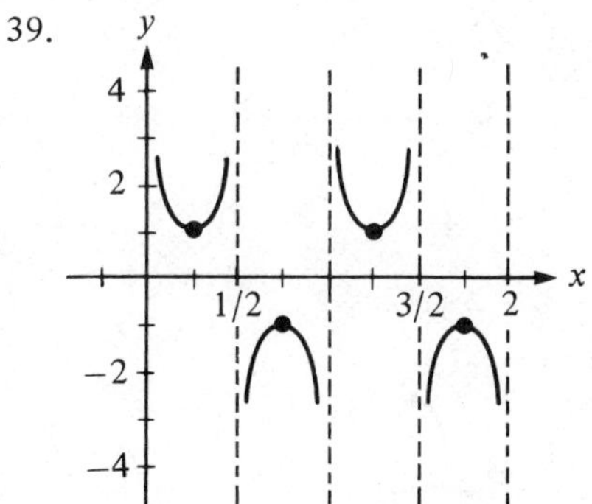

41. (a) $t = 6$ s
(b) 10
(c)

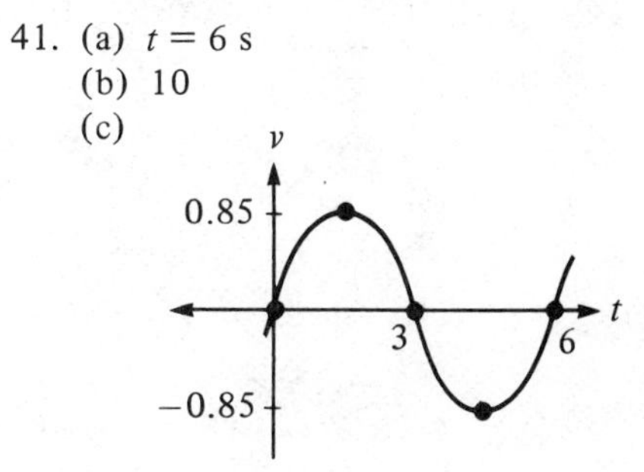

43. (a) $p = 1/440$
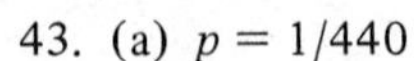
(b) $f = 440$
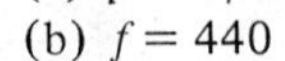
(c)

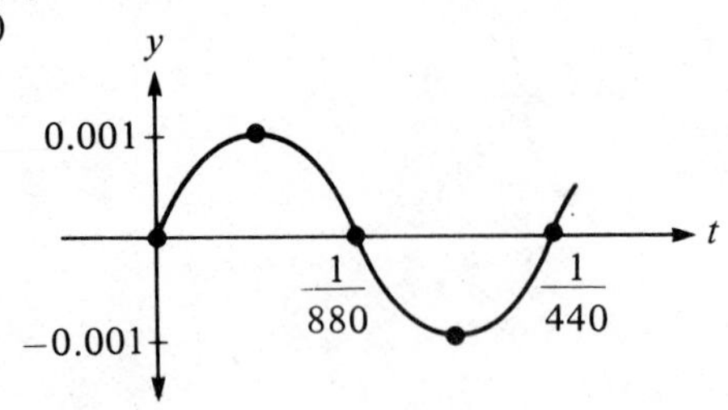

45. (a) $p = 6/5$
(b) 50
(c)

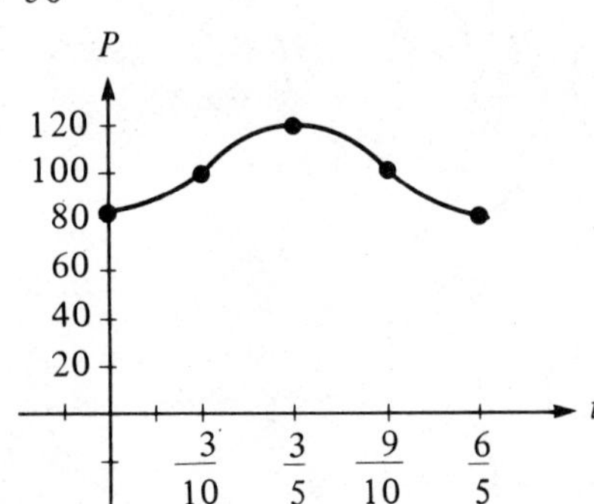

47.

x	−1.0	−0.8	−0.6	−0.4	−0.2	−0.1	−0.01	−0.001
$f(x)$	−0.4597	−0.3791	−0.2911	−0.1973	−0.0997	−0.0500	−0.0050	−0.0005

x	1.0	0.8	0.6	0.4	0.2	0.1	0.01	0.001
$f(x)$	0.4597	0.3791	0.2911	0.1973	0.0997	0.0500	0.0050	0.0005

$$\lim_{x \to 0} \frac{1 - \cos x}{x} = 0$$

Section 10.4

1. $4 \cos 4x$

3. $-\sin (x/2)$

5. $2x \sec (x^2) \tan (x^2)$

7. $\cos x - x \sin x$

9. $-\sin x$

11. $\dfrac{-x \sin x - \cos x - 1}{x^2}$

13. $-\csc 2x \cot 2x$

15. $\dfrac{\pi}{2} \sec^2 \left[\dfrac{\pi(x - 1)}{2}\right]$

17. $\sin \dfrac{1}{x} - \dfrac{1}{x} \cos \dfrac{1}{x}$

19. $2e^x \tan(e^x) \sec^2 (e^x)$

21. $-\sin 2x$

23. $-2 \sin 2x$

25. $-\csc^2 x$

27. $-\sec x \csc x$

29. $\tan^2 x$

31. $\csc x$

33. $\dfrac{1}{2}(x \sec^2 x + \tan x - \sec x \tan x)$

35. $3 \cos 3x(1 - 3 \sin^2 3x)$

41. 3

43. 2

45. 1

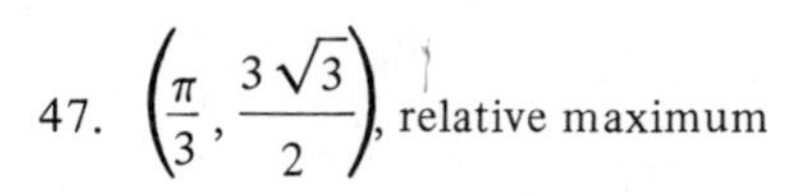

47. $\left(\frac{\pi}{3}, \frac{3\sqrt{3}}{2}\right)$, relative maximum

$(\pi, 0)$, point of inflection

$\left(\frac{5\pi}{3}, -\frac{3\sqrt{3}}{2}\right)$, relative minimum

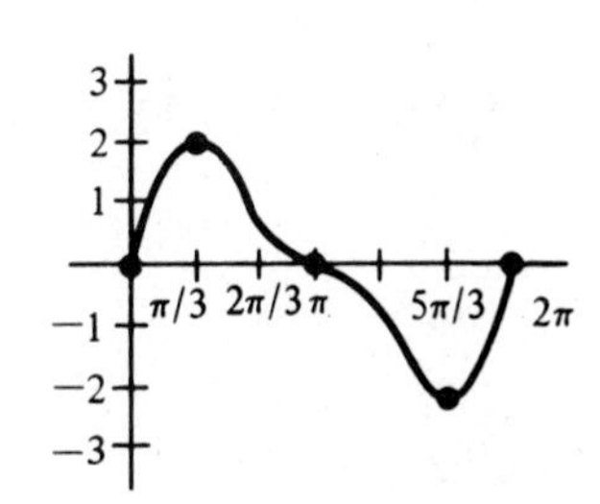

49. $\left(\frac{\pi}{3}, \frac{\pi}{3} - \sqrt{3}\right)$, relative minimum

$\left(\frac{5\pi}{3}, \frac{5\pi}{3} + \sqrt{3}\right)$, relative maximum

$\left(\frac{7\pi}{3}, \frac{7\pi}{3} - \sqrt{3}\right)$, relative minimum

$\left(\frac{11\pi}{3}, \frac{11\pi}{3} + \sqrt{3}\right)$, relative maximum

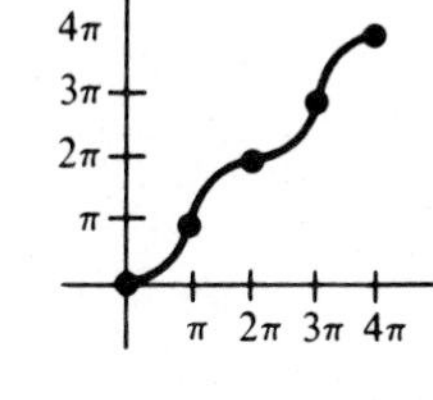

51. $\sum_{n=0}^{\infty} \frac{(-1)^n x^{2n+1}}{(2n+1)!}$; $R = \infty$

53. $\sum_{n=0}^{\infty} \frac{(-1)^n x^{2n}}{(2n+1)!}$; $R = \infty$

55. (a) $t = 214.5$, or August 2 and 3 are the warmest days
(b) $t = 32$, or February 1 is the coldest day

57. (a) $t = 1/2$, or January 15 and 16 are the days of highest consumption with a rate of 7.9 quadrillion Btus.
(b) $t = 4\frac{1}{2}$ and $8\frac{1}{2}$, or May 15, 16 and September 15 are the days of lowest consumption with a rate of 6.4 quadrillion Btus.

59. (a) midnight
(b) noon

Section 10.5

1. $-\frac{1}{2}\cos 2x + C$

3. $\frac{1}{2}\sin x^2 + C$

5. $2(\sqrt{3} - 1)$

7. $\ln \sqrt[3]{2}$

9. $\frac{1}{2}\ln|\sin x| + C$

11. $\sec 1 - 1 = 0.8508$

13. $\ln|\tan x| + C$

15. $-\frac{1}{3}\cot^3 x + C$

17. $\frac{1}{5}\tan^5 x + C$

19. 3/4

21. $\ln|\sec x - 1| + C$

23. $-\ln|1 + \cos x| + C$

25. $-\frac{2}{3}(1 + \cos\theta)^{3/2} + C$

27. $\sin e^x + C$

29. $2\sqrt{\sin x} + C$

31. $x \sin x + \cos x + C$

33. $x \tan x + \ln |\cos x| + C$

35. 2

37. $\ln \sqrt{2}$

39. 2

41. 1.3707

43. (a) 225.28 million barrels
 (b) 225.28 million barrels
 (c) 217 million barrels

45. 1.6234 liters

47. (a) $C = \$9.17$
 (b) $C = \$3.14$
 Savings = \$6.03

Index

Index

1 2 3 4 5 6 7 8 9 0